Salt Stress in Plants

Parvaiz Ahmad • M.M. Azooz
M.N.V. Prasad
Editors

Salt Stress in Plants

Signalling, Omics and Adaptations

Springer

Editors
Parvaiz Ahmad
Department of Botany
A.S. College
University of Kashmir
Srinagar, India

M.N.V. Prasad
Department of Plant Sciences
University of Hyderabad
Hyderabad, India

M.M. Azooz
Department of Botany
Faculty of Science
South Valley University
Qena, Egypt

Department of Biological Sciences
Faculty of Science
King Faisal University
Al Hassa, Saudi Arabia

ISBN 978-1-4614-6107-4 ISBN 978-1-4614-6108-1 (eBook)
DOI 10.1007/978-1-4614-6108-1
Springer New York Heidelberg Dordrecht London

Library of Congress Control Number: 2012956549

Printed on acid-free paper

Springer is part of Springer Science+Business Media (www.springer.com)

Preface

Anthropogenic pressures are implicated in the degradation of invaluable natural resources. Assessing the impacts of global soil salinization on plant growth and productivity and identifying approaches for mitigation salinization are subjects of global importance. It is reported that about 7% of the total land on Earth and 20% of the total arable area are affected by a high salt content. Plant productivity is also affected by the elevated levels of salt content in the soil. The reason for the low production is that various metabolic processes that work independently or in coordination with one another are affected by the deleterious effects of salt.

Poly-omics – namely, proteomics, genomics, micromics, transcriptomics, metablomics, inomics, metallomics, etc. – have emerged as a powerful tool for understanding the mechanism of plant response toward salinity stress. The exploitation of different genes and proteins involved in the regulation of various environmental stresses will be very useful in generating crops with enhanced food production under salinity stress. With the help of metabolomics, we will recognize different metabolic pathways the plant is rearranging during stress. Plants perceive both external and internal signals and use them to regulate various responses for their development.

During salinity stress, plants respond in various ways and can withstand the stress. Salinity stress is responsible for osmotic, ionic, and oxidative stresses, which lead to reduced growth and development of the plant. Plants can tolerate these stresses by the accumulation of osmolytes and osmoprotectants. Another machinery is the expression of different types of enzymatic and non-enzymatic antioxidants. Understanding the full mechanism of salt tolerance through different means is an enigmatic subject for scientists in general and plant biologists in particular.

The outline of this volume, "*Salt Stress in Plants: Omics, Signaling and Adaptations*," encompasses the following: Chapter 1 deals with advances of metabolomics to reveal plant response during salt stress. Chapter 2 narrates the role of microRNAs (micromics) in response to salt stress in plants. Chapter 3 sheds light on the role of proteomics in salt-stressed plants. Chapter 4 discusses improving salinity tolerance in plants through genetic approaches. Chapter 5 describes the role of LEA proteins in salinity tolerance in plants. Chapter 6 highlights the effect the salt stress on crop production and the role of omics in salinity tolerance. Chapters 7, 8 and 9 deal with the role of different kinds of signaling molecules in plants under salt stress. Chapters 10 and 11 examine the approaches to improve salt tolerance in rice and maize. Chapter 12 highlights the role of phytochromes in stress tolerance. Chapter 13 discusses alleviating salinity stress through arbuscular mycorrhiza. Chapter 14 deals with breeding approaches in stress tolerance in citrus. Chapter 15 highlights the effect of salt stress in photosynthesis under ambient and elevated atmospheric CO_2 concentrations. Chapter 16 deals with nitrogen-use-efficiency in plants under salt stress. Chapter 17 sheds light on the response of salt-affected plants to cadmium, and Chap. 18 highlights the role of plant tissue culture in screening the salt tolerance in plants.

This volume will provide valuable information about the omic approaches, signaling, and responses of plants under salt stress. We would like to thank all the authors of this volume for their contributions. We are also thankful to my colleagues who helped us directly or indirectly in completing this volume. We are also grateful to Hanna Smith (Associate Editor, Springer) and Margaret Burns (Developmental Editor, Springer) for their help, suggestions, and timely publication of the volume.

Srinagar, India Parvaiz Ahmad
Qena, Egypt M.M. Azooz
Hyderabad, India M.N.V. Prasad

Contents

Brief Biodata of Editor and Co-editors

Dr. Parvaiz Ahmad, Editor

Dr. Parvaiz Ahmad is an assistant professor in botany at A.S. College, Srinagar, Jammu and Kashmir, India. He completed his postgraduation in botany in 2000 from Jamia Hamdard, New Delhi, India. After receiving a doctorate degree from the Indian Institute of Technology (IIT) Delhi, India, he joined the International Centre for Genetic Engineering and Biotechnology, New Delhi, in 2007. His main research areas are stress physiology and molecular biology. He has published more than 30 research papers in peer-reviewed journals and 16 book chapters. He is also the editor of four book volumes (three with Springer Science + Business Media, NY, and one with Studium Press Pvt. India Ltd., New Delhi, India). He is the recipient of the Junior Research Fellowship and Senior Research Fellowship conducted by CSIR, New Delhi, India. Dr. Ahmad has been awarded the Young Scientist Award under Fast Track scheme, in 2007, by the Department of Science and Technology, Government of India, New Delhi, India. He is actively engaged in studying the molecular and physio-biochemical responses of different agricultural and horticultural plants under environmental stress.

Prof. Mohamed M. Azooz, Co-editor

Professor M.M. Azooz is a professor of plant physiology at the Department of Botany, Faculty of Science, South Valley University, Qena, Egypt. He received his B.Sc. in 1984 and M.Sc. in 1990 from Assiut University, Egypt. He received his Ph.D. in 1997 from Tübingen University, Germany. He has published more than 70 research articles in peer-reviewed journals and five book chapters. He has supervised many M.Sc. students and five Ph.D. scholars. He is an editorial board member and reviewer of many international journals. Professor Azooz has been included in *Marquis Who's in the World* (Biography, 28th Edition, 2011), and in the "Top 100 Scientists" list, International Biographical Centre, Cambridge,

UK, 2011. His research emphasizes exploring the physiological, biochemical, and molecular responses of medicinal and crop plants against various biotic and abiotic stresses and their tolerance mechanism under these stresses.

Prof. M.N.V. Prasad, Co-editor

Professor M.N.V. Prasad is a professor at the Department of Plant Sciences at the University of Hyderabad, India. He received his B.Sc. (1973) and M.Sc. (1975) degrees from Andhra University, India, and his Ph.D. degree (1979) in botany from the University of Lucknow, India. Prasad had published 141 articles in peer-reviewed journals and 84 book chapters and conference proceedings in the area of environmental botany and heavy metal stress in plants. He is the author and editor for 15 books (John Wiley and Sons Inc. New York; CRC Press, Boca Raton; Springer-Verlag, Heidelberg; Narosa Publishing House, New Delhi; Kluwer Academic Publishers, Dordrecht; Marcel Dekker, New York; Fizmatlit Publishers, Moscow; and Ministry of Environment and Forests, Govt. of India). He is the recipient of the Pitamber Plant National Environment Fellowship of 2007, awarded by the Ministry of Environment and Forests, Government of India. He has served as a visiting professor and guest researcher in many universities in different parts of the world, including Australia, Belgium, Canada, Finland, Kazakhstan, Poland, Portugal, and Sweden.

Contributors

Surekha Agarwal Directorate of Rice Research, Hyderabad, Andhra Pradesh, India

Parvaiz Ahmad Department of Botany, A.S. College, University of Kashmir, Srinagar, India

Gema Ancillo Instituto Valenciano de Investigaciones Agrarias (IVIA), Ctra Moncada-Náquera, Valencia, Spain

Junaid Aslam Department of Biotechnology, Jamia Hamdard (Hamdard University), New Delhi, India

R. Rusta Azad College of Chemical Engineering and Petroleum, Sharif University of Technology, Tehran, Iran

R.A. Azevedo Departamento de Genética, Escola Superior de Agricultura "Luiz de Queiroz", Universidade de São Paulo, Piracicaba, Brazil

M.M. Azooz Department of Botany, Faculty of Science, South Valley University, Qena, Egypt

Department of Biological Sciences, Faculty of Science, King Faisal University, Al Hassa, Saudi Arabia

N. Naresh Babu Division of Genetics, Indian Agricultural Research Institute, New Delhi, India

Monika Beszterda Department of Chemistry, Poznań University of Life Sciences, Poznań, Poland

Renu Bhardwaj Department of Botanical and Environmental Sciences, Guru Nanak Dev University, Punjab, India

Mukhtar A. Bhat Department of Botany, Degree College Beerwah, University of Kashmir, Srinagar, India

M.L. Campos Department of Energy, Plant Research Laboratory, Michigan State University, East Lansing, MI, USA

R.F. Carvalho Departamento de Biologia Aplicada à Agropecuária, Universidade Estadual Paulista "Júlio de Mesquita Filho", Jaboticabal, Brazil

Ruby Chandna National Institute for Plant Genome Research, New Delhi, India

Department of Botany, Jamia Hamdard, New Delhi, India

Sreekala Chellamma Dow AgroSciences LLC, Indianapolis, IN, USA

Joanna Dames Department of Biochemistry, Microbiology and Biotechnology, Rhodes University, Eastern Cape, South Africa

Maria Angeles Forner-Giner Instituto Valenciano de Investigaciones Agrarias (IVIA), Ctra Moncada-Náquera, Valencia, Spain

Masayuki Fujita Laboratory of Plant Stress Responses, Department of Applied Biological Science, Faculty of Agriculture, Kagawa University, Kagawa, Japan

Nicole Geissler Institute of Plant Ecology, Justus Liebig University Giessen, Giessen, Germany

Piotr Goliński Department of Chemistry, Poznań University of Life Sciences, Poznań, Poland

Aditi Gupta Centre for Rural Development and Technology, Indian Institute of Technology, New Delhi, Hauz Khas, India

R. Hajiboland Plant Science Department, University of Tabriz, Tabriz, Iran

Khalid Rehman Hakeem Faculty of Forestry, Universiti Putra Malaysia, Selangor, Malaysia

P.G Programme in Bioresources, Department of Botany, University of Kashmir, Srinagar, India

Neha Handa Department of Botanical and Environmental Sciences, Guru Nanak Dev University, Punjab, India

Mirza Hasanuzzaman Department of Agronomy, Faculty of Agriculture, Sher-e-Bangla Agricultural University, Dhaka, Bangladesh

Bernd Huchzermeyer Institute of Botany, Leibnitz University Hannover, Hannover, Germany

Muhammad Iqbal Department of Botany, Jamia Hamdard, New Delhi, India

Sumiya Jamshieed Department of Botany, Jamia Hamdard (Hamdard University), New Delhi, India

Mukesh Kanwar Department of Botanical and Environmental Sciences, Guru Nanak Dev University, Punjab, India

Dhriti Kapoor Department of Botanical and Environmental Sciences, Guru Nanak Dev University, Punjab, India

Harpreet Kaur Department of Botanical and Environmental Sciences, Guru Nanak Dev University, Punjab, India

K. Khajeh Department of Biochemistry, Faculty of Biological Science, Tarbiat Modares University, Tehran, Iran

Hans-Werner Koyro Institute of Plant Ecology, Justus Liebig University Giessen, Giessen, Germany

S. Gopala Krishnan Division of Genetics, Indian Agricultural Research Institute, New Delhi, India

Ashwani Kumar Department of Biochemistry, Microbiology and Biotechnology, Rhodes University, Eastern Cape, South Africa

M. Sheshu Madhav Directorate of Rice Research, Hyderabad, Andhra Pradesh, India

Mahmooduzzafar Department of Botany, Jamia Hamdard (Hamdard University), New Delhi, India

Satendra K. Mangrauthia Directorate of Rice Research, Hyderabad, Andhra Pradesh, India

M. Miransari Department of Biological Sciences, Faculty of Science, Tarbiat Modares University, Tehran, Iran

Abdul Mujib Department of Botany, Jamia Hamdard (Hamdard University), New Delhi, India

Behzad Murtaza Institute of Soil and Environmental Sciences, University of Agriculture, Faisalabad, Pakistan

Ghulam Murtaza Institute of Soil and Environmental Sciences, University of Agriculture, Faisalabad, Pakistan

F. Nagafi Department of Agriculture, Medicinal Plants Research Institute, Shaid Beheshti University, Tehran, Iran

M. Nagarajan Division of Genetics, Rice Breeding and Genetics Research Center, Indian Agricultural Research Institute, Aduthurai, India

Kamrun Nahar Department of Agricultural Botany, Faculty of Agriculture, Sher-e-Bangla Agricultural University, Dhaka, Bangladesh

Laboratory of Plant Stress Responses, Department of Applied Biological Science, Faculty of Agriculture, Kagawa University, Kagawa, Japan

Gabrijel Ondrasek University of Zagreb Faculty of Agriculture, Zagreb, Croatia

Munir Ozturk Department of Botany, Ege University, Bornova, Izmir, Turkey

V.Y. Patade Nuclear Agriculture and Biotechnology Division, Bhabha Atomic Research Centre, Mumbai, India

Bhinu V-S. Pillai Dow AgroSciences LLC, Indianapolis, IN, USA

Poonam Department of Botanical and Environmental Sciences, Guru Nanak Dev University, Punjab, India

M.N.V. Prasad Department of Plant Sciences, University of Hyderabad, Hyderabad, Andhra Pradesh, India

R. Rahnemaie Department of Soil Science, College of Agriculture, Tarbiat Modares University, Tehran, Iran

A.N. Rai Nuclear Agriculture and Biotechnology Division, Bhabha Atomic Research Centre, Mumbai, India

B. Rangbar Department of Biophysics, Faculty of Biological Sciences, Tarbiat Modares University, Tehran, Iran

Reiaz ul Rehman P.G. Programme in Bioresources, Department of Botany, University of Kashmir, Srinagar, India

B. Sailaja Directorate of Rice Research, Hyderabad, Andhra Pradesh, India

Muhammad Saqib Institute of Soil and Environmental Sciences, University of Agriculture, Faisalabad, Pakistan

Indu Sharma Department of Botanical and Environmental Sciences, Guru Nanak Dev University, Punjab, India

Resham Sharma Department of Botanical and Environmental Sciences, Guru Nanak Dev University, Punjab, India

Satyawati Sharma Centre for Rural Development and Technology, Indian Institute of Technology, New Delhi, India

A.K. Singh Division of Genetics, Indian Agricultural Research Institute, New Delhi, India

A.K. Srivastava Nuclear Agriculture and Biotechnology Division, Bhabha Atomic Research Centre, Mumbai, India

P. Suprasanna Nuclear Agriculture and Biotechnology Division, Bhabha Atomic Research Centre, Mumbai, India

Inayatullah Tahir P.G Programme in Bioresources, Department of Botany, University of Kashmir, Srinagar, India

M.M. Tehranchi Department of Physics, Faculty of Science, Shahid Beheshti University, Tehran, Iran

Yawar Usman Institute of Soil and Environmental Sciences, University of Agriculture, Faisalabad, Pakistan

K.K. Vinod Division of Genetics, Rice Breeding and Genetics Research Center, Indian Agricultural Research Institute, Aduthurai, India

S.R. Voleti Directorate of Rice Research, Hyderabad, Andhra Pradesh, India

Agnieszka Waśkiewicz Department of Chemistry, Poznań University of Life Sciences, Poznań, Poland

Chapter 1
Recent Advances of Metabolomics to Reveal Plant Response During Salt Stress

Ruby Chandna, M.M. Azooz, and Parvaiz Ahmad

Abstract Salt stress is the major limiting factor in agriculture and portraits a major challenge to food security. The adverse effect of salt stress is expressed on whole plant levels. Plants have acquired various processes that functions to balance cellular hyperosmolarity and ion disequilibrium in an effort to combat salt stress. These processes occur due to significant changes in the gene expression that in turn bring about changes in plant metabolism. These metabolic changes help the plant to adapt to disorganized metabolic homeostasis. It has been observed that adverse growth conditions have impact on the synthesis of secondary plant products or metabolites that help in plant defence. The diverse nature of these metabolites has lead to the development of 'Metabolomics'. The metabolite fingerprinting and profiling approaches provides accurate identification and quantification of stressed sample even before they can bring about change(s) in the transcriptome or proteome. Using metabolic profile changes as a marker for stress physiology, metabolic movements and factors can be analysed in combination with other 'omic' techniques, such as transcriptomics. Revealed analyses of salt acclimation effects and related stress factors to salinity stress may provide help in crop breeding programs to develop

R. Chandna
National Institute for Plant Genome research, New Delhi, India

Department of Botany, Jamia Hamdard, New Delhi, India

M.M. Azooz
Department of Botany, Faculty of science, South Valley University, 83523 Qena, Egypt

Department of Biological Sciences, Faculty of Science,
King Faisal University, Al Hassa, Saudi Arabia

P. Ahmad (✉)
Department of Botany, A.S. College, University of Kashmir, Srinagar, India
e-mail: parvaizbot@yahoo.com

P. Ahmad et al. (eds.), *Salt Stress in Plants: Signalling, Omics and Adaptations*,
DOI 10.1007/978-1-4614-6108-1_1,

salt tolerance varieties. In this review, we will focus on recent advancements and application of metabolomics in plants under salinity stress.

Keywords Salinity • Signaling • Phenolic compounds • Alkaloids

Salinity in agriculture has been a major restricting factor in food production. Soil salinity is known to restrict the land use and limits crop yield. The various environmental programs carried throughout the world estimates that approximately 20% of agricultural land and 50% of cropland in the world is salt-stressed (Munns and Tester 2008). Hence, the existing soil salinity is a large confront for food security. Increase in levels of water-soluble salts e.g. NaCl, Na_2CO_3 and $CaCl_2$ is mainly due to irrigation, results in soil salinity. Soil salinity results in reduced biomass production by affecting important physiological and biochemical processes of plant (Ahmad and John 2005; Ahmad and Sharma 2006, 2008, 2010; Ahmad 2010; Ahmad et al. 2010a, b, c, 2012).

The adverse effect of salt stress is expressed on whole plant level, and appears during all developmental stages including germination, seedling and vegetative stages. However, tolerance in respect to salt stress varies at different plant developmental stages and also from species to species. Salt stress occur as a calamitous episode, be imposed constantly or from time to time and then gradually becoming more severe at any stage during development. The plant respond to salt stress by various processes that functions in coordination to balance cellular hyperosmolarity and ion disequilibrium. Salt tolerance in plants and their yield stability are complex genetic traits that are complicated and difficult to establish in crops.

Plant's ability to tolerate salt is dependent on multiple biochemical pathways that lead to production of osmotically dynamic metabolites, free radicals and specific proteins to manage ion and water flux. Thus providing support to scavenging oxygen radicals and in turn maintaining ion homeostasis. Therefore there was need to determine the underlying biochemical mechanisms of salinity tolerance so as to provide plant breeders with appropriate indicators. But salt stress has showed to affect many intracellular substances, like nucleic acids, proteins, carbohydrates and amino acids (Ahmad and Sharma 2008, 2010; Ahmad et al. 2010a, b, c; Ahmad 2010). Thereby the introducing molecular biological techniques into plant stress physiology provided an enhanced effort that lead to the identification of stress-inducible genes (Bartels and Sunkar 2005; Umezawa et al. 2006). These studies succeeded in over-expression of genes that are known to be involved in stress responses, provided tolerance to abiotic stress. Another approach, quantitative trait locus (QTL) analysis brought about benefits in the enabled creation of stress-tolerant crops by combining QTLs for various stress tolerances traits (Takeda and Matsuoka 2008; Krasensky and Jonak 2012). Several studies related to QTL in various salt stress tolerances have been reported. For example, Ren et al. (2005) identified the SKC1 locus encoding a high affinity K^+ transporter (HKT)-type sodium transporter by analyzing a QTL for salinity tolerance using salt-tolerant and salt-susceptible rice varieties.

However, plant responses to salinity involve diversified changes in the activity of genes and proteins, which invariably lead to changes in plant metabolism. It is now known that plants have numerous metabolic pathways that direct to thousands of secondary products which are capable of effectively responding to salt stress situations. These pathways, often diverge from primary metabolic pathways upon initial gene duplication (Nascimento and Fett 2010; Mastrobuoni et al. 2012). So, the different type of growth conditions has a noteworthy impact on the synthesis and accretion of secondary plant products. Because of these extremely assorted chemical natures of metabolites, metabolome analyses can provide accurate identification and quantification from a single stressed sample (Birkemeyer et al. 2005). More precisely, the metabolite fingerprinting approach and its summary provides identification of the early compounds that signal the perception of stress even before they can bring about change(s) in the transcriptome or proteome could be detected to access the eventual biological information current connecting gene expression and metabolic phenotype. In this review, we will be summarizing some of the plant stress physiology and techniques used to study the metabolome of plants during salt stress.

Physiology of Plants Under Salt Stress

Mechanisms of plants towards salt tolerance occur by restricting the entry of salt into the plant (especially minimizing the accumulation of salt in photosynthetic tissues and cytoplasm) (Munns 2002). The plant follows two major adaptive strategies towards high environment salinity tolerance: firstly to avoid stress due to different physical and physiological barriers, secondly enhancing the adaptive mechanisms internally that will enable successful survival. Therefore, the Na^+ uptake and its transport regulation across the plasma membranes and tonoplast is one of the key factors that establish the plant cell response to salinity stress (Dajic 2006). Avoidance of salt uptake can take place by salt exclusion; it is a very efficient but complex way of reducing the permeability of massive ion in the root zone, especially sodium. This process enables a low uptake and accumulation of salts in the upper parts, especially in the transpiring organs of the plant. Many glycophytes are known to show better skills for Na^+ exclusion from the shoot and also for maintaining elevated levels of K^+ (Flowers and Hajibagheri 2001). Studies have revealed salt in-tolerant plants, such as beans and maize are known to be the most outstanding Na^+ excluders (Bayuelo-Jimenez et al. 2003) whereas salt tolerant crops like bread wheat has reduced speed of Na^+ transport to the shoots and high K^+/Na^+ intolerance (Gorham 1990).

A study done by Munns et al. (1988) and Jeschke and Hartung (2000) have shown salt exclusion to function at the cellular as well as at the whole plant level and to a greater extent is related to regulation of K/Na ion selection. In mangrove *Avicennia marina* is known to have 98% degree of salt exclusion property (Ball 1988). Whereas it was demonstrated by Munns et al. (1999) glycophyte or halophyte, has the property of restraining of Na^+ uptake and accumulation in the shoots. In some salt tolerant species, for example wheat have the property to exclude salts is achieved by changing sodium and calcium ions, rather than bringing about modification in osmotic potential, as adsorption on membranes of root cells of calcium ions directs towards reduced penetration of monovalent cations (Munns et al. 1999).

Salt excretion is another very efficient way of preventing excessive absorption and building up of salts in photosynthetic tissues. This mechanism is equipped with developed special features, which are mostly present in leaf epidermis, known as salt glands and salt hairs (bladders). These structures are commonly found in many halophytes such as *Spartina*, *Aeluropus* (Poaceae), *Limonium*, *Armeria* (Plumbaginaceae), *Atriplex* (Chenopodiaceae), *Glaux* (Primulaceae), *Tamarix*, *Reamuria* (Tamaricaceae), and in mangrove species, e.g. *Avicennia*, *Aegiceras* and *Acanthus* (Popp 1995). The glandular structures that are involved in salt excretion may vary in structure, position, mechanism and also in ecological significance. The simplest ones are two celled found in *Spartina* and *Aeluropus,* and three-celled types known to occur in *Chloris gayana*. There are also complex structures composed of 5–9 cells in *Avicennia*, 8 cells present in *Tamarix* and also 16 celled present in the family Plumbaginaceae (Crawford 1989). Glandular structures are present all over the surface area of the shoot, and are most abundant on the leaves.

Still excessive exposure of plants to salt stress can lead to the production of reactive oxygen species (ROS) such as H_2O_2 (hydrogen peroxide), $O_2^{\bullet-}$ (superoxide), $^1O_2$ (singlet oxygen) and $^{\bullet}OH$ (hydroxyl radical). Excess ROS is a source of toxic reactions like lipid peroxidation, protein degradation and DNA mutation (Vinocur and Altman 2005; Pitzschke et al. 2006; Ahmad et al. 2008, 2009, 2010b, 2011, 2012). In plant cells, H_2O_2, superoxide anion (O_2^-), and hydroxyl radical ($^{\bullet}OH$) are generated due to oxidative damage to cells during environmental stress in the cytosol, chloroplasts, mitochondria, and the apoplastic space and they have the potential to cause (Mittler 2002). Research work has revealed that ROS has an important role as signal transduction molecules in plants. It is involved in mediating responses to environmental stresses, pathogen infection, and programmed cell death (Torres and Dangl 2005). Increased production of ROS related to salt stress causes membrane injury (Shalata et al. 2001). This increase in levels of ROS results from closure of stomata that leads to reduced in CO_2 concentration within the chloroplasts and also decreases in $NADP^+$ concentration causing photoinhibition (Foyer and Noctor 2003).

Among all these synchronized physiological responses in plants, abscisic acid (ABA), the plant hormone, plays a essential role. ABA is a stress hormone as for its rapid accretion towards the response to stress and its intervention, help plant endurance over much stress. The first requirement is that ABA production should be rapidly triggered by the stress signals so that inhibition of physiological functions is avoided. Secondly, ABA should be quickly degraded and deactivated once the stress is reassured so that normal plant growth and functions can recommence.

Research on plant tolerance towards salt stress cover many portions of its influence on plant behaviour, which includes alterations at the morphological, physiological and molecular levels. Previously, stress studies are focused on: transgenic plants development, improvement of plant breeding and modification in the genetic structures of existing crops towards enhanced adaptation to salinity conditions.

Recently, the progress of research in 'omics' like proteomics and metablomics and has created a better platform for understanding of molecular mechanism in salinity stress (Tester and Davenport 2003).

Cell Signalling and Secondary Signal Molecules During Salt Stress

During environmental stress plant cells receive stress signals that are influenced by various signalling pathways. The secondary messengers, stress responsive plant hormones, signal transducers and regulatory transcriptional molecules together to induce signals via signalling pathway. Early response in plant cells towards salinity stress is Ca^{2+}, derived; leading to a sudden increase in its concentration in cytosol either from influx of Ca^{2+} from the apoplastic space or release from internal stores. The release of Ca^{2+} is further controlled by second messengers for example, cyclic ADP ribose, $NADP^{+}$ and inositol polyphosphates and that are present as ligand-sensitive Ca ion channels. These molecules enable the release of Ca^{2+} in plant cells particularly in guard cells (Schroeder et al. 2001). Hrabak et al. (2003) reported that two protein kinases are assumed to be the targets of the Ca^{2+} signal in plants. One is SnRK3-type kinases, whose action is reliant on the Ca^{2+}-binding calcineurin B-like (CBL) proteins (Krasensky and Jonak 2012; Sarwat et al. 2012). Work done by Luan (2009) in Arabidopsis has showed SnRK3 as one of the best characterized protein kinase which was eventually recognized as a vital factor in salt stress response. The other protein kinase concerned in stress response is Ca^{2+}-dependent kinase (CDPK).

Mitogen activated protein kinase (MAPK) mechanism in plant cell are as well accountable for the assembly of osmolytes and antioxidants. Receptors/sensors such as kinases (histidine kinases and protein tyrosine kinases) and G-protein are known to activate these MAPK pathways (Kong et al. 2011; Zhang et al. 2011).

Late embryogenesis–abundant (*LEA*)-type genes and the dehydration-responsive element (DRE)/C-repeat (CRT) among the class of important responsive genes towards stress play role in regulation of osmolyte production. Researchers showed that *LEA* -type genes represents damage repair pathways (Xiong and Zhu 2002). Since the activity of phospholipase C in plants that is regulated by G-proteins, and phosphoinositols organize the up-regulation of these *LEA* – like genes under salt stress. G-protein–related receptors may also provide with membrane-bound receptors for ABA response (Wang et al. 2001). Under salinity stress, ABA plays a very important role in bringing about a radical change in the expression profile of gene and cellular processes of plant (Park et al. 2009). Other plant hormones play direct or indirect substantial task during abiotic stress. ABA during abiotic stress is known to interconnect with Salicylic acid (SA), ethylene (ET) as well as jasmonic acid (JA) (Grant and Jones 2009).

Plasma membrane plays a role of barrier between living cells and the surrounding environments. It also has an important part in the insight and conduction of exterior details. Variation in phospholipid components occurs when osmotic stress is initiated and this change is detected in plants (Munnik et al. 1998). However, the major role of phospholipids is they form the backbone of cellular membranes and serve as precursor for the production of second-messenger molecules. The relevant enzymes involved in cleaving are the phospholipases (PIP) A2, C, and D. PIP besides being involved in signal transduction is also involved in several processes, like employing, assembly and transportation of signalling complexes to the specific membrane locations (Martin 1998). PIP 2 is also involved in cellular ion homeostasis. *PI5K*, is one of the gene that encodes a phosphatidylinositol 4-phosphate 5-kinase (PIPK) that functions in the production of PIP 2 (Mikami et al. 1998). As osmotic stress increases, so does the production of PIP 2 by upregulating the expression of *P15K* gene. Increase in PIPK isoforms expression contribute to increased conversion of PIP 2 to two important molecules, of diacylglycerol and inositol 1,4,5-trisphosphate (IP3). Diacylglycerol and IP3 are important secondary messengers that are able to stimulate protein kinase C and finally release trigger Ca^{2+} release.

Abiotic Stress Responses: Genome Wide Expression

Abiotic stresses causes increase of many intracellular substances, also affects nucleic acids, proteins, carbohydrates and amino acids. Molecular biology techniques introduction in plant biology enables an immense effort towards the identification of stress-inducible genes. Molecular studies thrive in isolating genes that were known to function in stress responses and tolerance. It is now known that transcriptional activation happens at different time points in response to stress stimuli. This suggests that abiotic stress responses are very complicated is controlled by a various signalling means and different transcription factors. Identification of many significant factors in the stress pathway had been able to use responsive genes as markers. For example, dehydration responsive element (CRE)/C-repeat (CRT) (A/GCCGAC) an abiotic stress-responsive cis-element, and its post-translational modifications were recognized, which in turn has increases the scope of research that play central role in identifying the transcriptional regulating factors (Shinozaki and Yamaguchi-Shinozaki 2007). Also the genetic screening for mutations affects the expression of stress inducible genes thus enabled the identification of novel components in the abiotic regulatory system (Chinnusamy et al. 2002).

Completions of the genomes of *Arabidopsis* and *Oryza sativa* have also added to the information available on stress physiology. The absolute genome sequence is now-a-days accessible and also has enabled genome-wide gene expression profiling to a variety of abiotic stresses (Kilian et al. 2007). Microarray technology has also enabled the knowledge of genes responding to abiotic stresses that have been identified more in detail than before. Complete transcriptome analysis have facilitated the relationships between stress-regulated transcripts, and their regulatory elements (Weston et al. 2008). The function of stress-inducible genes can also be determined by the reverse genetic approach, assisted by insertion mutation lines.

Vast microarray experiments have lead to the identification of the regulators for stress-inducible intracellular signalling and gene expression of various types of transcription factors (e.g. MAP kinases, phosphatises and metabolic phospholipid enzymes). This classification of inducible stress signal transducers augmented an thought that plants have developed transformable cellular response means to resourcefully react to various abiotic stresses.

Metabolic Profile Under Abiotic Stress

Metabolism reveals biological activities dependent on the environmental conditions. Study of metabolic profile under abiotic stress conditions has made possible the detection and recognition of metabolites. Under stress conditions, plants re-organize their metabolic pathways in order to adapt to changing conditions (Kaplan et al. 2004). Using these metabolic profile changes as a marker, metabolic movements and factors that regulate them were analysed in combination with other 'omic' analyses, such as transcriptome particularly through mass spectrometry-based analytical methods (Saito et al. 2008; Sawada et al. 2009). The metabolic pathways often employed from vital primary metabolism pathways, upon initial gene duplication, play main part in the plant and environment communication (Nascimento and Fett 2010). Accordingly, the diverse growth surroundings have a major impact on the synthesis and accrual of secondary plant products. Therefore, this production of secondary plant products acts as a surviving response for plants to manage the increasing stress. There are varieties of secondary products synthesized during these processes and are listed below.

Phenolic Compounds

A large variety of secondary products produced by plants contain a phenol group, a hydroxyl functional group on an aromatic ring called Phenol, a chemically heterogenous group. This group forms an important part of the plants defence system against biotic and abiotic stress condition. Increase in salt concentration increases the total phenolic content of leaves (Savirnata et al. 2010). Flavinoids form one of the largest classes of plant phenolics that carry out extremely dissimilar functions in plant coordination also including defence and pigmentation (Kondo et al. 1992). Like flavonoids, anthocyanins also have multiple biological activities as antioxidant component.

Isoflavonoids are derivatives of flavonone intermediate, naringenin that occurs universally in plants. They are known to be released by the legumes playing an important part in encouraging the creation of nitrogen-fixing nodules by symbiotic rhizobia (Sreevidya et al. 2006). Besides this they also participated in plant growth and defence responses. Studies done by Posmyk et al. (2009) observed that the production of these flavonoids is an efficient approach against ROS.

Alkaloids

About 20% of the species of vascular plants has substantial members of N-containing secondary metabolites. Most of alkaloids include pyrrolizidine alkaloids (PAs). They are considered to be toxic and primarily serve as defence against herbivoral attack (Schafer and Wink 2009). But under highly suppressive conditions these alkaloids play role as ROS scavengers. Studies have also shown that ROS production also regulates the alkaloid pathway occurring in undifferentiated cells. It seems they also have mechanisms for directing the alkaloid pathway in other parts of the plants (Sachan et al. 2010).

Influence of Salt Stress on the Synthesis of Secondary Plant Products

Salt stress is one of the serious factor that limits the efficiency of crops and especially quantity and quality of their metabolic (secondary plant products) products. Against the attack of pathogens, plants manufacture secondary plant products as a part of defence mechanism. Therefore, the concentrations of large amount of secondary products are totally dependent on the surrounding environmental circumstances. High salt concentrations in the soil are accountable for the production of the secondary plant products by making a major change in the metabolic enzymes/pathways. Volatile sulphur compounds, vitamins, carbonyl compounds, ascorbic acid and flavonoids are some of the active secondary metabolities that are stimulated under environmental stresses (Krasensky and Jonak 2012). Enzymes like Phenylalanine ammonium lyase (PAL) and Glutathione-S-transferase (GSTs) also get induced from unfavourable effects of stresses (Marrs 1996). PAL, in action with cinamates 4-hydroxylase forms essential group of enzymes that helps in biosynthesis of several important secondary metabolites from phenyl alanine (Singh et al. 2009). In a series of experimental observations, it could be shown that plants which are exposed to salt stress produce a greater amount of secondary plant products such as phenols, terpenes as well as N and S containing substances such as alkaloids, cyanogenic glucosides or gluco-sinolates (Singh et al. 2009).

Metabolomics: Recent Technology Developments and Applications

Plants are known to be nature's excellent chemists, as they have a huge variety of chemical substances that fit according to the needs of a highly variable and generally hostile environment conditions (Baxter and Borevitz 2006). Natural metabolic range and a lack of combination of principles require identification of compounds, forming major analytical challenge (Breitling et al. 2006). As, metabolomic applications in crop/plant analysis are constantly growing, therefore the use of liquid chromatography mass spectrometry (LC-MS), GCMS and NMR are greatly explored to get more full insight into the variation of compositions in the metabolities. Research in potato using flow injection mass spectrometry analysis (FIMS) of a varied range of genotypes, showed the correlation between genotypes with different traits in free amino acid content (Beckman et al. 2007). Metabolomic technologies facilitate the multivariant metabolic data using varied, chromatographic detection systems, such as GC-MS, Fourier-transformed infrared spectroscopy/NMR-based methods.

Techniques like Matrix-assisted laser desorption/ionization (MALDI-TOF) and GC-MS profiling revealed dissimilarity in metabolic composition like amino acids and organic acids in tomato cultivars even though these cultivars were closely related (Fraser et al. 2007). Thus, the excellence of crop/plants nutritional value is the expression of metabolite content (Memelink 2004). There is therefore great importance of using a metabolomics approach to know better what in particular has happened during stress encroachment and provide help to plan new ideas for crop improvement.

Metabolomics is entirely positioned to perceive the pathway that drive the expression of a trait and potentially enable breeders, to select the desired trait of superiority for high-yielding varieties also with tolerance to abiotic stress. Plant's response to salinity involves changes in the functionality of genes and proteins that consistently lead to changes in plant metabolism. Gas chromatography time-of-flight mass spectrometry outlines the novel details from the plant models, Rice, *Arabidopsis thaliana and Lotus japonicus* that demonstrated the power of metabolite profiling providing insight to disturbed cellular balance between amino acids and organic acids in response to salt stress.

Because of the highly assorted nature of metabolites, metabolome analyses are subjected to combination of technological and analytics. The most noteworthy advantage of metabolome analyses is the static chemical uniqueness of metabolite entities. In comparison to genomics, transcriptomics and proteomics analyses that enables the identity of genes and proteins, metabolomics provide highly appropriate investigations of metabolic, like the physiological responses caused by environmental perturbations (Desbrosses et al. 2005).

This data-rich analytical advancement have ignited the development of bioinformatic tools to sort through the complex fingerprints and profiles of data sets for relevant descriptive information. More specifically, bioinformatics with metabolite fingerprinting and profiling approaches grant access to the eventual biological information flow between gene expression and metabolic phenotype.

Some studies on higher plants upon exposure to salt stress utilize profiles of metabolic fingerprinting to explore changes in them. The metabolic impact of salt stress have been studied in crops like *Lycopersicom esculantum*, *Solanum lycopersicon*, *Oryza sativa*, *Vitis vinifera* and the model plant *Arabidopsis thaliana* (Cramer et al. 2007; Kim et al. 2007; Zuther et al. 2007). Comparisons of metabolite profiles have also been carried out in halophytic species, such as the *Populus euphratica* tree or the shrubs *Limonium latifolium* and *Thellungiella halophila* (Gong et al. 2005; Gagneul et al. 2007).

Analysis of the FT-IR spectra, provided information on compound classes specific that revealed signals from nitriles and amino radicals and some nitrogen containing compounds allowing the comparison between control samples and salt-treated fruits, leading to a clear classification of the investigated cultivars of *S. lycopersicon*. Another study on the salt-tolerant tree *P. euphratica* in combination with transcriptomics and GC-MS based metabolomic analyses revealed that within the natural habitat of plants, they are acclimated to the environment. It was observed that there was an increase in amino acid levels, specifically proline, valine and b-alanine, changes in sugar and polyol metabolism this may be due to high sodium concentration in the field. Increase in the levels of myo-inositol, glyceric acid and glycerol were reported while a decrease was observed in levels of fructose and mannitol (Brosché et al. 2005).

Study by Gong et al. (2005) where the transcriptional and metabolic profiles were investigated in the glycophyte *A. thaliana* to short term salt stress in comparison to halophyte *T. halophila.* An interspecies difference was demonstrated by GC-MS-based metabolite profiling that moderately amplified during response to 150 mM NaCl salt shock. Surprisingly, constant group of numerous metabolites and transcripts that are stress-responsive were found to be already changed in *T. halophila* even before exposure to salinity, signifying a continuous adjustment mechanism in halophytic species. Also, research have shown that sugars along with proline, citric acid, malic acid and succinic acids were reportedly higher in halophyte *T. halophila* than in *A. thaliana.*

Cramer et al. (2007) explored and compared the transcriptome and GC-MS-based metabolomic profile of drought and salt stress shoot tips from *V. vinifera* cv. Cabernet. It was revealed from metabolomics, sucrose, aspartic, succinic and fumaric acids levels reduced. The profile of proline, asparagine, malic acid and fructose showed increase in their levels under salt stress. Further, under water-limited conditions most metabolites exhibited similar trends, in contrast to glucose, malic acid and proline which increased noticeably. Recently, the role of the compatible solutes was also studied in the halophytic species *L. latifolium*, by means of untargeted and targeted metabolic profiles (Gagneul et al. 2007). It was noticed that sugars, inositols and proline acted as osmolytes balancing the cell environment, while organic acids decreased upon salt stress.

Investigations by Sanchez et al. (2008) during salt stress in legume plant *Lotus japonicus* used an combination of transcriptomics, ionomics and metabolomics technique, thereby found a vast enlargement in the constant levels of many amino acids, sugars and polyols, with a simultaneous reduction in accumulation of most of the organic acids. It was thus suggested that, metabolic responses during increased salinity showing changes in organic solute composition are there by guarded by adaptive developmental programs, that be inferred to metabolic anticipation of stress.

Conclusion and Future Perspective

It can be concluded from past and present metabolomic studies of plant response towards salinity is that; the changes occurring in metabolism are complex and therefore involve multiple pathways. Particularly during acclimation of salt stress the response is coupled with changes in metabolism of organic acid, amino acid and sugar. The changes that appear are in the form of primary metabolism that overlaps with the reaction of other interacting/regulatory stress factors. These traits are accompanied with a broad array of response at the whole-plant level (molecular and cellular). Metabolomics is coming up as one of the key tools in studying plant stress responses in the direction of gene expression individually/in group. Metabolic networks are highly dynamic particles and they keep on moving from one cellular compartment to another. Given that the metabolic profiling enables the better understanding of the unchanged level of metabolites, kinetics and flux analyses. It also adds to the knowledge of the unpredictable metabolic changes occurring towards stress. Metabolic examination at the plants-subcellular level in specific tissues plays a part for future challenge. Significant new discoveries in metablomics have enhanced the field. Combination of metabolomics, proteomics, transcriptomics and mathematical modeling in future will provide us an insight on how plants respond to salt stress and thus will enable us to develop strategies for enhancement towards the stress tolerance in plants.

References

Ahmad P (2010) Growth and antioxidant responses in mustard (*Brassica juncea* L.) plants subjected to combined effect of gibberellic acid and salinity. Arch Agro Soil Sci 56(5):575–588

Ahmad P, John R (2005) Effect of salt stress on growth and biochemical parameters of *Pisum sativum* L. Arch Agron Soil Sci 51:665–672

Ahmad P, Sharma S (2008) Salt stress and phyto-biochemical responses of plants. Plant Soil Environ 54(3):89–99

Ahmad P, Sharma S (2010) Physio-biochemical attributes in two cultivars of mulberry (*M. alba*) under $NaHCO_3$ stress. Int J Plant Produc 4(2):79–86

Ahmad P, Sarwat M, Sharma S (2008) Reactive oxygen species, antioxidants and signaling in plants. J Plant Biol 51(3):167–173

Ahmad P, Jeleel CA, Azooz MM, Nabi G (2009) Generation of ROS and non-enzymatic antioxidants during abiotic stress in plants. Bot Res Intern 2(1):11–20

Ahmad P, Jaleel CA, Salem MA, Nabi G, Sharma S (2010a) Roles of enzymatic and non-enzymatic antioxidants in plants during abiotic stress. Crit Rev Biotechnol 30(3):161–175

Ahmad P, Jaleel CA, Sharma S (2010b) Antioxidative defence system, lipid peroxidation, proline metabolizing enzymes and biochemical activity in two genotypes of *Morus alba* L. subjected to NaCl stress. Russ J Plant Physiol 57(4):509–517

Ahmad P, Umar S, Sharma S (2010c) Mechanism of free radical scavenging and role of phytohormones during abiotic stress in plants. In: Ashraf M, Ozturk M, Ahmad MSA (eds) Plant adaptation and phytoremediation. Springer, Dordrecht\Heidelberg\London\NY, pp 99–108

Ahmad P, Nabi G, Jeleel CA, Umar S (2011) Free radical production, oxidative damage and antioxidant defense mechanisms in plants under abiotic stress. In: Ahmad P, Umar S (eds) Oxidative stress: role of antioxidats in plants. Studium Press, New Delhi, pp 19–53

Ahmad P, Hakeem KR, Kumar A, Ashraf M, Akram NA (2012) Salt-induced changes in photosynthetic activity and oxidative defense system of three cultivars of mustard (*Brassica juncea* L.). Afr J Biotechnol 11(11):2694–2703

Ball MC (1988) Salinity tolerance in the mangroves *Aegiceras corniculatum* and *Avicennia marina*. I. Water use in relation to growth, carbon partitioning, and salt balance. Aust J Plant Physiol 15:447–464

Bartels D, Sunkar R (2005) Drought and salt tolerance in plants. Crit Rev Plant Sci 24:23–58

Baxter IR, Borevitz JO (2006) Mapping a plant's chemical vocabulary. Nat Genet 38:737–738

Bayuelo-Jimenez JS, Debouck DG, Lynch JP (2003) Growth, gas exchange, water relations and ion composition of *Phaseolus* species grown under saline conditions. Field Crops Res 80:207–222

Beckman M, Enot DP, Overy DP, Draper J (2007) Representation, comparison and interpretation of metabolome fingerprint data for total composition analysis and quality trait investigation in potato cultivars. J Agric Food Chem 55:3444–3451

Birkemeyer C, Luedemann A, Wagner C, Erban A, Kopka J (2005) Metabolome analysis: the potential of in vivo labeling with stable isotopes for metabolite profiling. Trend Plant Sci 23:28–33

Breitling R, Pitt AR, Barrett MP (2006) Precision mapping of the metabolome. Trends in Biotechnol 24:543–548

Brosché M, Vinocur B, Alatalo ER, Lamminmäki A, Teichmann T, Ottow EA, Djilianov D, Afif D, Bogeat-Triboulot MB, Altman A et al (2005) Gene expression and metabolite profiling of Populus euphratica growing in the Negev desert. Genome Biol 6:R101

Chinnusamy V, Stevenson B, Lee B, Zhu JK (2002) Screening for gene regulation mutants by bioluminescence imaging. Sci STKE 2002(140):PI10

Cramer GR, Ergul A, Grimplet J et al (2007) Water and salinity stress in grapevines: early and late changes in transcript and metabolite profiles. Funct Integr Genomics 7:111–134

Crawford RMM (1989) Studies in plant survival. Blackwell Scientific, Oxford\London\Edinburgh\Boston\Palo Alto\Melbourne

Dajic Z (2006) Salt stress. In: Madhava Rao KV, Raghavendra AS, Janardhan Reddy K (eds) Physiology and molecular biology of stress tolerance in plants. Springer, Dordrecht, pp 41–99

Desbrosses GG, Kopka J, Udvardi MK (2005) Lotus japonicus metabolic profiling. Development of gas chromatography–mass spectrometry resources for the study of plant-microbe interactions. Plant Physiol 137:1302–1318

Flowers TJ, Hajibagheri MA (2001) Salinity tolerance in *Hordeum vulgare*: ion concentrations in root cells of cultivars differing in salt tolerance. Plant Soil 231:1–9

Foyer CH, Noctor G (2003) Redox sensing and signaling associated with reactive oxygen in chloroplasts, peroxisomes and mitochondria. Physiol Plant 119:355–364

Fraser PD, Enfissi EMA, Goodfellow M, Eguchi T, Bramley PM (2007) Metabolite profiling of plant carotenoids using the matrix-assisted laser desorption ionization time-of-flight mass spectrometry. Plant J 49:552–564

Gagneul D, Ainouche A, Duhaze C, Lugan R, Lahrer FR, Bouchereau A (2007) A reassessment of the function of the so-called compatible solutes in the halophytic Plumbginaceae Limonium latifolium. Plant Physiol 144:1598–1611

Gong QQ, Li PH, Ma SS, Indu Rupassara S, Bohnert HJ (2005) Salinity stress adaptation competence in the extremophile Thellungiella halophila in comparison with its relative Arabidopsis thaliana. Plant J 44:826–839

Gorham J (1990) Salt tolerance in *Triticeae*: K/Na discrimination in synthetic hexaploid wheats. J Exp Bot 41:623–627

Grant MR, Jones JDG (2009) Hormone (dis)harmony molds plant health and disease. Science 324:750–752

Hrabak EM, Chan CW, Gribskov M et al (2003) The Arabidopsis CDPK– SnRK superfamily of protein kinases. Plant Physiol 132:666–680

Jeschke WD, Hartung W (2000) Root-shoot interactions in mineral nutrition. Plant Soil 226:57–69

Kaplan F, Kopka J, Haskell DW, Zhao W, Schiller KC, Gatzke N, Sung DY, Guy CL (2004) Exploring the temperature stress metabolome of Arabidopsis. Plant Physiol 136:4159–4168

Kilian J, Whitehead D, Horak J, Wanke D, Weinl S (2007) The AtGenExpress global stress expression data set: protocols, evaluation and model data analysis of UV-B light, drought and cold stress responses. Plant J 50:347–363

Kim JK, Bamba T, Harada K, Fukusaki E, Kobayashi A (2007) Time-course metabolic profiling in Arabidopsis thaliana cell cultures after salt stress treatment. J Exp Bot 58:415–424

Kondo T, Yoshida K, Nakagawa A, Kawai T, Tamura H, Goto T (1992) Commelinin, a highly associated metalloanthocyanin present in the blue flower petals of *Commelina communis*. Nature 358:515–517

Kong X, Pan J, Zhang M, Xing X, Zhou Y, Liu Y, Li D (2011) ZmMKK4, a novel group C mitogen-activated protein kinase kinase in maize (Zea mays), confers salt and cold tolerance in transgenic Arabidopsis. Plant Cell Environ 34:1291–1303

Krasensky J, Jonak C (2012) Drought, salt, and temperature stress-induced metabolic rearrangements and regulatory networks. J Exp Bot. doi:10.1093/jxb/err460

Luan S (2009) The CBL–CIPK network in plant calcium signaling. Trends Plant Sci 14:37–42

Marrs KA (1996) The function and regulation of glutathione-S-transferase in plants. Ann Rev Plant Physiol Plant Mol Biol 47:127–158

Martin TFJ (1998) Phosphoinositide lipids as signaling molecules: common themes for signal transduction, cytoskeletal regulation, and membrane trafficking. Annu Rev Cell Biol 14:231–264

Mastrobuoni G, Irgang S, Pietzke M, Wenzel M, Assmus HE, Schulze WX, Kempa S (2012) Proteome dynamics and early salt stress response of the photosynthetic organism Chlamydomonas reinhardtii. BMCGenomics 13:215

Memelink J (2004) Tailoring the plant metabolome without a loose stitch. Trends Plant Sci 7:305–307

Mikami K, Katagiri T, Luchi S, Yamaguchi-Shinozaki K, Shinozaki K (1998) A gene encoding phosphatidylinositol 4-phosphate 5-kinase is induced by water stress and abscisic acid in *Arabidopsis thaliana*. Plant J 15:563–568

Mittler R (2002) Oxidative stress, antioxidants and stress tolerance. Trends Plant Sci 7:405–410

Munnik T, Irvine RF, Musgrave A (1998) Phospholipid signaling in plants. Biochim Biophy Acta 1389:222–272

Munns R (2002) Comparative physiology of salt and water stress. Plant Cell Environ 25:239–250

Munns R, Tester M (2008) Mechanisms of salinity tolerance. Ann Rev Plant Biol 59:651–681

Munns R, Tonnet M, Shennan C, Gardner A (1988) Effect of high external NaCl concentrations on ion transport within the shoot of *Lupinus albus* II. ions in phloem sap. Plan Cell Environ 11:291–300

Munns R, Cramer GR, Ball MC (1999) Interactions between rising CO_2, soil salinity and plant growth. In: Luo Y, Mooney HA (eds) Carbon dioxide and environmental stress. Academic, London, pp 139–167

Nascimento NC, Fett-Neto AG (2010) Plant secondary metabolism and challenges in modifying its operation: an overview. Methods Mol Biol 643:1–13

Park S, Fung P, Nishimura N et al (2009) Abscisic acid inhibits type 2C protein phosphatases via the PYR/PYL family of START proteins. Science 324:1068–1071

Pitzschke A, Forzani C, Hirt H (2006) Reactive oxygen species signaling in plants. Antioxi Redox Signal 8:1757–1764

Popp M (1995) Salt resistance in herbaceous halophytes and mangroves. Prog Bot 56:416–429

Posmyk MM, Kontek R, Janas KM (2009) Antioxidant enzymes activity and phenolic compounds content in red cabbage seedlings exposed to copper stress. Ecotoxicol Environ Saf 72:596–602

Ren ZH, Gao JP, Li LG, Cai XL, Huang W, Chao DY et al (2005) A rice quantitative trait locus for salt tolerance encodes a sodium transporter. Nat Genet 37:1141–1146

Sachan N, Rogers DT, Yun KY, Littleton JM, Falcone DL (2010) Reactive oxygen species regulate alkaloid metabolism in undifferentiated N. tabacum cells. Plant Cell Rep 29:437–448

Saito K, Hirai MY, Yonekura-Sakakibara K (2008) Decoding genes with coexpression networks and metabolomics – 'majority report by precogs'. Trends Plant Sci 13:36–43

Sanchez DH, Siahpoosh MR, Roessner U, Udvardi M, Kopka J (2008) Plant metabolomics reveals conserved and divergent metabolic responses to salinity. Physiol Planta 132:209–219

Sarwat M, Ahmad P, Nabi G, Hu X (2012) Ca^{2+} signals: the versatile decoders of environmental cues. Crit Rev Biotechnol, 9 May 2012. [Epub ahead of print]

Savirnata NM, Jukunen-Titto R, Oksanen E, Karjalainen RO (2010) Leaf phenolic compounds in red clover (*Trfolium Pratense L.*) induced by exposure to moderately elevated ozone. Environ Pollut 158:440–446

Sawada Y, Akiyama K, Sakata A, Kuwahara A, Otsuki H, Sakurai T, Saito K, Hirai MY (2009) Widely targeted metabolomics based on large-scale MS/MS data for elucidating metabolite accumulation patterns in plants. Plant Cell Physiol 50:37–47

Schafer H, Wink MM (2009) Edicinally important secondary metabolites in recombinant microorganisms or plants: progress in alkaloid biosynthesis. Biotechnol J 4:1684–1703

Schroeder JI, Kwak JM, Allen GJ (2001) Guard cell abscisic acid signaling and engineering of drought hardiness in plants. Nature 410:327–330

Shalata A, Mittova V, Volokita M, Guy M, Tal M (2001) Response of cultivated tomato and its wild salt-tolerant relative Lycopersicon pennelli to salt-dependent oxidative stress: the root antioxidative system. Physiol Planta 122:487–494

Shinozaki K, Yamaguchi-Shinozaki K (2007) Gene networks involved in drought stress response and tolerance. J Exp Bot 58:221–227

Singh K, Kumar S, Rani A, Gulati A, Ahuja PS (2009) Phenylalanine ammonia-lyase (PAL) and cinnamate 4-hydroxylase (C4H) and catechins (flavan-3-ols) accumulation in tea. Funct Integr Genomics 9:125–134

Sreevidya VS, Srinivasa RC, Rao C, Sullia SB, Ladha JK, Reddy PM (2006) Metabolic engineering of rice with soyabean isoflavone synthase for promoting nodulation gene expression in rhizobia. J Exp Bot 57:1957–1969

Takeda S, Matsuoka M (2008) Genetic approaches to crop improvement: responding to environmental and population changes. Nat Rev Genet 9:444–457

Tester M, Davenport R (2003) Na^+ tolerance and Na+transport in higher plants. Ann Bot 91:503–507

Torres MA, Dangl JL (2005) Functions of the respiratory burst oxidase in biotic interactions, abiotic stress and development. Curr Opin Plant Biol 8:397–403

Umezawa T, Fujita M, Fujita Y, Yamaguchi-Shinozaki K, Shinozaki K (2006) Engineering drought tolerance in plants: discovering and tailoring genes to unlock the future. Curr Opin Biotechnol 17:113–122

Vinocur B, Altman A (2005) Recent advances in engineering plant tolerance to abiotic stress: achievements and limitations. Curr Opin Biotechnol 16:1–10

Wang XQ, Ullah H, Jones AM, Assmann SM (2001) G protein regulation of ion channels and abscisic acid signaling in Arabidopsis guard cells. Science 292:2070–2072

Weston DJ, Gunter LE, Rogers A, Wullschlerger SD (2008) Connecting genes, coexpression modules, and molecular signatures to environmental stress phenotypes in plants. BMC Syst Biol 2:16

Xiong L, Zhu JK (2002) Molecular and genetic aspects of plant responses to osmotic stress. Plant Cell Environ 25:131–139

Zhang L, Xi D, Li S, Gao Z, Zhao S, Shi J, Wu C, Guo X (2011) A cotton group C MAP kinase gene, GhMPK2, positively regulates salt and drought tolerance in tobacco. Plant Mol Biol 77:17–31

Zuther E, Koehl K, Kopka J (2007) Comparative metabolome analysis of the salt response in breeding cultivars of rice. In: Jenks MA, Hasegawa PM, Jain SM (eds) Advances in molecular breeding toward drought and salt tolerant crops. Springer, Berlin\Heidelberg\New York, pp 285–315

Chapter 2
MicroRNAs and Their Role in Salt Stress Response in Plants

Satendra K. Mangrauthia, Surekha Agarwal, B. Sailaja, M. Sheshu Madhav, and S.R. Voleti

Abstract MicroRNAs (miRNAs) are a subset of endogenous approximate 22 nucleotide (nt) small non-coding regulatory RNA molecules that regulate gene expression post-transcriptionally, by mediating mRNA degradation or translational repression in a sequence specific manner. These small regulatory molecules are involved in regulating the intrinsic normal growth of cells and development of organisms as well as in maintaining the integrity of genomes. The plant miRNA research gained momentum, 2002 onwards, which was accompanied by the discovery of plant proteins involved in miRNA biogenesis. Early discovery of miRNAs has been implicated in the regulation of developmental processes. Since then much has been discovered about their involvement in plant responses to adverse environmental conditions, including abiotic stress. Various approaches of miRNAs discovery such as cloning, deep sequencing and prediction using bioinformatic tools have been adapted to learn more about the miRNA expression patterns during stress. The master regulators such as miRNAs having important role in salt stress response are very much crucial to understand the molecular regulation of stress adaptation. Many target genes of miRNAs encode transcription factors, each of which further regulates a set of downstream genes and affect physiological responses. This chapter contains a concise account on historical importance of miRNAs discovery. The miRNA biogenesis pathway and the associated proteins are also discussed along with the tools of miRNAs prediction and identification. In addition, the role of plant miRNAs and their target in plant metabolism and in particular salt stress is elaborated. With the growing knowledge on salt responsive miRNAs, the efforts to develop salt stress tolerance using miRNAs are also given.

Keywords miRNA • Biogenesis • Target genes • Deep sequencing • Regulation

---

S.K. Mangrauthia (✉) • S. Agarwal • B. Sailaja • M.S. Madhav • S.R. Voleti
Directorate of Rice Research, Hyderabad 500 030, Andhra Pradesh, India
e-mail: skmdrr@gmail.com

P. Ahmad et al. (eds.), *Salt Stress in Plants: Signalling, Omics and Adaptations*, DOI 10.1007/978-1-4614-6108-1_2, © Springer Science+Business Media New York 2013

Plant development, metabolism and stress responses greatly rely on the correct and timely regulation of gene expression that ultimately affect the plant metabolites such as osmolytes and antioxidants (Koyro et al. 2012). A large class of plant genes is regulated by stresses such as diseases, insects, drought, cold, heat and soil salinity. Salinity is defined as the concentration of dissolved mineral salts present in the soils and waters. The dissolved mineral salts consist of the electrolytes of cations and anions. The major cations in saline soil solutions consist of Na^+, Ca^{2+}, Mg^{2+} and K^+ and the major anions are Cl^-, SO_4^{2-}, HCO_3^-, CO_3^{2-} and NO_3^-. Other electrolytes contributing to soil and water salinity include B, Sr^{2+}, SiO_2, Mo, Ba^{2+} and Al^{3+} (Hu et al. 2007). Salinity is generally a problem in arid and semi-arid areas where evapo-transpiration exceeds annual precipitation, and in such areas supply of irrigation water is therefore essential to meet water needs of crop plants. The plants get affected due to salinity by two major ways: the capacity of roots to extract water is disturbed due to high concentrations of salts in the soil, and high concentrations of salts within the plant cells can be toxic. These effects can ultimately result in an inhibition of many biochemical, molecular, and physiological processes, for instance, nutrient uptake, assimilation, cell signaling pathways including those that lead to synthesis of osmotically active metabolites, specific proteins, and certain free radical scavenging enzymes that control ion and water flux and support scavenging of oxygen radicals or chaperones. Salinity stress often activates reactive oxygen species (ROS) detoxification forms an important defense against salt stress (Monteiro et al. 2011). Furthermore, salt stress may cause water stress, ion toxicity, nutritional disorders, oxidative stress, alteration of metabolic processes, membrane disorganization, reduction of cell division and expansion, genotoxicity, etc. (Zhu 2007).

The salt stress response is a quantitative trait, involving activation of a large number of specific genes concomitant to the repression in activity of a large number of housekeeping genes (Sahi et al. 2006). Major changes in transcriptional and post-transcriptional events have been noticed to supplement response of plants to salt stress. In order to understand the transcriptional profiling of genes whose expression levels alter in response to salt stress, different EST/cDNA collections have been utilized in plants (Shiozaki et al. 2005). The different studies on genomic and proteomic suggest that overall tolerance to high salt concentration is due to effectors that directly modulate stress etiology or attenuate stress effects and due to key regulatory molecules which are involved in stress perception, signal transduction, and modulation of the effectors' functions (Hasegawa et al. 2000; Sahi et al. 2003). Furthermore, comparative genomics approaches have been employed to understand the progress on the transcript changes in response to high salinity. Multiple mechanisms involved in salt stress perception and tolerance, with perhaps the most important switch being exercised at the transcription level, the importance of post-transcriptional gene regulation is accelerated with the discovery of small RNAs such as siRNA (small interfering RNA) and miRNA (micro RNA).

Post-transcriptional regulation of genes through small RNAs is termed as RNA silencing or RNA interference (RNAi). It is a versatile, complex gene regulation and defense mechanism known to be operative in most, if not all eukaryotic organisms (Tomari and Zamore 2005). RNA silencing operates through a set of core reactions that are triggered by double stranded RNA (dsRNA), which is processed into 21–24 nt small RNA duplexes by the RNase III enzyme Dicer and its homologues (Bernstein et al. 2001). These small RNAs, in turn, mediate multiple regulatory and defense functions in cells (Brodersen and Voinnet 2006). MicroRNAs and small interfering RNAs are two major types of these small RNAs. Although miRNAs and siRNAs are similar in size (20–24 nucleotides), their biogenesis and modes of action are markedly different. In plants, post transcriptional gene regulation involves miRNAs, generated by Dicer-like 1 (DCL1) from miRNA precursors that are transcribed from miRNA genes (Agarwal et al. 2011).

miRNAs are known to down-regulate plant gene expression at the post-transcriptional level mainly by annealing to reverse complementary sequences, resulting in breakage or translational suppression of the target mRNAs. In addition plant miRNAs direct the methylation of target chromosomal loci to regulate the gene expression at transcription level. Genetic and biochemical studies further enhancing our understanding of the biogenesis and function of miRNAs in plant gene expression and regulation. The plant miRNAs have been identified and characterized to demonstrate that miRNAs play essential roles in growth and development in different plants species (Mallory and Vaucheret 2006). Role of miRNAs in stress responses could be linked from the discovery that biotic and abiotic stress can modulate miRNA levels, together with the alteration in expression level of stress-associated genes as miRNA targets. Now with various genomic tools, it is established that several plant miRNAs play vital roles in plant resistance to stresses. In this chapter we have provided different aspects of miRNAs and their role in salt stress response. The historical importance, enzymatic machinery and biogenesis of plant miRNAs have been covered along with the various tools of miRNA identification.

Historical Perspective of miRNAs

On the basis of the structural similarity between miRNAs and siRNAs, 21–23 nt miRNAs originally described in *Caenorhabditis elegans* that had less complementarity to their target genes (Lee et al. 1993; Wightman et al. 1993), it was then shown that miRNAs could also cause RNA silencing (Olsen and Ambros 1999; Reinhart et al. 2000) and that siRNAs were co-opting this endogenous machinery for RNA silencing. MicroRNAs were discovered by Ambros and colleagues in 1993 during the study of the role of *lin-14* gene in *C. elegans* developmental biology. They observed that LIN-14 protein abundance was regulated by a short RNA product encoded by another gene lin-4. Therefore, lin-4 was the first member of the miRNA family of plants and animals which was identified in *C. elegans* through a genetic screening for defects in the temporal control of post-embryonic development. Most of the genes identified from mutagenesis screens were protein-coding, but lin-4 encoded a 22-nucleotide non-coding RNA that was partially complementary to seven conserved sites located in the 3′-untranslated region (UTR) of the lin-14 gene. At that time it was thought to be a rarity of gene regulation mechanism. Further, these genetic and biochemical interactions inspired a series of studies showing that the direct, but imperfect, base pairing between lin-4 and the lin-14 3′UTR was essential for the ability of lin-4 to control LIN-14 expression through the regulation of protein synthesis. Next in the year 2000, a second miRNA called *let-7* was characterized using forward genetics in *C. elegans*, which repressed *lin-41* and *lin-57* expression during developmental stage transitions in this worm. *let-7* also encoded a temporally regulated 21-nucleotide short RNA that controls the developmental transition from the larval stage into the adult stage. Both the miRNAs (lin4 and let7) were key regulatory molecules in the pathway of the temporal regulation of larval development in the nematode. The term short-temporal RNA (stRNA) was used when this class of small RNA in *C. elegans* was described in 1993 (Lee et al. 1993; Wightman et al. 1993), but the term microRNA (miRNA) was only coined in 2001. The identification of these miRNAs in *C. elegans* not only provided another vivid example of developmental regulation by small RNAs, but also raised the possibility that such tiny RNAs might be present in other animals and plants species.

The discovery of let-7 miRNA in *C. elegans* in 2000 paved the way for identification of number of miRNAs in both plants and animals. The first miRNA of plant origin was identified in 2002 by cloning approach in *Arabidopsis thaliana* (Llave et al. 2002; Park et al. 2002; Reinhart et al. 2002). Since then, many miRNAs and their target genes have been identified and characterized in various other plants species including rice, Arabidopsis and maize etc. Most miRNA sequences are conserved among different plant species. With the availability of the complete genome sequence of rice, 138 miRNAs representing 20 families were predicted based on sequence conservation with Arabidopsis by using computational methods (Jones-Rhoades and Bartel 2004; Park et al. 2002; Reinhart et al. 2002). Later, many novel miRNAs have been revealed in other plants by direct cloning, traditional sequencing and deep sequencing tools.

It is believed that the miRNA mediated gene regulation is an ancient mechanism of controlling the gene expression and it occurred prior to the emergence of multicellularity. This also suggests that miRNAs must have a common ancestor in evolution (Zhang et al. 2005). Interestingly, two Arabidopsis miRNAs were found conserved in all lineages of land plants, including bryophytes, lycopods, ferns and seed plants. These miRNAs are known to be capable of regulating genes in HD-Zip gene family (Floyd and Bowman 2004). As for as role of miRNAs in plant biology is concerned, initially it was shown that plant miRNAs are involved mainly in developmental regulation such as leaf development, auxin signaling, phase transition and flowering etc. Later it was reported that these miRNAs also play a critical role in biotic as well as abiotic stress response. As it is known that different abiotic stresses may lead to alteration of expression of similar set of genes in plants similarly, different kinds of stresses have also been found to activate responses in similar sets of miRNAs. The first abiotic stress related miRNAs in plants was reported by Sunkar and Zhu (2004). A library of small RNAs was constructed from Arabidopsis seedlings exposed to different abiotic stresses including cold, dehydration, high salt, and abscisic acid (ABA) to identify several new miRNAs that are responsive to abiotic stress. After that, several abiotic stress related miRNAs were identified in different plant species such as *Oryza sativa*, *Brassica napus*, *Glycine max*, *Medicago truncatula*, *Physcomitrella patens*, *Populus trichocarpa*, *Saccharum officinarum*, *Sorghum bicolor*, and *Zea mays* etc.

Enzymatic Machinery of miRNAs Pathway

Dicer

Dicer is an enzyme that cleaves long double stranded RNA (dsRNA) into short dsRNA and it is the key step in the process of RNAi mediated gene regulation. Dicer represents the class of RNAse III endonucleases and posses five major domains for its dicing activity: (1) RBD (RNA Binding Domain) domain recognizes dsRNA structure; (2) two RNase III domains involved in cleavage of dsRNA into short RNA of ~21–24 nt; (3) PAZ (Piwi, Argonaute, Zwile) domain which binds to 3′-2 nt overhangs of cleaved RNA substrate; (4) C terminal helicase domain required for the dsRNA processing. Different forms of dicer enzymes are involved in the generation of different class of small RNAs. Plants have multiple dicers for instance Arabidopsis has four Dicer like proteins viz., DCL 1, 2, 3 and 4. DCL1 is responsible for miRNA generation while DCL2, DCL3 and DCL4 are required for the siRNA generation. Based on the observation that pri-miRNA levels increase and pre-miRNA levels decrease in the weak *dcl1-9* insertion mutant, DCL1 is believed to be the major player to cleave pri-miRNA to pre-miRNA (Song et al. 2007). Indeed, a dicer enzyme from *A. thaliana,* AtDCL1 is required at several steps in the maturation of the Arabidopsis miR163, and similarly, OsDCL1 appears to play an essential role in miRNA biogenesis in *Oryza sativa* (Kurihara and Watanabe 2004; Liu et al. 2005). Further, six DCLs have been reported in rice which are involved in recruiting different classes of small RNAs in plant gene expression and regulation (Liu et al. 2007).

Dawdle (DDL)

DAWDLE (DDL), an forkhead-associated domain (FHA) containing protein of Arabidopsis is involved in the production of miRNAs and endogenous siRNAs. DDL is an RNA binding protein and is able to interact with (DCL1) and participates in miRNA biogenesis by facilitating DCL1 to access or recognize pri-miRNAs. Unlike mutants of genes known to participate in the processing of miRNA precursors, *ddl* mutants show reduced levels of pri-miRNAs as well as mature miRNAs but the promoter activity of MIR genes is not affected by *ddl* mutations. SNIP1, the human homolog of DDL, is involved in miRNA biogenesis and interacts with Drosha (Yu et al. 2008).

Hyponastic Leaves 1 (HYL1) and Serrate (SE)

Several studies suggest that the dsRNA-binding protein (dsRBP) HYL1 and a C2H2 zinc-finger protein SE are DCL1 cofactors. HYL1 is involved in miRNA but not in siRNA biogenesis. This dsRNA binding protein is a part of a macromolecular complex involved in miRNA maturation (Vazquez et al. 2004). HYL1 has a function in assisting the efficient and precise cleavage of pri-miRNA through interaction with DCL1. Genetic studies with Arabidospsis indicate that three proteins, the RNase III DCL1, HYL1, and SE, are required for the accurate processing of microRNA (miRNA) precursors in the plant cell nucleus (Dong et al. 2008). It has been shown that in both *hyl1* and *se* mutants the mature miRNA levels are low (Vazquez et al. 2004), and pri-miRNAs accumulate (Song et al. 2007). DCL1 and HYL1 recombinant proteins form a complex in vitro and HYL1 has been reported to interact with SE (Dong et al. 2008).

Hua Enhancer1 (HEN1)

HEN1 has a putative dsRNA-binding motif and a C-terminal methyltransferase domain. It adds a 2′-O-methyl group on the 3′-terminal nucleotide of plant miRNAs and siRNAs to protect them against degradation (Huang et al. 2009b). It has been demonstrated that the *hen1-1, hen1-2,* and *hen1-4* mutations that compromise miRNA metabolism are all in the methyltransferase region. Purified HEN1 protein is able to methylate the miRNA/miRNA* duplex in vitro. HEN1 is highly selective of its substrate and miRNA/miRNA* of different primary sequences can serve as substrates, suggesting that HEN1 recognizes the structure (rather than the sequence) of the duplex produced by Dicer processing of pre-miRNA (Yang et al. 2006).

Argonaute

Argonaute protein (Ago) is the most important component of RISC (RNAi induced silencing complex) which is involved in cleavage or suppression of target mRNA. The small RNAs act like a guide of RISC to reach target mRNA. Argonaute protein possesses two domains viz., PAZ and PIWI domains. In plants and animals, different Ago proteins are members of si and mi RISC. Ago-1 is a part of si-RISC whereas Ago-2 for mi-RISC. Many other Ago proteins have been reported in Arabidopsis until now, Ago-4 generates repeat associated siRNAs (rasi-RNAs) which participate in RNA-directed DNA methylation, Ago-6 has role in DNA methylation and transcriptional gene silencing while Ago-7 has been shown to generate trans acting siRNA (tasi-RNA) and long siRNAs (Naqvi et al. 2009).

Biogenesis of Plant miRNAs

Plant miRNAs, primarily found in genomic regions not associated with protein coding genes (Reinhart et al. 2002), are produced from their own transcriptional units. miRNAs encoded by endogenous MIR genes are ~20–24 nt double stranded small

RNAs (dsRNA) with 5′-phosphate and 3′-hydroxyl groups with 2-nt overhangs. MicroRNA biogenesis requires multiple steps in order to form mature miRNAs from miRNA genes (Kurihara and Watanabe 2004). In a first step, a miRNA gene is transcribed to a primary miRNA (pri-miRNA), which is usually a long sequence of more than several hundred nucleotides. The RNA polymerase II is responsible for transcribing most plant miRNAs (Xie et al. 2005). There are several observations to strengthen the involvement of RNA polymerase II in processing of miRNAs; such as plant pri-miRNAs can be more than 1 kb in length, they are usually preceded by typical TATA box motifs, and that they can undergo canonical splicing, polyadenylation, and capping. The pri-miRNA is processed within the nucleus by a multiprotein complex called the Microprocessor. The second step involves cleavage of the pri-miRNA to a stem loop intermediate called miRNA precursor or pre-miRNA. This step is controlled by the Dicer-like 1 enzyme (DCL1) in plants (Kurihara and Watanabe 2004). The fork head-associated domain protein DAWDLE (DDL) is required for pri-RNA accumulation and also it has been proposed to play an important role in miRNA biogenesis by recruiting predominantly DICER-like protein 1 (DCL1) to pri-miRNA for downstream processing (Yu et al. 2008). The DCL1 together with HYL1 (HYPONASTIC LEAVES 1) and the zinc-finger protein SE (SERRATE) were required for processing of pre-miRNA into miRNA duplex. Most of the Arabidopsis miRNAs are matured in subnuclear bodies by DCL1, while a few appear to be DCL4 dependent (Rajagopalan et al. 2006).

The 2-nt 3′overhang, characteristic of RNase III–mediated cleavage gets methylated by HEN1 (HUA ENHANCER 1), that is recognized by Exportin (in animals) homolog, HST (HASTY), however, there are some reservations about whether miRNA duplexes or mature miRNAs are exported in plants by HASTY to the cytoplasm (Park et al. 2005). This whole process yields a precursor miRNA (pre-miRNA) and ultimately a mature miRNA/miRNA* duplex. In the cytoplasm, miRNAs are unwound into single strand mature miRNAs by helicase. The miRNA strand with relatively lower stability of base-pairing at its 5′ end act as guide molecule to reach the target mRNA and is incorporated into a ribonucleoprotein complex RISC, whereas the other miRNA strand is typically degraded (Du and Zamore 2005). Once incorporated into RISC, the miRNA directs AGO1 (or AGO10) containing RISCs to its target mRNA for cleavage or translational repression on the basis of sequence complementarity. In cases of perfect or near-perfect complementarity to the miRNA, target mRNA scan be cleaved (sliced) and degraded; otherwise, their translation is repressed (Martinez and Tuschl 2004). Therefore, miRNAs control gene expression by regulating mRNA stability and translation (Eulalio et al. 2008).

Identification of miRNAs

MicroRNAs can be identified either by bioinformatics tools, a computational method or by using experimental approaches.

Bioinformatics Tools of miRNAs Prediction

Several researchers have utilized computational approaches to predict miRNAs based on conserved sequence characteristics across the species. These approaches rely on a combination of RNA secondary structure analyses and conservation of the miRNA sequences among related plant genomes. Based on the sequence conservation and structure homologies of the genome across different species, it has paved the way to discover miRNAs in different organisms by using various bioinformatic tools with the help of known miRNAs of model plants such as Arabidopsis and *O. sativa*. Every day the number of computational methods for the identification of plant miRNAs is increasing. As the plant miRNAs regulate their target mRNAs based on sequence complementarity, it is also possible to predict the target genes in plants. There are many bioinformatic tools available for target gene prediction. Computational tools are also available for the prediction of secondary structure of precursor miRNAs (pre-miRNA). These computer based programs are highly efficient in identifying the miRNAs in plant species with the help of the data of experimentally validated miRNAs. These programs trained or validated using miRNA sequence and targets. Table 2.1 listed the bioinformatics tools for miRNA and target predication and miRNA database.

miRNA Prediction Tools

There are numerous computational methods implemented for miRNA gene prediction based on sequence conservation and/or structural similarity. A program for identification of miRNAs, called MiRscan was developed with 70% specificity at a sensitivity of 50% (Lim et al. 2003). MiRscan program utilized various miRNA features with associated weights to build a bioinformatic tool, which assigns scores to hairpin candidates. Many other researchers employed homology searches for revealing paralog and ortholog miRNAs (Weber 2005). Additionally, Wang and others (Wang et al. 2005) designed a program based on sequence and structure alignment for miRNA prediction. Another program, ProMiR (Nam et al. 2005) is based on machine learning for miRNA discovery that uses a highly specific probabilistic model (HMM) whose topology and states are handcrafted based on prior understanding and assumptions, and accumulated data is used to derive exact probabilities. RNAmicro (Hertel and Stadler 2006) is another miRNA prediction tool developed by Hertel and Stadler is based on comparative sequence analysis instead of structural features. MiRank (Xue et al. 2005) is a novel ranking algorithm based on a random walk through a graph consisting of known miRNA examples and unknown candidate sequences. There are many such tools to predict miRNAs in different plant species based on the data obtained from model plants. Some of the important programs are listed in Table 2.1.

Table 2.1 List of bioinformatics tools for miRNAs prediction, identification and characterization

miRNA prediction tools	
MiRseeker	
MiRscan	http://genes.mit.edu/mirscan/
miRank	MiRank is programmed in Matlab
ProMiR II	http://cbit.snu.ac.kr/~ProMiR2/
PalGrade	
mir-abela	http://www.mirz.unibas.ch/cgi/pred_miRNA_genes.cgi
triplet-SVM	http://bioinfo.au.tsinghua.edu.cn/mirnasvm/
Vmir	http://www.hpi-hamburg.de/fileadmin/downloads/VMir.zip
RNA micro	http://www.bioinf.uni-leipzig.de/~jana/software/index.html
mirCoS	Based on LIBSVM library package
BayesMiRNAfind	https://bioinfo.wistar.upenn.edu/miRNA/miRNA/login.php
One-ClassMirnaFind	http://wotan.wistar.upenn.edu/OneClassmiRNA/
miRFinder	http://www.bioinformatics.org/mirfinder/
Mireval	http://tagc.univ-mrs.fr/mireval
Target prediction tools	
TargetScanS	http://genes.mit.edu/targetscan
miRanda	http://www.microma.org
PicTar	http://pictar.bio.nyu.edu
RNAhybrid	http://bibiserv.techfak.uni-bielefeld.de/rnahybrid
Diana-microT	http://www.diana.pcbi.upenn.edu/cgi-bin/micro_t.cgi
Target Boost	https://demo1.interagon.com/demo
Rna22	http://cbcsrv.watson.ibm.com/rna22_targets.html
MicroTar	http://tiger.dbs.nus.edu.sg/microtar/
NbmiRTar	http://wotan.wistar.upenn.edu/NBmiRTar
miRTour	http://bio2server.bioinfo.uni-plovdiv.bg/miRTour/
miRecords	http://mirecords.umn.edu/miRecords/
miRU	http://bioinfo3.noble.org/miRNA/miRU.htm
TAPIR	http://bioinformatics.psb.ugent.be/webtools/tapir
Target-align	http//www.leonxie.com/targetAlign.php
miTarget	http://cbit.snu.ac.kr/~miTarget
microRNA.org	http://www.microrna.org/microrna/home.do
mirWIP	http://146.189.76.171/query.php
MicroCosm Targets	http://www.ebi.ac.uk/enright-srv/microcosm/htdocs/targets/v5/
miRNA database	
MiRBase	http://microrna.sanger.ac.uk/
TarBase	http://diana.cslab.ece.ntua.gr/tarbase/
Argonaute	http://www.ma.uni-heidelberg.de/apps/zmf/argonaute/
miRecords	http://mirecords.umn.edu/miRecords/
'miRNAMap 2.0	http://mirnamap.mbc.nctu.edu.tw/
PMRD	http://bioinformatics.cau.edu.cn/PMRD
CSRDB	http://sundarlab.ucdavis.edu/smrnas/
deepBase	http://deepbase.sysu.edu.cn/
miRNA secondary structure prediction tools	
RNA mFold	http://rna.tbi.univie.ac.at/cgi-bin/RNAfold.cgi
miRNA deep sequencing tools	
mirTools	http://59.79.168.90/mirtools/

(continued)

Table 2.1 (continued)

miRDeep	http://www.mdc-berlin.de/en/research/research_teams/systems_biology_of_gene_regulatory_elements/projects/miRDeep/index.html
deepBase	http://deepbase.sysu.edu.cn/
miRExpress	http://mirexpress.mbc.nctu.edu.tw/

Databases of miRNA/Gene Targets

There are very useful databases in public or private domain that provide a significant amount of information on miRNA and target gene predictions. The most extensive online database for both miRNA and target sequences is miRBase that contains both miRNA mature sequences, hairpin sequences of precursors and associated annotation. Release 18.0 of the database contains 18,226 hairpin precursor miRNA, expressing 21,643 mature miRNA products, and 1,929 novel mature miRNAS in 168 species including plants. MiRBase provides miRNA sequence data, annotation, references in the miRBase::Sequences, provides a gene naming and nomenclature facility in the miRBase:: Registry and contains predicted miRNA target genes in miRBase::Targets. TarBase contains a set of experimentally supported targets in different species that are collected manually from the literature. TarBase version 5 includes more than 1,300 experimentally validated miRNA target interactions. The database has information about the target site described by the duplex of miRNA and gene. The database also provides information on the experiments that were executed to validate the target genes, the efficacy of the site to induce translational repression or degradation of mRNA, and a complete reference to the article for in depth understanding.

PMRD (plant microRNA database) database includes all publicly known plant miRNA sequences including those in miRBase. It contains sequence information, secondary structure, target genes, expression profiles and a genome browser. The information is available for 121 plant species including model plants and major crops such as Arabidopsis, rice, wheat, soybean, maize, sorghum, barley, etc. Also for some crops like Arabidopsis, rice, poplar, soybean, cotton, medicago and maize the target genes for miRNAs with a predicted interaction site is available in the database. CSRDB (Cereal Small RNA Database), a database consisting of large scale datasets of maize and rice of small RNA sequences generated by high-throughput pyrosequencing. Small RNA sequences have been mapped to the rice genome and to the available maize genome sequence using the Generic Genome Browser.

miRNA Target Prediction

Recently developed bioinformatics tools have integrated analysis of expression profiles of microRNA and mRNA in conjunction with the predicted microRNA targets in order to minimize false positives and to detect the functional microRNA targets under a specific biological condition. Most of such computational tools rely on the simple principle that inverse relationships in the expression profiles of miRNAs and mRNAs should be held between a specific miRNA and its mRNA targets.

It should be noted that these tools are only reliable for the mRNA targets which are regulated by miRNA mediated degradation, and not sensitive to the targets that are regulated by microRNA mediated translational inhibition.

miRU is a potential plant mRNA target finder. Using this database, potential targets of miRNAs can be predicted. The mature miRNA sequence of interest has to be submitted to search potential targets in cDNA libraries hosted on the miRU server. Its updated version is psRNATarget, a Plant small RNA target analysis server (http://plantgrn.noble.org/psRNATarget/). RNAhybrid, using this database miRNA target prediction can be done. The features are disallowing of G:U base pairs in the seed region at nucleotides 12–18. Minimum free energy can also be calculated by submitting the mature miRNA sequence. miRTour, a Plant miRNA and target prediction tool. This database is mainly used for the detection of plant miRNA and their targets from sequencing datasets (EST, GSS, SRA, etc.). This database automates all the steps of miRNA similarity search, miRNA precursor selection, target prediction and annotation. Each of them can be performed with the same set of input sequences.

TAPIR provides a new feature for predicting novel interactions called 'target mimicry' between miRNAs and their imitated targets. This server is mainly used for the prediction of plant microRNA targets, including target mimics. By using a precise algorithm we can search for the plant miRNA target. The precise option is much slower but guarantees to find less perfectly paired miRNA – target duplexes. In addition, the precise option allows the prediction of target mimics, which are characterized by a miRNA – target duplex having a large loop, making them undetectable by traditional tools. Target-align, an alignment tool designed mainly for accurate prediction of miRNA targets. A score matrix can be build based on the complementarity of nucleotides in order to trace out the optimal local alignments. Important parameters, such as maximum mismatches and maximum consecutive mismatches between miRNAs and their targets, were also used for filtering the optimal local alignments. Target-align can identify multi-target sites as well potential for non-cleaved targets sites. Target-align integrates numerous factors influencing miRNA–target interactions and allows users to set a variety of parameters including alignment and maximum scores, number of consecutive mismatches, base site restrictions and numbers of G:U wobbles and gaps to refine the predictions

Secondary Structure Prediction

There are many programs for predicting the secondary structure of RNA molecule. Using RNAmFold secondary structures of single stranded RNA can be predicted. It is currently packaged in the Vienna RNA website, a collection of tools for folding, designing and analyzing of RNA sequences. This also provides additional analysis of folding parts using the barriers program and structural RNA alignments. The bioinformatics package contains basic programs such as RNAFold for structure prediction of single sequences and also for folding the minimum free energy (MFE) secondary structure, RNAalifold for consensus miRNA structure prediction on a set of aligned sequences, RNAinverse for sequence design, RNAcofold and RNAup for RNA-RNA interaction analysis, LocARNA for the generation of structural alignment and barriers.

Experimental Tools of miRNAs Identification

MicroRNAs can be identified experimentally using five approaches: genetic screening (Lee et al. 1993; Wightman et al. 1993), direct cloning after isolation of small RNAs (Lu et al. 2005a), expressed sequence tags (ESTs) analysis (Zhang et al. 2005), hybridization based approaches and next generation deep sequencing tools.

Genetic Screening

It was the classical approach to discover miRNAs role in gene regulation. In the year 1993, it was found that a small RNA (22 nt) derived from an endogenous gene called *lin-4* suppressed the expression of *lin-14*, which controls the timing of *C. elegans* larval development (Lee et al. 1993; Wightman et al. 1993). Method for identifying miRNAs by genetic screening was similar to methods for identifying other traditional genes through mutational analysis. The application of this method is limited because it is expensive, time consuming, and dominated by chance.

Direct Cloning and Sequencing

To overcome the limitations of genetic screening, approach involving direct cloning after isolation and purification of small RNAs (Fu et al. 2005; Lu et al. 2005a) was discovered. This approach requires isolation of small RNA molecules by followed by ligation of small RNAs to RNA adapters at their 5′ and 3′ ends. Finally, they are reverse transcribed into cDNA using reverse transcription, which is then amplified and sequenced (Lu et al. 2005a). Direct cloning is a more effective method to obtain miRNAs than general genetic screening as only small RNAs are isolated and screened by this method. This strategy was applied to identify stress associated miRNA of Arabidopsis seedlings (Sunkar and Zhu 2004). The big challenge with this approach is to screen the small RNA sequences and distinguish miRNAs from siRNAs and many of the small RNAs involved in different metabolic processes. Novel small RNAs identified by cloning approaches need to be annotated as putative miRNAs using experimental and computational filters (Ambros et al. 2003).

Expressed Sequence Tag (EST) Analysis

miRNAs are evolutionarily conserved from species to species. This feature is extremely useful in prediction of homologous or orthologues of previously known miRNAs. 481 homologues of miRNAs were identified based on previously known miRNAs of Arabidopsis. However, this method can be used to identify only conserved miRNAs (Zhang et al. 2005). The approach based on EST sequences can be well utilized as an attractive and complementary method of miRNA identification.

This approach is particularly useful in the cases where genome sequences are not known. The miRNAs which are species specific and are non-conserved, it is impossible to find these genes based on EST approach.

Microarray-Based Approaches

Various laboratories established that miRNA microarrays succeeded in evaluating miRNA expression on a global scale and empowered the expression profiling of hundreds of miRNA genes together. miRNA arrays are now being developed to explore the biogenesis of miRNAs, tissue distribution, differential miRNA expression between treated and non-treated tissues of plants. The microarrays use a high-density probe either synthetic oligonucleotides or cDNA fragments as capture probes set that can cover nearly every nucleotide in a genome or portion of a genome (Kapranov et al. 2002; Yamada et al. 2003). An ideal probe should have high specificity and high affinity. The microarray approach of expression analysis offers the potential to identify novel transcripts, including miRNA precursors. This approach has been successfully used in Arabidopsis for identification of polyadenylated RNAs from unannotated regions of the genome (Yamada et al. 2003). Several probe and array designs have been described specifically for the detection of known miRNAs using microarrays. In addition to probes that are complementary to the sense and antisense strands of miRNAs, different control probes are also required. They include exogenous and endogenous positive controls and negative controls. The miRNA microarrays allow for the detection of specific sequences, and can be used to assess miRNA differential expression between different tissues, growth stages, treatments or genotypes (Axtell and Bartel 2005; Kapranov et al. 2002; Yamada et al. 2003) and to analyze the expression profile of the corresponding targets genes. MicroRNA arrays can identify the expression of several 100 genes in the same sample at once while requiring only small amounts of total RNA. However, it is unlikely that the current microarray technologies will suffice the requirement for de novo identification of miRNAs, because of low signal owing to the low expression level of most endogenous miRNAs.

The Next Generation Sequencing (NGS) Approaches

The power of NGS technologies for miRNA identification and characterization has been particularly remarkable. The advent of high throughput sequencing methodologies has provided unprecedented opportunities to generate comprehensive sequencing data for the identification and quantification of known and novel miRNAs. These technological leaps forward pose new challenges for the biological interpretation of large sequencing data sets. With the help of NGS tools, miRNA family members, precursors as well as miRNA modifications can be easily identified. This approach of miRNA discovery involves the application of massively parallel signature sequencing (MPSS) to miRNA sequencing. The whole process is completed in various steps such as size fractionation of total RNA to get small RNAs, adaptor ligation and reverse transcription, size selection and sequencing. These sequencing methods generates high coverage of sequence and sets of more than 350,000 individual sequences per tissue or treatment representing substantial advance over existing methods for the identification of these RNA molecules. The expression profiling of miRNA can be performed by assessing the relative abundance of distinct sequences in each library. The small RNA MPSS data revealed that among the most abundant small RNAs are many miRNAs. Using this approach 77 of 92 available Arabidopsis miRNAs in the miRNA database identified (Lu et al. 2005b). This study was the first claim of a parallel, high-throughput sequencing methods for the identification and characterization of miRNAs.

There are many new high throughput sequencing approaches and chemistries that have recently been developed, and these are likely to deliver novel means of miRNA discovery. Two new sequencing technologies were introduced based on sequencing by synthesis (SBS). The 454 Life Sciences (http://www.454.com/), using pyrosequencing technology enables high-throughput, parallel sequencing of hundreds of thousands of DNA or cDNA fragments (Margulies et al. 2005). One major advantage of this technology is the longer read length of about 1 Kb that allows the full-length sequencing of miRNAs and its precursor. The other system Solexa http://www.illumina.com website, detects fluorescence signals that promises to generate tens of millions of 25-nt tags, potentially offering the richest source of small RNA data. Both the technologies execute millions of sequencing reactions in parallel, producing data at ultrahigh rates. In the past year, Applied Biosystems has introduced their SOLiD sequencer http://www3.appliedbiosystems.com webcite, another short-read 20–35 bp platform, with read lengths anticipated to be 50 bp in the upcoming SOLiD3 release. Similarly, miRNAs can be identified using Illumina platform also that offers good coverage and reasonable good read length. These sequencing platforms offer a variety of experimental approaches for characterizing a transcriptome, including single-end and paired-end cDNA sequencing, tag profiling, methylation assays, secondary structure analysis, miRNA sequencing, alternate polyadenylation, fusion gene analysis and splice variant analyses. These technologies offer advantages in both cost and throughput of several orders of magnitude over traditional sequencing methods. This would presumably require modifications for the sequencing of small RNAs, but this would appear to be straightforward based on the previous success with MPSS. The deep sequencing technologies have the advantage over microarray based approaches that novel sequences can be detected while microarrays can only identify expression profile of known miRNAs. Further, next generation sequencing technologies are highly sensitive and dynamic range can be gained by the high sequencing depth. Next generation sequencing is independent of predesigned probes, thus making it very suitable for the discovery of new miRNAs. Nevertheless, deep sequencing is a relatively novel approach and the associated computational analysis tools are still in their infancy and need to be improved to standardize normalization, mapping and thresholding (Motameny et al. 2010).

Role of miRNAs in Plant Metabolism

MicroRNA are said to be involved in various metabolic processes in plants like growth and development, morphogenesis, flowering, stress responses (biotic and abiotic stresses) (Khraiwesh et al. 2012), signal transduction (Meng et al. 2010; Zhang et al. 2006) and in feedback regulation of genes (Meng et al. 2010; Yanga et al. 2007). The miRNAs have been identified in many plant species like Arabidopsis, brassica, rice, wheat, barley etc. In various experiments, it has been demonstrated that miRNAs regulate various plant development processes, such as leaf morphogenesis and polarity (Kim et al. 2005), floral differentiation (Chen 2004), root initiation (Guo et al. 2005), vascular development (Kim et al. 2005), and transition of plant growth from vegetative growth to reproductive growth (Lauter et al. 2005). A majority of these miRNAs regulate expression of transcription factors that influence cell fate determination and ultimately affect plant traits (Mallory et al. 2004; Rhoades et al. 2002). Interestingly, about half of the identified miRNA targets are transcription factors involved in regulation of key metabolic processes of plants (Bartel 2004). Of these miRNAs, a great number of miRNA families target genes encoding transcriptional factors that regulate plant development (Allen et al. 2004). A number of studies demonstrated that a majority of miRNAs regulate plant development by controlling the levels of transcription factors.

MicroRNAs regulate key components of hormone signaling pathways also and further regulate hormone homeostasis and related developmental processes (Guo et al. 2005; Mallory et al. 2005). Several miRNAs (miR159, miR160, miR164, and miR167) were characterized from plant tissues induced by abscisic acid (ABA), Gibberellic acid (GA), jasmonic acid (JA), salicylic acid (SA), and other phytohormones (Zhang et al. 2005). In addition to miR160, miR164, and miR167, other miRNAs, such as miR393, are probably involved in signaling pathways by regulating TIR1 which is an important component of a SCF E3 ubiquitin ligase that degrades Aux/IAA proteins in response to auxin (Gray et al. 2001; Mallory et al. 2005). This study is important as demonstrates that miRNA can also regulate F-box proteins and affect the activity of the E3 enzyme. In conclusion, studies indicate that miRNAs are involved in plant metabolism through hormone-mediated signal transduction.

MicroRNA not only regulates the expression of genes controlling plant development but also regulates its own biogenesis and/or function. As of new, at least five miRNAs (miR162, miR168, miR173, miR390, and miR398) are known to regulate miRNA biogenesis or function. The mRNA encoding DCL1 is the target of miR162 (Xie et al. 2003). Similarly, miR168 guided cleavage of AGO1 mRNA is crucial for the post-transcriptional gene regulation of the ago1 transcript (Vazquez et al. 2004). The alteration in the nucleotides of AGO1 mRNA to make it less complementary with miR168 resulted in increasing levels of AGO1 mRNA and consequently caused developmental defects (Vaucheret et al. 2004).

In rice, the deep sequencing of developmental rice grains revealed the involvement of 20 different miRNA families which are expressed in organ specific manner (Zhu et al. 2008). Another study in rice, applying Massively Parallel Signature

Sequencing (MPSS) and integrative analysis identified 26 novel miRNA's and 12 miRNA candidates in rice seeds of which most of the identified miRNA's shown tissue specific expression (Sunkar et al. 2005). In maize also miRNA are detected in the early stages of the developmental processes. Using Solexa technology in maize, 106 novel miRNA has been identified (Wang et al. 2011). miRNA are also said to be in the development of roots and root architecture (Boualem et al. 2008) where auxin pathways plays an important role as shown in rice and Arabidopsis (Meng et al. 2010). The study in rice auxin resistant mutant discovered miRNA signal transduction pathway and also showed the feedback circuit between miRNA and auxin response factors (ARF). 133 miRNA has proved to be expressed tissue specifically and root zone specifically in Arabidopsis (Breakfield et al. 2011). Three miRNA's viz. miR482, miR1512, miR1515 were established for their function in root nodulation as identified in soybean (Li et al. 2010). miRNAs are also involved in meristem identity. Mutants defective in CARPEL FACTORY or Hen1, fail to produce miRNA and thus resulting in several developmental defects (Jones 2002). Zhao et al. using high through put sequencing technology, identified expression differences in miRNA between inferior and superior spikelets in rice. This study revealed the involvement of 43 novel miRNA's which are accumulated differentially in inferior and superior spikelets that are slow grain filling and low grain weight in rice (Peng et al. 2011).

MicroRNAs Involved in Stress Responses

Plants are sessile organisms that must endure stressful environments. Recently, there has been strong evidence leading to the proposal that miRNAs are hypersensitive to abiotic or biotic stress as well as to diverse physiological processes (Lu et al. 2005a; Sunkar and Zhu 2004). Drought, cold, and salinity are major abiotic stresses for plants; all of these conditions strongly induced miR402 over-expression. Other miRNAs, such as miR319, are induced by either cold or other stress (Sunkar and Zhu 2004). The first direct report linking miRNA and stress tolerance was miR398, expression of which is transcriptionally down-regulated by oxidative stresses. In Arabidopsis, miR398 was found to target two closely related Cu/Zn superoxide dismutase coding genes, cytosolic CSD1 and chloroplastic CSD2, and a reduced level of miR398 led to improved tolerance of transgenic lines compared with the wild-type plants under oxidative stress conditions (Sunkar et al. 2006). In rice, miR169g was showed as the only member induced by drought stress among the miR169 family (Zhao et al. 2007). Furthermore, to determine the role of miRNAs in stress response, 21 miRNAs of Arabidopsis were predicted to be associated under UV-B stress condition (Zhou et al. 2007). In a study using EST analyses, 25.8% of ESTs containing miRNAs were found in stress-induced plant tissues (Zhang et al. 2005). At low-sulfur conditions, the ATP sulfurylase APS4 and the sulfate transporter AST68 are accumulated and both of these genes are regulated by miR395 (Allen et al. 2005; Jones-Rhoades and Bartel 2004). Lu et al. (2005a) identified 48 miRNA sequences from the Populus genome. Majority of these Populus miRNAs showed their target as developmental and stress/defense related genes.

A recent study has identified the involvement of miRNA in relation to submergence tolerance (Zhang et al. 2008). In Arabidopsis, miR399 over expression resulted in the down regulation of the target mRNA transcript, thus, deciphering the role of miRNA to cope up the mineral nutrition fluctuations (Fujii et al. 2005). The study on the dehydration stress on barley identified 28 new miRNA's belonging to 18 families of which five miRNA have experimentally validated for their differential expression (Melda et al. 2010). Similarly, 21 maize miRNA are identified in drought stress of which 13 are proved to be specific for drought stress (Zhang et al. 2010). Many studies discovered numerous rice miRNA involved in different abiotic stresses viz., salinity (Zhao et al. 2009), cold (Lv et al. 2010), heavy metal (Ding et al. 2011; Huang et al. 2009a,), hydrogen peroxide (Li et al. 2011b), ABA (Liu et al. 2009) and drought (Jian et al. 2010; Zhao et al. 2007; Zhou et al. 2010).

MicroRNAs Expression Under Salt Stress

Salt stress is one of the most serious abiotic stresses of crop plants worldwide (Monteiro et al. 2011). Plant salt stress is a condition where excessive salts in soil solution cause inhibition of plant growth or plant death. High salt stress leads to disruption of homeostasis in water potential and ion distribution which occurs at both the cellular and the whole plant levels. Drastic changes in ion and water homeostasis lead to molecular damage, growth arrest and even death of plants (Zhu 2001). Salt stress furthermore induces the ABA synthesis which closes stomata when transported to guard cells that ultimately decreases photosynthesis activity and leads to oxidative stress.

To tolerate the high salt stress in their sessile lifestyle, plants have evolved a considerable degree of developmental plasticity, including adaptation via networks of molecular events. Numerous genes and gene products in plants are affected due to salt stress (Zhu 2002). Large number of gene transcripts gets up or down regulated during salt stress suggesting the tight regulation of transcription during stressed condition in plants. Therefore, post-transcriptional gene regulation plays a crucial role in the plant salt response (Fig. 2.1). MicroRNAs are now known as ubiquitous regulators of gene expression in eukaryotic organisms. In plants, functional analysis has demonstrated that several miRNAs play vital roles in plant resistance to abiotic and biotic stress (Navarro et al. 2006; Sunkar and Zhu 2004). Various studies on Arabidopsis, rice and other plants have revealed an important role for miRNAs in drought and salt responses. Recently in Arabidopsis, several differentially regulated miRNAs have been identified in salt-stressed tissues. In response to salt stress, miR156, miR158, miR159, miR165, miR167, miR168, miR169, miR171, miR319, miR393, miR394, miR396, and miR397 were up-regulated, while the miR398 was down-regulated in Arabidopsis, thus establishing a role for miRNAs in the adaptive response to salt stress (Liu et al. 2008).

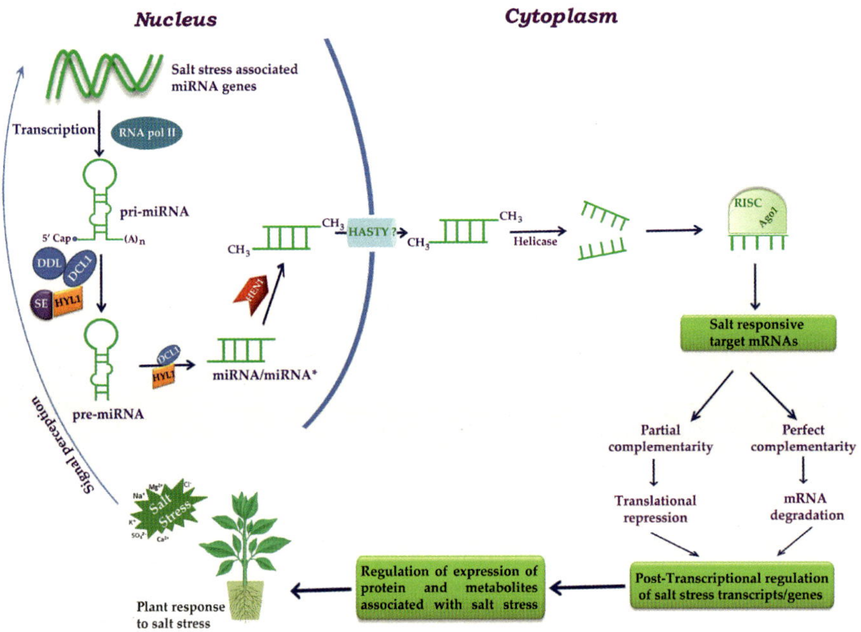

Fig. 2.1 A pathway showing miRNA mediated post-transcriptional regulation of salt stress responsive plant genes

Up-regulation of miRS1 and miR159.2 in response to salt stress was observed in *Phaseolous vulgaris*, (Arenas-Huertero et al. 2009). The expression of miR530a, miR1445, miR1446a-e, miR1447, and miR1711-n was increased, whereas miR482.2 and miR1450 was decreased during salt stress in *P. trichocarpa*, (Lu et al. 2008). Further, two members of miR169 family viz. miR169g and miR169n showed enhanced expression during salinity. Interestingly, a cis-acting ABA-responsive element (ABRE) was identified in the upstream region of miR169n, which gave an indication that miR169n may be regulated by stress hormone ABA. Both miR169g and miR169n showed its target mRNA as Nuclear factor Y subunit A (NF-YA) which was shown to be down-regulated in drought-affected leaves of wheat (Stephenson et al. 2007). Recently, the comparative study between salt tolerant maize genotype NC286 and salt sensitive maize genotypes Huangzao4 demonstrated that miR156, miR164, miR167, and miR396 families were down-regulated, while miR162, miR168, miR395, and miR474 families were up-regulated in salt-stressed maize roots. The study also proposed a gene model that regulates the abiotic stresses and gene networks coping with the stress (Ding et al. 2009). A microRNA miR393 is a conservative miRNA family present in a variety of different plant species. The two members of this miR393 family in rice are osa-MIR393 and osa-MIR393b. The expression level of osa-MIR393 altered, whereas that of osa-MIR393b did not alter under salinity and alkaline stress suggesting the precise regulation of salt associated genes. Soybean miRNAs associated with salt stress responses have been identified and analyzed by utilizing the next generation sequencing technology and bioinformatics tools. One hundred and thirty-three conserved miRNAs representing 95 miRNA families were differentially expressed in soybeans under different stress treatments along with 50 miRNAs differently expressed under salt stress (Li et al. 2011a). miR159 and 319 were up-regulated following saline treatment in artichoke tissues (De Paola et al. 2012). With the advancement of genomics tools and methods to identify novel miRNAs in various plant species, the number of miRNAs associated with salt stress response is increasing (Table 2.2). A better understanding of miRNA mediated gene regulation under salt stress will certainly help in elucidating the complex network of regulatory molecules, genes, proteins and metabolites.

Regulation of Target Genes of Salt Stress Associated miRNAs

Most miRNAs target multiple targets belonging to the same gene family in plants. Emerging data showed that miRNAs under specific conditions can selectively regulate the expression of specific target genes (Sunkar et al. 2012). Few times about target gene regulation via miRNAs detected in the maize salt stress response showed their target as many transcription factors (TFs) involved in plant development and organ formation which is in close agreement with previously reported data in model plants. The TFs Myb, NAC1 and homeodomain-leucine zipper protein (HD-ZIP) were predicted as the targets of zma-miR159a/b, zma-miR164a/b/c/d and zma-miR166l/m, respectively. The similar reports were made in arabidopsis and rice (Jones-Rhoades and Bartel 2004). Other transcription factors predicted as the miRNA targets include MADS-box proteins and zinc-finger proteins which have been reported as salt stress-responsive factors in plants (Fang et al. 2006; Xu et al. 2008). In addition to transcription factors, several miRNAs target genes that encode proteins in diverse metabolic pathways or are involved in various physiological processes of plants having important function. Many of the predicted targets of miRNAs, such as NADP-dependent malic enzyme (NADP-ME) and cytochrome oxidase, are known as salt stress-responsive plant genes (Cheng and Long 2007; Yan et al. 2005).

The targets sulfurylase and ASP1 genes are regulated by miR395 in salt induced soybean line under sulfate starvation conditions. Therefore the role of miR395 may be in non-specific salt stress responding pathways, such as the maintenance of energy supply (Ding et al. 2009; Jones-Rhoades and Bartel 2004). The targets F-box proteins and a basic-helix-loop-helix family protein induced during salt stress are regulated by miR393a (Jones-Rhoades and Bartel 2004). These stress responsive targets of miRNAs are regulated post-transcriptionally during stress responsive processes. Two putative targets for artichoke, cca-miR397 and 399 were homologous to members of laccase gene family, which has been demonstrated to be involved in salt stress response (De Paola et al. 2012). Laccases are

Table 2.2 List of reported miRNAs and their targets associated with salt stress in plants

S.No	miRNA	Plant species	Sequence	Predicted targets	References
1	miR156	*Arabidopsis thaliana*	UGACAGAAGAGAGUGAGCAC	Squamosa promoter-binding protein-like 11	Liu et al. (2008)
2	miR156	*Zea mays*	UGACAGAAGAGAGUGAGCAC	SBP-domain protein	Liu et al. (2008)
3	miR158	*Arabidopsis thaliana*	UCCCAAAUGUAGACAAAGCA	F-box family protein Pentatricopeptide (PPR) repeat-containing protein	Liu et al. (2008)
4	miR159	*Arabidopsis thaliana*	UUUGGAUUGAAGGGAGCUCUA	MYB and TCP transcription factors	Liu et al. (2008)
5	miR162	*Zea mays*	UCGAUAAACCUCUGCAUCCA	RNAseIII CAF protein Endoribonuclease Dicer Cytochrome P450	Ding et al. (2009)
6	miR164	*Zea mays*	UGGAGAAGCAGGGCACGUGCA	NAC domain protein NAC1	Ding et al. (2009)
7	miR165	*Arabidopsis thaliana*	UCGGACCAGGCUUCAUCCCCC	Class III HD-ZIP transcription factors	Liu et al. (2008)
8	miR166	*Zea mays*	UCGGACCAGGCUUCAUUCCCC	Homeo domain leucine Zipper protein (HD-ZIP)	Ding et al. (2009)
9	miR166b	*Glycine max*	UCGGACCAGGCUUCAUUCCCC	DNA binding protein Class III HD-Zip protein 4 and 8	Li et al. (2011b)
10	miR167	*Arabidopsis thaliana*	UGAAGCUGCCAGCAUGAUCUA	Auxin response factor 6 and 8	Liu et al. (2008)
11	miR167	*Zea mays*	UGAAGCUGCCAGCAUGAUCUA	Auxin response factor	Ding et al. (2009)
12	miR168	*Arabidopsis thaliana*	UCGCUUGGUGCAGGUCGGGAA	AGO1 (ARGONAUTE 1)	Liu et al. (2008)
13	miR168	*Zea mays*	UCGCUUGGUGCAGAUCGGGAC	PZE40 protein, Cytoplasmic aldolase, AGO1-1	Ding et al. (2009)
14	miR169	*Arabidopsis thaliana*	CAGCCAAGGAUGACUUGCCGA	CCAAT-binding transcription factor (CBF-B/NF-YA) family protein	Liu et al. (2008)
15	miR171	*Arabidopsis thaliana*	UGAUUGAGCCGCGCCAAUAUC	Scarecrow transcription factor family protein	Liu et al. (2008)
16	miR172	*Zea mays*	AGAAUCUUGAUGAUGCUGCA	Gamma-tubulin	Ding et al. (2009)
17	miR319	*Arabidopsis thaliana*	UUGGACUGAAGGGAGCUCCCU	TCP transcription factors	Liu et al. (2008)
18	miR393	*Arabidopsis thaliana*	UCCAAAGGGAUCGCAUUGAUCC	F-box protein; bHLH (basic helix–loop–helix) transcription factor	Liu et al. (2008)

19	miR393	*oryza sativa*	UCCAAAGGGAUCGCAUUGAUC	Phytosulfokine receptor precursor, Transport inhibitor response 1 protein Oxidoreductase,	
20	miR394	*Arabidopsis thaliana*	UUGGCAUUCUGUCCACCUCC	F-box family protein	Liu et al. (2008)
21	miR395	*Zea mays*	GUGAAGUGUUUGGGGGAACUC	ATP sulfurylase, L-Isoaspartyl methyltransferase, Beta-D-xylosidase, 8 NADP-dependent malic protein	Ding et al. (2009)
22	miR395	*Glycine max*	AUGAAGUGUUUGGGGGAACUC	Dehydration-responsive element binding protein 6 A7X2S6 sulfate transporter ATP sulfurylase Beta-glucosidase Disease resistance protein Glycoside hydrolase, family 17; Virulence factor, pectin lyase fold; Beta-glucosidase Phosphoadenylylsulfate reductase Dopamine beta-monooxygenase	Li et al. (2011b)
23	miR396	*Arabidopsis thaliana*	UUCCACAGCUUUCUUGAACUG	GRF2 transcription factor Rhodenase-like protein; Kinesin-like protein B	Liu et al. (2008)
24	miR396	*Zea mays*	UUCCACAGCUUUCUUGAACUG	Cytochrome oxidase subunit I	Ding et al. (2009)

(continued)

Table 2.2 (continued)

S.No	miRNA	Plant species	Sequence	Predicted targets	References
25	miR396c	*Oryza sativa*	UUCCACAGCUUUCUUGAACUU	heat shock 70 kDa protein 4, TBP-associated 59 kDa subunit protein, Leucine rich repeat family protein Ubiquitin-protein ligase COP1, At GRF 2 and 5 Growth regulating factor1 NBS-LRR type disease resistance protein IQ calmodulin-binding motif family protein Jasmonate O-methyltransferase	Gao et al. (2010)
26	miR397	*Arabidopsis thaliana*	UCAUUGAGUGCAGCGUUGAUG	LAC2 (laccases); ß-6 tubulin	Liu et al. (2008)
27	miR398	*Arabidopsis thaliana*	UGUGUUCUCAGGUCACCCCUU	InterPro domain Protein of unknown function DUF266, plant	Liu et al. (2008)
28	miR417	*Arabidopsis thaliana*	GAAGGUAGUGAAUUUGUUCGA	C2-domain containing protein SNF7family protein, contains Pfam domain Hydrolase Cell expansion protein RNA-directed RNA polymerase SNF domain/helicase domain protein Auxin response transcription factor	Jung and Kang (2007)
29	miR482.2	*Populus trichocarpa*	UCUUGCCUACUCCUCCCAUU	Disease resistance protein	
30	miR530a	*Populus trichocarpa*	UGCAUUUGCACCUGCACCUU	Zinc knuckle (CCHC-type) family protein Homeobox transcription factor KN3 Ribosomal protein L1 family protein	Lu et al. (2008)
31	miR1445	*Populus trichocarpa*	UCCCUUGUAGACUAGAAAAA	Dihydropyrimidinase	Lu et al. (2008)

32	miR1446a-e	*Populus trichocarpa*	UUCUGAACUCUCUCCCUCAA	GCN5-related N-acetyltransferase (GNAT) Family protein Gibberellin response modulator-like protein, Replication factor C-like Homeodomain transcription factor	Lu et al. (2008)
33	miR1447	*Populus trichocarpa*	CAGAAUUGCAGUGCCUUGAUU	Ankyrin repeat family protein, Beta-fructofuranosidase Oxidoreductase Leucine-rich repeat transmembrane protein kinase Disease resistance protein Translationally controlled tumor protein	Lu et al. (2008)
34	miR1450	*Populus trichocarpa*	UUCAAUGGCUCGGUCAGGUUAC	Leucine-rich repeat transmembrane protein kinase	Lu et al. (2008)
35	miR1507a	*Glycine max*	UCUCAUUCCAUACAUCGUCUG	Splicing factor yt521-B, NBS-LRR resistance protein RGH1 Cytosine-specific methyltransferase FAD linked oxidase, N-terminal	Li et al. (2011b)
36	miR 2001	*Oryza sativa*	CCCAGCUUGAGAAUCGGGCGGC	Hypothetical protein Protein GPR107 precursor	Jian et al. (2010)
37	miR2003	*Oryza sativa*	CCGGCCCCGAACCCGUCGGCU	HEAT repeat family protein, expressed Ribosomal protein S11 containing protein, Expressed protein BGGP Beta-1-3-galactosyl-O-glycosylglycoprotein, Hypothetical protein NAC domain-containing protein 90	Jian et al. (2010)

(continued)

Table 2.2 (continued)

S.No	miRNA	Plant species	Sequence	Predicted targets	References
38	miR2004	*Oryza sativa*	GACCGCAUAGCGCAGUGGAUU	EMB2745, Exonuclease, Lectin receptor-type protein kinase, FAD binding domain containing protein, Serine/threonine-protein kinase Hypothetical protein Oxidoreductase, expressed Peroxidase 52	Jian et al. (2010)
39	miR2006	*Oryza sativa*	GUGGCUGUAGUUUAGUGGUGA	Hypothetical protein Conserved hypothetical protein	Jian et al. (2010)

multicopper-containing glycoproteins, present in plants. It has been reported that the expression level of laccase genes is enhanced by high concentrations of NaCl in tomato, maize, and Arabidopsis roots (Cai et al. 2006; Liang et al. 2006; Wei et al. 2000). In artichoke, reduces expression of miR397a in roots during salt stress might possibly lead to enhanced expression of laccase (De Paola et al. 2012). A salt responsive target aspartic proteinase APA1 was predicted to be regulated by miRNA cca-novel-18. In Arabidopsis, a low decrease in the expression of target superoxide dismutase together with a slight increase in miR398 expression was observed under NaCl treatment (Attia et al. 2011). Another important target ARGONAUTE1 (AGO1) gene, which encodes the RNA slicer enzyme in the miRNA pathway, is regulated by miR168 (Wei et al. 2000). miR168 and AGO1 both are crucial in maintaining the balance between miRNAs and their target genes. The miR168 has also been found to be involved in salt stress in maize (Ding et al. 2009).

MicroRNAs Application in Development of Salt Stress Tolerant Plants

Understanding new RNA-guided stress regulatory networks should provide new way for the genetic improvement of plant stress tolerance. Indeed, it has been shown in few reports that manipulation of miRNA-guided gene regulation can help to engineer plants that will be more salt stress-tolerant. Differences in the expression level of miRNAs between inbred and hybrid lines of maize were studied. Under salt stress modest up-regulation of miR156 and miR166 in B73 and Mo17 lines was observed while an almost 2.5-fold up-regulation for miR156 and a 1.8-fold up-regulation for miR166 were observed in Mo17 × B73 hybrid. These results showed that the hybrid lines had drastic change in the miRNA expression and more flexible for salt stress when compared to the parents (Kong et al. 2010). Transgenic rice and *Arabidopsis thaliana* plants constitutively over-expressing osa-MIR396c showed reduced salt and alkali stress tolerance compared to that of wild-type plants (Gao et al. 2010). Similar efforts were made to overexpress osa-MIR393 in transgenic rice and Arabidopsis and it was shown that transgenics were sensitive to salt and alkali treatment as compare to wild-type plants. These results demonstrated that over-expression of osa-MIR393 can regulate rice salt-alkali stress tolerance in negative fashion (Gao et al. 2011). Transgenic Arabidopsis plants constitutively over-expressing miRNA417 showed that seed germination of the transgenic plants was retarded compared with the wild-type plants in the presence of high salt or ABA. These results also indicated that the miRNA417 plays a role of negative regulator of seed germination in Arabidopsis plants under salt stress conditions (Jung and Kang 2007). Recent study showed that specific downregulation of the bacterial-type PEPC (Phosphoenolpyruvate carboxylase) gene, Atppc4, by artificial microRNA improved salt tolerance in Arabidopsis. The improved salt tolerance may be related with the improved PEPC activity (Wang et al. 2012).

Conclusion and Future Perspective

Salinity is a significant problem affecting physiological, biochemical and molecular processes of plants and is predicted to become a larger problem in the coming decades. The detrimental effects of high salinity on plants can be observed at the whole-plant level or in the cellular level in terms of plant death and/or decrease in productivity. One way in which plants respond to salt stress is by modifying their gene expression through the post transcriptional gene regulation. MicroRNAs regulate the post-transcriptional expression of associated target genes involved in salt stress response. Hence, in addition to their roles in growth and development and maintenance of genome integrity, miRNAs are also important components in plant stress responses. Therefore, understanding how miRNAs regulate gene expression will enable researchers to explore the role of miRNAs in salt stress response in different plant species. Computational and experimental approaches have accelerated the efforts to discover large number of miRNAs and their targets in plants associated with various traits. The availability of complete genome information and next generation sequencing technologies has accelerated the efforts to understand the miRNA regulation during various abiotic and biotic stresses.

Identification of entire sets of miRNAs and their targets will lay the foundation that is needed to unravel the complex miRNA mediated regulatory networks controlling development and other physiological processes such as salt stress. The targets of plant miRNAs often belong to families of transcription factors involved in the control of genes associated with a particular trait. Given that miRNAs are crucial components in gene regulatory networks, it is certain that a complete understanding of the functions of miRNAs will greatly increase our understanding of plant tolerance to salt stress. Despite the existing voluminous data relating the miRNAs, a lot more remains to be known in terms of identification and characterization of the unknown miRNAs in diverse systems. Understanding miRNA-guided stress regulatory networks will provide new tools for the genetic improvement of salt tolerance in plants. Although this field is still in its infancy, the idea that miRNAs can be used in the therapy of plant stress is certain. If smart miRNAs can be used appropriately, a new avenue of biotechnology aimed at achieving enhanced salt tolerant plants will be opened. There are few reports showing that manipulation of miRNA guided gene regulation can help in the engineering of stress-resistant plants. Nevertheless, understanding of miRNA evolution is just at the starting point for elucidating their complex regulatory roles.

Studying stress-responsive miRNAs and their target gene expression in particular cell types will provide greater insights into miRNA target networks that operate in a cell- or tissue-specific manner during stress. Information generated from characterization of new small RNAs and the regulatory network of salt stress response needs to be scrutinized for development of tools to enhance plant tolerance to high salinity. It is a challenge to identify and characterize the catalogue of small RNAs that exhibit alteration in their expression level upon salt stress in various crop plants and to discover the target genes of those newly discovered miRNAs. Overcoming this challenge will allow rapid deciphering of new components in plant stress tolerance and lead to elucidation of the complex regulatory network of salt stress response. As our understanding of the roles of miRNAs during salt stress deepens, the possibilities for using miRNA-mediated gene regulation to enhance plant stress tolerance will become enormous.

Acknowledgments Authors thank Project Director (Directorate of Rice Research) for valuable support and encouragement. B. Sailaja acknowledges funding support and fellowship obtained from the NICRA (National Initiative on Climate Resilient Agriculture) project.

References

Agarwal S, Mohan M, Mangrauthia SK (2011) RNAi: machinery and role in pest and disease management. In: Bandi V, Shanker AK, Shanker C, Mandapaka M (eds) Crop stress and its management: perspectives and strategies. Springer, Netherlands, pp 447–469

Allen E, Xie Z, Gustafson AM, Sung GH, Spatafora JW, Carrington JC (2004) Evolution of microRNA genes by inverted duplication of target gene sequences in *Arabidopsis thaliana*. Nature Genet 36:1282–1290

Allen E, Xie Z, Gustafson AM, Carrington JC (2005) microRNA-directed phasing during trans-acting siRNA biogenesis in plants. Cell 121:207–221

Ambros V, Bartel B, Bartel DP, Burge CB, Carrington JC, Chen X, Dreyfuss G, Eddy SR, Griffiths-Jones S, Marshall M, Matzke M, Ruvkun G, Tuschl T (2003) A uniform system for microRNA annotation. RNA 9:277–279

Arenas-Huertero C, Pérez B, Rabanal F, Blanco-Melo D, De la Rosa C, Estrada-Navarrete G, Sanchez F, Covarrubias A, Reyes J (2009) Conserved and novel miRNAs in the legume *Phaseolus vulgaris* in response to stress. Plant Mol Biol 70:385–401

Attia H, Karray N, Msilini N, Lachaâl M (2011) Effect of salt stress on gene expression of superoxide dismutases and copper chaperone in *Arabidopsis thaliana*. Biologia Plantarum 55:159–163

Axtell MJ, Bartel DP (2005) Antiquity of microRNAs and their targets in land plants. Plant Cell 17:1658–1673

Bartel DP (2004) MicroRNAs: genomics, biogenesis, mechanism, and function. Cell 116:281–297

Bernstein E, Caudy AA, Hammond SM, Hannon GJ (2001) Role for a bidentate ribonuclease in the initiation step of RNA interference. Nature 409:363–366

Boualem A, Laporte P, Jovanovic M, Laffont C, Plet J, Combier JP, Niebel A, Crespi M, Frugier F (2008) MicroRNA166 controls root and nodule development in *Medicago truncatula*. Plant J 54(5):876–887

Breakfield NW, Corcoran DL, Petricka JJ, Shen J, Sae-Seaw J, Rubio-Somoza I, Weigel D, Ohler U, Benfey PH (2012) High-resolution experimental and computational profiling of tissue-specific known and novel miRNAs in Arabidopsis. Genome Res 22(1):163–176

Brodersen P, Voinnet O (2006) The diversity of RNA silencing pathways in plants. Trends Genet 22:268–280

Cai X, Davis EJ, Ballif J, Liang M, Bushman E, Haroldsen V, Torabinejad J, Wu Y (2006) Mutant identification and characterization of the laccase gene family in Arabidopsis. J Exp Bot 57:2563–2569

Chen X (2004) A microRNA as a translational repressor of APETALA2 in Arabidopsis flower development. Science 303:2022–2025

Cheng Y, Long M (2007) A cytosolic NADP-malic enzyme gene from rice (*Oryza sativa* L.) confers salt tolerance in transgenic Arabidopsis. Biotechnol Lett 29:1129–1134

De Paola D, Cattonaro F, Pignone D, Sonnante G (2012) The miRNAome of globe artichoke: conserved and novel micro RNAs and target analysis. BMC Genomics 13:41
Ding D, Zhang L, Wang H, Liu Z, Zhang Z, Zheng Y (2009) Differential expression of miRNAs in response to salt stress in maize roots. Ann Bot 103:29–38
Ding Y, Chen Z, Zhu C (2011) Microarray-based analysis of cadmium-responsive microRNAs in rice (*Oryza sativa*). J Exp Bot 62(10):3563–3573
Dong Z, Han MH, Fedoroff N (2008) The RNA-binding proteins HYL1 and SE promote accurate in vitro processing of pri-miRNA by DCL1. Proc Natl Acad Sci USA 105(29):9970–9975
Du T, Zamore PD (2005) microPrimer: the biogenesis and function of microRNA. Development 132:4645–4652
Eulalio A, Huntzinger E, Izaurralde E (2008) GW182 interaction with Argonaute is essential for miRNA-mediated translational repression and mRNA decay. Nat Struct Mol Biol 15:346–353
Fang Q, Xu Z, Song R (2006) Cloning, characterization and genetic engineering of FLC homolog in *Thellungiella halophila*. Biochem Biophys Res Commun 347:707–714
Floyd SK, Bowman JL (2004) Gene regulation: ancient microRNA target sequences in plants. Nature 428:485–486
Fu H, Tie Y, Xu C, Zhang Z, Zhu J, Shi Y, Jiang H, Sun Z, Zheng X (2005) Identification of human fetal liver miRNAs by a novel method. FEBS Lett 579:3849–3854
Fujii H, Chiou TJ, Lin SI, Aung K, Zhu JK (2005) A miRNA involved in phosphate-starvation response in Arabidopsis. Curr Biol 15(22):2038–2043
Gao P, Bai X, Yang L, Lv D, Li Y, Cai H, Ji W, Guo D, Zhu Y (2010) Over-expression of osa-MIR396c decreases salt and alkali stress tolerance. Planta 231:991–1001
Gao P, Bai X, Yang L, Lv D, Pan X, Li Y, Cai H, Ji W, Chen Q, Zhu Y (2011) osa-MIR393: a salinity- and alkaline stress-related microRNA gene. Mol Biol Rep 38:237–242
Gray WM, Kepinski S, Rouse D, Leyser O, Estelle M (2001) Auxin regulates SCF (TIR1)-dependent degradation of AUX/IAA proteins. Nature 414:271–276
Guo HS, Xie Q, Fei JF, Chua NH (2005) MicroRNA directs mRNA cleavage of the transcription factor NAC1 to downregulate auxin signals for Arabidopsis lateral root development. Plant Cell 17:1376–1386
Hasegawa PM, Bressan RA, Zhu JK, Bohnert HJ (2000) Plant cellular and molecular responses to high salinity. Annu Rev Plant Physiol Plant Mol Biol 51:463–499
Hertel J, Stadler PF (2006) Hairpins in a Haystack: recognizing microRNA precursors in comparative genomics data. Bioinformatics 22(14):e197–e202
Hu Y, Burucs Z, von Tucher S, Schmidhalter U (2007) Short-term effects of drought and salinity on mineral nutrient distribution along growing leaves of maize seedlings. Environ Exp Bot 60(2):268–275
Huang SQ, Peng J, Qiu CX, Yang ZM (2009a) Heavy metal-regulated new microRNAs from rice. J Inorg Biochem 103(2):282–287
Huang Y, Ji L, Huang Q, Vassylyev DG, Chen X, Ma JB (2009b) Structural insights into mechanisms of the small RNA methyltransferase HEN1. Nature 461:823–827
Jian X, Zhang L, Li G, Zhang L, Wang X, Cao X, Fang X, Chen F (2010) Identification of novel stress-regulated microRNAs from *Oryza sativa* L. Genomics 95(1):47–55
Jones L (2002) Revealing micro-RNAs in plants. Trends Plant Sci 7(11):473–475
Jones-Rhoades MW, Bartel DP (2004) Computational identification of plant micro- RNAs and their targets, including a stress-induced miRNA. Mol Cell 14:787–799
Jung HJ, Kang H (2007) Expression and functional analyses of microRNA417 in *Arabidopsis thaliana* under stress conditions. Plant Physiol Biochem 45:805–811
Kapranov P, Cawley SE, Drenkow J, Bekiranov S, Strausberg RL, Fodor SP, Gingeras TR (2002) Large-scale transcriptional activity in chromosomes 21 and 22. Science 296:916–919
Khraiwesh B, Zhu JK, Zhu J (2012) Role of miRNAs and siRNAs in biotic and abiotic stress responses of plants. Biochim Biophys Acta 1819(2):137–148
Kim J, Jung JH, Reyes JL, Kim YS, Kim SY, Chung KS, Kim JA, Lee M, Lee Y, Kim VN, Chua NH, Park CM (2005) microRNA directed cleavage of ATHB15 mRNA regulates vascular development in Arabidopsis inflorescence stems. Plant J 42:84–94

Kong YM, Elling AA, Chen B, Deng XW (2010) Differential expression of microRNAs in maize inbred and hybrid lines during salt and drought stress. Am J Plant Sci 1:69–76

Koyro HW, Ahmad P, Geissler N (2012) Abiotic stress responses in plants: an overview. In: Ahmad P, Prasad MNV (eds) Environmental adaptations and stress tolerance of plants in the era of climate change. doi: 10.1007/978-1-4614-0815-4_1

Kurihara Y, Watanabe Y (2004) Arabidopsis micro-RNA biogenesis through Dicer-like 1 protein functions. Proc Natl Acad Sci USA 101:12753–12758

Lauter N, Kampani A, Carlson S, Goebel M, Moose SP (2005) MicroRNA172 down-regulates glossy15 to promote vegetative phase change in maize. Proc Natl Acad Sci USA 102:9412–9417

Lee RC, Feinbaum RL, Ambros V (1993) The *C. elegans* heterochronic gene lin-4 encodes small RNAs with antisense complementarity to lin-14. Cell 75:843–854

Li H, Deng Y, Wu T, Subramanian S, Yu O (2010) Misexpression of miR482, miR1512, and miR1515 increases Soybean Nodulation. Plant Physiol 153:1759–1770

Li H, Dong Y, Yin H, Wang N, Yang J, Liu X, Wang Y, Wu J, Li X (2011a) Characterization of the stress associated microRNAs in Glycine max by deep sequencing. BMC Plant Biol 11:170

Li T, Li H, Zhang YX, Liu JY (2011b) Identification and analysis of seven H_2O_2-responsive miRNAs and 32 new miRNAs in the seedlings of rice (*Oryza sativa* L. ssp. indica). Nucleic Acids Res 39(7):2821–2833

Liang M, Haroldsen V, Cai X, Wu Y (2006) Expression of a putative laccase gene, ZmLAC1, in maize primary roots under stress. Plant Cell Environ 29:746–753

Lim LP, Glasner ME, Yekta S, Burge CB, Bartel DP (2003) Vertebrate microRNA genes. Science 299(5612):1540

Liu B, Li P, Li X, Liu C, Cao S, Chu C, Cao X (2005) Loss of function of OsDCL1 affects microRNA accumulation and causes developmental defects in rice. Plant Physiol 139:296–305

Liu B, Chen Z, Song X, Liu C, Cui X, Zhao X, Fang J, Xu W, Zhang H, Wang X, Chu C, Deng X, Xue Y, Cao X (2007) *Oryza sativa* Dicer-like4 reveals a key role for small interfering RNA silencing in plant development. Plant Cell 19:2705–2718

Liu HH, Tian X, Li YJ, Wu CA, Zheng CC (2008) Microarray-based analysis of stress-regulated microRNAs in *Arabidopsis thaliana*. RNA 14:836–843

Liu Q, Zhang YC, Wang CY, Luo YC, Huang QJ, Chen SY, Zhou H, Qu LH, Chen YQ (2009) Expression analysis of phytohormone-regulated microRNAs in rice, implying their regulation roles in plant hormone signaling. FEBS Lett 583(4):723–728

Llave C, Xie Z, Kasschau KD, Carrington JC (2002) Cleavage of Scarecrow-like mRNA targets directed by a class of Arabidopsis miRNA. Science 297:2053–2056

Lu C, Tej SS, Luo S, Haudenschild CD, Meyers BC, Green PJ (2005a) Elucidation of the small RNA component of the transcriptome. Science 309:1567–1569

Lu S, Sun YH, Shi R, Clark C, Li L, Chiang VL (2005b) Novel and mechanical stress-responsive microRNAs in *Populus trichocarpa* that are absent from Arabidopsis. Plant Cell 17:2186–2203

Lu SF, Sun YH, Chiang VL (2008) Stress-responsive microRNAs in Populus. Plant J 55:131–151

Lv DK, Bai X, Li Y, Ding XD, Ge Y, Cai H, Ji W, Wu N, Zhu YM (2010) Profiling of cold-stress-responsive miRNAs in rice by microarrays. Gene 459(1–2):39–47

Mallory AC, Vaucheret H (2006) Functions of microRNAs and related small RNAs in plants. Nat Genet 38(Suppl):S31–S36

Mallory AC, Dugas DV, Bartel DP, Bartel B (2004) MicroRNA regulation of NAC-domain targets is required for proper formation and separation of adjacent embryonic, vegetative, and floral organs. Curr Biol 14:1035–1046

Mallory AC, Bartel DP, Bartel B (2005) MicroRNA-directed regulation of Arabidopsis AUXIN RESPONSE FACTOR17 is essential for proper development and modulates expression of early auxin response genes. Plant Cell 17:1360–1375

Margulies M, Egholm M, Altman WE, Attiya S, Bader JS, Bemben LA, Berka J, Braverman MS, Chen YJ, Chen Z (2005) Genome sequencing in microfabricated high-density picolitre reactors. Nature 437:376–380

Martinez J, Tuschl T (2004) RISC is a 5′ phosphomonoester producing RNA endonuclease. Genes Dev 18:975–980

Melda K, Unver T, Budak H (2010) Regulation of barley miRNAs upon dehydration stress correlated with target gene expression. Funct Integr Genomics 10:493–507
Meng Y, Ma X, Chen D, Ping W, Chen M (2010) MicroRNA-mediated signaling involved in plant root development. Biochem Biophys Res Commun 393:345–349
Monteiro CC, Carvalho RF, Gratão PL, Carvalho G, Tezotto T, Medici LO, Peres LEP, Azevedo RA (2011) Biochemical responses of the ethylene-insensitive never ripe tomato mutant subjected to cadmium and sodium stresses. Environ Exp Bot 71:306–320
Motameny S, Wolters S, Nürnberg P, Schumacher B (2010) Next generation sequencing of miRNAs – strategies, resources and methods. Genes 1:70–84
Nam JW, Shin KR, Han J, Lee Y, Kim VN, Zhang BT (2005) Human microRNA prediction through a probabilistic co-learning model of sequence and structure. Nucleic Acids Res 33(11):3570–3581
Naqvi AR, Islam MN, Choudhury NR, Haq QMR (2009) The fascinating world of RNA interference. Int J Biol Sci 5(2):97–117
Navarro L, Dunoyer P, Jay F (2006) A plant miRNA contributes to antibacterial resistance by repressing auxin signaling. Science 312:436–439
Olsen PH, Ambros V (1999) The lin-4 regulatory RNA controls developmental timing in *Caenorhabditis elegans* by blocking LIN-14 protein synthesis after the initiation of translation. Dev Biol 216:671–680
Park W, Li J, Song R, Messing J, Chen X (2002) CARPEL FACTORY, a Dicer homolog, and HEN1, a novel protein, act in microRNA metabolism in *Arabidopsis thaliana*. Curr Biol 12:1484–1495
Park MY, Wu G, Gonzalez-Sulser A, Vaucheret H, Poethig RS (2005) Nuclear processing and export of microRNAs in Arabidopsis. Proc Natl Acad Sci USA 102:3691–3696
Peng T, Lv Q, Zhang J, Li J, Du Y, Zhao Q (2011) Differential expression of the microRNAs in superior and inferior spikelets in rice (*Oryza sativa*). J Exp Bot 62(14):4943–4954
Rajagopalan R, Vaucheret H, Trejo J, Bartel DP (2006) A diverse and evolutionarily fluid set of microRNAs in *Arabidopsis thaliana*. Genes Dev 20:3407–3425
Reinhart BJ, Slack FJ, Basson M, Pasquinelli AE, Bettinger JC (2000) The 21- nucleotide let-7 RNA regulates developmental timing in *Caenorhabditis elegans*. Nature 403:901–906
Reinhart BJ, Weinstein EG, Rhoades MW, Bartel B, Bartel DP (2002) MicroRNAs in plants. Genes Dev 16:1616–1626
Rhoades MW, Reinhart BJ, Lim LP, Burge CB, Bartel B, Bartel DP (2002) Prediction of plant microRNA targets. Cell 110:513–520
Sahi C, Agarwal M, Reddy MK, Sopory SK, Grover A (2003) Isolation and expression analysis of salt stress-associated ESTs from contrasting rice cultivars using a PCR-based subtraction method. Theor Appl Genet 106:620–628
Sahi C, Singh A, Blumwald E, Grover A (2006) Beyond osmolytes and transporters: novel plant salt stress tolerance-related genes from transcriptional profiling data. Physiol Plant 127:1–9
Shiozaki N, Yamada M, Yoshiba Y (2005) Analysis of salt-stressinducible ESTs isolated by PCR-subtraction in salt-tolerant rice. Theor Appl Genet 110:1177–1186
Song L, Han MH, Lesicka J, Fedoroff N (2007) Arabidopsis primary microRNA processing proteins HYL1 and DCL1 define a nuclear body distinct from the Cajal body. Proc Natl Acad Sci USA 104:5437–5442
Stephenson T, McIntyre C, Collet C, Xue GP (2007) Genome-wide identification and expression analysis of the NF-Y family of transcription factors in *Triticum aestivum*. Plant Mol Biol 65:77–92
Sunkar R, Zhu JK (2004) Novel and stress-regulated microRNAs and other small RNAs from Arabidopsis. Plant Cell 16:2001–2019
Sunkar R, Girke T, Jain PK, Zhu JK (2005) Cloning and characterization of microRNAs from rice. Plant Cell 17:1397–1411
Sunkar R, Kapoor A, Zhu JK (2006) Posttranscriptional induction of two Cu/Zn superoxide dismutase genes in Arabidopsis is mediated by downregulation of miR398 and important for oxidative stress tolerance. Plant Cell 18:2051–2065
Sunkar R, Li YF, Jagadeeswaran G (2012) Functions of microRNAs in plant stress responses. Trends Plant Sci 17(4):196–203

Tomari Y, Zamore PD (2005) Perspective: machines for RNAi. Genes Dev 19:517–529
Vaucheret H, Vazquez F, Crete P, Bartel DP (2004) The action of ARGONAUTE1 in the miRNA pathway and its regulation by the miRNA pathway are crucial for plant development. Genes Dev 18:1187–1197
Vazquez F, Gasciolli V, Crete P, Vaucheret H (2004) The nuclear dsRNA binding protein HYL1 is required for microRNA accumulation and plant development, but not posttranscriptional transgene silencing. Curr Biol 14:346–351
Wang X, Zhang J, Li F, Gu J, He T, Zhang X, Li Y (2005) MicroRNA identification based on sequence and structure alignment. Bioinformatics 21(18):3610–3614
Wang L, Liu H, Li D, Chen H (2011) Identification and characterization of maize microRNAs involved in the very early stage of seed germination. BMC Genomics 12:154
Wang F, Liu R, Wu G, Lang C, Chen J, Shi C (2012) Specific downregulation of the bacterial-type PEPC gene by artificial microRNA improves salt tolerance in Arabidopsis. Plant Mol Biol Rep. doi:10.1007/s11105-012-0418-6
Weber MJ (2005) New human and mouse microRNA genes found by homology search. FEBS J 272(1):59–73
Wei JZ, Tirajoh A, Effendy J, Plant AL (2000) Characterization of salt-induced changes in gene expression in tomato (*Lycopersicon esculentum*) roots and the role played by abscisic acid. Plant Sci 159:135–148
Wightman B, Ha I, Ruvkun G (1993) Posttranscriptional regulation of the heterochronic gene lin-14 by lin-4 mediates temporal pattern formation in *C. elegans*. Cell 75:855–862
Xie Z, Kasschau KD, Carrington JC (2003) Negative feedback regulation of Dicer-Like1 in Arabidopsis by microRNA-guided mRNA degradation. Curr Biol 13:784–789
Xie Z, Allen E, Fahlgren N, Calamar A, Givan SA, Carrington JC (2005) Expression of Arabidopsis MIRNA genes. Plant Physiol 138:2145–2154
Xu DQ, Huang J, Guo SQ (2008) Overexpression of a TFIIIA-type zinc finger protein gene ZFP252 enhances drought and salt tolerance in rice (*Oryza sativa* L.). FEBS Lett 582:1037–1043
Xue C, Li F, Tao H, Liu GP, Li Y, Zhang X (2005) Classification of real and pseudo microRNA precursors using local structure sequence features and support vector machine. BMC Bioinformatics 6(1):310
Yamada K, Lim J, Dale JM, Chen H, Shinn P, Palm CJ, Southwick AM, Wu HC, Kim C, Nguyen M (2003) Empirical analysis of transcriptional activity in the Arabidopsis genome. Science 302:842–846
Yan S, Tang Z, Su W, Sun W (2005) Proteomic analysis of salt stress responsive proteins in rice root. Proteomics 5:235–244
Yang Z, Yon WE, Yu B, Chen X (2006) HEN1 recognizes 21–24 nt small RNA duplexes and deposits a methyl group onto the 2′ OH of the 3′ terminal nucleotide. Nucleic Acids Res 34:667–675
Yanga T, Xuea L, An L (2007) Functional diversity of miRNA in plants. Plant Sci 172(3):423–432
Yu B, Bi L, Zheng B, Ji L, Chevalier D, Agarwal M, Ramachandran V, Li W, Lagrange T, John CW, Chen X (2008) The FHA domain proteins DAWDLE in Arabidopsis and SNIP1 in humans act in small RNA biogenesis. Proc Natl Acad Sci USA 105(29):10073–10078
Zhang BH, Pan XP, Wang QL, Cobb GP, Anderson TA (2005) Identification and characterization of new plant microRNAs using EST analysis. Cell Res 15:336–360
Zhang B, Pan X, Wang Q, Cobb GP, Anderson TA (2006) Computational identification of microRNAs and their targets. Comput Biol Chem 30(6):395–407
Zhang Z, Wei L, Zou X, Tao Y, Liu Z, Zheng Y (2008) Submergence-responsive microRNAs are potentially involved in the regulation of morphological and metabolic adaptations in maize root cells. Ann Bot 102(4):509–519
Zhang Z, Yu J, Li D, Zhang Z, Liu F, Zhou X, Wang T, Ling Y, Su Z (2010) PMRD: plant microRNA database. Nucleic Acids Res 38:D806–D813
Zhao B, Liang R, Ge L, Li W, Xiao H, Lin H, Ruan K, Jin Y (2007) Identification of drought-induced microRNAs in rice. Biochem Biophys Res Commun 354:585–590

Zhao B, Ge L, Liang R, Li W, Ruan K, Lin H, Jin Y (2009) Members of miR-169 family are induced by high salinity and transiently inhibit the NF-YA transcription factor. BMC Mol Biol 10:29

Zhou X, Wang G, Zhang W (2007) UV-B responsive microRNA genes in *Arabidopsis thaliana*. Mol Syst Biol 3:103

Zhou L, Liu Y, Liu Z, Kong D, Duan M, Luo L (2010) Genome-wide identification and analysis of drought-responsive microRNAs in *Oryza sativa*. J Exp Bot 61(15):4157–4168

Zhu JK (2001) Plant salt tolerance. Trends Plant Sci 6(2):66–71

Zhu JK (2002) Salt and drought stress signal transduction in plants. Annu Rev Plant Biol 53:247–273

Zhu JK (2007) Plant salt stress. Encyclopedia of life sciences. DOI: 10.1002/9780470015902.a0001300.pub2

Zhu QH, Andrew S, Matthew L, Fan L, Kennedy G, Gubler F, Helliwell C (2008) A diverse set of microRNAs and microRNA-like small RNAs in developing rice grains. Genome Res 18(9):1456–1465

Chapter 3
Unravelling Salt Stress in Plants Through Proteomics

Khalid Rehman Hakeem, Ruby Chandna, Reiaz ul Rehman, Inayatullah Tahir, Muhammad Sabir, and Muhammad Iqbal

Abstract Plants like other organisms are mostly under the threat of various stresses (both biotic as well as abiotic). Being sessile, plants lack the mechanisms to flee from these unfavourable situations. The development of exclusive and complicated responses to these environmental stresses in plants has fostered through evolution. Such alterations can influence plant growth, production and productivity in agriculture, plant nutritional potential and metabolic profile. Hence, plant abiotic stress has always been a matter of concern for the world economy and maintenance of human life on earth. Salinity stress, being one of the main abiotic stresses, may bring the morphological, anatomical, and physiological changes in plants. Distributed in both irrigated and non-irrigated areas of the world, around 6% of the world's total land area is affected by salt stress. So, it is a major concern to adopt the strategies against this great challenge by unravelling the mechanisms to overcome salt stress. In order to meet the challenges for biotechnological improvement of crop productivity; various steps involving genes, transcripts, proteins and metabolites, controlling the

K.R. Hakeem (✉)
Faculty of Forestry, Universiti Putra Malaysia, Serdang-43400, Selangor, Malaysia

P.G Programme in Bioresources, Department of Botany, University of Kashmir, Srinagar, India
e-mail: kur.hakeem@gmail.com; kur.hakeem@putra.upm.edu.my

R. Chandna
National Institute for Plant Genome Research, Aruna Asif Ali Road, New Delhi 110067, India

Department of Botany, Jamia Hamdard, New Delhi 110062, India

R. ul Rehman • I. Tahir
P.G Programme in Bioresources, Department of Botany, University of Kashmir, Srinagar, India

M. Sabir
Institute of Soil and Environmental Science, University of Agriculture, Faisalabad, Pakistan

M. Iqbal
Department of Botany, Jamia Hamdard, New Delhi 110062, India

P. Ahmad et al. (eds.), *Salt Stress in Plants: Signalling, Omics and Adaptations*,
DOI 10.1007/978-1-4614-6108-1_3,

stress resistance and/or architecture of crop plants in a wide array of environments needed to be recognized. Proteomics, the protein complement of genome, these days is one of the leading branches of research which enables the large-scale scanning of various substances, and offers great potential for post-genomics to elucidate the genotype-phenotype connections. The present chapter is an account of current knowledge in this regard. It focuses on effects of salt stress unrevealed by proteomics tools. It comprises information on recent advances in proteomics, which could be a new opportunity to comprehend abiotic responses and categorize genes responsible for significant crop traits.

Keywords Salt stress • Plant responses • Proteome • Metabolome • Tolerance

Abiotic stresses greatly affect the growth and development of crop plants. Among these stresses, salt stress is affecting world's over 6% of land area. As per the FAO report (2005), salt affected areas of the world accounts for around 800 million hectares (ha). The reason for such effect is either by salinity (397 million ha) or the associated condition of sodicity (434 million ha). Salinity of soil continues to be one of the world's most severe environmental predicaments in agriculture which limits agricultural production throughout the world. Salt stress induces various physiological and biochemical reactions in plants and influences almost all plant processes (Nemoto and Sasakuma 2002). Salinity stress can affect plant survival, biomass, plant height and plant form, where such changes in morphology alters the ability of a plant to collect light, water and nutrients (Locy et al. 1996).

Physiology, molecular genetics and functional genomics of salt stress studies have explored the molecular and physiological knowledge about salt stress tolerance. These approaches have also helped in encoding, cloning and characterizing important genes and proteins for osmolyte synthesis, ion channels, signalling factors and salt-responsive enzymes, thus, revealing the basic functions of the genes/proteins in plants' response and adaptation to salinity (Tuteja 2007). Besides this, high-throughput transcriptomics studies have also enabled to provide insight of gene expression at the mRNA level. However, occurrence of post-transcriptional and post-translational events such as phosphorylation and glycosylation, mRNA expression does not correlate with the levels of protein expressed, which are directly related to signalling and metabolic processes under salt stress conditions (Zhang et al. 2008). Thus, it is essential to understand salt stress response at the protein level. With the advancement in post genomic era, proteomics technologies have been coming up with global protein expression analysis. Recent research has been more focused on the salt-responsive proteomes in plants, since they yield more information in understanding the complex mechanisms of plant salt response and tolerance (Jiang et al. 2007). In the present chapter, we have integrated the salt-response by plants based on these proteomics studies. The integration of various 'omics' technologies and their breakthrough in understanding the plant processes during stress have also been discussed. This effort might provide a background for further investigation of salt response/tolerance networks and ultimate rational engineering of plants for enhanced stress tolerance.

Salt Stress in Plants

Soil degradation is largely associated with soil salinity. High conc. of salt in soil influences approximately 30% of the agricultural land in the world. The adverse effects of salinity on plant growth include the nutritional imbalance, low osmotic potential of the soil solution (water stress), specific ion effects (salt stress), or a combination of these factors (Munns and Tester 2008). Evelin et al. (2009) reported that soil salinity influences almost all the important physiological processes including growth, photosynthesis, protein and lipid metabolism. Munns and Tester (2008) explained the steps to cause the reduction of growth by salinity stress. They hypothesized that under saline conditions, nutritional disorders result from the effect of salinity on nutrient availability which promotes competitive uptake and transport, or partitioning within the plant and causes reduction in crop growth.

Salt stress generally disrupts the mineral nutrient acquisition in plants by two ways. Firstly, the ionic strength of the substrate can persuade nutrient uptake and translocation. Alternatively, by introducing competitive environment with major ions viz., Na^+ and Cl^-, in the substrate, salinity disrupts the mineral relations of plants. Such interactions often lead to Cl^- inhibition of NO_3^- absorption and Na^+-induced K^+ deficiency (Peuke and Jeschke 1999), which eventually generates imbalances and nutritional disorders and finally may confront the structure and composition of plant cells (Xu et al. 2010), injure macromolecules viz., chlorophyll, resulting in a loss of photosynthetic activity and consequently leading to senescence of the leaves (De Michele et al. 2009); and cause suppression of growth plants (Garg and Manchanda 2009).

Proteomics: An Evolution in Technology

The advancement of science is enabled by technologies; such process has been repeated many times throughout the history of science. One of these recent enabling technologies is proteomics. Proteomics can be applied to parallel protein-based analyses or to illustrate two-dimensional gel electrophoresis (2-DE) and mass spectrometric analyses of proteins of interest. This section focuses on the identification of protein components within system (profiling), their posttranslational modifications and/or quantitative analyses of the amounts of the proteins, and modifications. In 1975, first time 2- DE was reported by three groups which provided the first glance at the intricacy of the protein content of cells and, as such, can be thought of as marking the true beginning of proteomics (O'Farrell 1975), almost 20 years before the term "proteomics" was coined. Since then 2-DE and mass spectroscopy is still the largely utmost resolution protein separation method available that provides the first means of interrogating the levels of hundreds of proteins and their isoforms. However, even in its nascent form, proteomics provided discoveries such as determining the oscillating levels of proliferating cell nuclear antigens (PCNAs) during the cell cycle observed from cell biology experiments in the 1980s (Mathews et al. 1984). The interaction between protein and DNA were also revealed (Rauscher et al. 1988). The advances that enabled proteomics what it is today were the improvement in image analysis methods, as did the reproducibility of the 2-DE method, both of which continue to this day. In 1980s the advances in mass spectrometry (MS) improved the method for the effective detection of gel-separated proteins. Large-scale expressed sequence tag (EST) sequencing efforts provided access to the content of the sequence databases of both the public and private domains subsequently helpful in genomic sequencing efforts. Computational methods designed for the correlation of the results of mass spectrometric analyses with the content of sequence databases were developed. All these advances occurred in early 1990s and in 1994 at the first 2-DE meeting in Siena, Italy, the term "proteome" was coined, short for the "protein complement of the genome" (Wilkins et al. 1995). Figure 3.1 indicates the graphical representation of the proteomics technique used to unravel a stressful condition in an organism.

Large number of proteins in any biological system remained undefined; their size and complexity are also unknown. The number of potential proteins, including their isoforms, structural or functional (and often transient), are enormously greater than the number of genes. Extreme examples of the number of isoforms of individual proteins can be enormous, such as the neurexins, in which more than 1,000 isoforms have been defined from analyses of alternative splicing (Ullrich et al. 1995). Thus, by both genetic events (e.g., alternative splicing) and dynamic as also the static post-translational modifications add to the complexity of the proteome. Because these modifications are the results of other genes, a protein may be considered the phenotype of many genes (Klose 1999).

In addition to the dynamic and variable modification issues, the proteome is dynamic with respect to the turnover of the proteins within the system. Protein half-lives are under close regulatory control, from seconds to days. Many of the parallel protein-based analysis systems are currently unsuitable to the analysis of protein turnover. But with 2-DE this protein turnover can be measured using pulse-chase labelling with radioactive precursors and one can visualize the synthesis and degradation of a broad range of proteins.

The path of a benchmark proteomics experiment includes experimental plan, sampling, preparations of tissue/cell or organelle, protein extraction/fractionation/purification, labelling/modification, parting, Mass spectrometry (MS) analysis, protein identification, data analysis (statistically) and validation (Jorrín-Novo 2009). Extraction protocols mostly used for plant material may include buffer-based media or Inorganic-solvent media (TCA-acetone, phenol, precipitation protocols) for tissue homogenization. Rossignol (2006) concludes that highest possible assortment of proteins is important to which to achieve maximum efficiency in the extraction stage and this is often accomplished by combining different procedures (Rossignol 2006).

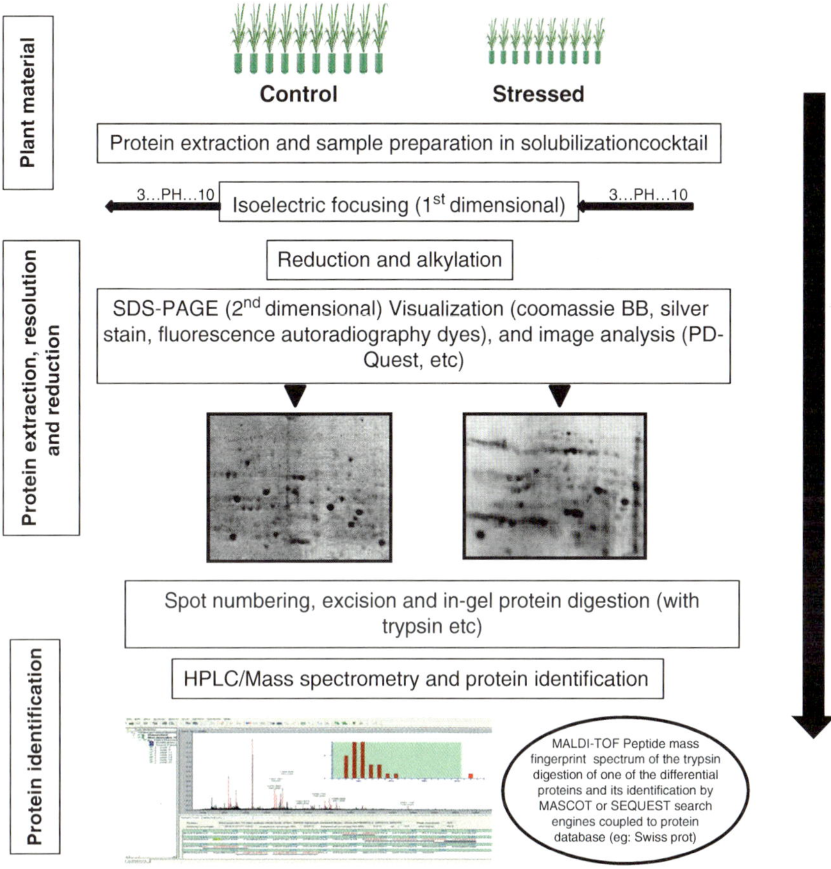

Fig. 3.1 Overall methodology used to perform proteomic analyses of control and stresses plants using 2-D electrophoresis and bioinformatics tools

Gel-based or gel-free approaches are the two commonest protein separation techniques. For gel-based studies, 1-DE and 2-DE are the preferred techniques used in amalgamation with MS (Görg 2009). Shotgun or Gel-free liquid chromatography tandem mass spectrometry (LC-MS/MS) analysis (Leitner and Lindner 2009) can raise the number of different proteins that can be recognized from complex samples, contrasted to more traditional gel-based approaches. There are numerous MS technologies, commonly known as 'Second generation MS technologies' for Quantitative proteomics that are now largely applied to cover the wide spectrum of plant proteomics. Some among these techniques may include isotope-coded affinity tags (ICAT), difference gel electrophoresis (DIGE), (Shiio and Aebersold 2006), isobaric tags for relative and absolute quantitation (iTRAQ) (Gan 2007) and stable isotope labelling by amino acids in cell culture (SILAC) (Palmblad et al. 2008). The gel-free multi-dimensional protein identification technology (MudPIT) is particularly well matched for the detection of hydrophobic proteins (Görg 2009).

In comparison with normal 2-DE, the gel-based 2-D DIGE technique is ample for quantitative proteomics. The reason for this is that it requires only a small amount of protein (0.025–0.050 mg) compared with 2-DE (ca. 0.7–1.0 mg) and therefore shuns the inadequacy of the existence of highly abundant proteins in the protein samples (Dunkley 2006). Presently, Label-free approaches (LC-MS or Fourier transform mass spectrometer) have been used to investigate highly complicated proteomics (Jorrín et al. 2007). Finally, some more advanced techniques like *in silico* proteomics is useful in both forecasting and validating experimental data (Heazlewood 2007). These innovations in the field of proteomics have influenced the resources to better discover and identify proteins.

Role of Proteomics in Enhancing Biological Discoveries

Proteomics has been used for systematic purposes viz., Analysing and evaluating the protein profiling and functional characterization of proteins present in the cells, tissues or organalles of an organism under particular conditions in a given time. Hence, proteomics serves a good contrivance to reveal the elements that are involved in stress perception and transduction. Most important role of proteomics is that it contributes to biological knowledge. This has been achieved through the integration of other molecular species (e.g., mRNA and metabolites) within a cell and organism together with proteomics analyses, thus enabling a systems biology approach. The upcoming impact that proteomics is likely to make is through the detection and substantiation of biomarkers of disease and disease progression or surrogate markers of drug efficacy, as it enables tracing of changes in the patterns of protein expression (Alterovitz et al. 2008). Interestingly, it has also been shown that peptides derived from proteolytic digests from tissue samples can be used in antibody-based arrays tests with a mass spectrometric readout (Whiteaker et al. 2007). The analysis of protein-protein interactions that has been demonstrated on a large scale biochemically were initially difficult to analyze but with respect to proteomics data analysis and validation of results provides substantive conclusions. As almost all proteomics methods use mass spectroscopy at some stage helping in generating large quantities of data. Finally, software aids in handling of these vast data sets and to meaningful conclusions (Anderson et al. 2004). The software tools which interpret the data still require some skill and experience on the part of the analyst and as such need to be improved. This can best be accomplished through biochemists/biologists and computer scientists working in concert. This is required not only for the analysis of mass spectrometric data but also for the intelligent collation of data from all aspects of scientific programs.

Various bioinformatics databases viz., the PODB, http://proteome.dc.affrc.go.jp/Soybean/; the Organellome, http://podb.nibb.ac.jp/Organellome (Mano et al. 2007); and the knowledge-based UniProt are now available to scientific community to evaluate proteomics information (Jorrín-Novo 2009). "Green Proteome", a project of Multinational Arabidopsis Steering Committee, Plant Proteomics in Europe

(COST Action FA0603) has been started to form a searchable database of MS/MS reference spectra (Weckwerth 2008). The database allows authentic protein recognition through a genome-independent approach.

Salt Stress Induced Proteome Alteration

Modern plant biologists are now focusing to improve plant salt stress tolerance. Generally, osmotic stress and ion toxicity is the creation of high salt concentration (Munns 2002). Increase in various salt concentrations, especially of Na^+ and Cl^- and also Ca^{2+}, K^+, $(CO_3)^{2-}$, $(NO_3)^-$ and $(SO_4)^{2-}$ decreases the water potential of soil, which in turn decreases the uptake of water by roots. Hence, elevated levels of salinity cause osmotic stress and disturb the ion homeostasis of the cell. During the osmotic stress, there occurs accumulation of several low-molecular weight osmolytes, like glycine betaine (GB), proline and raffinose-derived oligosaccharides, as well as of high molecular hydrophilic proteins from the LEA (late embryogenesis-abundant) superfamily (Azooz et al. 2011; Ahmad et al. 2012). Also a number of adverse processes like membrane disorganization, increase in amounts of toxic metabolites, ROS generation and nutrient uptake, inhibited photosynthesis and finally the death of cell and plant are known to occur during salt stress (Ahmad and Umar 2011). While analysing the effect of salinity on rice roots, Yan et al. (2006) observed 1,100 spots in the proteome of rice roots that were on exposure to high salt concentration (150 mmol/L), on 2D gels, with 34 protein spots up-regulated and 20 spots showed down regulation. Salinity generally reduces plant growth (Abbasi and Komatsu 2004). Yan et al. (2006) observed decrease in the levels of glutamate synthetase (GS), a key enzyme of nitrogen assimilation, during salt stress. It is interesting to examine that many proteins respond to salt stress and also to other abiotic stresses. For example, two proteins (putative nascent polypeptide-associated complex α-chain and peroxidase) respond to rice plants under salt stress and also show similar response to sugar beet leaves (*Beta vulgaris*) under drought stress (Hajheidari et al. 2005). Dooki et al. (2006) observed that 13 proteins in rice young panicles significantly changed their expression levels under salt stress. The MS of highly plentiful proteins of panicle led to the identification of proteins that are involved in several salt responsive mechanisms. The results indicating the recognition of proteins may help the plant to adapt under salinity stress by up-regulation of antioxidants and the enzymes involved in transcription, translation, signal transduction as well as ATP generation. Under salt stress, photosystem II (PII), oxygen-evolving enhancer protein precursor was up-regulated in rice leaf sheath (Abbasi and Komatsu 2004). Similarly, Caruso et al. (2008) observed that during salt stress ferredoxin $NADP^+$ oxidoreductase was up-regulated in wheat. Contrastingly, metabolism-related proteins in the roots of soybean seedlings were mainly down-regulated under salt stress. Also decline in some glycolytic enzymes (e.g. GAPDH) and a number of proteins involved in CO_2 assimilation were also observed as a result of salinity stress in crop plants (Sobhanian et al. 2010). Proteins with important functions, like membrane stabilization, proton homeostasis and signal transduction always show increased expression. Cheng et al. (2009) reported a new salt responding to leucine-rich-repeat type receptor-like protein kinase, OsRPK1. While studying the responses of Chinese common wheat cultivar Jinan 177 and its hybrid under salt stress with a salt-tolerant *Thinopyrum ponticum*, Wang et al. (2008) showed some remarkable results. Considerable change in expression of proteins involved in signal transduction (ethylene signalling, small G proteins, MAP kinase cascade) were observed in both the genotypes. Among the up-regulated proteins under salt stress, a wide range of transcription and translation factors were found. These includes DEAD box RNA helicase, involved in modulation of stress-inducible CBF/DREB transcriptional activators and putative transcription factor BTF3 from nascent polypeptide-associated NACA protein family. Also Up-regulation of several transporter proteins (vacuolar proton ATPase subunit E involved in Na^+/H^+ antiport, ABC transporters involved in transport of secondary metabolites, exocyst subunits SEC1a and EXO70) and several proteins with protective functions (HSP70 and other chaperones) was observed upon salt stress. Moreover, an increase in the level of transcription factor proteins by genetic engineering, for instance DREBs and NACs, considerably enhanced the tolerance of both the model and crop plants to a wide range of environmental adverse conditions, including cold, drought, salt even in the field conditions (Ahmad and Prasad 2011a, b; Ahmad and Umar 2011). Aghaei et al. (2009) reported that some proteases make their way to cytosol from injured vacuoles under salt stress in the hypocotyls and roots of soybean and were found to counteract the harmful effects of salt on cell proteins. Jelloulia et al. (2008) while working on grapevines, observed that under salt stress, the expression level of protein also showed significant change. A total of 48 proteins showed differential expression, of which 32 showed up-regulation, nine showed down-regulation and seven new protein spots were identified. Hence, modification of gene expression that alters the cellular machinery is a basic adaptation to salt stress. additionally, evaluating the plant responses to salt stress, quantitive analysis of gene expression at the protein level is necessary. Under stress conditions, expression profiling at the protein level of plant reflects the importance of proteomic approach (Aghaei et al. 2009). Hence, molecular tools are necessary for the detection of salt-stress- responsive genes and proteomics is the most reproducible tool to determine the functionality of identified gene (Aghaei et al. 2009). Evers et al. (2012) worked out that during cold and salt stresses in potato the majority of photosynthesis-related genes were down-regulated. On the other hand, cell rescue and transcription factor-related genes were mostly up-regulated. Recently, a study by Yang et al. (2012) in sugar beet line M14 displayed an interesting response towards salt stress tolerance. They reported proteomics analysis of M14 roots and leaves under 500 mM NaCl treatment for 7 days. Among the total protein spots analysed, 36 spots from root gels and 40 spots from leaf gels exhibited significant changes, respectively. Using mass spectrometry and database searching, they were able to identify 38 unique proteins in leaves and 29 unique proteins in roots. The proteins identified belonged to the category of photosynthesis, metabolism, protein folding and protein degradation. Thus revealing the new insights facilitated by the proteomics into the molecular mechanisms underlying salt tolerance for engineering crop plants towards enhanced tolerance. Therefore, proteomics proves to be a high-throughput approach that enables the study of sophisticated molecular networks in plants.

Metabolic Profile of Plant Responses to Abiotic Stresses

Stress prevention comprises a variety of protective mechanisms that hinder the adverse effect of a stress factor on a plant. Stress-inducible genes involve genes that play role in direct protection from such adverse conditions, including the production of osmoprotectants, detoxifying enzymes and transporters. Genes that programme regulatory proteins such as protein kinases, transcription factors and phosphatases also become active on the onset of stress stimuli. Proteins that are expressed grant a physiological benefit under stress conditions and thus are all together important objects for marker-assisted selection by studying the metabolite profile of plants.

Drought salinity and low temperature stress conditions encourage proteins related to detoxifying reactive oxygen species (ROS) (Hajheidari et al. 2007), as a consequence, photosynthetic inhibition, metabolic dysfunction and damage of cellular structures contribute to growth problems, reduced fertility and premature senescence.

Amino Acids

Accumulation of amino acids e.g. Proline and GABA has been observed in plants when exposed to abiotic stress (Lugan et al. 2010). This increase may be due to amino acid production and/or from increased stress-induced protein breakdown. This overall gathering of amino acids upon stress indicates cell damage in some species (Widodo et al. 2009), but in many other plants this increased levels of specific amino acids helps plant to cope up during stress acclimation.

Proline is a known osmolyte, a ROS scavenger, and a molecular chaperone that stabilizes the structure of proteins, hence protecting cells from the damage caused by stress (Szabados and Savoure 2010). In response to various environmental adverse conditions, like drought, high salinity and heavy metals, level of Proline increases significantly. c-aminobutyric acid (GABA), a non-protein amino acid, is also known to rapidly accumulate to greater levels under different unfavourable environmental conditions (Renault et al. 2010). Liu et al. (2011) observed that GABA metabolism has been connected with carbon–nitrogen balance and ROS scavenging. It is observed that during high salinity conditions, the activity of enzymes involved in GABA metabolism are increased (Renault et al. 2010).

Polyamines and Glycine Betaine

Polyamines (PA), small aliphatic molecules, are positively charged at cellular pH. Stresses, like drought, salinity and cold, increase the PA levels. The higher PA levels are, the plant shows increased stress tolerance (Alcazar et al. 2010). Spermidine, Putrescine and spermine are the common PAs in higher plants. Generally, PAs have been concerned in protecting membranes and alleviating oxidative stress (Hussain et al. 2011) but their exact function in stress tolerance is still unknown.

Glycine betaine (GB), a quaternary ammonium compound, normally occurs in a wide variety of plants. *Arabidopsis* and many crop species do not accumulate GB. Generally the plants that produce GB naturally, abiotic stress, such as drought, cold and salt stress, enhances GB accumulation (Chen and Murata 2011). Bansal et al. (2011) confirmed that introduction of the GB biosynthesis pathway genes into non-accumulators improved their ability to tolerate abiotic stress conditions, hence, indicating the advantageous role of GB in stress tolerance.

Integration of Metablomics, Transcriptomics and Proteomics for Understanding Plant Processes During Stress

To feed the teeming population and reduce food shortages, classical plant breeding methods alone did not attain the expected result. The increase in production must be accompanied by optimization of growing conditions, which nowadays are suboptimal for plant growth. It is esteemed that around 70% of the potential yield is lost because of unfavourable stress conditions.

Nowadays modern biotechnology serves as a facilitator in crop improvement by identifying and characterizing genes, transcripts, proteins and metabolites that control the architecture and/or stress resistance of crop plants in a wide range of environments. Moreover, now the focus of research has been shifted from just analysing a single gene or protein to whole analysis of genome, transcriptome, proteome as well as metabolome. There are number of questions which needed to be answered, as mentioned by Altman (2003) like what are the characteristics of the following events: stress perception ⇒ signal transduction ⇒ gene activation ⇒ protein expression ⇒ metabolite production ⇒ whole plant response?

In the Omics era, functional genomics is progressively attaining the heights of excellence by integrating various approaches to understand the full system biology. Systems biology, as a holistic approach, is the integration of data from various fields including physiology, genomics, transcriptomics, proteomics, and metabolomics into models that might, ultimately, symbolize and simulate the physiology of the organism. All these subjects combined in systems biology signify a new approach to discovering the genes and pathways that are decisive for stress responsiveness and tolerance (May 2011).

Both Proteomics as well as metabolomics, specifically, rely on label-free quantitative MS techniques, enabling absolutely important high throughput analyses. Results became more refined when result sets of integrated metabolites and protein data sets were examined than when they were studied individually (Morgenthal et al. 2007). To integrate these post-genomic platforms, computational tools are needed (Kitano 2002).

The amalgamation of different 'omics' tools, which rather than investigating a inadequate number of substances (Van Dam and Poppy 2008), enables the large-scale scanning of various substances, and offers a great deal of potential to elucidate the genotype-phenotype relationships (Wienkoop 2010).

For studying the overexpressing Arabidopsis lines for PAP1 gene that encodes for a Myb-like transcription factor by integrating transcriptome and LC-MS and FT-ICR MS analysis were employed (Tohge et al. 2005),which concluded that increased accumulation of anthocyanins is responsible for the changes in the metabolic profiles caused by PAP1 gene expression Thus, inferring that the expression of genes that were identified to be engaged in anthocyanin production was up-regulated and the other up-regulated genes could be assisting in a role for the development of anthocyanins, like members of the acyltransferase, glycosyltransferase and glutathione S-transferase families. Therefore, the functions of some of these unknown genes were also predicted experimentally through recombinant technologies. These approaches designate the feasibility of integrating transcriptomic and metabolomic analysis.

Research carried out by Barros et al. (2010) in two GE maize lines (MON810 and glyphosate tolerant) grown over three seasons in one location), used transcriptome, proteome, and metabolome profiling to contrast the respective control lines. It was observed that the environment affected gene expression, protein distribution, and metabolite content more strongly than the genetic modification. Additionally, to this, during one season, they also found distinct profiles for the three locations that were also part of their comparisons.

Studies by Goossens et al. (2003a) have showed that environmental factors also elicitate the production of metabolities. In their work, they have treated tobacco BY-2 cells with methyljasmonate. Using this elicitor, around 600 genes were reported to be differentially regulated as designated by cDNA-AFLP (AFLP) transcript profiling. This data was further linked to metabolite analysis possible genes that could be involved in structural or regulatory action with tobacco secondary metabolism, were deduced. This type of approach is becoming crucial now-a-days for working with rare and economically important (e.g. medicinal) plants or non-model plants for which the genome is unidentified, providing quantitative information on gene expression.

These advances in integration of omics (genomics, metabolomics, transcriptomics and proteomics) have brought biology together and appreciative the associations between different levels of biological systems have become really easy, providing a lead in realization of systems biology.

Conclusions and Future Perspective

The dynamic changes of the proteins under salinity conditions is a source of valuable information toward exploring and underlying complicated cellular and molecular processes in plant occurring in response to salt stress and tolerance. These processes include photosynthesis, energy metabolism, ROS scavenging, ion/osmotic homeostasis, signalling transduction, transcription regulation, translational regulation and cytoskeleton dynamics. Despite the marked progress in abiotic stress biology, large gaps remain in our knowledge about the basics of mechanism in combating stress by plants. Therefore, it is important to focus both on intracellular and intercellular molecular networks that plays role in interaction to stress. The extensive proteomics applications and advancement in technology such as multidimensional protein fractionation, absolute quantitation (iTRAQ), isobaric tags for relative, label free quantification mass spectrometry, phosphoprotein and glycoprotein enrichment and tagging, has enhanced the chance of discovery of low abundant proteins and novel regulatory mechanisms occurring during salt stress signalling and metabolic pathways. Integration of proteomics with transcriptomics, metabolomics and bioinformatics has further facilitated the insight of molecular networks underlying salt stress response and tolerance. This can now be used in future to foretell and certify how different components come together and generate responses to control divergent metabolic pathways.

References

Abbasi F, Komatsu S (2004) A proteomic approach to analyze salt-responsive proteins in rice leaf sheath. Proteomics 4:2072–2081

Aghaei K, Ehsanpour AA, Shah AH, Komatsu S (2009) Proteome analysis of soybean hypocotyl and root under salt stress. Amino Acids 36:91–98

Ahmad P, Prasad MNV (2011a) Environmental adaptations and stress tolerance in plants in the era of climate change. Springer Science+Business Media, LLC, New York

Ahmad P, Prasad MNV (2011b) Abiotic stress responses in plants: metabolism, productivity and sustainability. Springer Science + Business Media, LLC, New York

Ahmad P, Umar S (2011) Oxidative stress: role of antioxidants in plants. Studium Press, New Delhi

Ahmad P, Hakeem KR, Kumar A, Ashraf M, Akram NA (2012) Salt-induced changes in photosynthetic activity and oxidative defense system of three cultivars of mustard (*Brassica juncea* L.). Afr J Biotechnol 11:2694–2703

Alcazar R, Planas J, Saxena T, Zarza X, Bortolotti C, Cuevas J, Bitrian M, Tiburcio AF, Altabella T (2010) Putrescine accumulation confers drought tolerance in transgenic Arabidopsis plants overexpressing the homologous arginine decarboxylase 2 gene. Plant Physiol Biochem 48:547–552

Alterovitz G, Xiang M, Liu J, Chang A, Ramoni MF (2008) System-wide peripheral biomarker discovery using information theory. Pac Symp Biocomput 2008:231–242

Altman (2003) From plant tissue culture to biotechnology: scientific revolution. Abiotic stress tolerence and forestry. In vitro Biol Plant 39:75–84

Anderson NL, Anderson NG, Haines LR, Hardie DB, Olafson RW, Pearson TW (2004) Mass spectrometric quantitation of peptides and proteins using Stable Isotope Standards and Capture by Anti-Peptide Antibodies (SISCAPA). J Proteome Res 3:235–244

Azooz MM, Youssef AM, Ahmad P (2011) Evaluation of salicylic acid (SA) application on growth, osmotic solutes and antioxidant enzyme activities on broad bean seedlings grown under diluted seawater. Int J Plant Physiol Biochem 3:253–264

Bansal KC, Goel D, Singh AK, Yadav V, Babbar SB, Murata N (2011) Transformation of tomato with a bacterial codA gene enhances tolerance to salt and water stresses. J Plant Physiol 168:1286–1294

Barros ELS, Lezar S, Anttonen MJ, vanDijk JP, Röhlig RM, Kok EJ, Engel KH (2010) Comparison of two GM maize varieties with a near-isogenic non-GM variety using transcriptomics, proteomics and metabolomics. Plant Biotechnol J 8:436–451

Caruso G, Cavaliere C, Guarino C, Gubbiotti R, Foglia P, Laganà A (2008) Identification of changes in *Triticum durum* L. leaf proteome in response to salt stress by two-dimensional electrophoresis and MALDI-TOF mass spectrometry. Anal Bioanal Chem 391:381–390

Chen TH, Murata N (2011) Glycinebetaine protects plants against abiotic stress: mechanisms and biotechnological applications. Plant Cell Environ 34:1–20

Cheng Y, Qi Y, Zhu Q, Chen X, Wang N, Zhao X, Chen H, Cui X, Xu L, Zhang W (2009) New changes in the plasma-membrane-associated proteome of rice roots under salt stress. Proteomics 9(11):3100–3114

De Michele R, Formentin E, Schiavo FL (2009) Legume leaf senescence: a transcriptional analysis. Plant Signal Behav 4:319–320

Dong CH (2001) ADF proteins are involved in the control of flowering and regulate f-actin organization, cell expansion, and organ growth in arabidopsis. Plant Cell 13:1333–1346

Dooki AD, Mayer-Posner FJ, Askari H, Zaiee AA, Salekdeh GH (2006) Proteomic responses of rice young panicles to salinity. Proteomics 6:6498–6507

Dunkley TPJ (2006) Mapping the Arabidopsis organelle proteome. Proc Nat Acad Sci USA 103:6518–6523

Evelin H, Kapoor R, Giri B (2009) Arbuscular mycorrhizal fungi in alleviation of salt stress. Ann Bot 104:1263–1281

Evers D, Legay S, Lamoureux D, Hausman JF, Hoffmann L, Renaut J (2012) Towards a synthetic view of potato cold and salt stress response by transcriptomic and proteomic analyses. Plant Mol Biol 78:503–514

Gan CS (2007) Technical, experimental, and biological variations in isobaric tags for relative and absolute quantitation (iTRAQ). J Proteome Res 6:821–827

Garg N, Manchanda G (2009) Role of arbuscular mycorrhizae in the alleviation of ionic, osmotic and oxidative stresses induced by salinity in *Cajanuscajan* (L.) Millsp. (pigeonpea). J Agron Crop Sci 195:110–123

Goossens A, Häkkinen ST, Laakso I, Seppänen-Laakso T, Biondi S, De Sutter V, Lammertyn F, Nuutila AM, Soderlund H, Zabeau M (2003) A functional genomics approach toward the understanding of secondary metabolism in plant cells. Proc Natl Acad Sci USA 100: 8595–8600

Görg A (2009) 2-DE with IPGs. Electrophoresis 30:S122–S132

Hajheidari M, Abdollahian-Noghabi M, Askari A, Heidari M (2005) Proteome analysis of sugar beet leaves under drought stress. Proteomics 5:950–960

Hajheidari M, Eivazi A, Buchanan BB, Wong JH, Majidi I, Salekdeh GH (2007) Proteomics uncovers a role for redox in drought tolerance in wheat. J Proteome Res 6:1451–1460

Heazlewood JL (2007) SUBA: the arabidopsis subcellular database. Nucleic Acids Res 35:D213–D218

Hussain SS, Ali M, Ahmad M, Siddique KH (2011) Polyamines: natural and engineered abiotic and biotic stress tolerance in plants. Biotechnol Adv 29:300–311

Jelloulia N, Jouiraa HB, Skourib H, Ghorbela A, Gourgourib A, Mlikia A (2008) Proteomic analysis of Tunisian grapevine cultivar Razegui under salt stress. J Plant Physiol 165:471–481

Jiang Y, Yang B, Harris NS, Deyholos MK (2007) Comparative proteomic analysis of NaCl stress-responsive proteins in Arabidopsis roots. J Exp Bot 58:3591–3607

Jorrín JV, Maldonado AM, Castillejo MA (2007) Plant proteome analysis: a 2006 update. Proteomics 7:2947–2962

Jorrín-Novo JV (2009) Plant proteomics update (2007–2008): second-generation proteomic techniques, an appropriate experimental design, and data analysis to fulfill MIAPE standards, increase plant proteome coverage and expand biological knowledge. J Proteomics 72: 285–314

Kitano H (2002) Computational systems biology. Nature 420:206–210

Klose J (1999) Genotypes and phenotypes. Electrophoresis 20:643–652

Leitner A, Lindner W (2009) Chemical tagging strategies for mass spectrometry-based phospho-proteomics. In: de Graauw M (ed) Phospho-proteomics: methods and protocols. Humana Press, New York, pp 229–243

Liu C, Zhao L, Yu G (2011) The dominant glutamic acid metabolic flux to produce gamma-amino butyric acid over proline in Nicotiana tabacum leaves under water stress relates to its significant role in antioxidant activity. J Integr Plant Biol 53:608–618

Locy RD, Change C-C, Nielsen BL, Singh NK (1996) Photosynthesis in salt-adapted heterotrophic tobacco cells and regenerated plants. Plant Physiol 110:321–328

Lugan R, Niogret MF, Leport L, Guegan JP, Larher FR, Savoure A, Kopka J, Bouchereau A (2010) Metabolome and water homeostasis analysis of Thellungiella salsuginea suggests that dehydration tolerance is a key response to osmotic stress in this halophyte. Plant J 64:215–229

Mano S, Miwa T, Nishikawa S-I, Mimura T, Nishimura M (2007) The plant organelles database (PODB): a collection of visualizating plant organelles and protocols for plant organelle research. Nucleic Acids Res 36:D929–D937

Mathews MB, Bernstein RM, Franza BR, Garrels JI (1984) Identity of the proliferating cell nuclear antigen and cyclin. Nature 309:374–376

May P (2011) Integration of proteomic and metabolomic profiling as well as metabolic modeling for the functional analysis of metabolic networks. Methods Mol Biol (Clifton NJ) 694: 341–363

Morgenthal K, Wienkoop S, Wolschin F, Weckwerth W (2007) Integrative profiling of metabolites and proteins improving patternrecognition and biomarker selection for systems level approaches. Metabolom Methods Mol Biol 358:57–75

Munns R (2002) Comparative physiology of salt and water stress. Plant Cell Environ 25:239–250

Munns R, Tester M (2008) Mechanisms of salinity tolerance. Annu Rev Plant Biol 59:651–681

Nemoto Y, Sasakuma T (2002) Differential stress responses of early salt-stress responding genes in common wheat. Phytochemistry 61:129–133

O'Farrell PH (1975) High resolution two-dimensional gel electrophoresis of proteins. J Biol Chem 250:4007–4021

Palmblad M, Mills DJ, Bindschedler LV (2008) Heat-shock response in Arabidopsis thaliana explored by multiplexed quantitative proteomics using differential metabolic labeling. J Proteome Res 7:780–785

Peuke AD, Jeschke WD (1999) The characterization of inhibition of net nitrate uptake by salt in salt-tolerant barley (*Hordeum vulgare* L. cv. California Mariout). J Exp Bot 50:1365–1372

Rauscher FJ, Cohen DR, Curran T, Bos TJ, Vogt PK, Bohmann D, Tjian R, Franza BR (1988) Fos-associated protein p39 is the product of the jun proto-oncogene. Science 240:1010–1016

Renault H, Roussel V, El Amrani A, Arzel M, Renault D, Bouchereau A, Deleu C (2010) The Arabidopsis pop2-1 mutant reveals the involvement of GABA transaminase in salt stress tolerance. BMC Plant Biol 10:20

Rossignol M (2006) Plant proteome analysis: a 2004–2006 update. Proteomics 6:5529–5548

Schröder S (2008) Quantitative gel electrophoresis: sources of variation. J Proteome Res 7:1226–1234

Shiio Y, Aebersold R (2006) Quantitative proteome analysis using isotope-coded affinity tags and mass spectrometry. Nat Protocols 1:139–145

Sobhanian H, Razavizadeh R, Nanjo Y, Ehsanpour AA, Jazii FR, Motamed N (2010) Proteome analysis of soybean leaves, hypocotyls and roots under salt stress. Proteome Sci 8:19–25

Szabados L, Savoure A (2010) Proline: a multifunctional amino acid. Trends Plant Sci 15:89–97

Tohge T, Nishiyama Y, Hirai MY, Yano M, Nakajima J, Awazuhara M, Inoue E, Takahashi H, Goodenowe DB, Kitayama M (2005) Functional genomics by integrated analysis of metabo-

lome and transcriptome of Arabidopsis plants over-expressing an MYB transcription factor. Plant J 42:218–235
Tuteja N (2007) Mechanisms of high salinity tolerance in plants. Methods Enzymol 428:419–438
Ullrich B, Ushkaryov YA, Sudhof TC (1995) Cartography of neurexins: more than 1,000 isoforms generated by alternative splicing and expressed in distinct subsets of neurons. Neuron 14:497–507
Van Dam NM, Poppy GM (2008) Why plant volatile analysis needs bioinformatics – detecting signal from noise in increasingly complex profiles. Plant Biol 10:29–37
Wang MC, Peng ZY, Li CL, Li F, Liu C, Xia GM (2008) Proteomic analysis on a high salt tolerance introgression strain of *Triticum aestivum/Thinopyrum ponticum*. Proteomics 8:1470–1489
Weckwerth W (2008) The multinational Arabidopsis steering subcommittee for proteomics assembles the largest proteome database resource for plant systems biology. J Proteome Res 7:4209–4210
Whiteaker JR, Zhao L, Zhang HY (2007) Antibodybased enrichment of peptides on magnetic beads for massspectrometry-based quantification of serum biomarkers. Anal Biochem 362: 44–54
Widodo Patterson JH, Newbigin E, Tester M, Bacic A, Roessner U (2009) Metabolic responses to salt stress of barley (Hordeum vulgare L.) cultivars, Sahara and Clipper, which differ in salinity tolerance. J Exp Bot 60:4089–4103
Wienkoop S, Baginsky S, Weckwerth W (2010) Arabidopsis thaliana as a model organism for plant proteome research, 73:2239–2248
Wilkins MR, Sanchez JC, Gooley AA, Appel RD, Humphery-Smith I, Hochstrasser DF, Williams KL (1995) Progress with proteome projects: why all proteins expressed by a genome should be identified and how to do it. Biotechnol Genet Eng Rev 13:19–50
Xu F, Li L, Huang X, Cheng H, Wang Y, Cheng S (2010) Antioxidant and antibacterial properties of the leaves and stems of *Premna microphylla*. J Med Plant Res 4:2544–2550
Yan SP, Zhang QY, Tang ZC, Su WA, Sun WN (2006) Comparative proteomic analysis provides new insights into stress responses in rice. Mol Cell Prot 5:3–9
Yang L, Ma C, Wang L, Chen S, Li H (2012) Salt stress induced proteome and transcriptome changes in sugar beet monosomic addition line M14. J Plant Physiol 169:839–850
Zhang YY, Lai JB, Sun SH, Li Y, Liu YY, Liang LM, Chen MS, Xie Q (2008) Comparison analysis of transcripts from the halophyte Thellungiella halophila. J Integr Plant Biol 50:1327–1335

Chapter 4
Genetic Approaches to Improve Salinity Tolerance in Plants

Ashwani Kumar, Aditi Gupta, M.M Azooz, Satyawati Sharma, Parvaiz Ahmad, and Joanna Dames

Abstract Abiotic stress tolerance in plants is gaining importance day by day. Different techniques are being employed to develop salt tolerant plants that directly or indirectly combat global food problems. Advanced comprehension of stress signal perception and transduction of associated molecular networks is now possible with the development in functional genomics and high throughput sequencing. In plant stress tolerance various genes, proteins, transcription factors, DNA histone-modifying enzymes, and several metabolites are playing very important role in stress tolerance. Determination of genomes of *Arabidopsis*, *Oryza sativa* spp. japonica cv. Nipponbare and integration of omics approach has augmented our knowledge pertaining to salt tolerance mechanisms of plants in natural environments. Application of transcriptomics, metabolomics, bioinformatics, and high-through-put DNA sequencing has enabled active analyses of regulatory networks that control abiotic stress responses. To unravel and exploit the function of genes is a major challenge of the post genomic era. This chapter therefore reviews the effect

A. Kumar (✉)
Claude Leon Postdoctoral Fellow, Department of Biochemistry, Microbiology and Biotechnology, Rhodes University, P.O. Box 94, Grahamstown 6140, Eastern Cape, South Africa
e-mail: ashwaniiitd@hotmail.com

A. Gupta • S. Sharma
Centre for Rural Development and Technology, Indian Institute of Technology, New Delhi, Hauz Khas, India

M.M. Azooz
Department of Botany, Faculty of science, South Valley University, Qena, Egypt

Department of Biological Sciences, Faculty of Science, King Faisal University, Al Hassa, Saudi Arabia

P. Ahmad
Department of Botany, A.S. College, University of Kashmir, Srinagar, India

J. Dames
Department of Biochemistry, Microbiology and Biotechnology, Rhodes University, P.O. Box 94, Grahamstown 6140, Eastern Cape, South Africa

P. Ahmad et al. (eds.), *Salt Stress in Plants: Signalling, Omics and Adaptations*,
DOI 10.1007/978-1-4614-6108-1_4,

of salt stress on plants and the mechanism of salinity tolerance along with contributory roles of QTL, microRNA, microarray and proteomics.

Keywords Salt stress • Microarray • MicroRNA • Proteomics • Genomics

World population is increasing at a tremendous rate and is estimated to reach 9 billion by 2050 (http://www.fao.org/wsfs/world-summit/en/). Global warming, loss of resources and depletion of biodiversity is a consequential emerging hazard that can transform fertile conditions into extreme environments (Mittler 2006; Koyro et al. 2012). Such diverse environmental conditions can stimulate an array of stresses that can significantly alter plant metabolism, growth and development and can ultimately lead to plant death. Abiotic stress includes drought, salinity, temperature extremes, nutrient deficiencies and mineral toxicities that retard plant growth and therefore have a major impact on crop yield (Roy et al. 2011; Koyro et al. 2012). Problems of drought and salinity are widespread in many regions of the World, and are projected to cause serious salinization of more than 50% of all arable lands by the year 2050 (Ashraf 1994).

To meet the growing demands of increasing population, global food production of 44 million tons each year must be an increase, which is considerably more than the current annual average increase of 32 million tons (Tester and Langridge 2010). Therefore, there is an urgent need to develop eco-friendly techniques that controls salinization, use of salt tolerant species/halophytes or species that have high salt removing capacity and commercial value. It is known that most crop plants are hypersensitive to salt stress and showed stunted plant growth or even plant death at an increased salt level in the soil. Thus, soil salinity not only reduces the area of arable land available for crop production, but also reduces crop yield in affected area. Soil salinity occurs as a result of increased quantity of cations (Na^+, K^+, Mg^{2+}, Ca^{2+}) and anions (Cl^-, SO_4^{2-}, HCO_3^-) which are originated from more water-soluble salts such as $NaSO_4$, $NaHCO_3$, NaCl, and $MgCl_2$ as well as less water-soluble salts such as $CaSO_4$, $MgSO_4$, and $CaCO_3$.

Most plants have evolved mechanisms to cope up with the extreme environmental stresses to which they are exposed regularly. But the pathways of signal transduction that elicit these responses, or even the way in which plants perceive these environmental stresses, are not clearly understood (Braam et al. 1997; Mittler 2002).

Studies on effect of salt stress on plant tolerance covers many aspects of plant response, including alterations at the morphological, physiological and molecular levels. Nowadays, different approaches are used to enhance salinity adaptation in crop plants like development of transgenic plants, improvement in breeding/screening methodologies and modification of the existing crop genes. Molecular genetic analysis of the model plant *Arabidopsis thaliana* (L.) has improved our understanding of abiotic stress responses. The availability of the whole genome sequence of *Arabidopsis* in 2000 (The Arabidopsis Genome Initiative 2000a), and the development of genomic tools for *Arabidopsis* has hastened the research on different aspects of plant biology, including abiotic stress responses (Weigel et al. 2000; Bohnert et al. 2006; Koiwa et al. 2006). Such useful information can be extended to other plant species as well. The modulation of plant salinity tolerance in various cultivars has been extensively carried out using different approaches (Nenova 2008). In this chapter, we are discussing the effect of salinity stress on plants, and physiological and genetical responses of plants to these stresses.

Plant Response to Salinity Stress and Salt Tolerant Plants

Plants acclimate to particular stress conditions by producing specific responses in nature (Mittler 2006). It is now known that plant tolerance enhancement to biotic or abiotic stress conditions by activating a stress response signal transduction pathway in transgenic plants is a powerful and promising approach (Kovtun et al. 2000; Mittler 2006). In natural environment, simultaneous exposure of a plant to different abiotic stress conditions will results in co-activation of stress response pathways. The specific salt tolerant mechanisms might differ for different halophyte species but the general response includes tight regulation of internal Na^+, K^+ and Cl content (Flowers and Colmer 2008). Among the halophytes, *Thellungiella halophila*, has been used as a model plant to study salt stress response because it is closely related to Arabidopsis. Its response towards the stress conditions include higher accumulation of proline, high selectivity for ions, and a sensitive stomatal regulation (Inan et al. 2004). Further, comparative analysis of the differences between glycophytes and halophytes can furnish detailed information on the molecular basis of salinity tolerance and other stress tolerance attributes of halophytes (Hirayama and Shinozaki 2010).

Metabolome analyses of *Thellungiella* and *Arabidopsis* revealed drastically different profiles between these two. Compared with *Arabidopsis*, *Thellungiella* maintains higher levels of metabolites (osmolytes, such as fructose, sucrose, complex sugars, malate, proline, etc.) irrespective of presence or absence of salt stress. Results of transcriptome analysis of *Thellungiella* revealed high expression of several stress-related genes, even in the absence of salt stress (Gong et al. 2005). These results suggest that *Thellungiella* harbours a constant state of stress-anticipatory preparedness.

Other important uses of salt-tolerant plants include the production of economically important compounds such as essentials oils, flavours, fragrances, gums, resins, oils, pharmaceuticals, and fibres. These plants may also be marketed for use in landscape gardening and for their foliage or flowers.

Salinity Stress: Physiological Response

Presence of sodium salt in high amount in the soils, influence plant growth by four mechanisms; (1) osmotic stress, (2) inhibit uptake of K^+, (3) induced toxicity to cytosolic enzymes and, (4) oxidative stress and cell death (Jamil et al. 2011). Most of the crop plants have sophisticated mechanisms to protect themselves against these stresses. The underlying principle includes controlled expression and functioning of genes that participate in a myriad of processes related to sodium influx, efflux, storage, recycling and the suppression of high Na^+-triggered oxidative stress and cell death (Wang et al. 2006). On entering the plant, Na^+ ion is transported to shoot via transpiration flux where it accumulates and lowers the K^+/Na^+ ratio. This causes imbalance in the ion-homeostasis which triggers the excessive production of reactive oxygen species (ROS) ultimately leading to cell death (Huh et al. 2002). Ren et al. (2005) reported a Na^+-selective transporter, *SKC1*, in rice, to be partially responsible for salt tolerance in the resistant variety, Nona Bokro, and suggested its vital role in recycling the Na^+ from shoots to roots. The genes with a potential role in increasing salt tolerance in plants generally suppress oxidative stress or inhibit cell death. Suppression of oxidative stress leading to enhanced salt tolerance is well supported by the overexpression of the vesicle trafficking protein, Rab7, and several antioxidant regulators (dehydroascorbate reductase (DHAR), glutathione peroxidase (GPX) and ascorbate oxidase (AAO)) (Yamamoto et al. 2005; Ushimaru et al. 2006; Ahmad et al. 2008, 2010, 2011).

For a plant, to be able to adapt to high salinity conditions, the coordination between salinity tolerance and related physiological processes appears to be critical. The process of sensing high Na^+ concentration in the soil and the initiated signal transduction to activate a set of stress-responsive genes is yet to be elucidated. Finding key components of signal transduction for salt tolerance may be the key for the genetic improvement of plant salt tolerance.

Plants responded to these stresses by overproduction of different types of compatible organic solutes (amino acids, glycine betaine, sugars, or sugar alcohols) (Serraj and Sinclair 2002; Ahmad and Sharma 2008; Koyro et al. 2012) that are low in molecular weight, highly soluble and usually nontoxic at high cellular concentrations. Their protection mechanisms include cellular osmotic adjustment, ROS detoxification, membrane integrity protection, and enzymes/proteins stabilization (Yancey et al. 1982; Bohnert and Jensen 1996). These solutes accumulate predominantly in the cytoplasm at high concentrations under osmotic stress without interfering normal metabolism. Although many efforts have been devoted in the past to genetically engineer plants for over-production of various osmoprotectants, little success could be achieved.

Physiological studies also suggested the accumulation of nonstructural carbohydrates (sucrose, hexoses, and sugar alcohols) under stress conditions to varying degree in different plant species (Bartels and Sunkar 2005). Under stress conditions, hydrolytic enzymes start breakdown of starch into sugars. Current hypothesis supports that sugars either act as osmotica or protect specific macromolecules and contribute towards the stabilization of membrane structures (Bartels and Sunkar 2005). Production and accumulation of cyclic polyols such as D-pinitol (1D-3-O-methyl-chiro-inositol) or D-ononitol (1D-4-O-methyl-myoinositol) along with the involvement of cyclitols (in transgenic tobacco plants) has frequently been reported in response to drought and salinity stress (Sheveleva et al. 1997; Streeter et al. 2001). Among all the solutes accumulating during salt stress, proline is one of the most widely distributed osmolytes and is produced as a common response during stress conditions. Proline is involved in controlling osmotic adjustment and maintaining plasma membrane integrity; it acts as a sink of energy or reducing power, a source of carbon and nitrogen and also a hydroxyl radicle scavenger (Mansour 1998; Hong et al. 2000; Bartels and Sunkar 2005).

Ornithine-dependent pathway and the glutamate dependent pathway (especially stress conditions) are responsible for proline accumulation in plants. L-proline is synthesized by two enzymes, P5C synthetase (P5CS) and P5C reductase (P5CR) from L-glutamic acid viaΔ^1-pyrroline-5-carboxylate (P5C). The second pathway of proline biosynthesis involves transamination of ornithine catalyzed by ornithine-δ-aminotransferase (OAT) yielding two possible intermediates P5C (Δ-pyrroline-2-carboxylate) and P2C (*Δ*-pyrroline-2-carboxylate), these two can be reduced to proline (Mestichelli et al. 1979; Ahmad and Sharma 2008; Koyro et al. 2012).

Glycine betaine is another major osmolyte (Hanson and Burnet 1994), synthesized by many plants in response to abiotic stresses. Accumulation of this osmolyte is proportional to the degree of salt tolerance in plants – being highest for highly tolerant plants like *Spartina* and *Distichlis*, and low/nil in sensitive species (Rhodes et al. 1989). It is known that the elevated level of glycine-betaine in transgenic plants confers significant tolerance to salt, cold and heat stress.

Plants are also adapted to salt stresses by production of polyamines. The major polyamines (PAs) include triamine spermidine (Spd) and tetraamine spermine (Spm), as well as their precursor the diamine putrescine (Put). Polyamines are organic low-weight molecules with straight-chained C_3-C_{15} aliphatic structure with at least two primary amino groups and one or more internal amino groups (Gill and Tuteja 2010). Polyamines possess properties for free scavenging of ROS and may confer tolerance to different biotic and abiotic stresses, thus playing a pivotal role in conferring stress tolerance under unfavorable environmental conditions (Ahmad et al. 2012a).

Hormones, particularly abscisic acid (ABA) and ethylene, play an important role as regulators in responses to abiotic stress in plants. This hormone is a key regulator of many plant responses to environmental stresses, mainly osmotic stresses (Hubbard et al. 2010; Kim et al. 2010; Chinnusamy et al. 2008). Its signaling is very fast and does not involve transcriptional activity e.g. opening and closing of stomatal aperture by ABA through the biochemical regulation of ion and water transport processes (Cramer et al. 2011). ABA signaling activates a cascade of responses resulting in plant tolerance enhancement to dehydration stress. Studies revealed the activation of both ABA-dependent as well as ABA-independent pathways during dehydration and salinity stresses. When cell undergoes dehydration under water deficit, it induces an increase in endogenous level of ABA that trigger downstream target genes by coding signaling factors, transcription factors, metabolic enzymes and others as discussed by Yamaguchi-Shinozaki and Shinozaki (2006). While in other case, a dehydration responsive cis-acting element, DRE/CRT sequence and its DNA binding ERF/AP2-type TFs, DREB1/CBF and DREB2A, are related to the ABA-independent dehydration and temperature responsive pathways. DREB1/CBFs function in cold-responsive gene expression (Fowler and Thomashow 2002; Maruyama et al. 2004), whereas DREB2s is involved in dehydration responsive and heat-responsive gene expression (Sakuma et al. 2006).

Other plant hormones are also involved in abiotic stress, but play substantial roles, either directly or indirectly. In some studies, salicylic acid, ethylene and jasmonic acid have been shown to affect abiotic stress responses through interplay with ABA in a complex manner (Grant and Jones 2009; Pieterse et al. 2009).

Genetic Approach to Improve Salt Tolerance in Crop Plants

Conventional breeding still proves to be an effective strategy for development of new cultivars but is beset with some limitations. Hybridization-breeding experiments provide advantage of transfer of gene-controlled trait from a valuable germplasm to a related crop species. However, hybridization-breeding is not much effective for the improvement of quantitative traits, due to the genetic segregation of quantitative trait loci (QTL), transfer of trait is usually incomplete, and the selection process is both time-consuming and labor-intensive. Biotechnological advancement has provided us the opportunities to transfer genes of useful traits between species. Significant progress has been achieved in last decade to improve salinity tolerance in crop plant using molecular approaches (Jamil et al. 2011). Development of transgenic plants with the aid of genetic engineering constitutes one of the major achievements of plant science. In spite of these developments, it is not easy to produce genetically engineer plants with traits of interest or improved performance against abiotic stresses. The identification of genes linked to salt stress tolerance is a prerequisite to the improvement of crop plants on saline soils.

The understanding of genomic information with the help of computational biology have led to the identification of signaling pathways and regulatory genes and networks that control complex traits related to abiotic stresses. Knowledge obtained from post-transcriptional and post translational regulation could be used to improved performance under challenging environments of transgenic plants. Post-transcriptional and post-translational regulators exert both positive and negative control on the activities of stress response. Mazzucotelli et al. (2008) discussed in their paper about fruitful improvement of stress tolerance in plants, by positive or negative modification of regulators of post-transcriptional and post-translational levels. The characterization of post-transcriptional and post-translational regulatory systems is crucial for the deeper understanding of the molecular mechanisms governing plant adaptation to environment stress.

Ashraf (2009) has discussed the role of antioxidants in the alleviation of salt stress. Tolerance to a multitude of abiotic stresses has been correlated with the production of antioxidant enzymes and increased levels of non-enzymatic metabolites. This gives an opportunity to engineer plants that over-express transferred antioxidant genes and hence show an increased salt tolerance. Wang et al. (2004) observed that the activity of Mn-SOD was over twofold as compared to that of wild type. Engineered *Arabidopsis* not only showed higher salt tolerance but it also enhanced activities of other antioxidative enzymes such as Cu/Zn-SOD, Fe-SOD, CAT and POD compared to the wild type plants. Engineering of these genes for producing antioxidative enzymes has thus provided novel information about the role of these enzymes in counteracting stress-induced ROS in plant cells. Amaya et al. (1999) tested the germination ability of seeds of transgenic tobacco plants in over-expressing a cell wall peroxidase gene. They observed that the seeds of transgenic line were highly tolerant to salt stress during germination process, might have been due to the improved ability of transgenic seeds to keep considerable amount of water for germination by physically modifying water permeability of the wall. Previous reports indicated that, changes in ROS scavenging enzymatic systems may cause significant modification in oxidative stress tolerance mechanism and that leads to changes in tolerance to abiotic stresses (Pastori and Foyer 2002). Little success has been achieved in improving oxidative stress tolerance by manipulating a single antioxidant gene. This is because a balanced interaction of antioxidative protective enzymes as well as other metabolites may be vital to achieve a substantial improvement in plant stress tolerance (Tseng et al. 2007). Plants respond to salinity stress by perceiving external and internal signals and use these signals to regulate various responses. These signals are primarily perceived by the membrane and operate independent of each other and also modulate other pathways (Tuteja and Sopory 2008). Different types of molecules have been reported previously to have a role in abiotic stress signaling such as calcium (Ca^{2+}), nitric oxide (NO), sugars, ethylene, abscisic acid (ABA), brassinosteroids (BRs), jasmonates (JA), salicylic acid (SA), and auxins (Ahmad et al. 2012b). Rapid developments in the field of functional genomics and bioinformatics have enabled researchers to identify and characterize many stress-related functional and regulatory genes that are responsible in abiotic stress signaling pathways in several plant species (Nakashima et al. 2009). Among the regulatory proteins, transcription factors (TFs) are important and play crucial roles in signal transduction pathway by receiving the upstream signal and activating the expression of downstream stress-inducible genes (Xin et al. 2012).

Genetic engineering of crop plants with signaling components and transcription factors (TFs) leads to the expression of several genes involved in stress adaptation in plants (Ahmad et al. 2011; Varshney et al. 2011).

QTL and Salt Tolerance

In plants, adaptation to high salt levels involves several processes like osmotic adjustment, toxic ions compartmentalization and oxidative stress tolerance. Such plant responses to salt stress are therefore documented as multigenic in nature (Turkan and Demiral 2008). Apart from a few success stories of relation of QTLs with salinity, that have been published recently, genetic selection of stress tolerance traits is quite difficult in general, because these traits are dispersed in various quantitative trait loci (QTLs). Nevertheless, the use of molecular markers and powerful bioinformatics tools have led to the identification of genes with major contributions to Na^+ and K^+ homeostasis using the QTL approach (Pardo 2010).

Molecular isolation of salt tolerance-associated QTLs/effective gene located in a QTL would be highly valuable to understand the quantitative nature of salinity tolerance in plants. Such intense studies are currently feasible with the aid of high density genetic maps and availability of completely/partially sequenced genomes of several plant species. Frary et al. (2000) successfully cloned a QTL that controlled the fruit size in tomato illustrating an early booming example of map-based

QTL. It has been previously shown that 8 QTLs are associated with salt tolerance in Nona Bokra and of them two major QTLs are responsible for the variations of K^+/Na^+ ratio between salt tolerant and susceptible varieties (Lin et al. 2004). Such successes demonstrate the feasibility to identify and isolate major factors of a complex trait using molecular tools. However, it is still a big challenge to determine either up- or down-regulation of gene expression to amplify their functional performance without molecular manipulation of their function. A QTL-based approach has major benefits, that they may enable production of stress-tolerant crops by combining QTLs for various stress tolerances. This is well exemplified with the work of Ren et al. (2005) who identified the SKC1 locus encoding a high affinity K^+ transporter (HKT)-type sodium transporter by analyzing a QTL to improve the salinity tolerance of rice.

Role of Small RNA

There are evidences of changes occurring in the level of several small RNAs in *Arabidopsis* during the abiotic stress, indicating their involvement in stress response (Sunkar and Zhu 2004). RNA interference (RNAi) research increased considerably in recent years and studies are conducted in plants, fungi and eukaryotic organism. These RNA molecules act as a signal molecule that triggers gene regulation. Main categories of sRNAs have been identified on the basis of their origin, structure and functional identifications: short interfering RNAs (siRNAs), microRNAs (miRNAs), and piwi-interacting RNAs (piRNAs) (Guleria et al. 2012).

Comprehensive transcriptome studies have shown production of more than 7,000 transcriptional units that have the potential to produce endogenous siRNA (natural antisense transcript-associated siRNA, natsiRNA) (Yamada et al. 2003). Previous research showed the accumulation of SRO5 transcripts and 24-nt nat-siRNAs is up-regulated, whereas that of P5CDH is down-regulated. In this model, the 24-nt nat-siRNAs guide the initial cleavage of the P5CDH transcript, which generates secondary 21-nt nat-siRNAs by phasing cleavage. The secondary nat-siRNAs also suppress P5CDH transcripts and, after accumulation of the osmo-protectant proline, the plant acquires salt stress tolerance. It is not clearly known whether this nat-siRNA functions only in tolerance to salt stress in *Arabidopsis*. Involvement of microRNA (miRNA) in plant stress tolerance has led to a search of noncoding RNA and their role in salt stress mechanisms. These miRNA are endogenous, 21–24 nucleotide, single stranded, non-protein coding RNA. They have recently emerged as regulator of gene expression (Mantri et al. 2012) by catalyzing post-translational gene silencing or translational repression (Palatnik et al. 2003; Chen 2004). Shukla et al. (2008), reported the either up- or down- regulation of miRNA during stress response of plants. Several studies have been conducted previously on role of miRNA in stress such as dehydration, salinity and cold (Sunkar and Zhu 2004) in Arabidopsis, drought stress (Zhao et al. 2007) in rice, and short and long-term salt stress (Patade and Suprasanna 2010) in sugarcane. Various miRNAs responding to

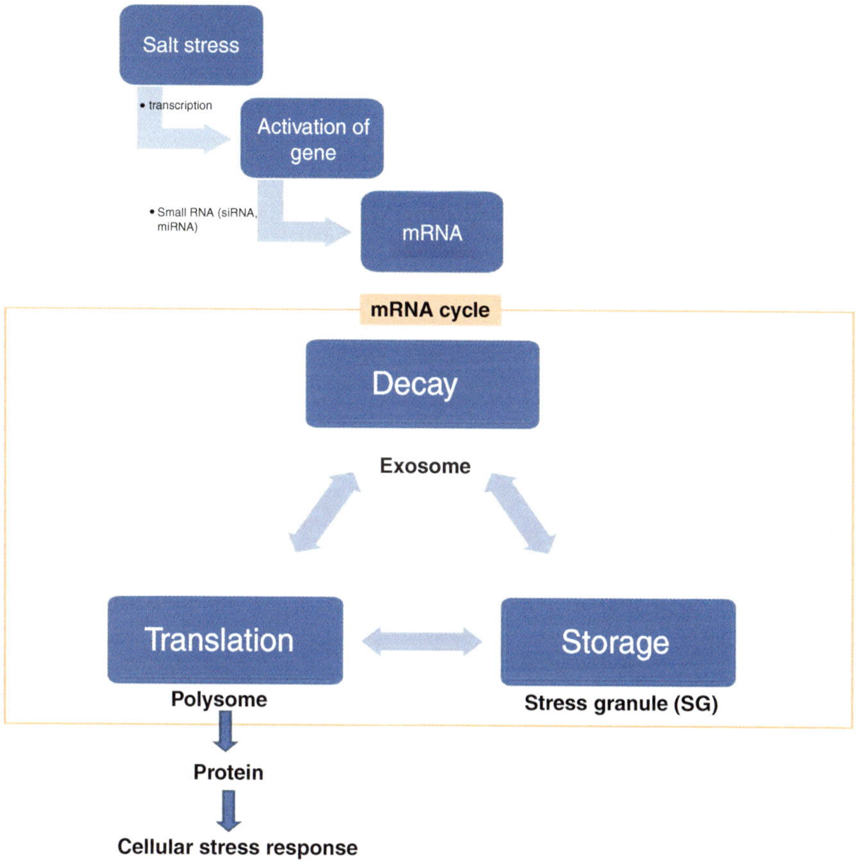

Fig. 4.1 Regulatory network of transcriptome profile in abiotic stress response

abiotic stresses in a specific manner have been reported from several plants such as *Arabidopsis*, *Oryza*, *Nicotiana*, *Brassica*, *Gossypium*, etc. Prolonged efforts would still be required to identify the complete set of miRNAs/other small RNAs that are involved in stress regulation pathways and help us in designing new strategies for improving salt stress tolerance in plants. Figure 4.1 showed regulatory network of transcriptome profile in abiotic stress response.

Microarray Analysis of Gene Expression

Intensive efforts have been made in the past to unveil both the constitutive as well as inducible mechanisms related with the plant survival under stress using molecular techniques, stress-related protein profiles, genome libraries/gene expression studies (Dubey and Grover 2000). Several major classes of genes are altered in response to

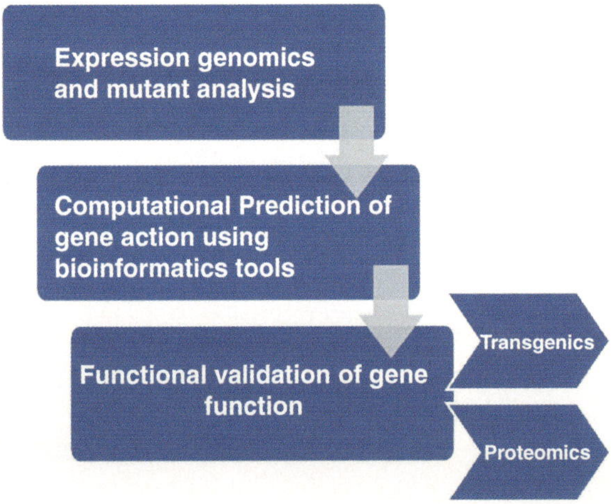

Fig. 4.2 Brief strategies to study salt tolerance in plants using functional genomics

stresses, particularly, the genes involved in signaling and regulation and gene products supporting cellular adaptation during stress (http://www.plantstress.com). Brief strategies are shown in Fig. 4.2 to study salt tolerance mechanism in plants.

A significant proportion of genes are induced by cold and dehydration stresses (Khan et al. 2007). Microarray studies, however, provide no information about changes in the corresponding protein expression related to salt stress tolerance. Most of the genes are identified by expression studies such as those involved in salicylic acid, jasmonic acid and ethylene signaling pathways. Classification of gene products of salt-stress response are classified and presented in Fig. 4.3. The first group salt-stress response includes functional proteins, are mainly proteins responsible for stress tolerance e.g. HSPs, LEAs., proteins involved in repair and protection from damages, membrane proteins such as transporters, protein synthesis-related proteins, proteins involved in osmoprotectants synthesis (proline, sugars and raffinose family oligosaccharides (RFO)), proteins involved in cellular metabolic processes (carbohydrate metabolism, secondary metabolism, biosynthesis of plant hormones (ABA, ethylene and IAA)), senescence-related proteins, proteins regulated by plant hormones (ABA and JA), RNA-binding proteins, cytochrome P450, alcohol and aldehyde dehydrogenase. The second group consists of regulatory proteins, related to signal transduction and gene expression that probably function in stress response (Khan et al. 2007). A list of inducible genes at high salinity stress identified by the cDNA microarray analysis is also available at http://pfgweb.gsc.riken.go.jp. Gygi et al. (2000) presented a marked difference between the relative mRNA expression levels and those of their proteins. Studies showed that the functions of proteins are considerably depend on post-translational modifications (PTMs) and their interactions with other proteins presented in cell, this information cannot be deduced from microarray analysis data (Rose et al. 2004). The proteome of any organism is highly dynamic and large number of variations is possible. For example, the rice genome contains 32,000–50,000 genes and each gene may give rise to multiple proteins by means of alternative splicing or PTMs (Yan et al. 2006).

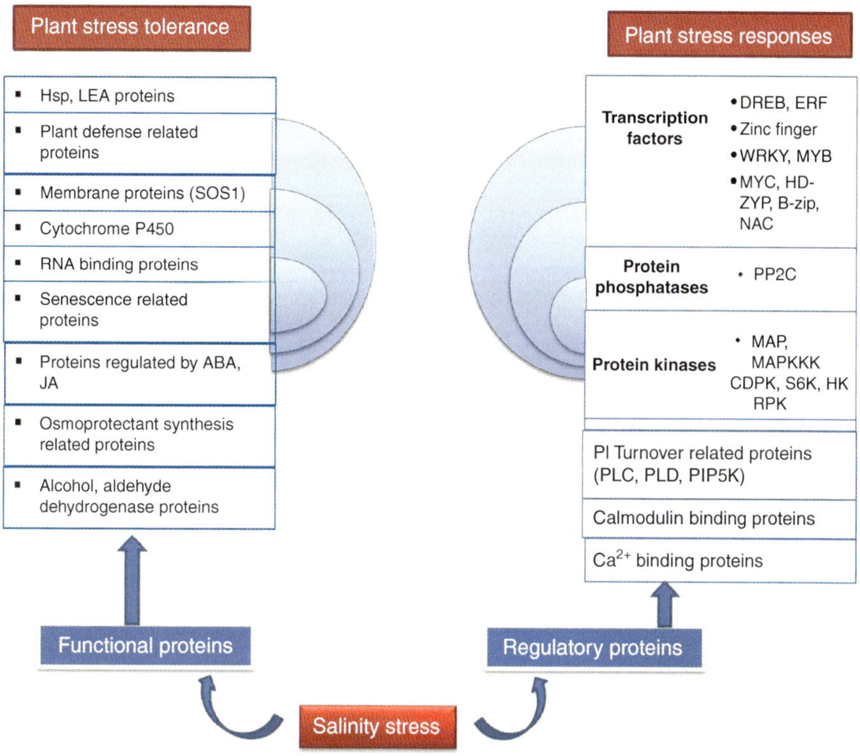

Fig. 4.3 Proteins induced under salt stress and their possible role in salt tolerance and response

Proteomic Approach to Salinity Tolerance

Development of proteomics is becoming a powerful tool for understanding the mechanism of plant response towards different stress conditions and provides a direct link between gene sequence and the biological activity. Proteomic experiments consist of four basic steps: sample preparation, protein separation, identification and function analysis. Several studies to characterize the expressed plant proteins (whole plant/organ proteomes) and/or the subcellular proteomes of organelles (chloroplast mitochondrion, peroxisome, nucleus) have been performed earlier (Lilley and Dupree 2007).

Development in "omic" tools have created major interests in scientists all over the world and opened new perspectives in stress biology of plants. The term proteomics was coined by Wilkins et al. (1996) and it include studies on systematic and detailed analysis of the proteins in a cellular compartment, tissue and the whole organism. The proteomics of several plants have already been studied in response to different environmental stresses such as drought (Salekdeh et al. 2002), heat (Majoul et al. 2003), cold (Cui et al. 2005) and salt (Abbasi and Komatsu 2004; Yan et al. 2005). However, proteomic analysis data of very few plants under salt stress is available.

Claes et al. (1990) observed the accumulation of lectin like protein with increasing salt treatments. Moons et al. (1995) identified production of high amount of a novel histidine-rich protein and two types of LEA in roots of tolerant rice varieties compared to those of sensitive varieties. 2-D gel electrophoresis revealed up-regulation of a 33 kDa protein with pI 5.2, as a result of NaCl treatment in the leaf extract of the mangrove plant, *Bruguiera gymnorhiza*.

Two MS platforms, Matrix-assisted laser desorption ionization time-of-flight (MALDI-TOF-MS) and sequencing of individual peptides by electrospray ionization tandem mass spectrometry (ESI-MS/MS) represent powerful tools for proteomic studies (Mann et al. 2001). The protein identification process by using these MS paltforms is well supported by a wide variety of non-redundant protein and translated nucleic acid databases in stress biology (Dubey and Grover 2000). The large protein databases include the SWISS-PROT and TrEMBL, NCBInr and pdEST and their corresponding genes from The *Arabidopsis* Information Resource (TAIR) (http://www.arabidopsis.org), International Rice Genome Consortium (IRGC) Databases (RGDs) (http://rgp.dna.affrc.go.jp) and TIGR (rice genome index databases). FASTA, BLAST and WUBLAST are the widely used programs for protein identification all over the world. The US National Center for Biotechnology Information (NCBI) database (http://www.ncbi.nlm.nih.gov) contains translated protein sequences from the entire collection of DNA sequences and are available at GeneBank (Rakwal and Agarawal 2003) and widely utilized by researchers.

Conclusions and Future Perspectives

Salinity tolerance in plants is a complex quantitative trait that is governed by a number of environmental and genetic interactions. Perfect correlation between molecular markers and their linking with phenotypic scores to predict a genetic region of DNA that contains genes influencing the plant response to salinity stress will be very much helpful in determining genetic basis of salinity tolerance. Emerged understanding of complexity of the physiological and the genetical responses of salt tolerance in plants in the present time enables the researchers to generate salt-tolerant crops. A variety of approaches that have been used to achieve these goals include conventional breeding, wide crossing, use of physiological traits and, more recently, marker-assisted selection and the use of transgenic plants. Although, phenotypically large number of plants population is a limiting component, in recent years, number of high-throughput assays for gene identification and molecular marker generation are coming online and increasing the pace of quality research. Continuous development of functional tools will enable researchers to dissect many of the fundamental questions related to the salt tolerance mechanism in crop plants and rapid development of salt tolerant high yielding varieties to solve global food problem.

Acknowledgments Ashwani Kumar acknowledges the Claude Leon Foundation and National Research Foundation (NRF), South Africa for providing Postdoctoral support.

References

Abbasi F, Komatsu S (2004) A proteomic approach to analyze salt responsive proteins in rice leaf sheath. Proteomics 4:2072–2081

Ahmad P, Sharma S (2008) Salt stress and phyto-biochemical responses of plants. Plant Soil Environ 54(3):89–99

Ahmad P, Jhon R, Sarwat M, Umar S (2008) Responses of proline, lipid peroxidation and antioxidative enzymes in two varieties of *Pisum sativum* L. under salt stress. Int J Plant Production 2(4):353–366

Ahmad P, Jaleel CA, Salem MA, Nabi G, Sharma S (2010) Roles of enzymatic and non-enzymatic antioxidants in plants during abiotic stress. Crit Rev Biotechnol 30(3):161–175

Ahmad P, Nabi G, Jeleel CA, Umar S (2011) Free radical production, oxidative damage and antioxidant defense mechanisms in plants under abiotic stress. In: Ahmad P, Umar S (eds) Oxidative stress: role of antioxidats in plants. Studium Press, New Delhi, pp 19–53

Ahmad P, Bhardwaj R, Tuteja N (2012a) Plant signaling under abiotic stress environment. In: Ahmad P, Prasad MNV (eds) Environmental adaptations and stress tolerance of plants in the era of climate change. 10.1007/978-1-4614-0815-4_12, © Springer Science+Business Media, LLC 2012

Ahmad P, Kumar A, Gupta A, Hu X, Hakeem KR, Azooz MM, Sharma S (2012b) Polyamines: role in plants under abiotic stress. In: Ashraf M, Ozturk M, Ahmad MSA, Aksoy A (eds) Crop production for agricultural improvement. pp 490–512, © Springer Science+Business Media, LLC 2012

Amaya I, Rotella MA, Calle M, Medina MI, Heredia A, Bressan RA et al (1999) Improved germination under osmotic stress of tobacco plants over-expressing a cell wall peroxidase. FEBS Lett 457:80–84

Ashraf M (1994) Breeding for salinity tolerance in plants. Crit Rev Plant Sci 13:17–42

Ashraf M (2009) Biotechnological approach of improving plant salt tolerance using antioxidants as markers. Biotechnol Adv 27:84–93

Bartels D, Sunkar R (2005) Drought and salt tolerance in plants. Crit Rev Plant Sci 24:23–58

Bohnert HJ, Jensen RG (1996) Strategies for engineering water-stress tolerance in plants. Trends Biotechnol 14:89–97

Bohnert HJ, Gong Q, Li P, Ma S (2006) Unraveling abiotic stress tolerance mechanisms – getting genomics going. Curr Opin Plant Biol 9:180–188

Braam J, Sistrunk ML, Polisensky DH, Xu W, Purugganan MM, Antosiewicz DM et al (1997) Plant responses to environmental stress: regulation and functions of the Arabidopsis TCH genes. Planta 203:35–41

Chen X (2004) A microRNA as translational repressor of APETALA2 in Arabidopsis flower development. Science 303:2022–2025

Chinnusamy V, Gong Z, Zhu JK (2008) Abscisic acid-mediated epigenetic processes in plant development and stress responses. J Integr Plant Biol 50(10):1187–1195

Claes B, Dekeyser R, Villarroel R, Bulcke VM, Bauw G, Montagu MV (1990) Characterization of rice gene showing organ specific expression in response to salt stress and drought. Plant Cell 2:19–27

Cramer GR, Urano K, Delrot S, Pezzotti M, Shinozaki K (2011) Effects of abiotic stress on plants: a systems biology perspective. BMC Plant Biol 11:163

Cui S, Huang F, Wang J, Ma X, Cheng Y, Liu J (2005) A proteomic analysis of cold stress responses in rice seedlings. Proteomics 5:3162–3172

Dubey H, Grover A (2000) Current initiatives in proteomic research: the plant perspective. Curr Sci 80:262–269

Flowers TJ, Colmer TD (2008) Salinity tolerance in halophytes. New Phytol 179:945–963

Fowler S, Thomashow MF (2002) *Arabidopsis* transcriptome profiling indicates that multiple regulatory pathways are activated during cold acclimation in addition to the CBF cold response pathway. Plant Cell 14:1675–1690

Frary A, Nesbitt TC, Frary A, Grandillo S, Van der Knaap E, Cong B, Liu J, Meller J, Elber R, Alpert KB, Tanksley SD (2000) *fw*-2.2: a quantitative trait locus key to the evolution of tomato fruit size. Science 289:85–88

Gill SS, Tuteja N (2010) Polyamines and abiotic stress tolerance in plants. Plant Signal Behav 5(1):26–33

Gong Q, Li P, Ma S, Indu Rupassara S, Bohnert HJ (2005) Salinity stress adaptation competence in the extremophile Thellungiella halophila in comparison with its relative *Arabidopsis thaliana*. Plant J 44:826–839

Grant MR, Jones JDG (2009) Hormone (dis)harmony moulds plant health and disease. Science 324:750–752

Guleria P, Goswami D, Mahajan M, Kumar V, Bhardwaj J, Kumar SY (2012) MicroRNAs and their role in plants during abiotic stresses. In: Ahmad P, Prasad MNV (eds) Environmental adaptations and stress tolerance of plants in the era of climate change. 10.1007/978-1-4614-0815-4_12, © Springer Science+Business Media, LLC 2012

Gygi SP, Rist B, Aebersold R (2000) Measuring gene expression by quantitative proteome analysis. Curr Opin Biotechnol 11:396–401

Hanson AD, Burnet M (1994) Evolution and metabolic engineering of osmoprotectant accumulation in higher plants. In: Cherry JH (ed) Cell biology: biochemical and cellular mechanisms of stress tolerance in plants, NATO ASI series H. Springer, Berlin, pp 291–302

Hirayama T, Shinozaki K (2010) Research on plant abiotic stress responses in the post-genome era: past, present and future. Plant J 61:1041–1052

Hong Z, Lakhineni K, Zhang Z, Verma DPS (2000) Removal of feedback inhibition of Ä1-pyrroline-5-carboxylate synthetase results in increased proline accumulation and protection of plants from osmotic stress. Plant Physiol 122:1129–1136

Hubbard KE, Nishimura N, Hitomi K, Getzoff ED, Schroeder JI (2010) Early abscisic acid signal transduction mechanisms: newly discovered components and newly emerging questions. Genes Dev 24(16):1695–1708

Huh G-H, Damsz B, Matsumoto TK, Reddy MP, Rus AM, Ibeas JI, Narasimhan ML, Bressan RA, Hasegawa PM (2002) Salt causes ion disequilibrium-induced programmed cell death in yeast and plants. Plant J 29:649–659

Inan G, Zhang Q, Li P et al (2004) Salt cress. A halophyte and cryophyte Arabidopsis relative model system and its applicability to molecular genetic analyses of growth and development of extremophiles. Plant Physiol 135:1718–1737

Jamil A, Riaz S, Ashraf M, Foolad MR (2011) Gene expression profiling of plants under salt stress. Crit Rev Plant Sci 30:435–458

Khan MR, Mohiddin FA, Khan MM (2007) Effect of low levels of SO_2 on the growth and yield of indigenous germplasm of black mustard. Environ Biol Conservat 12:53–57

Kim TH, Bohmer M, Hu H, Nishimura N, Schroeder JI (2010) Guard cell signal transduction network: advances in understanding abscisic acid, CO_2, and Ca^{2+} signaling. Annu Rev Plant Biol 61:561–591

Koiwa H, Bressan RA, Hasegawa PM (2006) Identification of plant stress-responsive determinants in *Arabidopsis* by large scale forward genetic screens. J Exp Bot 57:1119–1128

Kovtun Y, Chiu WL, Tena G, Sheen J (2000) Functional analysis of oxidative stress activated mitogen-activated protein kinase cascade in plants. Proc Natl Acad Sci USA 97:2940–2945

Koyro HW, Ahmad P, Geissler N (2012) Abiotic stress responses in plants: an overview. In: Ahmad P, Prasad MNV (eds) Environmental adaptations and stress tolerance of plants in the era of climate change. Springer, Science + business media, New York, pp 1–28

Lilley KS, Dupree P (2007) Plant organelle proteomics. Curr Opin Plant Biol 10:594–599

Lin HX, Zhu MZ, Yano M, Gao JP, Liang ZW, Su WA et al (2004) QTLs for Na+and K+uptake of the shoots and roots controlling rice salt tolerance. Theor Appl Genet 108:253–260

Majoul T, Bancel E, Triboi E, Hamida B, Branlard G (2003) Proteomic analysis of the effect of heat stress on hexaploid wheat grain: characterization of heat-responsive proteins from total endosperm. Proteomics 3:175–183

Mann M, Hendrickson RC, Pandey A (2001) Analysis of proteins and proteomes by mass spectrometry. Annu Rev Biol Chem 70:437–473

Mansour MMF (1998) Protection of plasma membrane of onion epidermal cells by glycine betain and proline against NaCl stress. Plant Physiol Biochem 36:767–772

Mantri N, Patade V, Penna S, Ford R, Pang E (2012) Abiotic stress responses in plants: present and future. In: Ahmad P, Prasad MNV (eds) Abiotic stress responses in plants: metabolism, productivity and sustainability. 10.1007/978-1-4614-0634-1, © Springer Science+Business Media, LLC 2012
Maruyama K, Sakuma Y, Kasuga M, Ito Y, Seki M, Goda H, Shimada Y, Yoshida S, Shinozaki K, Yamaguchi-Shinozaki K (2004) Identification of cold inducible downstream genes of the arabidopsis DREB1A/CBF3 transcriptional factor using two microarray systems. Plant J 38(6):982–993
Mazzucotelli E, Mastrangelo AM, Crosatti C, Guerra D, Stanca AM, Cattivelli L (2008) Abiotic stress response in plants: when post-transcriptional and post-translational regulations control transcription. Plant Sci 174:420–431
Mestichelli LJJ, Gupta RN, Spenser ID (1979) The biosynthetic route from ornithine to proline. J Biol Chem 254:640–647
Mittler R (2002) Oxidative stress, antioxidants, and stress tolerance. Trends Plant Sci 9:405–410
Mittler R (2006) Abiotic stress, the field environment and stress combination. Trends Plant Sci 11:15–19
Moons A, Bauw G, Prinsen E, Montagu MV, Van der Straeten D (1995) Molecular and physiological responses to abscisic acid and salts in roots of salt-sensitive and salt-tolerant indica rice varieties. Plant Physiol 107:177–186
Nakashima K, Ito Y, Yamaguchi-Shinozaki K (2009) Transcriptional regulatory networks in response to abiotic stresses in Arabidopsis and grasses. Plant Physiol 149:88–95
Nenova V (2008) Growth and mineral concentrations of pea plants under different salinity levels and iron supply. Gen Appl Plant Physiol 34(3–4):189–202
Palatnik JF, Allen E, Wu X, Schommer C, Schwab R, Carrington JC, Weigel D (2003) Control of leaf morphogenesis by microRNAs. Nature 425:257–263
Pardo JM (2010) Biotechnology of water and salinity stress tolerance. Curr Opin Biotechnol 21:185–196
Pastori GM, Foyer CH (2002) Common components, networks, and pathways of cross tolerance to stress. The central role of 'redox' and abscisic acid-mediated controls. Plant Physiol 129:460–468
Patade VY, Suprasanna P (2010) Short-term salt and PEG stresses regulate expression of MicroRNA, *miR159* in sugarcane leaves. J Crop Sci Biotechnol 13(3):177–182
Pieterse CMJ, Leon-Reyes A, Van der Ent S, Van Wees SCM (2009) Networking by small-molecules hormones in plant immunity. Nat Chem Biol 5:308–316
Rakwal R, Agarawal GK (2003) Rice proteomics: current status and future perspectives. Electrophoresis 24:3378–3389
Ren ZH, Gao JP, Li LG, Cai XL, Huang W, Chao DY, Zhu MZ, Wang ZY, Luan S, Lin HX (2005) A rice quantitative trait locus for salt tolerance encodes a sodium transporter. Nat Genet 37:1141–1146
Rhodes D, Rich PJ, Brunk DG, Ju GC, Rhodes JC, Pauly MH, Hansen LA (1989) Development of two isogenic sweet corn hybrids differing for glycine betaine content. Plant Physiol 9:1112–1121
Rose JKC, Bashir S, Giovannoni JJ, Jahn MM, Saravanan RS (2004) Tackling the plant proteome: practical approaches, hurdles and experimental tools. Plant J 39:715–733
Roy SJ, Tucker EJ, Tester M (2011) Genetic analysis of abiotic stress tolerance in crops. Curr Opin Plant Biol 14:232–239
Sakuma Y, Maruyama K, Osakabe Y, Qin F, Seki M, Shinozaki K, Yamaguchi-Shinozaki K (2006) Functional analysis of an Arabidopsis transcription factor, DREB2A, involved in drought-responsive gene expression. Plant Cell 18(5):1292–1309
Salekdeh GH, Siopongco J, Wade LJ, Ghareyazie B, Bennett J (2002) Proteomic analysis of rice leaves during drought stress and recovery. Proteomics 2:1131–1145
Serraj R, Sinclair TR (2002) Osmolyte accumulation: can it really help increase in crop yield under drought conditions. Plant Cell Environ 25:333–341
Sheveleva E, Chmara W, Bohnert HJ, Jensen RG (1997) Increased salt and drought tolerance by D-ononitol production in transgenic *Nicotiana tabacum* L. Plant Physiol 115:1211–1219
Shukla LI, Chinnusamy V, Sunkar R (2008) The role of microRNAs and other endogenous small RNAs in plant stress responses. Biochim Biophys Acta 1779:743–748

Streeter JG, Lohnes DG, Fioritto RJ (2001) Patterns of pinitol accumulation in soybean plants and relationships to drought tolerance. Plant Cell Environ 24:429–438

Sunkar R, Zhu JK (2004) Novel and stress-regulated microRNAs and other small RNA from Arabidopsis. Plant Cell 16:2001–2019

Tester M, Langridge P (2010) Breeding technologies to increase crop production in a changing world. Science 327:818–822

The Arabidopsis Genome Initiative (2000) Analysis of the genome sequence of the flowering plant Arabidopsis thaliana. Nature 408:796–815

Tseng MJ, Liu CW, Yiu JC (2007) Enhanced tolerance to sulfur dioxide and salt stress of transgenic Chinese cabbage plants expressing both superoxide dismutase and catalase in chloroplasts. Plant Physiol Biochem 45:822–833

Türkan I, Demiral T (2008) Salinity tolerance mechanisms of higher plants. In: Khan NA, Singh S (eds) Abiotic stress and plant responses. I.K. International, New Delhi, pp 106–123

Tuteja N, Sopory SK (2008) Chemical signaling under abiotic stress environment in plants. Plant Signal Behav 3:525–536

Ushimaru T, Nakagawa T, Fujioka Y, Daicho K, Naito M, Yamauchi Y, Nonaka H, Amako K, Yamawaki K, Murata N (2006) Transgenic *Arabidopsis* plants expressing the rice dehydroascorbate reductase gene are resistant to salt stress. J Plant Physiol 163:1179–1184

Varshney RK, Bansal KC, Aggarwal PK, Datta S, Craufurd PQ (2011) Agricultural biotechnology for crop improvement in a variable climate: hope or hype? Trends Plant Sci 1 6(7):363–371

Wang J, Zuo K, Wu W, Song J, Sun X, Lin J et al (2004) Expression of a novel antiporter gene from Brassica napus resulted in enhanced salt tolerance in transgenic tobacco plants. Biol Plantarum 48:509–515

Wang B, Davenport RJ, Volkov V, Amtmann A (2006) Low unidirectional sodium influx into root cells restricts net sodium accumulation in Thellungiella halophila, a salt-tolerant relative of Arabidopsis thaliana. J Exp Bot 57:1161–1170

Weigel D, Ahn JH, Blazquez MA, Borevitz JO, Christensen SK, Fankhauser C, Ferrandiz C, Kardailsky I, Malancharuvil EJ, Neff MM, Nguyen JT, Sato S, Wang ZY, Xia Y, Dixon RA, Harrison MJ, Lamb CJ, Yanofsky MF, Chory J (2000) Activation tagging in *Arabidopsis*. Plant Physiol 122:1003–1013

Wilkins MR, Sanchez JC, Gooley AA, Appel RD, Humphry-Smith I, Hochstrasser DF, Williams KL (1996) Progress with proteome projects: why all proteins expressed by a genome should be identified and how to do it. Biotechnol Genet Eng Rev 13:19–50

Xin H, Qin F, Tran Lam-Son P (2012) Transcription factors involved in environmental stress responses in plants. In: Ahmad P, Prasad MNV (eds) Environmental adaptations and stress tolerance 297 of plants in the era of climate change. 10.1007/978-1-4614-0815-4_12, © Springer Science+Business Media, LLC 2012

Yamada K, Lim J, Dale JM, Chen HM, Shinn P, Palm CJ, Southwick AM, Wu AC, Kim C, Nguyen M et al (2003) Empirical analysis of transcriptional activity in the arabidopsis genome. Science 302:842–846

Yamaguchi-Shinozaki K, Shinozaki K (2006) Transcriptional regulatory networks in cellular responses and tolerance to dehydration and cold stresses. Annu Rev Plant Biol 57:781–803

Yamamoto A, Bhuiyan MN, Waditee R, Tanaka Y, Esaka M, Oba K, Jagendorf AT, Takabe T (2005) Suppressed expression of the apoplastic ascorbate oxidase gene increases salt tolerance in tobacco and Arabidopsis plants. J Exp Bot 56:1785–1796

Yan S, Tang Z, Su W, Sun W (2005) Proteomic analysis of salt stress-responsive proteins in rice roots. Proteomics 5:235–244

Yan SP, Zhang QY, Tang ZC, Su WA, Sun WN (2006) Comparative proteomic analysis provides new insights into chilling stress responses in rice. Mol Cell Proteomics 5:484–496

Yancey PH, Clark ME, Hand SC, Bowlus RD, Somero GC (1982) Living with water stress: evolution of osmolyte systems. Science 217:1214–1222

Zhao BT, Liang RQ, Ge LF, Li W, Xiao HS, Lin HX, Ruan KC, Jin YX (2007) Identification of drought induced microRNAs in rice. Biochem Biophys Res Commun 354:585–590

Chapter 5
LEA Proteins in Salt Stress Tolerance

Renu Bhardwaj, Indu Sharma, Mukesh Kanwar, Resham Sharma, Neha Handa, Harpreet Kaur, Dhriti Kapoor, and Poonam

Abstract In late embryogenesis, the water content of living cell is reduced tremendously that leads to a state of dehydration and thus, might impose severe irreparable damage to cellular and macromolecular structures. However, the mature orthodox seeds can withstand severe desiccation due to role of osmoprotectants viz., reducing sugars, prolines, glycinebetaines or *L*ate *E*mbryogenesis *A*bundant (LEA) proteins. These operate on the virtue of intrinsic molecular mechanisms that alleviate multiple abiotic stresses in plants such as protein desiccation, membrane degradation, salt stress and cold and chilling stress. The LEA proteins are a group of versatile, adaptive, hydrophilic proteins considerably defined as 'molecular shields' for their anti-stress properties attributable to partial or complete structural randomness. On the basis of their amino acid composition and sequencing, LEA proteins have been clubbed into seven groups that are further sub divided into a number of protein sub families. Out of these, Group 2 LEA proteins called the 'Dehydrins' are of prime importance in the plant kingdom. The latent and unique stress remediating characteristics of this class of proteins has been further enhanced by transgenic studies, wherein the target LEA genes have been identified, sequenced to understand their molecular role in plants. Further investigations into the behavior of LEA proteins and mode of their regulation in stressed plants will facilitate in elucidating the function of LEA proteins. The present chapter reviews the versatility and role of LEA proteins in plant stress protection.

Keywords Dehydrins • Protein *aggrėgation* • *Desiccation* intrinsically unstructured proteins (IUPs) • Osmotic stress

---

R. Bhardwaj (✉) • I. Sharma • M. Kanwar • R. Sharma • N. Handa
• H. Kaur • D. Kapoor • Poonam
Department of Botanical and Environmental Sciences, Guru Nanak Dev University,
Amritsar 143005, Punjab, India
e-mail: dr.renubhardwaj@yahoo.in; renu_bhardwaj@rediffmail.com

P. Ahmad et al. (eds.), *Salt Stress in Plants: Signalling, Omics and Adaptations*,
DOI 10.1007/978-1-4614-6108-1_5,

*L*ate *E*mbryogenesis *A*bundant proteins (LEA proteins) are present in both animals and plants which protect other proteins from aggregation, desiccation or osmotic stresses associated with low temperature (Goyal et al. 2005a; Hand et al. 2011; Hundertmark et al. 2012; Furuki et al. 2012; Liu et al. 2011a). The LEA proteins were termed as LEA proteins due to their discovery in maturing plant seeds however these are also reported to be expressed either constitutively or in response to stress in plant tissues (Hand et al. 2011; Liu et al. 2011a). These proteins were initially discovered in cotton seeds and wheat during late embryogenesis (Cuming 1999) and are produced in abundance during seed development, comprising up to 4% of cellular proteins (Roberts et al. 1993). Although the expression of LEA proteins is linked to the acquisition of desiccation tolerance in orthodox seeds, pollen and anhydrobiotic plants, yet an array of LEA proteins are induced by cold or osmotic stress, by exogenous abscisic acid, or are even expressed constitutively, e.g. *dhnX* from *Arabidopsis thaliana* (Welin et al. 1994; Liu et al. 2011a). Burgeoning studies have revealed that LEA proteins have a protective role in a variety of organisms, including the bacterium *Deinococcus radiodurans*, nematode *Caenorhabditis elegans*, *Artemia* (Brine shrimp), and rotifers against desiccation, cold, or high salinity respectively (Amara et al. 2012; Goyal et al. 2005a; Tunnacliffe and Wise 2007). LEA proteins function by mechanisms which are distinct from those displayed by heat shock molecular chaperones (Goyal et al. 2005a). LEA proteins are particularly protective of mitochondrial membranes against dehydration damage. However, various functions of LEA proteins have been proposed, their precise role have not been defined. They are assumed to protect cellular or molecular structures from the damaging effects of water loss; a number of putative mechanisms have been proposed, including hydration buffering, ion sequestration, direct protection of other proteins or membranes, or renaturation of unfolded proteins (Goyal et al. 2005b; Zhang et al. 2000). Recent bioinformatics studies suggest that LEA proteins might behave as molecular chaperones (Wise 2003).

Heat-stress experiments with citrate synthase, which is susceptible to aggregation at high temperatures, suggested that LEA proteins do not behave as classical molecular chaperones, but they do exhibit a protective, synergistic effect in the presence of the so-called chemical chaperone, trehalose (Goyal et al. 2005a; Viner and Clegg 2000). In contrast, both LEA proteins can independently protect citrate synthase from aggregation due to desiccation and freezing. Besides this, there seems to be some effect on stress tolerance, since tomato, wheat and barley LEA proteins confer increased resistance to osmotic or freeze stresses when introduced into yeast. A barley LEA protein improves tolerance to water deficit in transgenic rice and wheat; furthermore, in vitro, an algal LEA protein diminished freeze damage of the enzyme lactate dehydrogenase (Bray 1993; Cuming 1999; Goyal et al. 2005a).

LEA proteins are a family of hydrophilic proteins, which get aggregated to high levels during the last stage of seed maturation and during onset of water deficiency in vegetative organs (Garay-Arroyo et al. 2000; Hoekstra et al. 2001). On the basis of sequence arrangements and similarity, LEA proteins have been clustered into different groups (Colmenero-Flores et al. 1999; Cuming 1999).

A great deal of structural similarity has not been identified between the various members of these groups. However, a common binding factor for most of them is their high degree of hydrophilic nature and increased levels in content of amino acids such as glycine, alanine and serine (Dure 1993). Genes of LEA proteins have been identified in many plant species, and at least six different groups of LEA proteins have been defined on the basis of expression pattern and sequence. The major categories are group 1, group 2 and group 3 LEA proteins. The Group 1 LEA proteins are found in plants and have been further subdivided into two super families by Wise (2003). The wheat Em protein is the type sequence of Group 1 LEA proteins. They are unstructured in solution, but contain a conserved 20-residue amino acid motif, most often in one copy. Group 2 LEA proteins are subdivided into three super families and are also known as 'dehydrins'. Group 2 proteins are characterized by up to three sequence motifs, known as the K-domain (lysine-rich), the Y-domain (DEYGNP) and the S-segment (poly-serine stutter). Again, they are largely unstructured, although they show some α-helical content.

Group 3 LEA proteins have amino acid sequences with characteristic 11-mer motifs and are known to reduce aggregation of proteins during dehydration (Chakrabortee et al. 2007; Goyal et al. 2005a; Furuki et al. 2012). Also, the homologues of this group in the organisms other than plants, including the nematodes *Caenorhabditis elegans*, *Steinernema feltiae* and *Aphelenchus avenae*, and the prokaryotes *Deinococcus radiodurans*, *Bacillus subtilis* and *Haemophilus influenzae* have been discovered. The group 3 LEA protein from anhydrobiotic nematode *A. avenae*, and a putative example from bullrush, are natively unfolded in solution, but seem to become more structured on drying. The genome of *Arabidopsis thaliana* contains 51 genes encoding LEA proteins and similar to other LEA proteins, three of these proteins have been shown to be unstructured in solution, while they fold into α-helices upon drying (Hundertmark et al. 2012). Previous studies indicate that the proteins ATEM6 and ATEM1 of *A. thaliana* group 1 LEA proteins protect maturing embryos against desiccation (Manfre et al. 2009). Several recent reports indicate the engagement of LEA proteins in desiccation stress tolerance in maturing seeds (Hand et al. 2011; Leprince and Buitink 2010; Shih et al. 2010).

Properties and Diversity of LEA Proteins

LEA proteins are considered as the highly hydrophilic and small but exception in both of the properties exist. In *Arabdiopsis thaliana* nine different groups of LEA proteins has been projected based on the amino acid sequence analysis (Bies-Etheve et al. 2008; Hundertmark and Hincha 2008). Due to the high hydrophilic nature of the LEA protein, they are predicted as the unstructured in the hydrated state (Uversky et al. 2000). Amara et al. (2012) has discovered 20 LEA proteins in maize under plant stress.

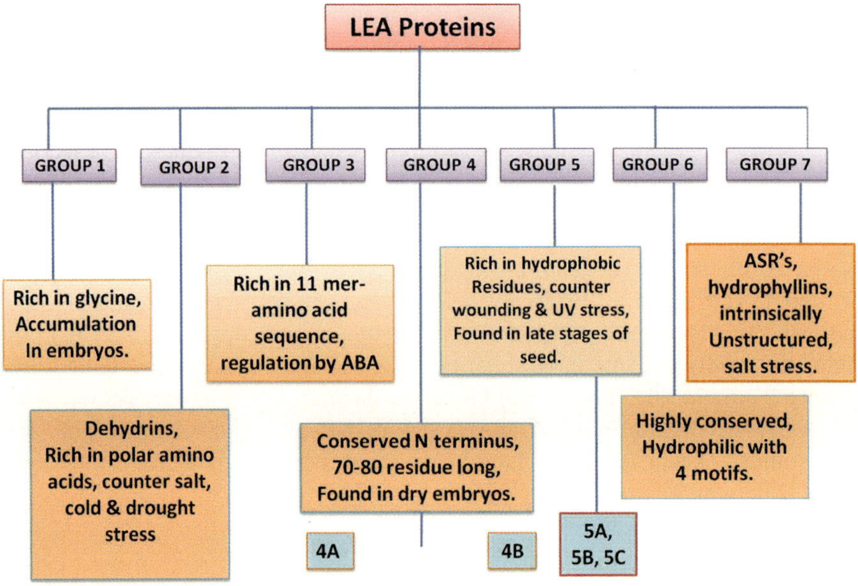

Fig. 5.1 Classification and occurrence of LEA proteins (*ABA* abscisic-acid, *ASR* abscisic-acid stress ripening)

The Versatility of LEA Proteins

Generally, LEA proteins were classified into seven subclasses classified according to expression pattern or protein characteristics on the basis of their amino acid sequences and conserved motifs (Fig. 5.1). Besides this, another criterion for classification was provided by a computer analysis, "Protein or Oligonucleotide Probability Profile" (POPP), that showed over or under-representation of particular amino acids in protein sequences (Wise 2003). These were reclassified into four sub-groups, after clustering by consensus POPP although the primary structures of most LEA proteins shared similar biophysical features (Battaglia et al. 2008; Wise 2003; Wise and Tunnacliffe 2004). In the present classification, Groups 1, 2, 3, 4, 6, and 7 correspond to the hydrophilic or "typical" LEA proteins (Battaglia et al. 2008), whereas Group 5 proteins show "atypical" or hydrophobic characteristics (Fig. 5.1). Bioinformatics approaches have been used in building a database known as LEAPdb, dedicated to the identification and *in silico* analysis of LEA proteins (Hunault and Jaspard 2010). LEAdb inculcates710 non-redondant and curated sequences of LEA proteins from all organisms, and is equipped with many analytical tools. Such a computational analysis led to a detailed and robust classification of LEA proteins.

Recent studies have revealed a number of unusual examples of LEA proteins. For instance two Group 3 (Pfam LEA-4) LEA proteins are described in the bdelloid rotifer *Adineta ricciae* which are rather less hydrophilic having a grand average hydropathy (GRAVY; Consensus scale, Eisenberg 1984) score of −0.46 in comparison to the average score for Group 3 LEA proteins is −0.97 (Wise 2003), which is analogous to that of many "normal" globular proteins such as serum albumin (−0.43). The first group of these LEA proteins, ArLEA1A shows characteristics of other group 3 LEA proteins that are unstructured in solution and gaining structure on drying. It reduces aggregation activity of Citrate Synthase in vitro desiccation assays. On the other hand the second group ArLEA1B is structured in solution largely folded as alpha-helix. It does not have anti-aggregation activity. In fact, it itself gets aggregated on drying. ArLEA1B shows a higher tendency to associate with phospholipid bilayers than ArLEA1A (Pouchkina-Stantcheva et al. 2007).

Group 1 Proteins

Group 1 proteins are prevalent in plants, but rare in metazoa (Sharon et al. 2009). This group of LEA proteins is represented by the D-19 and D-132 proteins from developing cotton seeds, (Baker et al. 1995; Galau et al. 1992; Tunnacliffe et al. 2010). They contain a 20-residue amino acid motif family and are unshaped in solution (Browne et al. 2002; McCubbin et al. 1985). This group contains large number of charged residues which lead to the high hydrophilicity of these proteins. This group also contains a high content of Gly residues and hence it is thought that they exist largely as random coils or unstructured in aqueous solution (Eom et al. 1996; Russouw et al. 1997; Soulages et al. 2002). Group 1 proteins accumulate during embryo development particularly in dry seeds and they also have been detected in the organs that undergo dehydration for example pollen grains (Espelund et al. 1992; Hollung et al. 1994; Prieto-Dapena et al. 1999; Ulrich et al. 1990; Vicient et al. 2000; Williams and Tsang 1994; Wurtele et al. 1993). Group 1 Lea protein express in water stress in horsegram (Veeranagamalliaiah et al. 2011).

Group 2 Proteins (Dehydrins)

Group 2 LEA proteins or dehydrins, were identified as D-11 family. These are mainly found in plants and in general accumulate in dehydrating plant tissues such as developing seeds during maturation, or in vegetative tissues which are subject to the environmental stress such as drought, low temperature and salinity (Allagulova et al. 2003; Close 1996). The main characteristic of dehydrins is a conserved, lysine-rich 15-amino acid domain, named k-segment which is mostly present near the C-terminus (Close 1996). Other features of dehydrins include lack of cysteine and tryptophan, presence of charged and polar amino acids and they have the ability to remain in solution after boiling (Close et al. 1989, 1996; Segrest et al. 1990). This group of LEA proteins is highly hydrophilic, contain a high proportion of charged and polar amino acids and have a low fraction of nonpolar, hydrophobic residues (Garay-Arroyo et al. 2000; Tunnacliffe et al. 2010).

Group 3 Proteins

The group 3 LEA proteins are distributed ubiquitously in the plant kingdom. They have also been found in nonvascular plants (Hellwege et al. 1996), in seedless vascular plants (Salmi et al. 2005), in all seed plants and in algae (Joh et al. 1995). These proteins have characteristics of a repeated 11-mer amino acid motif whose consensus sequence has been broadly defined as ΦΦE/QXΦKE/QKΦXE/D/Q, where Φ represents a hydrophobic residue (Dure 2001). A the study on bullrush *Typha latifolia* that the group 3 LEA proteins are natively unfolded in solution but looks more structural after drying (Wolkers et al. 2001). According to Dure (2001) there may be variability in the 11-mer motif which leads to the sub-classification of group 3 LEA proteins. The two subgroups of LEA proteins are 3A, represented by cotton D-7 LEA protein and 3B represented by cotton D-29 LEA protein. Based on the information available from transcriptomic projects and expression analysis, it has been found that these proteins accumulate in mature seeds, and also to the response of some abiotic factors like dehydration, salinity or low temperature (Cattivelli and Bartels 1990; Harada et al. 1989; Hsing et al. 1995; Romo et al. 2001). Structure and available regulation information of Group 3 genes/proteins in *Arabidopsis* have been reviewed (Zhao et al. 2011). The function of Group 3 proteins in response to desiccation and the relationship between protein structure and functions have also been discussed. It has been suggested that expression of Group 3 LEA proteins may be regulated by ABA during specific developmental stages or upon stress conditions. (Curry et al. 1991; Curry and Walker-Simmons 1993; Dehaye et al. 1997; Dong and Dunstan 1997; Piatkowski et al. 1990; Amara et al. 2012).

Group 4 Proteins

Group 4 proteins generally occur in the plant kingdom including vascular and nonvascular plants. These proteins are conserved in N-terminal portion and are approximately 70–80 residues long (Dure 1993). On the basis of motifs, Group 4 LEA proteins are sub classified into two subgroups. Group 4a contains small proteins of 80–124 residues long with motif 2 or 3 adjoining motif 1. The second group 4B has longer proteins of 108–180 residues containing motif 4 or 5 at C-terminus region. Cotton protein D-113 belongs to group 4B. Kyte and Doolittle (1982) have suggested that first 70–80 residues could adopt α-helix structure while other proteins assume a random coil conformation. Group 4 proteins were originally found in the dry embryos. Cotton D-113 proteins were found in the all embryo tissues in a concentration of 300 mM (Roberts et al. 1993). Similar proteins were found in the tomato plant leaves in response to water deficiency and ABA (Cohen et al. 1991). Carrillo et al. (2010) has obtained accumulation of group 4 LEA proteins in *Arabdiopsis* under water stress during different developmental stages. Absence of group 4 LEA proteins result in the seed dormancy in *arabdiopsis* plants (Carrillo et al. 2011).

Group 5 Proteins

This group of proteins contains a significantly higher proportion of hydrophobic residues. Since the first proteins of this group were D-34, D-73 and D-95, these were named as subgroups 5A, 5B and 5C respectively. These subgroup proteins are insoluble after boiling signifying that they adopt a globular conformation (Baker et al. 1988; Cuming 1999; Galau et al. 1993). While there is little knowledge about this group of proteins in the existing data indicates that these proteins accumulate during the late stage of seed development. Also these proteins have been reported to accumulate in response to the stress conditions like drought; salinity, wounding, cold and UV light (Kim et al. 2005; Kiyosue et al. 1992; Maitra and Cushman 1994; Park et al. 2003; Stacy et al. 1999; Zegzouti et al. 1997). Group 5 LEA protein genes are isolated in citrus plant (Kim et al. 2011).

Group 6 Proteins

PvLEA18 protein was the first protein found from bean (*Phaseolus vulgaris*) in this group (Colmenero-Flores et al. 1997). Till date, about 36 members of Group 6 protein family have been identified from various vascular plants. This group of proteins are highly conserved and have small size of about 7–14 kD. Four motifs help in distinguishing out of which motif 1 and 2 are highly conserved. Generally this group of proteins lack cysteine and tryptophan residues is highly hydrophilic and does not show coagulation on high temperature. They are predicted as intrinsically unstructured on the basis of their physicochemical properties and in silico analysis (Garay-Arroyo et al. 2000). The PvLea 18 protein levels are high in dry seeds and pollen grains. They also respond to water deficiency and ABA treatments. The expression of this gene is regulated during development in normal growth conditions (Colmenero-Flores et al. 1999).

Group 7 Proteins

The Group 7 proteins are also known as *A*bscisic-acid *S*tress *R*ipening (ASR) proteins and are considered as the members of hydrophilins. They are small, heat stable and intrinsically unstructured (Frankel et al. 2006; Goldgur et al. 2007; Silhavy et al. 1995). This group share physiochemical properties with other LEA proteins. Group 7 proteins accumulate in the seeds during late embryogenesis and also in response to the water deficiency (Maskin et al. 2008). These ASR genes contain three highly conserved regions (Silhavy et al. 1995). These genes transcripts are accumulated during senescence, seed and pollen maturation and fruit ripening. These are also responsive to the abiotic stress conditions such as salinity, water deficit, cold and limited light conditions (Doczi et al. 2005; Padmanabhan et al. 1997; Silhavy et al. 1995). The tissue specification of LEA proteins is variable. In some plants like tomato, melon, pomelo, apricot and grape gene transcripts are found in fruits (Cakir et al. 2003; Canel et al. 1995; Hong et al. 2002; Iusem et al. 1993; Mbeguie-A-Mbeguie et al. 1997) while in other plants these are located in other organs for example in potato tubers (Frankel et al. 2007), in roots of rice (Yang et al. 2004), in leaves or stems of tomato, rice, and maize (Amitai-Zeigerson et al. 1994; Maskin et al. 2008; Riccardi et al. 1998; Vaidyanathan et al. 1999), in pollens of lily (*Lilium longiflorum*; Wang et al. 1998), and in developing tomato seeds (Maskin et al. 2008). In tobacco plants the over-expression of tomato ASR1 protein resulted in increased salt tolerance (Kalifa et al. 2004).

LEA Proteins: Intrinsically Unstructured Proteins

As discussed in prior sections, LEA proteins are intrinsically disordered proteins (IDPs) with several physiological roles focused on desiccation tolerance or anhydrobiosis, water and salt stress abatement, membrane protection etc. The amino acids are specific for a group. For instance, Glycine is present in Group 1 whereas Alanine is characteristic of Group 2. However, it is known that glutamine is most important constituent amino acid, while others such as cysteine, phenylalanine, isoleucine, leucine and tryptophan are present, but in minority in all groups (Wise 2003). The first structural description of a LEA protein was reported for Group 1, Em in *Triticum aestivum* (McCubbin et al. 1985). It was seen that a lack of compactness existed in the protein configuration. Viscosity measurements indicated that the protein has an irregular or stretchy conformation, and for the secondary structure (i.e. α-helix or β-sheet), most of the protein appeared as a random coil. This unstructured nature of the protein reflected its high hydration potential, which in turn is obvious due to unique amino acid content, a high proportion of glycine and glutamine residues. There are approximately 78 Group 1 proteins in plants which have a 20 residue long unstructured motif (Dure 1993; Bray 1993). Group 1 LEA proteins such as those found in *Pisum sativum* were observed to be completely unstructured. Similarly, the Em homologue, EMB-1, from *Daucus carota* showed no specific secondary or tertiary structure (Eom et al. 1996). Group 2 LEA proteins in *Craterostigma* and *Arabidopsis* were observed to be predominantly unfolded and mainly active in phosphorylated forms (Lisse et al. 1996; Mouillon et al. 2006). Group 3 LEA proteins which form the largest group of all the major categories with about 124 plant proteins have largely repeated motifs, for instance those belonging to soybean (Shih et al. 2004; Haaning et al. 2008) show structural irregularity and unevenness. Several examples can be cited from recent research about the inconsistency in LEA protein structure through molecular characterization and functional analysis. He et al. (2012) has characterized *OsLEA5* gene, belonging to the atypical LEA group 5C from *Oryza sativa* L. The OsLEA5 polypeptide is rich in Leu, Ser and Asp while Cys, Trp and Gln residue contents are very low. A reference map of LEA proteins was established by proteomics on a hydrophilic protein fraction from mature *Medicago truncatula* seeds and identified 35 polypeptides encoded by 16 LEA genes. Five LEA polypeptides, representing 6% of the total LEA intensity, accumulated upon acquisition of desiccation tolerance (Chatelain et al. 2012). A cDNA clone encoding a group 7 late embryogenesis abundant proteins, termed *PgLEA*, has been isolated from *Pennisetum glaucum*. The *PgLEA* cDNA encodes a 176 amino acid polypeptide with a predicted molecular mass of 19.21 kDa (Reddy et al. 2012).

Reports have indicated that for most LEA protein groups, the entire molecule or regions of protein domains in the functional state are intrinsically unstructured/disordered (IUPs/IDPs) in LEA proteins (Dyson and Wright 2002). However, this very structural disorder confers special benefits to such proteins. These include an amplified speed of interaction, the ability to combine specificity with weak and reversible binding and the versatility to carry out more than one function (Tompa 2005). IDPs are widespread in plants and carry out mandatory functions coupled with cell signaling and regulation in stress physiology. IDPs, might support the folding of other proteins and prevent their aggregation, i.e., that they can act as chaperones though their molecular mode is different and more complicated as compared to the usual molecular chaperones (Tompa and Kovacs 2010). It is now known that such proteins have a folded configuration around a hydrophobic core, and this folding may be unsettled by the very typical "hydrophobic effect" (Scheef and Fink 2003). With such highly hydrophilic proteins like the LEA proteins, this effect is missing and hence the interactions with biological solvents such as water get to predominate. This may result in very little organised secondary structure (Tompa 2002; Bokor et al. 2005). Most importantly as it shall be further discussed, contrary to globular proteins, intrinsically disordered proteins (IDPs) lack a folded structure and due to this very elementary reason they do not lose solubility at elevated temperatures as well as low temperatures (Kosová et al. 2007; Tantos et al. 2009).

In addition to unstructured arrangements, majority of LEA proteins are low complexity proteins. Lack of elementary secondary structure means that members of the major LEA protein families are included in the large class of proteins which are "natively unfolded", "intrinsically disordered" or "intrinsically unstructured" (Dunker 2001; Uversky et al. 2000). As a consequence to this fact most attempts to isolate, crystallise and purify LEA proteins for X-ray crystallographic studies have failed incessantly (Goyal et al. 2003). LEA proteins from Groups 1, 2, 3 and the former group 4 are partially unfolded also; yet many smaller proteins are thought to be totally unfolded. Gaps in availability of ordered domain structure also affects LEA protein function. However, in some rare examples, secondary structure to a small extent has been revealed in the LEA proteins studied so far. It might be seen that many natively unfolded proteins are not simply highly mobile random coils but have some localised structural elements that are in equilibrium with unstructured states. Despite all this, many natively unfolded proteins are known to undergo increased folding under some specific conditions, usually when they get hooked to another molecule or cation (Dyson and Wright 2002; Uversky et al. 2000). Also, LEA proteins adopt a definite structure under certain conditions where they are required to become functional, i.e. during partial or complete dehydration brought about by freezing or drying or both. Many examples of plant LEA proteins that are unstructured in dilute solution but adapt to a folded configuration of alpha-helical structures in the dry state are reported (Hundertmark and Hincha 2008). A number of IUPs are integral components of protein complexes suggesting that molecular recognition is involved in their functional mechanism. For some IUPs, it is known that they are disordered under physiological conditions and acquire a defined conformation upon binding to their cellular targets. For plant IUPs, such as the typical LEA proteins, some results indicate that they acquire certain ordered conformation under low water availability. Environmental conditions such as variation in temperature, absence or presence of certain substances etc. influence folding, and several LEA proteins become more structured when subjected to high temperatures and dried. Group 3 LEA protein from *Typha latifolia* became largely α-helical when dried rapidly; slow drying resulted in intermolecular β-sheet formation, as well as α- helix (Wolkers et al. 2001). Only α-helix formation was observed in the presence of sucrose, however, regardless of rate of drying, which is perhaps significant, given the presence of non-reducing disaccharides in many desiccation tolerant systems. Similarly, the Group 3 LEA proteins LEAM from pea mitochondria also gain structure on drying. Some examples also indicate that LEA protein folding may be due to water stress and membrane stabilization. Group 3 LEA protein supports the fact that there are conformational shifts in LEA proteins on drying and in association with phospholipid vesicles (Tolleter et al. 2007). For instance, α-synuclein, ultrastructural investigations have shown its location close to, but not at, intracellular vesicle surfaces (Clayton and George 1999). Similarly, a Group 2 LEA protein has also been found close to membrane surfaces (Danyluk et al. 1998). More recently, the interplay and presence of LEA proteins across mitochondria possibly with a redox stabilising role have also been reported (Grelet et al. 2005; Salleh et al. 2011; Tolleter et al. 2010). LEA proteins are strewn across the entire plant cellular machinery and hence constantly adapt to specific structural changes in response to different stress signals. However, still their intrinsic irregularity and structural non confirmation at any level of organization remains elusive (Caramelo and Iusem 2009; Shih et al. 2008).

Functions of LEA Proteins

Despite structural irregularities, LEA Proteins are primarily known to play an array of important roles in plants. They protect cells against damage caused by cellular dehydration, hence countering any sort of water stress faced by the cells. This group is known to act as molecular chaperones against cold and saline stress as well as impart antioxidant and protein and membrane protection functions. In vitro studies have indicated that LEA proteins have cryoprotective properties (Hughes and Graether 2011), metal/lipid binding characteristics and antioxidative activities. These proteins are able to prevent conformational changes in reporter enzymes, induced by water limitation conditions. Liu et al. (2011b) tested the metal binding properties of two related soybean LEA4 proteins, GmPM1 and GmPM9, using immobilized metal ion affinity chromatography (IMAC). The metal ions Fe^{3+}, Ni^{2+}, Cu^{2+} and Zn^{2+} were observed to bind these two proteins to a considerable level. Similarly, accumulation patterns of *Arabidopsis thaliana* group 4 LEA proteins during different developmental stages and plant organs in response to water deficit were studied by Olvera-Carillo et al. (2011). This group of proteins conferred tolerance to severe drought in *Arabidopsis* plants and their deficiency leads to susceptible phenotypes upon water limitation, during germination, or in mature plants after recovery from severe dehydration. The lack of these proteins also indicated a reduced seed production despite optimal irrigation, supporting their role in fruit and/or seed development. These commendable roles essayed by LEA proteins further the prospects of breeding drought tolerant varieties by searching for anti-drought inducible LEA genes and their characterization. As mentioned, the LEA proteins are generally classified into six families according to their amino acid sequence and their corresponding mRNA homology, which are located primarily in cytoplasm and nuclear region. Their synthesis, expression and biological activities are regulated by many factors inclusive of developmental stages, hormones, ion changes, and cell signaling and *lea* genes. Recent research for the genomes of three plants, *Arabidopsis thaliana*, *Oryza sativa* and *Glycine max*, have lead to their sequencing and their many genes and promoters have been predicted. Transcription profiles of the cold- and dehydration-responsive genes were similar among these three species, showing representative up regulated (dehydrin/LEA) and down regulated (photosynthesis-related) genes (Maruyama et al. 2012). The study of the regulatory mechanism of *lea* gene expression is an important feature of modern plant molecular biology (Hong-Bo et al. 2005).

Protein Protection

Dehydrin mediated protein protection occurs via radical scavenging, protection against aggregation, desiccation and denaturation and stabilization. Moreover, dehydrins of KnS-type exhibit intense radical scavenging activity. Similarly, K3S-type CuCOR19 dehydrin from *Citrus unshiu* was seen to exhibit hydroxyl and peroxyl radical-scavenging activity in vitro and this was higher than that of mannitol and equal to that of albumin, which is known to be an antioxidative protein in mammals (Hara et al. 2005). It was also reported that CuCOR19 dehydrin purified from bacterial cells inhibited in vitro peroxidation of soybean (*Glycine max* L.) liposomes (Hara et al. 2004). The inhibitory activity of the CuCOR19 against liposome oxidation was higher than albumin, glutathione, proline, glycine betaine and sucrose. These studies indicated that LEA proteins might behave as or seem to mimic molecular chaperones. Recombinant forms of Em, a Group 1 LEA protein from wheat, have been subjected to functional analysis. Heat stress experiments with citrate synthase, suggest that LEA proteins do not behave as classical molecular chaperones, but they do exhibit a protective effect in the presence of the so-called chemical chaperone, trehalose.

In contrast, both LEA proteins alone protect citrate synthase from aggregation due to desiccation and freezing, in keeping with a role in water stress tolerance; similar results were obtained with lactate dehydrogenase (LDH). LEA proteins might act as "molecular shield" to help prevent the formation of damaging protein aggregates during water stress. LEA proteins were able to prevent this aggregation during multiple freeze-thaw cycles. LEA proteins also prevent LDH aggregate formation due to rapid freezing in liquid nitrogen, as determined in a light scattering assay (Goyal et al. 2005b). The SK (n)-and K-type seem to be directly involved in cold acclimation processes. Much recent in vitro data clearly indicates that dehydrins belonging to different subclasses exhibit distinct functions. Similar protective effects of LEA proteins were observed when LDH was subjected to slow freezing by placing at −20°C overnight (Momma et al. 2003; Sanchez-Ballesta et al. 2004). Freezing LDH alone eliminating enzyme activity, but the LEA proteins are able to prevent this inactivation, again in a concentration-dependent manner. Besides this, LEA protein groups are known to protect sensitive enzymes such as malate dehydrogenase, fumarase, CS etc. (Grelet et al. 2005; Reyes et al. 2005) against desiccation (without freeing). The LEA proteins analysed included the group 1 member, Em, from wheat and the nematode group 3 proteins AavLEA1. These did not stop citrate synthase aggregation due to heat stress, like classical heat shock proteins, but could be functioning as dehydration- specific molecular chaperones. However, chaperones not only prevent inappropriate protein aggregation but form specific, transient complexes with their partner proteins, through interaction of hydrophobic patches within the dehydrins (Ellis 2004). However, the involvement of LEA proteins in forming specific complexes with other partner proteins is unclear and it is difficult to show experimentally this in dry state. The well characterized *Em*-like gene, *OsLEA1a* of rice (*Oryza sativa*) encodes for OsLEA1which is similar to that of other plant Em proteins but lacks a 20-mer motif that is the most highlighted feature of a typical Em protein (Shih et al. 2010). Transcriptome analysis revealed that this protein is mainly expressed in embryos, with no or only a few transcripts in osmotic stress-treated vegetative tissues. Structural analysis revealed that the OsLEA1a protein adopts high amounts of disordered conformations in solution and undergoes desiccation-induced conformational changes.

LEA proteins, in an overly congested environment of a drying cytoplasm, serve to decrease the interaction between partially denatured polypeptides with the possibility to aggregate. Such molecular shielding functions are also similar to that of the entropic bristles of MAP2, tau and neurofilament side arms, which serve as spacers to prevent close association of microtubules and neurofilaments, except that shield proteins are not necessarily tethered to a surface (Mukhopadhyay et al. 2004). Association of molecular shields with the surface of other proteins and even membrane surfaces is also a possible mechanism for 'client' or 'partner' proteins. Finally, shield proteins might have a broader space-filling role and help to prevent collapse of the cell as it gets dehydrated.

Siminovitch and Briggs (1953) and Close (1996) correlated the accumulation of hydrophilic proteins with frost hardiness in the black locust tree and proposed their potential as "plasticizers or mechanical buffers in the cell". Still earlier work noted the contribution of these proteins to the "high non-solvent space" in frost-hardy cells (Asai 1943). Whereas recent reports indicate that this activity, termed molecular shield function, is somewhat distinct from that of a classical molecular chaperone, such as HSP70. While HSP70 diminishes aggregation of citrate synthase on heating, in comparison to AavLEA1 and Em; conversely, the LEA proteins reduce citrate synthase aggregation on desiccation, while HSP70 lacks this ability. In addition to this, there are differences in interaction with respective client proteins – HSP70 can be co-immunoprecipitated with a client containing polyglutamine, consistently with a firm complex formation, whereas the LEA proteins cannot do so to a great extent. Further revelations of analogous molecular shield function, it's seen that synthetic polysaccharides, like LEA proteins, are capable of reducing desiccation-induced aggregation of a water-soluble proteome (Chakrabortee et al. 2012).

Membrane Protection

A potential function of LEA protein, CuCOR19 in membrane protection has been documented by the expression of citrus CuCOR19 in transgenic tobacco plants (*Nicotiana tobacum* L.). Antioxidative activity may be a crucial function of KnS dehydrins in conditions leading to generate hydroxyl radicals in a metal/H_2O_2 system in plants during cellular dehydration. In membrane unsaturated fatty acids, hydroxyl radicals induce a process leading to the formation of lipid radicals. Peroxidation causes an immense loss of unsaturated fatty acids and creates membrane dysfunction through modifications in fluidity, thus affecting ion transport, selective permeability, enzyme activity and receptor availability. By binding the metal ions, the CuCOR19 reduces the potential to form hydroxyl radicals under water-stressed conditions. The activity of KnS-type dehydrins to scavenge hydroxyl radicals makes these proteins an important antioxidative factor in cells under cellular dehydration stress (Asghar et al. 1994).

Yet the exact function of dehydrin isn't specified, but in vitro experiments have exposed that some YSK(n)-type dehydrins bind to lipid vesicles that contain acidic phospholipids, and others such as K(n)S were revealed to bind to metal ions and have the capacity to scavenge hydroxyl radicals, protect lipid membranes against peroxidation and demonstrate cryoprotective activity towards freeze-sensitive enzymes (Asghar et al. 1994). Group 1 LEA proteins in soyabean protect organelle membranes against freezing, desiccation and osmotic stresses (Soulages et al. 2002). Chandra Babu et al. (2004) tested the barley LEA 4 group gene, *hva1* and found it functional in protecting cell membranes from injury during drought stress. Another report from *Zea mays* dehydrin DHN1, tells that this Group 2 member binds to liposomes containing anionic phospholipids, and this in return stimulated an increase in helicity, which led to the known membrane-stabilizing function (Koag et al. 2003, 2009). Importantly, LEAM (a mitochondrial LEA protein) found in plant cells reversibly gets folded into alpha helices upon desiccation and interacts with membranes upon drying. Further, Tolleter et al. (2007, 2010) proposed that LEAM protects inner mitochondrial membrane during desiccation. Coming back to the model plant *Arabidopsis*, two dehydrins (ERD10 and ERD14) were also shown to have affinity for anionic phospholipid vesicles, without inducing modifications in the fluidity of membranes in the fully hydrated state, indicating binding through peripheral electrostatic interactions (Kovacs et al. 2008). LEA proteins are characterized by high hydrophilic and thermal stabilities, and are known to stabilize the cell membrane structure and prevent oxidation. In another study, *TaLEA4*, a Group III member from the LEA family, was cloned from a cDNA library of stress-treated wheat seedlings. The full length clone of *TaLEA4* is 1,084 bp and contains a 570 bp open reading frame (ORF) encoding a 189-amino-acid protein. The prediction of protein-sorting signals and localization sites in amino acid sequences (PSORT) showed that TaLEA4 has a nuclear localization signal (NLS) in the amino acid C-terminal sequence. Expression profile analysis showed that *TaLEA4* was highly induced by drought, and low and high temperatures. Isolation of the *TaLEA4* promoter revealed a core promoter element and some cis-acting elements responding to abiotic stresses (Min et al. 2012). Hence on this basis, it is clear that LEA proteins balance out and protect membrane lipids against any free radical species, dehydration, desiccation or conformational changes.

LEA Proteins in Salinity: Dehydrins

As already known dehydrins (DHNs) are included in a large group of highly hydrophilic proteins known as LEA. They were originally identified as Group 2 of the LEA proteins. The distinctive feature of all DHNs is a highly conserved, lysine-rich 15-amino acid domain, named the K-segment. It is usually present near the C-terminus. They do not display a well-defined secondary structure. The number and order of the Y-, S-and K-segments define different DHN sub-classes: Y(n)SK(n), Y(n) Kn, SK(n), K(n) and K(n)S. Dehydrins are distributed across a wide range of organisms including the higher plants, algae, yeast and cyanobacteria. DHNs are localized in different cell compartments, such as the cytosol, nucleus, mitochondria, vacuole, and the vicinity of the plasma membrane; however, they are primarily localized to the cytoplasm and nucleus (Rorat 2006). Many studies reported a positive correlation between the accumulation of LEA transcripts or proteins and the tolerance to freezing, drought, and salinity. Transgenic plants and heterologous expression in yeast over expressing *Dhn* genes have also been used to elucidate the potential role of DHN proteins in stress tolerance. A total of 34 rice lea (Oslea) genes were identified, microarray data and semiquantitative reverse transcription PCR analysis revealed that the expressions of these genes are very diverse and some appear to be related to salt stress tolerance (Wang et al. 2006). LEA proteins got accumulated during the salinity-triggered growth arrest of young *Oryza sativa* cv Bura Rata seedlings and are mobilised during the recovery of seedlings from salinity stress (Chourey et al. 2003). Various parameters inclusive of the expression of late embryogenesis abundant proteins (LEA; group 1, 2, 3 and 4) under different levels of salt stress (0, 1.0, 1.5 and 2.0% NaCl) were investigated in *Morus alba* L. cultivars (S1 and ATP) with contrasting salt tolerance by Jyothsnakumari et al. (2009). The maximum content of LEA (group 3 and 4) was detected in S1 at 2.0% NaCl, which correlates with its salt tolerance. Several other reports and currently pursued investigations reveal that excess salt levels are countered in plants by accumulation of specific LEA protein groups. A 30 kDa dehydrin in *Chenopodium quinoa* embryos was reported to accumulate in both 300 and 500 mM NaCl growth conditions as revealed by densitometric analyses. Western blot analysis detected at least four dehydrins (55, 50, 34, and 30 kDa) in these seeds harvested under a wide range of salinities (Burrieza et al. 2012). As for the Group 2 LEA proteins in tomato, TAS14 dehydrin was characterized in and its expression was induced by NaCl stress, however its function became clearer, by using transgenic tomato plants over expressing *tas14* gene under the control of the 35SCaMV promoter in drought and salinity tolerance (Muñoz-Mayor et al. 2012). This field has several open possibilities and the mechanism probably is the ability of the proteins to bind to the surplus Na, Ca, Cl, K ions present in the soil or culture media of the plants.

Dehydrins: Cellular and Sub-cellular Localization in Plants

Godoy et al. (1994) isolated and characterized mRNA and dehydrin-TAS14 that is induced after giving ABA (abscisic acid) treatment and osmotic stress in tomato plants and that share with other dehydrin genes the sequence and expression features in different species. To study the expression of TAS14 protein, its subcellular localization and distribution in tissues, affinity purified antibodies against TAS14 protein were used both in seedlings and mature plants. TAS14 showed accumulation after NaCl, ABA or mannitol treatments and did not any presence in 4-day old seedlings. Some protein level was detected after 6 h of treatment in the seedlings which were treated by NaCl and maximum level was detected between 24 and 48 h. Similar levels of the protein were detected after NaCl treatment with concentration between 5 and 12.5 g/l, TAS14 showed abundant expression in aerial parts, but only some expression in roots, in salt – stressed mature plants. TAS14 showed accumulation in adventitious root primordial in salt stressed plants. With the help of Immunogold electron microscopy TAS14 protein was detected associated with the nucleolus and euchromatin in the nucleus and also in the cytosol. Freezing, drought, salinity and seed germination are the conditions which affect the plant cell's water status and dehydrins are the proteins which show presence in these conditions. Dehydrins function is not much clear but it is supposed that they stabilize macromolecules and membranes during cellular dehydration. *Cornus sericea* L. (Red-osier dogwood) is a woody plant which can tolerate extreme freezing conditions, during cold-acclimation it shows the accumulation of dehydrin like proteins and so these proteins are correlated with tolerance to extreme freezing conditions (Karlson 2001; Karlson et al. 2003; Sarnighausen et al. 2002).

Romanenko et al. (2010) studied the subcellular localization of dehydrins in stem cells tissues of winter wheat seedlings (*Triticum aestivum* L., cult Irkutskaya ozimaya) by using immunoelectron microscopy. When compared with control conditions (22°C), the dehydrin quantity doubled in the cold hardened cells at 4°C for 7 days. Minimum increase (fourfold) was found in chloroplasts and dehydrin maximum increase was found in the rough endoplasmic reticulum (3.8) fold, then in mitochondria (3.0 fold), followed by cell walls and intercellular spaces (2.8 fold). Low temperature resulted in increased dehydrin quantity near membranes and the inter membrane space of the membrane compartments (rough endoplasmic reticulum, mitochondria and chloroplasts). Nucleus showed increased amount of dehydrin (2.5 fold) under low temperature conditions. Hence, the translocation and accumulation of dehydrin to the regions of intracellular and intercellular compartments was induced by the cold-hardening of the plants. During the maturation drying phase, orthodox seeds are produced dehydrins as a part of their developmental program (Kermode 1997). Dehydrins have also been found in certain recalcitrant seeds (Finch-Savage et al. 1994; Farrant et al. 1996). Layton et al. (2010) performed experiment on desiccation – tolerant fern *Polypodium polypodioides*, they found extensive cell wall folding when it was dried to less than 15% RWC (relative water content) and showed rapid dehydration on exposure to water and high humidity. With the help of western blotting the expression of a 31-kDa putative dehydrin polypeptide in fully and partially dried tissues was detected. Expression was observed only during drying and rapidly dissipated upon tissue rehydration. No significant strain was found in the dry vascular tissue by atomic force microscopy of tracheal scalariform perforations. Adaxial and abaxial leaf surfaces showed differential hydrophobicity, also the reversible deformation as revealed by environmental scanning electron microscopy. The ability of some desiccation- tolerant species was for avoiding the cell wall damage for cell wall localization of dehydrin that enabled large and reversible cell-wall deformation. According to Kosová et al. (2011), when wheat and barley plants are exposed to cold temperature, both freezing-tolerant and freezing-susceptible plants starts accumulating dehydrin. Accumulation shows correlation with plant acquired frost tolerance.

Brini et al. (2007) performed experiment on two Tunisian durum wheat (*Triticum durum* Desf.) varieties and found that DHN-5 protein accumulated differentially in the two varieties with large differences in drought and salt tolerance. They proteins from mature embryos in two varieties were extracted by using dehydrin antibodies and 2D immunoblot analysis, DHN-5 differential phosphorylation patterns were observed. In sensitive variety (S) weakly detectable acidic spots were observed whereas in resistant variety (R), a series of acidic spots were detectable besides a basic protein spot. It was further proposed that these acidic forms correspond to phosphorylated forms of DHN-5, which accumulated mainly in the R variety thereby suggesting that P-DHN-5 plays a role in the preservation of cell integrity during desiccation and late embryogenesis. DHN-5: GFP fusion protein sub cellular localization indicates that DHN-5 is mainly nuclear and it suggests a nuclear role in wheat osmotic stress response. To confirm the presence of dehydrins in the leaves of *Pisum sativum*, Mueller and co-workers (2003) probed total protein extracts with the antidehydrin antibody. Immunoblot analysis indicated the presence of a 35-kD and a ca-31-kD dehydrin in all treatments. Less intensity bands were observed in seed protein extracts. Highest production of 31-kD dehydrin was detected in drought-stressed tissue, per equal total leaf protein, in comparison with other treatments. For the determination of chloroplast localization of dehydrins, chloroplasts from leaves of control, cold-stressed, drought-stressed, ABA treated and NaCl – stressed *Pisum sativum* plants were isolated. Detergent soluble protein immunoblots from the chloroplasts showed the presence of a ca-31kD dehydrin in control leaf tissue, like the whole leaf protein extract immunoblot. Thylakoid fraction showed the higher levels of chloroplast dehydrin and hence indicated most usual association of dehydrins with the thylakoid membrane. Using K-segment antibody, Panza et al. (2007) immunolocalized dehydrins in the cytoplasm and chromatin of *Euterpe edulis* seeds.

Membrane Association, Folding and Stabilization

Steponkus et al. (1998) suggested that LEA proteins could stabilize the membranes by associating with them. Danyluk et al. (1998) demonstrated the accumulation of acidic Group 2 LEA protein CO410_WHEAT (WCOR410) at plasma membrane during cold acclimation and particularly in tissues which were more sensitive to freeze damage. It had been suggested that LEA proteins interact more deeply with membranes and possibly provide protection in the dry state (Tunnacliffe et al. 2010). As described previously, interaction of Group 3 LEA protein (LEAM) from pea seed mitochondria with the membrane during the dry state provides protection (Tolleter et al. 2007). LEAM is an intrinsic polypeptide present in matrix of mitochondria but is able to fold into amphipathic alpha-helices during drying. K-segment, of dehydrins which is composed of 15-mer Lys-rich sequence, has been suggested to be involved in membrane binding (Close 1996). It is predicted to form an amphipathic class A helix, a motif which is involved in apolipoprotein binding to low density lipoprotein vesicles (Segrest et al. 1994). Further, Koag et al. (2003) reported that DHN1, a maize dehydrin, which was shown to bind liposomes containing anionic phospholipids and this binding stimulated an increase in helicity. It is expected that on folding during desiccation, or perhaps freezing, specific LEA proteins could indeed be targeted to membranes and contributed to their protection thus maintaining the integrity of cellular membranes during acute stress.

Ion Binding Activity

When cell undergoes dehydration, the concentration of intracellular components increases and so does the concentration of ions which causes damage to macromolecular structure and functions of the cell. It has been proposed that LEA proteins, because of their many charged amino acid residues, might act to sequester ions (Danyluk et al. 1998; Dure 1993). It was further suggested that Ca^{2+} binding might be a possible property of both Groups 2b and 3a proteins (Wise 2003; Wise and Tunnacliffe 2004). In some proteins, ion binding activity has also been observed. Svensson et al. (2000) reported that various dehydrins from *Arabidopsis* can be purified by immobilized metal ion affinity chromatography and particularly, bind with Ca^{2+} and Ni^{2+} ions. In group 2 LEA proteins, there proteins, there is no metal binding motif. The ability was confirmed due to the presence of high content of His residues, some of which are disposed as His-His pairs having strong metal-binding affinity. According to Hara et al. (2005) a citrus (*Citrus unshiu*) dehydrin binds Cu^{2+}, Fe^{3+}, Co^{2+}, Ni^{2+} and Zn^{2+} through specific sequence rich in His residues. Vacuole associated dehydrins of celery (*Apium graveolens*) and *Arabidopsis* possess calcium-binding properties, which is positively modulated by phosphorylation (Alsheikh et al. 2003, 2005; Heyen et al. 2002). A protein related to the dehydrins was described by Heyen et al. (2002) in celery (*Apium graveolens*) which is reported to bind Ca^{2+} when phosphorylated and is located in the vacuole of seedlings and its expression is increased by water stress. Another citrus dehydrin, CuCOR15, was shown to have metal binding activity, and the specific metal-binding domain in the protein sequence was identified (Hara et al. 2005). Fe^{3+}, Co^{2+}, Ni^{2+}, Cu^{2+} and Zn^{2+} bound to CuCOR15, but Mg^{2+}, Ca^{2+} and Mn^{2+} did not and highest affinity was detected for Cu^{2+}. Group 2 LEA proteins are also known to bind a number of other metal ions. Krüger et al. (2002) used an iron-binding protein from castor bean (*Ricinus communis*) with some similarities to Group 2 LEA proteins. This study revealed that binding affinities across a range of transition metal ions were highest for Fe^{3+} followed by Cu^{2+}, Zn^{2+}, Mn^{2+} and Fe^{2+}. ERD14 from *Arabidopsis* and VBA45 dehydrin from celery are reported to be involved in calcium binding and this activity is dependent on phosphorylation of these proteins (Alsheikh et al. 2003; Heyen et al. 2002). Apart from Group 2, two Group 4 LEA proteins were found to possess antioxidant activity (Liu et al. 2011b). GmPM1 and GmPM9 from soybean were tested for metal ion binding ability and it was found that these two proteins bound effectively to Fe^{3+}, Ni^{2+}, Cu^{2+} and Zn^{2+}. Alternatively, metal binding leads to detoxification required under stress conditions, where metal toxicity is related to the generation of reactive oxygen species (Hara et al. 2004).

Antioxidant Activity

The binding of metal ions by Group 2 LEA proteins might be linked to the antioxidant properties of these proteins. LEA protein from citrus CuCOR19 is reported to protect liposomes from peroxidation, as well as aided in reducing electrolyte leakage in transgenic tobacco seedlings during chilling stress (Hara et al. 2003). Experiments with the same protein also showed scavenging activity for hydroxyl and peroxyl radicals (Hara et al. 2004). Hydroxyl radicals and singlet oxygen are reactive oxygen species (ROS) that are generated by catalytic metals and that these are highly toxic to the organic molecules. Deficiency of water can also lead to accumulation of catalytic metal ions in plants thereby causing oxidative stress (Iturbe-Ormaetxe et al. 1998). Therefore, LEA proteins can be the probable molecules that can scavenge ROS directly or sequester metal ions that generate ROS and hence reduce the oxidative damage of the stressed plants.

Cryoprotective Activity

It was shown that dehydrins are more effective in controlling damage caused by freezing and tahwing in LDH than other molecules like sucrose or bovine serum albumin (Lin and Thomashow 1992). Certain dehydrins are also reported to possess the property of providing protection to freezing sensitive enzymes. CuCOR19 dehydrin from *Citrus unshiu* was reported to protect catalase and lactate dehydrogenase (LDH) against freezing inactivation. It was also found to be more effective than compatible solutes such as sucrose, glycine betaine and proline, or Bovine Serum Albumin (Hara et al. 2001). Study has revealed that the secondary structure of CuCOR19 in a solution is a random coil. Hence, it is suggested that the random coil structure of dehydrins might have an important role in the cryoprotection of freezing-sensitive enzymes. Randomly coiled structure may prevent the dissociation of oligomeric protein subunits. Antifreeze activity has also been reported in *Prunus persica* (Wisniewski et al. 1999) and the *Betula pubescens* (Rinne et al. 1999). PCA60 from *Prunus persica* was showed to protect the activity of LDH and dehydrins from *Betula pubescens* were shown to enhance α-amylase activity in the presence of polyethylene glycol. Lactate dehydrogenase has also been reported to be protected from freeze-thaw damage in the presence of K_2 dehydrin from *Vitis reparia* (Hughes and Graether 2011).

Transgenic Studies

A number of studies revealed the increase in stress tolerance property of plants or micro-organisms when heterologous LEA protein genes have been introduced. The tomato Group 4 protein LE25 when introduced and expressed in yeast, showed increased tolerance to salinity as well as chilling stress (Imai et al. 1996). Similar results were observed for Group 3 LEA proteins from *Chlorella vulgaris* when made to express in yeast (Honjoh et al. 1999). However, wheat group 1 protein, Em, when introduced in yeast showed no protection against freeze stress (Swire-Clark and Marcotte 1999). Transgenic tobacco having Group 2 LEA protein from *Citrus unshiu* revealed better growth and germination rates than controls at low temperature (15°C). Freeze tolerance of strawberry leaves was found to increase when wheat COR410 was introduced in it. Houde et al. (2004) introduced wheat COR410 into strawberry and observed a 5°C improvement in freeze tolerance of leaves expressing the protein compared to controls.

Osmotic stress tolerance was reported to increase in *Arabidopsis thaliana* when maize DHN1/Rab17 was made to overexpress (Figueras et al. 2004). Potato (*Solanum sogarandinum*) Group 2 proteins, DHN24 (Q7Y1A0_SOLSG), when introduced in cucumber and hence enhancement in chilling tolerance freeze tolerance was observed (Yin et al. 2006). The barley HVA1 gene (encoding Group 3 LEA protein sequence LEA1_HORVU) conferred enhanced tolerance of water stress and salt stress in transgenic rice (Rohila et al. 2002; Xu et al. 1996). Such transgenic plants were able to maintain higher relative water content in their leaves than non-transgenic controls and suffered less electrolyte leakage from cells (Babu et al. 2004; Rohila et al. 2002), suggesting that the HVA1 protein might protect cell membranes from injury during drought. Groups 1 and 2 LEA protein genes from wheat were also able to offer rice similar protection against dehydration stress (Cheng et al. 2002) and the same barley HVA1 gene introduced into wheat endowed better growth and water use under water stress conditions (Sivamani et al. 2000). Another Group 3 LEA protein, from rapeseed (*Brassica napus*), was used to make transgenic lines of Chinese cabbage (*B. campestris*), resulting in improved salt and drought tolerance (Park et al. 2005). Many studies also reported about slight or no effect of LEA proteins introduced into plants. Introduction of two dehydrins and a group 3 LEA protein from *Craterostigma plantagineum* could not improve drought tolerance of transgenic tobacco (Iturriaga et al. 1992). Also overexpression of RAB18 (DHR18_ARATH) in Arabidopsis *thaliana* could not improve freeze or drought tolerance (Lång 1993). The ability of LEA proteins to facilitate recombinant expression of recalcitrant and intrinsic membrane proteins was demonstrated by Singh et al. (2009). Two Group 3 LEA proteins from *Brassica napus* (BN115m and a truncated fragment of BNECP63) were fused to two target proteins identified as recalcitrant to overexpression in soluble form or outside of inclusion bodies. Fusion of a truncated peptide of BNECP63 was found to be sufficient to provide soluble and high levels of recombinant overexpression of BNPsbS protein of *B. napus* which is an intrinsic membrane chlorophyll-binding protein of photosystem II light harvesting complex and also to a peptide of the Hepatitis C viral polyprotein. Furthermore, fusion of the recombinant target proteins to BNECP63 or BN115 prevented irreversible heat- and freeze-induced precipitation.

Regulation of Dehydrins Under Stress (Post-translational Modifications)

Dehydrin proteins are modified mostly after translation and phosphorylation is the most common mechanism of modification. This mechanism was first reported in maize DHN1 (Rab17; YSK2-type) (Vilardell et al. 1990). The phosphorylated protein segment was found to be the S-segment (Plana et al. 1991). The amino acid region between position 66 and 96 contains a serine cluster followed by three acidic amino acids (EEE) as a putative consensus site for protein kinase2 (CK2) phosphorylation (Jensen et al. 1998). Also, there is a stretch of basic amino acids (RRKK) resembling a nuclear localization signal-binding domain. This region was shown to be necessary for the targeting of RAB17 (DHN1) to the nucleus. DHNs in certain plant systems undergo glycosylation viz. DHN-like proteins from blueberry and *Pistacia vera* (Golan-Goldhirsh et al. 1998; Levi et al. 1999). Phosphorylated form of DHN1 in isolated nuclei of transgenic *Arabidopsis* was found to be absent in the in vitro conditions when mutation occurred in consensus site for CK2 which further led to a strong decrease in the content of the protein. Hence, the nuclear location of the DHN1 is either facilitated by binding to specific proteins or as a direct part of the nuclear targeting apparatus and the above results imply that phosphorylation by CK2 is the relevant step for this (Jensen et al. 1998). Other DHNs such as tomato TAS14 (YSK2) (Godoy et al. 1994), DSP16 from *C. plantagineum* (Lisse et al. 1996), VCaB45 from *Apium graveolens* (Heyen et al. 2002), and ERD14 (SK2) from *Arabidopsis* (Alsheikh et al. 2003) also reported to show phosphorylation.

Role of LEA Proteins in Salt Stress Tolerance

LEA proteins are associated with the salt stress tolerance in many studies (Cuming 1999; Tunnacliffe and Wise 2007). Plants possess different mechanisms against salt stress (Melgar et al. 2008; Niknam et al. 2006; Sotiropoulos 2007). Exposure of plants to high salt stress stimulates the expression of gene for protective proteins like osmotin, LEA proteins, pathogenesis related (PR) proteins, ion transporters and SALT proteins (De Souza et al. 2003; Jyothsnakumari 2005; Moons et al. 1997a; Rorat 2006). Among different groups of LEA proteins, Group 2 LEA proteins are synthesized in response to salt stress (Allagulova et al. 2003; Cherian et al. 2006; Svensson et al. 2002; Wahid and Close 2007). According to Bishnoi et al. (2006), protein synthesis is altered in different plant species due to salinity stress and these changes are induced by the changes in the gene expression.

As previously mentioned, Jyothsnakumari et al. (2009) analyzed the cross reactivity of antigroup LEA 1,2,3 and 4 proteins under the salinity stress in the mulberry leaf. LEA 1 and LEA 2 expression was observed in tolerant cultivar S1 and completely absent in ATP susceptible cultivar. Expression of wheat LEA Group 1 and 2 protein genes were also reported in two transgenic rice plants under salinity stress (Cheng et al. 2002). Similarly, Moons et al. (1997b) observed Group 2 LEA and DHN proteins in the roots of rice under salinity stress. Higher levels of gene expression of LEA 3 proteins were reported by Moons et al. (1995) in the roots of salt tolerant rice genotypes as compared to salt sensitive genotypes. There are several groups which have used the *Saccharomyces cerevisiae* to check the functions of LEA protein. Group 4 protein LE25 in the tomato (*Solanum lycopersicum*/*Lycopersicon esculentum*) helped in improving the salt tolerance, when exposed in yeast (Wise 2003). According to Park et al. (2005), from the rape seed *Brassica napus* plants, 3 LEA proteins were used to prepare transgenic lines of Chinese cabbage *Brassica campestris*, which resulted in improved salt tolerant variety. More recently, Group 3 LEA proteins have been investigated for high salinity much recently by Zhao and co-workers (2011).

LEA Proteins and Other Hydrophilins in Stress

LEA proteins are a part of hydrophilins (glycine rich molecules), which are widespread group of proteins (Battaglia et al. 2008). Hydrophilins participate in acclimation and in the adaptive response to stress. In plants and yeast, ecotopic expression of some plant hydrophilins (LEA proteins) provide tolerance to water deficit conditions (Imai et al. 1996; Swire-Clark and Marcotte 1999; Xu et al. 1996; Zhang et al. 2000) and these also possess chilling tolerance (Danyluk et al. 1994, 1998; Ismail et al. 1999a, b; Nakayama et al. 2007; Puhakainen et al. 2004). LEA proteins directly protect plant cells against all types of stresses (Todaka et al. 2012).

Recently, the role of yeast hydrophilic genes and their inception into transgenic plants for increasing water stress tolerance has come to light (Dang and Hincha 2011; López-Martínez et al. 2012). Deletion of RMF hydrophilin gene in *E. coli* (Garay-Arroyo et al. 2000) and absence of LEA protein in the moss *Physcomitrella patens* (Saavedra et al. 2006) cause osmosensitive phenomenon. Mostly these proteins accumulate in all tissues in water deficit conditions imposed by salinity, drought or low temperature and these proteins respond to particular stress conditions. Certain dehydrins accumulate during low temperature conditions, but not at the time of drought or salinity stress (Rorat et al. 2006). It was reported that ABA-responsive element was first described for Group 2 LEA gene from *Oryza sativa* (Mundy and Chua 1988). ABA helps in mediating the expression of those genes of this group of proteins during seed development or in response to stress (Nylander et al. 2001). Response to stress is mediated by more than one pathway called dual regulation and one of which may be ABA dependent (Welling et al. 2004). Like Group 1 LEA proteins, Group 2 LEA proteins also accumulated during the seed desiccation in response to water deficit conditions induced by drought, low temperature or salinity (Bartels 2005; Ismail et al. 1999b; Nylander et al. 2001).

All hydrophilins from various phyla possess higher expression under the water-deficit conditions induced by different environmental or by developmental programs. Under partial dehydration, these prevent the enzyme inactivation (Goyal et al. 2005b; Grelet et al. 2005; Lin and Thomashow 1992; Reyes et al. 2005). Gradual decrease in water availability to the level as those detected in plant tissues subjected to drought, targets the conformational changes in the enzymes associated with inactivation. These changes do not occur if the hydrophilins are present before the dehydration treatment. Hydrophilins possess a protective activity that mitigates the effect of water deficit conditions exert on conformation and functions of proteins. Goyal et al. (2005b) reported that hydrophilins play protective role in the absence of energy source; it confirms that they act as typical molecular chaperones (Reyes et al. 2005). STF2, a yeast hydrophilin is known to participate in the stabilization of complex formed between F_1F_0 –ATPase and a protein that reduces the enzyme activity upon the cessation of phosphorylation (Yoshida et al. 1990). Other LEA proteins have been tested for protection of the enzyme activity during heat stress (Goyal et al. 2005b; Reyes et al. 2005). Some recent reports confirmed that two Group 2 LEA proteins from *Arabidopsis* (ERD10 and ERD14) prevent the aggregation or inactivation of various substrates during heat stress (Kovacs et al. 2008).

Twenty LEA proteins from maize dry embryos were analyzed by mass spectrometry out of which three major LEA proteins, Emb564, Rab17 and Mlg3, belonging to groups 1, 2 and 3, respectively were studied for differences and affinities in dehydration protection. Overall, the results highlighted differences and suggested functional diversity among maize LEA groups (Amara et al. 2012). Intrinsic flexible nature of hydrophilins makes them able to adjust their conformation to a particular environment. Different conformations are induced in the same proteins by different water availability levels, which results in exposure of particular motifs for recognition or interaction with specific target molecules to preserve their functions and helps in their assembly with partners (Battaglia et al. 2008; Eriksson et al. 2011).

Conclusions and Future Prospects

*L*ate *E*mbryogenesis *A*bundant (LEA) proteins play a pivotal role in plant stress tolerance, but the detailed underlying mechanisms of plant stress protection remains to be explored. Though the mode of regulation of LEA proteins is being studied in *Arabidopsis thaliana* yet studies in animal species would contribute to establish the role of LEA proteins in the long term-survival of dormant animals as well as plants. In future, the meticulous investigations needs to be carried out for understanding post-transcriptional and post-translational regulations of LEA genes in order to enhance their expression in plants that usually do not exhibit anti-stress properties beyond a certain limit. The various molecular and physiological aspects need further exploration to illuminate more stress countering roles for this protein family.

References

Allagulova CR, Gilamov FR, Shakirova FM, Vakhitov VA (2003) The plant dehydrins: structure and functions. Biochem 68:945–951

Alsheikh MK, Heyen BJ, Randall SK (2003) Ion binding properties of the dehydrin ERD14 are dependent upon phosphorylation. J Biol Chem 278:40882–40889

Alsheikh MK, Svensson JT, Randall SK (2005) Phosphorylation regulated ion-binding is a property shared by the acidic subclass dehydrins. Plant Cell Environ 28:1114–1122

Amara I, Odena A, Eliandre Oliveira E, Moreno A, Masmoudi K, Montserrat Pagès M, Goday A (2012) Insights into maize LEA proteins: from proteomics to functional approaches. Plant Cell Physiol 53(2):312–329

Amitai-Zeigerson H, Scolnik PA, Bar-Zvi D (1994) Genomic nucleotide sequence of tomato Asr2, a second member of the stress/ripeninginduced Asr1 gene family. Plant Physiol 106:1699–1700

Asai GN (1943) A study of frost injury and frost resistance in garden roses. Ph.D. thesis, Cornell University

Asghar R, Fenton R, DeMason D, Close T (1994) Nuclear and cytoplasmic localization of maize embryo and aleurone dehydrin. Protoplasma 177:87–94

Babu RC, Zhang J, Blum A, Ho T-HD, Wu R, Nguyen HT (2004) HVA1, a LEA gene from barley confers dehydration tolerance in transgenic rice (*Oryza sativa* L.) via cell membrane protection. Plant Sci 166:855–862

Baker J, Steele C, Dure L (1988) Sequence and characterization of 6 LEA proteins and their genes from cotton. Plant Mol Biol 11:277–291

Baker EH, Bradford KJ, Bryant JA, Rost TL (1995) A comparison of desiccation-related proteins (dehydrin and QP47) in pea (*Pisum sativum*). Seed Sci Res 5:185–193

Bartels D (2005) Desiccation tolerance studied in the resurrection plant *Craterostigma plantagineum*. Integr Comp Biol 45(5):696–701

Battaglia M, Olvera-Carrillo Y, Garciarrubio A, Campos F, Covarrubias AA (2008) The enigmatic LEA proteins and other hydrophilins. Plant Physiol 148:6–24

Bies-Etheve N, Gaubier-Comella P, Debures A, Lasserre E, Jobet E, Raynal M, Cooke R, Delseny M (2008) Inventory, evolution and expression profiling diversity of the LEA (late embryogenesis abundant) protein gene family in *Arabidopsis thaliana*. Plant Mol Biol 67:107–124

Bishnoi SK, Kumar B, Rani C, Datta KS, Kumari P, Sheoran IS, Angrish R (2006) Changes in protein profile of pigeonpea genotypes in response to NaCl and boron stress. Biol Plant 50:135–137

Bokor M, Csizmok V, Kovacs D, Banki P, Friedrich P, Tompa P, Tompa K (2005) NMR relaxation studies on the hydrate layer of intrinsically unstructured proteins. Biophys J 88:2030–2037

Bray EA (1993) Molecular responses to water deficit. Plant Physiol 103:1035–1040

Brini F, Hanin M, Lumbreras V, Irar S, Pages M, Masmondi K (2007) Functional characterization of DHN-5, a dehydrin showing a differential phosphorylation pattern in two Tunisian durum wheat (*Triticum durum* Desf.). Plant Sci 172:20–28

Browne J, Tunnacliffe A, Burnell A (2002) Plant desiccation gene found in a nematode. Nature 38:416

Burrieza HP, Koyro HW, Tosar LM, Kobayashi K, Maldonado S (2012) High salinity induces dehydrin accumulation in Chenopodium quinoa Willd. cv. Hualhuas embryos. Plant Soil 555:1–11

Cakir B, Agasse A, Gaillard C, Saumonneau A, Delrot S, Atanassova R (2003) A grape ASR protein involved in sugar and abscisic acid signaling. Plant Cell 15:2165–2180

Canel C, Bailey-Serres JN, Roose ML (1995) Pummelo fruit transcript homologous to ripening-induced genes. Plant Physiol 108:1323–1324

Caramelo JJ, Iusem ND (2009) When cells lose water: lessons from biophysics and molecular biology. Prog Biophys Mol Biol 99:1–6

Carrillo Y, Campos F, Reyers J, Garciarrubio A, Covarrubias AA (2010) Functional analysis of the group 4 late embryogenesis abundant proteins reveals their relevance in the adaptive response during water deficit in *Arabdiopsis*. Plant Physiol 154:373–390

Carrillo Y, Reyes JR, Covarrubias AA (2011) Late embryogenesis abundant proteins versatile players in the plant adaptation to water limiting environments. Plant Signal Behav 6(4):586–589

Cattivelli L, Bartels D (1990) Molecular cloning and characterization of cold-regulated genes in barley. Plant Physiol 93:1504–1510

Chakrabortee S, Boschetti C, Walton LJ, Sarkar S, Rubinsztein DC, Tunnacliffe A (2007) Hydrophilic protein associated with desiccation tolerance exhibits broad protein stabilization function. Proc Natl Acad Sci USA 104:18073–18078

Chakrabortee S, Tripathi R, Watson M, Schierle GS, Kurniawan DP, Kaminski CF, Wise MJ, Tunnacliffe A (2012) Intrinsically disordered proteins as molecular shields. Mol Biosyst 8(1):210–219

Chandra Babu R, Zhang JX, Blum A, Ho DT-H, Wu R, Nguyen HT (2004) HVA1, aLEAgene from barley confers dehydration tolerance in transgenic rice (Oryza sativa L.) via cell membrane protection. Plant Sci 166:855–862

Chatelain E, Hundertmark M, Leprince O, Gall SL, Satour P, Penninck SD, Rogniaux H, Buitink J (2012) Temporal profiling of the heat-stable proteome during late maturation of *Medicago truncatula* seeds identifies a restricted subset of late embryogenesis abundant proteins associated with longevitypce. Plant Cell Environ. doi:10.1111/j.1365-3040.2012.02501.x

Cheng Z, Targolli J, Huang X, Wu R (2002) Wheat LEA genes, PMA80 and PMA1959, enhance dehydration tolerance of transgenic rice (*Oryza sativa* L.). Mol Breed 10:71–82

Cherian S, Reddy MP, Ferreira RB (2006) Transgenic plants with improved dehydration stress tolerance: progress and future prospects. Plant Biol 50:481–495

Chourey K, Ramani S, Apte SK (2003) Accumulation of LEA proteins in salt (NaCl) stressed young seedlings of rice (*Oryza sativa* L.) cultivar Bura Rata and their degradation during recovery from salinity stress. Plant Physiol 160:1165–1174

Clayton DF, George JM (1999) Synucleins in synaptic plasticity and neurodegenerative disorders. J Neurosci Res 58:120–129

Close TJ (1996) Dehydrins: emergence of a biochemical role of a family of plant dehydration proteins. Plant Physiol 97:795–803

Close TJ, Kortt AA, Chandler PM (1989) A cDNA-based comparison of dehydration-induced proteins (dehydrins) in barley and corn. Plant Mol Biol 13:95–108

Cohen A, Plant AL, Moses MS, Bray EA (1991) Organ-specific and environmentally regulated expression of two abscisic acid-induced genes of tomato: nucleotide sequence and analysis of the corresponding cDNAs. Plant Physiol 97:1367–1374

Colmenero-Flores JM, Campos F, Garciarrubio A, Covarrubias AA (1997) Characterization of *Phaseolus vulgaris* cDNA clones responsive to water deficit: identification of a novel late embryogenesis abundant-like protein. Plant Mol Biol 35:393–405

Colmenero-Flores JM, Moreno LP, Smith C, Covarrubias AA (1999) PvLEA-18, a member of a new late-embryogenesis-abundant protein family that accumulates during water stress and in the growing regions of well-irrigated bean seedlings. Plant Physiol 120:93–103

Cuming AC (1999) LEA proteins. In: Shewry PR, Casey R (eds) Seed proteins. Kluwer, Dordrecht, pp 753–780

Curry J, Walker-Simmons MK (1993) Unusual sequence of group 3 LEA (II) mRNA inducible by dehydration stress in wheat. Plant Mol Biol 21:907–912

Curry J, Morris CF, Walker-Simmons MK (1991) Sequence analysis of a cDNA encoding a group 3 LEA mRNA inducible by ABA or dehydration stress in wheat. Plant Mol Biol 16:1073–1076

Dang NX, Hincha DK (2011) Identification of two hydrophilins that contribute to the desiccation and freezing tolerance of yeast (*Saccharomyces cerevisiae*) cells. Cryobiology 62: 188–193

Danyluk J, Houde M, Rassart E, Sarhan F (1994) Differential expression of a gene encoding an acidic dehydrin in chilling sensitive and freezing tolerant gramineae species. FEBS Lett 344: 20–24

Danyluk J, Perron A, Houde M, Limin A, Fowler B, Benhamou N, Sarhan F (1998) Accumulation of an acidic dehydrin in the vicinity of the plasma membrane during cold acclimation of wheat. Plant Cell 10:623–638

De Souza Filho GA, Ferreira BS, Dias JM, Queiroz KS, Bressan-Smith RE, Oliveira JG, Garcia AB (2003) Accumulation of SALT protein in rice plants as a response to environmental stresses. Plant Sci 164:623–628

Dehaye L, Duval M, Viguier D, Yaxley J, Job D (1997) Cloning and expression of the pea gene encoding SBP65, a seed-specific biotinylated protein. Plant Mol Biol 35:605–621

Doczi R, Kondrak M, Kovacs G, Beczner F, Banfalvi Z (2005) Conservation of the drought-inducible DS2 genes and divergences from their ASR paralogues in solanaceous species. Plant Physiol Biochem 43:269–276

Dong JZ, Dunstan DI (1997) Characterization of cDNAs representing five abscisic acid-responsive genes associated with somatic embryogenesis in *Picea glauca*, and their responses to abscisic acid stereostructure. Planta 203:448–453

Dunker AK (2001) Intrinsically disordered protein. J Mol Graph Model 19:26–59

Dure L III (1993) Structural motifs in Lea proteins. In: Close TJ, Bray EA (eds) Plant responses to cellular dehydration during environmental stress. The American Society of Plant Physiologists, Rockville, pp 91–103

Dure L III (2001) Occurrence of a repeating 11-mer amino acid sequence motif in diverse organisms. Protein Pept Lett 8:115–122

Dyson HJ, Wright PE (2002) Coupling of folding and binding for unstructured proteins. Curr Opin Struct Biol 12:54–60

Eisenberg D (1984) Three-dimensional structure of membrane and surface proteins. Annu Rev Biochem 53:595–623

Ellis RJ (2004) From chloroplasts to chaperones: how one thing led to another. Photosynth Res 80:333–343

Eom J, Baker W, Kintanar A, Wurtele ES (1996) The embryo-specific EMB-1 protein of *Daucus carota* is flexible and unstructured. Plant Sci 115:17–24

Eriksson SK, Kutzer M, Procek J, Gröbner G, Harryson P (2011) Tunable membrane binding of the intrinsically disordered dehydrin Lti30, a cold-induced plant stress protein. Plant Cell 23:2391–2404

Espelund M, Saeboe-Larssen S, Hughes DW, Galau GA, Larsen F, Jakobsen KS (1992) Late embryogenesis-abundant genes encoding proteins with different numbers of hydrophilic repeats are regulated differentially by abscisic acid and osmotic stress. Plant J 2: 241–252

Farrant JM, Pammenter NW, Berjak P, Farnsworth EJ, Vertucci CW (1996) Presence of dehydrin-like proteins and levels of abscisic acid in recalcitrant (desiccation sensitive) seeds may be related to habitat. Seed Sci Res 6:175–182

Figueras M, Pujal J, Saleh A, Savé R, Pagès M, Goday R (2004) Maize Rab17 overexpression in Arabidopsis plants promotes osmotic stress tolerance. Ann Appl Biol 144:251–257

Finch-Savage WE, Pramanik SK, Bewley JD (1994) The expression of dehydrin proteins in desiccation-sensitive (recalcitrant) seeds of temperate trees. Planta 193:478–485

Frankel N, Carrari F, Hasson E, Iusem ND (2006) Evolutionary history of the Asr gene family. Gene 378:74–83

Frankel N, Nunes-Nesi A, Balbo I, Mazuch J, Centeno D, Iusem ND, Fernie AR, Carrari F (2007) ci21A/Asr1 expression influences glucose accumulation in potato tubers. Plant Mol Biol 63(5):719–730

Furuki T, Shimizu T, Chakrabortee S, Yamakawa K, Hatanaka R, Takahashi T, Kikawada T, Okuda T, Mihara H, Tunnacliffe A, Sakurai M (2012) Effects of group 3 LEA protein model peptides on desiccation-induced protein aggregation. Biochim Biophys Acta – Proteins and Proteomics. doi:10.1016/j.bbapap. 2012.04.013

Galau GA, Wang HY, Hughes DW (1992) Cotton Lea4 (D19) and LeaA2 (D132) group 1 Lea genes encoding water stress-related proteins containing a 20-amino acid motif. Plant Physiol 99:783–788

Galau GA, Wang HY, Hughes DW (1993) Cotton Lea5 and Lea74 encode a typical late embryogenesis-abundant proteins. Plant Physiol 101:695–696

Garay-Arroyo A, Colmenero-Flores JM, Garciarrubio A, Covarrubias AA (2000) Highly hydrophilic proteins in prokaryotes and eukaryotes are common during conditions of water deficit. J Biol Chem 275:5668–5674

Godoy JA, Lunar R, Torres-Schumann S, Moreno J, Rodrigo RM, Pintor-Toro JA (1994) Expression, tissue distribution and subcellular localization of dehydrin TAS14 in salt-stressed tomato plants. Plant Mol Biol 26(6):1921–1934

Golan-Goldhirsh A, Peri I, Birk Y, Smirnoff P (1998) Inflorescence bud proteins of *Pistacia vera*. Trees-Struct Funct 12:415–419

Goldgur Y, Rom S, Ghirlando R, Shkolnik D, Shadrin N, Konrad Z, Bar-Zvi D (2007) Desiccation and zinc binding induce transition of tomato abscisic acid stress ripening 1, a water stress- and salt stress-regulated plant-specific protein, from unfolded to folded state. Plant Physiol 143:617–628

Goyal K, Tisi L, Basran A, Browne J, Burnell A, Zurdo J (2003) Transition from natively unfolded to folded state induced by desiccation in an anhydrobiotic nematode protein. J Biol Chem 278:12977–12984

Goyal K, Browne JA, Burnell AM, Tunnacliffe A (2005a) Dehydration-induced *tps* gene transcripts from an anhydrobiotic nematode contain novel spliced leaders and encode atypical GT-20 family proteins. Biochimie 87:565–574

Goyal K, Walton LJ, Tunnacliffe A (2005b) LEA proteins prevent protein aggregation due to water stress. Biochem J 388:151–157

Grelet J, Benamar A, Teyssier E, Avelange-Macherel MH, Grunwald D, Macherel D (2005) Identification in pea seed mitochondria of a late embryogenesis abundant protein able to protect enzymes from drying. Plant Physiol 137:157–167

Haaning S, Radutoiu S, Hoffmann SV, Dittmer J, Giehm L, Otzen DE, Stougaard J (2008) An unusual intrinsically disordered protein from the model legume *Lotus japonicus* stabilizes proteins in vitro. J Biol Chem 283:31142–31152

Hand SC, Menze MA, Toner M, Boswell L, Moore D (2011) LEA proteins during water stress: not just for plants anymore. Annu Rev Physiol 73:115–134

Hara M, Terashima S, Kuboi T (2001) Characterization and cryoprotective activity of cold-responsive dehydrin from *Citrus unshiu*. J Plant Physiol 158:1333–1339

Hara M, Terashima S, Fukaya T, Kuboi T (2003) Enhancement of cold tolerance and inhibition of lipid peroxidation by citrus dehydrin in transgenic tobacco. Planta 217:290–298

Hara M, Fujinaga M, Kuboi T (2004) Radical scavenging activity and oxidative modification of citrus dehydrin. Plant Physiol Biochem 42:657–662

Hara M, Fujinaga M, Kuboi T (2005) Metal binding by *citrus* dehydrin with histidine-rich domains. J Exp Bot 56:2695–2703

Harada JJ, DeLisle A, Baden C, Crouch M (1989) Unusual sequence of an abscisic acid inducible mRNA which accumulates late in *Brassica napus* seed development. Plant Mol Biol 12:395–401

He S, Tan L, Hu Z, Chen G, Wang G, Hu T (2012) Molecular characterization and functional analysis by heterologous expression in *E. coli* under diverse abiotic stresses for *OsLEA5*, the atypical hydrophobic LEA protein from *Oryza sativa* L. Mol Genet Genomics 287: 39–54

Hellwege EM, Dietz KJ, Hartung W (1996) Abscisic acid causes changes in gene expression involved in the induction of the landform of the liverwort *Riccia fluitans* L. Planta 198: 423–432

Heyen BJ, Alsheikh MK, Smith EA, Torvik CF, Seals DF, Randall SK (2002) The calcium-binding activity of a vacuole-associated, dehydrin like protein is regulated by phosphorylation. Plant Physiol 130:675–687

Hoekstra FA, Golovina EA, Tetteroo FA, Wolkers WF (2001) Induction of desiccation tolerance in plant somatic embryos: how exclusive is the protective role of sugars? Cryobiology 43:140–150

Hollung K, Espelund M, Jakobsen KS (1994) Another Lea B19 gene (group 1 LEA) from barley containing a single 20 amino acid hydrophilic motif. Plant Mol Biol 25:559–564

Hong SH, Kim IJ, Yang DC, Chung WI (2002) Characterization of an abscisic acid responsive gene homologue from *Cucumis melo*. J Exp Bot 53:2271–2272

Hong-Bo S, Zong-Suo L, Ming-An S (2005) Lea proteins in higher plants: structure, function, gene expression and regulation. Colloid Surf B 45:131–135

Honjoh K-I, Oda Y, Takata R, Miyamoto T, Hatano S (1999) Introduction of the hiC6 gene, which encodes a homologue of a late embryogenesis abundant (LEA) protein, enhances freezing tolerance of yeast. J Plant Physiol 155:509–512

Houde M, Dallaire S, N'Dong D, Sarhan F (2004) Overexpression of the acidic dehydrin WCOR410 improves freezing tolerance in transgenic strawberry leaves. Plant Biotech J 2:381–387

Hsing YC, Chen ZY, Shih MD, Hsieh JS, Chow TY (1995) Unusual sequences of group 3 LEA mRNA inducible by maturation or drying in soybean seeds. Plant Mol Biol 29:863–868

Hughes S, Graether SP (2011) Cryoprotective mechanism of a small intrinsically disordered dehydrin protein. Protein Sci 20:42–50

Hunault G, Jaspard E (2010) LEAPdb: a database for the late embryogenesis abundant proteins. BMC Genomics 11:221–230

Hundertmark M, Hincha DK (2008) LEA (late embryogenesis abundant) proteins and their encoding genes in *Arabidopsis thaliana*. BMC Genomics 9:118

Hundertmark M, Popova AV, Rausch S, Seckler R, Hincha DK (2012) Influence of drying on the secondary structure of intrinsically disordered and globular proteins. Biochem Biophys Res Commun 417:122–128

Imai R, Chang L, Ohta A, Bray EA, Takagi M (1996) A LEA-class gene of tomato confers salt and freezing tolerance when expressed in *Saccharomyces cerevisiae*. Gene 170:243–248

Ismail AM, Hall AE, Close TJ (1999a) Allelic variation of a dehydrin gene cosegregates with chilling tolerance during seedling emergence. Proc Natl Acad Sci USA 96:13566–13570

Ismail AM, Hall AE, Close TJ (1999b) Purification and partial characterization of a dehydrin involved in chilling tolerance during seedling emergence of cowpea. Plant Physiol 120:237–244

Iturbe-Ormaetxe I, Escuredo PR, Arrese-Igor C, Becana M (1998) Oxidative damage in pea plants exposed to water deficit or paraquat. Plant Physiol 116:173–181

Iturriaga G, Schneider K, Salamini F, Bartels D (1992) Expression of desiccation-related proteins from the resurrection plant *Craterostigma plantagineum* in transgenic tobacco. Plant Mol Biol 20:555–558

Iusem ND, Bartholomew DM, Hitz WD, Scolnik PA (1993) Tomato (*Lycopersicon esculentum)* transcript induced by water deficit and ripening. Plant Physiol 102:1353–1354

Jensen AB, Goday A, Figueras M, Jessop AC, Pages M (1998) Phosphorylation mediates the nuclear targeting of the maize RAB17 protein. Plant J 13:691–697

Joh T, Honjoh K, Yoshimoto M, Funabashi J, Miyamoto T, Hatano S (1995) Molecular-cloning and expression of hardening-induced genes in *Chlorella-vulgaris* C-27: the most abundant clone encodes a late embryogenesis abundant protein. Plant Cell Physiol 36:85–93

Jyothsnakumari G (2005) Studies on biochemical responses and proteome analysis of two high yielding genotypes of mulberry (*Morus alba* L.) with differential salt sensitivity. Ph.D. thesis, Sri Krishnadevaraya University, Anantapur

Jyothsnakumari G, Thippeswamy G, Veeranagamallaiah G, Sudhakar C (2009) Differential expression of LEA proteins in two genotypes of mulberry under salinity. Biol Plant 3(1):145–150

Kalifa Y, Perlson E, Gilad A, Konrad Z, Scolnik PA, Bar-Zvi D (2004) Over-expression of the water and salt stress-regulated Asr1 gene confers an increased salt tolerance. Plant Cell Environ 27:1459–1468

Karlson DT (2001) Characterization and environmental regulation of a 24-kDa *Cornus sericea* dehydrin-like protein and its relationship to freeze-tolerance. Ph.D. thesis, Purdue University, West Lafayette

Karlson DT, Zeng Y, Stirm VE, Joly RJ, Ashworth EN (2003) Photoperiodic regulation of a 24-kDa dehydrin-like proteinin red-osier dogwood (*Cornus sericea* L.) xylem with relation to freeze-tolerance. Plant Cell Physiol 44:25–34

Kermode AR (1997) Approaches to elucidate the basis of desiccation-tolerance in seeds. Seed Sci Res 7:75–95

Kim HS, Lee JH, Kim JJ, Kim CH, Jun SS, Hong YN (2005) Molecular and functional characterization of CaLEA6, the gene for a hydrophobic LEA protein from *Capsicum annuum*. Gene 344:115–123

Kim IJ, Lee J, Han JA, Kim CS, Hur Y (2011) Citrus LEA promoter confers fruit proferrential and stress- inducible gene expression in *Arabdiopsis*. Can J Plant Sci 91:459–466

Kiyosue T, Yamaguchi-Shinozaki K, Shinozaki K, Higashi K, Satoh S, Kamada H, Harada H (1992) Isolation and characterization of a cDNA that encodes ECP31, an embryogenic-cell protein from carrot. Plant Mol Biol 19:239–249

Koag MC, Fenton RD, Wilkens S, Close TJ (2003) The binding of maize DHN1 to lipid vesicles. Gain of structure and lipid specificity. Plant Physiol 131:309–316

Koag MC, Wilkens S, Fenton RD, Resnik J, Vo E, Close TJ (2009) The K-segment of maize DHN1 mediates binding to anionic phospholipid vesicles and concomitant structural changes. Plant Physiol 150:1503–1514

Kosová K, Vítámvás P, Prásil IT (2007) The role of dehydrins in plant response to cold. Plant Biol 51(4):601–617

Kosová K, Vítámvás P, Prášil IT (2011) Expression of dehydrins in wheat and barley under different temperatures. Plant Sci 180:46–52

Kovacs D, Kalmar E, Torok Z, Tompa P (2008) Chaperone activity of ERD10 and ERD14, two disordered stress-related plant proteins. Plant Physiol 147:381–390

Krüger C, Berkowitz O, Stephan UW, Hell R (2002) A metal-binding member of the late embryogenesis abundant protein family transports iron in the phloem of *Ricinus communis* L. J Biol Chem 277:25062–25069

Kyte J, Doolittle RF (1982) A simple method for displaying the hydropathic character of a protein. J Mol Biol 157:105–132

Lång V (1993) The role of ABA and ABA-induced gene expression in cold acclimation of *Arabidopsis thaliana*. Ph.D. thesis, Swedish University of Agricultural Sciences, Uppsala

Layton BE, Boyd MB, Tripepi MS, Bitonti BM, Dollahon MNR, Balsamo RA (2010) Dehydration-induced expression of a 31-KDA dehydrin in *Polypodium polypdioides* (polypodiaceae) may enable large, reversible deformation of cell walls. Am J Bot 97(4):535–544

Leprince O, Buitink J (2010) Desiccation tolerance: from genomics to the field. Plant Sci 179:554–564

Levi A, Panta GR, Parmentier CM, Muthalif MM, Arora R, Shanker S, Rowland LJ (1999) Complementary DNA cloning, sequencing and expression of an unusual dehydrin from blueberry floral buds. Plant Physiol 107:98–109

Lin C, Thomashow MF (1992) A cold-regulated *Arabidopsis* gene encodes a polypeptide having potent cryoprotective activity. Biochem Biophys Res Commun 183:1103–1108

Lisse T, Bartels D, Kalbitzer HR, Jaenicke R (1996) The recombinant dehydrinlike desiccation stress protein from the resurrection plant *Craterostigma plantagineum* displays no defined three-dimensional structure in its native state. Biol Chem 377:555–561

Liu Y, Chakrabortee S, Li R, Zheng Y, Tunnacliffe A (2011a) Both plant and animal LEA proteins act as kinetic stabilizers of polyglutamine-dependent protein aggregation. FEBS Lett 585:630–634

Liu G, Xu H, Zhang L, Zheng Y (2011b) Fe binding properties of two soybean (*Glycine max* L.) LEA4 proteins associated with antioxidant activity. Plant Cell Physiol 52(6):994–1002

López-Martínez G, Rodríguez-Porrata B, Margalef-Català M, Cordero-Otero R (2012) The *STF2p* hydrophilin from *Saccharomyces cerevisiae* is required for dehydration stress tolerance. PLoS One 7:e33324

Maitra N, Cushman JC (1994) Isolation and characterization of a drought induced soybean cDNA encoding a D95 family late-embryogenesis abundant protein. Plant Physiol 106:805–806

Manfre AJ, LaHatte GA, Climer CR, Marcotte WR (2009) Seed dehydration and the establishment of desiccation tolerance during seed maturation is altered in the *Arabidopsis thaliana* mutant atem6-1. Plant Cell Physiol 50:243–253

Maruyama K, Todaka D, Mizoi J, Yoshida T, Kidokoro S, Matsukura S, Ttakasaki H, Sakurai T, Yamamoto YY, Yoshiwara K, Kojima M, Sakakibara H, Shinozaki K, Shinozaki KY (2012) Identification of Cis-acting promoter elements in cold- and dehydration- induced transcriptional pathways in *Arabidopsis*, rice, and soybean. DNA Res 19:37–49

Maskin L, Maldonado S, Iusem ND (2008) Tomato leaf spatial expression of stress-induced Asr genes. Mol Biol Rep 35:501–505

Mbeguie-A-Mbeguie D, Gomez RM, Fils-Lycaon B (1997) Molecular cloning and nucleotide sequence of a protein from apricot fruit (accession no. U82760) homologous to LEC14B protein isolated from *Lithospermum* gene expression during fruit ripening (PGR 97–161). Plant Physiol 115:1287–1289

McCubbin WD, Kay CM, Lane B (1985) Hydrodynamic and optical properties of the wheat germ Em protein. Can J Biochem Cell Biol 63:803–811

Melgar JC, Syvertsen JP, Martínez V, García-Sánchez F (2008) Leaf gas exchange, water relations, nutrient content and growth in citrus and olive seedlings under salinity. Plant Biol 52:385–390

Min DH, Zhang XH, Xu XH, Zhao Y, Chen Y, Li LC, Chen M, Ma YZ (2012) Induction kinetics of a novel stress-related *LEA* gene in wheat. Plant Mol Biol Rep. doi:10.1007/s11105-012-0446-2

Momma M, Kaneko S, Haraguchi K, Matsukura U (2003) Peptide mapping and assessment of cryoprotective activity of 26/27-kDa dehydrin from soybean seeds. Biosci Biotech Biochem 67:1832–1835

Moons A, Bauw G, Prinsen E, Van Montagu M, Van der Straeten D (1995) Molecular and physiological responses to abscisic acid and salts in roots of salt-sensitive and salt tolerant Indica rice varieties. Plant Physiol 107:177–186

Moons A, Gielen J, Van der Kerckhove J, Van der Straeten D, Gheysen G, Van Montagu M (1997a) An abscisic acid and salt stress responsive rice cDNA from a novel plant gene family. Planta 202:443–454

Moons A, Prinsen E, Bauw G, Van Montagu M (1997b) Antagonistic effects of abscisic acid and jasmonates on salt stress inducible transcripts in rice roots. Plant Cell 9:2243–2259

Mouillon JM, Gustafsson P, Harryson P (2006) Structural investigation of disordered stress proteins: comparison of full-length dehydrins with isolated peptides of their conserved segments. Plant Physiol 141:638–650

Mueller JK, Heckathorn SA, Fernando D (2003) Identification of a chloroplast dehydrin in leaves of mature plants. Int J Plant Sci 164:535–542

Mukhopadhyay A, Vij S, Tyagi AK (2004) Overexpression of a zinc-finger protein gene from rice confers tolerance to cold, dehydration and salt stress in transgenic tobacco. Proc Natl Acad Sci USA 101:6309–6314

Mundy J, Chua NH (1988) Abscisic acid and water-stress induce the expression of a novel rice gene. EMBO J 7:2279–2286

Muñoz-Mayor A, Pineda B, Garcia-Abellán JO, Antón T, Garcia-Sogo B, Sanchez-Bel P, Flores FB, Atarés A, Angosto T, Pintor-Toro JA, Moreno V, Bolarin MC (2012) Overexpression of dehydrin tas14 gene improves the osmotic stress imposed by drought and salinity in tomato. J Plant Physiol 169(5):459–468

Nakayama K, Okawa K, Kakizaki T, Honma T, Itoh H, Inaba T (2007) *Arabidopsis* Cor15am is a chloroplast stromal protein that has cryoprotective activity and forms oligomers. Plant Physiol 144:513–523

Niknam V, Razavi N, Ebrahimzadeh H, Sharifizadeh B (2006) Effect of NaCl on biomass, protein and proline contents, and antioxidant enzymes in seedlings and calli of two *Trigonella* species. Plant Biol 50:594–596

Nylander M, Svensson J, Palva ET, Welin BV (2001) Stress-induced accumulation and tissue-specific localization of dehydrins in *Arabidopsis thaliana*. Plant Mol Biol 45:263–279

Olvera-Carrillo Y, Reyes JS, Covarrubia AA (2011) Late embryogenesis abundant proteins: versatile players in the plant adaptation to water limiting environments. Plant Signal Behav 6(4):586–589

Padmanabhan V, Dias DM, Newton RJ (1997) Expression analysis of a gene family in loblolly pine (*Pinus taeda* L.) induced by water deficit stress. Plant Mol Biol 35:801–807

Panza V, Distéfano AJ, Carjuzaa P, Láinez V, del Vas M, Maldonado S (2007) Detection of dehydrin-like proteins in embryos and endosperm of mature *Euterpe edulis* seeds. Protoplasma 231:1–5

Park JA, Cho SK, Kim JE, Chung HS, Hong JP, Hwang B, Hong CB, Kim WT (2003) Isolation of cDNAs differentially expressed in response to drought stress and characterization of the Ca-LEAL1 gene encoding a new family of atypical LEA-like protein homologue in hot pepper (*Capsicum annuum* L. cv. Pukang). Plant Sci 165:471–481

Park B-J, Liu ZC, Kanno A, Kameya T (2005) Genetic improvement of Chinese cabbage for salt and drought tolerance by constitutive expression of a *B. napus* LEA gene. Plant Sci 169:553–558

Piatkowski D, Schneider K, Salamini F, Bartels D (1990) Characterization of five abscisic acid-responsive cDNA clones isolated from the desiccation- tolerant plant *Craterostigma plantagineum* and their relationship to other water-stress genes. Plant Physiol 94:1682–1688

Plana M, Itarte E, Eritja R, Goday A, Pages M, Martinez MC (1991) Phosphorylation of maize RAB-17 protein by casein kinase-2. J Biol Chem 266:22510–22514

Pouchkina-Stantcheva NN, McGee BM, Boschetti C, Tolleter D, Chakrabortee S, Popova AV, Meersman F, Macherel D, Hincha DK, Tunnacliffe A (2007) Functional divergence of former alleles in an ancient asexual invertebrate. Science 318:268–271

Prieto-Dapena P, Almoguera C, Rojas A, Jordano J (1999) Seed-specific expression patterns and regulation by ABI3 of an unusual late embryogenesis-abundant gene in sunflower. Plant Mol Biol 39:615–627

Puhakainen T, Hess MW, Makela P, Svensson J, Heino P, Palva ET (2004) Overexpression of multiple dehydrin genes enhances tolerance to freezing stress in *Arabidopsis*. Plant Mol Biol 54:743–775

Reddy PS, Reddy GM, Pandey P, Chandrasekhar K, Reddy MK (2012) Cloning and molecular characterization of a gene encoding late embryogenesis abundant protein from *Pennisetum glaucum*: protection against abiotic stresses. Mol Biol Rep 39:7163–7174

Reyes JL, Rodrigo MJ, Colmenero-Flores JM, Gil JV, Garay-Arroyo A, Campos F, Salamini F, Bartels D, Covarrubias AA (2005) Hydrophilins from distant organisms can protect enzymatic activities from water limitation effects in vitro. Plant Cell Environ 28:709–718

Riccardi F, Gazeau P, Vienne DD, Zivy M (1998) Protein changes in response to progressive water deficit in maize: quantitative variation and polypeptide identification. Plant Physiol 117:1253–1263

Rinne PLH, Kaikuranta PLM, van der Plas LHW, van der Schoot C (1999) Dehydrins in cold-acclimated apices of birch (*Betula pubescens* Ehrh.): production, localization and potential role in rescuing enzyme function during dehydration. Planta 209:377–388

Roberts JK, DeSimone NA, Lingle WL, Dure L (1993) Cellular concentrations and uniformity of cell-type accumulation of two LEA proteins in cotton embryos. Plant Cell 5:769–780

Rohila JS, Jain RK, Wu R (2002) Genetic improvement of Basmati rice for salt and drought tolerance by regulated expression of a barley Hva1 cDNA. Plant Sci 163:525–532

Romanenko AS, Borovskii GB, Ukolova IV, Lomovatskaya LA (2010) Subcellular localization of dehydrins in wheat plant seedlings subjected to low-temperature adaptation. Biochem (Moscow) suppl ser A: Membr Cell Biol. doi:10.1134/S 1990747810020066

Romo S, Labrador E, Dopico B (2001) Water stress-regulated gene expression in *Cicer arietinum* seedlings and plants. Plant Physiol Biochem 39:1017–1026

Rorat T (2006) Plant dehydrins: tissue location, structure and function. Cel Mol Biol Lett 11:536–556

Rorat T, Szabala BM, Grygorowicz WJ, Wojtowicz B, Yin Z, Rey P (2006) Expression of SK3-type dehydrin in transporting organs is associated with cold acclimation in *Solanum* species. Planta 224:205–221

Russouw PS, Farrant J, Brandt W, Lindsey GG (1997) The most prevalent protein in a heat-treated extract of pea (*Pisum sativum*) embryos is an LEA group I protein: its conformation is not affected by exposure to high temperature. Seed Sci Res 7:117–123

Saavedra L, Svensson J, Carballo V, Izmendi D, Welin B, Vidal S (2006) A dehydrin gene in *Physcomitrella patens* is required for salt and osmotic stress tolerance. Plant J 45: 237–249

Salleh MF, Evans K, Goodall B, Machin H, Mowla S, Mur L, Runions J, Theodoulou F, Foyer C, Rogers H (2011) A novel function for a redox-related LEA protein (SAG21/AtLEA5) in root development and biotic stress responses. Plant Cell Environ 35:418–429

Salmi ML, Bushart TJ, Stout SC, Roux SJ (2005) Profile and analysis of gene expression changes during early development in germinating spores of *Ceratopteris richardii*. Plant Physiol 138:1734–1745

Sanchez-Ballesta MT, Rodrigo MJ, Lafuente MT, Granell A, Zacarias L (2004) Dehydrin from citrus, which confers in vitro dehydration and freezing protection activity, is constitutive and highly expressed in the flavedo of fruit but responsive to cold and water stress in leaves. J Agric Food Chem 52:1950–1957

Sarnighausen E, Karlson D, Ashworth E (2002) Seasonal regulation of a 24-kDa protein from red-osier dogwood (*Cornus sericea* L.) xylem. Tree Physiol 22:423–430

Scheef ED, Fink JL (2003) Fundamentals of protein structure. In: Bourne PE, Weissig H (eds) Structural bioinformatics. Wiley- Liss, Hoboken, pp 15–39

Segrest JP, Deloof H, Dohlman JG, Brouilette CG, Anantharamaiah GM (1990) Amphipathic helix motif: classes and properties. Proteins Struct Funct Genet 8:103–117

Segrest JP, Garber DW, Brouillette CG, Harvey SC, Anantharamaiah GM (1994) The amphipathic alpha helix: a multifunctional structural motif in plasma apolipoproteins. Adv Protein Chem 45:303–369

Sharon MA, Kozarova A, Clegg JS, Vacratsis PO, Warner AH (2009) Characterization of a group 1 late embryogenesis abundant protein in encysted embryos of the brine shrimp *Artemia franciscana*. Biochem Cell Biol 87(2):415–430

Shih MD, Linm SC, Hsieh JS, Tsou CH, Chow TY, Lin TP (2004) Gene cloning and characterization of a soybean (*Glycine max* L.) LEA protein, GmPM16. Plant Mol Biol 56:689–703

Shih M, Hoekstra F, Hsing Y (2008) Late embryogenesis abundant proteins. Adv Bot Res 48:211–255

Shih MD, Huang LT, Wei FJ, Wu MT, Hoekstra FA, Hsing YC (2010) OsLEA1a, a new Em-like protein of cereal plants. Plant Cell Physiol 51(12):2132–2144

Silhavy D, Hutvagner G, Barta E, Banfalvi Z (1995) Isolation and characterization of a water-stress-inducible cDNA clone from *Solanum chacoense*. Plant Mol Biol 27:587–595

Siminovitch D, Briggs DR (1953) Studies on the chemistry of the living bark of the black locust tree in relation to frost hardiness. IV. Effects of ringing on translocation, protein synthesis and the development of hardiness. Plant Physiol 28:177–200

Singh J, Whitwill S, Lacroix G, Douglas J, Dubuc E, Allard G, Keller W, Schernthaner JP (2009) The use of group 3 LEA proteins as fusion partners in facilitating recombinant expression of recalcitrant proteins in *E. coli*. Protein Expr Purif 67(1):15–22

Sivamani E, Bahieldin A, Wraith JM, Al-Niemi T, Dyer WE, Ho T-H D, Qu R (2000) Improved biomass productivity and water use efficiency under water deficit conditions in transgenic wheat constitutively expressing the barley HVA1 gene. Plant Sci 155:1–9

Sotiropoulos TE (2007) Effect of NaCl and $CaCl_2$ on growth and contents of minerals, chlorophyll, proline and sugars in the apple rootstock M4 cultured in vitro. Plant Biol 51:177–180

Soulages JL, Kim K, Walters C, Cushman JC (2002) Temperature-induced extended helix/random coil transitions in a group 1 late embryogenesis-abundant protein from soybean. Plant Physiol 128:822–832

Stacy RAP, Nordeng TW, Culianez-Macia FA, Aalen RB (1999) The dormancy-related peroxiredoxin anti-oxidant, PER1, is localized to the nucleus of barley embryo and aleurone cells. Plant J 19:1–8

Steponkus PL, Uemura M, Joseph RA, Gilmour SJ, Thomashow MF (1998) Mode of action of the COR15a gene on the freezing tolerance of *Arabidopsis thaliana*. Proc Natl Acad Sci USA 95:14570–14575

Svensson J, Palva ET, Welin B (2000) Purification of recombinant *Arabidopsis thaliana* dehydrins by metal ion affinity chromatography. Protein Expr Purif 20:169–178

Svensson J, Ismail AM, Palva ET, Close TJ (2002) Dehydrins. In: Storey KB, Storey JM (eds) Sensing, signaling and cell adaptation. Elsevier Science, Amsterdam, pp 155–171

Swire-Clark GA, Marcotte WR Jr (1999) The wheat LEA protein Em functions as an osmoprotective moledule in *Saccharomyces cerevisiae*. Plant Mol Biol 39:117–128

Tantos A, Friedrich P, Tompa P (2009) Cold stability of intrinsically disordered proteins. FEBS Lett 583:465–469

Todaka D, Nakashima K, Shinozaki K, Shinozaki KY (2012) Toward understanding transcriptional regulatory networks in abiotic stress responses and tolerance in rice. Rice 5:6

Tolleter D, Jaquinod M, Mangavel C, Passirani C, Saulnier P, Manon S, Teyssier E, Payet N, Avelange-Macherel MH, Macherel D (2007) Structure and function of a mitochondrial late embryogenesis abundant protein are revealed by desiccation. Plant Cell 19:1580–1589

Tolleter D, Hincha DK, Macherel D (2010) A mitochondrial late embryogenesis abundant protein stabilizes model membranes in the dry state. Biochim Biophys Acta 1798:1926–1933

Tompa P (2002) Intrinsically unstructured proteins. Trends Biochem Sci 27:527–533
Tompa P (2005) The interplay between structure and function in intrinsically unstructured proteins. FEBS Lett 579:3346–3354
Tompa P, Kovacs D (2010) Intrinsically disordered chaperones in plants and animals. Biochem Cell Biol 88(2):167–174
Tunnacliffe A, Wise MJ (2007) The continuing conundrum of the LEA proteins. Naturwissenschaften 94:791–812
Tunnacliffe A, Hincha DK, Leprince O, Macherel M (2010) LEA proteins: versatility of form and function. In: Lubzens E, Cerda J, Clark MS (eds) Dormancy and resistance in harsh environments. Springer, Berlin/Heidelberg, pp 91–108
Ulrich TU, Wurtele ES, Nikolau BJ (1990) Sequence of EMB-1, an mRNA accumulating specifically in embryos of carrot. Nucleic Acids Res 18:2826
Uversky VN, Gillespie JR, Fink AL (2000) Why are "natively unfolded" proteins unstructured under physiologic conditions. Proteins 41:415–427
Vaidyanathan R, Kuruvilla S, Thomas G (1999) Characterization and expression pattern of an abscisic acid and osmotic stress responsive gene from rice. Plant Sci 140:21–30
Veeranagamalliaiah G, Prasanthi J, Reddy KE, Pandurangaiah M, Babu OS, Sudhakar C (2011) Group 1 and 2 LEA protein expression correlates with a decrease in water stress induced protein aggregation in horsegram during germination and seedling growth. J Plant Physiol 168(7):671–677
Vicient CM, Hull G, Guilleminot J, Devic M, Delseny M (2000) Differential expression of the *Arabidopsis* genes coding for Em-like proteins. J Exp Bot 51:1211–1220
Vilardell J, Goday A, Freire MA, Torrent M, Martinez MC, Torne JM, Pages M (1990) Gene, sequence, developmental regulation and protein phosphorylation of RAB17 in maize. Plant Mol Biol 14:423–432
Viner RI, Clegg JS (2000) Influence of trehalose on the molecular chaperone activity of p26, a small heat shock/alpha-crystallin protein. Cell Stress Chaperones 6:126–135
Wahid A, Close TJ (2007) Expression of dehydrins under heat stress and their relationship with water relations of sugarcane leaves. Plant Biol 51:104–109
Wang CS, Liau YE, Huang JC, Wu TD, Su CC, Lin CH (1998) Characterization of a desiccation-related protein in lily pollen during development and stress. Plant Cell Physiol 39(12):1307–1314
Wang Y, Jiang J, Zhao X, Liu G, Yang C, Zhan L (2006) A novel LEA gene from *Tamarix androssowii* confers drought tolerance in transgenic tobacco. Plant Sci 171:655–662
Welin BV, Olson A, Nylander M, Palva ET (1994) Characterisation and differential expression of DHN/LEA/RAB-like genes during cold-acclimation and drought stress in *Arabidopsis thaliana*. Plant Mol Biol 26:131–144
Welling A, Rinne P, Vihera-Aarnio A, Kontunen-Soppela S, Heino P, Palva ET (2004) Photoperiod and temperature differentially regulate the expression of two dehydrin genes during overwintering of birch (*Betula pubescens* Ehrh.). J Exp Bot 55:507–516
Williams BA, Tsang A (1994) Analysis of multiple classes of abscisic acid responsive genes during embryogenesis in *Zea mays*. Dev Genet 15:415–424
Wise MJ (2003) LEAping to conclusions: a computational reanalysis of late embryogenesis abundant proteins and their possible roles. BMC Bioinformatics 4:52
Wise MJ, Tunnacliffe A (2004) POPP the question: what do LEA proteins do? Trends Plant Sci 9:13–17
Wisniewski M, Webb R, Balsamo R, Close TJ, Yu XM, Griffith M (1999) Purification, immunolocalization, cryoprotective, and antifreeze activity of PCA60: a dehydrin from peach (*Prunus persica*). Plant Physiol 105:600–608
Wolkers WF, McCready S, Brandt WF, Lindsey GG, Hoekstra FA (2001) Isolation and characterization of a D-7 LEA protein that stabilizes glasses in vitro. Biochim Biophys Acta 1544:196–206
Wurtele ES, Wang H, Durgerian S, Nikolau BJ, Ulrich TH (1993) Characterization of a gene that is expressed early in somatic embryogenesis of *Daucus carota*. Plant Physiol 102:303–312
Xu D, Duan X, Wang B, Hong B, Ho T-HD, Wu R (1996) Expression of a late embryogenesis abundant protein gene, HVA1, from barley confers tolerance to water deficit and salt stress in transgenic rice. Plant Physiol 110:249–257

Yang L, Zheng B, Mao C, Qi X, Liu F, Wu P (2004) Analysis of transcripts that are differentially expressed in three sectors of the rice root system under water deficit. Mol Genet Genomics 272:433–442

Yin Z, Rorat T, Szabala BM, Ziólkowska A, Malepszy S (2006) Expression of a *Solanum sogarandinum* SK3-type dehydrin enhances cold tolerance in transgenic cucumber seedlings. Plant Sci 170:1164–1172

Yoshida Y, Sato T, Hashimoto T, Ichikawa N, Nakai S, Yoshikawa H, Imamoto F, Tagawa K (1990) Isolation of a gene for a regulatory 15-kDa subunit of mitochondrial F_1F_0-ATPase and construction of mutant yeast lacking the protein. Eur J Biochem 192:49–53

Zegzouti H, Jones B, Marty C, Lelievre JM, Latche A, Pech JC, Bouzayen M (1997) ER5, a tomato cDNA encoding an ethylene-responsive LEA like protein: characterization and expression in response to drought, ABA and wounding. Plant Mol Biol 35:847–854

Zhang L, Ohta A, Takagi M, Imai R (2000) Expression of plant group 2 and group 3 lea genes in *Saccharomyces cerevisiae* revealed functional divergence among LEA proteins. J Biochem 127:611–616

Zhao PS, Liu F, Chang G, Zheng C, Liu H (2011) Group 3 late embryogenesis abundant protein in *Arabidopsis*: structure, regulation, and function. Acta Physiol Plant 3:1063–1073

Chapter 6
Enhancing Plant Productivity Under Salt Stress: Relevance of Poly-omics

Mirza Hasanuzzaman, Kamrun Nahar, Masayuki Fujita, Parvaiz Ahmad, Ruby Chandna, M.N.V. Prasad, and Munir Ozturk

Abstract At present more than 20% of all the irrigated land in the world is estimated as affected by salinity and this trend is increasing with the rapid climate changes as well as the excess use of irrigation water. Salt stress is one of the most devastating abiotic stresses which severely affects the agricultural productivity in various ways. High concentration of salt in the soil or in the irrigation water can have a overwhelming effect on plant metabolism, disrupting cellular homeostasis and uncoupling major physiological and biochemical processes. Salinity cause both osmotic stress and ionic toxicity which hamper the plant productivity by inhibiting or altering the plant growth,

M. Hasanuzzaman (✉)
Department of Agronomy, Faculty of Agriculture, Sher-e-Bangla Agricultural University, Sher-e-Bangla Nagar, Dhaka 1207, Bangladesh
e-mail: mhzsauag@yahoo.com

K. Nahar
Department of Agricultural Botany, Faculty of Agriculture, Sher-e-Bangla Agricultural University, Sher-e-Bangla Nagar, Dhaka 1207, Bangladesh

Laboratory of Plant Stress Responses, Department of Applied Biological Science, Faculty of Agriculture, Kagawa University, Kagawa, Japan

M. Fujita
Laboratory of Plant Stress Responses, Department of Applied Biological Science, Faculty of Agriculture, Kagawa University, Kagawa, Japan

P. Ahmad
Department of Botany, A.S. College, University of Kashmir, Srinagar, India

R. Chandna
National Institute for Plant Genome Research, 110067 New Delhi, India

Department of Botany, Jamia Hamdard, New Delhi, India

M.N.V. Prasad
Department of Plant Sciences, University of Hyderabad, Hydrabad, Andhra Pradesh, India

M. Ozturk
Department of Botany, Ege University, Bornova, Izmir, Turkey

P. Ahmad et al. (eds.), *Salt Stress in Plants: Signalling, Omics and Adaptations*, DOI 10.1007/978-1-4614-6108-1_6,

dry matter partitioning, seed germination, photosynthesis and yield. Considering the devastating effect of salt stress on plants, one of the important tasks for plant biologists is to explore the approaches that are able to develop salt tolerance in crop plants. In fact, salt tolerance is a multigenic trait which is governed by various morphological and physiological factors. Thus omics approaches therefore, come in forefront to develop salt tolerance as a part of different strategies of conventional plant breeding. Transcriptomics, proteomics, metabolomics, ionomics and micromics together have been a bloom in revealing plant stress responses and the mechanisms that underlie these responses. These techniques have been playing important part in discovering new genes, proteins and secondary plant metabolites those are responsible for plants adaptation to stress. In this review, we have focused on the causes and effects of salinity on crop plants and possible mechanisms of salt tolerance including the possible use of omics in conferring salt tolerance.

Keywords Salinity • Sodicity • Plant responses • Plant tolerance • Omics • Salt stress in plants

Introduction

In agrarian region of the world soil salinity and sodicity are among the major problems limiting plant growth and productivity. These problems are especially of great concern for countries whose economies rely to a great extent on agriculture (Pessarakli and Szabolcs 2010). Currently more than 100 countries are more or less affected by salinity and sodicity where irrigation water is an important causal factor (Glenn et al. 2009; de Souza et al. 2012). Worldwide about 831 million hectares (M ha) is salt-affected of which saline and sodic soils occupy 397 and 434 M ha, respectively (FAO 2000). The lower productivity of crops grown under saline soils is an obvious phenomenon which is mainly due to the osmotic stress and excess ion toxicity (Katerji et al. 1998; van Hoorn et al. 2001; Ahmad and Sharma 2008). The responses to salt stress comprise an array of changes at the molecular, biochemical and physiological levels (Garg and Manchanda 2008a, b). The common effects of salt stress in plants are decrease in seed germination and seedling vigor, stunted growth of plants, impaired photosynthesis and dry matter partitioning which rendered poor yield. In addition, excess salinity may cause ionic toxicity which led to retardation of many essential plant nutrients (Abrol et al. 1988; Hasanuzzaman et al. 2011a; 2011b).

As salt stress has become major concern for crop production, exploring suitable techniques to develop salt resistant plants are important tasks for plant scientists. Under salt stress condition plants exhibit various mechanisms for surviving. Salt tolerance in plants is a multifarious feature which is governed by several factors. In the last few decades salt tolerance mechanisms has been studied by many researchers and it was explained to some extent by stress adaptation effectors that mediate ion homeostasis, osmolyte biosynthesis, free radicals scavenging, water transport and long distance response co-ordination (Hasegawa et al. 2000; Ahmad and Prasad 2012a, b).

Omics (proteomics, genomics, metabolomics, micromics, ionomics etc) are essential to understand the molecular systems that underlie various plant functions (Mochida and Shinozaki 2011). For the improvement of crops, plant biologists need to understand mechanism of plant responses to environmental stress. Salt stress is responsible for the up and down regulation of genes is reported in many plants. A number of salt-responsive genes have been isolated and characterized and are used in conferring salt tolerance in salt sensitive plants (Debnath et al. 2011. Proteomics has appeared as an important tool in the field of plant science, enabling us to interpret the stress responses occurring in plants. The vast range of applications of proteomics in biological fields has greatly increased its use over the last decade (Bindschedler et al. 2008; Chen et al. 2011, Evers et al. 2012; Kaufmann et al. 2011; Nanjo et al. 2011; Thelen and Peck 2007; Yang et al. 2011; Yokthongwattana et al. 2012; Zhang et al. 2012; Zheng et al. 2012). Metabolomics involves a comprehensive non-biased analysis of metabolites in a given cell at a specific time. These metabolomic studies have demonstrated that active reconfiguration of the metabolome is regulated in part by changes in gene expression initiated by salt-stress-activated signaling and stress-related transcription factors (Bhalla et al. 2005; Guy et al. 2008). Unraveling additional stress-associated gene resources, from both crop plants and highly salt- and drought tolerant model plants, will enable future molecular dissection of salt-tolerance mechanisms in important crop plants (Vinocur and Altman 2005). Micomics have emerged as one of the important and interesting discoveries in past decades. MicroRNAs (miRNAs) a type of ssRNAs, 18-24 nucleotides long molecule of the genome plays a pivotal role in endogenous regulation of gene expression. miRNAs have been reported to regulate various stress responsive genes, proteins and transcription factors, thus helping to counteract adverse conditions. Various stress-inducible miRNAs have been identified and well characterized. Most of these miRNAs have been conserved among plants (Guleria et al. 2012).

In this chapter, we briefly describe the impact of salinity and sodicity on plant productivity. Mainly we focus on the responses of plants to salt stress and the adaptation of plants to salt stress by reviewing the recent findings on these aspects. Finally, we review the recent advancement of using various omics approaches in developing salt tolerance in crop plants.

Salinity and Sodicity

The term "salt-affected" refers to both saline and sodic soils. However, some people get confused between salinity and sodicity. Although sodium chloride (NaCl) is one of the dominant species in saline soils or water other salts may also be present in different amounts. Sodium sulphate (Na_2SO_4), calcium sulphate ($CaSO_4.2H_2O$, gypsum), calcium chloride ($CaCl_2$), magnesium chloride ($MgCl_2$) and sodium bicarbonate [$Ca(HCO_3)_2$] are found in acidic, neutral and alkaline soils in different proportions. For all of the cases, they dissociate into positively charged cations and negatively charged anions in aqueous solution (van de Graaff and Patterson 2001).

Based on the nature, characteristics and plant growth relationships in salt affected soils, two main types of soils have been classified as (a) saline soil and (b) sodic soil (Szabolcs 1974).

In case of saline soil, the soluble salts are chiefly NaCl and Na_2SO_4 and sometimes also contain appreciable quantities of chloride (Cl^-) and sulphate (SO_4^{2-}) of calcium (Ca^{2+}) and magnesium (Mg^{2+}). Sodium chloride dissolves easily when added to water, and produces a very concentrated electrolyte solution. Calcium sulphate is only sparingly soluble while calcium carbonate is only slightly soluble in water, and becomes even less soluble as the soil becomes more alkaline. These soils contain sufficient neutral soluble salts to pose negative effect on growth of most crop plants. The term 'salinity' is broadly known and refers to the amount of soluble salt in a soil. The salinity in a soil solution is generally measured as Electrical conductivity (EC) which is based on the fact that the electrical current transmitted between two electrodes (i.e. with standardised solution, temperature and electrodes areas that usually equal to unity) increases with an increase of soluble ionic salts and vice versa (Ezlit et al. 2010). The basic SI unit of EC is Siemens per metre (S m^{-1}). In agriculture, EC is often low; thus deciSiemens per metre (dS m^{-1}) is widely used. The unit (mmhos cm^{-1}) used in the past is numerically equal to dS m^{-1}.

Soil sodicity is expressed the amount of Na^+ in soils which develops through a process whereby Na^+ build up in preference to other soil cations (particularly Ca^{2+}) on the exchange complex of the soil. Previously these soils have also been termed as 'Alkali'. A sodic soil has excessive amount of Na^+ associated with the negatively charged clay particles in soils. Too much Na^+ leads to excessive swelling of the soil, which may result in a structural collapse which causes the separation of soil aggregates into the component sand, silt and clay particles (McKenzie and Orange 2003). Thus, soils become very sensitive to waterlogging which can hamper the crop production. In contrary, when soil becomes dry, dispersive soil tends to be too hard for roots and seedlings to penetrate. Sodic soils are high in exchangeable Na^+ compared to calcium and magnesium. EC is less than 4 dS m^{-1} and often less than 2 dS m^{-1}. Soil pH usually is greater than 8.5 and can be as high as 10 or even 11 in extreme cases.

The sodicity of water or soil is usually described in terms of the relative proportion of sodium cations (Na^+), compared to the divalent cations (i.e. Ca^{2+} and Mg^{2+}) in solution. The sodium adsorption ratio (SAR) of water (including soil solutions) is calculated as:

$$\mathrm{SAR} = \frac{\mathrm{Na}^{+}}{\sqrt{\frac{\mathrm{Ca}^{2+} + \mathrm{Mg}^{2+}}{2}}}$$

where the cation concentrations are expressed in mmolc L^{-1}. However, sodicity in soils is expressed by the exchangeable sodium percentage (ESP) and calculated as:

$$\mathrm{ESP} = \frac{\text{Exchangeable Na}}{\mathrm{CEC}} \times 100$$

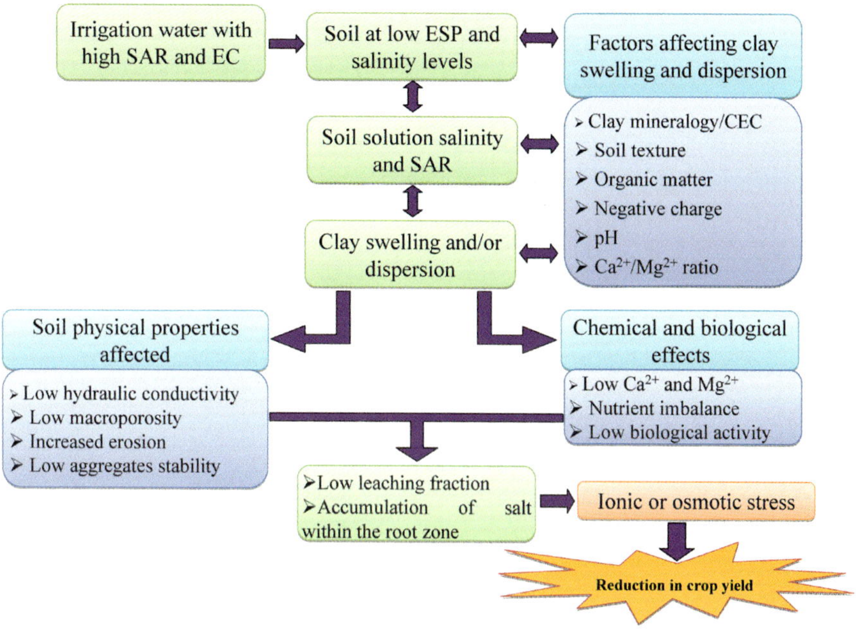

Fig. 6.1 Different problems in soils caused by soil salinity and sodicity

where CEC is the cation exchange capacity. The CEC is the sum of exchangeable cations such as Na^+, Ca^{2+}, Mg^{2+} and K^+ (as well as Al^{3+} in low pH soils) expressed in mmolc/100 g.

There are another type of soil called Saline-sodic soils which are high in Na^+ and other salts. This soil generally have EC greater than 4 dS m^{-1} (mmhos cm^{-1}), SAR greater than 13, and/or ESP greater than 15. Soil pH can be above or below 8.5. They may possess the characteristics both saline or sodic soil, depending on whether Na^+ or Ca^{2+} dominates. Both of the saline and sodic condition creates several problems in soils which affect the crop production (Fig. 6.1).

Causes of Salinity or Sodicity

Salinity and sodicity in soil caused due to excess accumulation of salts. The process of salinization may be occurred directly or indirectly. As salt is a natural element of soils and water, the ions responsible for salinization may accumulate in soil surface or near root zoon by different means. There are two major causes of salinity viz. (1) natural causes, and (2) anthropogenic causes (Fig. 6.2). Primary salinity is occurred due to the long-term natural accumulation of salts in the soil or surface water. This is a natural process which is caused mainly by weathering of parent materials containing soluble salts through break down of rocks containing Cl^- of Na^+, Ca^{2+} and Mg^{2+} and sometimes SO_4^{2-} and CO_3^{2-}. In addition, deposition of sea salt carried in wind and rain

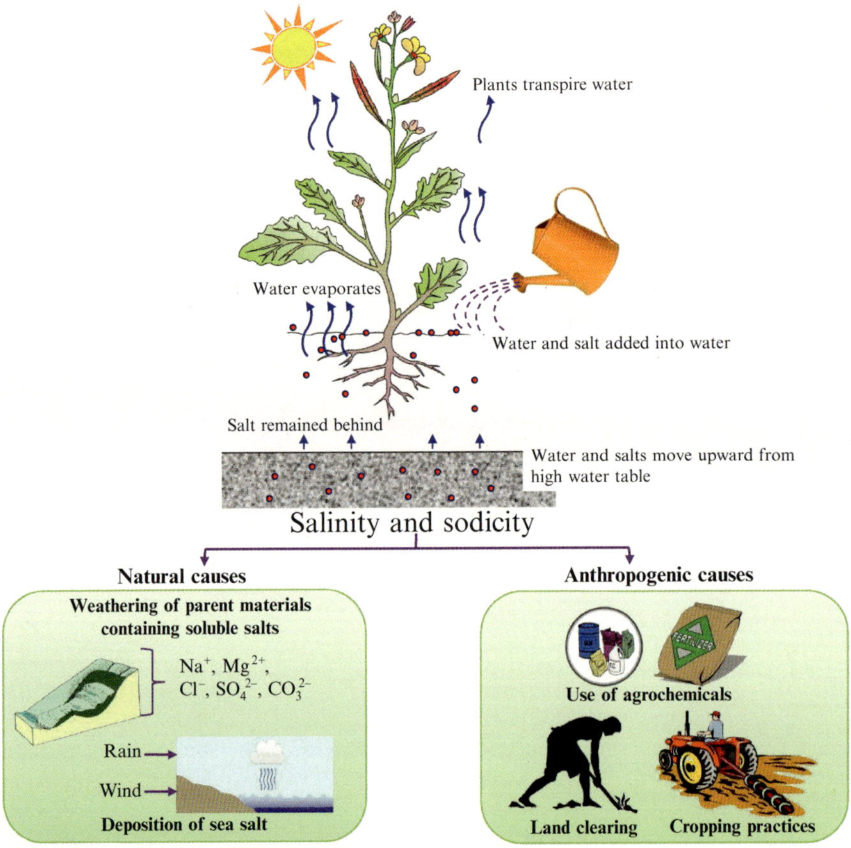

Fig. 6.2 Causes of salinity and sodicity in soil and water

also a reason which varies with the types of soil. The most important source of the salt in soils is the microscopic Aeolian salt which is disseminated to the land from the oceans that may account up to 13 kg ha^{-1} year^{-1}. Salinity may also caused by human activities (secondary salinization) such as land clearing, fertilization etc. which may accumulate salt up to 2.3 kg ha^{-1} year^{-1}. In nature, volcanic eruptions contribute 0.2 kg of salts ha^{-1} year^{-1} and contrary to the popular belief, the weathering of rocks contributes the least (0.04 kg ha^{-1} year^{-1}) (Waisel 1972). The extent to which each of these factors can affect growth depends on plant genotype and on environmental conditions (Munns 2002). Secondary salinity occurs due to anthropogenic activities that disrupt the hydrologic balance of the soil between water applied (irrigation or rainfall) and water used by crops (transpiration) (Munns 2005; Garg and Manchanda 2008b). The ions responsible for salinization are: Na^+, K^+, Ca^{2+}, Mg^{2+} and Cl^-. As the Na^+ predominates, soils can become sodic. Sodic soils are important because they create very poor soil structure due to disaggregation which impairs water movement. In nature, rocks and minerals containing salts are continuously weathering into small particles and over a long time these particles flushed or leached out of the soils. Additionally, salts are also deposited via dust and precipitation.

When access accumulation of salt occurs in ground water, these can be moved upward in the soil surface due to capillary transport from a salt laden water table and then accumulate due to evaporation. These salts may also transported into soils due to anthropogenic activities such as the use of fertilizer containing salt ions. However, salinization is a process that results from high levels of salt in the soils; landscape feature that allows salts to become mobile (movement of water table) and climatic trend that favor accumulation (Fig. 6.2). In dry and irrigated areas salinization processes are different. In dry areas, salinity may occur in soil having water table of 2–3 m. In such case, salts from the groundwater are raised by capillary action to the surface of the soil. Agricultural practices of dry areas are also different, e.g. in those areas most of the plant species are deep rooting and if groundwater is saline, and is favored by land use practices allowing more rainwater to enter the aquifer than it could accommodate. In irrigated areas, where water contains some salts, when uptaken by plants salinity can occur over time (ILRI 1989). In agricultural soils, the plants use the water as a natural process but the salts are left behind in the soil which gradually accumulate in the soil. Excessive irrigation and improper drainage increase the water table that allows the salty groundwater to reach in the upper soil layers and rhizosphere. In urban areas, where irrigation water is abundantly used in gardens and recreation areas salinity often results from the combination of irrigation and groundwater processes.

Problem of Saline Water Irrigation

Irrigated agriculture has been facing the challenge of sustainability of production for centuries. Due to the natural and geochemical factors, as well as excessive irrigation and poor drainage make the problem of salinity more severe in agricultural soils (Tanji and Wallender 2011). The salt damage in agricultural field is not a recent event. Historical records for past 6,000 years revealed that numerous civilizations based on irrigated agriculture have been suffered from salt damage. Over the time, salts may concentrate to such an extent that they hinder plant growth, yield and quality. Irrigation in crop field affects soil-water content equilibrium in various ways (Fig. 6.3).

Based on the salt concentration of irrigation water is classified into different categories (Rhoades et al. 1992; Table 6.1). However, due to the suitability of saline water for irrigation is dependent on several factors such as plant species, climatic condition, soil types, methods of irrigation and management practices. Hence, water quality classifications are always not advised for assessing suitability of water for irrigation.

Whatever the extent of salinity in irrigation water, it hampers the crop production in various ways depending on the levels of salinity. In moderately saline irrigation water, the absorption of water from the soil become difficult for plants, and, as a result, the plant soon begins to wilt, and growth is slowed, with reduced yields. In case of highly saline irrigation water, the process of osmosis can become reversed. In such cases, the concentration of salt outside the plant roots is higher than that of the root cells which rendered movement of water from the roots into the surrounding solution. Ultimately, the plant loses moisture, and so suffers stress. The general

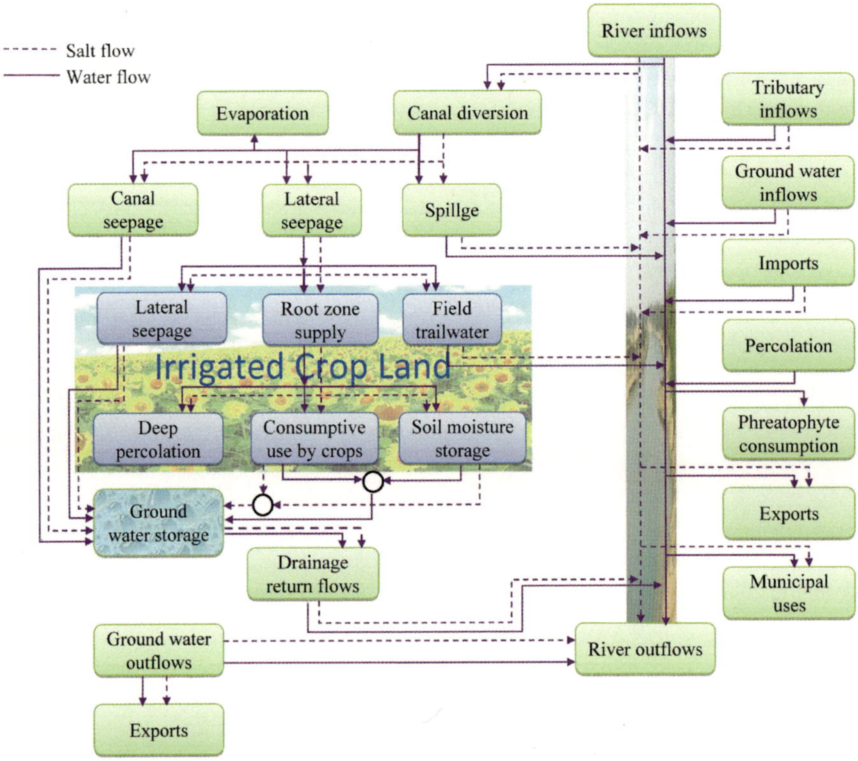

Fig. 6.3 Soil-water content equilibrium in irrigated crop field

symptoms of the crops grown under high salinity are: leaf tip dieback, yellowing of leaf, scorched and turning brown or black and finally death of leaves. When water is applied as sprinkler irrigation (spraying), it can cause salt scorch and leaf damage even at lower salinities. Besides these, any levels of salinity can cause ionic toxicity. Therefore, before using saline water for irrigation it is necessary to understand how salinity affects a crop. It is also necessary to monitor salinity levels constantly to ensure they remain within the acceptable range.

Plant Responses to Salt Stress

Salt stress affects the plants in different ways. Reduction of water potential, Na^+ and Cl^- phytotoxicity, disruption of nutrient transportation process are the three main physiological disorders due to salt stress as described by Munns (2002). These physiological disorders cause water deficit and nutrient imbalance and thus salt affected plants are suffered from their basic needs of water and nutrient. Increase of osmotic potential of soil caused by salinity is also another reason of reduced water availability of plants. Plant itself with the salinity stress are determinant factors for plant responses to salt stress and in these cases genotype and growth stage of plant, concentration and extent of salt stress imposition are considered. Upon imposition of salt stress plants

Table 6.1 Classification of irrigation water based on salinity levels

Water class	EC (dS m^{-1})	Salt concentration (mg L^{-1})	Type of water
Non-saline	<0.7	<500	Drinking and irrigation water
Slightly saline	0.7–2	500–1,500	Irrigation water
Moderately saline	2–10	1,500–7,000	Primary drainage water and groundwater
Highly saline	10–25	7,000–15,000	Secondary drainage water and groundwater
Very highly saline	25–45	15,000–35,000	Very saline groundwater
Brine	>45	>45,000	Seawater

Fig. 6.4 Major effects of salt stress on plants

are suffered in diversified ways. Salt stress may cause osmotic and ionic stresses, nutrient imbalance, disruption of hormone and enzymatic balance, and antioxidative defense mechanism. These phenomenon result in disintrigation of cell membrane, malfunctioning in photosynthesis, respiration and other physiological activities, generation of toxic and oxidative products causing serious damage to cell and even to death (Mahajan and Tuteja 2005). The major effects of salinity are the inhibition of germination, growth retardation, inhibition of photosynthesis, altered water uptake, oxidative stress, and yield reduction (Fig. 6.4). Excessive salt concentration in soil also hampers nutrient uptake and metabolism.

Germination

Salinity, as an abiotic exposure, provokes many disorders in seeds and propagules during germination (Koyro and Eisa 2008; Rasheed 2009). Every year salt stress is hampering the germination of number of species of crops which includes cereal, pulse, oilseed, vegetables, medicinal plants, trees. Adverse effect of germination was found in *Oryza sativa* (Xu et al. 2011), *Sorghum halepense* (Sinha and Gupta 1982), *Triticum aestivum* (Al-Harbi et al. 2008; Zheng et al. 2009), *Zea mays* (Carpıcı et al. 2009; Khodarahmpour et al. 2012), *Brassica* spp. (Ulfat et al. 2007), *Glycine max* (Essa 2002), *Vigna* spp., (Jabeen et al. 2003), *Solanum lycopersicum* (Kaveh et al. 2011), *Pinus pinea* (Sidari et al. 2008), *Lactuca sativa* (Zapata et al. 2004), *Anacardium occidentale* (Voigt et al. 2009), *Catharanthus roseus* (Jaleel et al. 2007), *Pistacia vera* (Benmahioul et al. 2009) and *Psoralea corylifolia* (Katare et al. 2012).

Salinity substantially reduces seed germination because of its osmotic effects and mineral toxicity during germination metabolism (Wahid et al. 2010). The most common examples of salt stress are delayed germination or maturity (Yadav et al. 2011). The four major events of germination are imbibition, active metabolism, emergence and elongation of embryonic tissues, and establishment of seedlings, all are affected by salinity (Wahid et al. 2010). The first step of germination is imbibition of water by seed and the osmotic component of salinity has a strong inhibitory effect on hydration of the embryo, cotyledon/s, and endosperm (Wahid et al. 1998) and it is true for all salinity type and the result is the reduced water uptake (Poljakoff-Mayber et al. 1994; Wahid et al. 1998). Upon salinization the second step active metabolism can adversely affected in several ways, viz. inter-membrane particles disintegrate for which leakage of solutes occur, mobilization of reserves reduced because of inhibition of carbohydrates and fatty acid metabolism enzyme activities, de novo protein synthesis which is a major event is also altered, osmotically active solutes produce (Dell'Aquila and Spada 1993; Yupsanis et al. 1994; Wahid et al. 1998); for all these unwanted events the cotyledon dry weight do not convert much into useful product for germinating seed (Ruffino et al. 2009). During seedling emergence salt stress delays and reduces emergence of radicle and plumule; reduced elongation of embryonic tissues, delayed the percentage and rate of seedling emergence (Turhan and Ayaz 2004; Sinha and Gupta 1982; Katerji et al. 1994). At last during seedling establishment increased salinity levels cause seedling mortality; reduced seedling growth, vigor and establishment (Rogers et al. 1995; Al-Mutawa 2003). Several reports can be presented as for example. Bordi (2010) found in his experiment that the germination percentage in *Brassica napus* significantly reduced at 150 and 200 mM NaCl. Nahar and Hasanuzzaman (2009) found to decrease the germination to 55% at 250 mM NaCl. Cuartero and Fernandez-Munoz (1999) reported that seeds need 50% more days to germinate at 80 mM NaCl (EC = 1.4) than in a medium without salt and almost 100% more days at 190 mM NaCl (EC = 3.4). Without these there are various internal (plant) and external (environmental) factors that affect seed germinability under saline conditions (Wahid et al. 2011).

Plant Growth

Salt stress affects the most crucial processes of cellular function including cell division, differentiation and expansion and these components have substantial impact on plant growth and development (Hasegawa et al. 2000; Zhu 2001; Ahmad 2010).

Soil salinity reduces the soil water potential and the ability of plants to take up water, and this reduces the rate of cell expansion in growing tissues. This may results in slower formation of photosynthetic leaf area in turns reduces the flow of dry matter or assimilates to the productive tissues of the plant. This effects are common in both roots and shoots which leads to reduced growth (Munns and Sharp 1993).

Accumulation of salts in the root zone beyond the tolerance level can lead to growth inhibition, leaf necrosis, accelerated senescence, wilting and death (Neumann 2011). Greenway and Munns (1980) studied with several crops which responded differently in same level of salt of 200 mM. They found 20% reduction in dry weight of salt-tolerant species such as *Beta vulgaris*, 60% reduction of moderately tolerant species such as *Gossypium hirsutum* and a sensitive species such as *Glycine max* might be dead, whereas a halophyte such as *Suaeda maritima* might be growing at its optimum rate at the same salinity level (Flowers et al. 1986). Salinity was also reported to increase cortex width and diameter of cortical cells in the stem cotton (Zhong and Läuchli 1993). It was observed that when salinity rises to 100 mM NaCl (about 10 dS m^{-1}), *Oryza sativa* died before maturity, while *Triticum aestivum* produced a reduced yield as a salt tolerant crop (Maas and Hoffman 1977). However, in *Hordeum vulgare* – the most-tolerant cereal, died only after treatment with more than 250 mM NaCl for an extended periods (USDA-ARS 2005). In different salinity levels all emergence and seedling growth characters were significantly affected in different lines of *Solanum lycopersicum* except seedling length was mentioned by Kaveh et al. (2011). In another experiment it was found that *Solanum lycopersicum* seedlings exposed to 300 mM NaCl for up to 9-h salt treatment rapidly and uniformly induced a complete wilting of the shoots and it was 100% lethal (Shalata et al. 2001; Shalata and Neumann 2001). Hasanuzzaman et al. (2009) observed a remarkable reduction in plant height and tiller number and leaf area index in *Oryza sativa* plants grown under saline soils. Root is the first plant organ to encounter the saline medium and is the first site of damage under salt stress. Salinity affects these root developmental processes differently. Root extension growth of many plants is severely inhibited by high concentrations of NaCl (Zhong and Läuchli 1993; Neumann et al. 1994), lateral root formation is less affected (Waisel and Breckle 1987), or stimulated (Kramer 1980). Salinization was shown to cause thicker but shorter roots in barley (Huang and Redman 1995; cotton: Zhong and Läuchli 1993a) and to reduce cotton root elongation and change cell shape (Kurth et al. 1986). A sorghum (Koryo 1997) root was shortened under saline conditions.

Excessive salinity in the root zone can also cause hydraulic limitations to leaf growth by decreasing in root hydraulic conductivity (Evlagon et al. 1990; Azaizeh and Steudle 1991; Chazen et al. 1995; Lu and Neumann 1999). Chemical signals coming from roots in dry or saline soil also reduce leaf growth (Munns 2011). Salinity led to decrease plant metabolic activities and finally decrease plant growth (Hussein et al. 2007). When Na^{+} or Cl^{-} is highly accumulated in plant it may limit the flow of carbon compounds to meristems and growing zones in leaves which led to the death of plant before maturity and seed production. The injury occurs when the salt load exceeds the ability of the plant cells to compartmentalize salts into the vacuole (Munns 2011).

Yield

Grain yield is a product of an organized interaction of several factors, which are highly subjected to environmental and genetic makeup of plants. That is why upon imposition of various environmental factors the yield of a plant varies widely, plant type, that is the species, or variety or intervarietal genetic makeup variation can alter the yield to a great extent. Global food production will need to increase by 38% by 2025 and by 57% by 2050 (Wild 2003) to supply food to the growing world population at the current level. The land of the world is not increasing. So, to increase the cultivable area is not wise. Rather, it is wise to increase in yield per unit even under adverse environmental condition (Rengasamy 2006).

Salinity is one of the major yield limiting factors for crop plants principally in arid and semiarid regions of the world (Munns 2005). Salinity stress not only affect vegetative growth, but also effect reproductive structures and translocation of C and N to developing seeds and fruits. High Soil salinity level can be a key environmental restriction to crop productivity (Kaveh et al. 2011). Most crops are vulnerable to salt stress and they either die or reduced the yield (Scholberg and Locascio 1999).

Levels of salinity affect different the yield and yield components. A low level of salinity may not reduce grain yield where the leaf area and shoot biomass is reduced, it may happen for a crop or plant that the yield may not decrease until a given 'threshold' salinity is reached (Rengasamy 2006). A comprehensive survey of US Salinity Laboratory (Maas and Hoffman 1977; USDA-ARS 2005) on salt tolerance of crops and pasture species presents that each species has a threshold salinity below which there is no reduction in yield, and then a linear reduction in yield with increasing salinity. In that survey, the yield of rice starts to decrease at 3 dS m^{-1} (30 mM NaCl) as compared to 6–8 dS m^{-1} for wheat (60–80 mM NaCl), and the subsequent linear decrease of rice yield with increasing concentration of salt is double that of wheat. Yield is the cumulative of yield components. Cullu (2003) reported that increasing EC values up to 13.4 dS m^{-1} caused diminishes the cotton and wheat yield of 29.6% and 35.4%, respectively. This reduction was due primarily to a decrease in grain weight/ear rather than to a reduction in ear formation (Francois et al. 1984). In our study we observed that at a level of 250 mM NaCl yield of *V. radiata* cv. BARI mung-2, BARI mung-5 and BARI mung-6 was reduced by 77%, 73% and 66%, respectively over control and it was the result of reduction of number pod per plant, seeds per pod and seed weight with the raised salinity level (Nahar and Hasanuzzaman 2009). Another experiment was conducted by Ahmed (2009) with five mungbean genotypes with the aim of ascertaining the effect of salt stress on the yield and its component. In that study the decrease in seed yield per plant under salt stress was more pronounced which was associated with decrease in number of seed per pod and seed weight. Salt stress have been found to be less effective at developmental stage than vegetative and flowering stages and delayed maturity due to salt stress caused desiccation stress and shriveled seeds as well (Ahmed 2009). Salinity significantly reduced the yield of lentil and chickpea (Katerji et al. 2001a, b). Lentil showed a yield reduction of about 20% at an ECe of 2 dS m^{-1} and 90–100% at an ECe of 3.2 dS

m^{-1} (van Hoorn et al. 2001). In annual plant salinity has negative impact on reproductive organs and it also alters the flowering time in cereals hence maturity (Munns and Rawson 1999). After studying with 18 advanced rice genotypes Ali et al. (2004) reported that yield per plant, morphological and chlorophyll content of all the genotypes were reduced by salinity. In *Vitis vinifera* L. yield reduction above the threshold was 8.9 ± 1.2% per 1 dS m^{-1} increase in soil electrical conductivity (Zhang et al. 2002).

Photosynthesis

Like other physiological processes salinity induced photosynthesis alteration also varied depending upon plant type, age of plant, dose and duration of stress. The rate of photosynthesis of salt tolerant variety may be less affected in higher salinity levels where the same levels of salinity significantly reduce the rate in salt sensitive varieties. At 342 mM salinity the photosynthetic rate is higher in *Salicornia which is* considered as the most salt-tolerant C_3 vascular plant (Abdulrahman and Williams 1981). In *Suaeda* plant photosynthesis was lower at 170 mM than at 340 or 680 mM salinity (Hajibagheri et al. 1984a). On the other hand, Kuramoto and Brest did experiment with 0–450 mM levels of salt stress and several plants viz. *Batis maritima, Spartina foliosa,* and *Distichlis spicata* (Kuramoto and Brest 1979). They observed the reduction in photosynthesis for all levels of salinity. Sometimes it happens that low salinity level increases photosynthesis where higher level decreases it (Parida et al. 2004). The photosynthetic pigments are altered in salt stress which is another major reason for diminished photosynthesis (Maxwell and Johnson 2000). The reduction of chl *a* and *b* contents in salt stress was found in *Oryza sativa* (Amirjani 2011), *Morus alba* (Ahmad et al. 2010b), *Brassica juncea* (Ahmad et al. 2012) and *Vicia faba* (Azooz et al. 2011). Chutipaijit et al. (2011) observed the reduction in chl *a*, chl *b* and carotenoids (Car) contents in the same plant upon exposure to 100 mM salinity.

The photosynthesis process not only affected due to the alteration in biochemical reactions under salt stress but also affected due to salinity induced morphological and physiological changes like reduction in leaf area, stomatal conductance, etc. and ambient environment like temperatures, sunlight are also important. Salinity creates a physiological drought in plant and which reduces water uptake ability of plant, then the cell expansion is reduced and this reduces the rate of leaf expansion, and closes stomata and thereby reduces photosynthesis (Rahnama et al. 2010). Salt stress reduces stomatal conductance in the older leaves, which limits their photosynthetic rate. In C_3 and C_4 plants, stomatal closure is generally the main cause of reduced photosynthesis under high salinity (Chaves et al. 2009; Delfine et al. 1999; Hura et al. 2006; Ghannoum 2009). Low mesophyll conductance in high salinity can reduce the partial pressure of CO_2 at the site of carboxylation and limit photosynthesis (Niinemets et al. 2009). The similar results were obtained by Flexas et al. (2004, 2008) and Niinemets et al. (2005, 2009).

Plant grown under salinity, accumulate high concentrations of Na^+ in leaves which impose an additional barrier to growth by reducing the longevity of photosynthetic tissues (Munns 2002). The toxic effect of salinity causes premature death of older leaves and can be seen by premature yellowing (Munns 2011). Variations in photosynthesis are also found in different temperatures but same salinity level. Photosynthesis was higher in *Lepochloa fusca at* 250 mM salt as compared to control when the temperature was 32°C or 39°C but it was lower in 19°C with the same treatments (Gorham 1987). Salinity brings on physiological drought that is why plant closes stomata to reduce water loss which causes rise in temperature by reducing transpiration. Leaf temperatures often rising up to 5°C or 6°C in relation to air temperature (Maria et al. 2011). Thus more adverse affect is created. The detrimental effects on leaf physiology were shown to depend on exposure to sun light also. Remorini et al. (2009) found more negative impacts for sunny than for the shade sites of the canopy of *Olea europea* trees.

Metabolic limitations occurring under high salinity are related with the high concentrations of Na^+ and Cl^- in leaf tissue (in general above 250 mM) (Munns et al. 2006). The depletion of organic acids accompanies stomatal closure and decreased assimilation. At higher concentrations, NaCl may directly inhibit photosynthesis due to oxidative stress (Chaves et al. 2009). In Sorghum under salinity, Netondo et al. (2004) reported a significant decrease in maximum quantum yield of photosystem II. Moradi and Ismail (2007) also mentioned sensitivity to salt stress in cereals might be due to reduction in PSII photochemical efficiency.

Ionic Toxicity and Osmotic Stresses

Osmotic effects of salt on plants are as a result of lowering of the soil water potential due to increase in solute concentration in the root zone. Salt induced osmotic imbalance starts from rhizosphere of a plant and then gradually spreads to plant internal cellular structure and causes ionic toxicity also. Increase of salt in the root medium can lead to a decrease in leaf water potential and may affect plant physiological processes (Sohan et al. 1999; Romero-Aranda et al. 2001). Very low soil water potentials is induced by high salt content in the soil water, this condition interferes with plants ability to extract water from the soil (Sohan et al. 1999) and one of the results of this is lack of osmotic adjustment. The plant then face osmotic stress resulting in reduced water absorption and physiological drought which are major causes of salinity injury in plants (Bernstein and Hayward 1958; Levitt 1980; Harivandi et al. 1992). During osmotic stress plants tend to increase their own cell osmotic potential so that they can compensate with that stress which is known as osmotic regulation or osmotic adjustment but this often cause ion toxicity (Hellebust 1976; Levitt 1980; Yeo 1983; Gorham et al. 1993). Ion toxicity is considered as the secondary effects of salt stress hampering the plant growth and survival as well.

The primary effects are ionic toxicity and disequilibrium, and hyperosmolality. Both Na^+ and Cl^- are inhibitory to cytosolic and organellar processes (Niu et al. 1995;

Zhu et al. 1998; Serrano et al. 1999). However, Na^+ appears to reach a toxic concentration before Cl^- does (Munns and Tester 2008). Excess Na^+ and Cl^- ion disrupt enzyme structure, cause cell proliferation, diminish energy production and hamper the physiological processes (Larcher 1980). To maintain the appropriate protein and enzyme structures a static hydrophobic–electrostatic balance is required and because of salt stress when the Na^+ exceeds 0.4 M of its concentration it disturbs that balance and disrupts the protein structure (Wyn Jones and Pollard 1983). High Na^+ also causes chlorosis, necrosis and premature senescence of leaves. Excessive salt concentration also restricts the cell expansion and growth by reducing the osmotic potential and turgidity of cell (Munns and Termaat 1986; Hasegawa et al. 2000; Zhu 2002).

Production of Reactive Oxygen Species

Like other abiotic stresses salt stress generate a secondary oxidative stress caused by the accumulation of reactive oxygen species (ROS) including singlet oxygen ($^1O_2$), superoxide ($O_2{\cdot}^-$), hydrogen peroxide (H_2O_2), and hydroxyl radicals ($OH{\cdot}$) (Ahmad et al. 2010a, c, 2011; Ahmad and Umar 2011; Hasanuzzaman et al. 2012b, c; Hasanuzzaman et al. 2013). When plants are exposed to salinity, the level of ROS production reaches as high as 720 $\mu M\ s^{-1}$ (a threefold increase) and H_2O_2 level can be as high as 15 μM (a 30-fold increase). It is reported that a H_2O_2 concentration of 10 μM reduces the net photosynthesis rate by 50% (Singh and Flowers 2010). Enhanced production of ROS resulted in an increase in lipid peroxidation, as documented by a more than fivefold increase of malondialdehyde (MDA) production in wheat (Price and Handry 1991). Salt induced ROS production was studied by many scientists after years in different crop plants such as *Oryza sativa*, *Lycopersicon esculentum*, *Triticum aestivum*, *Brassica napus*, *Citrus* sp., *Calendula officinalis* L., *Cicer arietinum*, *Hordeum vulgare* (Gueta-Dahan et al. 1997; Dionisio-Sese and Tobita 1998; Mittova et al. 2004; Ardic et al. 2009; Ali et al. 2011; Hasanuzzaman et al. 2011a, b; Sedghi et al. 2012). If once the superoxide and H_2O_2 toxicity have been created then they attributed to a cascade of reactions that result into the production of hydroxyl radicals and other destructive species such as lipid peroxidases those hamper plant processes and cellular components (Buxton et al. 1988; Halliwell and Gutteridge 1990; Asada 1994; Mittler 2002; Vaidyanathan et al. 2003; Miller et al. 2010). Especially these ROS are more toxic in the absence of protective mechanism in the plant and can cause protein denaturation, lipid peroxidation, hormonal imbalances, oxidation of nucleic acids and DNA, even mutation (Halliwell and Gutteridge 1989; Vaidyanathan et al. 2003; Munns and Tester 2008; Miller et al. 2010).

Generally ROS are rapidly removed by antioxidative mechanisms, but this removal can be impaired by salt stress (Allan and Fluhr 1997; Foyer and Noctor 2003). Increased levels of ROS also have the potential roles in stress signal transduction (Halliwell and Gutteridge 1989; Foyer and Noctor 2003; Kacperska 2004;

Table 6.2 Major crops showing different salt-tolerance levels

Sensitive	*Oryza sativa* L., *Sesamum indicum* L., *Phaseolus vulgaris* L., *Vigna radiata* (L.) R. Wilcz., *Daucus carota* L., *Vigna mungo* (L.) Hepper, *Allium cepa* L., *Fragaria* x *Ananassa* Duch., *Malus sylvestris* Mill., *Prunus armeniaca* L., *Musa acuminata* Colla, *Prunus avium* L., *Mangifera indica* L.
Moderately sensitive	*Cicer arietinum* L., *Zea mays* L., *Linum usitatissimum* L., *Arachis hypogaea* L., *Saccharum officinarum* L., *Medicago sativa* L., *Vicia faba* L., *Brassica oleracea* L. Botrytis, *Cucumis sativus* L., *Solanum melongena* L. var*esculentum* Nees., *Allium sativum* L., *Lactuca sativa* L., *Cucumis melo* L., *Abelmoschus esculentus* (L.) Moench, *Pisum sativum* L., *Capsicum annuum* L., *Solanum tuberosum* L., *Cucurbita pepo* L. var *Pepo, Spinacia oleracea* L., *Lycopersicon lycopersicum* (L.) Karst., *Vitis vinifera* L., *Carica papaya* L.
Moderatelt tolerant	*Hibiscus sabdariffa* L., *Carthamus tinctorius* L., *Sorghum bicolor* (L.) Moench, *Glycine max* (L.) Merrrill, *Helianthus annuus* L., *Triticum aestivum* L., *Vigna unguiculata* (L.) Walp., *Cocos nucifera* L., *Ziziphus mauritiana* Lam.
Tolerant	*Hordeum vulgare* L., *Brassica campestris* L., *B. napus*, *Gossypium hirsutum* L., *Hibiscus cannabinus* L., *Avena sativa* L., *Secale cereale* L., *Beta vulgaris* L., *Asparagus officinalis* L.
Highly tolerant	*Phoenix dactylifera, Allenrolfea occidentals, Pinus pinea, Distichlis spicata*

Hong-bo et al. 2008; Miller et al. 2008, 2010; Jaspers and Kangasjärvi 2010; Pardo 2010). ROS signaling has been shown to be an integral part of acclimation response to salinity. Therefore, ROS plays a dual role both as toxic compounds and signaling molecules under salt stress (Foyer and Noctor 2005; Miller et al. 2008, 2010; Ahmad et al. 2008, 2010a; Ahmad and Umar 2011).

Plant's Tolerance to Salinity

Excess salt concentration in soil or water adversely affects the plant growth and yield. However, there will be some variation in how salinity affects the plant, depending several factors such as plant genotypes, growth stage, and environmental factors. Some of the major crops showing different tolerance to salt are also presented in Table 6.2. Tolerance of crops changes with the age and growth stage. Sodicity tolerance of rice increased with age in the initial growth stages and it was found beneficial to transplant somewhat older rice seedlings, 35–40 days of age, in sodic soils instead of the usually recommended 30 day old seedlings (Abrol et al. 1988). Other factors include irrigation method (surface or flood, overhead sprinkler, drip), stage of plant growth and irrigation management. The salt tolerance of a crop can best be described by plotting its relative yield as a continuous function of soil salinity. Salt tolerance is the capacity to persist in the presence of increasing degree of salinity (Hayward and Wadleigh 1949). A plant species may make little or no growth at higher salinity levels

Table 6.3 The salinity levels (EC_e) at which 10%, 25%, and 50% yield reduction can be expected for major crops irrigated with saline water

Crop	% Yield reduction		
	10	25	50
Oryza sativa L.	5.1	5.9	8.0
Triticum aestivum L.	7.1	10.0	14.0
Zea mays L.	5.1	5.9	8.0
Hordeum vulgare L.	11.9	15.8	17.0
Phaseolus vulgaris L.	1.1	2.1	3.0
Sorghum bicolor (L.) Moench	5.9	9.0	11.9
Gossypium hirsutum L.	9.9	11.9	16.0
Carthamus tinctorius L.	7.0	11.0	14.0
Lycopersicon lycopersicum (L.) Karst	4.0	6.6	8.0
Solanum tuberosum L.	2.5	4.0	6.0
Spinacia oleracea L.	5.7	6.9	8.0
Lactuca sativa L.	2.0	3.0	4.8
Allium cepa L.	2.0	3.4	4.0
Daucus carota L.	1.3	2.5	4.2
Medicago sativa L.	3.0	4.9	8.2

EC*e* the electrical conductivity of a saturation extract

but does survive. Some scientists defined salt tolerance as "the degree to which osmotic adjustment can be made without sacrifice in growth" (Bernstein 1961) or "the absence of negative effects on growth in plants that accumulate salts in their tissues" (Greenway and Munns 1980). In regards to growth and yield, salt tolerance is defined as "the sustained growth of plants in an environment of excess salts in the growth medium" (Shannon 1984) or "yield decrease expected for a given level of soluble salts in the root medium as compared with yield under non-saline conditions" (Maas and Hoffman 1977). The yield reduction under different salinity levels greatly varies depending on the plant's ability to tolerate salt stress (Table 6.3). Salt tolerance rating is made based on the plants' ability to sustain against the EC value of a soil extract. US Salinity Laboratory classified the plants based on the capacity of tolerance to salt (Fig. 6.5). It has been demonstrated that most of the cereal crops are tolerant to soil salinity and the only exception is rice and corn (Maas 1990).

Salt Tolerance Strategies

Mechanisms of salt tolerance, not yet completely clear, can be explained to some extent by stress adaptation effectors that mediate ion homeostasis, osmolyte biosynthesis, toxic radical scavenging, water transport and long distance response co-ordination (Hasegawa et al. 2000). However, attempts to improve yield under stress conditions by plant improvement have been largely unsuccessful, primarily

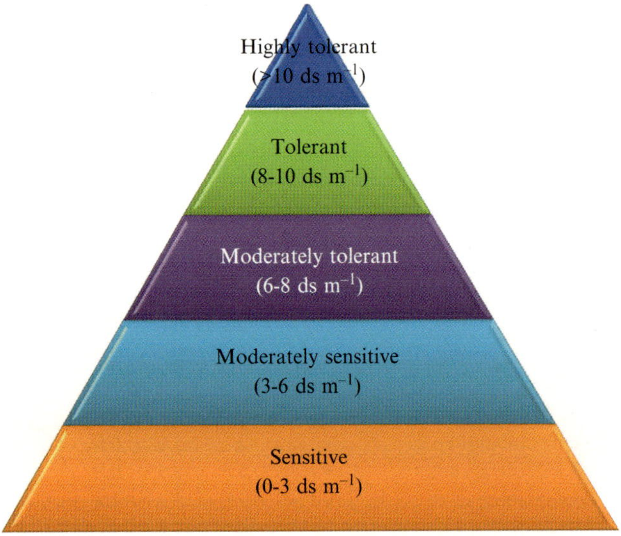

Fig. 6.5 Different categories of salt tolerance in plants (The EC values shown in salt tolerance classification must be determined in the soil saturation extract made from soil samples collected from the main root zone)

due to the multigenic origin of the adaptive responses. Therefore, a well-focused approach combining the molecular, physiological, biochemical and metabolic aspects of salt tolerance is essential to develop salt-tolerant crop varieties. Attempts to improve the salt tolerance of crops through conventional breeding program have met with very limited success, due to the complexity of the trait: salt tolerance is complex genetically and physiologically.

To achieve salt tolerance, three interconnected aspects of plant activities need to be investigated, these are: (1) damage must be prevented or alleviated, (2) homeostatic conditions must be re-established in the new, stressful environment, and (3) conducive conditions must be established for optimum growth or crop yield. Plants growing naturally on saline soils have developed various mechanisms to withstand the salinity stress Fig. 6.6 gives examples of control mechanisms in plants for maintaining a rather constant level of salt concentration in the living plant tissues. Plants show avoidance mechanism by performing their growth only in the favorable seasons (time niche) and on favorable sites (site niche). They also limit the root growth and absorption activity to distinct soil horizon. Most of the crop plants are relatively sensitive at early seedling and flowering stage. Rice, for example, being transplanted crop in most of the cases, can alleviate the salt stresses at seedling stage by management i.e. transplanting of aged seedlings but cannot avoid stress at flowering stage. However, under coastal saline conditions, salinity sometimes increase near the terminal growth stage of the plant. In that case, plants come to the maturity very quickly and complete their lifecycle. It is a typical case avoidance rather than

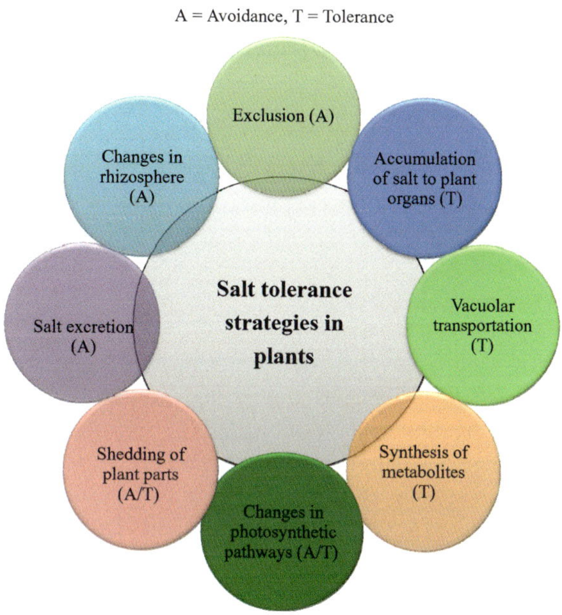

Fig. 6.6 Different salt tolerance strategies in plants

tolerance but it works as far as the productivity is concerned. Plants also develop salt tolerance in cells, tissues and organs employing a number of molecular and biochemical mechanisms. However, it is obvious that not only one of these mechanisms is active in different types of halophytes. One set can be active in one group, whereas another set of mechanisms is dominant in another group.

Salt Exclusion

In most of the cases, plants grown under saline condition plants are experienced with ionic stress and Na^+ appears to reach a toxic concentration before Cl^- does, and so most studies have concentrated on Na^+ exclusion and the control of Na^+ transport within the plant (Munns and Tester 2008). One of the mechanisms of salt tolerance involves minimum accumulation of Na^+ ions in the cytosol that in turn reduces the ionic stress in plants (Munns and Tester 2008; Carillo et al. 2011). Salt exclusion is a very efficient mechanism that prevents uptake of ions from the soil and helps the plants in less accumulation of salts especially in the transpiring organs like leaves (Dajic 2006). Lower permeability of plants root even under excessive concentration of soil salinity also actively helps in salt exclusion. Sometimes, glycophytes also found to exhibit Na^+ exclusion and hence, they perform better tolerance to salinity (Flowers and Hajibagheri 2001; Zhu 2001). There are plenty of evidences those indicated that Na^+ exclusion from leaves is associated with salt tolerance in cereal crops including rice, wheat, wheat and barley (James et al. 2011). Davenport et al. (2005) reported that exclusion of Na^+ from the leaves is due to low net Na^+ uptake by cells in the root cortex and the tight control of net loading of the xylem by parenchyma cells in the stele. Fortmeier and Schubert (1995) reported that reduction in growth of maize plants due to the higher accumulation of salt and the more salt tolerant cultivar was less affected due to the exclusion of Na^+ ions. Cereals are more or less proficient in salt exclusion; rice may be an exception, because of large rates of Na^+ influx into the roots under salinity stress, which was ascribed to leakage past the endodermis (Yeo et al. 1999).

Accumulation

Plant's ability to compartment salt ions is the most important mechanism conferring the salt tolerance. Plants transport the excess toxic ions to the older parts which soon die off and the young leaves or tissues are saved from the salt toxicity (Singh et al. 2002a, b). The ability of rice cultivars to compartmentalize ions in older leaves and structural tissues could crucially affect plant survival. Low salt concentration in the young leaves are the reasons that certain verities are able to survive in saline conditions. In case of grass family, compartmentation sometimes occurs in flag leaves. Rice, for example, salt tolerant cultivars maintain substantially lower salt concentration in the panicle and with concentration being lowest in grains compared to husks and rachis. Flag leaf health is also critical for higher yield under salt stress. Tolerant lines tend to maintain lower concentration of salt in the flag leaf (Singh et al. 2002a, b). During salt induced osmotic stress, Na^+ must be actively pumped into the vacuole from the cytoplasm due to the low concentration in the cytoplasm, whereas Cl^- might enter passively via anion channels to balance electrical charge differences across the membrane (Dupont 1992; Barkla et al. 1994; Rausch et al. 1996). This uptake of Na^+ into the vacuole appears to be mediated by Na^+/H^+-anitporters in the tonoplast, working in concert with H^+-ATPases and perhaps PPiases (Rea et al. 1992) that provides the proton motive force. Much more work has been done on the H^+ATPases, which have homology with enzymes from other organisms and therefore were easily cloned (Dupont 1992), than with Na^+/H^+-antiports, which have been cloned only recently (Apse et al. 1998; Darley et al. 1998). Accumulation of salt in the shoots of plants was reported as one of the most avoidance mechanisms of plant grown under saline condition and a number of research groups found higher accumulation of salts in salt-tolerant species (Ungar 1991; Dajic 1996, 2006). Although all of the halophytes exhibit better accumulation of salt, however, the level of total salt accumulation in the shoot is mostly species specific, depending on different adaptive strategies (Dajic 1996; Fig. 6.7).

Vacuolar Compartmentation

One of the common features of salt stress is the accumulation of toxic ions (Na^+ and Cl^-) in plants parts. However, in salt tolerant plants, one of the tolerance mechanisms is associated with the efficiency of such plants to deliver these toxic ions into

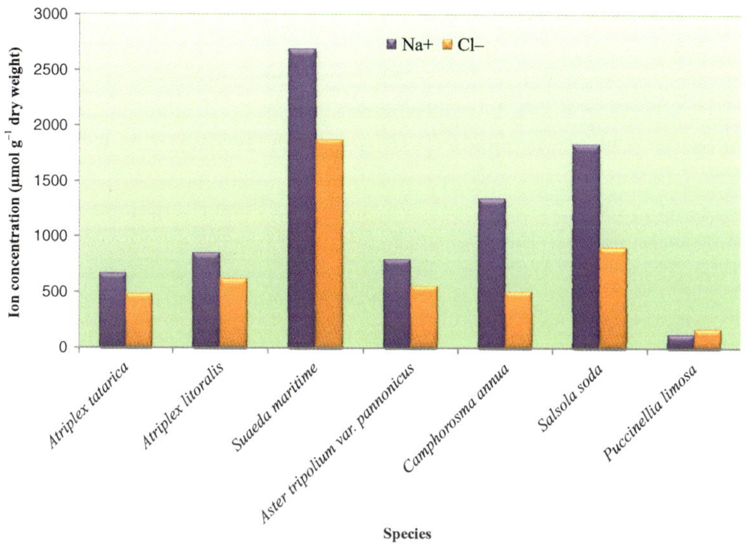

Fig. 6.7 Seasonal Na^+ and Cl^- ion concentration in the shoot of some halophytes grown in natural habitats

the vacuoles. However, this capacity of transfer ions into the vacuole is dependent on the proportion of highly vacuolated cells and tissues, as well as the activity of transport systems located at the tonoplast, which prevent excessive concentration of ions in the cytoplasm (Dajic 2006). This sequestration of toxic salt ions into the vacuole of aboveground parts of the plants sometimes coupled with other physiological adaptations, such as regulation of transpiration and performing of cell metabolism with low K^+ concentrations (Flowers and Dalmond 1992). In was noted that high salt tolerant (halophytes) species posses large vacuoles. For example, *Suaeda maritime*, a potential halophyte, occupy 77% of the mesophyll cells for vacuoles (Hajibagheri et al. 1984b) and are capable of accumulating salts to concentrations higher than 500 mM (Dracup and Greenway 1985). In *S. maritime*, Dajic et al. (1997) reported that, Na^+ concentration of the cell sap exceeded even 800 mM, while the total salt content contributed to the osmotic potential up to 91%.

Synthesis of Metabolites and Alteration of Metabolic Activities

Most of the plants species accumulate certain organic solutes (sugar, alcohol, proline (Pro), quarternary ammonium compounds) in response of osmotic stress. These organic solutes are known as compatible solutes or osmoprotectants because they don't interfere with enzymatic activities even in high concentrations (Johnson et al. 1968; Chen et al. 2007; Tavakkoli et al. 2012). These are localized in cytoplasm and the inorganic ions such as Na^+ and Cl^- are preferentially sequestered into vacuole, thus leads to the turgor maintenance for the cell under osmotic stress

(Flowers et al. 1977; Bohnert et al. 1995; Nounjan et al. 2012; Tavakkoli et al. 2012). Though osmoprotectant enable plants to tolerate more salinity but still a significant amount of Na^+ needs to be compartmentalized for better tolerance. Therefore, it is desirable that overproduction of osmoprotectant is governed by the pleiotropic control of vacuolar Na^+/H^+ antiporter activity. The accumulation of osmolytes such as Pro is a well-known adaptive mechanism in plants against salt stressed conditions. It has also been suggested that Pro accumulation can serve as a selection criterion for the tolerance of most species to stressed conditions (Parida and Das 2005; Ashraf and Foolad 2007). Since the first report on Pro accumulation in wilting perennial rye grass (Kemble and MacPherson 1954), a number of research works have been carried out concerning the role of Pro as a compatible osmolyte and osmoprotectant and its roles in salt stress tolerance. Glycinebetaine (GB) is a small organic metabolite soluble in water and non-toxic at high concentrations which can potentially play a protective role against salt stress (Ashraf and Foolad 2007; Chen and Murata 2008). The major role of GB in plants exposed to salt is probably protecting plant cells by osmotic adjustment (Gadallah 1999), protein stabilization (RuBisCo) (Mäkelä et al. 2000), photosynthetic apparatus protection (Allakhverdiev et al. 2003; Cha-Um and Kirdmanee 2010), and reduction of ROS (Ashraf and Foolad 2007). Trehalose (Tre), a non-reducing sugar, possess a unique feature of reversible water storage capacity to protect biological molecules from desiccation damages. Recently there has been growing interest of utilization of Tre metabolism to ameliorate the effects of abiotic stresses. Nounjan et al. (2012) reported that exogenous Tre treatment under salt stress condition reduced the Na^+/K^+ ratio and strongly decreased endogenous Pro in *O. sativa* seedlings.

Salt stress in plants induce higher concentration of ROS/intermediate such as O_2^-, H_2O_2 and OH due to the impaired election transport processes in chloroplast, mitochondria and photorespiration pathway. Several reports showed the overproduction of ROS in plants under saline conditions and ROS-induced membrane damage is a major cause of cellular toxicity by salinity (Mittova et al. 2004; Hasanuzzaman et al. 2011a, b). As salt stress is complex and imposes a water deficit because of osmotic effects on a wide variety of metabolic activities (Greenway and Munns 1980; Cheeseman 1988; Hasanuzzaman and Fujita 2011). This water deficit leads to the formation of ROS (Halliwell and Gutteridge 1985; Elstner 1987). ROS are highly reactive and may cause cellular damage through oxidation of lipids, proteins and nucleic acids (Pastori and Foyer 2002; Apel and Hirt 2004). If there is a serious imbalance in any cell compartment between the production of ROS and antioxidant defense, oxidative stress and damage occurs (Mittler 2002). Enhanced production of ROS under salinity stress induces phytotoxic reactions such as lipid peroxidation, protein degradation, and DNA mutation (Hefny and Abdel-Kader 2009; Tanou et al. 2009; Ahmad et al. 2010a, 2011; Ahmad and Umar 2011; Ahmad and Prasad 2012a, b). Plants possess an efficient non-enzymatic (ascorbate, AsA; glutathione, GSH; α-tocopherol; phenolic compounds, alkaloids and non-protein amino acids) and enzymatic (superoxide dismutase, SOD; catalase, CAT; ascorbate peroxidase, APX; monodehydroascorbate reductase, MDHAR; dehydroascorbate reductase, DHAR; glutathione reductase, GR; glutathione peroxidase, GPX; glutathione *S*-transferase,

GST and peroxidases, POX) antioxidant defense systems which work in concert to control the cascades of uncontrolled oxidation and protect plant cells from oxidative damage by scavenging ROS (Mittler et al. 2004; Gill and Tuteja 2010; Hasanuzzaman et al. 2012a, b; Hasanuzzaman et al. 2013). There is variability among rice genotypes for the enzymatic and non-enzymatic scavenging system hence it is possible to tag the genes coding for both enzymatic and non-enzymatic ROS scavenging agents and use them in engineering the desired plants of MAB.

Change in Photosynthetic Pathway

The reduction in photosynthetic rates in plants under salt stress is one of the common effects which is mainly due to the reduction in water potential. Photosynthesis is also inhibited when high concentrations of Na^+ and/or Cl^- accumulated in chloroplasts. It is noteworthy that some of the halophytes (especially facultative halophytes) such as *M. crystallinum* shift their C_3 mode of photosynthesis to CAM (Cushman et al. 1989). This change helps the plant able to open the stomata at night which reduce water loss, thus decreasing transpiration loss under saline conditions (Parida and Das 2005). In another study, Zhu and Meinzer (1999) noticed a shift from the C_3 to the C_4 pathway in response to salinity in *Atriplex lentiformis* (a salt-tolerant plant).

Salt Excretion

Salt tolerant plants often posses a number of special characteristics which are directly linked to physiological adaptation to excessive salt. Among them, salt excretion is also a very efficient way of preventing excessive concentrations of salts building up in photosynthetic tissues. Some of the halophytes possess multicellular salt glands and salt hairs (bladders) (Popp 1995). These structures are common for many halophytic genera such as *Cressa* (Convolvulaceae), *Frankenia* (Frankeniaceae), *Spartina*, *Chloris*, *Aeluropus* (Poaceae), *Atriplex* (Chenopodiaceae), *Statice*, *Limonium*, *Plumbago*, *Armeria* (Plumbaginaceae), *Glaux* (Primulaceae), *Tamarix*, *Reamuria* (Tamaricaceae), as well as, some mangrove species, e.g. *Avicennia*, *Aegialitis*, *Aegiceras* and *Acanthus* (Waisel 1972; Crawford 1989; Popp 1995). Salt glands is composed with set of epidermal cells complexes which remove salt from the mesophyll cells beneath them, to which they are connected by numerous plasmadesmata, and secrete it at the leaf surface, where a layer of salt crystals is formed (Mohr and Schopfer 1995; Fig. 6.8). However, glandular structures involved in salt excretion vary in structure, position, physiological mechanism and related ecological significance (Waisel 1972). This an energy-dependent process and this energy (ATP) required for ion pump is provided by the active respiration of the glandular cells. However, there is still a dilemma about the salt glands, whether they functions to excrete, secrete, or recrete (Freitas and Breckle 1992; Marcum and Murdoch 1992). There are a lot of similarities in the metabolic principles of ion transport and functioning of the proton pumps in salt-excreting structures and cells of other tissues.

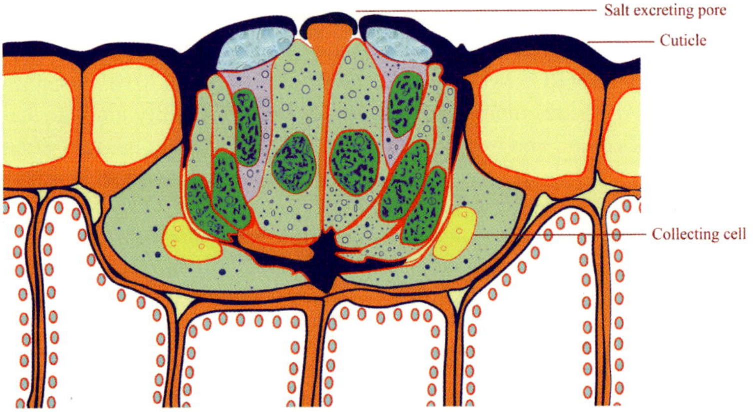

Fig. 6.8 Cross section of a salt gland

Salt Tolerance in Plants Using Omic Approaches

Crop plants, are particularly sensitive to salinity and sodic soils. Though within the crop plants, there are genotypes which have shown tolerance to sodicity and salinity. But, little progress has been made in analysis of changes in gene expression in sensitive and tolerant plants exposed to salt stress and sodic soils. Work done in the direction of evaluating the importance of processes those are involved in tolerance that limits cellular damage enhances the for bearance to soils where NaCl is the leading salt is the present need. This knowledge would enable us to plan successful approach towards genetic engineering of improved salt tolerance in crop plants.

Enabling Salt Tolerance in Plants Using Proteomics Approach

Proteins are known to undergo significant post-translational modification of their primary sequences when readily targeted to proteolysis. Thus, quantitative analysis of gene expression at the protein level is indispensable for evaluating response to salt stress (Parker et al. 2006). The expression profiling of proteins at subsequent levels symbolize the core of proteome analysis that is performed these days. Two-dimensional gel electrophoresis (2-DE) is the most common tool that is being used for revealing the expression of intact proteins. Still, resolution of the range of protein separated depends upon the sample preparation and heterogeneous physiochemical nature of protein mixtures (Nam et al. 2012). Once the sample preparation is optimal, experimental design and quantification leads to the identification of biologically relevant markers with the help of mass spectroscopy. In this growing era of research, improvement in proteomics technology has also made this technique more sensitive. Protein separation and detection as well as identification of proteins using mass spectroscopy has brought about increasing impact on the study of salinity stress in plants (Caruso et al. 2008). Databases were constructed like www.matrix.com containing all expressed proteins from plant organs and cell organelles from various species. Many new insights have been reported with the use of this technology in studying salt stress in rice and *Arabidopsis* (Yan et al. 2005; Jiang et al. 2007) such as novel candidate proteins were characterized, detection of protein phosphorylation sites and location of stress responsive proteins in rice root apoplast (Chitteti and Peng 2007). Signal transduction function of some of the putative proteins identified also indicated the importance of ion uptake and its transport regulation in root was also revealed by proteomics study. Besides these studies, various researchers have also analyzed rice leaves (Salekdeh et al. 2002), rice leaf lamina (Parker et al. 2006), and wheat roots (Wang et al. 2008) in response to salt stress. These studies have revealed valuable insight into species- as well as tissue-specific stress responses and proved that 2-D gel electrophoresis is an adequate tool for large-scale analysis of protein expression.

Recently, Nam et al. (2012) studied the proteome profile of wild type and *OSRK1* transgenic rice roots on being exposed to 150 mM NaCl for couple of 7 h. They reported 52 salt responsive proteins that were indentified from wild rice roots. Most of the up-regulated proteins belonged to the category of energy regulation, methylglyoxal detoxification, amino acid metabolism and redox regulation. Down-regulated proteins were found to be fructose bisphosphate aldolase and methylmalonate semialdehyde dehydrogenase. Where as in transgenic *OSRK1* rice roots, 43 salt responsive proteins were identified, mostly involved in amino acid catabolism, dnaK-type molecular chaperones, calcium binding proteins, Sal T and glyoxalase revealing the fact that they are the part of broad spectrum of proteins involved in fighting salt stress.

Besides identification of proteins through proteomics, characterization of post-transcriptional and post-translational regulatory systems also enhance the understanding of the molecular mechanisms lying behind plant adaptation to salt stress (Asano et al. 2010). Further, improvement in proteomics technology has evolved the concept of phosphoproteome or proteome changes that are related to specific protein kinase signalling. These days, phosphorylation is the important post-translational protein modifications that affects conformation, activity, localization and stability of target proteins (Mazzucotelli et al. 2008). During salt stress and increased sodicity in soil, phosphorylation cascades gets activated and play a critical role in stress tolerance. Protein kinases are known to be involved as key regulators of salt stress and ABA signalling in plants. Mitogen-activated protein kinases (MAPK), calcium-dependent protein kinases (CDPK), SNF1-related protein kinases (SnRK) and receptor like kinases (RLKs) are some of the known protein kinases that were found to be activated by ABA and diverse stress signals (Jossier et al. 2009; Ouyang et al. 2010). Thus, phosphoproteomics enables the identification of core kinase family that might be involved in ABA signal transduction during hyperosmotic stress response in plants (Alvarez et al. 2011). Further analysis of the identified kinase family might provide insight in metabolic regulation during stress.

Transcriptomic Approach

Research has been carried in the direction to identify genes conferring tolerance to stress using a gene-by-gene approach, since stress tolerance is a multigenic trait; with this approach stress resistant networks are not understood entirely (Brinker et al. 2010). Therefore, complete genome sequences and large-scale EST sequencing projects were facilitated on model and crop plants that assisted in the determination of gene functions. Many genes play an important role in adapting plants to grow and yield well in saline soil. To better understand the genetic control of salt tolerance and to use this knowledge to increase the salt tolerance of important crops and pasture species. Various approaches can be explored to discover genes for salinity tolerance. The upcoming advancement in revealing stress tolerance through exploring the genome resulted in the generation of techniques, like as microarrays and yeast two-hybrid, for high-throughput analysis of the transcriptome. Hatzimanikatis et al. (1999) revealed in his research that to understand how protein abundance is related to mRNA levels is necessary for infer gene expression and protein function involved in stress response. Recent study done by Yang et al. (2012) in sugar beet monosomic addition line M14 under salt stress showed the differential expression of proteins and genes using proteomics and transcriptomics approach. Comparison between gene transcription data and protein data, they were able to identified *S*-adenosyl-L-methionine synthetase (SAMS) which showed increased expression under salt stress. Together with proteomics and transcriptomics many interesting genes and proteins such as GS, CPS, BADH, ferritin, HSPs, FtsH, proteasome proteins, GST and 14-3-3s were identified. These identified proteins and genes play important role in combating stress. Thus, protein dynamics and expression patterns together can throw light onto the molecular mechanisms underlying salt tolerance. Another interesting study by Brinker et al. (2010) carried out in salt-resistant Populus species in an effort to identify important genes for stress acclimation using transcriptomics. By using comparative in silico analysis with *Arabidopsis* orthologs they were able to identify a lipocalin-like gene (AtTIL) and a gene encoding a protein with previously unknown functions (AtSIS), which showed to play roles in salt tolerance. *Populus euphratica* is a model plant for salt tolerance in trees (Chen and Polle 2010). Number of previous studies has also been carried out to unreveal its mechanism of tolerance using transcriptomics. Gu et al. 2004 used microarray method with 315 cDNAs of transcripts that were obtained by subtractive hybridization from *P. euphratica* was used to analyze salt responses. An EST-based microarrays was constructed by Brosché et al. (2005) for *P. euphratica* that contained 7,342 different ESTs corresponding to about 6,300 different genes. Using microarrays with this collection of stress-responsive genes, they were able to find that leaves of mature desert-grown *P. euphratica* trees were tolerant to high salinity.

Therefore, transcriptome mapping of plant has currently gained popularity for studying the probable mechanisms revealing response to salt stress thereby illuminating the pathways involved in signal transduction during salt stress (Sanchez et al. 2011a, b). Yao et al. (2011) also used microarray based technology to reveal changes of transcriptome during salt stress in cotton. The signal transduction pathways associated to salt stress in NaCl exposed cotton roots were studied. Yao et al. (2011) used GeneChip® Cotton Genome Array that represents 21,854 cotton transcripts, and they were able to monitor the salt responsive expression profiles of genes. The biological processes that are involved in roots of cotton in response to salt stress were studied using network analysis for exposing the up- or down-regulated genes, gene ontology (GO) enrichment techniques and MapMan. Their study provided much constructive knowledge revealing detailed salt-mediated signal transduction pathways in roots, also offering a account of candidate genes acting as potential markers of tolerance to salt stress.

In, NCBI GEO data sets available at www.ncbi.nlm.nih.gov, microarray-based analyses in response to NaCl of many plant species is available (Huang et al. 2012). Thus, transcriptome map provides major information related to the underlying molecular mechanisms that direct the response of plants to salt stress. The upregulated and downregulated genes act as potential candidates for genetic engineering of salt tolerance in plants.

Metabolomics

Metabolomics is the science that has made possible the detection and recognition of metabolites that are expressed depending on the environmental conditions. During stress conditions, plants rearrange their metabolic pathways in an effort to adapt to changing conditions (Widodo et al. 2009). Apparently, the different environmental conditions make a major impact on the synthesis and accumulation of secondary plant products. This metabolic profile acts as a marker for measuring metabolic movements and factors that regulate them. In combination with other 'omic' techniques, such as transcriptome particularly through mass spectrometry-based analytical methods (Sawada et al. 2009) provide a deeper insight of the stress molecular responses. During stress, the plant synthesises a variety of secondary products like phenolic compounds and alkaloids. Secondary products containing a phenol group and a hydroxyl functional group on an aromatic ring, belongs to the category of phenolics. Phenolics, forms an important part of the plants defence system against salt stress. It is observed that with increase in salt concentration, phenolic content of leaves also increases (Savirnata et al. 2010). In plant phenolics, flavinoids form one of the largest classes that play role in plant defence (Kondo et al. 1992). Isoflavonoids are derivatives of flavonone intermediate e.g. naringenin, secreted by the legumes play a critical role in plant development and defense response and also in supporting the configuration of nitrogen-fixing nodules by symbiotic rhizobia (Sreevidya et al. 2006). Posmyk et al. (2009) reported that synthesis of these flavonoids is an effective strategy against ROS.

Family of N-containing secondary metabolites known as alkaloids that are found in 20% of the total species of vascular plants e.g. Pyrrolizidine alkaloids (PAs). Schafer and Wink (2009) described them as ROS scavengers that played role in herbivoral attack.

During high salt concentrations in the soil the metabolic pathways that play part in synthesis and accretion of the secondary products of plants become active. Sulphur compounds, vitamins and flavonoids are the secondary metabolites that have central role in overcoming different environmental stresses. Besides this, ascorbic acid and carbonyl compounds also play important role in regulating stress (Krasensky and Jonak 2012). Glutathione-*S*-transferases and phenylalanine ammonium lyase (PAL) also known to get increase during salt stress thus providing help to plant from unfavorable effects of stresses (Krasensky and Jonak 2012).

With the advancement in metabolomic applications in plants, the use of GCMS/ liquid chromatography mass spectrometry (LC-MS) and NMR based techniques are profoundly used to get imminent insight into the disparity of metabolities compositions (Gavaghan et al. 2011). Plants in response to salinity involved multiple differentiation/expression in the function of genes/proteins that constantly lead to alteration in plant metabolism. An important advantage of metabolome analyses is that it minutely distinguishes the chemical nature of metabolite entities. Further analytical advancement in metabolomics have introduced bioinformatic tools that help in analysis of complex fingerprints and profiles of data sets to the eventual flow of biological information connecting gene and metabolic expression.

The study of metabolite changes due to of salt stress have been studied in model plants like *Oryza sativa*, *Arabidopsis thaliana*; also in crops plants such as tomato, Soybean, *Solanum lycopersicon* and *Vitis vinifera* (Cramer et al. 2007; Kim et al. 2007; Zuther et al. 2007; Guo et al. 2011). Gong et al. (2005) and Gagneul et al. (2007) studied the metabolite profiles of halophytic tree *Populus euphratica* and the shrubs *Thellungiella halophila* and *Limonium latifolium*. A comparison study by Cramer et al. (2007) in drought and salt stressed shoot tips from grapevine, *V. vinifera* cv. Cabernet was carried out using transcriptomic and GC-MS-based metabolomic profiling. Metabolomics revealed the reduced levels of sucrose, aspartic, succinic and fumaric acids whereas proline, asparagine, malic acid and fructose showed increase in their levels under salt stress. From this study they suggested that, metabolic responses during increased salinity showed changes in organic solute composition. Thus, analyses of metabolic adjustments in plants during salt stress have offered significant evidences for improved understanding of the role of played by various metabolites in regulating harsh environment conditions.

Ionomics

Ionome defines a broad range of biological events that include electrophysiology, signaling, enzymology, osmoregulation, and transport. It involves the simultaneous measurement of the elemental composition of living organisms and changes that occur in this composition in response to environment conditions (Salt et al. 2008). Ionomic study promises to yield new and significant biological insight. Ionome was first described by Lahner et al. (2003) to categorize all the metals, metalloids, and non-metals present in an organism. The margins between the ionome, metabolome, and proteome are unclear. Compound that contain the nonmetals phosphorus, sulfur, or nitrogen fills the category of ionome and metabolome, and metals such as zinc, copper, manganese, and iron in metalloproteins lies within the proteome, or metalloproteome (Szpunar 2004). The three most commonly used methods for elemental analysis are atomic absorption spectroscopy, ICP (Inductivity coupled plasma)-optical emission spectroscopy (ICPOES), and ICP-MS. ICP is a reasonable choice for an ionomics screening, as it is more sensitive in identifying trace elements (Salt et al. 2008; Baxter et al. 2012).

Plants that get exposed to salts have difficulty in survival not only because of dehydration, ion toxicity and oxidative stress, but also of metabolic and nutrient misbalance (Tester and Davenport 2003). Interestingly, plant have strategies to cope with saline environments by mechanisms like salt exclusion and sequestration, limitation of K^+ loss, osmotic adjustment and control of water homeostasis (Sanchez et al. 2008b). Sanchez et al. (2008a) investigated the change in metablome using combination of ionomics, transcriptomics and metabolomics in legume plant *Lotus japonicus* under the effect of salt stress. They found a constant increase in the levels of many amino acids, sugars and polyols, and same time a decrease in concentration of most organic acids were observed. Recently, Sanchez et al. (2011b) also studied complete shoots (pooling leaves, petioles and stems) of extremophile *Lotus creticus* two cultivated glycophytic grassland forage species, *Lotus corniculatus* and *Lotus tenuis* using the combination of ionomic and gas chromatography mass spectrometry (GC-MS)-based metabolite profiling. They found that *L. creticus* exhibited better survival after exposure to long term lethal salinity and was more efficient at excluding salt from the shoots than the glycophytes. With help of ionomics they confirmed a differential rearrangement of shoot nutrient levels in the extremophile upon salt exposure. Baxter et al. (2010) encoded sodium (Na^+) transporter, that played major role in controlling the natural variation in leaf Na^+ accumulation capacity in *A. thaliana* population using genome wide association mapping and ionomics approach. Finally, the changes in ionome and understanding the mechanism through which it relate and interacts with other components of cell systems like the genome, proteome and metabolome, also the cell environment influence the how plants assimilate their organic and inorganic metabolisms.

Micromics

MicroRNAs (miRNAs) plays role in transcriptional or post-transcriptional gene silencing and constitute the part of Micromics. They are genome encoded 20–24 nucleotide long non-coding small RNAs (Paul et al. 2011). Recent advances in high throughput sequencing methods have made possible the classification of non-conserved miRNAs which are low in number as observed in *Arabidopsis* (Axtell et al. 2007). The identification of small RNA-mediated regulatory networks initialize the exploring of complex responses by plants during growth, response to nutrients and biotic and abiotic stress tolerance (Tuteja et al. 2012).

The mechanism of RNA induced silencing starts when mature miRNA gets integrated into the RNA induced silencing complex (RISC) directing the complex for targeted mRNA degradation (Lin et al. 2005). Previous researches have specified that miRNAs plays role in plant development, signaling, organ morphogenesis and regulate the gene expression at the level of post-transcriptional during stress induced protein changes (Chen et al. 2004; Yoshikawa et al. 2005; Kidner and Martienssen 2005; Ding et al. 2009). Researchers have been able to identify 2,043 plant miRNAs with in the plants including 106 miRNAs from *Brachypodium distachyon* and 93 miRNAs in wheat (Wei et al. 2009), 27 miRNAs were detected from *Citrus aestivum* (Song et al. 2009), 29 miRNAs from *Phaseolus vulgaris* (Arenas-Huertero et al. 2009).

Plant development and adaptation during stress are different but strongly linked processes. During the condition of stress, miRNAs expression among most of the plants shows alternation that is essential for development of plant tolerance (Kantar et al. 2010). The transformed levels of miRNAs regulate the expression patterns of the targeted genes and indirectly restrain the development and morphogenesis of plants during stress. Thereafter, this leads to the development of adaptive responses towards stress.

Conclusion and Future Perspectives

It is clear that salinity and sodicity problem is increasing rapidly throughout the world. Because of this, about half a billion hectares of land have not being appropriately used for the purpose of crop production. Observing the change in soil composition and increase in salinity, approaches to improve saline soils in support of highly productive and meaningful land-use systems needs to be uplifted so that challenges of global food security are met. In addition, the crop adaptability to saline conditions should also be improved. Since the responses of plant and the tolerance of plants to salt stress are still unclear, the physiological and molecular studies to reveal the underlying mechanisms of these processes are important. In addition, discovering the induction of signaling cascades leading to profound changes in specific gene expression is also considered an important salt stress adaptation. Molecular knowledge that enables tolerance mechanisms finally will lead the way for manufacturing stress tolerant plants, thereby forming the basis for crop production and increase in economic yield. In recent years phytoremediation of saline soils have been studied by researchers and it was observed that the use of some halophytes could remove salt from soil. Phytoremediation could become a cost-effective and environmentally sound technology for remediation of salt-impacted sites if it can be properly developed. There are certain limitations that must be overcome for this plant-based remediation system to come into common usage. Phytoremediation can be time-consuming because it requires several growing seasons to lower the level of contaminants in soil. It is also limited to soil depths that are in the rooting zone. It is necessary to find the plants that are having ability to remove the build-up quantity of salts thereby producing high biomass having economic importance are largely chosen for phytoremediation, the selected plant species should tolerate high salt concentration. The forthcoming challenge for using halophytes to remediate soil salinity is to develop a plant with diverse salt accumulating capacity in a cost effective way. Identification of novel genes with high biomass yield characteristics and the subsequent development of transgenic plants with superior remediation capacities would be crucial for further research.

With all the efforts in building salinity tolerance in the plants, still many challenges lie to overcome. Advancement in the technologies has introduced proteomics, transcriptomics and metabolomics, and when associated together they act as powerful tools for enhancing the gene information related to understanding their function and regulatory networks. Progressive studies have shown that combination of metabolic fluxes and physiological changes of plants provide accurate predictions about possible mechanism required for adaptation to stress. Response to salt stress being a multi-facet trait remains more elusive as some novel proteins will eventually need complementary approaches to be identified. Functional genomics till now have been used to unravel genes that have been limited to the model plants like Arabidopsis, rice and tobacco, further input is required to focus on regulatory genes in other important crop plants also. Consequently, advancement in biotechnology facilitating discovery of new stress related genes and how they work to confer stress tolerance will be an addition to the knowledge in engineering crop plants. Therefore, attempts to reveal the undiscovered facts about tolerance to stress using the 'omics' approach together will provide deep insight as to how manage the up-coming challenges.

Acknowledgments We wish to thank Mr. Md. Mahabub Alam, Laboratory of Plant Stress Responses, Faculty of Agriculture, Kagawa University, Japan for his vital assistance during the preparation of the manuscript. We also express our sincere thanks to Prof. Dr. Kamal Uddin Ahamed, Department of Agricultural Botany, Sher-e-Bangla Agricultural University, Dhaka, Bangladesh for his continuous encouragement and constructive suggestion during the manuscript preparation. We apologize to all researchers for those parts of their work that were not cited in the manuscripts because of the page limitation.

References

Abdulrahman FS, William GJ (1981) Temperature and salinity regulation of growth and gas exchange of *Saliconia fruticosa* L. Oecologia 48:346–352

Abrol IP, Yadav JSP, Massoud FI (1988) Salt-affected soils and their management, vol 39, FAO soils bulletin. Food and Agriculture Organization of the United Nations, Rome

Ahmad P (2010) Growth and antioxidant responses in mustard (*Brassica juncea* L.) plants subjected to combined effect of gibberellic acid and salinity. Arch Agron Soil Sci 56:575–588

Ahmad P, Prasad MNV (2012a) Environmental adaptations and stress tolerance in plants in the era of climate change. Springer, LLC, New York

Ahmad P, Prasad MNV (2012b) Abiotic stress responses in plants: metabolism, productivity and sustainability. Springer, New York

Ahmad P, Sharma S (2008) Salt stress and phyto-biochemical responses of plants. Plant Soil Environ 54:89–99

Ahmad P, Umar S (2011) Oxidative stress: role of antioxidants in plants. Studium Press, New Delhi

Ahmad P, Sarwat M, Sharma S (2008) Reactive oxygen species, antioxidants and signaling in plants. J Plant Biol 51:167–173

Ahmad P, Jaleel CA, Salem MA, Nabi G, Sharma S (2010a) Roles of enzymatic and non-enzymatic antioxidants in plants during abiotic stress. Crit Rev Biotechnol 30:161–175

Ahmad P, Jaleel CA, Sharma S (2010b) Antioxidative defence system, lipid peroxidation, proline metabolizing enzymes and biochemical activity in two genotypes of *Morus alba* L. subjected to NaCl stress. Russ J Plant Physiol 57:509–517

Ahmad P, Umar S, Sharma S (2010c) Mechanism of free radical scavenging and role of phytohormones during abiotic stress in plants. In: Ashraf M, Ozturk M, Ahmad MSA (eds) Plant adaptation and phytoremediation. Springer, Dordrecht/Heidelberg/London/New York, pp 99–108

Ahmad P, Nabi G, Jeleel CA, Umar S (2011) Free radical production, oxidative damage and antioxidant defense mechanisms in plants under abiotic stress. In: Ahmad P, Umar S (eds) Oxidative stress: role of antioxidants in plants. Studium Press, New Delhi, pp 19–53

Ahmad P, Hakeem KR, Kumar A, Ashraf M, Akram NA (2012) Salt-induced changes in photosynthetic activity and oxidative defense system of three cultivars of mustard (*Brassica juncea* L.). Afr J Biotechnol 11:2694–2703

Ahmed S (2009) Effect of soil salinity on the yield and yield components of mungbean. Pak J Bot 41:263–268

Al-Harbi AR, Wahb-Allah MA, Abu-Muriefah SS (2008) Salinity and nitrogen level affects germination, emergence, and seedling growth of tomato. Int J Veg Sci 14:380–392

Ali Y, Aslam Z, Ashraf MY, Tahir GR (2004) Effect of salinity on chlorophyll concentration, leaf area, yield and yield components of rice genotypes grown under saline environment. Int J Environ Sci Technol 1:221–225

Ali S, Zeng F, Cai S, Qiu B, Zhang G (2011) The interaction of salinity and chromium in the influence of barley growth and oxidative stress. Plant Soil Environ 57:153–159

Allakhverdiev SI, Hayashi H, Nishiyama Y, Ivanov AG, Aliev JA, Klimov VV, Murata N, Carpemtier R (2003) Glycinebetaine protects the D1/D2/*Cyt b 559* complex of photosystem II against photo-induced and heat-induced inactivation. J Plant Physiol 160:41–49

Allan AC, Fluhr R (1997) Two distinct sources of elicited reactive oxygen species in tobacco epidermal cells. Plant Cell 9:1559–1572

Al-Mutawa MM (2003) Effect of salinity on germination and seedling growth of chickpea (*Cicer arietinum* L.) genotypes. Int J Agric Biol 5:226–229

Alvarez S, Hicks LM, Pandey S (2011) ABA-dependent and -independent Gprotein signaling in Arabidopsis roots revealed through an iTRAQ proteomics approach. J Proteome Res 10:3107–3122

Amirjani MR (2011) Effect of salinity stress on growth, sugar content, pigments and enzyme activity of rice. Int J Bot 7:73–81

Apel K, Hirt H (2004) Reactive oxygen species: metabolism, oxidative stress and signal transduction. Annu Rev Plant Biol 55:373–399

Apse MP, Aharon GS, Snedden WA, Blumwald E (1998) Cloning and characterization of plant sodium/proton antiports. In: Proceedings of the 11th international workshop on plant membrane biology, Cambridge, p 84

Ardic M, Sekmen AH, Tokur S, Ozdemir F, Turkan I (2009) Antioxidant responses of chickpea plants subjected to boron toxicity. Plant Biol 11:328–338

Arenas-Huertero C, Pérez B, Rabanal F, Blanco-Melo D, De la Rosa C, Estrada Navarrete G, Sanchez F, Covarrubias AA, Reyes JL (2009) Conserved and novel miRNAs in the legume *Phaseolus vulgaris* in response to stress. Plant Mol Biol 70:385–401

Asada K (1994) Production and action of active oxygen species in photosynthetic tissues. In: Foyer CH, Mullineaux PM (eds) Causes of photooxidative stress and amelioration of defense systems in plants. CRC Press, London, pp 77–104

Asano T, Hakata M, Nakamura H, Aoki N, Komatsu S, Ichikawa H, Hirochika H, Ohsugi R (2010) Functional characterisation of *OsCPK21*, a calciumdependent protein kinase that confers salt tolerance in rice. Plant Mol Biol 75:179–191

Ashraf M, Foolad MR (2007) Roles of glycinebetaine and proline in improving plant abiotic stress resistance. Environ Exp Bot 59:206–216
Axtell MJ, Snyder JA, Bartel DP (2007) Common functions for diverse small RNAs of land plants. Plant Cell 19:1750–1769
Azaizeh H, Steudle E (1991) Effects of salinity on water transport of excised maize (*Zea mays* L.) roots. Plant Physiol 99:1136–1145
Azooz MM, Youssef AM, Ahmad P (2011) Evaluation of salicylic acid (SA) application on growth, osmotic solutes and antioxidant enzyme activities on broad bean seedlings grown under diluted seawater. Int J Plant Physiol Biochem 3:253–264
Barkla B, Apse MP, Manolson MF, Blumwald E (1994) The plant vacuolar Na^+/H^+ antiport. In: Blatt M, Leigh R, Sanders D (eds) Membrane transport in plants and fungi: molecular mechanisms and control. SEB, London, pp 141–153
Baxter I, Brazelton JN, Yu D, Huang YS, Lahner B (2010) A coastal cline in sodium accumulation in *Arabidopsis thaliana* is driven by natural variation of the sodium transporter *AtHKT1*;1. PLoS Genet 6(11):e1001193
Baxter I, Hermans C, Lahner B, Yakubova E, Tikhonova M, Verbruggen N, Chao DY, Salt DE (2012) Biodiversity of mineral nutrient and trace element accumulation in *Arabidopsis thaliana*. PLoS One 7:e35121
Benmahioul B, Daguin F, Kaid-Harche M (2009) Effects of salt stress on germination and in vitro growth of pistachio (*Pistacia vera* L.). C R Biol 332:752–758
Bernstein L (1961) Osmotic adjustment of plants to saline media. I. Steady state. Am J Bot 48:909–918
Bernstein L, Hayward HE (1958) Physiology of salt tolerance. Annu Rev Plant Physiol 9:25–46
Bhalla R, Narasimhan K, Swarup S (2005) Metabolomics and its role in understanding cellular responses in plants. Plant Cell Rep 24:562–571
Bindschedler LV, Palmblad M, Cramer R (2008) Hydroponic isotope labelling of entire plants (HILEP) for quantitative plant proteomics; an oxidative stress case study. Phytochem 69:1962–1972
Bohnert HJ, Nelson DE, Jensen RG (1995) Adaptations to environmental stresses. Plant Cell 7:1099–1111
Bordi A (2010) The influence of salt stress on seed germination, growth and yield of canola cultivars. Not Bot Hort Agrobot Cluj 38:128–133
Brinker M, Brosche M, Vinocur B, Abo-ogiala A, Fayyaz P, Janz D, Ottow EA, Cullmann AD, Saborowski J, Kangasjärvi J, Altman A, Polle A (2010) Linking the salt transcriptome with physiological responses of a salt-resistant *Populus* species as a strategy to identify genes important for stress acclimation. Plant Physiol 154:1697–1709
Brosché M, Vinocur B, Alatalo ER, Lamminmäki A, Teichmann T, Ottow EA, Djilianov D, Afif D, Bogeat-Triboulot MB, Altman A (2005) Gene expression and metabolite profiling of *Populus euphratica* growing in the Negev desert. Genome Biol 6:R101
Buxton GV, Greenstock CL, Helman WP (1988) Critical review of rate constants for reactions of hydrated electrons, hydrogen atoms and hydroxyl radicals in aqueous solution. J Phys Chem Ref Data 17:512–579
Carillo P, Grazia Annunziata M, Pontecorvo G, Fuggi A, Woodrow P (2011) Salinity stress and salt tolerance. In: Shanker AK, Venkateswarlu B (eds) Abiotic stress in plants – mechanisms and adaptations. InTech, Rijeka, pp 21–38
Carpıcı EB, Celık N, Bayram G (2009) Effects of salt stress on germination of some maize (*Zea mays* L.) cultivars. Afr J Biotechnol 8:4918–4922
Caruso GC, Cavaliere C, Guarino R, Gubbiotti PF, Lagana A (2008) Identification of changes in *Triticum durum* L. Leaf proteome in response to salt stress by two-dimenstional electrophoresis and MALDI-TOF mass spectrometry. Anal Bioanal Chem 391:381–390
Cha-Um S, Kirdmanee C (2010) Effect of glycinebetaine on proline, water use, and photosynthetic efficiencies, and growth of rice seedlings under salt stress. Turk J Agric For 34:517–527
Chaves MM, Flexas J, Pinheiro C (2009) Photosynthesis under drought and salt stress: regulation mechanisms from whole plant to cell. Ann Bot 103:551–560

Chazen O, Hartung W, Neumann PM (1995) The different effects of PEG 6000 and NaCl on leaf development are associated with differential inhibition of root water transport. Plant Cell Environ 18:727–735

Cheeseman JM (1988) Mechanism of salinity tolerance in plants. Plant Physiol 87:547–550

Chen Q, Yang L, Ahmad P, Wan X, Hu X (2011) Proteomic profiling and redox status alteration of recalcitrant tea (*Camellia sinensis*) seed in response to desiccation. Planta 233:593–609

Chen THH, Murata N (2008) Glycinebetaine: an effective protectant against abiotic stress in plants. Trends Plant Sci 13:499–505

Chen SL, Polle A (2010) Salinity tolerance of Populus. Plant Biol 12:317–333

Chen J, Li WX, Xie D, Peng JR, Ding SW (2004) Viral virulence protein suppresses RNA silencing-mediated defense but upregulates the role of microrna in host gene expression. Plant Cell 16:1302–1313

Chen Z, Cuin TA, Zhou M, Twomey A, Naidu BP, Shabala S (2007) Compatible solute accumulation and stress-mitigating effects in barley genotypes contrasting in their salt tolerance. J Exp Bot 58:4245–4255

Chitteti BR, Peng ZH (2007) Proteome and phosphor proteome differential expression under salinity stress in rice (*Oryza sative*) roots. J Proteome Res 6:1718–1727

Chutipaijit S, Cha-um S, Sompornpailin K (2011) High contents of proline and anthocyanin increase protective response to salinity in *Oryza sativa* L. spp. indica. Aust J Crop Sci 5:1191–1198

Cramer GR, Ergul A, Grimplet J et al (2007) Water and salinity stress in grapevines: early and late changes in transcript and metabolite profiles. Funct Integr Genomics 7:111–134

Crawford RMM (1989) Studies in plant survival. Blackwell Scientific, Oxford/London/Edinburgh/Boston/Palo Alto/Melbourne

Cuartero J, Fernandez-Munoz R (1999) Tomato and salinity. Sci Hortic 78:83–125

Cullu MA (2003) Estimation of the effect of soil salinity on crop yield using remote sensing and geographic information system. Turk J Agric For 27:23–28

Cushman JC, Meyer G, Michalowski CB, Schmitt JM, Bohnert HJ (1989) Salt stress leads to differential expression of two isogenes of phosphoenolpyruvate carboxylase during crassulacean acid metabolism induction in the common ice plant. Plant Cell 1(7):715–725

Dajic Z (1996) *Ekoloska studija halofitske zajednice Puccinellietum limosae (Rapcs.) Wend.* (Ecological study of halophytic community Puccinellietum limosae (Rapcs.) Wend.). Unpublished doctoral dissertation. Faculty of Biology, University of Belgrade, FRY

Dajic Z (2006) Salt stress. In: Madhava Rao KV, Raghavendra AS, Janardhan Reddy K (eds) Physiology and molecular biology of salt tolerance in plant. Springer, Dordrecht, pp 41–99

Dajic Z, Stajkovic M, Jakovljevic M (1997) An ecophysiological study of *Suaeda maritime* (Chenopodiaceae) in Serbia. Bocconea 5:11–516

Darley CP, van Wuyt Swinkel D, van der Woude K, Mager P, de Goer B (1998) *ANA1* a Na^+/H^+ antiport from *Arabidopsis*? In: Proceedings of the international workshop on plant membrane biology. Cambridge, p 8

Davenport R, James R, Zakrisson-Plogander A, Tester M, Munns R (2005) Control of sodium transport in durum wheat. Plant Physiol 137:807–818

de Souza ER, dos Santos Freire MBG, da Cunha KPV, do Nascimento CWA, Ruiz HA, Teixeira Lins CM (2012) Biomass, anatomical changes and osmotic potential in *Atriplex nummularia* Lindl. cultivated in sodic saline soil under water stress. Environ Exp Bot 82:20–27

Debnath M, Pandey M, Bisen PS (2011) An omics approach to understand the plant abiotic stress. OMICS J Integ Biol 15:739–762

Delfine S, Alvino A, Villani MC, Loreto F (1999) Restrictions to carbon dioxide conductance and photosynthesis in spinach leaves recovering from salt stress. Plant Physiol 119:1101–1106

Dell'Aquila A, Spada D (1993) The effect of salinity stress upon protein synthesis of germinating wheat embryos. Ann Bot 72:97–101

Ding D, Zhang L, Wang H, Liu Z, Zhang Z, Zheng Y (2009) Differential expression of miRNAs in response to salt stress in maize roots. Ann Bot 103:29–38

Dionisio-Sese ML, Tobita S (1998) Antioxidant responses of rice seedlings to salinity stress. Plant Sci 135:1–9
Dracup MNH, Greenway H (1985) A procedure for isolating vacuoles from leaves of the halophyte *Suaeda maritima*. Plant Cell Environ 8:149–154
Dupont F (1992) Salt-induced changes in ion transport: regulation of primary pumps and secondary transporters. In: Cooke D, Clarkson D (eds) Transport and receptor proteins of plant membranes. Plenum, New York, pp 91–100
Elstner EF (1987) Metabolism of activated oxygen species. In: Davies DD (ed) The biochemistry of plants, vol II, Biochemistry of metabolism. Academic, San Diego, pp 252–353
Essa TA (2002) Effect of salinity stress on growth and nutrient composition of three soybean (*Glycine max* L. Merrill) cultivars. J Agron Crop Sci 88:86–93
Evers D, Legay S, Lamoureux D, Hausman JF, Hoffmann L, Renaut J (2012) Towards a synthetic view of potato cold and salt stress response by transcriptomic and proteomic analyses. Plant Mol Biol 78:503–514
Evlagon D, Ravina I, Neumann PM (1990) Interactive effects of salinity and calcium on osmotic adjustment, hydraulic conductivity and growth in primary roots of maize seedlings. Isr J Bot 39:239–247
Ezlit YD, Smith RJ, Raine SR (2010) A review of salinity and sodicity in irrigation, vol 01/10, Irrigation matters series. CRC for Irrigation Futures, Toowoomba
FAO (2000) Global network on integrated soil management for sustain-able use of salt-affected soils. Rome. http://www.fao.org/ag/agl/agll/spush
Flexas J, Bota J, Loreto F, Cornic G, Sharkey TD (2004) Diffusive and metabolic limitations to photosynthesis under drought and salinity in C-3 plants. Plant Biol 6:269–279
Flexas J, Ribas-Carbo M, Diaz-Espejo A, Galmes J, Medrano H (2008) Mesophyll conductance to CO2: current knowledge and future prospects. Plant Cell Environ 31:602–621
Flowers TJ, Dalmond D (1992) Protein synthesis in halophytes: the influence of potassium, sodium and magnesium in vitro. Plant Soil 146:153–161
Flowers TJ, Hajibagheri MA (2001) Salinity tolerance in *Hordeum vulgare*: ion concentrations in root cells of cultivars differing in salt tolerance. Plant Soil 231:1–9
Flowers TJ, Troke PF, Yeo AR (1977) The mechanisms of salt tolerance in halophytes. Annu Rev Plant Physiol 28:89–121
Flowers TJ, Hajibagheri MA, Clipson NJW (1986) Halophytes. Q Rev Biol 61:313–337
Fortmeier R, Schubert S (1995) Salt tolerance of maize (*Zea mays* L.): the role of sodium exclusion. Plant Cell Environ 18:1041–1047
Foyer CH, Noctor G (2003) Redox sensing and signalling associated with reactive oxygen in chloroplasts, peroxisomes and mitochondria. Physiol Plant 119:355–364
Foyer CH, Noctor G (2005) Redox homeostasis and antioxidant signaling a metabolic interface between stress perception and physiological responses. Plant Cell 17:1866–1875
Francois LE, Donovan T, Maas EV (1984) Salinity effects on seed yield, growth, and germination of grain sorghum. Agron J 76:741–744
Freitas H, Breckle SW (1992) Importance of bladder hairs for the salt tolerance of field grown *Atriplex* species from a Portuguese salt marsh. Flora 187:283–297
Gadallah MAA (1999) Effect of proline and glycinebetaine on *Vicia faba* responses to salt stress. Biol Plant 42:249–257
Gagneul D, Ainouche A, Duhaze C, Lugan R, Lahrer FR, Bouchereau A (2007) A reassessment of the function of the so-called compatible solutes in the halophytic Plumbginaceae *Limonium latifolium*. Plant Physiol 144:1598–1611
Garg N, Manchanda G (2008a) Effect of arbuscular mycorrhizal inoculation on salt-induces nodule senescence in *Cajanus cajan* (Pigeonpea). J Plant Growth Regul 27:115–124
Garg N, Manchanda G (2008b) Salinity and its effects on the functional biology of legumes. Acta Physiol Plant 30:595–618

Gavaghan CL, Li JV, Hadfield ST, Hole S, Nicholson JK, Wilson ID, Howe PWA, Stanley PD, Holmes E (2011) Application of NMR-based metabolomics to the investigation of salt stress in maize (*Zea mays*). Phytochem Anal 22:214–224

Ghannoum O (2009) C-4 photosynthesis and water stress. Ann Bot 103:635–644

Gill SS, Tuteja N (2010) Polyamines and abiotic stress tolerance in plants. Plant Signal Behav 51:26–33

Glenn E, Miyamoto M, Moore D, Brown JJ, Thompson TL, Brown P (1997) Water requirements for cultivating *Salicornia bigelovii* Torr. with seawater on sand in a coastal desert environment. J Arid Environ 36:711–730

Gong QQ, Li PH, Ma SS, Indu Rupassara S, Bohnert HJ (2005) Salinity stress adaptation competence in the extremophile *Thellungiella halophila* in comparison with its relative *Arabidopsis thaliana*. Plant J 44:826–839

Gorham J (1987) Photosynthesis, transpiration and salt fluxes through leaves of *Leptocloa fuska* L. Kunth. Plant Cell Environ 10:191–196

Gorham J, Randall PJ, Delhaize E, Richards RA, Munns R (1993) Genetics and physiology of enhanced K/Na discrimination. Genetic aspects of plant mineral nutrition. Dev Plant Soil Sci 50:151–158

Greenway H, Munns R (1980) Mechanisms of salt tolerance in non halophytes. Annu Rev Plant Physiol 31:149–190

Gu R, Fonseca S, Puskás LG, Hackler L Jr, Zvara A, Dudits D, Pais MS (2004) Transcript identification and profiling during salt stress and recovery of *Populus euphratica*. Tree Physiol 24:265–276

Gueta-Dahan Y, Yaniv Z, Zilinskas BA, Ben-Hayyim G (1997) Salt and oxidative stress: similar and specific responses and their relation to salt tolerance in citrus. Planta 204:460–469

Guleria P, Goswami D, Mahajan M, Kumar V, Bhardwaj J, Yadav SK (2012) MicroRNAs and their role in plants during abiotic stresses. In: Environmental adaptations and stress tolerance in plants in the era of climate change. Springer, New York, pp 265–278

Guo Y, Wang X-B, He W, Zhou G, Guo B-F, Zhang L, Liu Z-X, Luo Z, Wang L, Qiu L (2011) Soybean omics and biotechnology in China. Plant Omics J 4:318–328

Guy C, Kaplan F, Kopka J, Selbig J, Hincha DK (2008) Metabolomics of temperature stress. Physiol Plant132:220–235

Hajibagheri MA, Hall JL, Flowers TJ (1984a) Stereological analysis of leaf cell of the halophyte *Suaeda maritima* (L.) Dum. J Exp Bot 35:1547–1557

Hajibagheri MA, Harvey DMR, Flowers TJ (1984b) Photosynthetic oxygen evolution in relation to ion contents in the chloroplast of *Suaeda maritima*. Plant Sci Lett 34:353

Halliwell B, Gutteridge JMC (1985) Free radicals in biology and medicine. Clarendon, Oxford

Halliwell B, Gutteridge JMC (1989) Protection against oxidants in biological systems: the super oxide theory of oxygen toxicity. In: Halliwell B, Gutteridge JM (eds) Free radicals in biology and medicine. Clarendon, Oxford, pp 86–123

Halliwell B, Gutteridge MC (1990) Role of free radicals and catalytic metal ions in human disease: an overview. Methods Enzymol 186:1–85

Harivandi MA, Butler JD, Wu L (1992) Salinity and turfgrass culture. In: Waddington DV, Carrow RN, Shearman RC (eds) Turfgrass. ASA, CSSA, and SSSA, Madison, WI, pp 207–229

Hasanuzzaman M, Fujita M (2011) Selenium pretreatment upregulates the antioxidant defense and methylglyoxal detoxification system and confers enhanced tolerance to drought stress in rapeseed seedlings. Biol Trace Elem Res 143:1758–1776

Hasanuzzaman M, Fujita M, Islam MN, Ahamed KU, Nahar K (2009) Performance of four irrigated rice varieties under different levels of salinity stress. Int J Integr Biol 6:85–90

Hasanuzzaman M, Hossain MA, Fujita M (2011a) Selenium-induced up-regulation of the antioxidant defense and methylglyoxal detoxification system reduces salinity-induced damage in rapeseed seedlings. Biol Trace Elem Res 143:1704–1721

Hasanuzzaman M, Hossain MA, Fujita M (2011b) Nitric oxide modulates antioxidant defense and the methylglyoxal detoxification system and reduces salinity-induced damage of wheat seedlings. Plant Biotechnol Rep 5:353–365

Hasanuzzaman M, Hossain MA, da Silva JAT, Fujita M (2012a) Plant responses and tolerance to abiotic oxidative stress: antioxidant defenses is a key factors. In: Bandi V, Shanker AK, Shanker C, Mandapaka M (eds) Crop stress and its management: perspectives and strategies. Springer, Berlin, pp 261–316

Hasanuzzaman M, Hossain MA, Fujita M (2012b) Exogenous selenium pretreatment protects rapeseed seedlings from cadmium-induced oxidative stress by upregulating the antioxidant defense and methylglyoxal detoxification systems. Biol Trace Elem Res 149:248–261

Hasanuzzaman M, Nahar K, Alam MM, Fujita M (2012c) Exogenous nitric oxide alleviates high temperature induced oxidative stress in wheat (*Triticum aestivum*) seedlings by modulating the antioxidant defense and glyoxalase system. Aust J Crop Sci 6:1314–1323

Hasanuzzaman M, Nahar K, Fujita M (2013) Plant response to salt stress and role of exogenous protectants to mitigate salt-induced damages. In: Ahmad P, Azooz MM, Prasad MNV (eds) Ecophysiology and responses of plants under salt stress. Springer, New York. pp 25–87

Hasegawa PM, Bressan RA, Zhu JK, Bohnert HJ (2000) Plant cellular and molecular responses to high salinity. Annu Rev Plant Physiol Plant Mol Biol 51:463–499

Hatzimanikatis V, Choe LH, Lee KH (1999) Proteomics: theoretical and experimental considerations. Biotechnol Prog 15:312–318

Hayward HE, Wadleigh CH (1949) Plant growth on saline and alkali soils. Adv Agron 1:1–35

Hefny M, Abdel-Kader DZ (2009) Antioxidant enzyme system as selection criteria for salt tolerance in forage sorghum genotypes (*Sorghum bicolor* L. Moench). In: Ashraf M, Ozturk M, Athar HR (eds) Salinity and water stress. Springer, Amsterdam, pp 25–36

Hellebust JA (1976) Osmoregulation. Annu Rev Plant Physiol 27:485–505

Hong-bo S, Li-ye C, Ming-an S, Abdul Jaleel C, Hong-mei M (2008) Higher plant antioxidants and redox signaling under environmental stresses. C R Biol 331:433–441

Huang J, Redman RE (1995) Response of growth, morphology and anatomy to salinity and calcium supply in cultivated and wild barley. Can J Bot 73:1859–1866

Huang J, Lu X, Yan H, Chen S, Zhang W, Huang R, Zheng Y (2012) Transcriptome characterization and sequencing-based identification of salt-responsive genes in *Millettia pinnata*, a semi-mangrove plant. DNA Res19:1–13

Hura T, Grzesiak S, Hura K, Grzesiak M, Rzepka A (2006) Differences in the physiological state between triticale and maize plants during drought stress and followed rehydration expressed by the leaf gas exchange and spectrofluorimetric methods. Acta Physiol Plant 28:433–443

Hussein MM, Balbaa LK, Gaballah MS (2007) Salicylic acid and salinity effects on growth of maize plants. Res J Agric Biol Sci 3(4):321–328

ILRI (1989) Effectiveness and social/environmental impacts of irrigation projects: a review, In: Annual report 1988 of the international institute for land reclamation and improvement (ILRI), Wageningen, pp 18–34

Jabeen M, Ibrar M, Azim F, Hussain F, Ilahi I (2003) The effect of sodium chloride salinity on germination and productivity of mung bean (*Vigna mungo* Linn.). J Sci Technol Univ Peshawar 27:1–5

Jaleel CA, Gopi R, Sankar B, Manivannan P, Kishorekumar A, Sridharan R, Panneerselvam R (2007) Studies on germination, seedling vigour, lipid peroxidation and proline metabolism in *Catharanthus roseus* seedlings under salt stress. S Afr J Bot 73:190–195

James RA, Blake C, Byrt CS, Munns R (2011) Major genes for Na^+ exclusion, ^{Nax1} and ^{Nax2} (wheat HKT1;4 and HKT1;5), decrease Na^+ accumulation in bread wheat leaves under saline and waterlogged conditions. J Exp Bot. doi:10.1093/jxb/err003

Jaspers P, Kangasjärvi J (2010) Reactive oxygen species in abiotic stress signaling. Physiol Plant 138:405–413

Jiang Y, Yang B, Harris NS, Deyholos MK (2007) Comparative proteomics analysis of NaCl stress responsive proteins in Arabidopsis roots. J Exp Bot 58:3591–3607

Johnson MK, Johnson EJ, MacElroy RD, Speer HL, Bruff BS (1968) Effects of salts on the halophilic alga *Dunaliella viridis*. J Bacteriol 95:1461–1468

Jossier M, Bouly JP, Meimoun P, Arjmand A, Lessard P, Hawley S, Grahame Hardie D, Thomas M (2009) SnRK1 (SNF1-related kinase 1) has a central role in sugar and ABA signalling in *Arabidopsis thaliana*. Plant J 59:316–328

Kacperska A (2004) Sensor types in signal transduction pathways in plant cells responding to abiotic stressors: do they depend on stress intensity? Physiol Plant 122:159–168

Kantar M, Unver T, Budak H (2010) Regulation of barley miRNAs upon dehydration stress correlated with target gene expression. Funct Integr Genomics 10:493–507

Katare DP, Nabi G, Azooz MM, Aeri V, Ahmad P (2012) Biochemical modifications and enhancement of psoralen content in salt-stressed seedlings of *Psoralea corylifolia* Linn. J Funct Environ Bot 2:65–74

Katerji N, van Hoorn ZW, Hamdy A, Karam F, Mastrorilli M (1994) Effect of salinity on emergence and on water stress and early seedling growth of sunflower and maize. Agric Water Manag 26:81–91

Katerji N, van Hoorn JW, Hamdy A, Mastrorilli M (1998) Response of tomatoes, a crop of determinate growth to soil salinity. Agic Water Manag 38:59–68

Katerji N, Van Hoorn JW, Hamdy A, Mastrorelli M (2001a) Salt tolerance of crops according to three classifications methods and examination of some hypothesis about salt tolerance. Agric Water Manag 47:1–8

Katerji N, Van Hoorn JW, Hamdy A, Mastrorilli M, Oweis T, Erskine W (2001b) Response of two varieties of lentil to soil salinity. Agric Water Manag 47:179–190

Kaufmann K, Smaczniak C, de Vries S, Angenent GC, Karlova R (2011) Proteomics insights into plant signaling and development. Proteomics 11:744–755

Kaveh H, Nemati H, Farsi M, Jartoodeh SV (2011) How salinity affect germination and emergence of tomato lines. J Biol Environ Sci 5:159–163

Kemble AR, MacPherson HT (1954) Liberation of amino acids in perennial ray grass during wilting. Biochem J 58:46–59

Khodarahmpour Z, Ifar M, Motamedi M (2012) Effects of NaCl salinity on maize (*Zea mays* L.) at germination and early seedling stage. Afr J Biotechnol 11:298–304

Kidner CA, Martienssen RA (2005) The developmental role of microRNA in plants. Curr Opin Plant Biol 8:38–44

Kim JK, Bamba T, Harada K, Fukusaki E, Kobayashi A (2007) Time-course metabolic profiling in *Arabidopsis thaliana* cell cultures after salt stress treatment. J Exp Bot 58:415–424

Kondo T, Yoshida K, Nakagawa A, Kawai T, Tamura H, Goto T (1992) Commelinin, a highly associated metalloanthocyanin present in the blue flower petals of *Commelina communis*. Nature 358:515–517

Koryo HW (1997) Ulstrastructural and physiological changes in root cells of sorghum plants (*Shorghum bicolor* S. Sudanensis cv. Sweet Sioux) induced by NaCl. J Exp Bot 308:693–706

Koyro HW, Eisa SS (2008) Effect of salinity on composition, viability and germination of seeds of *Chenopodium quinoa* Willd. Plant Soil 302:79–90

Kramer D (1980) Transfer cells in the epidermis of roots. In: Spanswick RM, Lucas WJ, Dainty J (eds) Plant membrane transport: current conceptual issues. Elsevier, Amsterdam, pp 393–394

Krasensky J, Jonak C (2012) Drought, salt, and temperature stress-induced metabolic rearrangements and regulatory networks. J Exp Bot. doi:10.1093/jxb/err460

Kuramoto RT, Brest DE (1979) Physiological responses to salinity by four salt marsh plants. Bot Gaz 140:295

Kurth E, Cramer GR, Läuchli A, Epstein E (1986) Effects of NaCl and $CaCl_2$ on cell enlargement and cell production in cotton roots. Plant Physiol 82:1102–1106

Lahner B, Gong J, Mahmoudian M, Smith EL, Abid KB, Rogers EE, Guerinot ML, Harper JF, Ward JM, McIntyre L et al (2003) Genomic scale profiling of nutrient and trace elements in *Arabidopsis thaliana*. Nat Biotechnol 21:1215–1221

Larcher W (1980) Physiological plant ecology, 2nd edn. Springer, Berlin, pp 303

Levitt J (1980) Responses of plants to environmental stresses, vol II. Academic, New York, pp 35–50

Lin SL, Chang D, Ying SY (2005) Asymmetry of intronic pre-miRNA structures in functional RISC assembly. Gene 356:32–38

Lu Z, Neumann PM (1999) Water stress inhibits hydraulic conductance and leaf growth in rice seedlings but not transport of water via mercury sensitive water channels in the root. Plant Physiol 120:143–152

Maas EV (1990) Crop salt tolerance. In: Tanji KK (ed) Agricultural salinity assessment and management, vol 71, ASCE manuals and reports on engineering practice. American Society of Civil Engineers, New York, pp 262–304
Maas EV, Hoffman GJ (1977) Crop salt tolerance–current assessment. J Irrig Drain Div 103:115–134
Mahajan S, Tuteja N (2005) Cold, salinity and drought stresses: an overview. Arch Biochem Biophys 444:139–158
Mäkelä P, Kärkkäinen J, Somersalo S (2000) Effect of glycine betaine on chloroplast ultrastructure, chlorophyll and protein content, and RuBPCO activity in tomato grown under drought or salinity. Biol Plant 43:471–475
Marcum KB, Murdoch CL (1992) Salt tolerance of the coastal salt marsh grass, *Sporobulus virginicus* (L.) Kunth. New Phytol 120:281–288
Maria MC, Costa JM, Saibo NJM (2011) Recent advances in photosynthesis under drought and salinity. Adv Bot Res 57:50–83
Maxwell K, Johnson GN (2000) Chlorophyll fluorescence-a practical guide. J Exp Bot 51:659–668
Mazzucotelli E, Mastrangelo AM, Crosatti C, Guerra D, Stanca AM, Cattivelli L (2008) Abiotic stress response in plants: when post-transcriptionaland post-translational regulations control transcription. Plant Sci 174:420–431
McKenzie D, Orange PLM (2003) Salinity and sodicity – what's the difference. The Australian Cottongrower 24:28
Miller G, Shulaev V, Mittler R (2008) Reactive oxygen signaling and abiotic stress. Physiol Plant 133:481–489
Miller G, Suzuki N, Ciftci-Yilmaz S, Mittler R (2010) Reactive oxygen species homeostasis and signaling during drought and salinity stresses. Plant Cell Environ 33:453–467
Mittler M (2002) Oxidative stress, antioxidants and stress tolerance. Trends Plant Sci 7:405–410
Mittler R, Vanderauwera S, Gollery M, Breusegem FV (2004) Reactive oxygen gene network of plants. Trends Plant Sci 9:490–498
Mittova V, Guy M, Tal M, Volokita M (2004) Salinity upregulates the antioxidative system in root mitochondria and peroxisomes of the wild salt-tolerant tomato species *Lycopersicon pennellii*. J Exp Bot 55:1105–1113
Mochida K, Shinozaki K (2011) Advances in omics and bioinformatics tools for systems analyses of plant functions. Plant Cell Physiol 52:2017–2038
Mohr H, Schopfer P (1995) Metabolism of water and inorganic ions. In: Mohr H, Schopfer P (eds) Plant physiology. Springer, Berlin, p 265
Moradi F, Ismail AM (2007) Responses of photosynthesis, chlorophyll fluorescence and ROS-scavenging systems to salt stress during seedling and reproductive stages in rice. Ann Bot 99:1161–1173
Munns R (2002) Comparative physiology of salt and water stress. Plant Cell Environ 25:239–250
Munns R (2005) Genes and salt tolerance: bringing them together. New Phytol 167:645–663
Munns R (2011) Plant adaptations to salt and water stress: differences and commonalities. Adv Bot Res 57. doi:10.1016/B978-0-12-387692-8.00001-1, Elsevier Ltd
Munns R, Rawson HM (1999) Effect of salinity on salt accumulation and reproductive development in the apical meristem of wheat and barley. Aust J Plant Physiol 26:459–464
Munns R, Sharp RE (1993) Involvement of abscisic acid in controlling plant growth in soils of low water potential. Aust J Plant Physiol 20:425–437
Munns R, Termaat A (1986) Whole-plant responses to salinity. Aust J Plant Physiol 13:143–160
Munns R, Tester M (2008) Mechanisms of salinity tolerance. Annu Rev Plant Biol 59:651–681
Munns R, James RA, Lauchli A (2006) Approaches to increasing the salt tolerance of wheat and other cereals. J Exp Bot 57:1025–1043
Nahar K, Hasanuzzaman M (2009) Germination, growth, nodulation and yield performance of three mungbean varieties under different levels of salinity stress. Green Farming 2:825–829
Nam MH, Huh SM, Kim KM, Park WJ, Seo JB, Cho K, Kim DY, Kim BG, Yoon IS (2012) Comparative proteomic analysis of early salt stress-responsive proteins in roots of ^{SnRK2} transgenic rice. Proteome Sci 10:25

Nanjo Y, Nouri MZ, Komatsu S (2011) Quantitative proteomics analyses of crop seedlings subjected to stress conditions; a commentary. Phytochemistry 72:1263–1272

Netondo GW, Onyango JC, Beck E (2004) Sorghum and salinity: II. Gas exchange and chlorophyll fluorescence of sorghum under salt stress. Crop Sci 44:806–811

Neumann PM, Azaizeh H, Leon D (1994) Hardening of root cell walls: a growth inhibitory response to saliny stress. Plant Cell Environ 17:303–309

Neumann PM (2011) Recent advances in understanding the regulation of whole-plant growth inhibition by salinity, drought and colloid stress. In: Kader JC, Delseny M (eds) Adv Bot Res 57:33–48

Niinemets U, Cescatti A, Rodeghiero M, Tosens T (2005) Leaf internal diffusion conductance limits photosynthesis more strongly in older leaves of Mediterranean evergreen broad-leaved species. Plant Cell Environ 28:1552–1566

Niinemets U, Diaz-Espejo A, Flexas J, Galmes J, Warren CR (2009) Role of mesophyll diffusion conductance in constraining potential photosynthetic productivity in the field. J Exp Bot 60:2249–2270

Niu X, Bressan RA, Hasegawa PM, Pardo JM (1995) Ion homeostasis in NaCl stress environments. Plant Physiol 109:735–742

Nounjan N, Nghia PT, Theerakulpisut P (2012) Exogenous proline and trehalose promote recovery of rice seedlings from salt-stress and differentially modulate antioxidant enzymes and expression of related genes. J Plant Physiol 169:596–604

Ouyang SQ, Liu YF, Liu P, Lei G, He SJ, Ma B, Zhang WK, Zhang JS, Chen SY (2010) Receptor-like kinase OsSIK1 improves drought and salt stress tolerance in rice (*Oryza sativa*) plants. Plant J 62:316–329

Pardo JM (2010) Biotechnology of water and salinity stress tolerance. Curr Opin Biotechnol 21:185–196

Parida AK, Das AB (2005) Salt tolerance and salinity effects on plants: a review. Ecotoxicol Environ Saf 60:324–349

Parida AK, Das AB, Mohanty P (2004) Investigations on the antioxidative defense responses to NaCl stress in a mangrove, *Bruguiera parviflora*: differential regulations of isoforms of some antioxidative enzymes. Plant Growth Regul 42:213–226

Parker R, Flowers TJ, Moore AL, Harpham VJ (2006) An accurate and reproducible method for proteome profiling of the effects of salt stress in the rice leaf lamina. J Exp Bot 57:1109–1118

Pastori GM, Foyer CH (2002) Common components, networks, and pathways of cross-tolerance to stress. The central role of "Redox" and abscisic acid-mediated controls. Plant Physiol 129:460–468

Paul S, Kundu A, Pal A (2011) Identification and validation of conserved microRNAs along with their differential expression in roots of *Vigna unguiculata* grown under salt stress. Plant Cell Tissue Organ Cult 105:233–242

Pessarakli M, Szabolcs I (2010) Soil salinity and sodicity as particular plant/crop stress factors. In: Pessarakli M (ed) Handbook of plant and crop stress, 3rd edn. CRC Press, Boca Raton, pp 3–21

Poljakoff-Mayber A, Somers GF, Werker E, Gallagher JL (1994) Seeds of *Koteletzkya virginica* (Malvaceae): their structure, germination and salt tolerance. Am J Bot 81:54–59

Popp M (1995) Salt resistance in herbaceous halophytes and mangroves. Prog Bot 56:415–429

Posmyk MM, Kontek R, Janas KM (2009) Antioxidant enzymes activity and phenolic compounds content in red cabbage seedlings exposed to copper stress. Ecotoxicol Environ Saf 72:596–602

Price AH, Handry GAF (1991) Iron-catalysed oxygen radical formation and its possible contribution to drought damage in nine native grasses and three cereals. Plant Cell Environ 14:477–484

Rahnama A, Poustini K, Munns R, James RA (2010) Stomatal conductance as a screen for osmotic stress tolerance in durum wheat growing in saline soil. Funct Plant Biol 37:255–263

Rasheed R (2009) Salinity and extreme temperature effects on sprouting buds of sugarcane (*Saccharum officinarum* L.): some histological and biochemical studies. Ph.D. thesis, Department of Botany, University of Agriculture, Faisalabad

Rausch T, Kirsch M, Low R, Lehr A, Viereck R, Zhigang A (1996) Salt stress responses of higher plants: the role of proton pumps and Na^+/H^+-antiporters. J Plant Physiol 148:425–433
Rea PA, Yongcheol K, Sarafian V, Poole RJ, Davies JM, Sanders D (1992) Vacuolar H+ –translocating pyrophosphatases: a new category of ion translocase. TIBS 17:348–353
Remorini D, Melgar JC, Guidi L, Degl'Innocenti E, Castelli S, Traversi ML, Massai R, Tattini M (2009) Interaction effects of root-zone salinity and solar irradiance on the physiology and biochemistry of *Olea europaea*. Environ Exp Bot 65:210–219
Rengasamy P (2006) World salinization with emphasis on Australia. J Exp Bot 57:1017–1023
Rhoades JD, Kandiah A, Mashali AM (1992) The use of saline waters for crop production, vol 48, FAO irrigation and drainage paper. Food and Agriculture Organization of the United Nations, Rome
Rogers ME, Noble CL, Halloran GM, Nicolas ME (1995) The effect of NaCl on the germination and early seedling growth of white clover (*Trifolium repens* L.) populations selected for high and low salinity tolerance. Seed Sci Technol 23:277–287
Romero-Aranda R, Soria T, Cuartero S (2001) Tomato plant-water uptake and plant-water relationships under saline growth conditions. Plant Sci 160:265–272
Ruffino AMC, Rosa M, Hilal M, González JA, Prado FE (2009) The role of cotyledon metabolism in the establishment of quinoa (*Chenopodium quinoa*) seedlings growing under salinity. Plant Soil 326:213–224
Salekdeh GH, Siopongco J, Wade LJ, Ghareyazie B, Bennet J (2002) Proteomic analysis of rice leaves during drought stress and recovery. Proteomics 2:1131–1145
Salt DE, Baxter I, Lahner B (2008) Ionomics and the study of the plant ionome. Annu Rev Plant Biol 59:709–733
Sanchez DH, Lippold F, Redestig H, Hannah M, Erban A, Krämer U, Kopka J, Udvardi MK (2008a) Integrative functional genomics of salt acclimatization in the model legume *Lotus japonicus*. Plant J 53:973–987
Sanchez DH, Redestig H, Krämer U, Udvardi MK, Kopka J (2008b) Metabolome–ionome–biomass interactions: what can we learn about salt stress by multiparallel phenotyping? Plant Signal Behav 3:598–600
Sanchez D, Pieckenstain FL, Escaray F, Erban A, Kraemer U, Udvardi MK, Kopka J (2011a) Comparative ionomics and metabolomics in extremophile and glycophytic Lotus species under salt stress challenge the metabolic pre-adaptation hypothesis. Plant Cell Environ 34: 605–617
Sanchez DH, Pieckenstain FL, Szymanski J, Erban A, Bromke M (2011b) Comparative functional genomics of salt stress in related model and cultivated plants identifies and overcomes limitations to translational genomics. PLoS One 6(2):e17094
Savirnata NM, Jukunen-Titto R, Oksanen E, Karjalainen RO (2010) Leaf phenolic compounds in red clover (*Trfolium Pratense L.*) induced by exposure to moderately elevated ozone. Environ Pollut 158:440–446
Sawada Y, Akiyama K, Sakata A, Kuwahara A, Otsuki H, Sakurai T, Saito K, Hirai MY (2009) Widely targeted metabolomics based on large-scale MS/MS data for elucidating metabolite accumulation patterns in plants. Plant Cell Physiol 50:37–47
Schafer H, Wink MM (2009) Edicinally important secondary metabolites in recombinant microorganisms or plants: progress in alkaloid biosynthesis. Biotechnol J 4:1684–1703
Scholberg JMS, Locascio SJ (1999) Growth response of snap bean and tomato as affected by salinity and irrigation method. HortScience 34:259–264
Sedghi M, Seyed Sharifi R, Pirzad AR, Amanpour-Balaneji B (2012) Phytohormonal regulation of antioxidant systems in petals of drought stressed pot marigold (*Calendula officinalis* L.). J Agric Sci Technol 14:869–878
Serrano R, Mulet JM, Rios G, Marquez JA, de Larriona IF, Leube MP, Mendizabal I, Pascual-Ahuir A, Proft M, Ros R, Montesinos C (1999) A glimpse of the mechanism of ion homeostasis during salt stress. J Exp Bot 50:1023–1036
Shalata A, Neumann PM (2001) Exogenous ascorbic acid (vitamin C) increases resistance to salt stress and reduces lipid peroxidation. J Exp Bot 52:2207–2211

Shalata A, Mittova V, Volokita M, Guy M, Tal M (2001) Response of the cultivated tomato and its wild salt-tolerant relative *Lycopersicon pennellii* to salt-dependent oxidative stress: the root antioxidative system. Physiol Plant 112:487–494

Shannon MC (1984) Breeding, selection, and the genetics of salt tolerance. In: Staples RC, Toenniessen GH (eds) Salinity tolerance in plants: strategies for crop improvement. Wiley, New York, pp 231–254

Sidari M, Mallamaci C, Muscolo A (2008) Drought, salinity and heat differently affect seed germination of *Pinus pinea*. J For Res 13:326–330

Singh RK, Flowers TJ (2010) Physiology and molecular biology of the effects of salinity on rice. In: Handbook of plant and crop stress. CRC Press, Boca Raton, pp 899–939. doi: 101201/b10329-44

Singh RK, Mishra B, Chauhan MS, Yeo AR, Flowers SA, Flowers TJ (2002a) Solution culture for screening rice varieties for sodicity tolerance. J Agric Sci Camb 139:327–333

Singh RK, Singh NK, Mishra B (2002b) Identification of genes for physiological mechanisms for salinity tolerance and pyramiding through marker assisted selection. Paper presented during "Rice functional genomics workshop" held at National Research Centre on plant biotechnology, IARI, New Delhi, 20–21 May 2002. Abstract Proceeding page 27

Sinha A, Gupta SR (1982) Effect of osmotic tension and salt stress on germination of three grass species. Plant Soil 69:13–19

Sohan D, Jason R, Zajcek J (1999) Plant-water relations of NaCl and calcium-treated sunflower plants. Environ Exp Bot 42:105–111

Song C, Fang J, Li X, Liu H, Chao CT (2009) Identification and characterization of 27 conserved microRNAs in citrus. Planta 230:671

Sreevidya VS, Srinivasa RC, Rao C, Sullia SB, Ladha JK, Reddy PM (2006) Metabolic engineering of rice with soyabean isoflavone synthase for promoting nodulation gene expression in rhizobia. J Exp Bot 57:1957–1969

Szabolcs I (1974) Salt affected soils in Europe. Martinus Nijhoff, The Hague, p 63

Szpunar J (2004) Metallomics: a new frontier in analytical chemistry. Anal Bioanal Chem 378:54–56

Tanji KK, Wallender WW (2011) Nature and extent of agricultural salinity and sodicity. In: Wallender WW, Tanji KK (eds) Agricultural salinity assessment and management, 2nd edn. American Society of Civil Engineers, New York

Tanou G, Molassiotis A, Diamantidis G (2009) Induction of reactive oxygen species and necrotic death-like destruction in strawberry leaves by salinity. Environ Exp Bot 65:270–281

Tavakkoli E, Fatehi F, Rengasamy P, McDonald GK (2012) A comparison of hydroponic and soil-based screening methods to identify salt tolerance in the field in barley. J Exp Bot. doi:10.1093/jxb/ers085

Tester M, Davenport R (2003) Na^+ tolerance and Na^+ transport in plants. Ann Bot 91:503–527

Thelen JJ, Peck SC (2007) Quantitative proteomics in plants: choices in abundance. Plant Cell 19:3339–33346

Turhan H, Ayaz C (2004) Effect of salinity on seedling emergence and growth of sunflower (*Helianthus annuus* L.) cultivars. Int J Agric Biol 6:149–152

Tuteja N, Gill SS, Tiburcio AF, Tuteja R (2012) The micromics revolution: microRNA-mediated approaches to develop stress-resistant crops. In: López C, Pérez-Quintero AL (eds) Improving crop resistance to abiotic stress volume 1 and volume 2. Wiley-VCH Verlag GmbH & Co. KGaA, Weinheim. doi:10.1002/9783527632930.ch25

Ulfat M, Athar H, Ashraf M, Akram NA, Jamil A (2007) Appraisal of physiological and biochemical selection criteria for evaluation of salt tolerance in canola (*Brassica napus* L.). Pak J Bot 39:1593–1608

Ungar IA (1991) Ecophysiology of vascular halophytes. CRC Press, Boca Raton

USDA-ARS (2005) George E. Brown Jr salinity laboratory, Riverside. http://www.ars.usda.gov/Services/docs.htm? docid=8908

Vaidyanathan H, Sivakumar P, Chakrabarty R, Thomas G (2003) Scavenging of reactive oxygen species in NaCl-stressed rice (*Oryza sativa* L.)—differential response in salt-tolerant and sensitive varieties. Plant Sci 165:1411–1418 van de Graaff R, Patterson RA (2001) Explaining the mysteries of salinity, sodicity, SAR and ESP in on-site practice. In: Patterson RA, Jones MJ (eds) Proceedings of on-site '01 conference: advancing on-site wastewater systems. Lanfax Laboratories, Armidale, pp 361–368

van Hoorn JW, Katerji N, Hamdy A, Mastrorilli M (2001) Effect of salinity on yield and nitrogen uptake of four grain legumes and on biological nitrogen contribution from the soil. Agric Water Manage 51:87–98

Vinocur B, Altman A (2005) Recent advances in engineering plant tolerance to abiotic stress: achievements and limitations. Curr Opin Biotechnol 16:123–132

Voigt EL, Almeida TD, Chagas RM, Ponte LFA, Viégas RA, Silveira JAG (2009) Source–sink regulation of cotyledonary reserve mobilization during cashew (*Anacardium occidentale*) seedling establishment under NaCl salinity. J Plant Physiol 166(1):80–89

Wahid A, Javed IUH, Ali I, Baig A, Rasul E (1998) Short term incubation of sorghum caryopses in sodium chloride levels: changes in some pre and post germination physiological parameters. Plant Sci 139:223–232

Wahid A, Farooq M, Basra SMA, Rasul E, Siddique KHM (2010) Germination of seeds and propagules under salt stress. In: Handbook of plant and crop stress. CRC Press, Boca Raton. doi: 10.1201/b10329-16

Wahid A, Farooq M, Basra SMA, Rasul E, Siddique KHM (2011) Germination of seeds and propagules under salt stress. In: Pessarakli M (ed) Handbook of plant and crop stress, 3rd edn. CRC Press, Boca Raton, pp 321–337

Waisel Y (1972) Biology of halophytes. Academic, New York/London

Waisel Y, Breckle SW (1987) Differences in responses of various radish roots to salinity. Plant Soil 104:191–194

Wang MC, Peng ZY, Li CL, Li F, Liu C, Xia GM (2008) Proteomic analysis on a high salt tolerance introgression strain of *Triticum aestivum/Thinopyrum ponticum*. Proteomics 8: 1470–1489

Wei B, Cai T, Zhang R, Li A, Huo N, Li S, Gu QY, Vogel J, Jia J, Qi Y, Mao L (2009) Novel microRNAs uncovered by deep sequencing of small RNA transcriptomes in bread wheat (*Triticum aestivum* L.) and *Brachypodium distachyon* (L.) Beauv. Funct Integr Genomics 9:499–511

Widodo JJ, Patterson JH, Newbigin E, Tester M, Bacic A, Roessner U (2009) Metabolic responses to salt stress of barley (*Hordeum vulgare L.*) cultivars, Sahara and Clipper, which differ in salinity tolerance. J Exp Bot 60:4089–4103

Wild A (2003) Soils, land and food: managing the land during the twenty-first century. Cambridge University Press, Cambridge, UK

Wyn Jones RG, Pollard A (1983) Proteins, enzymes and inorganic ions. In: Laüchli A, Pirson A (eds) Encyclopedia of plant physiology, new series, vol 15B. Springer, Berlin, pp 528–562

Xu S, Hu B, He Z, Ma F, Feng J, Shen W, Yan J (2011) Enhancement of salinity tolerance during rice seed germination by presoaking with hemoglobin. Int J Mol Sci 12:2488–2501

Yadav S, Irfan M, Ahmad A, Hayat S (2011) Causes of salinity and plant manifestations to salt stress: a review. J Environ Biol 32:667–685

Yan SZ, Tang WS, Sun W (2005) Proteomic analysis of salt stress responsive proteins in rice roots. Proteomics 5:235–244

Yang L, Bai X, Yang Y, Ahmad P, Yang Y, Hu X (2011) Deciphering the protective role of nitric oxide against salt stress at the physiological and proteomic levels in maize. J Proteom Res 10:4349–4364

Yang L, Ma C, Wang L, Chen S, Li H (2012) Salt stress induced proteome and transcriptome changes in sugar beet monosomic addition line M14. J Plant Physiol 169:839–850

Yao D, Zhang X, Zhao X, Liu C, Wang C, Zhang Z, Zhang C, Wei Q, Wang Q, Yan H, Li F, Su Z (2011) Transcriptome analysis reveals salt-stress-regulated biological processes and key pathways in roots of cotton (*Gossypium hirsutum L.*). Genomics 98:47–55

Yeo AR (1983) Salinity resistance: physiologies and prices. Physiol Plant 58:214–222

Yeo AR, Flowers SA, Rao G, Welfare K, Senanayake N, Flowers TJ (1999) Silicon reduces sodium uptake in rice (*Oryza sativa* L.) in saline conditions and this is accounted for a reduction in the transpirational bypass flow. Plant Cell Environ 22:559–565

Yokthongwattana C, Mahong B, Roytrakul S, Phaonaklop N, Narangajavana J, Yokthongwattana K (2012) Proteomic analysis of salinity-stressed *Chlamydomonas reinhardtii* revealed differential suppression and induction of a large number of important housekeeping proteins. Planta 235:649–659

Yoshikawa M, Peragine A, Park MY, Poethig RS (2005) A pathway for the biogenesis of trans-acting siRNAs in Arabidopsis. Genes Dev 19:2164–2175

Yupsanis T, Moustakas M, Domiandou K (1994) Protein phosphorylation-dephosphorylation in alfalfa seeds germinating under salt stress. J Plant Physiol 143:234–240

Zapata PJ, Serrano M, Pretel MT, Amoros A, Botella MA (2004) Polyamines and ethylene changes during germination of different plant species under salinity. Plant Sci 167:781–788

Zhang X, Walker RR, Stevens RM, Prior LD (2002) Yield-salinity relationships of different grapevine (*Vitis vinifera* L.) scion-rootstock combinations. Aust J GrapeWine Res 8:150–156

Zhang H, Han B, Wang T, Chen S, Li H, Zhang Y, Dai S (2012) Mechanisms of plant salt response: Insights from proteomics. J Proteome Res 11:49–67

Zheng C, Jiang D, Liu F (2009) Exogenous nitric oxide improves seed germination in wheat against mitochondrial oxidative damage induced by high salinity. Environ Exp Bot 67:222–227

Zheng M, Wang Y, Liu K, Shu H, Zhou Z (2012) Protein expression changes during cotton fiber elongation in response to low temperature stress. J Plant Physiol 169:399–409

Zhong H, Läuchli A (1993) Spatial and temporal aspects of growth in the primary root of cotton seedlings: effect of NaCl and $CaCl_2$. J Exp Bot 44:763–771

Zhu JK (2001) Plant salt tolerance. Trends Plant Sci 6:66–71

Zhu JK (2002) Salt and drought stress signal transduction in plants. Annu Rev Plant Biol 53:247–273

Zhu J, Meinzer FC (1999) Efficiency of C-4 photosynthesis in *Atriplex lentiformis* under salinity stress. Aust J Plant Physiol 26:79–86

Zhu JK, Liu J, Xiong L (1998) Genetic analysis of salt tolerance in arabidopsis. Evidence for a critical role of potassium nutrition. Plant Cell 10:1181–1191

Zuther E, Koehl K, Kopka J (2007) Comparative metabolome analysis of the salt response in breeding cultivars of rice. In: Jenks MA, Hasegawa PM, Jain SM (eds) Advances in molecular breeding toward drought and salt tolerant crops. Springer, Berlin/Heidelberg/New York, pp 285–315

Chapter 7
Salt Stress and MAPK Signaling in Plants

M. Miransari, B. Rangbar, K. Khajeh, M.M. Tehranchi, R. Rusta Azad, F. Nagafi, and R. Rahnemaie

Abstract Plants are subjected to different kinds of stress, which can adversely affect their performance. Under stress, plant's response can determine its survival by altering its morphological and physiological properties. Salt stress negatively affects plant growth and yield production in different parts of the world. Hence, it is important to evaluate plant response under the stress so that the production of more tolerant plants may be possible. Among such responses the activation of different signaling pathways, which can enhance plant ability to tolerate the stress, is the most important one. Under stress, plant genes, which are responsible for plant resistance, are activated resulting in the production of signaling pathways, which can increase plant tolerance to stress. The cross-talk between different signaling pathways during stress can significantly influence plant performance. There are a range of important signaling pathways such as mitogen–activated protein kinase (MAPK), which can trigger plant response to biotic and abiotic stresses. MAPK components

M. Miransari (✉)
Department of Biological Sciences, Faculty of Science, Tarbiat Modares University, Tehran, Iran
e-mail: miransari1@gmail.com

B. Rangbar
Department of Biophysics, Faculty of Biological Sciences, Tarbiat Modares University, Tehran, Iran

K. Khajeh
Department of Biochemistry, Faculty of Biological Sciences, Tarbiat Modares University, Tehran, Iran

M.M. Tehranchi
Department of Physics, Faculty of Science, Shahid Beheshti University, Tehran, Iran

R. Rusta Azad
College of Chemical Engineering and Petroleum, Sharif University of Technology, Tehran, Iran

F. Nagafi
Department of Agriculture, Medicinal Plants Research Institute, Shaid Beheshti University, Tehran, Iran

R. Rahnemaie
Department of Soil Science, College of Agriculture, Tarbiat Modares University, Tehran, Iran

P. Ahmad et al. (eds.), *Salt Stress in Plants: Signalling, Omics and Adaptations*,
DOI 10.1007/978-1-4614-6108-1_7,

are a set of enzymes, causing plant response to the stimuli, for example by activating the antioxidant enzymes, resulted by different stresses. MAPK molecules are found in cytosol and nucleus, interacting with transcription components and phosphatase enzymes, regulate cell polarization, division and morphology by affecting microtubules, plant growth and development, and result in different signaling pathways under stress. In this chapter, some of the important findings regarding the MAPK signaling component under salinity stress are presented. Such details can be used for the production of plants with higher tolerance to such stresses.

Keywords Mitogen-activated protein kinase (MAPK) • Plant response • Stress • Signaling pathways

Plants must be able to survive under stress. Hence, with time they have evolved different mechanisms to be able to grow under such conditions. A wide range of mechanisms are used by plants among which the production of signaling pathways and the exchange of signal molecules is of great significance. Signal molecules are biochemical products, which are received by plant receptors, usually located in the cellular membrane and result in the expression of different genes including stress genes. Such signal molecules help the plant to be able to tolerate and survive under the stress (Hamel et al. 2006; Colcombet and Hirt 2008; Miransari and Smith 2007, 2008, 2009).

Salt stress is common in different parts of the world, adversely affecting plant growth by causing osmotic stress and cellular homeostasis. Hence, for the plant to survive under salt stress, the ratio of ions such as potassium and sodium (K^+/Na^+) must be kept constant in the cytosol (Tuteja 2007; Munns and Tester 2008). Accordingly, the amount of K^+ and Na^+ uptake as well as their transfer in the plant can significantly affect plant tolerance to salt stress (Fig. 7.1). Plants use different mechanisms to keep the cellular homeostasis of K^+ and Na^+ and alleviate salt stress including the exclusion of Na^+ from the cell or its localization in the vacuole by Na^+/H^+ activity (Tuteja 2007). Hence, the over expression of vacuolar or plasma membrane Na^+/H^+ can significantly enhance plant tolerance to salinity (Shi et al. 2003; Mehlmer et al. 2010).

During stress, different signalling pathways are activated including the activation of mitogen-activated protein kinases (MAPK) signalling cascade (Fig. 7.1). MAPK components are a set of enzymes, causing plant response to the stimuli resulted by different stresses. Accordingly, MAPK components are able to trigger different plant responses to the stress such as the activity of antioxidant enzymes. MAPK components are activated by the phosphorylation of the activation loop through the upstream kinases. Such kind of activation can be inhibited by the activity of phosphatases (Lee et al. 2009). Arabidopsis genome is able to activate five MKP molecules including AtMKP1, AtMKP2, DsPTP1, PHS1 and IBR5. Among such molecules, AtMKP2 and DsPTP1 are able to dephosphorylate *Arabidopsis* MAPK molecules (MPK3, 4 and 6) *in vitro* (Lee and Ellis 2007). Under stress the activation of different MAPK components and subsequent expression of stress genes may enhance plant response to stress.

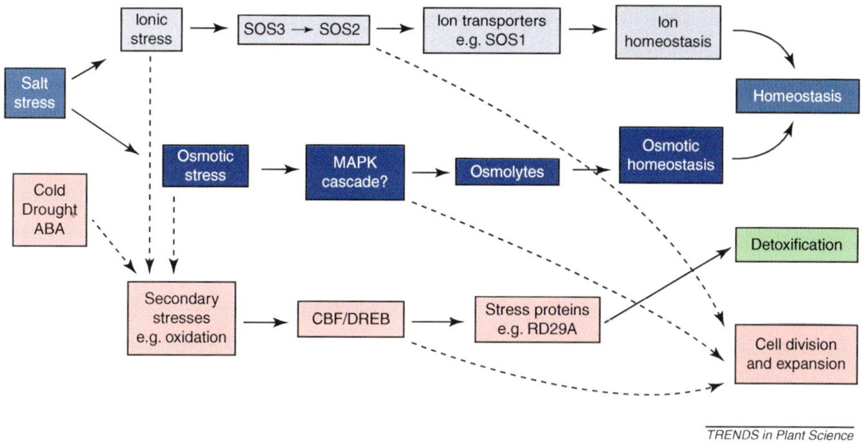

Fig. 7.1 The three different aspects of salt stress in plants including homeostasis, detoxification and growth control (From Zhu (2001))

According to Betsuyaku et al. (2011) in *Arabidopsis* the population size of stem cell in the apical meristem of shoot is regulated by the CLAVATA pathway in which the activation of a little peptide adversely affect the activity of the related transcription factors by different receptor kinases. In addition, the activation of a recently known RECEPTOR-LIKE PROTEIN KINASE 2 is also activated in the pathway by phospho-signalling. They accordingly found that such kinase receptors in the pathway can regulate the activity of MAPK signalling cascade and hence regulate the homeostasis of stem cell in the apical meristem.

In brief, MAPK molecules are found in cytosol and nucleus, interacting with transcription components and phosphatase enzymes (Schweighofer et al. 2007; Boudsocq et al. 2010), regulate cell polarization, division and morphology by affecting microtubules, plant growth and development, and result in different signaling pathways under stress (Samaj et al. 2004; Komis et al. 2010; Wurzinger et al. 2011). The transduction of MAPK signals from the membrane to the nucleus or cytosolic space is determined by the localization of the related kinases (Marmagne et al. 2007; Benetka et al. 2008; Yang et al. 2012).

MAPK Signalling Cascade and Salt Stress

The first response of plant to salt stress is by increasing the concentration of cytosolic Ca^{2+} resulting in the regulation of different stress signalling pathways (Zhu 2002). To adjust the cellular concentration of Na^+, the proton pumps localized on the plasma membrane direct the extrusion of Na^+ form the cell by the Na^+/H^+ SOS1 (Shi et al. 2003). The localization of Na^+ into the vacuoles in *Arabidopsis* is directed by

Na^+/H^+, NHX1 (Yu et al. 2010). Transporters and channels are activated and mediated by a set of signalling pathways in plant (Hirayama and Shinozaki 2010). Under salinity stress the increased concentration of Ca^{2+} is perceived by a calcineurin B-like protein, SOS3, which activates a MAPK molecule SOS2 regulating the Na^+/H^+ SOS1 localized on the plasma membrane (Zhu 2003).

There are about 120 kinases, however only a few of them have been so far recognized. Protein phosphorylation, which is performed by protein kinases, is a common mechanism used by different organisms. In a eukaryotic cell, about 30% of proteins are phosphorylated. Accordingly, 5% of plant genome results in the activation of protein kinases (The Arabidopsis Genome Initiative 2000; IRGSP 2005).

Protein phosphorylation is a way by which the activity of enzyme is regulated. For example, under stress the production of the stress enzyme ethylene is induced by different signaling pathways. Accordingly, 1-aminocyclopropane-1-carboxylate (ACC) synthase is activated, which can trigger the production of the enzyme by regulating transcription factors (Teige et al. 2004) or directly phosphorylating proteins (Liu and Zhang 2004) resulting in protein regulation and accumulation (Joo et al. 2008). Ethylene is able to regulate the crosstalk between Ca^{2+} signaling and MAPK signaling (Ludwig et al. 2005; Lindberg et al. 2012; Miransari 2012a, b).

It has been indicated that MAPK molecules become activated under stress. Under stress, the important task of MAPK pathways is to make the extracellular stimuli understandable by the cells and hence resulting into an appropriate response by plant. Such kind of cellular responses include the expression of different genes, the activation of enzymes and the channel proteins (Proft and Struhl 2004; Morris 2010). The main MAPK molecules, activated during stress are MPK3, MPK4 and MPK6. Using cytoplasm expression method, Colcombet and Hirt (2008) indicated that under salt and cold stress M2K2 activates MPK4. There are some indications that MAPK molecules are able to influence ABA signaling. For example, use of MAP2K inhibitor deceased ABA activity during stomatal closure (Burnett et al. 2000; Mehlmer et al. 2010).

Under different stresses such as heat, cold and salt, the activation of MAPK molecules results in the remodeling of microtubules. Among the first identified MAPK molecules, regulating cytoskeletal arrangement is a MMK2 related to *Medicago sativa*, being activated during the cell cycle stages, late anaphase/early telophase (Jonak et al. 1995; Calderini et al. 1998; Borge et al. 1999).

MAPK components are serine/threonine kinases phosphorylating a high number of substrates such as other kinases, transcription factors and cytoplasmic substrates. MAPK molecules are of significant importance in the transduction of different extracellular and intracellular pathways including the stress related signaling pathways. Among the other important functions of MAPK cascade is the cytoskeletal regulation (Tuteja 2007; Komis et al. 2010; Rodriguez et al. 2010; Johnson 2011) (Table 7.1). The way by which MAPK cascade is able to regulate microtubules structure and functioning has been recently investigated in tobacco. In MAPK mutants the processes of microtubules phosphorylation is influenced (Holmfeldt et al. 2009).

Table 7.1 Classifying MAPK, MAP2K and MAP3K molecules according to their functions

Molecule	Biological function
MAPK	
AtMPK3	Bacterial signalling, oxidative stress
AtMPK4	Drought, salinity, cold, pathogenic resistance
AtMPK6	Drought, salinity, cold, bacterial and fungal signalling, oxidative stress
MsSAMK	Drought, cold, fungal signalling
MsSIMK	Drought, salinity, cold, fungal signalling
MsMMK2	Fugal signalling
MsMMK3	Cytokinesis, Fugal signalling
NtWIPK	Salinity, fungal signalling, viral infection
NtSIPK	Salinity, SA, bacterial and fungal signalling, viral infection
NtNTF6	Cytokinesis
MAP2K	
AtMKK1	Drought, salinity, cold
AtMKK4,5	Bacterial signalling
MsPRKK	Fungal signalling
MsSIMKK	Salinity, fungal signalling
NtMEK1	Cytokinesis
NtMEK2	Hyper sensitive response
MAP3k	
AtMEKK1	Salinity, cold, bacterial signalling
AtANP1,2,3	Cytokinesis, auxin signalling, oxidative stress
AtCTR1	Ethylene signalling
AtEDR1	Pathogenic response
NtNPK1	Salinity, cold, heat, cytokinesis, oxidative stress

Data from Jonak et al. (2002); Hirayama and Shinozaki (2010)

Ten percent of plant kinases regulate the pathways of MAPK signal molecules, which are highly preserved in eukaryotes (Xu et al. 2008). The three signal molecules, forming the structure of MAPK are MAP3K (MAPK2K kinase), MAP2K (MAPK kinase), and MAPK affecting the activity of receptors (Tena et al. 2001; Pedley and Martin 2005). In a cascade the MAPK molecules interact with each other through the help of other proteins. Such kind of MAPK activity is regulated by MKP molecules (MAPK phosphatases).

Under stress the phosphorylation and dephosphorylation of proteins can activate different signalling pathways. For example, the expression of sucrose non-fermenting 1-related protein kinase2 *(TaSnRK2.4)* under different stresses in *Arabidopsis*, rice and maize can significantly enhance plant resistance under different stresses and hence can be used for the production of tolerant plants (Mao et al. 2010).

The genome sequencing of *Arabidopsis*, rice (*Oryza sativa*) and grapevine (*Vitis vinifera*) has indicated that there are a large number of genes, which are able to activate the MAPK related kinases (Hamel et al. 2006; Colcombet and Hirt 2008;

Mehlmer et al. 2010). For example, the *Arabidopsis* genome results in the expression of different MAPK genes, which are expressed differently under different conditions. Plant response to stress, plant growth and development (cell division), the regulation of plant response to plant hormones and soil microbes as well as stomatal developmental pathway are controlled by MAPK cascade signalling (Takahashi et al. 2007; Doczi et al. 2007).

Under salt stress the MAPK cascade is related to ME2K, which upregulates M2K2 and downregulates MPK4 and MPK6 (Teige et al. 2004). MPK1 adversely affects the pathway related to salt stress signaling by MAPK molecules (MPK4 and MPK6) (Ulm et al. 2002). A 46 kDa MAPK was induced by salt stress in alfalfa (Munnik et al. 1999). Salt stress resulted in the expression of AtMPK3, AtMPK4 and AtMPK6 in *Arabidopsis* (Droillard et al. 2002, 2004). Similarly, the expression of MAPKs, ZmMPK3, ZmMAPK5 and ZmSIMK1 under salt stress have been indicated in corn (*Zea mays* L.) (Ding et al. 2009; Wang et al. 2010c). Under the stresses such as salinity, drought and cold, the DNA sequence of AtMPK3 promoter has been recently indicated, which can contribute to our knowledge regarding the molecular pathways related to AtMPK3 expression under stress (Sinha et al. 2011).

Maize is a very important crop plant feeding a large number of people in the world. Under stresses such as salinity and cold, maize growth and yield is severely affected (Kizis and Pages 2002). There are six MAPK molecules, which have been so far recognized in maize. Under salinity stress or treating plant with ethylene, salicylic acid (SA), H_2O_2 or abscisic acid (ABA), ZmMPK3 RNA is accumulated in the plant (Wang et al. 2010c). ZmMPK5 and ZmMPK7 can be up-regulated through treating plant with ABA and H_2O_2 (Lin et al. 2009; Zong et al. 2009). Expression of *ZmSIMK* in *Arabidopsis* increased plant tolerance to salinity (Kong et al. 2011). Under salinity stress and other abiotic stresses different MAPK molecules such as AtMEKK1 are transcriptionally activated (Jonak et al. 2002). They are able to mediate plant response to stress (Kovtun et al. 2000). Asai et al. (2002) indicated that the signalling of flagellin is mediated by AtMEKK1 as it activates AtMKK4/AtMKK5 and AtMPK3/AtMPK6. However, yeast analyses have shown that AtMEKK1 functioning is upstream of AtMKK1 and AtMPK4 (Ichimura et al. 1998). As a response to abiotic signalling of stresses such as salinity, the activation of AtMKK1 results in the phosphorylation of AtMPK4 (Matsuoka et al. 2002).

A MAPKK gene, *ZmMKK4*, whose protein is found in the nucleolus, was isolated from maize by Kong et al. (2011). The gene was upregulated under stresses such as salinity, exogenous H_2O_2 and cold, but exogenous ABA resulted in the down regulation of gene. As a result of the gene over-expression, plant tolerance to stresses such as salinity and cold was enhanced (increased proline and sugar content in the cells). Because under the stress the rate of germination, root growth, the activity of antioxidant enzymes (peroxidase and catalse), soluble sugars, proline and chlorophyll increased related to the control treatment (Kong et al. 2011). Under salt and cold stress the activity of a 37-kDa kinase increased by ZmMKK4. According to RT-PCR analysis, in plants that the gene was over-expressed, higher expression of transcripts levels and functional genes was indicated. Hence, the above mentioned details indicate that ZmMKK4 can confer the plant the tolerance to cold and stress (Brock et al. 2010).

Under salt stress the dependent and independent ABA signalling pathways regulate the expression of salt genes in plant, however this is not the case for cold stress (Xiong et al. 2002; Zhu 2002). Accordingly, the research by Kong et al. (2011) indicated that ABA is able to down regulate *ZmM2K4*. MAPK components are localized to different plant organelles (Rodriguez et al. 2010). The expression analysis of MAPK components in *Arabidopsis* protoplasts indicated that M2K9 is mostly found in the nucleus activating MPK3 and MPK6, however, M2K4 is more found in the nucleus and cytoplasm activating MPK6 (Kong et al. 2011). If *AtM2K3* gene is overexpressed in *Arabidopsis*, plant tolerance to salinity as well as its sensitivity to ABA increases indicating the signalling role of *AtM2K3* in ABA activity and plant resistance under salinity (Hwa and Yang 2008). Similarly, overexpression of *AtMKK2* in plant can enhance plant resistance to cold and salinity stress (Teige et al. 2004).

Under the stress of heavy metals including cadmium (Cd^{2+}) and copper (Cu^{2+}), MAPK signalling pathway were activated in rice. Accordingly, Cd^{2+} stress resulted in the activation of a 42 kDa MAPK molecule. Diphenylene iodonium (DPI), which is an NADPH oxidase inhibitor, suppressed the activities of MAPK signalling cascade activated by Cd^{2+} but not by Cu^{2+}. This may indicate that NADPH oxidases results in the activation of MAPK components under Cd^{2+} stress.

In addition, for the activation of MAPK molecules by Cd^{2+} and Cu^{2+}, the presence of a Ca^{2+} dependent protein kinase and phosphatidylinositol 3-kinase was necessary (Yeh et al. 2007). They also indicated that the activation of MAP kinase by Cd^{2+} is dependent on the conditions of mitochondria. The higher activities of the 42 kDa MAPK molecule in the Cd^{2+} tolerant varieties may suggest that MAP kinase can induce plant resistance to the stress of Cd^{2+}.

The effects of salinity on the suspension cells of tobacco was investigated using in-gel kinase assays, which indicated the phosphorylating activities of multiple proteins including MAPKs SIPK and WIPK. Another protein kinase was also activated under salt stress, which is related to the class of SNF1 (SUCROSE-NONFERMENTING 1) (Mikolajczyk et al. 2000).

Under high salinity stress, the response of alfalfa cells was through the activation of a 38-kDa kinase, from the class of SNF1, however under moderate salinity the SIMK pathway was activated (Munnik et al. 1999). SIMKK can function as cell response to salt stress and as an elicitor during signalling. While SIMKK results in the activation of SIMK just under salinity stress, SIMK and MMK3 are activated in the presence of elicitor. Accordingly, exactly similar MAPK molecules can function in different signalling pathways (Whitmarsh and Davis 1998).

MAPK Signalling Pathway and Plant Hormones

Plant hormones are vital for different plant activities and development as well as plant response to stress and hence plant immunity. Salicylic acid (SA), jasmonic acid (JA) and ethylene (ET) are important hormones for plant immunity when interacting with different microbes such as pathogens (Spoel and Dong 2008). Pathogenic microbes are able to alter plant hormonal homeostasis. For example, Cui et al. (2010) indicated that production of AvrB protein by *Pseudomonas syringae* can affect plant hormonal signalling such as the upregulation of JA response genes resulting in the enhanced plant susceptibility. However, for the AvrB protein to act effectively the *Arabidopsis* MPK4, chaperon components and interactive proteins are necessary. The interaction between such parameters results in the activation of AvrB signalling pathway, and hence the subsequent alteration of plant hormonal signalling for the benefit of the bacteria and enhanced plant susceptibility.

MAPK cascades regulate and are regulated by plant hormones. One of the endoplasmic-reticulum ET receptors can target the MAP3K cascade (Kieber et al. 1993; Clark et al. 1998). ET activated a 47 kDa MAPK like kinase (Novikova et al. 2000) identified as MPK6. It was recently that M2K9 was identified as the activator of both MPK3 and MPK6 during ET signaling by relocalizing from the cytoplasm to the nucleus (Hamel et al. 2006; Colcombet and Hirt 2008). Muller et al. (2010) indicated that AtMPK6 is localized at plasma membrane and trans-Golgi network.

MPK6 is activated by many biotic and abiotic stresses resulting in the induction of ET production (Liu and Zhang 2004). The indication of ACS6 [ACC (1-aminocyclopropane- 1-carboxylic acid) SYNTHASE 6] as the substrate for MPK6 verify such a hypothesis (Liu and Zhang 2004; Joo et al. 2008). The point that the activation of M2K3-dependent MPK6, can adversely regulate JA signaling was recently indicated (Takahashi et al. 2007). MeJA (methyl jasmonate) is also able to activate MPK1 and MPK2. MPK4 is able to regulate the balance between SA and JA/ET during plant interaction with microbes (Brodersen et al. 2006). ACC results in the re-localization of M2K9 from the cytoplasm to the nucleus and hence the activation of MKP3 and MPK6 (Kong et al. 2011).

SA is able to activate the ortholog of tobacco MPK6, and the MPK4 of *Arabidopsis* mutant can increase the SA level resulting in the enhancement of plant resistance to pathogen (Zhang and Klessig 1997; Petersen et al. 2000). The negative regulation of ethylene by a kinase like molecule, which controls the activation of MPK3/6 by a MAPK kinase 9, has recently verified one more time the interaction effect between MAPK signalling and ethylene (Bethke et al. 2009).

Although ACC, which is the precursor of JA and SA, is able to activate MPK6 in *Arabidopsis* (Takahashi et al. 2007), it is not able to activate the cascade in tobacco (Kumar and Klessig 2000). In addition to the effects of MAPK molecules on ACC in the cytoplasm, they can also influence the proteins in the nucleus including MPK4 and MKS1 substrates (Andreasson et al. 2005). MKS1 is the downstream of MPK4 regulating the SA dependent pathway but not the ET/JA pathway (Qiu et al. 2008). Accordingly, these researchers indicated that during the stress, plant response is by the activation of MPK4 and phosphorylation of MKS1 and hence the release of the substrate and the related transcription factor from the MPK4 molecule. The transcription factor affects the related promoter and hence an enzyme is produced as a response to the stress. It is hence indicated how the activation of plant MAP kinase may result in the regulation of gene expression by the release of transcription factors into the nucleus (Qiu et al. 2008).

MAPK signalling is able to significantly affect plant immunity. Although MAP kinase kinase 7 (M2K7) adversely influences the regulation of polar auxin transport, it positively upregulates plant systemic resistance. In the *bud1* mutant of *Arabidopsis* the highly activation of M2K7 increases the amount of SA and the expression of pathogenesis related genes resulting in the plant increased resistance to pathogens. In the wild type the pathogen results in the expression of M2K7 gene (Zhang et al. 2007).

Nitrous oxide (NO) as a biomolecule regulates different plant developmental and physiological processes including the development of plant lateral roots. The production of NO as a very important signal molecule is regulated by MPK6 in the presence of H_2O_2. The *mpk6* mutants produced little amounts of NO as well as nitrate reductase (NR), when subjected to H_2O_2. In addition, the isoform of NR was necessary for the activity (acting as the substrate) of MPK6. Phosphorylation of the isoform by MPK6 increased the activity of NR. Accordingly, the role of MPK6 in the production of NO by H_2O_2 and hence regulation of root development is verified by such kind of data (Wang et al. 2010a).

The mis-regulation of signalling pathways can significantly affect plant phenotype and hence are controlled by different mechanisms and processes such as the stability of mRNA and the proteins regulated by phosphorylation (Gutierrez et al. 2002; Bao et al. 2004). According to Zhang et al. (2010) *P. syringae* affects plant immunity by influencing the receptor-like cytoplasmic kinases. Wang et al. (2010b) also indicated that *P. syringae* is able to suppress *Arabidopsis* immunity by producing an effector, which can interact with the plant MAPK molecules including MAP kinase kinase 5 and the other ones. The bacteria is able to interrupt the phosphorylation of MAPK molecules.

Abscisic acid (ABA) is a very important plant hormone regulating different plant activities under different conditions including stress. Under stress ABA results in the production of H_2O_2, which, as a signal, is able to mediate plant response to stress. H_2O_2 acts as a signal molecule in different plant activities including root growth and development, evapotranspiration, cell cycle, plant-microbe interactions and plant response to stress (Vanderauwera et al. 2005; Pitzschke and Hirt 2006). Such kind of responses resulted by H_2O_2 can activate the MAPK signalling cascade (Samuel et al. 2000; Zhang et al. 2006).

The activity of ABA under stress also results in the activation of catalase by expressing *CAT1* as well as by the production of H_2O_2. This idea has been supported by the research work of Xing et al. (2008) who indicated that the induction of *CAT1* activity by ABA is regulated by the combined MAPK signaling of AtMKK1-AtMPK6. Different researchers have indicated that H_2O_2 regulation is performed by MAPK cascade (Zhang et al. 2006). For example, in *Arabidopsis* H_2O_2 activated two MAPK molecules including AtMPK6 and AtMPK3. During ABA signalling, the signalling cascade of MdM2K1 and MdMPK1 act by regulating ABI5 or ABI5-like transcription factors (Wang et al. 2010d).

Using *Arabidopsis* mutants the role of MAPK molecules in auxin signaling has been indicated (Dai et al. 2006). Auxin stimulated MAPK-like activities in *Arabidopsis* and tobacco (*Nicotiana tabacum*) (Mizoguchi et al. 1994). However, using maize (*Zea mays* L.) and tobacco, Kovtun et al. (1998) indicated that auxin signaling was adversely affected by MAPK cascade. The role of M2K cascade on auxin transport was recently indicated using *Arabidopsis* mutant by investigating different auxin related phenotypes such as root, leaf and hypocotyls morphology and physiology affected by auxin transport (Dai et al. 2006).

Although auxin is able to activate the MAPK signaling pathway, the activation of MAPK molecules under oxidative stress can adversely affect auxin activity. Accordingly, it has been suggested that IBR5 can positively affect auxin signaling pathway by the inactivation of MAPK components (Monroe-Augustus et al. 2003). Lee et al. (2009) indicated that MPK12 adversely regulates auxin signaling and is a substrate for IBR5. There is interaction between MPK12 and IBR5 and IBR5 phosphatase is able to dephosphorylate and hence deactivate the activated MPK12 verifying MKP activity in *Arabidopsis*. They also showed that under in vivo conditions MPK12 is activated by auxin and in transgenic plants, if MPK12 is suppressed, it results in the upregulation of the genes related to auxin activity affecting root morphology. In transgenic plants, such as *Arabidopsis thaliana* the overexpression of a MAPK related gene may enhance plant tolerance to stress (Wang et al. 2010a, b, c, d).

MAPK Signalling Cascade and Regulation of Plant Activities

The precise co-ordination of chromosome relocalization during cell divisions both in time and space is an important cell activity. The regulators of mitosis are the cyclin-dependent kinase (CDK) and mitotic B-type cyclins. The degradation of mitotic cyclins during the transition from methaphase to anaphase indicates that the regulation of mitotic is performed by other regulators. In plants, cytokinesis structures can localize CDK indicating the likely roles of CDK during cytokinesis (Weingartner et al. 2001). According to their localization and activation during cytokinesis two MAPK molecules in alfalfa, MMK3, and tobacco, Ntf6, have been recognized in cytokinesis (Calderini et al. 1998; Bogre et al. 1999).

By experimenting with taxol, as a microtubule stabilizer blocking the lateral expanding of phragmoplasts, the significance of microtubules during cytokinesis has been indicated. The related microtubule protein NtMAP65-1 with sites of MAPK and CDK phosphorylation localizes to the midplane of cell division. This molecule can be accordingly a likely protein in the MAPK signalling pathway. The *mpk4* mutants of *Arabidopsis* indicated deficiency in the structure of microtubules, which resulted in the cell abnormal growth including the growth of root hairs and epidermal cells. It is because the phosphorylation of MAPK and hence the microtubule-related protein (MAP65) is negatively affected. Such kind of signalling pathways is necessary for the right formation of microtubules structure in epidermal cells (Beck et al. 2010; Gardiner and Marc 2011).

Although deactivation of MAPK signaling cascade by dephosphorylation is an important process, it has not been investigated much compared with the activation process. *PHS1* mutants negatively affected the structure of microtubule (Naoi and Hashimoto 2004) and ABA hypersensitivity (Quettier et al. 2006). However, the responsiveness to auxin and ABA decreased in *ibr5* plants related to the wild-type ones (Monroe-Augustus et al. 2003). It has been previously indicated that there is interaction between MKP and MAPK for AtMPK1, which is interactive with Arabidopsis MKP6 as well as with AtMPK3 and AtMPK4 (Ulm et al. 2002). MPK12 and IBR5 are localized in the nucleus verifying their potential interaction.

Under *in vitro* and *in vivo* conditions IBR5 was able to dephosphorylate and inactivate phospho MPK12 while it did not influence a different phosphorylated MAPK (AtMPK3). Lee and Ellis (2007) showed that another MKP protein (AtMKP2) was not able to dephosphorylate and inactivate MPK12. According to the above mentioned details IBR5 is a MKP (for example, if an active MAPK is dephosphorylated) and its catalytic activity such as its interactions with other proteins can highly and specifically affect MPK12 related to the other Arabidopsis MPK components (Lee et al. 2009).

Coincubation with MPK3 or MPK6 significantly increased the activity of AtMKP2 (Lee and Ellis 2007). However, there was not any effect on the catalytic activity of IBR5 when incubated with MPK12, although IBR5 is able to effectively and significantly interact with MPK12 resulting in its dephosphorylation (Lee et al. 2009). The authors accordingly suggested that MPK12 and IBR5 can adversely and positively affect auxin signaling pathway, respectively (Monroe-Augustus et al. 2003). It has not yet been indicated if in *ibr5* mutants, MPK12 has a role in ABA signaling.

Stomatal development is a suitable way of evaluating the MAPK signaling cascade. In addition to affecting stomatal development, MAPK signaling can also influence stomatal response to environmental parameters by affecting its physiology. Accordingly, MAPK signaling can positively affect stomatal closure under stress and adversely influence stomatal development (Bergmann et al. 2004; Wang et al. 2007; Neill et al. 2008).

The MAPK cascade with MKK4/5 and MPK3/6 is able to affect the regulation of plant response to environmental parameters as well as stomatal development (Bergmann et al. 2004; Wang et al. 2007; Colcombet and Hirt 2008). Hence, inhibiting the MAPK signaling cascade with the use of null alleles or RNA interference results in the overproliferation clustering of stomata (Bergmann et al. 2004; Wang et al. 2007).

The regulation of stomatal development by MAPK signaling pathway can be of significance for plant under different conditions including stress. It is because water efficiency is important for plant survival under drought and salt stress. Accordingly, the role of MAPK molecules under water stress can also be reflected in such a way, which is related to a set of interactions with stomatal development and functioning as well as with plant hormones specially ABA, which can control plant water behavior. Hence, investigating the greater details related to how such kind of interactions can be controlled by MAPK signaling pathways under water deficient conditions may result in the development of more tolerant plants under stress. The MAPK signalling pathway is common in all eukaryotic organisms for their response and hence their survival under different conditions including stress. For example, Rispail et al. (2009) analysed the MAPK signalling pathway in different fungi species.

Conclusion and Perspectives for Future Research

Under abiotic stress, different signaling pathways are activated including the MAPK cascade. MAPK molecules are a set of proteins, which can handle different functions in plant including cell cycle, plant growth and development, plant response to stress, etc. Indication of different MAPK pathways, under the stress can be useful for the production of transgenic plants, which are more tolerant under stress. There is a cross talk between different signaling pathways during the stress as well as the interactions with plant hormones. Although such kind of cross talk and interactions are complicated, researchers have been able to identify some of such processes, which can be very important for handling plant under stress. Adjusting plant behavior, especially Na^+ cellular concentration, under salt stress is an important key issue, to make plant survive the stress. MAPK signaling can importantly affect such phenomenon by affecting the activity of proton pumps, Na^+ localization into vacuoles, and the regulation of microtubules arrangement (cell cycle). To make plants survive more efficiently under salinity stress, the signaling pathways including the MAPK cascade must be recognized in greater details and accordingly the tolerant plants be developed.

References

Andreasson E, Jenkins T, Brodersen P, Thorgrimsen S, Petersen N, Zhu S, Qiu J, Micheelsen P, Rocher A, Petersen M, Newman M, Nielsen H, Hirt H, Somssich I, Mattsson O, Mundy J (2005) The MAP kinase substrate MKS1 is a regulator of plant defense responses. EMBO J 24:2579–2589

Asai T, Tena G, Plotnikova J, Willmann MR, Chiu WL, Gomez-Gomez L, Boller T, Ausubel FM, Sheen J (2002) MAP kinase signalling cascade in *Arabidopsis* innate immunity. Nature 415:977–983

Bao MZ, Schwartz MA, Cantin GT, Yates JR 3rd, Madhani HD (2004) Pheromone-dependent destruction of the Tec1 transcription factor is required for MAP kinase signaling specificity in yeast. Cell 119:991–1000

Beck M, Komis G, Muller J, Menzel D, Samaj J (2010) *Arabidopsis* homologs of nucleus- and phragmoplast-localized kinase 2 and 3 and mitogen-activated protein kinase 4 are essential for microtubule organization. Plant Cell 22:755–771

Benetka W, Mehlmer N, Maurer-Stroh S, Sammer M, Koranda M, Neumuller R, Betschinger J, Knoblich JA, Teige M, Eisenhaber F (2008) Experimental testing of predicted myristoylation targets involved in asymmetric cell division and calcium-dependent signalling. Cell Cycle 7:3709–3719

Bergmann DC, Lukowitz W, Somerville CR (2004) Stomatal development and pattern controlled by a MAPKK kinase. Science 304:1494–1497

Bethkea G, Unthan T, Uhrig J, Poschla Y, Gustc A, Scheel D, Lee J (2009) Flg22 regulates the release of an ethylene response factor substrate from MAP kinase 6 in Arabidopsis thaliana via ethylene signaling. Proc Nat Acad Sci, USA 106:8067–8072

Betsuyaku S, Takahashi F, Kinoshita A, Miwa H, Shinozaki K, Fukuda H, Sawa S (2011) Mitogen-activated protein kinase regulated by the CLAVATA receptors contributes to shoot apical meristem homeostasis. Plant Cell Physiol 52:14–29

Bogre L, Calderini O, Binarova P, Mattauch M, Till S, Kiegerl S, Jonak C, Pollaschek C, Barker P, Huskisson NS, Hirt H, Heberle-Bors E (1999) A MAP kinase is activated late in plant mitosis and becomes localized to the plane of cell division. Plant Cell 11:101–113

Boudsocq M, Willmann MR, McCormack M, Lee H, Shan L, He P, Bush J, Cheng SH, Sheen J (2010) Differential innate immune signalling via Ca^{2+} sensor protein kinases. Nature 464:418–422

Brock A, Willmann R, Kolb D, Grefen L, Lajunen HM, Bethke G, Lee J, Nurnberger T, Gust A (2010) The *Arabidopsis* mitogen-activated protein kinase phosphatase PP2C5 affects seed germination, stomatal aperture, and abscisic acid-inducible gene expression. Plant Physiol 153:1098–1111

Brodersen P, Petersen M, Bjorn Nielsen H, Zhu S, Newman MA, Shokat KM, Rietz S, Parker J, Mundy J (2006) *Arabidopsis* MAP kinase 4 regulates salicylic acid- and jasmonic acid/ethylene-dependent responses via EDS1 and PAD4. Plant J 47:532–546

Burnett EC, Desikan R, Moser RC, Neill SJ (2000) ABA activation of an MBP kinase in *Pisum sativum* epidermal peels correlates with stomatal responses to ABA. J Exp Bot 51:197–205

Calderini O, Bogre L, Vicente O, Binarova P, Heberle-Bors E, Wilson C (1998) A cell cycle regulated MAP kinase with a possible role in cytokinesis in tobacco cells. J Cell Sci 111:3091–3100

Clark KL, Larsen PB, Wang X, Chang C (1998) Association of the Arabidopsis CTR1 Raf-like kinase with the ETR1 and ERS ET receptors. Proc Nat Acad Sci USA 95:5401–5406

Colcombet J, Hirt H (2008) Arabidopsis MAPKs: a complex signalling network involved in multiple biological processes. Biochem J 413:217–226

Cui H, Wang Y, Xue L, Chu J, Yan C, Fu J, Chen M, Innes R, Zhou J-M (2010) *Pseudomonas syringae* effector protein AvrB perturbs arabidopsis hormone signaling by activating MAP kinase 4. Cell Host Microbe 7:164–175

Dai Y, Wang H, Li B, Huang J, Liu X, Zhou Y, Mou Z, Li J (2006) Increased expression of MAP KINASE KINASE7 causes deficiency in polar auxin transport and leads to plant architectural abnormality in *Arabidopsis*. Plant Cell 18:308–320

Ding HD, Zhang XH, Xu SC, Sun LL, Jiang MY, Zhang AY, Jin YG (2009) Induction of protection against paraquat-induced oxidative damage by abscisic acid in maize leaves is mediated through mitogen-activated protein kinase. J Integr Plant Biol 51:961–972

Doczi R, Brader G, Pettko-Szandtner A, Rajh I, Djamei A, Pitzschke A, Teige M, Hirt H (2007) The Arabidopsis mitogen-activated protein kinase kinase MKK3 is upstream of group C mitogen-activated protein kinases and participates in pathogen signaling. Plant Cell 19:3266–3279

Droillard M, Boudsocq M, Barbier-Brygoo H, Lauriere C (2002) Different protein kinase families are activated by osmotic stresses in *Arabidopsis thaliana* cell suspensions. Involvement of the MAP kinases AtMPK3 and AtMPK6. FEBS Lett 527:43–50

Droillard MJ, Boudsocq M, Barbier-Brygoo H, Lauriere C (2004) Involvement of MPK4 in osmotic stress response pathways in cell suspensions and plantlets of *A. thaliana*: activation by hypoosmolarity and negative role in hyperosmolarity tolerance. FEBS Lett 574:42–48

Gardiner J, Marc J (2011) Arabidopsis thaliana, a plant model organism for the neuronal microtubule cytoskeleton? J Exp Bot 62:89–97

Gutierrez RA, Ewing RM, Cherry JM, Green PJ (2002) Identification of unstable transcripts in *Arabidopsis* by cDNA microarray analysis: rapid decay is associated with a group of touch- and specific clock-controlled genes. Proc Nat Acad Sci USA 99:11513–11518

Hamel LP, Nicole MC, Sritubtim S, Morency M-J, Ellis M, Ehlting J, Beaudoin N, Barbazuk B, Klessig D, Lee J, Martin G, Mundy J, Ohashi Y, Scheel D, Sheen J, Xing T, Zhang S, Seguin A, Ellis BE (2006) Ancient signals: comparative genomics of plant MAPK and MAPKK gene families. Trend Plant Sci 11:192–198

Hirayama T, Shinozaki K (2010) Research on plant abiotic stress responses in the post-genome era: past, present and future. Plant J 61:1041–1052

Holmfeldt P, Sellin ME, Gullberg M (2009) Predominant regulators of tubulin monomer–polymer partitioning and their implication for cell polarization. Cell Mol Life Sci 66:3263–3276

Hwa CM, Yang XC (2008) The AtMKK3 pathway mediates ABA and salt signaling in *Arabidopsis*. Acta Physiol Plant 30:277–286

Ichimura K, Mizoguchi T, Irie K, Morris P, Giraudat J, Matsumoto K, Shinozaki K (1998) Isolation of ATMEKK1 (a MAP kinase kinase kinase)-interacting protein and analysis of a MAP kinase cascade in *Arabidopsis*. Biochem Biophys Res Commun 253:532–543

International Rive Genome Sequencing Project (2005) The map-based sequence of the rice genome. Nature 436:793–800

Johnson GL (2011) Defining MAPK interactomes. ACS Chem Biol 6:18–20

Jonak C, Kiegerl S, Lloyd C, Chan J, Hirt H (1995) H: MMK2, a novel alfalfa MAP kinase, specifically complements the yeast MPK1 function. Mol Gen Genet 248:686–694

Jonak C, Ökrész L, Bögre L, Hirt H (2002) Complexity, cross talk and integration of plant MAP kinase signalling. Curr Opin Plant Biol 5:415–424

Joo S, Liu Y, Lueth A, Zhang S (2008) MAPK phosphorylation-induced stabilization of ACS6 protein is mediated by the non-catalytic C-terminal domain, which also contains the cis-determinant for rapid degradation by the 26 S proteasome pathway. Plant J 54:129–140

Kieber JJ, Rothenberg M, Roman G, Feldmann KA, Ecker JR (1993) CTR1, a negative regulator of the ET response pathway in *Arabidopsis*, encodes a member of the raf family of protein kinases. Cell 72:427–441

Kizis D, Pages M (2002) Maize DRE-binding proteins DBF1 and DBF2 are involved in rab17 regulation through the droughtresponsive element in an ABA-dependent pathway. Plant J 30:679–689

Komis G, Illes P, Beck M, Samaj J (2010) Microtubules and mitogen-activated protein kinase signaling. Curr Opin Plant Biol 14:650–657

Kong X, Pan J, Zhang M, Xing X, Zhou Y, Liu Y, Li D, Li D (2011) *ZmMKK4*, a novel group C mitogen-activated protein kinase kinase in maize (*Zea mays*), confers salt and cold tolerance in transgenic Arabidopsis. Plant Cell Environ 34:1291–1303

Kovtun Y, Chiu WL, Zeng W, Sheen J (1998) Suppression of auxin signal transduction by a MAPK cascade in higher plants. Nature 395:716–720

Kovtun Y, Chiu WL, Tena G, Sheen J (2000) Functional analysis of oxidative stress-activated mitogen-activated protein kinase cascade in plants. Proc Nat Acad Sci USA 97:2940–2945

Kumar D, Klessig DF (2000) Differential induction of tobacco MAP kinases by the defense signals nitric oxide, salicylic acid, ethylene, and jasmonic acid. Mol Plant Microbe Interact 13:347–351

Lee JS, Ellis BE (2007) *Arabidopsis* MAPK phosphatase 2 (MKP2) positively regulates oxidative stress tolerance and inactivates the MPK3 and MPK6 MAPKs. J Biol Chem 282:25020–25029

Lee JS, Wang S, Sritubtim S, Chen JG, Ellis BE (2009) Arabidopsis mitogen-activated protein kinase MPK12 interacts with the MAPK phosphatase IBR5 and regulates auxin signaling. Plant J 57:975–985

Lin F, Ding H, Wang J, Zhang H, Zhang A, Zhang Y, Tan M, Dong W, Jiang M (2009) Positive feedback regulation of maize NADPH oxidase by mitogen-activated protein kinase cascade in abscisic acid signalling. J Exp Bot 60:3221–3238

Lindberg S, Kader A, Yemelyanov V (2012) Calcium signalling in plant cells under environmental stress. In: Ahmad P, Prasad MNV (eds) Environmental adaptations and stress tolerance of plants in the era of climate change. Springer, New York, pp 325–360

Liu Y, Zhang S (2004) Phosphorylation of 1-aminocyclopropane-1-carboxylic acid synthase by MPK6, a stress-responsive mitogen-activated protein kinase, induces ET biosynthesis in *Arabidopsis*. Plant Cell 16:3386–3399

Ludwig AA, Saitoh H, Felix G, Freymark G, Miersch O, Wasternack C, Boller T, Jones JD, Romeis T (2005) Ethylene-mediated cross-talk between calcium-dependent protein kinase and MAPK signaling controls stress responses in plants. Proc Nat Acad Sci USA 102:10736–10741

Mao X, Zhang H, Tian S, Chang X, Jing R (2010) TaSnRK2.4, an SNF1-type serine/threonine protein kinase of wheat (*Triticum aestivum* L.), confers enhanced multistress tolerance in *Arabidopsis*. J Exp Bot 61:683–696

Marmagne A, Ferro M, Meinnel T, Bruley C, Kuhn L, Garin J, Barbier-Brygoo H, Ephritikhine G (2007) A high content in lipid-modified peripheral proteins and integral receptor kinases features in the Arabidopsis plasma membrane proteome. Mol Cell Proteomics 6:1980–1996

Matsuoka D, Nanmori T, Sato K, Fukami Y, Kikkawa U, Yasuda T (2002) Activation of AtMEK1, an *Arabidopsis* mitogen-activated protein kinase kinase, *in vitro* and in vivo: analysis of active mutants expressed in *E. coli* and generation of the active form in stress response in seedlings. Plant J 29:637–647

Mehlmer N, Wurzinger B, Stael S, Hofmann-Rodrigues D, Csaszar E, Pfister B, Bayer R, Teige M (2010) The Ca^{2+}-dependent protein kinase CPK3 is required for MAPK-independent salt-stress acclimation in *Arabidopsis*. Plant J 63:484–498

Mikolajczyk M, Awotunde OS, Muszynska G, Klessig DF, Dobrowolska G (2000) Osmotic stress induces rapid activation of a salicylic acid induced protein kinase and a homolog of protein kinase ASK1 in tobacco cells. Plant Cell 12:165–178

Miransari M (2012a) Role of phytohormone signaling during stress. In: Ahmad A, Prasad MNV (eds) Environmental adaptations and stress tolerance of plants in the era of climate change. Springer, New York, pp 381–393

Miransari M (2012b) Microbial products and soil stresses. In: Maheshwari DK (ed) Bacteria in agrobiology: stress management. Springer, Berlin\New York, pp 65–75

Miransari M, Smith DL (2007) Overcoming the stressful effects of salinity and acidity on soybean [*Glycine max* (L.) Merr.] nodulation and yields using signal molecule genistein under field conditions. J Plant Nutr 30:1967–1992

Miransari M, Smith DL (2008) Using signal molecule genistein to alleviate the stress of suboptimal root zone temperature on soybean-*Bradyrhizobium* symbiosis under different soil textures. J Plant Interact 3:287–295

Miransari M, Smith D (2009) Alleviating salt stress on soybean (*Glycine max* (L.) Merr.) – *Bradyrhizobium japonicum* symbiosis, using signal molecule genistein. Europ J Soil Biol 45:146–152

Mizoguchi T, Gotoh Y, Nishida E, Yamaguchi-Shinozaki K, Hayashida N, Iwasaki T, Kamada H, Shinozaki K (1994) Characterization of two cDNAs that encode MAP kinase homologues in *Arabidopsis thaliana* and analysis of the possible role of auxin in activating such kinase activities in cultured cells. Plant J 5:111–122

Monroe-Augustus M, Zolman BK, Bartel B (2003) IBR5, a dual-specificity phosphatase-like protein modulating auxin and abscisic acid responsiveness in Arabidopsis. Plant Cell 15:2979–2991

Morris P (2010) Integrating lipid signalling, mitogen-activated protein kinase cascades and salt tolerance. New Phytol 188:640–643

Muller J, Beck M, Mettbach U, Komis G, Hause G, Menzel D, Samaj J (2010) Arabidopsis MPK6 is involved in cell division plane control during early root development, and localizes to the pre-prophase band, phragmoplast, trans-golgi network and plasma membrane. Plant J 61:234–248

Munnik T, Ligterink W, Meskiene II, Calderini O, Beyerly J, Musgrave A, Hirt H (1999) Distinct osmo-sensing protein kinase pathways are involved in signaling moderate and severe hyper-osmotic stress. Plant J 20:381–388

Munns R, Tester M (2008) Mechanisms of salinity tolerance. Ann Rev Plant Biol 59:651–681

Naoi K, Hashimoto T (2004) A semidominant mutation in an Arabidopsis mitogen-activated protein kinase phosphatase-like gene compromises cortical microtubule organization. Plant Cell 16:1841–1853

Neill S, Barros R, Bright J, Desikan R, Hancock J, Harrison J, Morris P, Ribeiro D, Wilson I (2008) Nitric oxide, stomatal closure, and abiotic stress. J Exp Bot 59:165–176

Novikova GV, Moshkov IE, Smith AR, Hall MA (2000) The effect of ET on MAPKinase-like activity in *Arabidopsis thaliana*. FEBS Lett 474:29–32

Pedley KF, Martin GB (2005) Role of mitogen-activated protein kinases in plant immunity. Curr Opin Plant Biol 8:541–547

Petersen M, Brodersen P, Naested H, Andreasson E, Lindhart U, Johansen B, Nielsen HB, Lacy M, Austin MJ, Parker JE, Sharma SB, Klessig DF, Martienssen R, Mattsson O, Jensen AB, Mundy

J (2000) *Arabidopsis* MAP kinase 4 negatively regulates systemic acquired resistance. Cell 103:1111–1120
Pitzschke A, Hirt H (2006) Mitogen-activated protein kinases and reactive oxygen species signaling in plants. Plant Physiol 141:351–356
Proft M, Struhl K (2004) MAP kinase-mediated stress relief that precedes and regulates the timing of transcriptional induction. Cell 118:351–361
Qiu J-L, Fiil BK, Petersen K, Nielsen HB, Botanga CJ, Thorgrimsen S, Palma K, Suarez-Rodriguez CM, Sandbech-Clausen S, Lichota J, Brodersen P, Grasser KD, Mattsson O, Glazebrook J, Mundy J, Petersen M (2008) Arabidopsis MAP kinase 4 regulates gene expression through transcription factor release in the nucleus. EMBO J 27:2214–2221
Quettier AL, Bertrand C, Habricot Y, Miginiac E, Agnes C, Jeannette E, Maldiney R (2006) The phs1-3 mutation in a putative dual-specificity protein tyrosine phosphatase gene provokes hypersensitive responses to abscisic acid in *Arabidopsis thaliana*. Plant J 47:711–719
Rispail N, Soanes DM, Ant C, Czajkowski R, Grünler A, Huguet R, Perez-Nadales E, Poli A, Sartorel E, Valiante V, Yang M, Beffa R, Brakhage A, Gowi N, Kahmann R, Lebrun M-H, Lenasi H, Perez-Martin J, Talbot NJ, Wendland J, Di Pietro A (2009) Comparative genomics of MAP kinase and calcium–calcineurin signalling components in plant and human pathogenic fungi. Fungal Gen Biol 46:287–298
Rodriguez MC, Petersen M, Mundy J (2010) Mitogen-activated protein kinase signaling in plants. Ann Rev Plant Biol 61:621–649
Samaj J, Baluska F, Hirt H (2004) From signal to cell polarity: mitogen-activated protein kinases as sensors and effectors of cytoskeleton dynamicity. J Exp Bot 55:189–198
Samuel MA, Miles GP, Ellis BE (2000) Ozone treatment rapidly activates MAP kinase signalling in plants. Plant J 22:367–376
Schweighofer A, Kazanaviciute V, Scheikl E, Teige M, Doczi R, Hirt H, Schwanninger M, Kant M, Schuurink R, Mauch F, Buchala A, Cardinale F, Meskiene I (2007) The PP2C-type phosphatase AP2C1, which negatively regulates MPK4 and MPK6, modulates innate immunity, jasmonic acid and ethylene levels in *Arabidopsis*. Plant Cell 19:2213–2224
Shi H, Lee BH, Wu SJ, Zhu JK (2003) Overexpression of a plasma membrane Na^+/H^+ antiporter gene improves salt tolerance in *Arabidopsis thaliana*. Nat Biotechnol 21:81–85
Sinha A, Jaggi M, Raghuram B, Tuteja N (2011) Mitogen-activated protein kinase signaling in plants under abiotic stress. Plant Signal Behav 2:196–203
Spoel SH, Dong X (2008) Making sense of hormone crosstalk during plant immune responses. Cell Host Microbe 3:348–351
Takahashi F, Yoshida R, Ichimura K, Mizoguchi T, Seo S, Yonezawa M, Maruyama K, Yamaguchi-Shinozaki K, Shinozaki K (2007) The mitogen-activated protein kinase cascade MKK3-MPK6 is an important part of the jasmonate signal transduction pathway in *Arabidopsis*. Plant Cell 19:805–818
Teige M, Scheikl E, Eulgem T, Doczi R, Ichimura K, Shinozaki K, Dangl JL, Hirt H (2004) The MKK2 pathway mediates cold and salt stress signaling in Arabidopsis. Mol Cell 15:141–152
Tena G, Asai T, Chiu WL, Sheen J (2001) Plant mitogen-activated protein kinase signaling cascades. Curr Opin Plant Biol 4:392–400
The Arabidopsis Genome Initiative (2000) Analysis of the genome sequence of the flowering plant *Arabidopsis thaliana*. Nature 408:796–815
Tuteja N (2007) Mechanisms of high salinity tolerance in plants. Methods Enzymol 428:419–438
Ulm R, Ichimura K, Mizoguchi T, Peck SC, Zhu T, Wang X, Shinozaki K, Paszkowski J (2002) Distinct regulation of salinity and genotoxic stress responses by *Arabidopsis* MAP kinase phosphatase 1. EMBO J 21:6483–6493
Vanderauwera S, Zimmermann P, Rombauts S, Vandenabeele S, Langebartels C, Gruissem W, Inze D, Van Breusegem F (2005) Genome-wide analysis of hydrogen peroxide-regulated gene expression in *Arabidopsis* reveals a high light-induced transcriptional cluster involved in anthocyanin biosynthesis. Plant Physiol 139:806–821
Wang H, Ngwenyama N, Liu Y, Walker JC, Zhang S (2007) Stomatal development and patterning are regulated by environmentally responsive mitogen-activated protein kinases in Arabidopsis. Plant Cell 19:63–73

Wang J, Ding H, Zhang A, Ma F, Cao J, Jiang M (2010a) A novel mitogen-activated protein kinase gene in maize (*Zea mays*), ZmMPK3, is involved in response to diverse environmental cues. J Integr Plant Biol 52:442–452

Wang P, Du Y, Li Y, Ren D, Songa C-P (2010b) Hydrogen peroxide–mediated activation of MAP Kinase 6 modulates nitric oxide biosynthesis and signal transduction in *Arabidopsis*. Plant Cell 22:2981–2998

Wang XJ, Zhu SY, Lu YF, Zhao R, Xin Q, Wang XF, Zhang DP (2010c) Two coupled components of the mitogen-activated protein kinase cascade MdMPK1 and MdMKK1 from apple function in ABA signal transduction. Plant Cell Physiol 51:754–766

Wang Y, Li J, Hou S, Wang X, Li Y, Ren D, Chen S, Tang X, Zhou J-M (2010d) A *Pseudomonas syringae* ADP-ribosyltransferase inhibits *Arabidopsis* mitogen-activated protein kinase kinases. Plant Cell 22:2033–2044

Weingartner M, Binarova P, Drykova D, Schweighofer A, David JP, Heberle-Bors E, Doonan J, Bogre L (2001) Dynamic recruitment of Cdc2 to specific microtubule structures during mitosis. Plant Cell 13:1929–1943

Whitmarsh AJ, Davis RJ (1998) Structural organization of MAP-kinase signaling modules by scaffold proteins in yeast and mammals. Trends Biochem Sci 23:481–485

Wurzinger B, Mair A, Pfister B, Teige M (2011) Cross-talk of calcium-dependent protein kinase and MAP kinase signaling. Plant Signal Behav 1:8–12

Xing Y, Jia W, Zhang J (2008) AtMKK1 mediates ABA-induced CAT1 expression and H_2O_2 production via AtMPK6-coupled signaling in *Arabidopsis*. Plant J 54:440–451

Xiong L, Schumaker K, Zhu JK (2002) Cell signaling during cold, drought, and salt stress. Plant Cell 14:S165–S183

Xu J, Li Y, Wang Y, Liu H, Lei L, Yang H, Liu G, Ren D (2008) Activation of MAPK kinase 9 induces ethylene and camalexin biosynthesis and enhances sensitivity to salt stress in *Arabidopsis*. J Biol Chem 283:26996–27006

Yang L, Ji W, Gao P, Li Y, Cai H, Bai X, Chen Q, Zhu Y (2012) GsAPK, an ABA-activated and calcium-independent snrk2-type kinase from G. soja, mediates the regulation of plant tolerance to salinity and ABA stress. PLoS One 7:e33838

Yeh CM, Chien P, Huang H (2007) Distinct signalling pathways for induction of MAP kinase activities by cadmium and copper in rice roots. J Exp Bot 58:659–671

Yu L, Nie J, Cao C, Jin Y, Yan M, Wang F, Liu J, Xiao Y, Liang Y, Zhang W (2010) Phosphatidic acid mediates salt stress response by regulation of MPK6 in *Arabidopsis thaliana*. New Phytol 188(3):762–773

Zhang S, Klessig DF (1997) Salicylic acid activates a 48-kD MAP kinase in tobacco. Plant Cell 9:809–824

Zhang AY, Jiang MY, Zhang JH, Tan MP, Hu XL (2006) Mitogen-activated protein kinase is involved in abscisic acid- induced antioxidant defense and acts downstream of reactive oxygen species production in leaves of maize plants. Plant Physiol 141:475–487

Zhang X, Dai Y, Xiong DFC, Li J, Dong X, Mou Z (2007) Overexpression of Arabidopsis MAP kinase kinase 7 leads to activation of plant basal and systemic acquired resistance. Plant J 52:1066–1079

Zhang J, Li W, Xiang T, Liu Z, Laluk K, Ding X, Zou Y, Gao M, Zhang X, Chen S, Mengiste T, Zhang Y, Zhou J-M (2010) Receptor-like cytoplasmic kinases integrate signaling from multiple plant immune receptors and are targeted by a *Pseudomonas syringae* effector. Cell Host Microbe 7:290–301

Zhu JK (2001) Plant salt tolerance. Trends Plant Sci 6:66–71

Zhu JK (2002) Salt and drought stress signal transduction in plants. Ann Rev Plant Biol 53:247–273

Zhu JK (2003) Regulation of ion homeostasis under salt stress. Curr Opin Plant Biol 6:441–445

Zong XJ, Li DP, Gu LK, Li DQ, Liu LX, Hu XL (2009) Abscisic acid and hydrogen peroxide induce a novel maize group C MAP kinase gene, *ZmMPK7*, which is responsible for the removal of reactive oxygen species. Planta 229:485–495

Chapter 8
ABA: Role in Plant Signaling Under Salt Stress

Agnieszka Waśkiewicz, Monika Beszterda, and Piotr Goliński

Abstract Salt stress in soil and water is one of the primary abiotic stresses which limit plant growth and productivity, especially in arid and semi-arid regions. Salinity is responsible for other stresses such as ion toxicity, and nutrient imbalances. During the development of salt stress within the plant, all the major processes such as photosynthesis, protein synthesis, energy and lipid metabolisms are affected. In terms of salinity tolerance, plants are classified as halophytes, which can grow and reproduce under high salinity (>400 mM NaCl), and glycophytes, which cannot survive high salinity. Most of the grain crops and vegetables like bean, eggplant, corn, potato and sugarcane are natrophobic (glycophytes) and are highly susceptible to soil salinity.

Among physiological responses to abiotic stresses, the plant hormone abscisic acid (ABA) – a sesquiterpenoid with one (C-1) asymmetric carbon plays an important role. The accumulation of ABA in response to water or salt stress is a cell signaling process, encompassing initial stress signal perception, cellular signal transduction and regulation of expression of genes encoding key enzymes in ABA biosynthesis and catabolism. This phytohormone plays a dual roles in its physiological regulation. It exhibits inhibitive functions when it is accumulated in large amount under stress to help plant survival through inhibition of processes such as stomatal opening and plant size expansion. Moreover ABA is involved in the inhibition of ethylene production, which is a growth inhibitor under stress. At low concentration it exhibits promoting influence while at 'normal' conditions, the metabolite has been shown essential for vegetative growth in several organs, e.g., primary root growth and post-germination seedling development. Also, in seeds ABA modulates the biosynthesis of storage components such as lipids and proteins. The amount of active ABA can be regulated by synthesis, conjugation and catabolism. The present review will throw light on role of ABA in signal transduction under salt stress.

A. Waśkiewicz • M. Beszterda • P. Goliński (✉)
Department of Chemistry, Poznań University of Life Sciences, Wojska Polskiego 75, 60-625 Poznań, Poland
e-mail: piotrg@up.poznan.pl

P. Ahmad et al. (eds.), *Salt Stress in Plants: Signalling, Omics and Adaptations*,
DOI 10.1007/978-1-4614-6108-1_8,

Keywords Salt stress • ABA • Signaling • Halophytes • Glycophytes

Under optimal conditions plants grow and reproduce, but often face a changing environment that can cause unfavorable conditions. In such an environment plants are considered to be "stressed", it prevents them from expressing their full ability to reproduce. The consequence of stress in plants can vary from impeded growth to death. Among the various environmental stresses (like drought, salinity, high and low temperature and ultraviolet radiation stress) salt stress is one of the major abiotic stresses reducing the productivity and quality of agricultural crops, with unfavourable influence on germination, plant homeostasis and crop yield, especially in arid and semi-arid regions (Koca et al. 2007; Munns and Tester 2008; Ahmad and Sharma 2008).

Salinization of agricultural areas might caused negative global effects and currently includes over 6% of the world's land and 20% of the world's irrigated land (Saeedipour 2011). In such soils salt contents (NaCl) usually exceed 40 mM, and much higher levels are often found (Munns 2005), leading to reduce in water potential affecting to the water availability (Hasegawa et al. 2000). Salt stress is among the most harmful abiotic stresses affecting current crops. Salinity occurs frequently in areas where soils are naturally rich in salts and/or when significant amounts of water are delivered by irrigation, with no adequate provision of drainage for the leaching and deletion of salts, resulting in the soils becoming salty and unproductive (Zhang et al. 2006).

Salt Stress Damage to Plants

In plants under development of salt stress, all the major processes, such as photosynthesis, energy and lipid metabolisms, protein synthesis, are impaired (Anuradha and Rao 2001; Babu et al. 2012; Bazihizina et al. 2012). Salt stress leads to the membrane leakage, ion imbalance, enhanced lipid peroxidation and an increased of reactive oxygen species (ROS) production, like superoxide radicals ($^{\bullet}O_2^-$), hydrogen peroxide (H_2O_2) and hydroxy radicals ($^{\bullet}OH$) scavenged by enzymatic and non-enzymatic reactions (Roychoudury et al. 2008). Salinity is unfavorable for the plant growth and development, inhibiting seed germination (Ungar 1996; Dash and Panda 2001), seedling growth (Ashraf et al. 2002; Katembe et al. 1998), enzyme activity (Seckin et al. 2009), DNA, RNA and mitosis (Tabur and Demir 2010). Soil salinity imposes ion toxicity (mainly due to Na^+, Cl^- and SO_4^{2-}), osmotic stress and nutrient (N, Ca, K, P, Fe, Zn) deficiency (Zhu 2003). Salinity also indirectly limits plant productivity through its adverse effects on the growth of beneficial and symbiotic microbes (Munns 2002).

The mechanisms of tolerance to osmotic and ionic components of salinity stress are controlled at the cellular, organ, and whole-plant levels. There are three kinds of plant adaptations under high salt content: osmotic stress tolerance, Na^+ or Cl^- exclusion, and the tolerance of tissue to accumulated Na^+ or Cl^- (Munns and Tester 2008). At high concentration of Na^+ is disrupt the uptake of other nutrients by Na^+ interfering with transporters in the root plasma membrane, such as K^+-selective ion channels, and reduction in root growth (Tester and Davenport 2003).

Effect of salinity on plant growth is a two phase process. The first stage is a rapid, osmotic phase inhibiting growth of young leaves and the second is a slower, ionic phase accelerating senescence of mature leaves. Na^+-specific destruction is associated with the accumulation of Na^+ in leaf tissues and causes necrosis of older leaves (Parvaiz and Satyawati 2008). Also leaves are more vulnerable than roots to Na^+ for this reason that Na^+ and Cl^- are accumulated at higher levels in shoots than in roots (Tester and Davenport 2003). Although Na^+ is transported to shoots through the xylem, it can only return to roots *via* the phloem suggesting that Na^+ transport is strongly one-way and results in accumulation of Na^+ in leaves (Tester and Davenport 2003).

Plants can be classified according to the tolerance to salinity as halophytes and glycophytes. Halophytes can grow and reproduce under high salinity, where as glycophytes cannot tolerate average and high levels of salt (Sairam and Tyagi 2004; Sulian et al. 2012). Halophytes, plants having the ability to exist in high salinity conditions (about 200 mM NaCl or more), constitute about 1% of the world's flora. Tolerance of the photosynthetic apparatus of halophytes under salt stress is brought about ranging from biochemical adaptations to specialized morphologies (Lovelock and Ball 2002; Rajaravindran and Natarajan 2012; Shabala et al. 2012).

All halophytes need to regulate cellular Na^+, Cl^- and K^+ contents consisting in a controlled uptake and compartmentalization of ions, as they adaptation to the external water potential. There are some differences between halophytic monocotyledonae and dicotyledonae for their adaptation to the various salinity conditions. In dicotyledonae plants the growth rate, internal Na^+ to K^+ ratio and water content are higher at increased salinity conditions. Meanwhile monocotyledonous plants exhibit more effectively water use, which is associated with the growth meristem position in developing leaves (Flowers and Colmer 2008).

Abscisic Acid: Chemical Structure, Properties and Its Role in Plants Under Salt Stress

Plant hormones are classified as specific organic substances that at very low concentration acting on target tissues as regulators of growth and development (Kaya et al. 2009). These include abscisic acid, jasmonic acid, salicylic acid and ethylene modifying the physiological reaction of plants exposed to salt stress. The multiplicity of responses is an important aspect of the complexity of stress signaling. Abscisic acid (ABA) – a lipophilic plant hormone – is ubiquitous in lower and higher plants and participating in the complex processes throughout the lifecycle of plants (Javid et al. 2011).

Chemical Structure

Upon discovery in the early 1960s the chemical structure of abscisic acid, to identify the mechanisms of action and specific receptors binding ABA, numerous scientific institutions synthesize in laboratory and industrial scales the chemicals on the structure and properties similar to ABA-precursor and ABA-metabolites. These derived chemical compounds have agricultural and horticultural use as plant growth regulators. Likewise, the phenomenon of ABA and ABA analogs altering ABA metabolism can confound bioassay results and should be considered in biochemical investigations.

Abscisic acid (5-(1-hydroxy-2,6,6-trimethyl-4-oxocyclohex-2-en-1-yl)-3-methylpenta-2,4-dienoic acid) is a sesquiterpenoid ($C_{15}H_{20}O_4$), with one asymmetric, optically active carbon atom in position 1′ (Pilet 1998). In plants there is *S*-ABA and *R*-ABA – the mirror-image form has not been reported in nature (Zaharia et al. 2005). The side chain of the *S*-(+)-ABA structure is by definition 2-*cis*,4-*trans* (Fig. 8.1a). The 2-*trans*,4-*trans* form of ABA is biologically inactive and occurs in small amounts (Fig. 8.1b). Plants synthesize *S*-ABA by 2-C-methyl-D-erythritol-4-phosphate (MEP) and the carotenoid pathway and the precursor xanthoxal (Fig. 8.1f). In physiological studies the effects of *R*-ABA and racemic ABA (which is a 1:1 mixture of the mirror-image forms) need not be similar to the effects of *S*-ABA. Also the impact of the two forms can be various in inducing gene expression and physiological responses. *R*-ABA shows a low ABA-activity in stomatal aperture assays and in inducing expression of genes participating in lipid synthesis (Zaharia et al. 2005). Besides, the two forms are metabolized by plant enzymes to various products, what is important, because some of the *S*-ABA catabolites also have hormonal activity (Zhou et al. 2004). Delivered *R*-ABA can also stimulate the produce of *S*-ABA in plant tissues, was investigated when deuterated *R*-ABA was supplied to *Marsilea quadrifolia* plants, and *S*-ABA and its metabolites levels increased (Lin et al. 2005). Phaseic acid (Fig. 8.1c), dihydrophaseic acid (Fig. 8.1d) and 4′-desoxy-abscisic acid (Fig. 8.1e) are three metabolites of ABA. Phaseic acid – an inactive product of ABA metabolism – is applied as a negative control in biochemical investigations (Todoroki et al. 1995). There are several metabolic pathways, by which ABA can be removed or degraded in plant tissues as a means of further regulating ABA concentrations. The principal oxidation pathway is through hydroxylation of the 8′-carbon atom of ABA, mediated by a cytochrome P450 (Finkelstein and Rock 2002).

Properties of ABA

In the early phase of a plant's life, the hormone abscisic acid regulates seed maturation and the maintenance of embryo dormancy (Himmelbach et al. 1998). Later, during ontogenesis, it acts as a negative growth regulator in order to adapt plants to

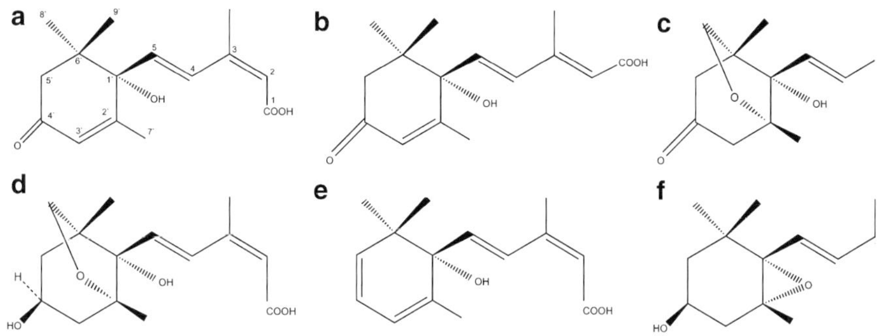

Fig. 8.1 Chemical structures of abscisic acid and related compounds: (**a**) *S*(+)-ABA, (**b**) *trans*-ABA, (**c**) phaseic acid, (**d**) dihydrophaseic acid, (**e**) 4′-desoxy-ABA, (**f**) (−)-xanthoxal. (Data from Pilet (1998))

adverse environmental conditions like drought, cold, pathogen infection, and salinity (Ueno 1998; Albinsky et al. 1999; Mahajan and Tuteja 2005; Tao et al. 2011; Saeedipour 2012; Zhang et al. 2012). The accumulation of abscisic acid under salinity is a cell signaling process, including initial stress signal perception, cellular signal transduction and regulation of genes expression encoding key enzymes in ABA biosynthesis and catabolism (Zhang et al. 2006). This phytohormone plays dual roles in physiological regulation. It exhibits inhibitive functions when it is accumulated in large amounts under salinity to protect plant development through the inhibition of processes such as stomatal opening and plant size expansion (Mauch-Mani and Mauch 2005; Rai et al. 2011). Moreover, ABA participates in the inhibition of ethylene production – a growth inhibitor under stress. At low concentrations it exhibits a promoting influence, while under ‘normal’ conditions the metabolite has been demonstrated to be crucial for vegetative growth in several tissues, such as root growth and post-germination seedling development. What is more, in seeds ABA modulates the biosynthesis of storage components such as lipids and proteins. The amount of active ABA may be regulated by synthesis, conjugation and hydroxylation. A lot of aspects of ABA metabolism may be attend in a homeostasis mechanism that prevents excess ABA accumulation and matching ABA concentration to the kind and the severity of the stress in exposed plants (Verslues and Bray 2006).

Role of Abscisic Acid Signaling Under Salt Stress in Various Plants

Halophytes

Abscisic acid plays a key protection role in all halophyte species under salt stress conditions. *Atriplex leucoclada*, *Suaeda fruticosa* and *Salicornica verginica* exhibit higher concentrations of ABA under salt stress in comparison with indole acetic acid (IAA), which plays a major role in regulating plant growth (Bano and Bano 2011). In a Mediterranean xero-halophyte *Atriplex halimus* exogenous ABA at 50 μM improved osmotic stress resistance in seedlings from a coastal salt-resistant population, through an improvement in the efficiency of stomatal conductance regulation. Abscisic acid also enhanced NaCl resistance in seedlings from an inland water-stress-resistant population, through growth in sodium excretion through the external bladders (Hassine et al. 2009). In another halophyte *Salicornia bigelovii* under salinity stress from 0.005 to 500 mM NaCl abscisic acid concentration in the shoots was negatively related to growth and water content, which suggests that ABA induced by water deficit may inhibit growth at both the suboptimal and supraoptimal salinities (Ohori and Fujiyama 2011). Yoshida et al. (2003) indicated that exogenously added ABA plays a certain physiological role in *Chlamydomonas reinhardtii* to increase growth in this alga.

Salinity stress is a major factor inducing carotenogenesis in some species of halotolerant green algae *Dunaliella* and also induces *Dunaliella* to produce high levels of ABA. There has been little information concerning the mechanism through which ABA responds to environmental stress in algae. Levels of endogenous ABA were examined in two *Dunaliella* species with different capacity of *β*-carotene synthesis during salt stress (*Dunaliella salina* and *D. viridis*, as higher and lower *β*-carotene accumulator species, respectively) in order to enable appropriate comparisons (Sarmad et al. 2007). In controls (1.5 M NaCl) of both species, low levels of endogenous ABA remained nearly constant, whereas ABA levels in stressed cells (3.5 M NaCl) increased 2–3 fold after 6 h of hypersaline conditions and reached 11 and 65 ng/mg DW for *D. salina* and *D. viridis*, respectively. These data suggests that *D. salina* (containing higher amounts of *β*-carotene) responds to salt stress conditions by producing significantly higher amounts of ABA when compared to *D. viridis*.

Glycophytes

Most agriculturally important crops are glycophytes what means that cannot tolerate salt stress. They have developed mechanisms that allow them to monitor the incoming stresses and rapidly regulate their physiology and metabolism to deal with them (Sairam and Tyagi 2004). The root and shoot growth reduces abruptly in glycophytes and this effect need not depend on salt levels in the growing tissues (Parvaiz and Satyawati 2008). In contrast to stress-induced ABA accumulation, there is a little information on the role and function of ABA in plant growth and development under unstressed conditions (Nambara and Marion-Poll 2005).

Salinity damage in wheat and other kinds of cereals is commonly due to excessive sodium and chloride uptake by plants. Among cereals, wheat is considered to be moderately resistant to salt stress. The effects of salinity at the seedling stage of wheat ranges from reduction in germination rate, fresh and dry weight of shoots and roots to the uptake of different nutrient ions (Javid et al. 2011). It is thought that the negative effect of salinity on germination could be associated with the reduction of endogenous hormones levels (Debez et al. 2001). Afzal et al. (2006) investigated an alleviation of salt stress in spring wheat by hormonal priming with abscisic acid (10, 30 and 50 ppm), salicylic acid (50, 100 and 200 ppm) and ascorbic acid (50, 100 and 200 ppm), which play a key role in plant responses under salinity. In this experiment salt stress caused both a significant reduction in germination rates and seedling growth in fresh and dry weights of roots and shoots in wheat. In turn, hormonal priming with ascorbic acid and salicylic acid (50 ppm each of them) increase the capacity of wheat to cope with salinity, whereas abscisic acid was not efficient in inducing salt tolerance. Similar studies concerning the effect of abscisic acid (10^{-5} M for 24 h) on growth and ion accumulation in two spring wheat cultivars under salinity conditions were conducted by Gurmani et al. (2007). ABA-treated plants showed a significant decrease in Na^+ concentration, but an increased K^+ concentration in flag leaf and a reduced plant height, but an increased number of kernels per spike and crop yield of both wheat cultivars.

Nitric oxide (NO) is a short-life very bioactive chemical involved in many processes and diverse biological pathway, which induced increase in plant tolerance or resistance against various abiotic and biotic stresses, like pathogen attack (Delledonne et al. 1998), drought (Gracia-Mata and Lamatina 2001), salt stress (Zhang et al. 2007; Ruan et al. 2002, 2004a; Uchida et al. 2002) and UV-radiation (Mackerness et al. 2001). Studies demonstrated that NO stimulated the production of proline, the – probable – protective action of NO against salt-induced oxidative damage to wheat seedlings. Ruan et al. (2004b) investigated the relationship of NO and ABA in proline accumulation under salinity. It was shown that nitric oxide might activate the synthesis of endogenous ABA and the proline content was significantly higher by exogenously supplemented abscisic acid in wheat seedling leaves under 150 mmol/L NaCl salt stress after 4 days of treatment.

Comparing tolerant and sensitive cultivars within the same species, some reports have shown higher ABA levels in tolerant cultivars (Bravo et al. 1998; Moons et al. 1995; Chen et al. 2002). In physiological studies, characteristics such as low shoot Na^+ concentrations, tolerance of leaves to salinity, compartmentation of salt in older rather than younger leaves can affect to deal better with salt conditions. Some of the most important environmental factors limiting rice yields include soil salinization and sodification during seedling and reproductive phases. The magnitude of the salt effect in rice depends on the duration of salt treatment and the age of plants. Rice (*Oryza sativa* L.) is moderately sensitive to soil salinity. A long-term effect of growing rice in saline soils is that excess salts accumulate in leaves. Saeedipour (2011) tested two cultivars (sensitive and tolerant) of indica rice under salinity conditions (0 and 100 mM NaCl) with respect to the non-stressed controls. The tolerant variety may have a higher ability to produce ABA than the sensitive cultivar and an increased ABA content in leaves of the tolerant variety plays a positive role in reducing stress effects. In other studies the effectiveness of ABA in alleviating the toxic effects of salt stress (0, 50 and 75 mM NaCl) in rice was demonstrated by Gurmani et al. (2011). It was found that ABA was the most efficient plant growth regulator in decreasing Na^+ and Cl^- contents and Na^+/K^+ ratio, increasing K^+ and Ca^{2+} concentrations, proline accumulation, soluble sugar content and grain yield.

In turn, Bano (2010) monitored variability in ABA content in xylem sap and leaves in three rice cultivars (salt-sensitive, moderately tolerant and tolerant) under salinity conditions (1.2 dS/m NaCl). The concentration of xylem ABA increased significantly, reaching a peak in the salt-sensitive cultivar much earlier (24 h of salt treatment) than in other cultivars. The ABA accumulation was delayed and the magnitude of ABA accumulation was greater in both moderately tolerant and tolerant cultivars. The xylem flux of ABA followed a similar pattern. On exposure to salt stress the free ABA content in leaves was increased in all the varieties, but the magnitude of this increase was greater (about two times of the basal level) in the salt-sensitive cultivar at 48 h after salt treatment. The basal levels of ABA and cytokinin appear to play an important role in determining the response of a variety to salt stress.

The molecular and physiological effects of jasmonic acid (1, 3, 5, 9, 10, 20 and 40 μM) and abscisic acid (1, 3, 5, 9, 10, 20 and 40 μM) in roots of rice under salt conditions (60, 80, 100, 125, 150, 175 and 210 mM NaCl) were investigated by Moons et al. (1997). Endogenous root levels of ABA and methyl jasmonate (methyl ester of jasmonic acid) showed a differential increase with the dose and the duration of salt stress. The presented results demonstrate that ABA and jasmonates antagonistically regulate the expression of salt stress-inducible proteins associated with water deficit or defense responses.

Sorghum – the fifth most important cereal grown worldwide – is well-adapted to semiarid and arid tropics, where salinity is the major problem because of water deficit. The effect of ABA on germination and growth in *Sorghum bicolor* L. Moench seeds under salt stress was investigates by Sharma et al. (2004). Imposition of ABA and NaCl treatments showed decrease in germination rate, a very significant reduction in dry matter content of embryos and a significantly higher acid phosphatase activity.

In soybean (generally considered to be salt-sensitive), salinity stress reduced seed germination and seedling growth, inhibits nodulation, and decreases biomass accumulation and yield (Essa 2002). These effects are induced by osmotically mediated interference with water and nutrient uptake (Brady and Weill 2002). Salinity stress can also lead severe leaf chlorosis, leaf bleaching and necrosis, and ultimately plant death. Hamayun et al. (2010) tested the impact of NaCl (70 and 140 mM) induced salt stress on growth attributes and endogenous levels of abscisic, jasmonic and salicylic acids in soybean. Abscisic acid was found in much higher amounts in soybean as compared to other endogenous plant hormones. It was observed that the amount of ABA followed the growth of plants and the highest ABA contents were found at later stages of soybean growth and development. The ABA contents in leaves significantly increased with the exposure of soybean plants to elevated NaCl stress. Under salt stress, similarly to ABA, the endogenous jasmonic acid content significantly increased, while a significant decrease in salicylic acid content was observed. The results showed that salinity stress drastically reduces growth and yield components of soybean by affecting endogenous growth hormones.

Beans – a good sources of proteins, vitamins and minerals, are major foodstuffs especially in tropical and subtropical areas, due to unfavorable climatic and agronomic conditions facing drought and salinity (Camacho-Barron and Gonzalez de Mejia 1998). Salinity during any stage of beans development is a factor that significantly disrupting the growth productivity. Similar to many other crops, discrepancies in physiological responses on drought and salinity have also been noted among the various species of the genus *Phaseolus* (Lazcano-Ferrat and Lovatt 1999). Hereof its glycophytic nature limits its development in salt stress conditions and enforces the need for selection of more resistant bean varieties that may be recommended for tillage under hostile environmental conditions. ABA levels in two bean species, *Phaseolus vulgaris* L. (sensitive to water scarcity and salt stress) and *P. acutifolius* (tolerant to high temperature and drought) were tested by Yurekli et al. (2004) under different stress conditions (50, 100 and 150 mM NaCl). The abscisic acid concent in the leaves of control plants of *P. acutifolius* and *P. vulgaris* did not change significantly after 24, 48 and 72 h. Whereas, in leaves of salinity-stressed *P. vulgaris* plants, abscisic acid content level at 24, 48 and 72 h after use of the additive all three NaCl levels showed large increases comparison to the control leaves, while there were no significant differences in abscisic acid concentration in leaves of *P. acutifolius* plants. On the basis of experiments it is suggested that leaf development under short-term salt stress action is verified by the water status of the root through a root-derived signal, which in certain plants, might be abscisic acid-mediated (Montero et al. 1997, 1998).

The role of ABA in other adaptive reactions, like leaf growth retardation, remains controversial (Voisin et al. 2006). Although numerous scientific works about the importance of abscisic acid in controlling ion fluxes in guard cells, the subject of abscisin acid participation in the verification of uptake and distribution of toxic ions like sodium is still the topic of many studies. Previous results (Montero et al. 1998; Sibole et al. 1998) indicated that in long-term salt-treated beans (1, 25, 50 and 75 mM NaCl) there is a high positive dependence between leaf sodium and xylem and leaf abscisic acid, however no relationship was found between leaf abscisic acid and leaf chloride ions content. Plants exposed to 75 mM sodium chloride dose demonstrated an increase in sodium leaf level with an accompanying decrease in growth and photosynthesis as salt exposure progressed.

Cabot et al. (2009) investigated the ABA signaling properties for the Na^+ exclusion mechanism of salinity-treated bean plants using various physiological combinations, like addition of Ca^{2+}, increased nutrient strength, various isosmotic variants, appendix of exogenous abscisic acid, fluridone (an abscisic acid inhibitor) and aminooxiacetic acid (an ethylene inhibitor). Sodium chloride and potassium chloride equally increased leaf abscisic acid content and decreased transpiration rates, while the addition of calcium and an increased strength of the nutrient solution reduced leaf abscisic acid and leaf sodium concentration. These results explained a non-ion-specific growth in abscisic acid that presumably signals the osmotic element of salt, and grown abscisic acid contents that effected in higher leaf sodium level due to lower sodium exclusion or grown root to shoot sodium translocation. At present a considerable amount of experimental evidence is available showing that the physiological results induced by salt stress may be regulated by abscisic acid. The effect of salt stress (100 mM NaCl) and various doses of exogenous ABA

(1 and 10 μM) on two bean cultivars different in salinity tolerance were investigated by Khadri et al. (2007). The effect of salt stress on shoot and root dry weights were investigated in the salinity-sensitive cultivar (Coco) and the salinity-resistant cultivar (Africa). The less intense root growth than shoot growth was observed, and the extension of root to shoot ratio with sodium chloride was upper in Coco cultivar. Concerning abscisic acid, a decrease of shoot dry weight was demonstrated in both cultivars and of root dry weight in Africa cultivar with the addition of phytohormone, ipso facto demonstrating that abscisic acid is a growth inhibitor. The phytohormone at 1 μM doses seems to strengthen the response or effectiveness the adaptation of common bean to salinity conditions, at least in the salinity-sensitive cultivar. Previous results from that researcher group indicated that plant dry weight, nodule dry weight, nitrogen fixation, and majority enzymes of the ammonium and ureide metabolism were influenced by abscisic acid and sodium chloride (Khadri et al. 2006).

In other studies the effect of three levels of salt stress (0, 100 and 400 mM NaCl) and four levels of ascorbic acid (0, 25, 50 and 100 mM) on ABA content was investigated in bean plants (Dolatabadian and Jouneghani 2009). The results demonstrated that salinity caused increase of abscisic acid concentration compared to unstressed plants. Ascorbic acid appendix prevented the enhancement of abscisic acid in plants exposed to sodium chloride. Under stress-free conditions no changes were reported in abscisic acid concentration level due to addition of ascorbic acid. Plant hormones like abscisic acid and ethylene may entail the aging processes under salinity terms. Ascorbic acid reduced abscisic acid concentration via lagging of senescence by scavenging of reactive oxygen species as promoters of this process. During development under salt stress conditions, endogenous abscisic acid level grows in different tissues, including nodules. Jebara et al. (2006) analysed the influence of salinity (50 and 100 mM NaCl) and abscisic acid (100 μM) on O_2 uptake by N_2-fixing nodules of bean plants. It was shown that sodium chloride influenced more unfavorable on shoot and nodule growth than on root development. Also, the short-term sodium chloride application (lasting 35 days) was less destroying than a long-term application (lasting 42 days). It was also found, that an external applying of ABA inhibited nodule respiration. Nodule conductance was 30 and 12% lower for ABA and NaCl, respectively, than for the control (4.4 μm/s).

He and Cramer (1996) demonstrated that abscisic acid application on *Brassica* can reinforce salt tolerance and the application of an adequate abscisic acid doses in salinized plants causes the stimulation of tissues development. In this experiment *B. carinata* and *B. napus* were tested under salt stress (8 dS/m) by various exogenous ABA concentrations levels (0, 1, 10, 80 and 200 μM). Around 40% of plant growth with or without salt was retarded by abscisic acid in amount of 1 μg/g DW. Increasing the concentration of endogenous abscisic acid ranging from 1 to 7 μg/g DW, the decrease in plant growth was between 40% and 90%.

The death of plants follows at high abscisic acid concentration (average 24 μg/g DW). In the majority cases stress-induced abscisic acid restrains shoot development; however the response of roots varies, being unaffected, stimulated and inhibited (Lovelli et al. 2012). Depending on the intensity of salinity the inhibition differentiation of shoot and root growth may be positive or negative for salt tolerance. Under low to moderate salt stress conditions, growing relative root mass to shoots may increase the requirements for carbon, resulting in competition for carbon necessary to shoot development, which would eventually lead to a fall in the total growth of salt-stressed plants. However, at high salt concentration conditions ascending root to shoot ratio may enhance water conduction to the shoots, disrupting previously adopted schemes growth. Under these conditions the increasing root mass allows to plants overcome a hard water deficit terms, leads to beneficial consequence salinity tolerance. It is also known that abscisic acid concentration level in plants tissues is significantly associated with degree to which stressor modifies plant water status as reported by fluctuation in turgor and relative water content (Verslues and Bray 2006). At present more is known about ABA-specific effects on the tissues responsible for plants water economy, stomatal conductance and hence transpiration (Freundl et al. 2000; Hose et al. 2000; Yang et al. 2012) in response to abiotic stresses that commonly coincide in the environment, through changes in cell wall extensibility (Dodd and Davies 1996; Thompson et al. 1997; Cramer et al. 1998; Bacon 1999) or apoplastic pH (Bacon et al. 1998). To the efficient operation of ABA in the plants tissues are needed Ca^{2+} stores, activation of mitogen-activated protein kinase and the phosphoinositol cascade affecting on ion channels and changes in protein phosphorylation and gene expression (Xiong et al. 2002; Swiatek et al. 2003). For many years the subject of researches is also the potential role of abscisic acid in the initiation of the plant growth after the physiological shock associated with unfavorable conditions for plant development (Thompson et al. 1997).

Among glycophytes, vegetables and fruit are economically important world crops, but their tillage is effectively limited by their lack of tolerance to salinity. Most experiences on abscisic acid deals with the application a single dose. This procedure may not reflect the prolonged and gradual increase in bioactive substance in plant tissue, what occurs in natural conditions of exposure to various negative environment factors.

The biochemical mechanism of plant response to abiotic stress conditions were analyzed inter alia on the example of potatoes seedlings. In the experiment consisting of abscisic acid application used four potato genotypes with various characteristic responses to salinity: conventional cultivar, ABA-deficient mutant, resistant genotype line 9506, and cultivar 'Norland' of average resistance (Etehadnia et al. 2008). Single or multiple growing doses of abscisic acid at the range of 0, 50, 75, or 100 mM were directly given to root by drench. Application a single dose of abscisic acid caused enhanced vertical plant growth rate under salinity conditions in all genotypes taking part in the experiment, while applicable multiple abscisic acid doses remained stable lateral shoot growth, excluding the ABA-deficient mutant, in whom was not observed upward trend (Etehadnia et al. 2008). Mulholland et al. (2003) examined the effect of varied sodium chloride doses (in the range of 0–120 mM) on biochemical and physiological stability of two genotypes of tomato: wild type (Ailsa Craig) and abscisic acid-deficient mutant (notabilis). There were not noticed changes in the leaves turgor of the analyzed genotypes of tomatoes under controlled salinity conditions, however it was observed an increase ratio between root and shoot weight. More information about importance of the abscisic acid concentration in plants leaves and the root system expanding were obtained by breeding in average concentrations of NaCl terms in relation to extreme doses. Importantly, the decrease of stomatal conductance may inform that a except hydraulic element, some non hydraulic factors may participate in root medium reaction to abiotic stress. Endogenous abscisic acid content were grown in reply to salt stress and statistic analysis revealed occurrence a negative dependence between stomatal conductance and growing xylem abscisic acid concentration for both: Ailsa Craig and notabilis genotypes (Mulholland et al. 2003).

Among these regulated physiological responses, one of the physiological ABA responses is connected with stomatal closure that prevents excessive transpiration. As a consequence, availability of CO_2 for fixation of the C3 cycle was limited, which may, in turn, enhance ROS production and chloroplasts. Despite this well-known link between ABA and ROS, little attention has been drawn to the relationship between these two signaling molecules and their pathways (Bhattacharjee 2005).

Citrus species are very salinity-sensitive crops. Due to the economic impact it evokes citrus crop area reduction by soil salinity, to analyze the effect of different sodium chloride concentration in the ground on sensitive citrus plants biochemical response (Arbona et al. 2006). Gómez-Cadenas et al. (2003) and Mengual et al. (2003) examined ABA content under salinity (100 mM NaCl) on citrus plants, i.e. Salustiana scion (*Citrus sinensis* (L) Osbeck) grafted onto Carrizo citrange (*Citrus sinensis* [L.] Osbeck x *Poncirus trifoliata* [L.] Raf). Plants non treated with growth regulator showed a significant decrease in the photosynthetic activity in response to salt stress, an increase in leaf production of ethylene and a high abscission rate as a result of a massive leaf chloride accumulation. Use of the additive of 10 μM abscisic acid to the nutrient medium 10 days before the threat to salinity decreased ethylene secretion and leaf abscission – commonly induced by the chloride build-up through a mechanism that stimulates leaf aminocyclopropane-1-carboxylic acid synthesis and further conversion to ethylene (Gómez-Cadenas et al. 1998, 2003). These changes were probably the result of a decrease in the content of toxic Cl^- ions in leaves.

To continue, the response of the same plant to 200 mM NaCl was also determined (Gómez-Cadenas et al. 1998). The salt shock reduced stomatal conductance, increased proline and aminocyclopropane-1-carboxylic acid content in roots, xylem fluid and leaves, intensified ethylene production and showed massive leaf abscission (Gómez-Cadenas et al. 1998). As a conclusion of both experiments, it may be conclude that to reduce the negative effects of high NaCl concentrations on young citrus plants can added abscisic acid to the irrigation solution (Mengual et al. 2003).

The effect of abscisic acid in cucumber (*Cucumis sativus* L.) F1 hybrid Zozulya seedlings to combined exposure to salinity and high temperature were examined (Talanova et al. 2006). Seedlings were consecutively or simultaneously exposed to 38°C and 120 or 154 mM NaCl. Prior to the exposure, a fraction of seedlings were placed in ABA solutions of 0.1 or 0.05 mM. Seedlings pretreated with ABA demonstrated a more considerable increase in salt resistance as compared to untreated ones.

Consecutive exposure of seedlings to high temperature and NaCl both with and without ABA temporarily increased their salt resistance. Likewise, cucumber seeds treated with exogenous abscisic acid protected cucumber seedlings from water loss damage and increased their resistance or dehydration tolerance, especially improving their water use efficiency by an improved photosynthesis rate rather than by a reduced water uptake under dehydration (Wang et al. 2010).

In kiwifruit vines (*Actinidia deliciosa* var. *deliciosa*) was shown a positive relationship between ABA concentration and rapid fruit growth what suggests that ABA may have a role in the allocation of assimilates to fruit (Smith et al. 1995). In leaves from grapevines (*Vitis vinifera* L.) subjected to NaCl the endogenous abscisic acid concentration increased for the 50 and 100 mM NaCl doses, while phaseic acid showed kinetics consistent with its being derived from ABA. In another experiment, under determined salinity (0, 50, 100 and 250 mM NaCl), the endogenous ABA content also increased in grape rootstocks of Dogridge, 1613, St. George and Salt Creek, the root to shoot dry mass ratio in all rootstocks grown in case of up to 100 mM sodium chloride and with growing sodium chloride content putrescine, spermine and spermidine concentration demonstrated a consistent increasing trend (Upreti and Murti 2010).

ABA concentrations reaching up to 10 μM and different levels of salt content conditions were summarized in terms of their impact on shoot growth in salinity-sensitive and salinity-tolerant clones of jojoba (*Simmondsia chinensis*) (Mills et al. 2001). A low ABA concentration of 1.0 μM does not induce structural lesions in salinity-tolerant clones, whereas disturbs dry mass synthesis and shoot expansion in sensitive clones. Externally applied ABA at a low concentration stimulated also the accumulation of K^+, Na^+, Ca^{2+} and Cl^- in shoots of salt-sensitive clones. Abscisic acid at 10 μM content level caused an increment of epicuticular wax collecting on leaves of the salinity non resistant clones by two fold, but it had no influence on leaves of the salinity resistant clone. The quoted results confirm the assumption that responses to salt stress in jojoba shoots such as the abrupt interruption of elongation, growing ion concentration and accumulation of epicuticular wax, participates abscisic acid (Mills et al. 2001). Wang et al. (2001) investigated the effects of short- (0, 100, 200 and 400 mM NaCl for 24 and 48 h) and long-term (15, 50 and 100 mM NaCl for 19 months) salinity on endogenous abscisic acid in *Iris hexagona*. The central and apical portions of mature flower stalks contained the highest ABA levels, establishing a basipetal gradient. Therefore, these areas are likely sites of ABA biosynthesis. Salinity also increased ABA contents in both young and mature *I. hexagona* leaves (Wang et al. 2001). For comparison, in *Chrysanthemum morifolium* L. cv. Puritan stressing by irrigation with 100 mM NaCl before harvest reduced following water loss and stomatal aperture, but had no significant effect on ABA concentration in leaves.

In similar studies the effects of short- (24 h, 100 mM NaCl) and long-term (4 weeks, NaCl increased weekly from 100 to 400 mM) salinity on leaf abscisic acid were tested on 1-year-old seedlings of two mangrove species, *Kandelia candel* and *Bruguiera gymnorrhiza* (Li et al. 2009). When compared with *B. gymnorrhiza*, *K. candel* showed a higher capacity to exclude salt under increasing salinity, which may partially may be effects by the marked ABA accumulation over the duration of salt exposure, since ABA limits Na^+ and Cl^- concentrations in leaves. Leaf ABA content in *K. candel* increased significantly at the beginning of salt treatment, but it remained unchanged in *B. gymnorrhiza*, indicating that *K. candel* is more sensitive to soil salinity when compared to *B. gymnorrhiza*. A similar trend was found in a longer-term salinity, in which salt-induced ABA synthesis is more evident in *K. candel*.

The efficiency of abscisic acid analogs as half measures against salt stress in glycophytes is increasingly the subject of experiments. The influence of ABA and its analogs, i.e. 8′-methylene ABA, 8′-acetylene ABA, ABA methyl ester, 8′-methylene ABA methyl ester, and 8′-acetylene ABA methyl ester, was tested on *Citrus sinensis* (L) Osbeck and *Citrus reticulata* Blanco under salt stress conditions (Arbona et al. 2006). Among all, 8′-methylene abscisic acid was characterized by the highest effectiveness in preventing of disruptive effects of salt stress in citrus plants. Application of this analogue in the cultivation of plants exposed to moderate and high concentrations of sodium chloride in the soil, would prevent such a biochemical disorder as a high chlorine content in the leaves, excessive production of ethylene and consequently reduction of the leafage. Similarly, abscisic acid and 8′-acetylene abscisic acid supplementation also decreased salinity stress induced damages in citrus plants, but at a lower intensity. Verified experimentally abscisic acid methyl ester and 8′-C modified analogs do not exhibit a similar protective activity as the earlier analogues (Arbona et al. 2006).

Stimulative role of abscisic acid in relation to intensity polypeptide synthesis in tomato (*Lycopersicon esculentum* Mill. cv. Ailsa Craig) roots under salt stress conditions was analyzed (Chen and Plant 1999). In this experiment endogenous abscisic acid concentration grew twofold in salt-treated Ailsa Craig roots and 14-fold in salt-treated ABA-deficient mutant, *flacca* roots. In Alisa Craig and flacca roots the protein profiles under salinity do not change significantly, which imply that on synthesis of novel polypeptides in roots under salinity conditions do not affect the concentration of endogenous abscisic acid. Under unstressed conditions the exogenous abscisic acid caused the accumulation of few polypeptides that were characteristic of this proceeding, but also a subset of those biosynthesized in salt-treated roots (Chen and Plant 1999). Meanwhile auto- and hetero-grafting was applied to analyze the influence of rootstock and abscisic acid on the expression of the Ca^{2+}-storage protein calreticulin (CR) and salinity tolerance in potato (*Solanum tuberosum*) (Shaterian et al. 2005). External addition of abscisic acid did not cause upper leaf osmotic potential in sensitive on salinity abscisic acid-deficient clone, however not at the ABA-normal sibling line clone. Calreticulin expression seems to participate in abscisic acid-induced salinity resistance and both salinity tolerance and calreticulin expression may be limited by the roots (Shaterian et al. 2005). In another studies related to biochemical functions associated with salt tolerance in citrus plants, a salt-stress-associated protein (Cit-SAP), which is induced by NaCl in *Citrus sinensis* cultured cells and citrus plants (Etrog citron and Cleopatra mandarin), was identified, purified and partially characterized (Holland et al. 1993). Pretreatment with abscisic acid can significantly reduce the membrane damage caused by freezing stress in trifoliate orange (*Poncirus trifoliata* (L.) Raf.) and lemon (*Citrus limonia* Osbeck.) (Yang et al. 2011).

Maintaining of internal stability of biochemical reactions requires regulating or controlling the most important parameters of the internal body environment. Thus, under soil salinity conditions the plant cells trigger the mechanisms of regulating the osmotic pressure. Plants under the influence of stress reduce the osmotic potential by collecting osmolytes – osmotically active chemicals of low molecular weight, such as proline and higher polyamines (Roychoudury et al. 2008; Shafi et al. 2011). Proline is accumulated in many plants at high levels of in response to osmotic stress, and plays an adaptive role during osmotic stress. Plants can monitor the proline content inter alia by transcriptional regulation of key enzymes, such as Δ^1*-pyrroline-carboxylate synthetase 1*, proline dehydrogenase, ornithine amino transferase and Δ^1*-pyrroline-carboxylate dehydrogenase*, and it is entered into a small area in the cells cytoplasm at concentrations that may affect on the process of osmotic adjustment (Sharma and Verslues 2010; Summart et al. 2010). It can operate as an enzyme protectant, strengthening membranes and cellular structures under unfavorable conditions, detoxifies free radicals by creating stable adducts and furthermore influences solubility of different proteins by interacting with their hydrophobic residues (Hong et al. 2000). Proline can be used in the plants as organic nitrogen deposit as a tool to maintain of amino acids and polypeptides synthesis (Sairam and Tyagi 2004). The notability of proline in salt tolerance has been examined in agriculturally important crop plants such as alfalfa, wheat, soybean, rice, beans, and tobacco (Kong et al. 2001; Basu et al. 2002). The regulatory role of the abscisic acid in proline biosynthesis was tested by the expression of the At-P5S and At-P5R genes in an *Arabidopsis thaliana* wild type, in various mutants including abscisic acid-deficient aba1-1, abscisic acid-insensitive abi1-1 and abi2-1under different kind of abiotic stress conditions (Savouré et al. 1997). Results indicate that the endogenous abscisic acid concentration level can influence the proline accumulation upon salinity conditions, implying post-transcriptional verification of proline synthesis in reaction to sodium chloride presence (Savouré et al. 1997). Verslues and Bray (2006) also used an abscisic acid-deficient mutant (aba2-1) to assess the role of phytohormone in low water potential induced proline collecting and osmotic processes of *Arabidopsis thaliana* seedlings. Osmotic adjustment followed irrespectively of abscisic acid accumulation in aba 2–1. Verslues and Bray (2006) also indicated the presence of alternative arrangements of proline homeostasis that retain seedling proline concentration adequately to the stress exposure, even if abscisic acid level is increased more than normal concentration. Abscisic acid influences on proline content and water status were also investigated in two *Helianthus annuus* genotypes (cv. Nantio F1 and cv. Özdemirbey (TR3080)) exhibited to drought and excess water stress. Drought stress and subjected to abscisic acid resulted grow in proline level in relation to non-subject leaves (Ünyayar et al. 2004). Recent studies (Sharma and Verslues 2010) to determine if there are tissue-specific differentiations in gene expression that can operate to fine-tune proline metabolism to meet differentiation metabolic demands.

ABA is related with vulnerability of plants to bacteria, fungi, and oomycetes, but comparatively not much is known about its function in abiotic stress propensity to root pathogens attack. Many observations argue for abscisic acid as a predominant factor in the salt stress-induced predisposition to *Phytophthora* spp. contagion. DiLeo et al. (2010) demonstrated that exogenous ABA can mimic the salinity-induced susceptibility of tomato roots to *Phytophthora* spp. and that endogenous root ABA levels are strongly associated with the onset of salinity-induced susceptibility of tomato to root and crown rot caused by *Phytophthora capsici*. The implications of salinity and abscisic acid on the interplay of tomato (*Lycopersicon esculentum*) with a biotrophic fungus *Oidium neolycopersici* and a necrotrophic fungus *Botrytis cinerea* were tested (Achuo et al. 2006). Basal endogenous abscisic acid contents undermine resistance of tomato to *Oidium neolycopersici* and *Botrytis cinerea*, while increasing the abscisic acid concentration by the addition of exogenous doses of phytohormone, not increased susceptibility to these pathogens. The salinity noticeably decreased contagion of *O. neolycopersici*, but has no influence on contagion of *B. cinerea*, with no clear increase in endogenous abscisic acid concentration. The lower susceptibility to infection of *O. neolycopersici* and *B. cinerea* characterized by the abscisic acid-deficient *sitiens* mutant with respect to the wild type. Exogenous abscisic acid caused in an increased susceptibility of sitiens to both pathogens, but did not enhance the basal susceptibility of wild-type plants (Achuo et al. 2006). Tomato (*Solanum lycopersicum*) *abscisic acid-induced myb1* (*SlAIM1*) RNA interference (RNAi) plants with reduced *SlAIM1* gene expression show an increased susceptibility to the necrotrophic fungus *Botrytis cinerea* and increased sensitivity to salt and oxidative stress (Abu Qamar et al. 2009). When exposed to high root-zone salinity levels, *SlAIM1* RNAi plants accumulate more Na^+, whereas the overexpression lines accumulate less Na^+ relative to wild-type plants, suggesting that *SlAIM1* regulates ion fluxes.

Conclusion and Future Perspective

High concentration of salt imposes on plants stressors inter alia ion toxicity, as a effect of ion penetration in excess of proper compartmentation, and disorder of nutrient balances, as generally seen in the displacement of monovalent ions (K^+ by Na^+). During the progress of salinity within the plant, all the primary biochemical reaction, as photosynthesis, protein synthesis, energy and lipid metabolisms, are disturbed. Salt stress leads to the leak of membrane, ion imbalance or disequilibrium, reinforcement lipid peroxidation and an formation of reactive oxygen species (ROS), such as superoxide radicals ($^{\bullet}O_2^-$), hydrogen peroxide (H_2O_2) and hydroxy radicals ($^{\bullet}OH$), which are scavenged by enzymatic and non-enzymatic reactions. Salinity adversely affects plant growth and development, hindering seed germination, seedling growth, enzyme activity, DNA, RNA, protein synthesis and mitosis. Soil salinity imposes ion toxicity (mainly due to Na^+, $Cl^-$and SO_4^{2-}), osmotic stress and alimentary components (N, Ca, K, P, Fe, Zn) deficit. Salt stress likewise not directly reduces plant productivity through its damaging consequences on the increase of beneficial and symbiotic microbes.

Plant hormones are classified as specific organic substances that at very low concentration acting on target tissues as regulators of growth and development. These include abscisic acid, jasmonic acid, salicylic acid and ethylene modifying the physiological reaction of plants exposed to salt stress. The multiplicity of responses is an important aspect of the complexity of stress signaling. Abscisic acid (ABA) – a lipophilic plant hormone – is ubiquitous in lower and higher plants and participating in the complex processes throughout the lifecycle of plants.

It is easy to understand that improvement of – up to now – achieved results, as well as development of knowledge in the field of role of the plant hormones in salt stress is of prime concern and are impossible to be over-evaluated at present time and at up-dated knowledge level and stage.

References

Abu Qamar S, Luo H, Laluk K, Mickelbart MV, Mengiste T (2009) Crosstalk between biotic and abiotic stress responses in tomato is mediated by the *AIM1* transcription factor. Plant J 58(2):347–360

Achuo EA, Prinsen E, Höfte M (2006) Influence of drought, salt stress and abscisic acid on the resistance of tomato to *Botrytis cinerea* and *Oidium neolycopersici*. Plant Pathol 55:178–186

Afzal I, Basra SMA, Farooq M, Nawaz A (2006) Alleviation of salinity stress in spring wheat by hormonal priming with ABA, salicylic acid and ascorbic acid. Int J Agr Biol 8:23–28

Ahmad P, Sharma S (2008) Salt stress and phyto-biochemical responses of plants. Plant Soil Environ 54:89–99

Albinsky D, Masson JE, Bogucki A, Afsar K, Vass I, Nagy F, Paszkowski J (1999) Plant responses to genotoxic stress are linked to an ABA-salinity signalling pathway. Plant J 17:73–82

Anuradha S, Rao SSR (2001) Effect of brassinosteroids on salinity stress induced inhibition of seed germination and seedling growth of rice (*Oryza sativa* L.). J Plant Growth Regul 33:151–153

Arbona V, López-Climent MF, Mahouachi J, Pérez-Clemente RM, Abrams SR, Gómez-Cadenas A (2006) Use of persistent analogs of abscisic acid as palliatives against salt-stress induced damage in citrus plants. J Plant Growth Regul 25:1–9

Ashraf MY, Sarwar G, Ashraf M, Afaf R, Sattar A (2002) Salinity induced changes in α-amylase activity during germination and early cotton seedling growth. Biol Plantarum 45:589–591

Babu MA, Singh D, Gothandam KM (2012) The effect of salinity on growth, hormones and mineral elements in leaf and fruit of tomato cultivar PKM1. J Anim Plant Sci 22(1):159–164

Bacon MA (1999) The biochemical control of leaf expansion during drought. J Plant Growth Regul 29:101–112

Bacon MA, Wilkinson S, Davies WJ (1998) pH-regulated leaf cell expansion in droughted plants is abscisic acid dependent. Plant Physiol 118:1507–1515

Bano A (2010) Root-to-shoot signal transduction in rice under salt stress. Pak J Bot 42:329–339

Bano S, Bano A (2011) Physiological and biochemical analysis of the selected halophytes of district Mardan, Pakistan. J Biosci Biochem Bioinf 1:239–243

Basu S, Gangopadhyay G, Mukherjee BB (2002) Salt tolerance in rice *in vitro*: Implication of accumulation of Na^+, K^+ and proline. Plant Cell Tiss Org Cult 69:55–64

Bazihizina N, Barrett-Lennard EG, Colmer TD (2012) Plant growth and physiology under heterogeneous salinity. Plant Soil 354:1–19

Bhattacharjee S (2005) Reactive oxygen species and oxidative burst: roles in stress, senescence and signal transduction in plants. Cur Sci 89:1113–1121
Brady NC, Weill RR (2002) The nature and property of soils, 13th edn. Prentice Hall, Upper Saddle River, 960
Bravo LA, Zúñiga GE, Alberdi M, Corcuera LJ (1998) The role of ABA in freezing tolerance and cold acclimation in barley. Physiol Plant 103:17–23
Cabot C, John C, Sibole V, Balcero J, Poschenrieder C (2009) Abscisic acid decreases leaf Na^+ exclusion in salt-treated *Phaseolus vulgaris* L. J Plant Growth Regul 28:187–192
Camacho-Barron M, Gonzalez de Mejia E (1998) Comparative study of enzymes related to proline metabolism in tepary bean (*Phaseolus acutifolius*) and common bean (*Phaseolus vulgaris*) under drought and irrigated conditions, and various urea concentrations. Plant Foods Human Nutr 52:119–132
Chen CCS, Plant AL (1999) Salt-induced protein synthesis in tomato roots: the role of ABA. J Exp Bot 50:677–687
Chen W, Provart NJ, Glazebrook J, Katagiri F, Chang HS, Eulgem T, Mauch F, Luan S, Zou G, Whitham SA, Budworth PR, Tao Y, Xie Z, Chen X, Lam S, Kreps JA, Harper JF, Si-Ammour A, Mauch-Mani B, Heinlein M, Kobayashi K, Hohn T, Dangl JL, Wang X, Zhu T (2002) Expression profile matrix of arabidopsis transcription factor genes suggests their putative functions in response to environmental stresses. Plant Cell 14:559–574
Cramer GR, Krishnan K, Abrams SR (1998) Kinetics of maize leaf elongation. IV. Effects of (+)- and (−)-absisic acid. J Exp Bot 49:191–198
Dash M, Panda SK (2001) Salt stress induced changes in growth and enzyme activities in germinating *Phaseolus muingo* seeds. Biol Plantarum 44:587–589
Debez A, Chaibi W, Bouzid S (2001) Effect du NaCl et de regulatoeurs de croissance sur la germination d' (*Atriplex halimus* L.). Cahiers Agricultures 10:135–138
Delledonne M, Xia Y, Dixon RA, Lamb C (1998) Nitric oxide functions as a signal in plant disease resistance. Nature 394:585–588
DiLeo MV, Pye MF, Roubtsova TV, Duniway JM, MacDonald JD, Rizzo DM, Bostock RM (2010) Abscisic acid in salt stress predisposition to phytophthora root and crown rot in tomato and *Chrysanthemum*. Plant Stress and Abiotic Disord 100:871–879
Dodd IC, Davies WJ (1996) The relationship between leaf growth and ABA accumulation in the grass leaf elongation zone. Plant Cell Environ 19:1047–1056
Dolatabadian A, Jouneghani S (2009) Impact of exogenous ascorbic acid on antioxidant activity and some physiological traits of common bean subjected to salinity stress. Not Bot Hort Agrobot Cluj 37:165–172
Essa TA (2002) Effect of salinity stress on growth and nutrient composition of three soybean (*Glycine max* (L.) Merrill) cultivars. J Agr Crop Sci 188:86–93
Etehadnia M, Waterer DR, Tanino KK (2008) The method of ABA application affects salt stress responses in resistant and sensitive potato lines. J Plant Growth Regul 27:331–341
Finkelstein RR, Rock CD (2002) Abscisic acid biosynthesis and response. In: Somerville CR, Meyerowitz EM (eds) The Arabidopsis book. American Society of Plant Biologists, Rockville, pp 1–52
Flowers TJ, Colmer TD (2008) Salinity tolerance in halophytes. New Phytol 179:945–963
Freundl E, Steudle E, Hartung W (2000) Apoplastic transport of abscisic acid through roots of maize: effect of the exodermis. Planta 210:222–231
Gómez-Cadenas A, Tadeo FR, Primo-Millo E, Talon M (1998) Involvement of abscisic acid and ethylene in the responses of citrus seedlings to salt shock. Physiol Plant 103:475–484
Gómez-Cadenas A, Arbona V, Jacas J, Primo-Millo E, Talon M (2003) Abscisic acid reduces leaf abscission and increases salt tolerance in citrus plants. J Plant Growth Regul 21:234–240
Gracia-Mata C, Lamatina L (2001) Nitric oxide induces stomatal closure and enhances the adaptive plant responses against drought stress. Plant Physiol 126:1196–1204
Gurmani AR, Bano A, Salim M (2007) Effect of abscisic acid and benzyladenine on growth and ion accumulation of wheat under salinity stress. J Bot 39:141–149

Gurmani AR, Bano A, Khan SU, Din J, Zhang JL (2011) Alleviation of salt stress by seed treatment with abscisic acid (ABA), 6-benzylaminopurine (BA) and chlormequat chloride (CCC) optimizes ion and organic matter accumulation and increases yield of rice (*Oryza sativa* L.). Austr J Crop Sci 5:1278–1285

Hamayun M, Khan SA, Khan AL, Shinwari ZK, Hussain J, Sohn E-Y, Kang S-M, Kim Y-H, Khan MA, Lee I-J (2010) Effect of salt stress on growth hormones of soybean cultivar Hwangkeumkong. Pak J Bot 42:3103–3112

Hasegawa PM, Bressan RA, Zhu JK, Bohnert HJ (2000) Plant cellular and molecular responses to high salinity. Ann Rev Plant Physiol Plant Mol Biol 51:463–499

Hassine AB, Ghanem ME, Bouzid S, Lutts S (2009) Abscisic acid has contrasting effects on salt excretion and polyamine concentrations of an inland and a coastal population of the Mediterranean xero-halophyte species *Atriplex halimus*. Ann Bot 104:925–936

He T, Cramer GR (1996) Abscisic acid concentrations are correlated with leaf area reductions in two salt-stressed rapidcycling *Brassica* species. Plant and Soil 179:25–33

Himmelbach A, Iten M, Grill E (1998) Signalling of abscisic acid to regulate plant growth. Philos Trans R Soc Lond B 353:1439–1444

Holland D, Ben-Hayyim G, Faltin Z, Camoin L, Strosberg AD, Eshdat Y (1993) Molecular characterization of salt-stress-associated protein in citrus: protein and cDNA sequence homology to mammalian glutathione peroxidases. Plant Mol Biol 21:923–927

Hong ZL, Lakkineni K, Zhang ZM, Verma DPS (2000) Removal of feed back inhibition of DELTA-1-pyrroline-5-carboxylate synthetase results in increased proline accumulation and protection of plants from osmotic stress. Plant Physiol 122:1129–1136

Hose E, Steudle E, Hartung W (2000) Abscisic acid and hydraulic conductivity of maize roots: a study using cell- and root-pressure probes. Planta 211:874–882

Javid MG, Sorooshzadeh A, Moradi F, Sanavy SAMM, Allahdadi I (2011) The role of phytohormones in alleviating salt stress in crop plants. Austr J Crop Sci 5:726–734

Jebara M, Harzalli-Jebara AL, Payre H, Aouani ME, Drevon JJ (2006) Influence of salinity and abscisic acid on the O_2 uptake by N_2-fixing nodules of common bean. Biol Plantarum 50:717–721

Katembe WJ, Ungar IA, Mitchell JP (1998) Effect of salinity on germination and seedling growth of two Atriplex species (Chenopodiaceae). Ann Bot 82:167–175

Kaya C, Tuna AL, Yokas I (2009) The role of plant hormones in plants under salinity stress. In: Ashraf M, Ozturk M, Athar HR (eds) Salinity and water stress: improving crop efficiency. Springer, Berlin, pp 45–50

Khadri M, Tejera NA, Lluch C (2006) Alleviation of salt stress in common bean (*Phaseolus vulgaris*) by exogenous abscisic acid supply. J Plant Growth Regul 25:110–119

Khadri M, Tejera NA, Lluch C (2007) Sodium chloride–ABA interaction in two common bean (*Phaseolus vulgaris*) cultivars differing in salinity tolerance. Environ Exp Bot 60:211–218

Koca M, Bor M, Ozdemir F, Turkan I (2007) The effect of salt stress on lipid peroxidation, antioxidative enzymes and proline content of sesame cultivars. Environ Exp Bot 60:344–351

Kong Y, Zhou G, Wang Y (2001) Physiological characteristics and alternative respiratory pathway under stress in two wheat cultivars differing in salt tolerance. Russ J Plant Physiol 48:595–600

Lazcano-Ferrat I, Lovatt CJ (1999) Relationship between relative water content, nitrogen pools, and growth of *Phaseolus vulgaris* L. and *P. acutifolius* A. Gray during water deficit. Crop Sci 39:467–475

Li C, Fang B, Yang C, Shi D, Wang D (2009) Effects of various salt-alkaline mixed stresses on the state of mineral elements in nutrient solutions and the growth of alkali resistant halophyte *Chloris virgata*. J Plant Nutr 32:1137–1147

Lin B-J, Wang H-J, Wang J-S, Zaharia LI, Abrams SR (2005) Abscisic acid regulation of heterophylly in *Marsilea guadrifolia*: Effects of R-(−) and S-(+) isomers. J Exp Bot 56: 2935–2948

Lovelli S, Scopa A, Perniola M, Tommaso TD, Sofo A (2012) Abscisic acid root and leaf concentration in relation to biomass partitioning in salinized tomato plants. J Plant Physiol 169(3):226–233

Lovelock C, Ball M (2002) Influence of salinity on photosynthesis of halophytes. In: Lauchli A, Luttge U (eds) Salinity: environment – plants – molecules, Kluwer Academic Publishers, Dordrecht, pp. 315–339
Mackerness SAH, John CF, Jordan B, Thomas B (2001) Early signaling components in ultraviolet-B responses: distinct roles for different reactive oxygen species and nitric oxide. FEBS Lett 489:237–242
Mahajan S, Tuteja N (2005) Cold, salinity and drought stresses: an overview. Arch Biochem Biophys 444:139–158
Mauch-Mani B, Mauch F (2005) The role of abscisic acid in plant-pathogen interactions. Curr Opin Plant Biol 8:409–414
Mengual VA, Serra MLF, Marín PE, Casanova AJM, Miret JAJ, Cadenas AG (2003) Influence of abscisic acid and other plant growth regulators on citrus defence mechanism to salt stress. Spanish J Agr Res 1:59–65
Mills D, Zhang G, Benzioni A (2001) Effect of different salts and of ABA on growth and mineral uptake in Jojoba shoots grown *in vitro*. J Plant Physiol 158:1031–1039
Montero E, Cabot C, Barcelo J, Poschenrieder C (1997) Endogenous abscisic acid levels are linked to decreased growth of bush bean treated NaCl. Physiol Plant 101:17–22
Montero E, Cabot C, Poschenrieder C, Barcelo J (1998) Relative importance of osmotic stress and ion-specific effects on ABA-mediated inhibition of leaf expansion growth in *Phaseolus vulgaris*. Plant Cell Environ 21:54–62
Moons A, Bauw G, Prinsen E, Van Montagu M, Van Der Straeten D (1995) Molecular and physiological responses to abscisic acid and salt in roots of salt-sensitive and salt-tolerant indica rice varieties. Plant Physiol 107:177–186
Moons A, Prinsen E, Bauw G, Van Montagu M (1997) Antagonistic effects of abscisic acid and jasmonates on salt stress inducible transcripts in rice roots. Plant Cell 9:2243–2259
Mulholland BJ, Taylor IB, Jackson AC, Thompson AJ (2003) Can ABA mediate responses of salinity stressed tomato. Environ Exp Bot 50:17–28
Munns R (2002) Comparative physiology of salt and water stress. Plant Cell Environ 25:239–250
Munns R (2005) Gene and salt tolerance: bringing them together. New Phytol 167:645–663
Munns R, Tester M (2008) Mechanisms of salinity tolerance. Ann Rev Plant Biol 59:651–681
Nambara E, Marion-Poll A (2005) Abscisic acid biosynthesis and catabolism. Ann Rev Plant Biol 56:165–185
Ohori T, Fujiyama H (2011) Water deficit and abscisic acid production of *Salicornia bigelovii* under salinity stress. Soil Sci Plant Nutr 57:566–572
Parvaiz A, Satyawati S (2008) Salt stress and phyto-biochemical responses of plants – a review. Plant Soil Environ 54:89–99
Pilet PE (1998) Some cellular and molecular properties of abscisic acid: its particular involvement in growing plant roots. Cell Mol Life Sci 54:851–865
Rai M, Harish NS, Gupta A, Ram MPK, Jaiswal U (2011) The role of abscisic acid in plant tissue culture: a review of recent progress. Plant Cell Tiss Org 106:179–190
Rajaravindran M, Natarajan S (2012) Effects of salinity stress on growth and antioxidant enzymes of the halophyte Sesuvium portulacastrum. Int J Res Plant Sci 2(1):23–28
Roychoudury A, Basu S, Sarkar SN, Sengupta DN (2008) Comparative physiological and molecular responses of a common aromatic indica rice cultivar to high salinity with non-aromatic indica rice cultivars. Plant Cell Rep 27:1395–1410
Ruan H-H, Shen W-B, Ye M-B, Xu L-L (2002) Protective effects of nitric oxide on salt stress-induced oxidative damages to wheat (*Triticum aestivum* L.) leaves. Chin Sci Bull 47:677–681
Ruan H-H, Shen W-B, Xu L-L (2004a) Nitric oxide modulates the activities of plasma membrane ATPase and PPase in wheat seedling roots and promotes the salt tolerance against salt stress. Acta Bot Sin 46:415–422
Ruan H-H, Shen W-B, Xu L-L (2004b) Nitric oxide involved in the abscisic acid induced proline accumulation in wheat seedling leaves nder salt stress. Acta Bot Sin 46:1307–1315
Saeedipour S (2011) Salinity tolerance of rice lines related to endogenous abscisic acid (ABA) level synthesis under stress. Afr J Plant Sci 5:628–633

Saeedipour S (2012) Is salinity tolerance of rice lines concerned to endogenous ABA content or to the cellular ability for ABA synthesis under stress? Afr J Biotechnol 11(49):10938–10943

Sairam RK, Tyagi A (2004) Physiology and molecular biology of salinity stress tolerance in plants. Cur Sci 86:407–421

Sarmad J, Shariati M, Haghjou MM (2007) Relationship between endogenous abscisic acid and β-carotene synthesis in unicellular green alga *Dunaliella*. American-Eurasian J Agric Environ Sci 2:559–564

Savouré A, Hua XJ, Bertauche N, Montagu MV, Verbruggen N (1997) Abscisic acid-independent and abscisic acid-dependent regulation of proline biosynthesis following cold and osmotic stresses in *Arabidopsis thaliana*. Mol Gen Genet 254:104–109

Seckin B, Sekmen AH, Turkan I (2009) An enhancing effect of exogenous mannitol on the antioxidant enzyme activities in roots of wheat under salt stress. J Plant Growth Regul 28:12–20

Shabala L, Mackay A, Tian Y, Jacobsen S-E, Zhou D, Shabala S (2012) Oxidative stress protection and stomatal patterning as components of salinity tolerance mechanism in quinoa (*Chenopodium quinoa*). Physiol Plant. doi:10.1111/j.1399-3054.2012.01599

Shafi M, Bakht J, Khan MJ, Khan MA, Raziudin M (2011) Role of abscisic acid and proline in salinity tolerance of wheat genotypes. Pak J Bot 43(2):1111–1118

Sharma S, Verslues PE (2010) Mechanisms independent of abscisic acid (ABA) or proline feedback have a predominant role in transcriptional regulation of proline metabolism during low water potential and stress recovery. Plant Cell Environ 33:1838–1851

Sharma S, Thakur M, Rana M, Singh K (2004) Effect of plant growth hormones and abiotic stresses on germination, growth and phosphatase activities in *Sorghum bicolor* (L.) Moench seeds. Afr J Biotechnol 3:308–312

Shaterian J, Georges F, Hussain A, Waterer D, De Jong H, Tanino KK (2005) Root to shoot communication and abscisic acid in calreticulin (*CR*) gene expression and salt-stress tolerance in grafted diploid potato clones. Environ Exp Bot 53:323–332

Sibole JV, Montero E, Cabot C, Poschenrieder C, Barcelo J (1998) Role of sodium in the ABA-mediated long-term growth response of bean to salt stress. Physiol Plant 104:299–305

Smith GS, Klages KU, Green TGA, Walton EF (1995) Changes in abscisic acid concentration, surface conductance, and water content of developing kiwifruit. Sci Hortic 61:13–27

Sulian LV, Jiang P, Chen X, Fan P, Wang X, Li Y (2012) Multiple compartmentalization of sodium conferred salt tolerance in *Salicornia europaea*. Plant Physiol Biochem 51:47–52

Summart J, Thanonkeo P, Panichajakul S, Prathepha P, McManus MT (2010) Effect of salt stress on growth, inorganic ion and proline accumulation in Thai aromatic rice, Khao Dawk Mali 105, callus culture. Afr J Biotechnol 9:145–152

Swiatek A, Azmi A, Witters E, Van Onckelen H (2003) Stress messengers jasmonic acid and abscisic acid negatively regulate plant cell cycle. Bulg J Plant Physiol 29:172–178

Tabur S, Demir K (2010) Role of some growth regulators on cytogenetic activity of barley under salt stress. J Plant Growth Regul 60:99–104

Talanova VV, Topchieva LV, Titov AF (2006) Effect of abscisic acid on the resistance of cucumber seedlings to combined exposure to high temperature and chloride. Biol Bull 33:619–622

Tao Z, Kou Y, Liu H, Li X, Xiao J, Wang S (2011) OsWRKY45 alleles play different roles in abscisic acid signalling and salt stress tolerance but similar roles in drought and cold tolerance in rice. J Exp Bot 1:1–12

Tester M, Davenport R (2003) Na^+ tolerance and Na^+ transport in higher plants. Ann Bot 91:503–527

Thompson DS, Wilkinson S, Bacon MA, Davies WJ (1997) Multiple signals and mechanisms that regulate leaf growth and stomatal behaviour during water deficit. Physiol Plant 100: 303–313

Todoroki Y, Hirai N, Koshimizu K (1995) 8′, 8′-Difluoro- and 8′, 8′, 8′ – trifluoroabscisic acids as highly potent, long-lasting analogs of abscisic acid. Phytochemistry 38:561–568

Uchida A, Jagendorf AT, Hibino T, Takabe T, Takabe T (2002) Effects of hydrogen peroxide and nitric oxide on both salt and heat stress tolerance in rice. Plant Sci 163:515–523

Ueno O (1998) Induction of Kranz anatomy and C%-like biochemical characteristics in a submerged amphibious plant by abscisic acid. Plant Cell 10:571–583
Ungar IA (1996) Effect of salinity on seed germination, growth, and ion accumulation of Atriplex patula (Chenopodiaceae). Am J Bot 83:604–607
Ünyayar S, Kele Y, Ünal E (2004) Proline and ABA levels in two sunflower genotypes subjected to water stress. Bulg J Plant Physiol 30:34–47
Upreti KK, Murti GSR (2010) Response of grape rootstocks to salinity: changes in root growth, polyamines and abscisic acid. Biol Plantarum 54:730–734
Verslues PE, Bray EA (2006) Role of abscisic acid (ABA) and Arabidopsis thaliana ABA-insensitive loci in low water potential-induced ABA and proline accumulation. J Exp Bot 57:201–212
Voisin AS, Reidy B, Parent B, Rolland G, Redondo E, Gerentes D, Tardieu F, Muller B (2006) Are ABA, ethylene or their interaction involved in the response of leaf growth to soil water deficit? An analysis using naturally occurring variation or genetic transformation of ABA production in maize. Plant Cell Environ 29:1829–1840
Wang Y, Mopper S, Hasenstein KH (2001) Effects of salinity on endogenous ABA, IAA, JA, AND SA in Iris hexagona. J Chem Ecol 27:327–342
Wang S, Sui X, Hu L, Sun J, Wei Y, Zhang Z (2010) Effects of exogenous abscisic acid pre-treatment of cucumber (*Cucumis sativus*) seeds on seedling growth and water-stress tolerance. New Zealand J Crop Hort Sci 38:7–18
Xiong L, Schumaker KS, Zhu J-K (2002) Cell signaling during cold, drought and salt stress. Plant Cell 14:165–183
Yang H, Li H, Rao L, Long G, Shi G, Peng G (2011) Effects of exogenous ABA on antioxidant enzymes in detached citrus leaves treated by rapid freezing. Afr J Biotech 10:9779–9785
Yang Z, Yu J, Merewitz E, Huang B (2012) Differential effects of abscisic acid and glycine betaine on physiological responses to drought and salinity stress for two perennial grass species. J Am Soc Hort Sci 137:96–106
Yoshida K, Igarashi E, Mukai K, Hirata K, Miyamoto K (2003) Induction of tolerance to oxidative stress in the green alga, *Chlamydomonas reinhardtii*, by abscisic acid. Plant Cell Environ 26:451–457
Yurekli F, Porgali B, Turkan I (2004) Variation in abscisic acid, indole-3-acetic acid, gibberellic acid and zeatin concentrations in two bean species subjected to salt stress. Acta Biol Cracov S Bot 46:201–212
Zaharia LI, Walker-Simmon MK, Rodríguez CN, Abrams SR (2005) Chemistry of abscisic acid, abscisic acid catabolites and analogs. J Plant Growth Regul 24:274–284
Zhang J, Jia W, Yang J, Ismail AM (2006) Role of ABA in integrating plant responses to drought and salt stress. Field Crop Res 97:111–119
Zhang F, Wang Y, Wang D (2007) Role of nitric oxide and hydrogen peroxide during the salt resistance response. Plant Signal Behav 2:473–474
Zhang HJ, Dong HZ, Li WJ, Zhang JDM (2012) Effects of soil salinity and plant density on yield and leaf senescence of field-grown cotton. Agr Crop Sci 198:27–37
Zhou R, Cutler AJ, Ambrose SJ, Galka MM, Nelson KM (2004) A new abscisic acid catabolic pathway. Plant Physiol 134:361–369
Zhu JK (2003) Regulation of ion homeostasis under salt stress. Curr Opin Plant Biol 6:441–445

Chapter 9
Calcium Signaling and Its Significance in Alleviating Salt Stress in Plants

A.K. Srivastava, A.N. Rai, V.Y. Patade, and P. Suprasanna

Abstract Environmental stresses such as salinity, temperature, drought and heavy metals negatively impact the agricultural productivity. Of these, salinity stands as a major problem mainly in the developing countries. Calcium is an essential nutrient that regulates the plant growth and development and it has evolved as a ubiquitous secondary messenger in mediating complex responses towards various developmental and environmental cues. Thus, understanding the calcium signaling and consequent calcium-dependent events is essential to improve plant productivity under extreme environment. The first step in calcium signaling is the induction of $[Ca^{2+}]_{cyt}$-transient/signatures which is defined as the repetitive oscillations or spiking of $[Ca^{2+}]_{cyt}$ level. This in turn activates a set of calcium binding proteins including Ca^{2+} sensors/decoders, protein kinases and transcription factors. The interplay between Ca^{2+} signatures and these proteins together contributes towards the stimulus specificity. Various efforts have been made to manipulate calcium signaling events either by exogenous calcium supplementation or by genetic modification of calcium signaling related genes in many plant species and considerable progress has been made in managing the plant responses toward salt stress. Additionally, these studies also help in understanding the effect of salt stress on the process of calcium signaling. The present review deals with the basic steps of calcium signaling process and its possible modulations that can lead to the enhanced salt tolerance in plants.

Keywords $[Ca^{2+}]_{cyt}$-transients/signature • Calcium sensor/decoder elements • Exogenous calcium supplementation • Genetic manipulation • SOS pathway • Transcription factors

---

A.K. Srivastava • A.N. Rai • V.Y. Patade • P. Suprasanna (✉)
Nuclear Agriculture and Biotechnology Division, Bhabha Atomic Research Centre, Mumbai 400085, India
e-mail: prasanna@barc.gov.in

P. Ahmad et al. (eds.), *Salt Stress in Plants: Signalling, Omics and Adaptations*,
DOI 10.1007/978-1-4614-6108-1_9,

The different abiotic stresses such as drought, salinity, cold, and heat negatively affect the survival, growth and yield of crop plants and hence, challenge the food security. Of these stresses, salinity is considered as one of the most important factors that affect the plant at all the developmental stages (Munns and Tester 2008). According to the Food and Agriculture Organization (FAO 2008; http://www.fao.org/ag/agl/agll/spush/), over 6% of the world's land area (400 million ha) is affected by salinity. Of this, 45 and 32 million ha land is associated with irrigated and dry land agriculture, respectively, and the remaining is non-cultivated. In India, 6.73 million ha of land is affected by salinity and 25% of ground water used for irrigation is saline. Salinity adversely affects the growth of several crop plants. It causes poor and spotty stands of crops, uneven and stunted growth and poor yields. The primary effect of excess salinity is that it renders water less available to plants although some is still present in the root zone. This is because the osmotic pressure of the soil solution increases with the increase in salt concentration. Apart from the osmotic effect of salt in soil solution, excessive concentration and absorption of individual ions may prove toxic to the plants and/or may retard the absorption of other essential plant nutrients. Besides, salinity also has an indirect effect through its negative effect on the growth of beneficial and symbiotic microbes.

To mitigate the negative effects of salt-stress, research world-wide is focused to design various strategies such as plant breeding (conventional and mutation breeding based approach); biotechnological (transgenic based approach) and agronomical (priming based approach) either for the development of salt-tolerant crops or for imparting salinity tolerance to available crop germplasm. Traditional breeding methods have played a significant role in integrating favorable genes to induce stress tolerance in crops. However, such attempts have met with limited success due to the complexity of salt tolerance trait (Yamaguchi and Blumwald 2005). The mutation breeding has generated a number of salt-sensitive types which became useful genetic material for understanding mechanisms of salt tolerance (Tester and Davenport 2003). The transgenic technology deals with alteration in the activity of one or two genes to achieve overall tolerance and has generated salt-tolerant lines in some crops. However none of the developed lines have been field tested and released commercially (Ashraf and Akram 2009; Mittler and Blumwald 2010). Thus, in order to ensure the success, it is necessary to understand the detailed molecular mechanisms responsible for the induction of plant tolerance against salt stress. The physiological basis for these mechanisms is the integration of various signal transduction events into a comprehensive network that finally leads to the activation of plant defense system.

In plants, calcium has been acknowledged as a ubiquitous secondary messenger, for environmental as well as developmental stimuli, because of its flexibility to exhibit different coordination numbers and complex geometries (Gilroy and Trewavas 2001; Sarwat et al. 2012). In recent years, the functions of calcium as an important signal transducer are subject to intensive investigations in different plant species and it is proposed that most of the abiotic stress causes the release of calcium from various intracellular stores that form the basis for the induction of stress-specific response (Kudla et al. 2010). A substantial progress has been made to understand the plant's response under salt stress in terms of SOS-pathway. As proposed by Zhu (2002), an increase in cytosolic calcium ($[Ca^{2+}]_{cyt}$) under salinity stress is read by SOS3. The SOS3 protein interacts with a SOS2 protein kinase and the SOS3-SOS2 complex then activates the SOS1 protein (a plasma membrane Na^+/H^+ antiporter) and thereby re-establishing Na^+ homeostasis in cells. The current research on calcium signaling focuses on the identification of salt-responsive calcium channels, sensor/decoders elements, calcium-dependent transcription factors their differential localization and post-translation modification under salt stress, both in model as well as crop plant. This will give more insight into the suitable candidate genes that can be used by plant biotechnologist for the purpose of crop improvement. In the present review, we have discussed different aspects of calcium as a signal transducer, approaches for the measurement of calcium dynamics and recent strategies to apply the calcium signaling for the crop improvement under salt stress conditions.

Concept of Calcium Signaling

In response to any external stimuli, one of the early response is a change in calcium signal in the form of transient increase in cytosolic free Ca^{2+} ($[Ca^{2+}]_{cyt}$) which arises because of the flux of Ca^{2+} into the cytosol, either from the external medium or from sub cellular compartments, where the concentration of Ca^{2+} is higher as compared to that of cytosol. The increase in $[Ca^{2+}]_{cyt}$ led Webb et al. (1996) to formulate the concept of "Ca^{2+} signatures" which is defined as the repetitive oscillations or spiking of $[Ca^{2+}]_{cyt}$. The frequency (period), amplitude and shape (e.g. sinusoidal, square-wave) of Ca^{2+} signature are determined by the nature and magnitude of the stimulus. The fact that Ca^{2+} serves as a messenger and regulator in so many different processes raises a fundamental question of how specificity in information processing and output determination are achieved. In this regard, the first clue emanates from the calcium signature/transient which enables the Ca^{2+} to encode stimulus-specific information (Dodd et al. 2010). An additional level of specificity is achieved by a set of calcium binding toolkit, which includes the Ca^{2+}-binding proteins functioning as Ca^{2+} sensors (CBLs, calcineurin B-like proteins) and decoders (CIPKs, CBL-interacting protein kinases) elements that together relay the information encoded within calcium signatures. An overview of the molecular components of calcium signaling process responsible for the salt-stress perception and tolerance is depicted in Fig. 9.1.

Major Determinants of Calcium Signature

Single-cell systems (guard cells, growing pollen tubes, or root hairs) represent an excellent model to understand the coding system and determinants of calcium signature. However, the recent research is progressing forward to elucidate the Ca^{2+} dynamics at the organ level because the response of plant to external stimuli is

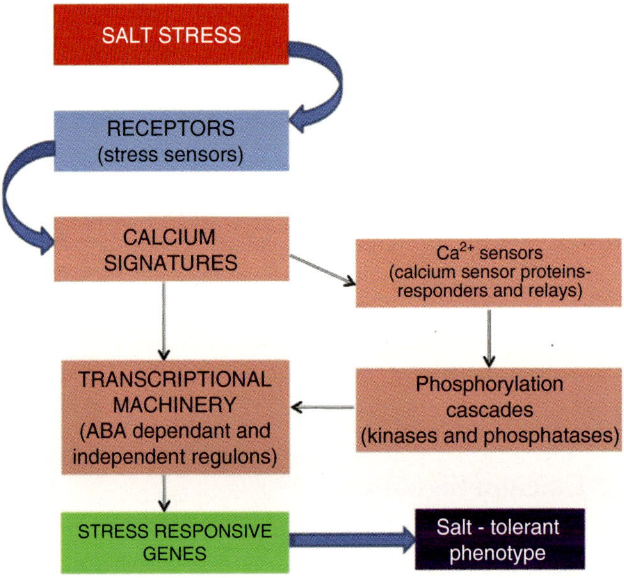

Fig. 9.1 Molecular components involved in calcium mediated salt stress responses. The first reaction after the onset of salt stress is the induction of $[Ca^{2+}]_{cyt}$-transients/signature. Such a signature is dependent upon the nature and dose of the salt and is responsible for the induction of salt-specific response. The calcium signature is sensed by a set of sensor/decoder elements and the appropriate downstream signaling is being activated in the terms of salt-responsive transcription factors (TFs). Additionally, the calcium alone or in combination with the calmodulin can also directly activate the stress-responsive TFs. These TFs in turn activates the various defense mechanisms such as Na^+-ion efflux, vacuole sequestration, synthesis of osmolytes and enhanced antioxidant metabolism leading to the amelioration of salt stress

mainly manifested by distinct organs. At the organ level, the induction of the calcium-transient is mainly in the form of single spike. The induction of $[Ca^{2+}]$-signature is dependent on the Ca^{2+} influx channels at the plasma membrane and endomembrane system (both mediate Ca^{2+} release into the cytosol) and Ca^{2+} efflux transporters (responsible for the removal of Ca^{2+} from cytosol). Although, the complete range of Ca^{2+} influx channels and efflux transporters along with their properties have been reviewed (McAinsh and Pittman 2009; Kudla et al. 2010), however, the specific-role of each components in shaping Ca^{2+} signature is still to be elucidated.

Ca^{2+} Influx Channels

Different Ca^{2+} channels/transporters can be either classified on the basis of their activation mechanism as voltage-, ligand- and stretch- activated or on the basis of their location as plasmamembrane- or endo-membrane located. On the electrophysiological basis, three distinct groups of Ca^{2+}-permeable channels have been characterized on the plasma membrane of plant cells. These are the mechanosensitive Ca^{2+} channel (MCC), the depolarization-activated Ca^{2+} channel (DACC) and the hyperpolarization-activated Ca^{2+} channel (HACC). All these three are together termed as nonselective cation channels (NSCC). The DACC and HACC in together are termed as voltage-dependent channels (Demidchik and Maathuis 2007). The MCC have been recorded in various cell types and are responsible for shaping mechanically induced $[Ca^{2+}]_{cyt}$-signature. However, despite the identification of 10 MCC-like genes in *Arabidopsis* genome, limited information is available on the specific functions of this channel type. The DACC is activated in response to stress-induced depolarization and contributes to the short transient influx of Ca^{2+}. In contrast, HACC contributes to a sustained Ca^{2+} influx. In *Arabidopsis*, the HACC activity is localized in the apical region and is down-regulated in sub-apical regions of growing root hairs and at the tips of mature hairs, suggesting its role in generating the root hair apical $[Ca^{2+}]_{cyt}$-gradient (Very and Davies 2000). In contrast to plasma membrane channels, the electrophysiological characterization of Ca^{2+} channels in the endomembrane is not possible in intact cells, imposing additional challenges when assigning the role of channel for the generation of specific Ca^{2+} signatures. Apart from plasmamembrane, at least four Ca^{2+}-permeable channels have been identified at the vacuolar membrane such as inositol 1,4,5-trisphosphate (InsP3)- and cyclic ADP-ribose (cADPR)-gated channels, voltage-gated Ca^{2+} (VVCa) and slow-activating vacuolar (SV) channels (Pottosin and Schonknecht 2007). Whether InsP3- and cADPR-gated Ca^{2+}-permeable channels reside solely in the vacuolar membrane or more widely distributed at the endoplasmic reticulum (ER) remains to be established. In *Arabidopsis*, SV channel is encoded by the AtTPC1 (two-pore channel 1) gene. The demonstration that AtTPC1 encodes the *Arabidopsis* SV channel has permitted the first functional analysis of endomembrane Ca^{2+}-permeable channel involvement in the generation of plant Ca^{2+} signatures at the molecular level (Peiter et al. 2005). In addition, Ca^{2+} release from the vacuole and ER in response to inositol hexakisphosphate (InsP6) and nicotinic acid adenine dinucleotide phosphate (NAADP), respectively also highlights the role of ligand-gated endomembrane Ca^{2+}-permeable channels in shaping the Ca^{2+} signatures.

Ca^{2+} Efflux Channels

Calcium is an essential nutrient, yet in all organisms Ca^{2+} is extremely toxic when present at high concentrations in the cytosol. This is because it readily forms insoluble complexes with ATP and makes the cell energy deficient. Thus transport mechanisms to rapidly remove Ca^{2+} from the cytosol developed early during evolution (Case et al. 2007). In addition, the Ca^{2+} efflux transporters are also important in generating Ca^{2+} signature. Plants have two main pathways for $[Ca^{2+}]_{cyt}$ removal, for instance high-affinity Ca^{2+}-ATPases and low-affinity Ca^{2+}-exchangers. The high-affinity Ca^{2+} efflux ATPases are basically a subgroup of P-type ATPases (the P2-ATPases) (Baxter et al. 2003). These are further divided into P2A-ATPases, which include sarcoplasmic/endoplasmic reticulum Ca^{2+}-ATPase (SERCA) in animals, and the ER-type Ca^{2+}-ATPase (ECA) in plants, and P2B-ATPases, including the animal CaM-regulated plasma membrane Ca^{2+}-ATPase (PMCA) and the autoinhibited Ca^{2+}-ATPase (ACA) in plants. These transporters are usually of lower Ca^{2+} affinity than the Ca^{2+} pumps but transport Ca^{2+} from the cytosol rapidly and at high capacity. *Arabidopsis* has six CAX genes plus five related genes, designated cation/Ca^{2+} exchanger (CCX) that are more similar to an animal Na^{+}/Ca^{2+} exchanger isoform (Shigaki et al. 2006). H^{+}/Ca^{2+} exchange activity has long been known to be a major route for Ca^{2+} removal from the cytosol into the vacuole (Schumaker and Sze 1985; Blumwald and Poole 1986), although exchange activity has also been detected at the plasma membrane (Kasai and Muto 1990). CAX genes encoding tonoplast H^{+}/Ca^{2+} exchangers have been subsequently identified from various plant species and have predominant roles in maintaining low $[Ca^{2+}]_{cyt}$ (Mei et al. 2007).

Decoding and Relay of Ca^{2+} Signals

The cytosolic Ca^{2+} changes are perceived by a large number of proteins termed as Ca^{2+}-sensors. A majority of such sensors have classical helix-loop-helix EF-hand motif responsible for Ca^{2+}-dependent conformational change. Upon binding with Ca^{2+}, these sensor proteins can either act as an activator (sensor relays) or else may bind with another group of proteins to activate the downstream signaling (sensor transducer). The entire group of calcium sensor proteins includes calmodulin (CaMs), calmodulin-like proteins (CMLs), Ca^{2+}-dependent protein kinase (CDPKs) and the calcineurin B-like proteins (CBLs). In CDPKs, the Ca^{2+} sensing (EF-motifs) and a responding function (protein kinase activity) is combined within a single protein and hence termed as sensor responders. In contrast, CaMs/CMLs have no enzymatic function and they alter the downstream target activities via Ca^{2+}-dependent protein-protein interaction. Therefore, they represent the bonafide sensor relay proteins. CBLs also belong to sensor relay proteins due to the lack of any enzymatic activity. However, CBLs specifically interact with a family of protein kinases designated as CBL-interacting protein kinases (CIPKs). Consequently, CBL-CIPK complexes could be considered as bimolecular sensor responders. In *Arabidopsis*, a total 10 CBLs and 26 CIPKs are reported and substantial progress has been made to understand the signaling between SOS3 (CBL4) and SOS2 (CIPK24) which is responsible for mediating the salt-stress response (Qiu et al. 2003). In response to salt, the $[Ca^{2+}]_{cyt}$ is increased which is perceived by SOS3 (CBL-4; a Ca^{2+} sensor). The SOS3 protein interacts with SOS2 (CIPK-24; a Ca^{2+} decoder) protein kinase and SOS3-SOS2 complex then activates the SOS1 protein (a plasma membrane Na^{+}/H^{+} antiporter) and thereby re-establishes the Na^{+} homeostasis in cells. The activated calcium sensor/decoder elements finally manifest their effect through a group of transcription factors such as CaM-Binding transcription factors (CAMTAs), GT-element-binding proteins (GTLs), MYBs, WRKYs and NACs.

Measurement of Calcium Signature in Plants

The entire research towards the calcium signaling is based on the ability to monitor the dynamic changes in the level of $[Ca^{2+}]_{cyt}$. Such measurements have been done using different probes which can be categorized into small molecule based fluorescent dyes (such as Fura-2) and genetically encoded Ca^{2+} sensors (Swanson et al. 2011). Although, both these methods are widely used, however; genetically encoded sensors (GECIs) have several advantages over fluorescent dyes because of the possibility to fuse the indicator with a protein or tag of interest to monitor Ca^{2+} in individual subcellular locations that are not accessible for loading with fluorescent dyes. The GECIs can be further divided in chemiluminescent Ca^{2+} sensors, based on the protein aequorin and fluorescent Ca^{2+} indicator proteins (FCIPs).

Small-Molecule Fluorescent Dyes

The most widely used Ca^{2+}-imaging dye interact with Ca^{2+} through carboxylic acid group that alters its property by changing either its fluorescence intensity or/both excitation maximum. For example, Calcium Green-1 shows an approximately 100-fold increase in fluorescence emission when binds with μM levels of calcium. However, the fluorescence property of this dye is affected by many factors such as photo bleaching that can reduce the probe signal intensity by destroying dye molecules; movement of the dye within the cell that leads to the accumulation of dye in some areas and depletion in others; changes in the optical properties of the cell in response to treatments and different cell types could accumulate or leak the reporter to different extents (Probes 2010). All of these factors led to the misinterpretation of the result. This is an important limitation of these single-wavelength probes (where a single emission or excitation wavelength of fluorescence is used to monitor the probe's behavior) which can be rectified by simultaneously measuring the Ca^{2+}-dependent fluorescence intensity and probe's concentration; however, in most of the time it is a not technically feasible. The other alternative is to opt the ratiometric dyes such as Indo-1 and Fura-2 that provides a robust approach to measure the in vivo change in the calcium level (Grynkiewicz et al. 1985). The fluorescence intensity of these dyes changes with some part of their fluorescence emission (Indo-1) or excitation (Fura-2) spectrum increasing as Ca^{2+} levels rise, whereas other parts of the spectrum show no change and others even decrease in signal. The ratio of the fluorescence intensity at the increasing wavelength to that of the invariant or decreasing wavelength provides a measure of Ca^{2+}-dependent response and is independent of the dye concentration or the optical changes which is the major limitation of using the single-wavelength dye. Another technical limitation is that of membrane impermeability which arises because of their highly charged groups. Thus, different techniques are being evolved to increase the loading efficiency of these dyes. One such technique is to breach the plasma membrane by electroporation or detergent solubilization and then perform the loading by microinjection or biolistic delivery. Acid and ester loading are also used. For acid loading, the medium pH is dropped to 4.5 to protonate the dye and thereby allowing it to cross the membrane in an uncharged form. Here, careful controls for adverse effects of incubating the plant at this low pH for the requisite 1–2 h are needed. For ester loading, the charged groups are chemically derivatized with ester groups, generally the acetoxymethyl ester, allowing them to cross the membrane where nonspecific hydrolazes cleave the ester bonds, releasing the free dye in the cytosol. For ester loading in plants, significant hydrolysis can occur in the wall, but this can be reduced by incubating the samples at low temperature. Cytosolic accumulation of uncleaved dye and accumulation within organelles of the free dye are potential problems; therefore, a widely used control is to also use microinjection of a dextran-conjugated version of the dye that cannot cross membranes and so remains localized to the cytosol. This is an important control but necessitates use of technically demanding loading approaches such as microinjection of the dextran into the cell.

Genetically Encoded Calcium Sensors

The genetically encoded sensors can be divided into the chemiluminescent bases Ca^{2+} sensors, based on the protein aequorin and fluorescent Ca^{2+} indicator proteins (FCIPs). The aequorin-based system is inherently different from the FCIPs, as the emitted light is generated by a chemical reaction that requires reconstitution of aequorin with a co-factor (McCombs and Palmer 2008). The aequorin method is well established in plants and is used substantially to advance the current understanding of Ca^{2+} signaling (Kaplan et al. 2006; Tanaka et al. 2010). Since light emission by aequorin does not depend upon the optical excitation, problems with chromophore bleaching and autofluorescence, observed when using FCIPs, do not occur. However, the main inconvenience of the aequorin system is the low light emission, which severely limits the spatiotemporal resolution during imaging as Ca^{2+} signals have to be detected from whole seedlings or tissues (Alonso et al. 2009). Mathematical simulation has shown that oscillatory single-cell Ca^{2+} signals cannot be resolved with aequorin, at least when using conventional light detection devices (Dodd et al. 2006). Thus, FCIPs are preferred to achieve the high temporal and spatial resolution Ca^{2+} imaging.

The FCIP probes are chimeric proteins that transfer the energy from a donor to an acceptor protein. If the acceptor excitation matches the donor emission and the two proteins are very close together, the energy of emission from the donor can excite the acceptor and cause it to fluoresce. The energy transfer is through resonance phenomena rather than the acceptor absorbing an emitted photon from the donor. Resonance is highly efficient but only over very short distances, Yellow cameleons (YC) is a well studied FCIP probes and is composed of a donor chromophore (CFP), calmodulin (CaM), a glycylglycine linker, the CaM-binding peptide of myosin light-chain kinase (M13), and an acceptor chromophore (YFP). The Ca^{2+}-binding to CaM initiates an intramolecular interaction between CaM and M13 which changes the protein from an extended to a more compact conformation resulting in an increased FRET-efficiency between CFP and YFP (Miyawaki et al. 1997). In plants, YC2.1 was successfully used to study guard cell-specific Ca^{2+}-dynamics. The substitution of the acceptor FP (YFP) in YC2.1 for a circularly permuted version named Venus yielded a fivefold increase in the dynamic range of the change in FRET signal upon binding Ca^{2+}, providing a significant improvement in the signal-to-noise ratio; this improved version is called YC3.6. The increased resolution afforded by this sensor has allowed imaging of both the temporal and spatial dynamics of cytoplasmic Ca^{2+} fluxes in growing *Arabidopsis* root hairs. To make ratiometric measurements with the YC sensors, the emission from CFP (donor) and the FRET emission (from Venus or YFP) are measured (both using excitation of the CFP donor). As Ca^{2+} levels increase, the donor emission falls as it transfers more of its emission energy to the acceptor, and thus FRET emission increases. The ratio of FRET/CFP therefore provides a measure of Ca^{2+} level compared to observation of YFP emission which only shows differences in absolute intensity and not differences in calcium concentration. The CFP and YFP signal can be monitored using a conventional epifluorescence microscope with appropriate filters, but these probes are also compatible with most confocal microscopes.

Alleviation of Salinity Stress Using Exogenous Calcium Supplementation

Considering the importance of calcium in mediating the process of salt tolerance, various efforts have been undertaken to use exogenous calcium supplementation for ameliorating the effect of salt stress (Table 9.1). The application of calcium can be done either as a pre-treatment or/and simultaneous with the stress. Although, the calcium can be supplied in different forms but in most of the cases calcium was supplemented in the form of $CaCl_2$. The exact mechanism of calcium mediated response is variable and dependent upon the plant system. For instance, in barley (*Hordeum vulgare*), the exogenous calcium significantly decreased the salt-induced efflux of H^+, K^+ and NH_4^+ from root epidermis which resulted in the improved growth rate and biomass accumulation (Shabala et al. 2003). In rice (*Oryza sativa* L.), calcium supplementation enhanced the net photosynthetic rate which was simultaneous with increased synthesis of osmolytes in leaves. This results in a significant reduction in the oxidative damage in NaCl + Ca^{2+} treated plant as compared to that of NaCl alone (Zhu et al. 2004). In a pot culture experiment with *Dioscorea rotundata* plants, the increased activities of antioxidant enzymes viz. superoxide dismutase (SOD), ascorbate peroxidase (APX) and catalase (CAT) was found to be responsible for the calcium mediated amelioration of the salt stress (Jaleel et al. 2008). Gobinathan et al. (2009) reported the enhanced level of osmoprotection, mainly through the glycine betaine, in response to calcium supplementation in salt stressed *Pennisetum typoidies* plants. The ameliorative effect of calcium supplementation was also demonstrated in broccoli plant and better response was seen in sensitive cultivar rather than the tolerant through enhanced K^+ and Ca^{2+} uptake and

Table 9.1 Exogenous calcium supplementation for alleviation of salinity stress in different plants

S. No.	Type of Ca^{2+} salt	Ca^{2+} level	Mode of application	Plant	References
1.	$CaCl_2$	10 mM	Along with salt stress	*Hordeum vulgare*	Shabala et al. (2003)
2.	$CaSO_4$, $CaCl_2$	10 mM	Along with salt stress	*Amaranthus tricolor*; *A. cruentus*	Omami (2005)
3.	$CaCl_2$	5 mM	Along with salt stress	*Vigna radiata*	Manivannan et al. (2007)
4.	$CaCl_2$	2.5, 5.0, 10.0, 50.0 mmol/L	Along with salt stress	*Urochondra setulosa*	Shaikh et al. (2007)
5.	$CaCl_2$	5 mM	Along with salt stress	*Dioscorea rotundata*	Jaleel et al. (2008)
6.	$CaCl_2$	10 mM	Before and along with salt stress	*Zea mays, Festuca arundinacea, Phalaris arundinacea*	Maeda and Nakazawa (2008)
7.	$CaCl_2$	5 mM	Along with salt stress	*Avena sativa* L.	Xu et al. (2008)
8.	$CaCl_2$	5 mM	Along with salt stress	*Pennisetum typoidies*	Gobinathan et al. (2009a, b)
9.	Ca_2SiO_4	1.4, 2.1, 2.8 mM	Along with salt stress	*Saccharum officinarum* L.	Ashraf et al. (2010)
10.	$CaCl_2$	10 mM	Along with salt stress	Broccolis (*Brassica oleracea)*	Nengfei et al. (2010)
11.	$CaCl_2$	10 mg/Kg sand	Along with GA_3 and salt	*Linum usitatissimum* L.	Khan et al. (2010)
12.	$CaCl_2$	10 mM	Before and along with salt stress	*Solanum lycopersicum*	Yang et al. (2010)
13.	$CaCl_2$	10 mM	Along with salt stress	*Vicia faba*	Barakat (2011)
14.	$Ca(NO_3)_2$	12 mM	Along with salt stress	*Cyclocarya paliurus*	Yao and Wang (2012)

reduced uptake of Na^+ by shoot and root (Nengfei et al. 2010). In rice, Ca application (4 mM Ca^{2+}) has improved the survival percentage in a salt sensitive cultivar and limited the cellular levels of Na^+ (Anil et al. 2005). Exogenous application of $CaCl_2$ has also been shown to alleviate stress in bean (Cabot et al. 2009), greengram (Manivannan et al. 2007), Jerusalem artichoke (Xue et al. 2008), soybean (Arshi et al. 2010), *Atriplex halimus* (Nedjimi and Daoud 2009), *Rumex* sp. (Chen et al. 2007), cowpea (Murillo-Amador et al. 2006) and linseed (Khan et al. 2010). Moreover, the gypsum ($CaSO_4.2H_2O$) has also been exogenously applied to the saline soil leading to enhance the crop productivity (Cha-um et al. 2011). Yang et al. (2010) studied the effect of calcium supplementation at different time of application in tomato (*Lycopersicon esculentum*) plant and concluded that co-existence of calcium and salt in growth medium is necessary for alleviation of salt stress. No ameliorative effect of calcium was observed only before the stress. The effect was also reduced when calcium was provided after few days of the onset of stress. Similar results have been obtained in maize, tall fescue and reed canary grass wherein maximum salt stress alleviation was achieved when calcium was applied at the same time of salt exposure (Maeda and Nakazawa 2008). These data together suggested that the calcium dependent amelioration is not only due to the Ca^{2+}:Na^+ ratio in the outside matrix but is also dependent upon the threshold external Ca^{2+} level which has to be reached for the optimum effect to be seen. A unified mechanism can also be proposed to explain the calcium mediated amelioration of salt stress where the Ca^{2+}-supplementation reduces the cell surface negativity that results into the lower accumulation of Na^+-ion. Owing to this, plants co-coordinately activate the salt-specific defense mechanisms such as enhanced synthesis of osmolytes, increased anti-oxidant capacitance and reduction in the K^+-efflux.

Genetic Manipulation of Calcium Signaling Related Gents for Enhancing the Plant Salt-Tolerance

An extensive research has been conducted where the expression level of different calcium signaling related genes have been altered for the purpose of increasing the plant salt-tolerance (Fig. 9.2). In the following sections, brief information about the nature of the gene (either calcium transporter, sensor, decoder or transcription factor), mode of change (either over-expression or repression) and final effect in terms of plant phenotype is described.

Calcium Transporters

The different calcium transporters present on the plasmamembrane and other endo-membranes are responsible for shaping the $[Ca^{2+}]_{cyt}$-transients under particular stimuli such as salt stress. The $[Ca^{2+}]_{cyt}$-transients induced under NaCl

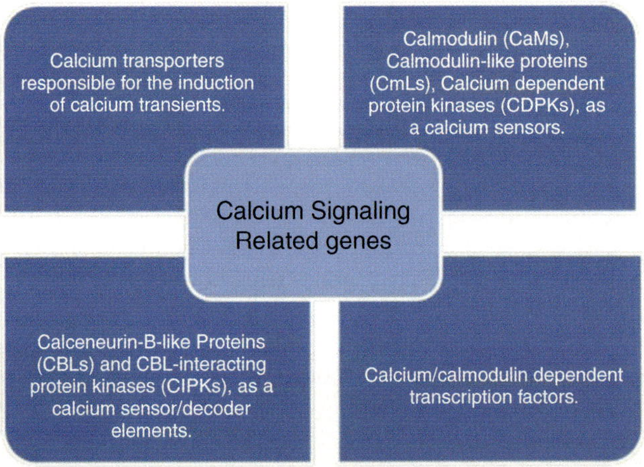

Fig. 9.2 Genetic manipulation of calcium signaling network for the modulation of salt tolerance. A range of calcium signaling related genes have been modulated either by overexpression or repression to evaluate their efficacy in ameliorating the impact of salt stress and to identify the positive and negative regulator of salt tolerance

stress is highly variable and is mainly restricted to the roots. In general, it is of a biphasic in nature and its duration and intensity is dependent upon the exact nature and dost of the salt (Tracy et al. 2008). Many calcium transporters have been genetically manipulated to evaluate their impact in the terms of salt tolerance (Bose et al. 2011). Two calcium-transporting ATPases (AtACA4 and AtACA2) were expressed in *Saccharomyces cerevisiae* to increase its tolerance against the salt stress. In case of AtACA2, the total response time of $[Ca^{2+}]_{cyt}$-transient was also reduced (Anil et al. 2008). The loss-of-function mutant of ACA type ATPase (PCA1) in *Physcomitrella patens* resulted in a sustained $[Ca^{2+}]_{cyt}$ elevation and never returned to resting level (Qudeimat et al. 2008). Apart from the genetic manipulation, attempts were also made to modulate the activity of calcium transporters using pharmacological approach. Romani et al. (2004) used EY (a P2B-type Ca^{2+}-ATPase inhibitor) to show its effect in preventing the increase in Ca^{2+}-efflux and the ROS-transient in *Egeria densa* subjected to ABA treatment. All these data point towards the significance of Ca^{2+}-ATPase in the re-establishment $[Ca^{2+}]_{cyt}$-transient to resting levels which is essential for the correct transduction of the signal. Apart from Ca^{2+}-ATPase, the calcium exchangers (CAX; another determinant for shaping the $[Ca^{2+}]_{cyt}$-transient) are also studied for their role under different abiotic stresses. The *Arabidopsis* mutant impaired in cax1 showed no significant differences with respect to the wild type when analyzed for dehydration, high-salt, chilling, or constitutive freezing tolerance. However, they exhibited increased freezing tolerance after cold acclimation, demonstrating that CAX1 plays an important role in this adaptive response (Catala et al. 2003). However, the plants impaired in cax3 showed increased sensitivity towards salt stress (Zhao et al. 2009). These data together suggests that induction of a particular isoforms of CAX is stimulus-specific.

Calcium Sensor/Decoder Elements

Salt stress is composed of both osmotic and ionic component and can be sensed either at the outer or inner surface of the plasma membrane by a trans-membrane protein, or within the cytosol by yet unidentified sensors (Zhu 2003). In the paradigm of information processing, stimulus perception is followed by the signal transduction and activation of appropriate physiological response. In plants, salt stress leads to an increase in $[Ca^{2+}]_{cyt}$, which initiates the stimulus-specific downstream signal transduction. This is dependent upon the array of calcium sensor/decoder elements. The sensor proteins have a typical Ca^{2+}-binding EF-hand motif that occur in pairs and facilitate the high-affinity binding with Ca^{2+}. The calcium sensors are categorized into CaMs (calmodulins), CMLs (CaM-like proteins), CDPKs (Ca^{2+}-dependent protein kinases), and CBLs (calcineurin B-like proteins). These sensor proteins may directly perform its function or interact with the corresponding decoder element (CIPKs: CBL-interacting protein kinase) for the same (refer Sect. 9.1). The description about the manipulation of each set of calcium sensor/decoder elements in different plant species is given below.

Calmodulins (CaMs) and CaM Like Proteins (CMLs)

The CaMs are known to play multiple regulatory roles in eukaryotes; however, direct function as transcriptional regulators is unknown. Kushwaha et al. (2008) showed that one of the four *Arabidopsis thaliana* CaM isoforms (CAM7) is a transcriptional regulator that directly interacts with the promoters of light-inducible genes and promotes photomorphogenesis. CAM7 overexpression caused hyperphotomorphogenic growth and an increase in the expression of light-inducible genes. Mutations in CAM7 produce no visible effects on photomorphogenic growth, indicating likely redundant gene functions. However, cam7 mutants displayed reduced expression of light-inducible genes, and cam7 hy5 double mutants showed an enhancement of the hy5 phenotype. The biological role of CAM7 under salt stress is yet to be elucidated. The roles of CaM like proteins (CMLs) have been studied under different abiotic stresses. The CML18 is thought to be involved in salinity tolerance through interaction with the tonoplastic Na^+/H^+ antiporter within the vacuolar lumen (Yamaguchi et al. 2005). CML18 interaction with AtNHX1 is pH-dependent and reduces its Na^+/H^+ exchange activity. Thus acidification of the vacuole under salinity stress could interrupt the CML18–AtNHX1 interaction, thereby increasing AtNHX1 activity to promote Na^+ sequestration under salinity stress (Yamaguchi et al. 2005). CML9 is a candidate for participation in abiotic stress responses, as *cml9*-knockout mutants showed enhanced tolerance to both salinity and drought stress. While germination of *cml9* seeds is not affected by medium containing mannitol, is delayed by NaCl, implying that CML9 may not involved in general osmotic stress responses. Furthermore, *cml9* seedlings are hypersensitive to ABA, and it was hypothesized that CML9 functions as a negative regulator of ABA-dependent salinity tolerance (Magnan et al. 2008). However, activity of the ABA- and salinity-responsive *RD29A* promoter is reduced in *cml9* mutants upon treatment with ABA or high salinity.

Calcium-Dependent Protein Kinases (CDPKs)

CDPKs are directly activated by the binding of Ca^{2+} to the calmodulin-like domain, and activated CDPKs regulate downstream components by themselves (Harmon et al. 2000). CDPKs constitute a large multigene family consisting of 34 and 29 genes in *Arabidopsis* and Rice, respectively and are involved in various physiological processes in plants (refer Sect. 9.1). In *Arabidopsis,* AtCDPK4/11 was found to phosphorylate a drought-responsive zinc-finger domain protein, AtDi19, which suggested that CDPKs regulate plant abiotic stress responses (Milla 2006). Zhu et al. (2007) studied the effect of genetic manipulation of CDPK4/11 in *Arabidopsis*. The loss-of-function mutations of CPK4 and CPK11 resulted in pleiotropic ABA-insensitive phenotypes in seed germination, seedling growth, and stomatal movement and led to salt insensitivity in seed germination and decreased tolerance of seedlings to salt stress. Double mutants of the two CDPK genes had stronger phenotypes than the single mutants. CDPK4- or CDPK11-overexpressing plants generally showed inverse ABA-related phenotypes relative to those of the loss-of-function mutants. Expression levels of many ABA-responsive genes were altered in the loss-of-function mutants and over expression lines. These data provide genetic evidence to show that CDPK4 and CDPK11 are the two important positive regulators of CDPK or calcium-mediated ABA signaling pathway. Franz et al. (2011) showed that CDPK21 from *Arabidopsis thaliana* is biochemically activated in vivo in response to hyperosmotic stress. Loss-of-function seedlings of cdpk21 are more tolerant to hyperosmotic stress and mutant plants show increased stress responses with respect to marker gene expression and metabolite accumulation. In transgenic *Arabidopsis* complementation lines in the cdpk21 mutant background, in which either CDPK21 wild-type, or a full-length enzyme variant carrying an amino-acid substitution were stably expressed, stress response was restored by CDPK21 but not with the kinase inactive variant. AtCDPK23 is another regulator of osmotic stress tolerance. A loss-of-function *cdpk23* plants show markedly increased drought and salinity tolerance, whereas overexpression lines display increased drought and salinity sensitivity. Although its targets remain unknown, AtCDPK23 may function in both positive regulation of stomatal opening and regulation of K^+-acquisition under salinity stress (Ma and Wu 2007).

In contrast to *Arabidopsis* CDPKs, little is known about the functions of rice CDPKs. OsCDPK7-overexpressing rice plants show enhanced tolerance to abiotic stresses, such as cold, drought and salinity (Saijo et al. 2000). Recently, Asano et al. (2012) identified OsCDPK12 as a co-regulator for both abiotic and biotic stresses. The OsCPK12-overexpressing plants exhibited increased tolerance to salt stress. Conversely, a retrotransposon (Tos17) insertion mutant, oscpk12, and plants transformed with an OsCPK12 RNA interference (RNAi) construct were more sensitive to high salinity than that of wild-type. In both the mutant lines, the H_2O_2 accumulation was higher than the wild-type which suggest that OsCDPK12 promotes tolerance to salt stress by reducing the ROS accumulation in plants. The OsCDPK12-overexpressed seedlings also had an increased sensitivity to abscisic acid (ABA) and increased susceptibility to blast fungus which is probably due to the repression of ROS production.

Calcium-Dependent Protein Kinases (CBLs)/CBL Interacting Protein Kinases (CIPKs)

The role of CBL/CIPKs in the salinity response has been well documented. The first CBL and CIPK described in plants were isolated during a screen for *Arabidopsis* mutants with altered responses to osmotic stress and were termed SOS3 (salt overly sensitive 3) and SOS2 respectively (Zhu 2003). Current nomenclature identifies them as CBL4 and CIPK24, respectively. A series of detailed studies demonstrated that CBL4 and CIPK24 interacts both in vitro and in vivo and function during salinity stress through the regulation of a plasma-membrane localized Na^+/H^+ exchanger (pmNHX or SOS1; Shi et al. 2000). The SOS-1 overexpressed lines of *Arabidopsis* showed the better growth under slat stress because of lower Na^+ ion accumulation (Shi et al. 2003). Because SOS1 requires the SOS2/SOS3 complex for its maximal activity, attempts have been made to co-express more than one SOS gene in parallel. Yang et al. (2009) developed different transgenic *Arabidopsis* plants that over expressed *SOS1, SOS3, SOS1+SOS3, SOS2+SOS3, or SOS1+SOS2+SOS3*. The transgenic plants over expressing *SOS3* and *SOS1* exhibited increased salt tolerance; however, the tolerance level of *SOS1+SOS3, SOS2+SOS3*, or *SOS1+SOS2+SOS3* lines were comparable to that of either *SOS1* or *SOS3* alone. Thus, it was suggested that no significant additive effect could be obtained by combining different *SOS* genes. Huertas et al. (2012) have identified and functionally characterized a gene encoding calcineurin-interacting protein kinase of SOS pathway (SlSOS2) from tomato. The SlSOS2 is the functional homolog of AtSOS2 and the transgenic plants with higher level of SlSOS2 were shown to grow faster than the wild-type under NaCl. The increased tolerance was also associated with higher sodium content in stems and leaves and with the induction and up-regulation of the plasma membrane Na^+/H^+ (SlSOS1) and endosomal-vacuolar Na^+/H^+ antiporters (LeNHX2 and LeNHX4), responsible for Na^+ extrusion out of the root, active loading of Na^+ into the xylem and Na^+ and K^+ compartmentalization. The expression of *Arabidopsis* SOS genes are also done in a heterologous system for increasing the salt-tolerance. For instance, Yue et al. (2012) expressed the AtSOS1 in tobacco and showed the improved phenotype of transgenic seedlings under salt stress. Apart from the SOS system, Kim et al. (2007) identified CBL10 which is as another regulator of salt-tolerance. CBL10 mutants exhibited the significant growth retardation under high-salt conditions. However, the Na^+ content of the cbl10 mutant, unlike cbl4, was significantly lower than that of the wild type under either normal or high-salt conditions, suggesting that CBL10 mediates a novel Ca^{2+}-signaling pathway for salt tolerance. The CBL10 protein physically interacts with CIPK24 (SOS2), but unlike CBL4/SOS2 complex (that is formed on plasmamembrane), the CBL10/SOS2 complex was associated with the vacuolar compartments that are responsible for salt sequestration. These findings suggested that CBL10 is mainly responsible for the Na^+-ion sequestration inside the vacuole and CBL4 is responsible for its efflux outside the cell. This concept is also supported by the fact that CBL4 is found mainly in the root cells while the CBL10 found mainly in the aerial tissues (Kim et al. 2007). Apart from the Na^+-ion efflux, the CBL4 also plays important role in K^+-nutrition. Held et al. (2011) showed that CBL4 interacts with the CIPK6 and mediates the translocation of K^+-specific channel (AKT2) to the plasmamembrane. The disruption of CIPK6 significantly reduced the expression of a number of genes involved in auxin transport and abiotic stress tolerance. On the contrary, CIPK6 overexpressing lines showed enhanced salt tolerance as compared to that of the wild-type (Tripathi et al. 2009). CBL1 is another member of calcium sensor that acts in ABA-independent manner to integrate the response of salt, drought and cold stress. The CBL1 is a positive regulator of salt and drought responses and negative regulator of cold response in plants. Thus, the CBL1 overexpressing *Arabidopsis* lines were tolerant towards the salt and drought stress but sensitive towards the cold stress (Cheong et al. 2003). The CBL1 interacts with the CIPK23 and functions to regulate the K^+ homeostasis under salt stress. The CBL1 has a redundant function with CBL9 which can also interact with CIPK23. Thus, cbl1 and cbl9 double mutant, but not the cbl1 or cbl9 single mutants, exhibit low-potassium sensitivity (Cheong et al. 2007). The CBL5 is also a positive regulator of high salt or drought tolerance. The CBL5 over-expression *Arabidopsis* lines displayed enhanced tolerance to high salt or drought stress but did not alter their response to ABA (Cheong et al. 2010). Recently, Tsou et al. (2012) have characterized a new CIPK (CIPK6) from *Arabidopsis*. Although the interacting CBL partner of CIPK6 is not yet known, but the null mutation of *cipk6* was found to abolish the salt-tolerance behavior of CBP (calcium binding peptide)-over expressing transgenic lines. A rice *OsCIPK03* is overexpressed in rice and the transgenic rice showed increased sensitivity towards salt stress at seed germination stage and seedling growth. However, RNAi lines that under-expressed *OsCIPK03* exhibited higher tolerance to NaCl stress than that of wild-type. These data suggested that *OsCIPK03* is a negative regulator of salt tolerance (Rao et al. 2011).

Calcium-Dependent Transcription Factors

There are a diverse set of transcription factors which are regulated by either the direct binding of Ca^{2+} or by Ca^{2+}/CaM complex with their promoter sequences or by Ca^{2+} or by Ca^{2+}/CaM complex-mediated post-translation modification (Kim et al. 2009). The AtNIG1 (*Arabidopsis thaliana* NaCl-inducible gene 1) is the first identified Ca^{2+}-binding transcription factor in plants. AtNIG1 is a basic helix-loop-helix-type transcription factor that contains an EF-hand motif and it gets targeted to the nucleus after binding with the calcium. In addition, AtNIG1 also binds with E-box-DNA sequence (CANNTG), which is found in the promoter regions of many salt stress-related genes. Functional analyses with an atnig1-1 knockout mutant revealed that the mutant plants are hypersensitive towards salt stress with significantly lower survival rate, fresh weight, chlorophyll content, and protein content as compared to that of the wild-type (Kim and Kim 2006). The MYB is another set of R2R3-type MYB transcription factor which is an upstream regulator of a set of salt- and dehydration-responsive genes. The MYB is generally regulated by the CaM4 isoform that binds with the R2R3 DNA-binding domain of AtMYB2 protein in a Ca^{2+}-dependent manner and in turn, enhances not only the DNA-binding activity of AtMYB2, but also the AtMYB2-mediated transcriptional activation. The overexpression of GmCaM4 in *Arabidopsis* leads to constitutive expression of salt- and dehydration-responsive genes, which include P5CS1, ADH1, and rd22, and confers salt tolerance (Yoo et al. 2005). A NAC domain-containing transcription factor, termed CBNAC (CaM-binding NAC protein) was identified as a CaM-binding protein via the screening of an *Arabidopsis* cDNA expression library. The function of the Ca^{2+}/CaM complex in CBNAC-mediated transcriptional regulation was characterized (Kim et al. 2007). The NAC protein family comprises a large group of plant-specific transcription factors. Ooka et al. (2003) reported the existence of 75 and 105 NAC genes in *Oryza sativa* and *Arabidopsis thaliana*, respectively. Hu et al. (2006) showed that overexpression of NAC1 significantly enhanced the drought resistance in transgenic rice in the field under severe drought stress conditions at the reproductive stage while showing no phenotypic changes or yield penalty. The transgenic rice also showed improved salt tolerance at the vegetative stage by closing more stomatal pores to prevent the salt-induced water loss. Mao et al. (2012) isolated TaNAC2 from wheat and then characterized it in *Arabidopsis thaliana*. The overexpressing TaNAC2 lines resulted in enhanced tolerance to drought, salt, and freezing stresses in *Arabidopsis*, which were associated with the increased expression of various abiotic stress-responsive genes.

Additionally, there are several proteins that binds with Ca^{2+} but do not contain EF-hand motifs. These include phospholipase D (PLD), annexins, calreticulin, calnexin and Pistil-expressed Ca^{2+}-binding protein (PCP). The activity of PLD, which cleaves membrane phospholipids into a soluble head group and phosphatidic acid, is regulated by $[Ca^{2+}]_{cyt}$ through a Ca^{2+}/phospholipids binding-site termed as the 'C2 domain' (Li et al. 2009). Plants have several PLD isoforms that can be modulated by calcium, free fatty acids or lipids. These biochemical modulators of PLD are the substrates or products of phospholipase C (PLC), which generates IP3 and diacylglycerol (DAG), both of which are regulated by CaM (Munnik and Vermeer 2009). Shen et al. (2011) demonstrated the function of one PLD isoforms of rice (OsPLDα) to regulate the activity of tonoplast (TP) and plasma membrane (PM) H^+-ATPase in response to salt stress. The OsPLDα1 knockdown cells developed using RNA interference (RNAi) showed significantly lower TP and PM H^+-ATPase activity under NaCl stress as compared to the wild-type. Knockdown OsPLDα was also shown to prevent the NaCl-induced increase in the transcript level of OsVHA-A (TP H^+-ATPase), OSA2 (PM H+-ATPase) and OsNHX1 (TP Na^+/H^+ antiporter) which suggested that OsPLDα is involved in salt tolerance in rice through the mediation of H^+-ATPase activity and transcription.

Conclusions and Future Perspective

Abiotic stress factors such as drought, salinity, heat and cold stress have imposed challenge to world agriculture creating a threat to food security. Over the past few decades, considerable progress has been made in understanding the plant responses towards these stresses at the molecular level. These studies have unfolded the central role of Ca^{2+}, as a key secondary messenger in regulating the plant growth and developmental under normal and stress condition. It is also known that externally supplied Ca^{2+} alleviates the adverse effects of salinity in many plant species. At the molecular front, a large number of calcium transporters, sensor/decoder elements and calcium-dependent transcription factors are known to be regulated by Ca^{2+} at different levels either by direct binding of Ca^{2+}, or calmodulin or by other kinases/phosphatases. The genetic manipulation for developing tolerant plants is underway and the findings substantiate the great potential in generating the salt-tolerant plants by changing the level of either one or more genes of calcium signaling pathway. In order to further strengthen these efforts, it is imperative to understand the possible crosstalk of calcium with redox, hormonal and MAP kinase based signaling. This includes the application of all possible 'omics' approaches with the computational tools, combined with high-resolution phenotypic analysis and investigation of spatial and temporal dynamics of sub-cellular Ca^{2+}-transients and their downstream receptors. Thus, the major focus of future research should be to apply such approaches, in an integrated manner, to identify and employ "novel" and "efficient" regulators/switches for the development of stress tolerant crops.

References

Alonso MT, Manjarres IM, Garcia-Sancho J (2009) Modulation of calcium signaling by intracellular organelles seen with targeted aequorins. Blackwell, Oxford, pp 37–49

Anil VS, Krishnamurthy P, Kuruvilla S, Sucharitha K, Thomas G, Mathew MK (2005) Regulation of the uptake and distribution of Na^+ in shoots of rice (*Oryza sativa*) variety Pokkali: role of Ca^{2+} in salt tolerance response. Physiol Plant 124:451–464

Anil VS, Rajkumar P, Kumar P, Mathew M (2008) A plant Ca^{2+} pump, ACA2, relieves salt hypersensitivity in yeast. J Biol Chem 283:3497–3506

Arshi A, Ahmad A, Aref IM, Iqbal M (2010) Calcium interaction with salinity-induced effects on growth and metabolism of soybean (*Glycine max* L.) cultivars. J Environ Biol 31:795–801

Asano T, Hayashi N, Kobayashi M, Aoki N, Miyao A, Mitsuhara I, Ichikawa H, Komatsu S, Hirochika H, Kikuchi S, Ohsugi R (2012) A rice calcium-dependent protein kinase OsCPK12 oppositely modulates salt-stress tolerance and blast disease resistance. Plant J 69(1):26–36

Ashraf M, Akram NA (2009) Improving salinity tolerance of plants through conventional breeding and genetic engineering: an analytical comparison. Biotechnol Adv 27:744–752

Ashraf M, Ahmad R, Bhatti AS, Afzal M, Sarwar A, Maqsood MA, Kanwal S (2010) Amelioration of Salt Stress in Sugarcane (*Saccharum officinarum* L.) by Supplying Potassium and Silicon in Hydroponics. Pedosphere 20:153–162

Barakat NAM (2011) Ameliorative effects of Ca^{2+} on the growth, metabolism, cationic status and cell wall degrading enzymes of induced salinity stress *Vicia faba* L. J Stress Physiol Biochem 7:369–386

Baxter I, Tchieu J, Sussman MR, Boutry M, Palmgren MG, Gribskov M, Harper JF, Axelsen KB (2003) Genomic comparison of P-type ATPase ion pumps in *Arabidopsis* and rice. Plant Physiol 132:618–628

Blumwald E, Poole RJ (1986) Kinetics of Ca^{2+}/H^+ antiport in isolated tonoplast vesicles from storage tissue of *Beta vulgaris* L. Plant Physiol 80:727–731

Bose J, Pottosin II, Shabala SS, Palmgren MG, Shabala S (2011) Calcium efflux systems in stress signaling and adaptation in plants. Front Plant Sci 2:85

Cabot C, Sibole JV, Barcelo J, Poschenrieder C (2009) Sodium-calcium interactions with growth, water, and photosynthetic parameters in salt-treated beans. J Plant Nutr Soil Sci 172:637–643

Case RM, Eisner D, Gurney A, Jones O, Muallem S, Verkhratsky A (2007) Evolution of calcium homeostasis: from birth of the first cell to an omnipresent signaling system. Cell Calcium 42:345–350

Catala R, Santos E, Alonso JM, Ecker JR, Martínez-Zapater JM, Salinas J (2003) Mutations in the Ca^{2+}/H^+ transporter CAX1 increase CBF/DREB1 expression and the cold-acclimation response in *Arabidopsis*. Plant Cell 15:2940–2951

Cha-um S, Pokasombat Y, Kirdmanee C (2011) Remediation of salt-affected soil by gypsum and farmyard manure importance for the production of Jasmine rice. Aust J Crop Sci 5:458–465

Chen HX, Li PM, Gao HY (2007) Alleviation of photoinhibition by calcium supplement in salt-treated *Rumex* leaves. Physiol Plant 129:386–396

Cheong YH, Kim KN, Pandey GK, Gupta R, Grant JJ, Luan S (2003) CBL1, a calcium sensor that differentially regulates salt, drought, and cold responses in *Arabidopsis*. Plant Cell 15:1833–1845

Cheong YH, Pandey GK, Grant JJ, Batisti O, Li L, Kim BG, Lee SC, Kudla J, Luan S (2007) Two calcineurin B-like calcium sensors, interacting with protein kinase CIPK23, regulate leaf transpiration and root potassium uptake in *Arabidopsis*. Plant J 52:223–239

Cheong YH, Sung SJ, Kim BG, Pandey GK, Cho JS, Kim KN, Luan S (2010) Constitutive overexpression of the calcium sensor CBL5 confers osmotic or drought stress tolerance in *Arabidopsis*. Mol Cells 29:159–165

Demidchik V, Maathuis FJM (2007) Physiological roles of nonselective cation channels in plants: from salt stress to signaling and development. New Phytol 175:387–404

Dodd AN, Jakobsen MK, Baker AJ, Telzerow A, Hou SW, Laplaze L, Barrot L, Scott Poethig R, Haseloff J, Webb AAR (2006) Time of day modulates low-temperature Ca^{2+} signals in *Arabidopsis*. Plant J 48:962–973

Dodd AN, Kudla J, Sanders D (2010) The language of calcium signaling. Annu Rev Plant Biol 61:593–620

FAO (2008) FAO land and plant nutrition management service. http://www.fao.org/ag/agl/agll/spush

Franz S, Ehlert B, Liese A, Kurth J, Cazale AC, Romeis T (2011) Calcium-dependent protein kinase CPK21 functions in abiotic stress response in *Arabidopsis thaliana*. Mol Plant 4:83–96

Gilroy S, Trewavas AJ (2001) Signal processing and transduction in plant cells: the end of the beginning? Nat Rev Mol Cell Biol 2:307–314

Gobinathan P, Murali PV, Panneerselvam R (2009) Intearactive effects of calcium chloride on salinity induced proline metabolism in *Pennisetum typoides*. Adv Biol Res 3:168–173

Grynkiewicz G, Poenie M, Tsien RY (1985) A new generation of Ca^{2+} indicators with greatly improved fluorescence properties. J Biol Chem 260:3440–3450

Harmon AC, Gribskov M, Harper JF (2000) CDPKs – a kinase for every Ca^{2+} signal? Trends Plant Sci 5:154–159

Held K, Pascaud F, Eckert C, Gajdanowicz P, Hashimoto K, Corratge-Faillie C, Offenborn JN (2011) Calcium-dependent modulation and plasma membrane targeting of the AKT2 potassium channel by the CBL4/CIPK6 calcium sensor/protein kinase complex. Cell Res 21(7):1116–1130

Hu H, Dai M, Yao J, Xiao B, Li X, Zhang Q, Xiong L (2006) Overexpressing a NAM, ATAF, and CUC (NAC) transcription factor enhances drought resistance and salt tolerance in rice. Proc Natl Acad Sci USA 103:12987–12992

Huertas R, Olías R, Eljakaoui Z, Galvez FJ, Li J, Morales PAD, Belver A, Rodriguez-Rosales MP (2012) Overexpression of SlSOS2 (SlCIPK24) confers salt tolerance to transgenic tomato. Plant Cell Environ 35(8):1467–1482

Jaleel CA, Gopi R, Gomathinayagam M, Panneerselvam R (2008) Effects of calcium chloride on metabolism of salt-stressed *Dioscorea rotundata*. Acta Biol Crac Series Bot 50:63–67

Kaplan B, Davydov O, Knight H, Galon Y, Knight MR (2006) Rapid transcriptome changes induced by cytosolic Ca^{2+} transients reveal ABRE-related sequences as Ca^{2+}-responsive cis elements in *Arabidopsis*. Plant Cell 18:2733–2748

Kasai M, Muto S (1990) Ca^{2+} pump and Ca^{2+}/H^{+} antiporter in plasma membrane vesicles isolated by aqueous two-phase partitioning from corn leaves. J Mem Biol 114:133–142

Khan MN, Siddiqui MH, Mohammad F, Naeem M, Masroor M, Khan A (2010) Calcium chloride and gibberellic acid protect linseed (*Linum usitatissimum* L.) from NaCl stress by inducing antioxidative defense system and osmoprotectant accumulation. Acta Physiol Plant 32:121–132

Kim J, Kim HY (2006) Functional analysis of a calcium binding transcription factor involved in plant salt stress signaling. FEBS Lett 580:5251–5256

Kim HS, Park BO, Yoo JH, Jung MY, Lee SM, Han HJ, Kim KE, Kim SH, Lim CO, Yun DJ, Lee SY, Chung WS (2007) Identification of a calmodulin-binding NAC protein as a transcriptional repressor in *Arabidopsis*. J Biol Chem 282:36292–36302

Kim MC, Chung WS, Yun DJ, Cho MJ (2009) Calcium and calmodulin-mediated regulation of gene expression in plants. Mol Plant 2:13–21

Kudla J, Batistic O, Hashimoto K (2010) Calcium signals: the lead currency of plant information processing. Plant Cell 22:541–563

Kushwaha R, Singh A, Chattopadhyay S (2008) Calmodulin7 plays an important role as transcriptional regulator in *Arabidopsis* seedling development. Plant Cell 20:1747–1759

Li M, Hong Y, Wang X (2009) Phospholipase D- and phosphatidic acid-mediated signaling in plants. Biochim Biophys Acta 1791:927–935

Ma SY, Wu WH (2007) AtCPK23 functions in *Arabidopsis* responses to drought and salt stresses. Plant Mol Biol 65:511–518

Maeda Y, Nakazawa R (2008) Effects of the timing of calcium application on the alleviation of salt stress in the maize, tall fescue and reed canarygrass seedlings. Biol Plantarum 52:153–156

Magnan F, Ranty B, Charpenteau M, Sotta B, Galaud JP, Aldon D (2008) Mutations in AtCML9, a calmodulin-like protein from *Arabidopsis thaliana*, alter plant responses to abiotic stress and abscisic acid. Plant J 56:575–589

Manivannan P, Jaleel CA, Sankar B, Somasundaram R, Murali PV, Sridharan R, Panneerselvam R (2007) Salt stress mitigation by calcium chloride in *Vigna radiata* (l.) Wilczek. Acta Biol Crac Series Bot 49:105–109

Mao X, Zhang H, Qian X, Li A, Zhao G, Jing R (2012) TaNAC2, a NAC-type wheat transcription factor conferring enhanced multiple abiotic stress tolerances in *Arabidopsis*. J Exp Bot 63(8):2933–2946

McAinsh MR, Pittman JK (2009) Shaping the calcium signature. New Phytol 181:275–294

McCombs JE, Palmer AE (2008) Measuring calcium dynamics in living cells with genetically encodable calcium indicators. Methods 46:152–159

Mei H, Zhao J, Pittman JK, Lachmansingh J, Park S, Hirschi KD (2007) In planta regulation of the *Arabidopsis* Ca^{2+}/H^{+} antiporter CAX1. J Exp Bot 58:3419–3427

Milla MAR, Uno Y, Chang IF, Townsend J, Maher EA, Quilici D, Cushman JC (2006) A novel yeast two-hybrid approach to identify CDPK substrates: characterization of the interaction between CPK11 and AtDi19, a nuclear zinc finger protein. FEBS Lett 580:904–911

Mittler R, Blumwald E (2010) Genetic engineering for modern agriculture: challenges and perspectives. Annu Rev Plant Biol 61:443–462

Miyawaki A, Llopis J, Heim R, McCaffery JM, Adams JA, Ikura M, Tsien RY (1997) Fluorescent indicators for Ca^{2+} based on green fluorescent proteins and calmodulin. Nature 388:882–887

Munnik T, Vermeer JM (2009) Osmotic stress-induced phosphoinositide and inositol phosphate signalling in plants. Plant Cell Environ 33:655–669

Munns R, Tester M (2008) Mechanisms of salinity tolerance. Annu Rev Plant Biol 59:651–681

Murillo-Amador B, Jones HG, Kaya C, Aguilar RL, Gárcia-Hernández JL, Troyo-Diéguez E, Ávila-Serrano NY, Rueda-Puente E (2006) Effects of foliar application of calcium nitrate on growth and physiological attributes of cowpea (*Vigna unguiculata* L. Walp.) grown under salt stress. Env Exp Bot 58:188–196

Nedjimi B, Daoud Y (2009) Ameliorative effect of $CaCl_2$ on growth, membrane permeability and nutrient uptake in *Atriplex halimus* subsp. *schweinfurthii* grown at high (NaCl) salinity. Desalination 249:163–166

Nengfei D, Qinglin F, Chen L, Yichen L, Bin G, Huifeng S (2010) Effect of exogenous calcium chloride on antioxidant enzyme activities and ions uptake of broccoli under salt stress. Chin Agri Sci Bull 26:133–137

Omami EN (2005) Response of Amaranth to salinity stress. Ph.D. thesis. Faculty of Natural and Agricultural Sciences. University of Pretoria

Ooka H, Satoh K, Doi K, Nagata T, Otomo Y, Murakami K, Matsubara K, Osato N, Kawai J, Carninci P (2003) Comprehensive Analysis of NAC Family Genes in *Oryza sativa* and *Arabidopsis thaliana*. DNA Res 10:239–247

Peiter E, Maathuis FJM, Mills LN, Knight H, Pelloux M, Hetherington AM, Sanders D (2005) The vacuolar Ca^{2+}-activated channel TPC1 regulates germination and stomatal movement. Nature 434:404–408

Pottosin II, Schonknecht G (2007) Vacuolar calcium channels. J Exp Bot 58:1559–1569

Probes M (2010) The handbook. Invitrogen, Carlsbad

Qiu QS, Barkla BJ, Vera-Estrella R, Zhu JK, Schumaker KS (2003) Na^+/H^+ exchange activity in the plasma membrane of *Arabidopsis*. Plant Physiol 132:1041–1052

Qudeimat E, Faltusz AMC, Wheeler G, Lang D, Brownlee C, Reski R, Frank W (2008) A PIIB-type Ca^{2+}-ATPase is essential for stress adaptation in *Physcomitrella patens*. Proc Natl Acad Sci USA 105:19555–19560

Rao XL, Zhang XH, Li RJ, Shi HT, Lu YT (2011) A calcium sensor-interacting protein kinase negatively regulates salt stress tolerance in rice (*Oryza sativa*). Funct Plant Biol 38:441–450

Romani G, Bonza MC, Filippini I, Cerana M, Beffagna N, De Michelis MI (2004) Involvement of the plasma membrane Ca^{2+}-ATPase in the short-term response of *Arabidopsis thaliana* cultured cells to oligogalacturonides. Plant Biol 6:192–200

Saijo Y, Hata S, Kyozuka J, Shimamoto K, Izui K (2000) Over-expression of a single Ca^{2+}-dependent protein kinase confers both cold and salt/drought tolerance on rice plants. Plant J 23:319–327

Sarwat M, Ahmad P, Nabi G, Hu X (2012) Ca(2+) signals: the versatile decoders of environmental cues. Crit Rev Biotechnol 2012 May 9. [Epub ahead of print]

Schumaker KS, Sze H (1985) A Ca^{2+}/H^+ antiport system driven by the proton electrochemical gradient of a tonoplast H^+-ATPase from oat roots. Plant Physiol 79:1111–1117

Shabala S, Shabala L, Volkenburgh EV (2003) Effect of calcium on root development and root ion fluxes in salinised barley seedlings. Funct Plant Biol 30:507–514

Shaikh F, Gul B, Li WQ, Jing LX, Khan MA (2007) Effect of calcium and light on the germination of Urochondra setulosa under different salts. J Zheiziang University Sci B 8:20–26

Shen P, Wang R, Jing W, Zhang W (2011) Rice Phospholipase Dα is involved in salt tolerance by the mediation of H^+-ATPase activity and transcription. J Int Plant Biol 53:289–299

Shi H, Ishitani M, Kim C, Zhu JK (2000) The *Arabidopsis thaliana* salt tolerance gene SOS1 encodes a putative Na^+/H^+ antiporter. Proc Natl Acad Sci USA 97:6896–6901

Shi H, Lee BH, Wu SJ, Zhu JK (2003) Overexpression of a plasma membrane Na^+/H^+ antiporter gene improves salt tolerance in *Arabidopsis thaliana*. Nat Biotechnol 21:81–85

Shigaki T, Hirschi KD (2006) Diverse functions and molecular properties emerging for CAX cation/H^+ exchangers in plants. Plant Biol 8:419–429

Swanson SJ, Choi WG, Chanoca A, Gilroy S (2011) In vivo imaging of Ca^{2+}, pH, and reactive oxygen species using fluorescent probes in plants. Annu Rev Plant Biol 62:273–297

Tanaka K, Swanson SJ, Gilroy S, Stacey G (2010) Extracellular nucleotides elicit cytosolic free calcium oscillations in *Arabidopsis*. Plant Physiol 154:705–719

Tester M, Davenport RJ (2003) Na^+ tolerance and Na^+ transport in higher plants. Ann Bot (Lon) 91:503–527

Tracy FE, Gilliham M, Dodd AN, Webb AAR, Tester M (2008) NaCl-induced changes in cytosolic free Ca^{2+} in *Arabidopsis thaliana* are heterogeneous and modified by external ionic composition. Plant Cell Environ 31:1063–1073

Tripathi V, Parasuraman B, Laxmi A, Chattopadhyay D (2009) CIPK6, a CBL-interacting protein kinase is required for development and salt tolerance in plant. Plant J 58:778–790

Tsou PL, Lee SY, Allen NS, Winter-Sederoff H, Robertson D (2012) An ER-targeted calcium-binding peptide confers salt and drought tolerance mediated by CIPK6 in *Arabidopsis*. Planta 235:539–552

Very AA, Davies JM (2000) Hyperpolarization-activated calcium channels at the tip of *Arabidopsis* root hairs. Proc Natl Acad Sci USA 97:9801–9806

Webb AAR, McAinsh MR, Taylor JE, Hetherington AM (1996) Calcium ions as intracellular second messengers in higher plants. Advances Bot Res 22:45–96

Xue YF, Liu L, Liu ZP, Mehta SK, Zhao GM (2008) Protective role of Ca^{2+} against NaCl toxicity in *Jerusalem artichoke* by up-regulation of antioxidant enzymes. Pedosphere 18:766–774

Yamaguchi T, Blumwald E (2005) Developing salt-tolerant crop plants: challenges and opportunities. Trends Plant Sci 12:1360–1385

Yamaguchi T, Aharon GS, Sottosanto JB, Blumwald E (2005) Vacuolar Na^+/H^+ antiporter cation selectivity is regulated by calmodulin from within the vacuole in a Ca^{2+}- and pH-dependent manner. Proc Natl Acad Sci USA 102:16107–16112

Yang Q, Chen ZZ, Zhou XF, Yin HB, Li X, Xin XF, Hong XH, Zhu JK, Gong Z (2009) Overexpression of SOS (salt overly sensitive) genes increases salt tolerance in transgenic *Arabidopsis*. Mol Plant 2:22–31

Yang FJ, Li TL, Zang ZJ, Lu SW (2010) Effects of timing of exogenous calcium application on the alleviation of salt stress in the tomato seedlings. Sci Agri Sin 43:1181–1188

Yoo JH, Park CY, Kim JC, Heo WD, Cheong MS, Park HC, Kim MC, Moon BC, Choi MS, Kang YH, Lee JH, Kim HS, Lee SM, Yoon HW, Lim CO, Yun DJ, Lee SY, Chung WS, Cho MJ (2005) Direct interaction of a divergent CaM isoform and the transcription factor, MYB2, enhances salt tolerance in *Arabidopsis*. J Biol Chem 280:3697–3706

Yue Y, Zhang M, Zhang J, Duan L, Li Z (2012) SOS1 gene overexpression increased salt tolerance in transgenic tobacco by maintaining a higher K^+/Na^+ ratio. J Plant Physiol 169:255–261

Zhao J, Shigaki T, Mei H, Guo Y, Cheng NH, Hirschi KD (2009) Interaction between *Arabidopsis* Ca^{2+}/H^+ exchangers CAX1 and CAX3. J Biol Chem 284:4605–4615

Zhu JK (2002) Salt and drought stress signal transduction in plants. Annu Rev Plant Biol 53:247–273

Zhu JK (2003) Regulation of ion homeostasis under salt stress. Curr Opin Plant Biol 6:441–445

Zhu XJ, Yang JS, Liang YC, Lou YS, Yang XY (2004) Effect of exogenous calcium on photosynthesis and its related physiological characteristics of rice seedlings under salt stress. Sci Agri Sin 37:1497–1503

Zhu SY, Yu XC, Wang XJ, Zhao R, Li Y, Fan RC, Shang Y, Du SY, Wang XF, Wu FQ, Xu YH, Zhang XY, Zhang DP (2007) Two calcium-dependent protein kinases, CPK4 and CPK11, regulate abscisic acid signal transduction in *Arabidopsis*. Plant Cell 19:3019–3036

Chapter 10
Improving Salt Tolerance in Rice: Looking Beyond the Conventional

K.K. Vinod, S. Gopala Krishnan, N. Naresh Babu, M. Nagarajan, and A.K. Singh

Abstract Several factors in the intensive cropping system have played significant role in deteriorating soil health in general. Soil salinization is one of the major issues threatening crop productivity in major irrigated rice growing areas of the world. Salinity is a serious issue in rice, the crop that feeds half the world, since it is sensitive to salt accumulation. With the world population growing incessantly, there is an urgent need to increase rice productivity especially in salinized lands as well as to reutilize lands that are rendered unproductive due to salt accumulation. It is therefore essential to develop varieties that are phenologically capable of sustaining excess salt throughout its life span and produce higher yield. Although there is sufficient variability in rice germplasm for salt tolerance, conventional breeding has been far less fruitful in addressing this complex problem. With the deeper understanding of the intricate mechanisms of salt tolerance and the array of genes and useable quantitative trait loci that are being discovered, the breeding scenario towards salt tolerant rice is poised to take a more productive turn in near future. This chapter outlines the latest developments in rice breeding towards salt tolerance through employment of modern molecular techniques in conjunction with the conventional breeding approaches.

Keywords Soil salinity • Rice • Molecular genetics • Salt tolerance • Molecular breeding

---

K.K. Vinod (✉) • M. Nagarajan
Division of Genetics, Rice Breeding and Genetics Research Center,
Indian Agricultural Research Institute, Aduthurai, India
e-mail: kkvinod@iari.res.in; drmnagarajan2000@yahoo.co.in

S.G. Krishnan • N.N. Babu • A.K. Singh
Division of Genetics, Indian Agricultural Research Institute,
New Delhi, India
e-mail: krish.icar@gmail.com; babun9832@gmail.com; aks_gene@yahoo.com

P. Ahmad et al. (eds.), *Salt Stress in Plants: Signalling, Omics and Adaptations*,
DOI 10.1007/978-1-4614-6108-1_10, © Springer Science+Business Media New York 2013

Salinity is one of the major factors limiting productivity of crops including rice around the world. Soil salinity is often accompanied by osmotic imbalance, mineral deficiency and toxicity that has adverse effects on crop growth (Parvaiz and Satyawati 2008, Ahmad 2010; Ahmad et al. 2010, 2011, 2012a; Ahmad and Prasad 2012a, b). Although the term salinity in general refers to adverse effects of salt accumulation in soils, in strict sense there are two broad types of salt problem soils (a) alkaline (sodic) and (b) saline. Alkaline or sodic soils have appreciable quantities of sodium and ions capable of alkaline hydrolysis such as carbonates with Exchangeable Sodium Percent (ESP) of more than 15% coupled with a pH above 8.2 and an average root-zone electrical conductivity of saturated soil extract (ECe) of less than 4 dS m^{-1} (approx. 40 mM NaCl) and poor soil structure. Whereas, saline soils are predominant with neutral salts of sulphates and chlorides of sodium, calcium and magnesium with ECe of 4 dS m^{-1} or more, pH less than 8.2 and ESP lower than 15% (Richards 1954). Soil salinization is gradual and occurs naturally, anthropogenically or by a combination of both (secondary salinization). Most common factors that influence soil salinization are, excessive irrigation without proper drainage in inlands (Borlaug and Dowswell 2005), underlying rocks rich in harmful salts, use of saline ground water, saline seeps triggered by deforestation and shift in cropping pattern, sea water ingression, water balance disturbance due to irrigation, localized redistribution of salts, excessive evaporation over precipitation, anaerobic reduction under high organic matter and climate change. Recent mega-tsunamis that affected many South and Southeast Asian countries and Japan had engrossed several million acres of land rendering them salt affected (Chandrasekharan et al. 2008; Abe et al. 2012).

Soil Salinity in Rice Growing Regions of the World

More than 6.2% of the total world area amounting to 837 million hectares (M ha) is salt affected (Fisher et al. 2002), of which roughly about 48% is saline and 52% is sodic (Bot et al. 2000) with saline soils predominating major rice areas of the world (Fig. 10.1). Area under salt stress is on the rise with an estimated 10–40 M ha becoming salinized every year due to secondary salinization that renders vast areas of land un(der) utilized (Pessarakli and Szabolcs 1999; Ahmed and Qamar 2004; Ansari et al. 2007). Published estimates show that more than 76 M ha of land worldwide has become salt-affected due to secondary salinization of which about 45 M ha is irrigated (Dregne et al. 1991; Oldeman et al. 1991).

Rice is cultivated in more than 115 countries, of which Asia's share is more than 91% of the world total. Majority of the Asian rice production zone is confined to South and South East Asia wherein severe salinity related problems are rampant in 20% of the area amounting to 47 M ha (Abbas et al. 1994) consisting of warm humid coastal regions and marshy inlands (Fig. 10.2). India and Pakistan has the largest share of area under salinity among the South Asian nations (Table 10.1).

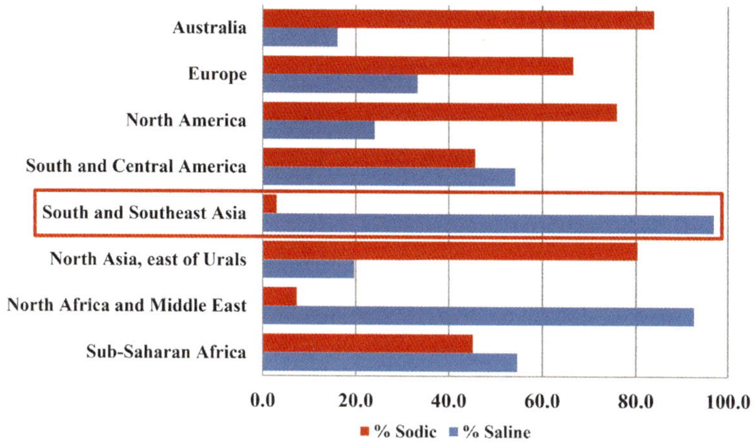

Fig. 10.1 Worldwide distribution of sodic and saline soils showing prominence of saline soils in major rice growing regions of South and Southeast Asia

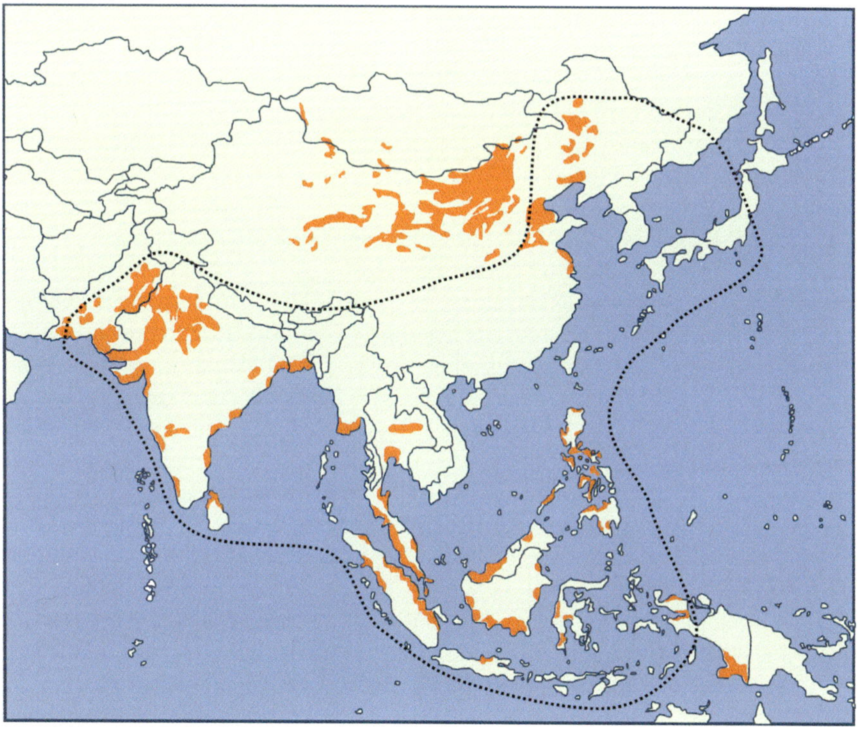

Fig. 10.2 Distribution of saline soils in Asia (*in colored patches*), with demarcation (*dotted*) showing 91% of the rice growing areas of the world confined to South and Southeast Asia

Table 10.1 Distribution of saline and sodic soils in major rice growing countries of South and Southeast Asia

	Area in million hectares (M ha)				
Country	Salinity	%	Sodicity	%	Total
Bangladesh	0.9	6.3	0.0	0.0	14.4
Cambodia	0.2	1.1	0.4	2.2	18.1
India	21.0	6.7	1.5	0.5	315.7
Indonesia	2.1	1.1	0.0	0.0	191.6
Laos	0.0	0.0	0.1	0.4	23.7
Malaysia	1.0	3.0	0.0	0.0	33.3
Myanmar	1.1	1.6	0.0	0.0	67.7
Pakistan	15.8	19.7	0.1	0.1	80.2
Papua New Guinea	0.4	0.9	0.0	0.0	46.2
Philippines	0.0	0.0	0.0	0.0	29.9
Sri Lanka	0.1	1.5	0.5	7.7	6.5
Thailand	1.0	1.9	0.4	0.8	51.3
Viet Nam	0.7	2.1	0.0	0.0	32.9
World	402.5	3.0	434.4	3.2	13,490.7

Data from Bot et al. (2000)

About 7% of India's land area amounting to 21 M ha and 20% of the land area (16 M ha) in Pakistan are salt affected. In India, salinity coupled with waterlogging is seriously threatening agricultural economy in Indo-Gangetic plains covering the states of Haryana, Punjab and Uttar Pradesh. In Pakistan, reports indicate that more than 1 M ha rice area is salt affected with 25–60% reduction in production (Haq et al. 2010) and more than 1.4 M ha land is abandoned due to salinity (World Bank 2006).

Salinity is a common problem along the coastal belts of rice growing countries, especially in India. Such areas are characterized by occasional or frequent sea water ingression during tides resulting in submergence that builds up salt accumulation. Rice is the only viable crop in these areas because it can withstand submergence and shows wide variability for salt tolerance. However, under salt prone conditions, yield is reduced from 10% to 80%, and coupled with erratic rainfall, loss can reach up to 100%. Further, low productivity can also be attributed to occasional flooding, rainfed rice cultivation, frequent water-deficit stress and continued cultivation of traditional low-yielding rice varieties and landraces (Singh et al. 2009). In near future, Asia requires more rice to feed its burgeoning population, since rice is the staple food for more than 90% of Asians. Ironically, millions of hectares of rice cultivable areas are either being rendered uncultivable or are grown with very low yields because of growing soil salinization. Two options to recover more out of such problem areas are reclaiming soil off the salinity and by cultivating salt tolerant varieties. The second option seems more feasible and sustainable, as there is enough variability in the rice germplasm for salt tolerance which could be utilized in rice breeding against soil salinity.

Salt Stress in Rice

Rice is sensitive to salinity, particularly during the seedling stage (Maas and Hoffman 1977) and the earlier set benchmarks (Maas and Grattan 1999; Hanson et al. 1999) indicate that rice yield decreases 12% for every unit (dS/m) increase in ECe above the threshold tolerance of 3.0 dS m^{-1}(Maas 1990). Salt sensitivity in rice is now revised to a much lower threshold tolerance of 1.8 dS m^{-1} with yield decline slope of 9.1% (Grattan et al. 2002). Salinity affects yield components such as panicle length, spikelet number per panicle, and grain yield (Zeng and Shannon 2000), besides delayed panicle emergence and flowering, and reduced seed set percentage due to lower pollen viability (Khatun and Flowers 1995).

As against the earlier belief that the salt induced damage in rice is caused by osmotic imbalance and accumulation of chloride (Cl^-) ions (Tagawa and Ishizaka 1963), it is now known that injury is primarily caused by sodium ion (Na^+) toxicity due to cellular ion imbalance (Mandhania et al. 2006). In contrast to adverse effect on root growth, presence of excess amounts of Na^+ results in greater reduction in shoot growth and yield (Esechie et al. 2002). When compared to other cereals, Cl^- ions are relatively well tolerated by rice at varying concentrations preempting them being toxic (Clarkson and Hanson 1980). Rice plants uptake excess levels of Na^+ under salt rich conditions, interfering with the uptake of potassium (K^+) and calcium (Ca^{2+}) inciting deficiency symptoms. Low K^+ uptake result in high Na^+/K^+ ratio within plants, which together with low Ca^{2+} uptake causes impairment of mineral transport resulting in reduced shoot growth. The rate of conversion of soluble sugars into starch is reduced concomitantly with the reduced uptake of K^+, as K^+ is needed for the catalytic activities of starch biosynthesis enzymes (Zhang et al. 2012). Further, decrease in carbohydrate accumulation under severe salt stress may also occur due to reduced carbon assimilation (Moradi and Ismail 2007; Pattanagul and Thitisaksakul 2008) resulting from the damage of photosynthetic machinery (Moradi and Ismail 2007). Photoinhibition along with salt stress can cause serious damage to photosynthesis, nutrient uptake, water absorption, root growth and cellular metabolism leading to yield loss (Hasegawa et al. 2000; Zeng and Shannon 2000; Zhu 2001). Besides, salt stress rapidly activates several lipid responses in rice leaves, however, whether these responses do have any role in salt tolerance is not clear (Darwish et al. 2009). Further, on long term exposure to salinity, especially during development, morphological modifications may be seen occurring in leaves by development of smaller and densely packed cells with thickened cell walls (Qiu et al. 2007).

Rice cultivars show variable sensitivity towards salinity at different phenological stages, with better tolerance during germination and tillering (Khan et al. 1997). Tolerance drops significantly at young seedling stage and especially during early reproductive stage becoming very sensitive during panicle initiation and fertilization, directly affecting the crop yield (Heenan et al. 1988; Zeng et al. 2001). A possible reason for this variability is the ability of rice plants to grow in standing water that can dilute and leach away excess salts in the soil (Bhumbla and Abrol 1978). Particularly due to the early seedling susceptibility, older seedlings are generally recommended for transplanting into saline soils. Young rice plants of susceptible varieties die after germination, while those of the adapted survive showing reduced growth, together with osmotic adjustments to avoid dehydration. Seedling biomass is now recognized as an important parameter for the survival of transplanted seedlings in saline soils (Summart et al. 2010). However, there are reports in which rice varieties show poor correlation of seedling and adult plant salt tolerance (Moradi et al. 2003), which can pose a major challenge because combinatorial expression of tolerance sustaining throughout the crop lifespan is essential for breeding tolerant varieties (Ismail et al. 2007). Therefore, most of these reports are only suggestive because the earlier investigations on salt tolerance are done on seedlings grown under hydroponics, tissue culture environment, and under artificial salinization of the culture media. Although artificial screening may simulate near perfect situations of salt sufficiency, it may fail to duplicate natural situations that are practically relevant for breeding.

Physiological Basis of Salt Tolerance in Rice

Salt tolerance in rice, is an integrated phenomenon contributed by several traits relating to water and mineral uptake, transpiration, osmotic balance, tissue tolerance, oxidant scavenging and growth vigor (Moradi et al. 2003). Physiological basis of salt tolerance in rice plants is primarily manifested by Na^+ exclusion from young tissues and flag leaves and developing panicles (Asch et al. 2000; Haq et al. 2010). High Na^+ concentration in the apoplastic solution results in increased accumulation of proline in the cytoplasm especially in tolerant rice genotypes, which helps in restoring the osmotic potential between the cytosol and apoplastic solution (Demiral and Turkan 2005). A comprehensive review of mechanisms of salt tolerance (Munns and Tester 2008) with major focus on rice (Ismail et al. 2007; Negrão et al. 2011) can be found elsewhere. Other mechanisms operating in rice include confining of toxic ions to older leaves and vacuoles, secondary responses such as scavenging reactive oxygen species (ROS), enhanced growth response to dilute salts and increased stomatal response. Salt stress was reported to increase chlorophyll concentration in leaves of tolerant and moderately tolerant rice genotypes, with significant levels of chlorophyll *a* concentration and high chlorophyll *a*/*b* ratio. Being the major photosynthetic pigment, reduction in chlorophyll *a* may be associated with reduced photosynthetic activity under salt stress (Moradi and Ismail 2007). Salt induced inhibition of conversion of soluble sugars into starch was less in salt tolerant genotypes (Zhang et al. 2012). Additionally, long-term reduction of mesophyll conductance under salt stress results in anatomical modifications in leaves (Chaves et al. 2009).

In a recent investigation to identify biochemical markers for salt tolerance, significantly reduced level of H_2O_2 activity was observed in the tolerant variety Pokkali, suggesting the existence of an efficient antioxidant defense system to cope up with salt-induced oxidative stress. Supporting this hypothesis, higher activities of antioxidant enzymes and isozyme patterns that are either directly or indirectly involved in the detoxification of ROS, were observed in Pokkali. Further, Pokkali exhibited a higher reduced versus oxidized glutathione ratio (GSH/GSSG) together with a higher ratio of reduced versus oxidized ascorbate ratio and higher activity of methylglyoxal detoxification system (glyoxalase I and II). As reduced glutathione is involved in the ascorbate–glutathione pathway as well as in the methylglyoxal detoxification pathway, the results suggest that both ascorbate and glutathione homeostasis, which is also modulated via glyoxalase enzymes, can be considered as biomarkers for salt tolerance in rice (El-Shabrawi et al. 2010). In salt tolerant genotypes of rice, activity of enzymes such as ascorbate peroxidase, catalase and peroxide dismutase, that are known to be involved in ROS scavenging was found to be either constitutively expressed or induced by salt stress (Moradi and Ismail 2007; Nagamiya et al. 2007; Ahmad et al. 2008).

Genetics of Salt Tolerance in Rice

Genetic Variability and Inheritance of Salt Tolerance

Rice genetic diversity harbors natural variability for salt tolerance. In a large scale screening of rice genotypes conducted at the International Rice Research Institute (IRRI), about 17% of the total 1,38,000 genotypes have been found to possess acceptable levels of salt tolerance (EC 10 dsm^{-1}) at seedling stage (De Datta et al. 1993). A basic understanding on the genetics and relationships between varietal groups and phenotypic variation for salt tolerance is vital for bioprospecting of genes and mining useful alleles. Till date, there have been no systematic studies for comparing within and between group variability for salt tolerance in rice. Several independent studies have reported many landraces and varieties tolerant to salt accumulation by one mechanism or another (Gregorio et al. 2002; Ismail et al. 2007; Mohammadi-Nejad et al. 2010). However, most of the traditional salt tolerant varieties from coastal regions of India such as *pokkali* rice types (Pokkali, Cheruvirippu, Bali, Orkayama, Eravapandy, Orpandy, Oorumundakan, Chettivirippu, Kuruka, Anakodan, Chottupokkali etc.) from Kerala and Nona Bokra, Getu, Kalarata 1–24, SR26B, Damodar, Dasal, Patnai and Nona Sail from West Bengal possess undesirable agronomic and grain quality characters. Other salt tolerant *indica* cultivars grown traditionally in the coastal areas of other countries include Kalimekri, Bhirpala, Kajalsail (Bangladesh), Ketumbar, Kuatik Putih (Indonesia), Khao Seetha (Thailand) and Soc Nau (Vietnam). Several breeding lines have been developed at the IRRI with specific characteristic features for salt tolerance namely IR4630-22-2-5-1-3 (a donor for leaf compartmentation), IR60167-129-3-4 (a donor for tissue tolerance), and IR66946-3R-178-1-1 (also known as FL478). Similarly, popular salt tolerant varieties such as CSR10, CSR13, CSR27 and CSR30 have been developed and released from the Central Soil Salinity Research Institute (CSSRI) at Karnal in India using traditional salt tolerant varieties as parent (Negrão et al. 2011).

Hardly little was known about inheritance of salt tolerance in rice until 1970s, when Akbar and Yabuno (1977) reported that panicle sterility under salt stress was a dominant trait controlled by a small number of genes. Later, overdominance of salt tolerance was demonstrated in crosses between tolerant and susceptible varieties, accompanied by dominance and a sizeable degree of fixable additive variance (Moeljopawiro and Ikehashi 1981). Subsequent genetic studies indicated that salt tolerance in rice was a complex trait under polygenic control (Flowers 2004) with large environmental effects and low heritability (Gregorio and Senadhira 1993). However, in crosses with moderately tolerant and susceptible parents, duplicate type of epistasis was also reported (Ray and Islam 2008). It is rather difficult to consolidate and quantify genetic effects of tolerant traits since screening methods for tolerance were different at different growth stages and hardly any relation exists between phenological tolerance expressions. Nevertheless, considerable variation for difference in survival traits and components of salt injury are reported coupled with significant genotype × environment interactions (Zhou et al. 2010; Ali et al. 2006). Many workers have attempted genetic component analysis on various traits, especially on Na^+/K^+ compensation, and reported both additive and dominant gene effects and overdominance (Gregorio and Senadhira 1993; Ray and Islam 2008). Agronomic performance under salinity showed preponderance of dominant gene action for yield components and additive gene action for morphological traits (Kalaiyarasi et al. 2002; Sankar et al. 2008). Recent mixed model genetic analysis (Wang et al. 2010) on seed germination under salinity, reported polygenic control of the component traits with preponderance of two to three major genes showing high heritability and accounting for 12.5–99.0% of the total phenotypic variation. A specific genetic model was fitted for each trait, that showed control of two major genes on imbibition rate, two major genes plus polygene on germination and vigor indices, three major genes plus polygene on germination rate and two major genes or two major genes plus polygene on shoot and root length. Significant dominant effects and absence of epistasis for salt tolerant traits have also been reported suggesting the possibility of hybrid rice development under saline situations (Ray and Islam 2008) and using modified bulk and single seed descent methods with later generation selections using larger population as breeding strategies (Gregorio and Senadhira 1993).

Molecular Genetics

Molecular mechanisms triggered in plants' defense against salt stress may be based on either avoidance or tolerance strategies. Selective ion uptake, dilution of excess ions, sequestration and extrusion are the major avoidance mechanisms, manifested at cellular, organellar or whole-plant levels. Moreover, plants have an array of tolerance mechanisms to sustain growth under unfavorable conditions, triggering a cascade of systemic reactions related to salt injury that jeopardize osmotic balances, photorespiration, mineral transport, membrane stability, cell division, cell architecture organization and survival. Therefore most of the defensive genes against salt injury may share common profile of other abiotic stresses (reviewed by Bartels and Sunkar 2005; Chaves et al. 2003; Munns and Tester 2008; Witcombe et al. 2008; Singh et al. 2008; Peleg et al. 2011; Negrão et al. 2011; Krasensky and Jonak 2012). In rice, several genes have been functionally validated for salt tolerance, many of which are sourced from other systems (Table 10.2), besides rice genes (Table 10.3).

Monovalent ion transport such as that of K^+, Na^+ and Cl^- are governed by ion channel and carrier proteins and their isoforms that are encoded by large multi-gene families. When encountered with salt stress, these transporters play crucial roles in uptake, efflux, translocation and sequestration. Main classes of ion channel proteins are non-selective cation channels (NSCC), two-pore K^+ channels (TPK), shaker proteins and voltage gated Cl^- channels (CLC). They include sub-forms such as cyclic nucleotide gated channel (CNGC), glutamate like receptor (GLR), and shaker type K^+ inward rectifying channels (KIRC) and outward rectifying channels (KORC). The carrier proteins include (a) symporters and their sub-forms such as cation chloride co-transporter (CCC), K^+ uptake permease (KUP) and high affinity K^+ transporters (HKT), and (b) antiporters that include Na^+/H^+ exchangers (NHX), cation/H^+ exchangers (CHX), K^+/H^+ exchangers (KHX) and salt overlay sensitive (SOS) gene families (Szczerba et al. 2009). Primary mechanism of salt tolerance in plants is now known to be manifested by ion exclusion mechanisms, especially of Na^+, in association with osmo-regulation, in which physiological friendly solutes such as sugars (trehalose, fructan), sugar alcohols (galactinol, trehalose and mannitol), amino acids (proline) and amines (glycine betaine) are accumulated in the plant systems to ward off the ill effects of accumulating solutes. Besides, several transcription factors (TF) and their cis-regulatory sequences act as molecular switches of stress response gene expression at temporal and spatial levels (Puranik et al. 2012).

The HKT Gene Family

Na^+-K^+ homeostasis governs the principal mechanisms of salt tolerance in plants such as ion uptake and transport, sequestration and extrusion. The major genes involved in Na^+ uptake under saline conditions are governed by HKTs, while vacuolar Na^+/H^+ antiporters regulate Na^+ sequestration in vacuoles and membrane Na^+/H^+ antiporters regulate Na^+ extrusion as seen in halophytes (Yamamoto and Yano 2008). Plant HKTs represent a class of xylem–parenchyma-expressed Na^+-permeable genes that govern primary mechanism mediating salt tolerance and Na^+ exclusion from leaves. HKT genes are the most widely studied genetic system for salt tolerance in *Arabidopsis*, and in rice they constitute a large gene family of nine genes consisting of two distinctly grouped sub-families of *OsHKT1* with five genes (Garciadeblás et al. 2003; Platten et al. 2006; Huang et al. 2008b) and *OsHKT2* containing four genes (Table 10.4). Although named after their relation with bacterial high affinity K^+ transport genes, HKT genes are primarily Na^+ transporters and

Table 10.2 External genes functionally tested in transgenic rice for imparting salt tolerance

Code	Encoded protein	Source organism	Promotor	Associated trait	References
HVA1	LEA protein	*Hordeum vulgare*	Rice Actin1	Stress response	Xu et al. (1996)
P5CS	Δ^1-pyrroline-5-carboxylate synthetase	*Vigna aconitifolia*	Stress-inducible	Proline accumulation	Zhu et al. (1998)
codA	Choline oxidase	*Arthrobacter globiformis*	CaMV 35S	Glycinebetaine synthesis	Sakamoto et al. (1998)
MnSOD	Superoxide dismutase	*Saccharomyces cerevisiae*	–	Antioxidants synthesis	Tanaka et al. (1999)
mtlD, gutD	Mannitol-1-phosphate dehydrogenase Glucitol-6-phosphate dehydrogenase	*Escherichia coli*	CaMV 35S	Sugar alcohol synthesis	Wang et al. (2000)
ADC	Arginine decarboxylase	*Avena sativa*	ABA inducible	Polyamine activity	Roy and Wu (2001)
SAMDC	S-adenosylmethionine decarboxylase	*Tritordeum*	ABA inducible	Polyamine synthesis	Roy and Wu (2002)
codA	Choline oxidase	*Arthrobacter globiformis*	CaMV 35S	Glycinebetaine synthesis	Mohanty et al. (2002)
AgNHX1	Na^+/H^+ antiporter1	*Atriplex gmelini*	CaMV 35S	Na^+ homeostasis	Ohta et al. (2002)
otsA, otsB	Trehalose biosynthetic genes	*Escherichia coli*	CaMV 35S, rice *rbcS*	Trehalose accumulation	Garg et al. (2002)
HVA1	LEA protein	*Hordeum vulgare*	Rice Actin1	Stress response	Rohila et al. (2002)
TPS, TPP	Trehalose biosynthetic genes	*Escherichia coli*	Ubiquitin1	Trehalose accumulation	Jang et al. (2003)
δ-OAT	Ornithine-δ-aminotransferase	*Arabidopsis thaliana*	CaMV 35S	Proline accumulation	Wu et al. (2003)
P5CS	Δ^1-pyrroline-5-carboxylate synthetase	*Vigna aconitifolia*	CaMV 35S	Proline accumulation	Su and Wu (2004)
CN Atr	Calcineurin	Mouse	CaMV 35S	Ion homeostasis	Ma et al. (2005)
AtMYB2	MYB transcription factor	*Arabidopsis thaliana*	AIPC	Stress response	Malik and Wu (2005)
CBF3, ABF3	ABA independent CBF3/DREB1A	*Arabidopsis thaliana*	Ubiquitin1	Stress response	Oh et al. (2005)
nhaA	Na^+/H^+ antiporter	*Escherichia coli*	CaMV 5S	Ion homeostasis	Wu et al. (2005)
DREB1A, DREB1B, DREB1C	Dehydration responsive element binding (DREB)	*Arabidopsis thaliana*	CaMV 5S , Ubiquitin	Dehydration response	Ito et al. (2006)

GST, CAT	Glutathione S-transferase, Catalase	*Suaeda salsa*	CaMV 35S	Antioxidant activity	Zhao and Zhang (2006)
SOD2	Sodium2	*Schizosaccharomyces pombe*	CaMV35S	Na^+ homeostasis	Zhao et al. (2006a)
SsNHX1, AVP1	Vacuolar membrane Na^+/H^+ antiporter, Vacuolar H^+ pyrophosphatase proton pump	*Suaeda salsa, Arabidopsis thaliana*	CaMV35S	Na^+ homeostasis	Zhao et al. (2006b)
codA	Choline oxidase	*Arthrobacter pascens*	ABA inducible	Glycinebetaine synthesis	Su et al. (2006)
*kat*E	Catalase	*Escherichia coli*	CaMV 35S	Antioxidant activity	Nagamiya et al. (2007)
HvCBF4	CBF transcription factor	*Hordeum vulgare*	Ubiquitin1	Cold tolerance	Oh et al. (2007)
PgNHX1	Vacuolar Na+/H+ antiporter	*Pennisetum glaucum*	ABA inducible	Na^+ homeostasis	Verma et al. (2007)
Cu/Zn SOD1	Superoxide dismutase	*Avicennia marina*	Ubiquitin	Antioxidative pathway	Prashanth et al. (2008)
TERF1	Tomato ethylene responsive factor	*Lycopercicum esculentum*	CaMV 35S	Stress regulatory	Gao et al. (2008)
NtOPBP1	Osmotin promoter binding protein 1	*Nocotiana tabacum*	Ubiquitin	Salt tolerance	Chen and Guo (2008)
TaSTRG	Salt tolerance-related gene	*Triticum aestivum*	Actin	Salt stress response	Zhou et al. (2009)
AtHKT1;1	High affinity K^+ transporter	*Arabidopsis thaliana*	Root specific	Na^+ homeostasis	Plett et al. (2010)
P5CSF129A	Δ^1-pyrroline-5-carboxylate synthetase	*Vigna aconitifolia*	CaMV 35S	Proline accumulation	Kumar et al. (2010)
PgNHX1	Vacuolar Na+/H+ antiporter	*Pennisetum glaucum*	CaMV 35S	Na^+ homeostasis	Islam et al. (2010)
P5CS	Δ^1-pyrroline-5-carboxylate synthetase	*Vigna aconitifolia*	CaMV 35S	Proline accumulation	Karthikeyan et al. (2011)
PDH45	DEAD-box helicase	*Pisum sativum*	CaMV 35S	Salt stress response	Amin et al. (2012)

Table 10.3 Rice genes that are functionally tested for improved salt tolerance through transgenesis

Gene	Encoded protein	Transgene candidate	Promotor	Associated trait	References
OsCDPK7	Calcium-dependent protein kinase	*Oryza sativa*	CaMV 35S	Cytosolic Ca^{2+} influx	Saijo et al. (2000)
GS2	Chloroplastic glutamine synthetase	*Oryza sativa*	CaMV35s	Photorespiration	Hoshida et al. (2000)
OsMAPK5a	Mitogen-activated protein kinase	*Oryza sativa*	Ubiquitin	Cytosolic Ca^{2+} influx	Xiong and Yang (2003)
OsDREB1A	DREB Transcription factor	*Arabidopsis thaliana*	CaMV35s	Dehydration response	Dubouzet et al. (2003)
OsNHX1	Na^+/H^+ antiporter	*Oryza sativa*	CaMV 35S	Na^+ homeostasis	Fukuda et al. (2004)
OsMAPK44	Mitogen-activated protein kinase	*Oryza sativa*	–	Cytosolic Ca^{2+} influx	Jeong et al. (2006)
OsDREB1A, *OsDREB1B*	Dehydration responsive element binding (DREB)	*Oryza sativa*	Ubiquitin	Dehydration response	Ito et al. (2006)
SNAC1	Stress-responsive NAC 1	*Oryza sativa*	CaMV 35S	Stress response	Hu et al. (2006)
OsSOS1	Salt overlay sensitive	*Arabidopsis thaliana*	CaMV 35S	Na+ homeostasis	Martinez-Atienza et al. (2007)
Rab16A	Responsive to ABA dehydrins	*Nicotiana tabacum*	Rab16A	Dehydration response	RoyChoudhury et al. (2007)
OsNHX1	Vacuolar type Na^+/H^+ antiporter	*Oryza sativa*	CaMV 35S	Na^+ homeostasis	Chen et al. (2007)
OsCIPK12	Calcineurin B-like protein-interacting protein kinase	*Oryza sativa*	Ubiquitin	Cytosolic Ca^{2+} influx	Xiang et al. (2007)
OsKAT1	Shaker family K^+ channel	*Oryza sativa*	CaMV 35S	K^+ uptake	Obata et al. (2007)
OsAPXa, OsAPXb	Ascorbate peroxidase	*Oryza sativa*	CaMV 35S	Antioxidant activity	Lu et al. (2007)
OsNAC6	NAC transcription factor	*Oryza sativa*	Ubiquitin	Stress response	Nakashima et al. (2007)
glyII	Glyoxalase II	*Oryza sativa*	CaMV 35S	Salt tolerance	Singla-Pareek et al. (2008)
OsTOP6A1	Meiotic recombination protein	*Arabidopsis thaliana*	CaMV 35S	Multiple stress response	Jain et al. (2008)
OsiSAP8	Stress associated protein	*Nicotiana benthamiana*	CaMV 35S	Multiple stress response	Kanneganti and Gupta (2008)
OsiSAP8	Stress associated protein	*Oryza sativa*	CaMV 35S	Multiple stress response	Kanneganti and Gupta (2008)

OsHsfA2e	Heat shock transcription factor	*Arabidopsis thaliana*	CaMV 35S	Stress response	Yokotani et al. (2008)
ZFP252	TFIIIA-type zinc finger protein	*Oryza sativa*	CaMV 35S	Proline accumulation	Xu et al. (2008)
OsTPP1	Trehalose-6-phosphate phosphatase	*Oryza sativa*	CaMV 35S	Trehalose accumulation	Ge et al. (2008)
SNAC2	Stress-responsive NAC	*Oryza sativa*	Ubiquitin, *Ubi1*	Stress response	Hu et al. (2008)
ONAC063	NAC transcription factor	*Arabidopsis thaliana*	CaMV 35S		Yokotani et al. (2009)
ONAC045	Stress-responsive NAC	*Oryza sativa*	CaMV 35S	LEA gene expression	Zheng et al. (2009)
AP37	APETALA 2 transcription factor	*Oryza sativa*	*OsCC1*	Stress regulatory	Oh et al. (2009)
DST	Ethylene zinc finger protein	*Oryza sativa*	CaMV 35S	Stomatal control	Huang et al. (2009)
OsSIK1	Receptor-like kinase	*Oryza sativa*	*OsSIK1*	Antioxidant activity	Ouyang et al. (2010)
OsNAC10	Stress-responsive NAC	*Oryza sativa*	*GOS2*, *RCc3*	Stress response	Jeong et al. (2010)
OsNAC5	Stress-responsive NAC	*Oryza sativa*	Ubiquitin	LEA gene expression	Takasaki et al. (2010)
ZFP179	Cys2/His2-type zinc finger protein	*Oryza sativa*	CaMV 35S	Proline accumulation	Sun et al. (2010)
OsNHX1	Na^+/H^+ exchanger	*Oryza sativa*	CaMV 35S	Osmoregulation system	Liu et al. (2010b)
OsVP1	H^+-pyrophosphatase in tonoplasts	*Oryza sativa*	CaMV 35S	Na^+ homeostasis	Liu et al. (2010b)
OsTPKb	Two pore K^+ channel	*Oryza sativa*	CaMV 35S	K^+ homeostasis	Mian (2010)
OsAKT1	K^+ inward rectifying channel	*Oryza sativa*	CaMV 35S	K^+ uptake	Mian (2010)
OsTPS1	Trehalose-6-phosphate synthase	*Oryza sativa*	*Actin1*	Trehalose accumulation	Li et al. (2011)
OsHAK5	Sodium-insensitive potassium transporter	*Nicotiana tabacum*	CaMV 35S	K^+ transport	Horie et al. (2011b)
OsNAC5	Stress-responsive NAC	*Oryza sativa*	CaMV 35S	Proline accumualtion	Song et al. (2011)
OsHsfC1b	Heat shock factors	*Oryza sativa*	Ubiquitin	Stress response	Schmidt et al. (2012)

Table 10.4 High affinity potassium transporters (HKT) on rice reference genome cv. Nipponbare

Gene	Chromosome	Genome location	Other names	Length (bp)	No. of transcripts
OsHKT1;1	4	LOC_Os04g51820	*OsHKT4*	2,443, 2,224, 2,097	3
OsHKT1;2	4	–	*OsHKT5*	Pseudogene	–
OsHKT1;3	2	LOC_Os02g07830	*OsHKT6*	1,733	1
OsHKT1;4	4	LOC_Os04g51830	*OsHKT7*	2,269	1
OsHKT1;5	1	LOC_Os01g20160	*OsHKT8, SKC1*	2,164	1
OsHKT2;1	6	LOC_Os06g48810	*OsHKT1*	1,881	1
OsHKT2;2	?	–	*OsHKT2*	–	–
OsHKT2;3	1	LOC_Os01g34850	*OsHKT3*	1,628	1
OsHKT2;4	6	LOC_Os06g48800	*OsHKT9*	1,557	1

regulates a variety of cellular mechanisms such as Na^+ sequestration, extrusion and exclusion (Hauser and Horie 2010) and play a key role in regulation of Na^+ homeostasis (Rodríguez-Navarro and Rubio 2006). Evidences show that *OsHKT1* genes distinctly act as Na^+ uniporters and *OsHKT2* as Na^+-K^+ symporters or uniporters depending on the ionic conditions (Huang et al. 2008b; Pardo 2010). A detailed review of HKT transporter-mediated salt tolerance mechanisms can be found at Horie et al. (2009) and Hauser and Horie (2010).

HKT transporters fulfill distinctive roles at the whole plant level in rice, each system playing decisive roles in different cell types. *OsHKT1;5* (*OsHKT8*), was the first among the HKT genes to be mapped, identified initially as a quantitative trait locus (QTL), *shoot* K^+ *content 1* (*SKC1*; Lin et al. 2004). Functional analysis later identified *SKC1* to code for a transporter that unloads Na^+ from the root xylem and preferentially expressed in the parenchyma cells surrounding xylem vessels. It was postulated that relative salt tolerance of rice landraces Pokkali and Nona Bokra is due to the presence of *OsHKT1;5* (Ren et al. 2005). *OsHKT1;1* and *OsHKT1;3* are exclusively permeable to Na^+ and are expressed in roots and leaves, suggesting more wider functional role other than ion transport. These transporters may be involved in ion fluxes triggering turgor changes in bulliform cells that control leaf rolling and unrolling, allowing them to regulate leaf folding in response to environmental conditions (Jabnoune et al. 2009). Although *OsHKT1;4* sequences has a structural synteny with *Na*$^+$ *exclusion 1* (*Nax1*) gene conferring salt tolerance in durum wheat (Huang et al. 2006), that is preferentially expressed in shoot (Garciadeblás et al. 2003), no QTL for salt tolerance was detected on its locus on rice chromosome 4, suggesting that it may either be silenced or unexpressed in rice (James et al. 2011). *OsHKT1;2* is identified as a pseudogene in Nipponbare genome and no functional properties are reported for this gene so far.

Depending on the ionic conditions, members of HKT2 transporter subfamily were found to mediate Na^+-K^+ symport under normal concentrations and Na^+-selective transport under high Na^+ concentrations. *OsHKT2;1* was a highly conserved protein (Oomen et al. 2012) that is expressed strongly in roots, and weakly in mesophyll cells of mature leaves, displaying three conducting modes depending on external Na^+ and K^+, K^+-Na^+ symport, Na^+ and K^+ uniport (Jabnoune et al. 2009). The second member of the family, *OsHKT2;2* is demonstrated to act as Na^+-K^+ symporter in tobacco cells

(Yao et al. 2010). Recently an isoform of *OsHKT2;2* (No-*OsHKT2;2/1*) that is likely to have originated from a deletion in chromosome 6, producing a chimeric gene is identified in Nona Bokra, a highly salt-tolerant cultivar (Oomen et al. 2012). It has a 5′ region corresponds to that of *OsHKT2;2,* as found in Pokkali but with a 3′ region corresponds to that of *OsHKT2;1.* In contrast to *OsHKT2;1,* No-*OsHKT2;2/1* is essentially expressed in roots and displays a significant level of permeability to Na^+ and K^+ even at high external Na^+ concentrations. No-*OsHKT2;2/1* perhaps contributes to the salt tolerance of Nona Bokra by enabling high root K^+ uptake under saline conditions. Other genes of the sub-family 2, *OsHKT2;3* and *OsHKT2;4* are structurally 93% similar at the amino acid sequence level and traceable on the reference rice genome of cv. Nipponbare (Horie et al. 2011a). They retain the four selectivity filter Gly residues typical of class II HKT transporters (Horie et al. 2009; Hauser and Horie 2010). Recently, *OsHKT2;4* was shown to possess atypical Na^+ transport properties and show dominant selectivity for K^+ under competition over Mg^{2+} and Ca^{2+} ions, however, *OsHKT2;3* failed to complement a high-affinity K^+ uptake-deficient mutant of yeast strain (Horie et al. 2011a).

Other Genes for Salt Tolerance

Vacuolar sequestration by ion antiporters plays a significant role in maintaining osmotic balance in plants. Apart from these, several genes coding for osmotic homeostasis such as protein kinases, aquaporins and enzymes for osmolyte biosynthesis, enzymes for damage prevention and repair pathways such as antioxidant biosynthesis, Late Embryogenesis Abundant (LEA) proteins, dehydrins, antitoxic enzymes, chaperons, proteases, ubiquitination-related enzymes (Abogadalla 2010; Peleg et al. 2011), stress signaling pathways and a variety of transcription factors also regulate temporal and spatial gene expression. Polyamines (PA), the small aliphatic molecules positively charged at cellular pH, are modulated by salt stress and high PA levels have been positively correlated with stress tolerance (Alcázar et al. 2006; Kusano et al. 2008; Krasensky and Jonak 2012; Ahmad et al. 2012b). The induction of two *OsLEA* genes (*OsLEA3*, *OsLEA21*) by salinity has been clearly demonstrated through semi-quantitative RT-PCR analysis (Wang et al. 2007; Hu 2008). The membrane protein Na^+/H^+ antiporters catalyze the exchange of Na^+ for H^+ thereby helping plants to tolerate high salt levels through internal distribution of ions for osmotic adjustment and it has been shown that the expression of antiporter gene, *OsNHX1* is increased in rice roots and shoots with salt stress (Fukuda et al. 2004). They further demonstrated that transgenic rice plants overexpressing the gene showed improved salt tolerance. Transcription Factors (TFs) defined as proteins which can activate or repress the gene expressions with its affinity for sequence specific DNA binding, has been shown to be involved in plant responses to stress including salinity. More than 40 TFs belonging to APETALA2/ethylene-responsive element binding proteins (AP2/EREBP), basic leucine zipper (bZIP), homeodomain proteins (HD), zinc finger proteins (ZFP), myeloblastosis (MYB), heteromeric CCAAT-box-binding heme-activator protein complex (CCAAT-HAP2), heat shock factor (HSF) and No apical meristem [NAM]-Arabidopsis Transcription Activation Factor [ATAF] - Cup shaped cotyledon

[CUC] (NAC) gene families have been shown to be involved in rice responses to high salinity (Negrão et al. 2011). Three genes encoding ZFPs namely *ZFP179, ZFP182* and *SRZ1* have been reported as responsive to high salinity (Huang et al. 2007). The overexpression of *ZFP179* (Sun et al. 2010) and *ZFP182* and repression of *SRZ1* improves salt tolerance (Huang et al. 2008a). Four genes encoding bZIPs (*OSABF1, OsAB15, OsbZIP23* and *OSBZ8*) have been reported to be associated with salt stress responses (Hossain et al. 2010; Nakagawa et al. 1996; Xiang et al. 2008; Zou et al. 2008). Overexpression of *OsbZIP23* improves salt tolerance while *OSAB15* is a negative regulator of salt stress response and repression of this gene helps improving salt tolerance (Zou et al. 2008). TFs belonging to different families, such as MYB, HSF, Trihelix or CCAAT-HAP2 (*OsMYB3R-2, OsGTγ-1* and *OsHsfA2e)* have also been found to be induced by high salinity and all improve salt tolerance (Dai et al. 2007; Fang et al. 2010; Liu et al. 2010a). More recently, Schimdt et al. (2012) showed that expression of an HSF, *OsHsfC1b* was induced by salt and overexpression of this gene improves salt tolerance in rice. One of the largest TF superfamily in plants is the NAC, which is known to play a significant role in abiotic stress tolerance in plants including salt tolerance. Rice has 151 NAC genes that are being shown to impart resistance to various stresses (Puranik et al. 2012).

Salt overly sensitive (SOS) pathway has been demonstrated to play a remarkable role in salt tolerance in *Arabidopsis*. A calcium signal elicited by salt stress is picked up by *SOS3* protein and sends a downstream signal that activates *SOS2*, a serine/threonine protein kinase. *SOS3* together with *SOS2* regulate *SOS1*, a salt tolerance effector gene that encodes for a plasma membrane antiporter Na^+/H^+ (Yamamoto and Yano 2008). Rice has a conservative system of SOS pathway, with *OsSOS1, OsSOS2/OsCBL4* and *OsSOS3/OsCIPK24* showing functional similarity to their *Arabidopsis* counterparts *AtSOS1*, *AtSOS2* and *AtSOS3* (Martínez-Atienza et al. 2007; Kumar et al. 2012). Differential expression *OsSOS2* has been established in salt tolerant (Pokkali) and sensitive (IR64) genotypes with tissue specific expression in field grown mature plants (Kumar et al. 2012).

Quantitative Trait Loci and Markers

Extending the DNA based molecular technology to classical linkage analysis, mapping of QTLs is the simplest way of identifying trait-related genomic regions that are otherwise difficult to identify due to several interfering factors such as polygenes, linkage, and low heritability. To date, several QTLs have been mapped for salt tolerance related traits in rice (Table 10.5), especially for the vegetative stage tolerance. Significant QTLs for reproductive stage tolerance are yet to be identified in rice (Jena and Mackill 2008). However, excepting few significant ones (Table 10.6) most of the QTLs identified so far are small effect QTLs and many of those reported from populations of early generations probably may remain in reports.

Earliest attempt to map QTL for salt tolerance in rice was reported by Zhang et al. (1995), in which a QTL was mapped on chromosome 7 in an F_2 population

Table 10.5 Quantitative trait loci (QTLs) mapped for salt tolerance in rice

Cross	Population	No. of QTL	Method	Marker system	References
M20/77–170	F_2	1	–	RFLP	Zhang et al. (1995)
IR29/Pokkali	RIL	10	IM	AFLP	Gregorio (1997)
Tesanai 2/CB	RIL	1	SF-ANOVA	RFLP	Lin et al. (1998)
M20/77–170	F_2	1	SMA	RAPD	Ding et al. (1998)
Zhaiyeqing 8/Jingxi 17	DHL	8	SIM, CIM	RFLP	Gong et al. (1999)
IR64/Azucena	DHL	7	SIM	RFLP	Prasad et al. (2000)
IR 59462[a]	F_7	16	–	AFLP	Flowers et al. (2000)
IR4630/IR15324	RIL	25	SMA	AFLP, RFLP, SSR	Koyama et al. (2001)
Zhaiyeqing 8/Jingxi 17	DHL	24	SIM	RFLP	Gong et al. (2001)
IR29/Pokkali	RIL	1	–	AFLP, SSLP	Bonilla et al. (2002)
Tesanai 2/CB	F_8	31	SMA	RFLP	Masood et al. (2004)
Nipponbare/Kasalath	BIL	28	IM	RFLP	Takehisa et al. (2004)
Nona Bokra/Koshihikari	F_3	8	SIM	RFLP	Lin et al. (2004)
IR64/ Tarom Molaii	BIL	??	CIM	SSR	Fotokian et al. (2005)
Milyang 23/Gihobyeo	RIL	3	IM	RFLP	Lee et al. (2007)
CSR 27/MI 48	F_2	6	SIM	STMS	Ammar et al. (2007)
IR64/Binam	BIL	13–22	SIM	SSR	Zang et al. (2008)
AS996/IR50404	RIL	1	Regression	SSR	Lang et al. (2008)
Tarommahali/Khazar	F_3	32	CIM	SSR	Sabouri and Sabouri (2008)
Tarommahali/Khazar	F_2	14	CIM	SSR	Sabouri et al. (2009)
Tarommahali/Khazar	F_3	12	CIM	SSR	Sabouri and Biabani (2009)
Ilpumbyeo/Moroberekan	BIL	8	CIM	SSR	Kim et al. (2009)
CSR 27/MI 48	RIL	18	CIM	SSR, SNP	Pandit et al. (2010)
IR29/Pokkali	RIL	17	CIM	SSR	Thomson et al. (2010)
Co39/Moroberekan	RIL	Many	IM	RFLP	Ul Haq et al. (2010)

(continued)

Table 10.5 (continued)

Cross	Population	No. of QTL	Method	Marker system	References
IR26/Jiucaiqing	F_9	16	MIM	SSR	Wang et al. (2011)
Tarome-Molaei/Tiqing	BIL	14	CIM	SSR	Ahmadi and Fotokian (2011)
Pokkali/IR29	BIL	13	SMA, Regression	SSR	Alam et al. (2011)
Pokkali/Shaheen Basmati	F_3	22	SMA	SSR	Javed et al. (2011)
BRRI Dhan40/ IR61920-3B-22-2-1	F_2	3	SMA, CIM	SSR	Islam et al. (2011)
Teqing/*Oryza rufipogon*	IL	15	SMA	SSR	Tian et al. (2011)

[a]IR 59462 = Nona Bokra/Pokkali//IR 4630-22-2-5-1-3/IR 10167-129-3-4

Table 10.6 Prominent quantitative trait loci mapped under salt tolerance screening

Trait	QTL	Chromosome	Flanking markers	R^2 (%)	References
Salt tolerance[a]	*Saltol*	1	P3/M9-8 – P1/M-9-3	81.0	Gregorio (1997)
Seedling survival days	–	1	RG612 – C131	14.3	Gong et al. (1999)
Seedling root length	*qSRTL-6*	6	RG162 – RG653	18.9	Prasad et al. (2000)
Salt tolerance	*Saltol*	1	C52903S – C1733S	39.2	Bonilla et al. (2002)
Salt tolerance	*Saltol*	1	RM23 – RM140	43.2	Bonilla et al. (2002)
Salt tolerance	*Saltol*	1	CP03970 – CP06224	–	Niones (2004)
Shoot K^+ concentration	*qSKC-1*	1	C1211-S2139	48.5	Lin et al. (2004)
Shoot Na^+ concentration	*qSNC-7*	7	C1057-R2401	40.1	Lin et al. (2004)
Seedling salt tolerance	*qST-1*	1	Est12 – RZ569A	27.8	Lee et al. (2007)
Shoot Na–K ratio	*qSNK1(Saltol)*	1	RM1287 – RM10825	20.0	Thomson et al. (2010)
Imbibition rate	*qIR-6*	6	RM3687 – RM3306	33.6	Wang et al. (2011)
Imbibition rate	*qIR-9*	9	RM276 – RM5531	33.7	Wang et al. (2011)
Germination percentage	*qGP-2*	2	RM8254 – RM5804	36.5	Wang et al. (2011)
Germination percentage	*qGP-9*	9	RM219 – RM7048	43.7	Wang et al. (2011)
Relative root dry weight	*qRDW-10*	10	RM273	22.7	Tian et al. (2011)
Relative shoot dry weight	*qRSW-10*	10	RM273	17.3	Tian et al. (2011)
Relative total dry weight	*qRTW-10*	10	RM273	18.5	Tian et al. (2011)

[a]Salt tolerance = high K^+ uptake + low Na^+ uptake + low Na^+/K^+ ratio

derived of the cross M-20 × 77–170. M-20 was a stable mutant of 77–170 derived by *in vitro* selection. Later a large effect QTL was mapped on chromosome 1 in an IR29/Pokkali derived recombinant inbred line (RIL) population significantly influencing three salt tolerant traits viz., high K^+ uptake, low Na^+ uptake and low Na^+/K^+ ratio (Gregorio 1997). Named *Saltol*, this remains as the most prominent QTL mapped so far for salt tolerance in rice. Subsequently, two large effect QTLs were mapped for shoot concentration of Na^+ and K^+, *qSNC-7* on chromosome 7 and *qSKC-1* on chromosome 1, respectively in an $F_{2:3}$ population between Nona Bokra, a salt tolerant *indica* landrace and Koshihikari (Lin et al. 2004), with *qSNC-7* explaining 49% and *qSKC-1* explaining 40% of the phenotypic variation. Later, *qSKC-1* was cloned and found to encode a member of HKT-type transporters, *OsHKT1;5*, that is preferentially expressed in the parenchyma cells surrounding the xylem vessels. *SKC1* protein functions as a Na^+-selective transporter, involved in regulating K^+/Na^+ homeostasis under salt stress (Ren et al. 2005).

Map based cloning has become a useful tool in identifying genes that are responsible for the desired trait expression. It is now known that the best route forward to use a QTL is to identify it at the molecular level and to check their expression in each QTL combination. Developing of chromosome segment substitution lines (CSSLs) through backcross procedure will provide a route to accelerate this process (Yamamoto et al. 2009). Although a larger proportion of genes cloned by map-based cloning belongs to simple Mendelian traits, quantitative genes are also attempted to be cloned, with an objective of pyramiding genes targeting quantum improvement in the salt tolerance.

Recently a novel technique called MutMap-a method which combines DNA sequencing and EMS induced mutagenesis-has been developed for rapid gene isolation using a cross of the mutant to wild-type parental line (Abe et al. 2012). A mutant is crossed once with the wild type used for mutagenesis, followed by subsequent selfing. Gene identification is realized faster using MutMap with the help of next generation sequencing following the unequivocal segregation between the mutant and wild-type phenotypes. A programme for screening for salt tolerant genes from the cultivar Hitomebore has been initiated using MutMap, aiming at developing rice cultivars suitable for cultivation in the 2011 tsunami-hit of paddy fields of the Northern Japan coast (Abe et al. 2012).

Massive parallel sequencing of mRNA using RNA-Sequencing in Nipponbare has enabled the identification and annotation of genes which are differentially expressed under salt stress. Among the unannotated genes, 213 genes from shoot and 436 genes from root were differentially expressed in response to salinity stress (Mizuno et al. 2010).

The Saltol Region

The *Saltol* was mapped on the short arm of rice chromosome 1 derived from the tolerant parent Pokkali by AFLP genotyping (Gregorio 1997). The QTL had a LOD score of 14.5 and explained up to 81% of the phenotypic variation. Subsequently,

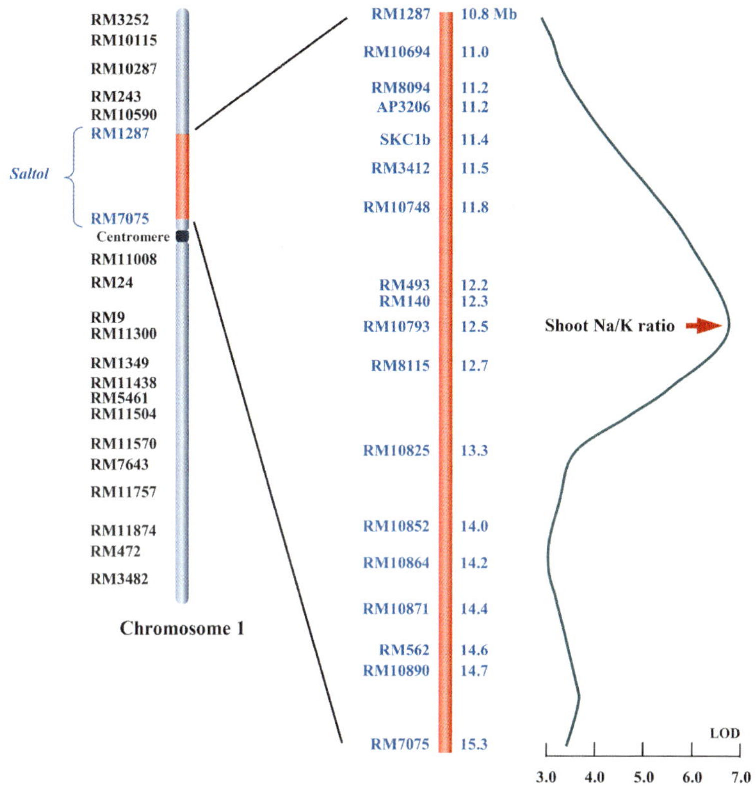

Fig. 10.3 The *Saltol* region on chromosome 1 showing 18 polymorphic SSR markers, and peak of the QTL position for shoot Na/K ratio by simple interval mapping

Bonilla et al. (2002) integrated RFLP and SSR markers to the *Saltol* map, and in a hydroponic screen at the seedling stage using 54 RILs remapped this QTL that explained 43% of the phenotypic variation for shoot Na–K ratio. Further, many workers remapped and fine mapped *Saltol* in other mapping populations (Niones 2004; Lin et al. 2004; Elahi et al. 2004; Thomson et al. 2010). This QTL region is now confined within a 1.2 kb region (Niones 2004). Recent confirmative mapping of *Saltol* locus (Fig. 10.3), shows that *Saltol* contributes to Na^+/K^+ homeostasis with an LOD of 7.6 and R^2 of 27% across the 140 RILs and a 30% decrease in the shoot Na–K ratio (Thomson et al. 2010). *Saltol* is flanked between microsatellite markers, RM1287 and RM7075 at physical position between 10.8 and 15.3 Mb on chromosome 1 (Alam et al. 2011).

SalT, another important gene, co-localized with *Saltol* on chromosome 1 was first isolated and characterized from the roots of salt-treated rice plants (Claes et al. 1990). Its expression is correlated with osmoprotectants, such as trehalose and proline. The treatment of rice with trehalose improved salt tolerance but suppressed *SalT* upregulation, while proline treatment increased growth inhibition of salt-treated rice plants and upregulated *SalT* (Garcia et al. 1997). *SalT* transcripts were found to accumulate with wounding and heat treatment and the gene is also induced by fungal elicitors, jasmonic acid and abscisic acid (de Souza et al. 2003; Kim et al. 2004; Moons et al. 1997), which suggests that the role of *SalT* protein may be involved in a broader response/ sensor mechanism to the imposed stress (de Souza et al. 2003).

FL478 (IR 66946-3R-178-1-1), a highly salt tolerant RIL otherwise similar to IR29 (Bonilla et al. 2002), was subsequently used as donor for salt tolerance breeding worldwide. Contrary to the expectations, investigations revealed that *Saltol* region of FL478 was indeed contributed by the sensitive parent IR29, but activated to trigger high salt tolerance in presence of other positive alleles form Pokkali (Walia et al. 2005). *Saltol* region of FL478 is very complex, and now poised to contain many Pokkali QTLs including that of *SKC1* (Thomson et al. 2010) and a <1 Mb Pokkali DNA fragment at 10.6–11.5 Mb flanked by IR29 alleles (Kim et al. 2009). The fact that *Saltol* affected the Na–K ratio predominantly, the causal gene underlying this effect could be the sodium transporter *SKC1* (*OsHKT1;5*) (Thomson et al. 2010).

Comparative genomic investigations reveal that *Saltol* region seemed to contain an array of homologous sequences of known genes, viz., transcription factors, signal transduction components, cell wall components, and membrane transporters (Walia et al. 2005, 2007). The membrane transporter genes included those coding for carriers and channels involved in transporting cations, anions and organic substrates such as sugar transporters (Senadheera et al. 2009). Specific genes identified so far are root tissue and membrane transporters such as cation-proton exchanger (*OsCHX11*) and Cyclic nucleotide-gated ion channel (*OsCNGC1*) (Senadheera et al. 2009), HKT1 (high affinity potassium transporter), ABC1 (ATP-binding cassette transporter) genes (Walia et al. 2005). Various other genes identified near *Saltol* are, salt stress-induced protein (EF576533) and tetracopeptide repeat domain containing protein (EF575991) showing salt induced activation (Kumari et al. 2009).

Improving Salt Tolerance in Rice

Cultivation of salt tolerant rice in India perhaps had begun much earlier than anywhere else in the world. Although there is no historical record available of its beginning, a traditional organic rice-shrimp farming system known as *pokkali* still exists in the coastal saline areas of Kerala that is characterized by daily ingression of tidal waves causing partial flooding of rice fields and seasonal shrimp farming in the rice fallows during high saline phase (Pillai 1999; Shylaraj and Sasidharan 2005). In fact, there were many *pokkali* rice varieties in use, all of which were salt tolerant. In India, however, organized research on breeding salt tolerant rice was begun *circa* 1940, especially in the states of Maharashtra and Madras. In 1943, two salt tolerant varieties Kala Rata 1–24 and Bhura Rata 4–10 were released in Maharashtra

(Shendge et al. 1959). Around this time, in 1939, a salt-tolerant landrace called 'Pokkali' was introduced to Sri Lanka, which was later recommended for cultivation in saline areas in 1945 (Fernando 1949).

However, varietal development program for salt affected areas met with low success due to many reasons such as (a) lack of understanding of the complex nature of inheritance of salt tolerance, (b) lack of sufficient sources of resistance, (c) complexity and diversity of salt affected areas, (d) lack of precise and reliable screening techniques, and (e) lack of sufficient research backing. Conventional breeding methods such as introduction and selection of landraces, pedigree method, modified bulk pedigree method, mutation and shuttle breeding were used for development of new varieties in India (Reviewed by Singh et al. 2009). Shuttle breeding under an IRRI-India collaborative project had resulted in development of two salt-tolerant rice varieties, CSR23 and CSR27 (Mishra 1994).

Despite decade long breeding efforts, potential yield gap between the salt affected areas and normal regions remains wide, and there is an immediate need to harness resources to develop target-specific, locally adapted high-yielding varieties. Thanks to the modern approaches such as improved screening technique for phenotyping, in vitro and marker assisted selection, gap between potential and actual yield within these coastal areas is narrowing because of the development of salt-tolerant, fertilizer responsive and intermediate stature high-yielding varieties.

Screening for Salt Tolerance

Success of a target specific varietal development programme such as salt tolerance depends on reliable screening techniques that translates the results to reality in the field. Based on the target traits, methods of screening can be either phenological or physiological. While phenological screens included germination, survival, injury, morphology, yield and index such as mean tolerance index, physiological screening was done for Na^+, K^+, Cl^- concentrations and their derived ratios. Several screening techniques have been developed in rice such as hydroponics, pot culture, microplots and field evaluation, besides specialized solution culture screening methods such as bread boxes with perforated lids, seedling float technique and adult plant screening system (reviewed by Singh et al. 2010). Among all, field screening is the best because it is the only method that could accommodate salt tolerance in its holistic form with entire temporal and spatial variability. Notwithstanding, field screening is the most cumbersome of all the methods and hence limits the number of genotypes/progenies to be handled per screening. However, augmented designs allow screening of large number of varieties than conventional complete designs. Artificially created soil plots that resembles mini in situ fields but devoid of soil heterogeneity and maintains gradient levels of salinity in each designated plots are used for microplot screening. Such microplots are utilized for screening early generation materials. For precise individual plant studies pot culture experiments are employed, which facilitate closer observations. Hydroponics screens have been very popular

Table 10.7 Salt tolerance classification based on the modified standard evaluation score of visual salt injury at seedling stage

Phenotype	Score	Tolerance
Normal growth, no leaf symptoms	1	Highly tolerant
Nearly normal growth, but leaf tips or few leaves whitish and rolled	3	Tolerant
Growth severely retarded; most leaves rolled; only a few are elongating	5	Moderately tolerant
Complete cessation of growth; most leaves dry; some plants dying	7	Susceptible
Almost all plants dead or dying	9	Highly susceptible

because of its simplicity in setting up, precise control over salt concentration and it helps in creating a water-plant interface to which rice is adapted. Furthermore, specialized laboratory screening such as prolonged soaking in high salt concentration for 9 days prior to germination test was proven to be effective in delineating salt tolerant varieties (Abeysiriwardena 2004). Salt tolerance at juvenile screening are classified based on a modified standard evaluation score (Gregorio et al. 1997) of visual salt injury at seedling level (Table 10.7).

Most of these methods except field screening are limited to seedling stage, and therefore could not account for adult plant salt tolerance. Notwithstanding, artificial screens remain different from natural soil-water-plant interface limiting their direct application in crop improvement research. Therefore, an integrated approach is desirable that uses different screens so that the selected variety performs well under all stages of growth and therefore can be feasible for commercial cultivation.

In Vitro Techniques and Transgenesis

Since 1980s, cell and tissue culture techniques, have been recognized as powerful tools augmenting conventional breeding for the development of plants with increased tolerance to stresses such as salt stress. Later in vitro technology found their applications in molecular linkage mapping through Doubled Haploid (DH) lines and as an integral part of genetic engineering for the development of transgenic plants. In vitro techniques such as anther and pollen culture, somaclonal variation and protoplast fusion were used to develop salt tolerant lines in rice (Ram and Nabors 1985; Lynch et al. 1991). Among these, anther culture was used extensively in deriving salt tolerant lines, because of its advantage of faster development and efficiency in handling large number of progeny lines.

Anther Culture

The success of anther culture derived salt tolerant lines was established by the release of PSBRc50 (Bicol) targeted for saline-prone areas. Developed at IRRI, from the *indica-indica* cross IR5657-33-2/IR4630-22-2-5-1-3 (Zapata et al. 1991) this variety, originally known as IR51500-AC11-1, was the first ever anther culture derived variety to be released in the Philippines (Senadhira et al. 2002) and also the first cultivar recommended for adverse environments (Datta et al. 2009). Two other anther culture derived lines from IRRI, IR51500-AC17 and IR51485-AC6534-4 were released as commercial cultivars CSR21 and CSR28, respectively, for cultivation in saline-alkaline soils of India. Several anther derived DH lines were developed at IRRI, most of which had been used as a donor parents in breeding programs in various rice growing nations (Datta et al. 2009).

Although anther culture has limitations of reduced success, anther derived lines are still being developed for salt tolerance. Recently in Bangladesh, Rahman et al. (2010) generated 25 salt tolerant DH lines from a cross IR52724/ BR36 with line AC 1 showing excellent seedling stress survival coupled with moderately low Na/K ratio close to that of the tolerant control Pokkali, besides producing good yield in field trials conducted in a saline zone. Similarly many anther derived lines are under testing in Vietnam (Tam and Lang 2004) and Thailand (Cha-um et al. 2008).

Somaclonal Variation

Successful exploitation of somaclonal variants, the mutant cell lines that survive in salt rich selective media and regeneration of whole plants from such variants, stimulated many attempts for the development of salt-tolerant plants (Reddy and Vaidyanath 1986; Kavi-Kishor 1988). Although several salt tolerant lines have been reported, in many cases regenerants either failed to inherit the trait effectively, or showed developmental defects or in extreme cases showed complete reversal of tolerance. Such failure are now attributed to lack of distinction between mutant and adapted cell lines, distinct driving mechanisms for cellular and whole plant tolerance, multigenicity of salt tolerance and loss of regeneration capacity during selection (Tal 1993; Oono 1984). In an earlier reported attempt from IRRI, Pokkali cell lines were subjected to in vitro induction of somaclonal variants, with the objective of improving agronomic traits. A variant, TCCP 266-2-49-B-B-3 had improved agronomic performance coupled with good salt tolerance (Senadhira et al. 1994), showing vigorous growth, semi-dwarf nature, white pericarp and better cooked rice quality, features that were distinctly different from the original Pokkali line, which was tall with red pericarp and poor cooking quality. TCCP266-2-49-B-B-3 has later become a popular donor for producing new high-yielding salt-tolerant lines, some of which were released as varieties (Datta et al. 2009).

Transgenesis

Development of transgenic plants, by introducing new genes from external sources is an ideal tool to test the expression of orthologous genes. With the advancement in molecular mapping of QTLs together with transcriptome and whole genome profiles, map-based cloning has been successfully used to isolate and clone candidate genes and QTLs of biological and/or agricultural importance (Senadheera et al. 2009). However, before putting them into use it is essential to functionally validate the genes for trait expression. This helps to target the gene precisely and develop markers for marker assisted selection (MAS) programmes. In the modern biology, information on useful genes accumulate from different directions such as whole-genome information of both eukaryotic and prokaryotic model organisms, expressed sequence tag (EST) libraries, QTL mapping, microarrays etc. and transgenic system is the most handy tool in testing the expression of target genes under given environments such as high salinity. The testing can either be on the same or different organisms.

Several rice genes have already been positively tested in plant systems such as Arabidopsis, maize, tobacco and within rice itself, besides testing of foreign genes in rice for their role in imparting salt tolerance through transgenic approaches (Tables 10.2 and 10.3). In many cases *Agrobacterium* mediated gene transfer has been used to generate transgenic plants, proving the usefulness of transgenic approach in gene validation.

Molecular Breeding

With the availability of several molecular markers and saturated molecular genetic map of rice, MAS has now become feasible both for traits controlled by major genes as well as QTLs. Molecular breeding, a generic term now includes different MAS approaches, such as marker assisted backcrossing (MABC), marker assisted recurrent selection (MARS) and MAS based diallel selective mating system (MAS-DSMS). MAS has two distinct advantages translating to significant monetary and time benefits, namely, (a) it reduces the product delivery time considerably and (b) it reduces the genetic load or linkage drag associated commonly with backcross breeding programmes (Alpuerto et al. 2008; Gopalakrishnan et al. 2008; Singh et al. 2012). Furthermore, it is now possible to defer early generation phenotyping of segregating populations, by foreground selecting for stable QTLs and genes that have already been validated to confer stable salt tolerance under varying situations. This can accelerate breeding cycle as well as helps in handling large number of individuals per population, channeling to a successful salt tolerant variety. MAS as a tool to augment breeding programme, is widely used in targeted transfer of specific genes/ QTLs into popular cultivars through indirect selection based on gene or QTL linked based markers for foreground selection (Singh et al. 2011).

Marker Assisted Backcrossing

MABC is of great practical interest in applied breeding programmes, because it is done almost in the conventional way, but without or minimized phenotype testing in the early generations, rapidly advancing to the target genotype by following the inheritance of simple molecular tags that segregate in classical Mendelian fashion. Given the information available, molecular markers can be successfully deployed for foreground as well as background selection in order to confirm the presence of resistance gene(s) and speedy recovery of recurrent parent genome (RPG) and phenome (Singh et al. 2011). Detailed reviews on usefulness of MABC in rice are now available (Collard and Mackill 2008; Singh et al. 2011).

MABC for transferring salt tolerance QTL, *Saltol* is currently practiced in many rice growing nations, viz., Philippines, India, Thailand, Vietnam and Bangladesh (Elahi et al. 2004; Singh et al. 2011; Lang et al. 2011). The procedure involves crossing of the *Saltol* donor (preferably FL478) with the recipient in three backcrosses, selecting for the markers flanking the *Saltol* (foreground selection) and selecting against the donor markers for other regions (background selection) within each backcross generations. At the end of the programme recombinants are selected in which *Saltol* alleles are fixed, and show salt tolerance as against the original recipient. Various MABC procedures are in practice that combines phenotype selection for faster background recovery (Singh et al. 2011), together with stepwise transfer or simultaneous transfer or simultaneous and stepwise transfer for QTL pyramiding (Joshi and Nayak 2010).

In India, FL478 is being used as a donor to transfer *Saltol* into the recurrent parents Pusa Basmati 1121 and Pusa Basmati 6 through MABC in two independent backcross programs. Three *Saltol* linked markers RM8094, RM3412 and RM493 that are polymorphic between the recurrent and donor parents are used for foreground selection in each backcross generation, coupled with stringent phenotypic selection for rapid recovery of RPG and phenome with salt tolerance (Singh et al. 2011; Babu et al. 2012). Furthermore, various institutions across India are working on transferring *Saltol* in the backgrounds of popular rice varieties such as Sarjoo 52, Pusa 44, PR114, Gayatri, Savithri, MTU 1010, White Ponni and ADT45. IRRI in collaboration with national institutes under Stress-Tolerant Rice for Africa and South Asia (STRASA) project is gearing up with MABC for transferring *Saltol* into popular varieties such as BRRI dhan 28, IR64, BR11 and Swarna (STRASA 2011).

Marker Assisted Recurrent Selection and Diallel Selective Mating System

MARS and its variant DSMS-MAS are recent introductions in molecular breeding protocols and are not widely being practiced. MARS targets the identification and selection of several genomic regions involved in the expression of complex traits to 'assemble' the best-performing genotype within a single, or across related populations (Ribaut et al. 2010), while DSMS-MAS uses a multi-parent crossing strategy to develop recombinants using partial or full diallel crossings and select those recombinants with desired alleles and employ them into intensive crossing programmes so that the ultimate genotypes accumulate as many desired genes as possible so the probability of selection of desired combinations become maximized (Singh et al. 2010). These methods, although long term, has the advantage of transferring multiple stress resistance, are gaining importance as permanent breeding strategy in institutes like IRRI. Selection for *Saltol* is now routinely done at IRRI for selection of desirable recombinants for selective mating (Singh et al. 2008; Gautam et al. 2009).

Association Mapping and Genomic Selection

Association mapping (AM) and genomic selection (GS) are latest technologies that have found applications in plant molecular breeding (Abdurakhmonov and Abdukarimov 2008; Heffner et al. 2009). Although both differ in their applications, these techniques promise identification and use of target genomic regions that are otherwise difficult to map using QTL mapping approach. Both these techniques use more accurate and ubiquitous single nucleotide polymorphisms (SNPs) and genome wide variations. AM is based on the evolutionary linkage disequilibrium, that is conserved in the haplotype blocks and identification of the precise marker-trait association on an unstructured population, providing better precision of the associated markers to the target trait. Emerging scientific reports indicate that many laboratories are currently working on AM towards identifying novel QTLs associated with salt tolerance in rice (Courtios et al. 2011; Li et al. 2012). AM in the European rice core collection (ERCC) has recently identified 19 distinct loci associated with salt tolerant traits in the temperate *japonica* background, divulging that no accession carried all favorable alleles suggesting the potential for further improvement. The effective strategy for the accumulation of the favorable alleles would be marker-assisted population improvement (Ahmadi et al. 2011).

Rapid advances in the development of novel high-throughput DNA sequencing has drastically reduced the cost of whole genome sequencing, providing low cost coverage of any genome. Whole genome sequence resources are useful in the development of high density molecular markers that could be used in molecular marker assisted breeding (Subbaiyan et al. 2012). Genomic selection (GS) is the selection based on genomic estimated breeding value (GEBV) determined from simultaneous estimation of all locus effects across the genome in a set of training population without significance testing and without identifying *a priori* a subset of markers associated with the trait (Heffner et al. 2009). GS is gaining popularity in molecular breeding because it eliminates the need of QTL mapping and provides renewed promise in breeding for difficult traits such as salt tolerance. GS is still at the exploratory stage for plants, however, many laboratories have already started using the technique in rice, and one of the target traits set is salt tolerance.

Achievements, Impact, and Prospects

Breeding for salt tolerance in rice has taken a leap forward during last 15 years. In India, remarkable progress was achieved in developing improved salt tolerant genotypes at the CSSRI under the IRRI-India collaborative project. So far nine varieties (CSR21 to CSR29) were developed under this project and recommended for cultivation at various salt-affected ecologies (Singh et al. 2009). MABC derived IRRI-bred salt tolerant variety carrying *Saltol*, IR63307-4B-4-3, has been recently released in Bangladesh as BRRI dhan47 (Salam et al. 2007), and in the Philippines for cultivation. Many popular varieties are in the pipeline of MABC based improvement. Reason for the success of these programmes can be attributed to the practicable developments in the molecular biology together with better phenotyping screens augmented with tremendous developments in computing, communication and automation. Although conventional breeding efforts for the past 70 years had amassed wealth of information on salt tolerance behavior in rice, in addition, today we know more about the intricate mechanisms of salt stress signaling and responses in plants. The novel information in tandem with the age-old wisdom can now be channelized towards precise development of novel rice cultivars with improved tolerance.

For having promoted for better performance than the existing stock on a contemporary scale, majority of the rice varieties developed through conventional approaches worldwide have varying levels of salt tolerance. Many of them are low to medium yielders with poor quality grains. The present day understanding of the genetics of various agronomic traits and salt tolerance can steer molecular breeding towards an integrated crop improvement approach for high yielding good quality salt tolerant rice genotypes in the future.

Several reports that are available on screening of germplasm for salt tolerance, like that of screening of more than 25,000 rice germplasm at IRRI identifying 1,495 salt tolerant types with varying levels of tolerance (Ponnamperuma and Bandyopadhya 1980) and similar efforts in other countries indicate that massive breeding efforts are required to bring together desired genes into limelight. Notwithstanding, the key to success lies in the unexplored germplasm, especially in the landraces, that were traditionally grown in saline lowlands. Excepting Pokkali and Nona Bokra, not many traditional saline tolerant varieties have been subjected to serious investigations. In India, besides *pokkali* rice types there are several salt tolerant rice varieties that remains to studied in detail, such as Assgo, Bello, Damgo, Kalo Damgo, Kalo Korgut, Kalo Novan, Khochro, Korgut, Muno and Shiedi from the khazan lands of Goa (Bhonsle and Krishnan 2011). It is unlikely that all the traditional salt tolerant varieties would be harboring variants of the major QTL, *Saltol*. Having long term programmes aimed at allele mining from salt tolerant rice germplasm, mapping QTLs, cloning of tolerance QTLs/genes, pyramiding of multiple genes and QTLs coupled with yield and quality would be the ideal strategic solution for developing varieties suited for saline environments. Recent developments of AM, GS and high throughput techniques such as MutMap (Abe et al. 2012), are promise towards future for novel gene mapping and allele mining technologies (Henry et al. 2012).

Recent developments in salt tolerance research with accomplishments on the discovery of several functional genes and QTLs is poised to take new challenges for the future. Soon the MAS attempts will reflect in significant improvements in tolerance of salt-sensitive varieties.

Conclusion and Future Perspective

Growing concerns of land salinization is threatening crop productivity in major rice growing areas of the world. The issue is more serious because rice, the crop that feeds half the world, is sensitive to salinity. Varying sensitivity at different phenological stages of crop growth, particularly the high sensitivity during seedling and flowering stages strongly compromises plant survival and yield. With the world population growing incessantly, there is an urgent need to produce more grains from the salinized lands as well as to reutilize lands that are rendered unproductive due to salt accumulation. Although variability exists in rice germplasm towards salt tolerance, conventional breeding has been far less fruitful in addressing this complex problem. Probable reasons are lesser understanding of the nature of salt tolerance genetics, poorer screening facilities, biased screening towards juvenile tolerance, poor exploitation of the germplasm and lack of integrated breeding approaches towards an 'ideal' cultivar. It is essential to develop varieties that are phenologically capable of sustaining excess salt throughout its life span and produce higher yield and better quality. With the deeper understanding of the intricate mechanisms of salt tolerance and the array of genes and useable QTLs that are being discovered every day, the breeding scenario towards salt tolerant rice is poised to take a more productive turn in near future. Since QTLs are measurable and identifiable in the genome, it is essential to identify the genes underlying the trait so that genetic load or linkage drag from the undesirable genes can be minimized. Association mapping and genome wide selections are already making their footprints in salt tolerance research in rice. Complimentary validation of candidate genes using transgenic and transcriptomic approaches are supplementary tools in designing precise MAS strategies. Since MAS is going to be an integral part of the future breeding plans, it should be made much simpler so that breeders can perform the selection using a minimum genotype screening setup, which will boost the success of participatory and shuttle breeding approaches that promise sustainable cultivars for the future.

References

Abbas ST, Quraishi A, Chughtai MID (1994) Salt affected soils- problems and prospects. Pak J Agric Res 15:176–184

Abdurakhmonov IY, Abdukarimov A (2008) Application of association mapping to understanding the genetic diversity of plant germplasm resources. Int J Plant Genomic. doi:10.1155/2008/574927

Abe A, Kosugi S, Yoshida K, Natsume S, Takagi H, Kanzaki H, Matsumura H, Yoshida K, Mitsuoka C, Tamiru M, Innan H, Cano L, Kamoun S, Terauchi R (2012) Genome sequencing reveals agronomically important loci in rice using MutMap. Nat Biotechnol. doi:10.1038/nbt.2095

Abeysiriwardena DS (2004) A simple screening technique for salinity tolerance in rice: germination rate under stress. Int Rice Res Note 29:78–79

Abogadallah GM (2010) Antioxidative defense under salt stress. Plant Signal Behav 5:369–374

Ahmad P (2010) Growth and antioxidant responses in mustard (*Brassica juncea* L.) plants subjected to combined effect of gibberellic acid and salinity. Arch Agro Soil Sci 56(5):575–588

Ahmad P, Prasad MNV (2012a) Environmental adaptations and stress tolerance in plants in the era of climate change. Springer Science + Business Media, LLC, New York
Ahmad P, Prasad MNV (2012b) Abiotic stress responses in plants: metabolism, productivity and sustainability. Springer Science + Business Media, LLC., New York
Ahmad P, Sarwat M, Sharma S (2008) Reactive oxygen species, antioxidants and signaling in plants. J Plant Biol 51(3):167–173
Ahmad P, Jaleel CA, Salem MA, Nabi G, Sharma S (2010) Roles of enzymatic and non-enzymatic antioxidants in plants during abiotic stress. Crit Rev Biotechnol 30(3):161–175
Ahmad P, Nabi G, Jeleel CA, Umar S (2011) Free radical production, oxidative damage and antioxidant defense mechanisms in plants under abiotic stress. In: Ahmad P, Umar S (eds) Oxidative stress: role of antioxidats in plants. Studium Press, New Delhi, pp 19–53
Ahmad P, Ashraf M, Younis M, Hu X, Kumar A, Akram NA, Al-Qurainy F (2012a) Role of transgenic plants in agriculture and biopharming. Biotechnol Adv 30(3):524–540
Ahmad P, Kumar A, Gupta A, Hu X, Hakeem KR, Azooz MM, Sharma S (2012b) Polyamines: role in plants under abiotic stress. In: Ashraf M, Ozturk M, Ahmad MSA, Aksoy A (eds) Crop production for agricultural improvement. Springer, Dordrecht/Heidelberg/London/New York, pp 490–512
Ahmadi J, Fotokian MH (2011) Identification and mapping of quantitative trait loci associated with salinity tolerance in rice (*Oryza sativa*) using SSR markers. Iran J Biotechnol 9:21–30
Ahmadi N, Negrão S, Katsantonis D, Frouin J, Ploux J, Letourmy P, Droc G, Babo P, Trindade H, Bruschi G, Greco R, Oliveira MM, PiVanelli P, Courtois B (2011) Targeted association analysis identified japonica rice varieties achieving Na^+/K^+ homeostasis without the allelic make-up of the salt tolerant indica variety Nona Bokra. Theor Appl Genet 123:881–895
Ahmed M, Qamar I (2004) Productive rehabilitation and use of salt-affected land through afforestation (a review). Q Sci Vis 19:1–14
Akbar M, Yabuno T (1977) Breeding for saline-resistant varieties of rice inheritance of delayed-type panicle sterility induced by salinity. Japan J Breed 27:237–240
Alam R, Rahman MS, Seraj ZI, Thomson MJ, Ismail AM, Tumimbang-Raiz E, Gregorio GB (2011) Investigation of seedling-stage salinity tolerance QTLs using backcross lines derived from *Oryza sativa* L. Pokkali. Plant Breed 130:430–437
Alcázar R, Marco F, Cuevas JC, Patron M, Ferrando A, Carrasco P, Tiburcio AF, Altabella T (2006) Involvement of polyamines in plant response to abiotic stress. Biotechnol Lett 28:1867–1876
Ali Y, Sarwar G, Aslam Z, Hussain F (2006) Genotypic and environmental interaction in advanced lines of rice under salt-affected soils of Punjab. Int J Environ Sci Tech 3:191–195
Alpuerto V, Norton GW, Alwang J (2008) Economic impact analysis of marker-assisted breeding in rice. Paper presented at American agricultural economics association annual meeting, Orlando, 28p
Amin M, Elias SM, Hossain A, Ferdousi A, Rahman MS, Tuteja N, Seraj ZI (2012) Over-expression of a DEAD-box helicase, *PDH45*, confers both seedling and reproductive stage salinity tolerance to rice (*Oryza sativa* L.). Mol Breed 30:345–354
Ammar MHM, Singh RK, Singh AK, Mohapatra T, Sharma TR, Singh NK (2007) Mapping QTLs related to salinity tolerance of rice at the young seedling stage. In: African Crop Science Conference Proceedings, vol 8, El-Minia, Egypt, pp 617–620
Ansari R, Khan MA, Gul B (2007) Gainful utilization of salt affected lands: prospects and precautions. In: Kafi M, Khan MA (eds) Crop and forage production using saline waters, vol X., pp 103–108, NAM S&T Centre
Asch F, Dingkuhn M, Dörffling K, Mièzan K (2000) Leaf K/Na ratio predicts salinity induced yield loss in irrigated rice. Euphytica 113:109–118
Babu NN, Sharma SK, Ellur RK, Singh VK, Pal M, Pathania S, Singh VK, Singh D, Gopalakrishnan S, Bhowmick PK, Nagarajan M, Vinod KK, Singh NK, Prabhu KV, Singh AK (2012) Marker assisted improvement of Pusa Basmati 1121 for salinity tolerance. International conference on plant biotechnology for food security: new frontiers, national agricultural science centre, New Delhi, pp 72–73
Bartels D, Sunkar R (2005) Drought and salt tolerance in plants. Crit Rev Plant Sci 24:23–58
Bhonsle SJ, Krishnan S (2011) Traditionally cultivated salt tolerant rice varieties grown in khazan lands of Goa, India and their grain quality characteristics. J Phytol 3:11–17

Bhumbla DR, Abrol IP (1978) Saline and sodic soils. In: Soils and rice. International Rice Research Institute, Los Baños, pp 719–738

Bonilla P, Dvorak J, Mackill DJ, Deal K, Gregorio GB (2002) RFLP and SSLP mapping of salinity tolerance genes in chromosome 1 of rice (*Oryza sativa* L.) using recombinant inbred lines. Philipp Agric Sci 85:68–76

Borlaug NE, Dowswell CR (2005) Feeding a world of ten billion people: a 21st century challenge. In: Tuberosa R., Phillips RL, Gale M (eds) Proceedings of the international congress "In the wake of the double helix: from the green revolution to the gene revolution". Bologna, pp 3–23, ©2005 Avenue media, Bologna, 27–31 May 2003

Bot AJ, Nachtergaele FO, Young A (2000) Land resource potential and constraints at regional and country levels, vol 90, World soil resources reports. Food and Agriculture Organization of the United Nations, Rome, 113p

Chandrasekharan H, Sarangi A, Nagarajan M, Singh VP, Rao DUM, Stalin P, Natarajan K, Chandrasekaran B, Anbazhagan S (2008) Variability of soil–water quality due to Tsunami-2004 in the coastal belt of Nagapattinam district, Tamilnadu. J Environ Manage 89:63–72

Cha-um S, Srithandon S, Wanitchananan P, Kirdmanee C (2008) A progressive on salt-tolerant rice improvement using plant biotechnology in Thailand. In: Proceedings for the 5th international crop science congress, Jeju

Chaves MM, Maroco JP, Pereira JS (2003) Understanding plant responses to drought – from genes to the whole plant. Funct Plant Biol 30:239–264

Chaves MM, Flexas J, Pinheiro C (2009) Photosynthesis under drought and salt stress: regulation mechanisms from whole plant to cell. Ann Bot 103:551–560

Chen X, Guo Z (2008) Tobacco OPBP1 enhances salt tolerance and disease resistance of transgenic rice. Int J Mol Sci 9:2601–2613

Chen H, An R, Tang JH, Cui XH, Hao FS, Chen J, Wang XC (2007) Over-expression of a vacuolar Na^+/H^+ antiporter gene improves salt tolerance in an upland rice. Mol Breed 19:215–225

Claes B, Dekeyser R, Villarroel R, Bulcke MV, Bauw G, Montagu MV, Caplan A (1990) Characterization of a rice gene showing organ-specific expression in response to salt stress and drought. Plant Cell 2:19–27

Clarkson DT, Hanson JB (1980) The mineral nutrition of higher plants. Annu Rev Plant Physiol 31:239–298

Collard BCY, Mackill DJ (2008) Marker-assisted selection: an approach for precision plant breeding in the twenty-first century. Phil Trans R Soc B 363:557–572

Courtois B, Greco R, Bruschi G, Frouin J, Ahmadi N, Droc G, Hamelin C, Ruiz M, Evrard JC, Katsantonis D, Oliveira M, Negrão S, Cavigiolo S, Lupotto E, Piffanelli P (2011) Molecular characterization of the European rice collection in view of association mapping. Plant Genetic Resour 9:233–235

Dai X, Xu Y, Ma Q, Xu W, Wang T, Xue Y, Chong K (2007) Overexpression of an R1R2R3 MYB gene, *OsMYB3R-2*, increases tolerance to freezing, drought, and salt stress in transgenic *Arabidopsis*. Plant Physiol 143:1739–1751

Darwish E, Testerink C, Khalil M, El-Shihy O, Munnik T (2009) Phospholipid signaling responses in salt-stressed rice leaves. Plant Cell Physiol 50:986–997

Datta SK, Torrizo LB, Gregorio GB, Moon HP (2009) Haploid breeding in rice improvement. In: Datta SK (ed) Rice improvement in the genomics era. CRC Press/Taylor & Francis Group, Boca Raton, pp 71–104

De Datta SK, Neue HU, Senadhira D, Quijano C (1993) Success in rice improvement for poor soils. In: Proceedings of the workshop on adaptation of plants to soil stress. University of Nebraska, Lincoln, Nebraska, pp 248–268

de Souza GA, Ferreira BS, Dias JM, Queiroz KS, Branco AT, Bressan-Smith RE, Oliveira JG, Garcia AB (2003) Accumulation of SALT protein in rice plants as a response to environmental stresses. Plant Sci 164:623–628

Demiral T, Turkan I (2005) Comparative lipid peroxidation, antioxidant defense systems and proline content in roots of two rice cultivars differing in salt tolerance. Environ Exp Bot 53:247–257

Ding H, Zhang G, Guo Y, Chen S, Chen S (1998) RAPD tagging of a salt tolerant gene in rice. Chinese Sci Bull 43:330–332

Dregne H, Kassas M, Razanov B (1991) A new assessment of the world status of desertification. Desertif Control Bull (U N Environ Programme) 20:6–18

Dubouzet JG, Sakuma Y, Ito Y, Kasuga M, Dubouzet EG, Miura S, Seki M, Shinozaki K, Yamaguchi-Shinozaki K (2003) *OsDREB* genes in rice, *Oryza sativa* L., encode transcription activators that function in drought-, high-salt- and cold-responsive gene expression. Plant J 33:751–763

Elahi CMF, Seraj ZI, Rasul NM, Das KC, Biswas K, Salam MA, Gomosta AR, Tumimbang E, Adorada D, Gregorio G, Bennett J (2004) Breeding rice for salinity tolerance using the Pokkali allele: finding a linked DNA marker. In: Islam AS (ed) In vitro culture, transformation and molecular markers for crop improvement. Science, Enfield, pp 157–170

El-Shabrawi H, Kumar B, Kaul T, Reddy MK, Singla-Pareek SL, Sopory SK (2010) Redox homeostasis, antioxidant defense, and methylglyoxal detoxification as markers for salt tolerance in Pokkali rice. Protoplasma 245:85–96

Esechie HA, Al-Barhi B, Al-Gheity S, Al-Khanjari S (2002) Root and shoot growth in salinity-stressed alfalfa in response to nitrogen source. J Plant Nutr 25:2559–2569

Fang Y, Xie K, Hou X, Hu H, Xiong L (2010) Systematic analysis of GT factor family of rice reveals a novel subfamily involved in stress responses. Mol Genet Genomics 283:157–169

Fernando LH (1949) The performance of salt resistant paddy, Pokkali in Ceylon. Trop Agric 105:124–126

Fischer G, van Velthuizen H, Shah M, Nachtergaele F (2002) Global agro-ecological assessment for agriculture in the 21st century: methodology and results. International Institute for Applied Systems Analysis, Laxenburg

Flowers TJ (2004) Improving crop salt tolerance. J Exp Bot 55:307–319

Flowers TJ, Koyama ML, Flowers SA, Sudhakar C, Singh KP, Yeo AR (2000) QTL: their place in engineering tolerance of rice to salinity. J Exp Bot 51:99–106

Fotokian M, Taleie A, Ghareyazie B, Postini K, Bushehri AAS, Li ZK (2005) QTL mapping of genes affecting salt tolerance in rice (*Oryza sativa* L.) using microsatellite markers. Iran J Crop Sci 6:13

Fukuda A, Nakamura A, Tagiri A, Tanaka H, Miyao A, Hirochika H, Tanaka Y (2004) Function, intracellular localization and the importance in salt tolerance of a vacuolar Na^+/H^+ antiporter from rice. Plant Cell Physiol 45:146–159

Gao S, Zhang H, Tian Y, Li F, Zhang Z, Lu X, Chen X, Huang R (2008) Expression of *TERF1* in rice regulates expression of stress-responsive genes and enhances tolerance to drought and high-salinity. Plant Cell Rep 27:1787–1795

Garcia AB, Engler JD, Iyer S, Gerats T, VanMontagu M, Caplan AB (1997) Effects of osmoprotectants upon NaCl stress in rice. Plant Physiol 115:159–169

Garciadeblás B, Senn ME, Bañuelos MA, Rodríguez-Navarro A (2003) Sodium transport and HKT transporters: the rice model. Plant J 34:788–801

Garg AK, Kim JK, Owens TG, Ranwala AP, Choi YD, Kochian LV, Wu RJ (2002) Trehalose accumulation in rice plants confers high tolerance levels to different abiotic stresses. Proc Natl Acad Sci U S A 99:15898–15903

Gautam RK, Singh RK, Qadar A (2009) Incorporating salt tolerance in rice with more precision-status and prospects. J Soil Salin Water Qual 1:73–84

Ge LF, Chao DY, Shi M, Zhu MZ, Gao JP, Lin HX (2008) Overexpression of the trehalose-6-phosphate phosphatase gene *OsTPP1* confers stress tolerance in rice and results in the activation of stress responsive genes. Planta 228:191–201

Gong J, He P, QLAN Q, Shen L, Zhu L, Chen S (1999) Identification of salt-tolerance QTL in rice (*Oryza sativa* L.). Chinese Sci Bull 44:68–71

Gong J, Zheng X, Du B, Qian Q, Chen S, Zhu L, He P (2001) Comparative study of QTLs for agronomic traits of rice (*Oryza sativa* L.) between salt stress and nonstress environment. Sci in China (Ser C) 44:73–82

Gopalakrishnan S, Sharma RK, Rajkumar KA, Joseph M, Singh VP, Singh AK, Bhat KV, Singh NK, Mohapatra T (2008) Integrating marker assisted background analysis with foreground selection for identification of superior bacterial blight resistant recombinants in Basmati rice. Plant Breeding 127:131–139

Grattan SR, Zeng L, Shannon MC, Roberts SR (2002) Rice is more sensitive to salinity than previously thought. Calif Agric 56:189–195

Gregorio GB (1997) Tagging salinity tolerance genes in rice using amplified fragment length polymorphism (AFLP). Ph.D. thesis, University of the Philippines, Los Baños, 118p

Gregorio GB, Senadhira D (1993) Genetic analysis of salinity tolerance in rice (*Oryza sativa* L.). Theor Appl Genet 86:333–338

Gregorio GB, Senadhira D, Mendoza RD (1997) Screening rice for salinity tolerance, vol 22, IRRI discussion paper series. International Rice research Institute, Manila, 30p

Gregorio GB, Senadhira D, Mendoza RD, Manigbas NL, Roxas JP, Guerta CQ (2002) Progress in breeding for salinity tolerance and associated abiotic stresses in rice. Field Crops Res 76:91–101

Hanson B, Grattan SR, Fulton A (1999) Agricultural salinity and drainage. University of California Pub 3375, Division of Agricultural and Natural Resources, University of California, Davis, 160 p

Haq TU, Gorham J, Akhtar J, Akhtar N, Steele KA (2010) Dynamic quantitative trait loci for salt stress components on chromosome 1 of rice. Funct Plant Biol 37:634–645

Hasegawa PM, Bressan RA, Zhu JK, Bohnert HJ (2000) Plant cellular and molecular responses to high salinity. Annu Rev Plant Physiol Plant Mol Biol 51:463–499

Hauser F, Horie T (2010) A conserved primary salt tolerance mechanism mediated by HKT transporters: a mechanism for sodium exclusion and maintenance of high K^+/Na^+ ratio in leaves during salinity stress. Plant Cell Environ 33:552–565

Heenan DP, Lewin LG, McCaffery DW (1988) Salinity tolerance in rice varieties at different growth stages. Aust J Exp Agric 28:343–349

Heffner EL, Sorrells ME, Jannink JL (2009) Genomic selection for crop improvement. Crop Sci 49:1–12

Henry RJ, Edwards M, Waters DLE, Krishnan GS, Bundock P, Sexton TR, Masouleh AR, Nock CJ, Pattiemore J (2012) Molecular markers for plants derived from large scale sequencing. J Biosci 37:829–841

Horie T, Hauser F, Schroeder JI (2009) HKT transporter-mediated salinity resistance mechanisms in *Arabidopsis* and monocot crop plants. Trends Plant Sci 14:660–668

Horie T, Brodsky DE, Costa A, Kaneko T, Schiavo FL, Katsuhara M, Schroeder JI (2011a) K^+ transport by the *OsHKT2;4* transporter from rice with atypical Na^+ transport properties and competition in permeation of K^+ over Mg^{2+} and Ca^{2+} ions. Plant Physiol 156:1493–1507

Horie T, Sugawara M, Okada T, Taira K, Kaothien-Nakayama P, Katsuhara M, Shinmyo A, Nakayama H (2011b) Rice sodium-insensitive potassium transporter, *OsHAK5*, confers increased salt tolerance in tobacco BY2 cells. J Biosci Bioeng 111:346–356

Hoshida H, Tanaka Y, Hibino T, Hayashi Y, Tanaka A, Takabe T (2000) Enhanced tolerance to salt stress in transgenic rice that overexpresses chloroplast glutamine synthetase. Plant Mol Biol 43:103–111

Hossain MA, Lee Y, Cho JI, Ahn CH, Lee SK, Jeon JS, Kang H, Lee CH, An G, Park PB (2010) The bZIP transcription factor *OsABF1* is an ABA responsive element binding factor that enhances abiotic stress signaling in rice. Plant Mol Biol 72:557–566

Hu TZ (2008) *OsLEA3*, a late embryogenesis abundant protein gene from rice, confers tolerance to water deficit and salt stress to transgenic rice. Russian J Plant Physiol 55:530–537

Hu H, Dai M, Yao J, Xiao B, Li X, Zhang Q, Xiong L (2006) Overexpressing a NAM, ATAF, and CUC (NAC) transcription factor enhances drought resistance and salt tolerance in rice. Proc Natl Acad Sci USA 103:12987–12992

Hu H, You J, Fang J, Zhu X, Qi Z, Xiong L (2008) Characterization of transcription factor gene *SNAC2* conferring cold and salt tolerance in rice. Plant Mol Biol 67:169–181

Huang S, Spielmeyer W, Lagudah ES, James RA, Platten JD, Dennis ES, Munns R (2006) A sodium transporter (HKT7) is a candidate for *Nax1*, a gene for salt tolerance in durum wheat. Plant Physiol 142:1718–1727

Huang J, Yang X, Wang MM, Tang HJ, Ding LY, Shen Y, Zhang HS (2007) A novel rice C2H2-type zinc finger protein lacking DLNbox/ EAR-motif plays a role in salt tolerance. Biochim Biophys Acta 1769:220–227

Huang J, Wang MM, Jiang Y, Wang QH, Huang X, Zhang HS (2008a) Stress repressive expression of rice *SRZ1* and characterization of plant SRZ gene family. Plant Sci 174:227–235

Huang S, Spielmeyer W, Lagudah ES, Munns R (2008b) Comparative mapping of HKT genes in wheat, barley, and rice, key determinants of Na^+ transport, and salt tolerance. J Exp Bot 59:927–937

Huang XY, Chao DY, Gao JP, Zhu MZ, Shi M, Lin HX (2009) A previously unknown zinc finger protein, DST, regulates drought and salt tolerance in rice via stomatal aperture control. Genes Dev 23:1805–1817

Islam SMT, Tammi RS, Singla-Pareek SL, Seraj ZI (2010) Enhanced salinity tolerance and improved yield properties in Bangladeshi rice Binnatoa through *Agrobacterium*-mediated transformation of PgNHX1 from *Pennisetum glaucum*. Acta Physiol Plant 32:657–663

Islam MR, Salam MA, Hassan L, Collard BCY, Singh RK, Gregorio GB (2011) QTL mapping for salinity tolerance at seedling stage in rice. Emir J Food Agric 23:137–146

Ismail AM, Heuer S, Thomson MJ, Wissuwa M (2007) Genetic and genomic approaches to develop rice germplasm for problem soils. Plant Mol Biol 65:547–570

Ito Y, Katsura K, Maruyama K, Taji T, Kobayashi M, Seki M, Shinozaki K, Yamaguchi-Shinozaki K (2006) Functional analysis of rice DREB1/CBF-type transcription factors involved in cold-responsive gene expression in transgenic rice. Plant Cell Physiol 47:141–153

Jabnoune M, Espeout S, Mieulet D, Fizames C, Verdeil JL, Conéjéro G, Rodríguez-Navarro A, Sentenac H, Guiderdoni E, Abdelly C, Véry AA (2009) Diversity in expression patterns and functional properties in the rice HKT transporter family. Plant Physiol 150:1955–1971

Jain M, Tyagi AK, Khurana JP (2008) Constitutive expression of a meiotic recombination protein gene homolog, *OsTOP6A1*, from rice confers abiotic stress tolerance in transgenic *Arabidopsis* plants. Plant Cell Rep 27:767–778

James RA, Blake C, Byrt CS, Munns R (2011) Major genes for Na^+ exclusion, *Nax1* and *Nax2* (wheat *HKT1;4* and *HKT1;5*), decrease Na^+ accumulation in bread wheat leaves under saline and waterlogged conditions. J Exp Bot 62:2939–2947

Jang IC, Oh SJ, Seo JS, Choi WB, Song SI, Kim CH, Kim YS, Seo HS, Choi YD, Nahm BH, Kim JK (2003) Expression of a bifunctional fusion of the *Escherichia coli* genes for trehalose-6-phosphate synthase and trehalose-6-phosphate phosphatase in transgenic rice plants increases trehalose accumulation and abiotic stress tolerance without stunting growth. Plant Physiol 131:516–524

Javed MA, Huyop FZ, Wagiran A, Salleh FM (2011) Identification of QTLs for morph-physiological traits related to salinity tolerance at seedling stage in indica rice. Procedia Environ Sci 8:389–395

Jena KK, Mackill DJ (2008) Molecular markers and their use in marker-assisted selection in rice. Crop Sci 48:1266–1276

Jeong MJ, Lee SK, Kim BG, Kwon TR, Cho WS, Park YT, Lee JO, Kwon HB, Byun MO, Park SC (2006) A rice (*Oryza sativa* L.) MAP kinase gene, *OsMAPK44*, is involved in response to abiotic stresses. Plant Cell Tissue Organ Cult 85:151–160

Jeong JS, Kim YS, Baek KH, Jung H, Ha SH, Do Choi Y, Kim M, Reuzeau C, Kim JK (2010) Root-specific expression of *OsNAC10* improves drought tolerance and grain yield in rice under field drought conditions. Plant Physiol 153:185–197

Joshi RK, Nayak S (2010) Gene pyramiding-a broad spectrum technique for developing durable stress resistance in crops. Biotechnol Mol Biol Rev 5:51–60

Kalaiyarasi R, Palanisamy GA, Vaidyanathan P (2002) The potentials and scope of utilizing TGMS lines in inter-subspecies crosses of rice (*Oryza sativa* L.). J Genet Breed 56:137–143

Kanneganti V, Gupta AK (2008) Overexpression of OsiSAP8, a member of stress associated protein (SAP) gene family of rice confers tolerance to salt, drought and cold stress in transgenic tobacco and rice. Plant Mol Biol 66:445–462

Karthikeyan A, Pandian SK, Ramesh M (2011) Transgenic indica rice cv. ADT 43 expressing a Δ^1-pyrroline-5-carboxylate synthetase (P5CS) gene from *Vigna aconitifolia* demonstrates salt tolerance. Plant Cell Tissue Organ Cult 107:383–395

Kavi-Kishor PB (1988) Effect of salt stress on callus cultures of *Oryza sativa* L. J Exp Bot 39:235–240

Khan MSA, Hamid A, Karim MA (1997) Effect of sodium chloride on germination and seedling characters of different types of rice (*Oryza sativa* L.). J Agron Crop Sci 179:163–169

Khatun S, Flowers TJ (1995) Effects of salinity on seed set in rice. Plant Cell Environ 18:61–67

Kim ST, Kim SG, Hwang DH, Kang SY, Koo SC, Cho MJ, Kang KY (2004) Expression of a salt-induced protein (SALT) in suspension-cultured cells and leaves of rice following exposure to fungal elicitor and phytohormones. Plant Cell Rep 23:256–262

Kim DM, Ju HG, Kwon TR, Oh CS, Ahn SN (2009) Mapping QTLs for salt tolerance in an introgression line population between *japonica* cultivars in rice. J Crop Sci Biotechnol 12:121–128

Koyama ML, Levesley A, Koebner RMD, Flowers TJ, Yeo AR (2001) Quantitative trait loci for component physiological traits determining salt tolerance in rice. Plant Physiol 125:406–422

Krasensky J, Jonak C (2012) Drought, salt, and temperature stress-induced metabolic rearrangements and regulatory networks. J Exp Bot 63:1593–1608

Kumar V, Shriram V, Kishor PBK, Jawali N, Shitole MG (2010) Enhanced proline accumulation and salt stress tolerance of transgenic indica rice by over-expressing *P5CSF129A* gene. Plant Biotechnol Rep 4:37–48

Kumar G, Kushwaha HR, Purty RS, Kumari S, Singla-Pareek SL, Pareek A (2012) Cloning, structural and expression analysis of *OsSOS2* in contrasting cultivars of rice under salinity stress. Genes Genomes Genomics 6:34–41

Kumari S, Sabharwal VP, Kushwaha HR, Sopory SK, Singla-Pareek SL, Pareek A (2009) Transcriptome map for seedling stage specific salinity stress response indicates a specific set of genes as candidate for saline tolerance in *Oryza sativa* L. Funct Integr Genomics 9:109–123

Kusano T, Berberich T, Tateda C, Takahashi Y (2008) Polyamines: essential factors for growth and survival. Planta 228:367–381

Lang N, Buu BC, Ismail A (2008) Molecular mapping and marker-assisted selection for salt tolerance in rice (*Oryza sativa* L.). Omonrice 16:50–56

Lang N, Buu BC, Ismail AM (2011) Enhancing and stabilizing the productivity of salt- affected areas by incorporating genes for tolerance of abiotic stresses in rice. Omonrice 18:41–49

Lee SY, Ahn JH, Cha YS, Yun DW, Lee MC, Ko JC, Lee KS, Eun MY (2007) Mapping QTLs related to salinity tolerance of rice at the young seedling stage. Plant Breed 126:43–46

Li HW, Zang BS, Deng XW, Wang XP (2011) Overexpression of the trehalose-6-phosphate synthase gene OsTPS1 enhances abiotic stress tolerance in rice. Planta 234:1007–1018

Li G, Zhao WG, Kwon TR, Kwon SW, Park YJ (2012) Genome-wide association mapping of salt tolerance in rice seedlings. In: Plant and animal genome XX conference, San Diego

Lin H, Yanagihara S, Zhuang J, Senboku T, Zheng K, Yashima S (1998) Identification of QTL for salt tolerance in rice via molecular markers. Chinese J Rice Sci 12:72–78

Lin HX, Zhu MZ, Yano M, Gao JP, Liang ZW, Su WA, Hu XH, Ren ZH, Chao DY (2004) QTLs for Na^+ and K^+ uptake of the shoots and roots controlling rice salt tolerance. Theor Appl Genet 108:253–260

Liu AL, Zou J, Zhang XW, Zhou XY, Wang WF, Xiong XY, Chen LY, Chen XB (2010a) Expression profiles of class a rice heat shock transcription factor genes under abiotic stresses. J Plant Biol 53:142–149

Liu S, Zheng L, Xue Y, Zhang Q, Wang L, Shou HJ (2010b) Overexpression of *OsVP1* and *OsNHX1* increases tolerance to drought and salinity in rice. Plant Biol 53:444–452

Lu Z, Liu D, Liu S (2007) Two rice cytosolic ascorbate peroxidases differentially improve salt tolerance in transgenic Arabidopsis. Plant Cell Rep 26:1909–1917

Lynch PT, Finch RP, Davey MR, Cocking EC (1991) Rice tissue culture and its application. In: Khush GS, Toenniessen GH (eds) Rice biotechnology. CAB International, Wallingford, pp 135–155

Ma X, Qian Q, Zhu D (2005) Expression of a calcineurin gene improves salt stress tolerance in transgenic rice. Plant Mol Biol 58:483–495
Maas EV (1990) Agricultural salnity assessment and management. In: Tanji KK (ed) ASCE manuals and reports on engineering. American Society of Civil Engineers, New York, pp 262–304
Maas EV, Grattan SR (1999) Crop yields as affected by salinity. In: Skaggs RW, van Schilfgaarde J (eds) Agricultural drainage, vol 38, Agronomy monograph. American Society of Agronomy: Crop Science Society of America: Soil Science Society of America, Madison, pp 55–108
Maas EV, Hoffman GJ (1977) Crop salt tolerance – current assessment. J Irrig Drain Div ASCE 103:115–134
Malik V, Wu R (2005) Transcription factor *AtMyb2* increased salt-stress tolerance in rice *Oryza sativa* L. Rice Genet Newslett 22:63–67
Mandhania S, Madan S, Sawhney V (2006) Antioxidant defense mechanism under salt stress in wheat seedlings. Biol Plantarum 227:227–231
Martínez-Atienza J, Jiang X, Garciadeblas B, Mendoza I, Zhu JK, Pardo JM, Quintero FJ (2007) Conservation of the salt overly sensitive pathway in rice. Plant Physiol 143:1001–1012
Masood MS, Seiji Y, Shinwari ZK, Anwar R (2004) Mapping quantitative trait loci (QTLs) for salt tolerance in rice (*Oryza sativa*) using RFLPs. Pak J Bot 36:825–834
Mian AA (2010) Improving salt stress resistance in cereals. Ph.D. thesis, University of York, Heslington, 120p
Mishra B (1994) Breeding for salt tolerance in crops. In: Rao DLN, Singh NT, Gupta RK, Tyagi NK (eds) Salinity management for sustainable agriculture: 25 years of research at CSSRI. Central Soil Salinity Research Institute, Karnal, pp 226–259
Mizuno H, Kawahara Y, Sakai H, Kanamori H, Wakimoto H, Yamagata H, Oono Y, Wu J, Ikawa H, Itoh T, Matsumoto T (2010) Massive parallel sequencing of mRNA in identification of unannotated salinity stress-inducible transcripts in rice (*Oryza sativa* L.). BMC Genomics 11:683
Moeljopawiro S, Ikehashi H (1981) Inheritance of salt tolerance in rice. Euphytica 30:291–300
Mohammadi-Nejad G, Singh RK, Arzani A, Rezaie AM, Sabouri H, Gregorio GB (2010) Evaluation of salinity tolerance in rice genotypes. Int J Plant Prod 4:199–207
Mohanty A, Kathuria H, Ferjani A, Sakamoto A, Mohanty P, Murata N, Tyagi AK (2002) Transgenics of an elite indica rice variety Pusa Basmati 1 harbouring the *codA* gene are highly tolerant to salt stress. Theor Appl Genet 106:51–57
Moons A, Prinsen E, Bauw G, Van Montagu M (1997) Antagonistic effects of abscisic acid and jasmonates on salt stress-inducible transcripts in rice roots. Plant Cell 9:2243–2259
Moradi F, Ismail AM (2007) Responses of photosynthesis, chlorophyll fluorescence and ROS-Scavenging systems to salt stress during seedling and reproductive stages in rice. Ann Bot 99:1161–1173
Moradi F, Ismail AM, Egdane A, Gregorio GB (2003) Salinity tolerance of rice during reproductive development and association with tolerance at the seedling stage. Indian J Plant Physiol 8:105–116
Munns R, Tester M (2008) Mechanisms of salinity tolerance. Annu Rev Plant Biol 59:651–681
Nagamiya K, Motohashi T, Nakao K, Prodhan SH, Hattori E, Hirose S, Ozawa K, Ohkawa Y, Takabe T, Takabe T, Komamine A (2007) Enhancement of salt tolerance in transgenic rice expressing an Escherichia coli catalase gene, *katE*. Plant Biotechnol Rep 1:49–55
Nakagawa H, Ohmiya K, Hattori T (1996) A rice bZIP protein, designated *OSBZ8*, is rapidly induced by abscisic acid. Plant J 9:217–227
Nakashima K, Tran LS, Van Nguyen D, Fujita M, Maruyama K, Todaka D, Ito Y, Hayashi N, Shinozaki K, Yamaguchi-Shinozaki K (2007) Functional analysis of a NAC-type transcription factor *OsNAC6* involved in abiotic and biotic stress responsive gene expression in rice. Plant J 51:617–630
Negrão S, Courtois B, Ahmadi N, Abreu I, Saibo N, Oliveira MM (2011) Recent updates on salinity stress in rice: from physiological to molecular responses. Crit Rev Plant Sci 30:329–377
Niones JM (2004) Fine mapping of the salinity tolerance gene on chromosome 1 of rice (*Oryza sativa* L.) using near-isogenic lines. M.Sc. thesis, University of the Philippines, Los Baños, 78p

Obata T, Kitamoto HK, Nakamura A, Fukuda A, Tanaka Y (2007) Rice shaker potassium channel OsKAT1 confers tolerance to salinity stress on yeast and rice cells. Plant Physiol 144:1978–1985

Oh SJ, Song SI, Kim YS, Jang HJ, Kim SY, Kim M, Kim YK, Nahm BH, Kim JK (2005) Arabidopsis CBF3/DREB1A and ABF3 in transgenic rice increased tolerance to abiotic stress without stunting growth. Plant Physiol 138:341–351

Oh SJ, Kwon CW, Choi DW, Song SI, Kim JK (2007) Expression of barley HvCBF4 enhances tolerance to abiotic stress in transgenic rice. Plant Biotechnol J 5:646–656

Oh SJ, Kim YS, Kwon CW, Park HK, Jeong JS, Kim JK (2009) Overexpression of the transcription factor AP37 in rice improves grain yield under drought conditions. Plant Physiol 150:1368–1379

Ohta M, Hayashi Y, Nakashima A, Hamada A, Tanaka A, Nakamura T, Hayakawa T (2002) Introduction of a Na^+/H^+ antiporter gene from *Atriplex gmelini* confers salt tolerance to rice. FEBS Lett 532:279–282

Oldeman LR, Van Englen VWP, Pulles JHM (1991) The extent of human-induced soil degradation. In: Oldeman LR, Hakkeling RTA, Sombroek WG (eds) World map of status of human-induced soil degradation: an explanatory note. International Soil Reference and Information Centre (ISRIC), Wageningen, pp 27–33

Oomen RJ, Benito B, Sentenac H, Rodríguez-Navarro A, Talón M, Véry AA, Domingo C (2012) *HKT2;2/1*, a K^+-permeable transporter identified in a salt tolerant rice cultivar through surveys of natural genetic polymorphism. Plant J. doi:10.1111/j.1365-313X.2012.05031.x

Oono K (1984) Tissue culture and genetic engineering in rice. In: Tsanoda S, Takahashi N (eds) Biology of rice. Japan Science Society Press, Japan, pp 339–358

Ouyang SQ, Liu YF, Liu P, Lei G, He SJ, Ma B, Zhang WK, Zhang JS, Chen SY (2010) Receptor-like kinase *OsSIK1* improves drought and salt stress tolerance in rice (*Oryza sativa*) plants. Plant J 62:316–329

Pandit A, Vandna R, Subhashis B, Shikha S, Vinod K, Chauhan M, Gautam RK, Singh R, Sharma PC, Singh AK, Gaikwad K, Sharma TR, Mohapatra T, Singh NK (2010) Combining QTL mapping and transcriptome profiling of bulked RILs for identification of functional polymorphism for salt tolerance genes in rice (*Oryza sativa* L.). Mol Genet Genomic 284:121–136

Pardo JM (2010) Biotechnology of water and salinity stress tolerance. Curr Opin Biotechnol 21:185–196

Parvaiz A, Satyawati S (2008) Salt stress and phyto-biochemical responses of plants. Plant Soil Environ 54:89–99

Pattanagul W, Thitisaksakul M (2008) Effects of salinity stress on growth and carbohydrate metabolism in three rice (*Oryza sativa* L.) cultivars differing in salinity tolerance. Indian J Exp Biol 46:736–742

Peleg Z, Apse MP, Blumwald E (2011) Engineering salinity and water-stress tolerance in crop plants: getting closer to the field. Adv Bot Res 57:405–443

Pessarakli M, Szabolcs I (1999) Soil salinity and sodicity as particular plant/crop stress factors. In: Pessarakli M (ed) Handbook of plant and crop stress. Marcel Dekker, New York, pp 1–16

Pillai SM (1999) Traditional and improved traditional shrimp farming in the Pokkali fields of Kerala. J Indian Soc Coast Agric Res 17:171–181

Platten JD, Cotsaftis O, Berthomieu P, Bohnert H, Davenport RJ, Fairbairn DJ, Horie T, Leigh RA, Lin HX, Luan S, Mäser P, Pantoja O, Rodríguez-Navarro A, Schachtman DP, Schroeder JI, Sentenac H, Uozumi N, Véry AA, Zhu JK, Dennis ES, Tester M (2006) Nomenclature for HKT transporters, key determinants of plant salinity tolerance. Trends Plant Sci 11:372–374

Plett D, Safwat G, Gilliham M, Skrumsager Møller I, Roy S, Shirley N, Jacobs A, Johnson A, Tester M (2010) Improved salinity tolerance of rice through cell type-specific expression of *AtHKT1;1*. PLoS One 5:e12571

Ponnamperuma FN, Bandyopadhya AK (1980) Soil salinity as a constraint to food production in the humid tropics. In: Priorities for alleviating soil-related constraints to food production in the tropics. International Rice Research Institute, Manila

Prasad SR, Bagali PG, Hittalmani S, Shashidhar HE (2000) Molecular mapping of quantitative trait loci associated with seedling tolerance to salt stress in rice (*Oryza sativa* L.). Curr Sci 78:162–164

Prashanth SR, Sadhasivam V, Parida A (2008) Over expression of cytosolic copper/zinc superoxide dismutase from a mangrove plant *Avicennia marina* in indica rice var Pusa Basmati-1 confers abiotic stress tolerance. Transgenic Res 17:281–291

Puranik S, Sahu PP, Srivastava PS, Prasad M (2012) NAC proteins: regulation and role in stress tolerance. Trends Plant Sci 17:369–381

Qiu D, Lin P, Guo SZ (2007) Effects of salinity on leaf characteristics and CO_2/H_2O exchange of *Kandelia candel* (L.) Druce seedlings. J Forest Sci 53:13–19

Rahman MS, Das KC, Dipok KD, Kuntal BM, Badrul HC, Nilufer HK, Salam MA, Seraj ZI (2010) Breeding and anther derived lines of rice (*Oryza sativa* L.) for saline coastal areas of Bangladesh. Bangladesh J Bot 39:71–78

Ram NVR, Nabors MW (1985) Plant regeneration from tissue cultures of Pokkali rice is promoted by optimizing callus to medium volume ratio and by a medium conditioning factor produced by embryogenic callus. Plant Cell Tissue Organ Cult 4:241–248

Ray PKS, Islam A (2008) Genetic analysis of salinity tolerance in rice. Bangladesh J Agric Res 33:519–529

Reddy PJ, Vaidyanath K (1986) In vitro characterization of salt stress effects and the selection of salt tolerant plants in rice *Oryza sativa*. Theor Appl Genet 71:757–760

Ren ZH, Gao JP, Li LG, Cai XL, Huang W, Chao DY, Zhu MZ, Wang ZY, Luan S, Lin HX (2005) A rice quantitative trait locus for salt tolerance encodes a sodium transporter. Nat Genet 37:1141–1146

Ribaut J-M, de Vicente MC, Delannay X (2010) Molecular breeding in developing countries: challenges and perspectives. Curr Opin Plant Biol 13:1–6

Richards LA (1954) Diagnosis and improvement of saline and alkali soils, vol 60, USDA handbook. U.S. Department of Agriculture, Washington, DC

Rodríguez-Navarro A, Rubio F (2006) High-affinity potassium and sodium transport systems in plants. J Exp Bot 57:1149–1160

Rohila JS, Jain RK, Wu R (2002) Genetic improvement of basmati rice for salt and drought tolerance by regulated expression of a barley *HVA1* cDNA. Plant Sci 163:525–532

Roy M, Wu R (2001) Arginine decarboxylase transgene expression and analysis of environmental stress tolerance in transgenic rice. Plant Sci 160:869–875

Roy M, Wu R (2002) Overexpression of S-adenosylmethionine decarboxylase gene in rice increases polyamine level and enhances sodium chloride-stress tolerance. Plant Sci 163:987–992

RoyChoudhury A, Roy C, Sengupta DN (2007) Transgenic tobacco plants overexpressing the heterologous lea gene Rab16A from rice during high salt and water deficit display enhanced tolerance to salinity stress. Plant Cell Rep 26:1839–1859

Sabouri H, Biabani A (2009) Toward the mapping of agronomic characters on a rice genetic map: quantitative trait loci analysis under saline condition. Biotechnology 8:144–149

Sabouri H, Sabouri A (2008) New evidence of QTLs attributed to salinity tolerance in rice. Afr J Biotechnol 7:4376–4383

Sabouri H, Rezai AM, Moumeni A, Kavousi A, Katouzi M, Sabouri A (2009) QTLs mapping of physiological traits related to salt tolerance in young rice seedlings. Biol Plantarum 53:657–662

Saijo Y, Hata S, Kyozuka J, Shimamoto K, Izui K (2000) Over-expression of a single Ca2+-dependent protein kinase confers both cold and salt/drought tolerance on rice plants. Plant J 23:319–327

Sakamoto A, Alia, Murata N (1998) Metabolic engineering of rice leading to biosynthesis of glycinebetaine and tolerance to salt and cold. Plant Mol Biol 38:1011–1019

Salam MA, Rahman MA, Bhuiyan MAR, Uddin K, Sarker MRA, Yasmeen R, Rahman MS (2007) BRRI dhan 47: a salt-tolerant variety for the boro season. Int Rice Res News 32:42–43

Sankar PD, Subbaraman N, Narayanan SL (2008) Heterosis, combining ability and gene action studies in TGMS based rice hybrids under normal and salt affected environments. Indian J Agric Res 42:177–182

Schmidt R, Schippers JH, Welker A, Mieulet D, Guiderdoni E, Mueller-Roeber B (2012) Transcription factor *OsHsfC1b* regulates salt tolerance and development in *Oryza sativa* ssp. japonica. AoB Plants: pls011, doi:10.1093/aobpla/pls011

Senadheera P, Singh RK, Maathuis FJM (2009) Differentially expressed membrane transporters in rice roots may contribute to cultivar dependent salt tolerance. J Exp Bot 60:2553–2563
Senadhira D, Neue HU, Akbar M (1994) Development of improved donors for salinity tolerance in rice through somaclonal variation. SABRAO J 26:12–25
Senadhira D, Zapata-Arias FJ, Gregorio GB, Alejar MS, de la Cruz HC, Padolina TF, Galvez AM (2002) Development of the first salt-tolerant rice cultivar through indica/indica anther culture. Field Crop Res 76:103–110
Shendge PY, Chavan VM, Deshpande JD (1959) Breeding of saline resistant varieties in Bombay state. Rice News Lett 8:18–19
Shylaraj KS, Sasidharan NK (2005) VTL 5: a high yielding salinity tolerant rice variety for the coastal saline ecosystems of Kerala. J Tropic Agric 43:25–28
Singh AK, Ansari MW, Pareek AH, Singla-Pareek S (2008) Raising salinity tolerant rice: recent progress and future perspectives. Physiol Mol Biol Plants 14:137–154
Singh RK, Mishra B, Ismail AM, Gregorio GB (2009) Breeding rice for salt-affected areas of India. In: Hossain M, Bennett J, Mackill D, Hardy B (eds) Progress in crop improvement research. International Rice Research Institute, Los Baños, pp 78–90
Singh RK, Redoña E, Refuerzo L, Govindjee (2010) Varietal improvement for abiotic stress tolerance in crop plants: special reference to salinity in rice. In: Pareek A, Sopory SK, Bohnert HJ (eds) Abiotic stress adaptation in plants: physiological, molecular and genomic foundation. Springer, Dordrecht/London, pp 387–415
Singh AK, Gopalakrishnan S, Singh VP, Prabhu KV, Mohapatra T, Singh NK, Sharma TR, Nagarajan M, Vinod KK, Singh D, Singh UD, Chander S, Atwal SS, Seth R, Singh VK, Ellur RK, Singh A, Anand D, Khanna A, Yadav S, Goel N, Singh A, Shikari AB, Singh A, Marathi B (2011) Marker assisted selection: a paradigm shift in Basmati breeding. Indian J Genet Plant Breed 71:120–128
Singh VK, Singh A, Singh SP, Ellur RK, Singh D, Gopalakrishnan S, Nagarajan M, Vinod KK, Singh UD, Rathore R, Prasanthi SK, Agrawal PK, Bhatt JC, Mohapatra T, Prabhu KV, Singh AK (2012) Incorporation of blast resistance gene in elite basmati rice restorer line PRR78, using marker assisted selection. Field Crop Res 128:8–16
Singla-Pareek SL, Yadav SK, Pareek A, Reddy MK, Sopory SK (2008) Enhancing salt tolerance in a crop plant by overexpression of glyoxalase II. Transgenic Res 17:171–180
Song SY, Chen Y, Chen J, Dai XY, Zhang WH (2011) Physiological mechanisms underlying *OsNAC5*-dependent tolerance of rice plants to abiotic stress. Planta 234:331–345
STRASA (2011) Salt-tolerant rice. http://irri.org. Accessed 19 June 2012
Su J, Wu R (2004) Stress-inducible synthesis of proline in transgenic rice confers faster growth under stress conditions than that with constitutive synthesis. Plant Sci 166:941–948
Su J, Hirji R, Zhang L, He C, Selvaraj G, Wu R (2006) Evaluation of the stress-inducible production of choline oxidase in transgenic rice as a strategy for producing the stress-protectant glycine betaine. J Exp Bot 57:1129–1135
Subbaiyan GK, Waters DLE, Katiyar SK, Sadananda AR, Vaddadi S, Henry RJ (2012) Genome-wide DNA polymorphisms in elite indica rice inbreds discovered by whole-genome sequencing. Plant Biotechnol J 10:623–634
Summart J, Thanonkeo P, Panichajakul S, Prathepha P, McManus MT (2010) Effects of salt stress on growth, inorganic ion and proline accumulation in Thai aromatic rice, Khao Dawk Mali 105, callus culture. Afr J Biotechnol 9:145–152
Sun SJ, Guo SQ, Yang X, Bao YM, Tang HJ, Sun H, Huang J, Zhang HS (2010) Functional analysis of a novel Cys2/His2-type zinc finger protein involved in salt tolerance in rice. J Exp Bot 61:2807–2818
Szczerba MW, Britto DT, Kronzucker HJ (2009) K^+ transport in plants: Physiology and molecular biology. J Plant Phy 166:447–466
Tagawa T, Ishizaka N (1963) Physiological studies on the tolerance of rice plants to salinity. Proc Crop Sci Soc Jpn 31:249–252
Takasaki H, Maruyama K, Kidokoro S, Ito Y, Fujita Y, Shinozaki K, Yamaguchi-Shinozaki K, Nakashima K (2010) The abiotic stress-responsive NAC-type transcription factor *OsNAC5* regulates stress-inducible genes and stress tolerance in rice. Mol Genet Genomics 284:173–183

Takehisa H, Shimoda Y, Fukuta Y, Ueda T, Yano M, Yamaya T, Kameya T, Sato T (2004) Identification of quantitative trait loci for plant growth of rice in paddy field flooded with salt water. Field Crops Res 89:85–95

Tal M (1993) In vitro methodology for increasing salt tolerance in crop plants. Acta Hort 336:69–78

Tam DM, Lang NT (2004) Selection of salt tolerance genotypes from doubled haploids in rice. Omonrice 12:33–37

Tanaka Y, Hibin T, Hayashi Y, Tanaka A, Kishitani S, Takabe T, Yokota S, Takabe T (1999) Salt tolerance of transgenic rice overexpressing yeast mitochondrial Mn-SOD in chloroplasts. Plant Sci 148:131–138

Thomson MJ, Ocampo M, Egdane J, Rahman MA, Sajise AG, Adorada DL, Raiz ET, Blumwald E, Seraj ZI, Singh RK, Gregorio GB, Ismail AM (2010) Characterizing the Saltol quantitative trait locus for salinity tolerance in rice. Rice 3:148–160

Tian L, Tan L, Liu F, Cai H, Sun C (2011) Identification of quantitative trait loci associated with salt tolerance at seedling stage from *Oryza rufipogon*. J Genet Genom 38:593–601

Verma D, Singla-Pareek SL, Rajagopal D, Reddy MK, Sopory SK (2007) Functional validation of a novel isoform of Na^+/H^+ antiporter from *Pennisetum glaucum* for enhancing salinity tolerance in rice. J Biosci 32:621–628

World Bank (2006) Pakistan strategic country environmental assessment. Vol II technical annex: the cost of environmental degradation in Pakistan – An analysis of physical and monetary losses in environmental health and natural resources. South Asia Environment and Social Development Unit, South Asia Region. 42p

Walia H, Wilson C, Condamine P, Liu X, Ismail AM, Zeng L, Wanamaker SI, Mandal J, Xu J, Cui X, Close TJ (2005) Comparative transcriptional profiling of two contrasting rice genotypes under salinity stress during the vegetative growth stage. Plant Physiol 139:822–835

Walia H, Wilson C, Zeng L, Ismail AM, Condamine P, Close TJ (2007) Genome-wide transcriptional analysis of salinity stressed japonica and indica rice genotypes during panicle initiation stage. Plant Mol Biol 63:609–623

Wang H, Huang D, Lu R, Liu J, Qian Q, Peng X (2000) Salt tolerance of transgenic rice (*Oryza sativa* L.) with *mtlD* gene and *gutD* gene. Chinese Sci Bull 45:1685–1690

Wang XS, Zhu HB, Jin GL, Liu HL, Wu WR, Zhu J (2007) Genome-scale identification and analysis of LEA genes in rice (*Oryza sativa* L.). Plant Sci 172:414–420

Wang ZF, Wang JF, Bao YM, Wu YY, Su X, Zhang HS (2010) Inheritance of rice seed germination ability under salt stress. Rice Sci 17:105–110

Wang Z, Wang J, Bao Y, Wu Y, Zhang H (2011) Quantitative trait loci controlling rice seed germination under salt stress. Euphytica 178:297–307

Witcombe JR, Hollington PA, Howarth CJ, Reader S, Steele KA (2008) Breeding for abiotic stresses for sustainable agriculture. Phil Trans Roy Soc B 363:703–716

Wu L, Fan Z, Guo L, Li Y, Zhang W, Qu LJ, Chen Z (2003) Over-expression of an Arabidopsis δ-OAT gene enhances salt and drought tolerance in transgenic rice. Chinese Sci Bull 48:2594–2600

Wu L, Fan Z, Guo L, Li Y, Chen ZL, Qu LJ (2005) Over-expression of the bacterial *nhaA* gene in rice enhances salt and drought tolerance. Plant Sci 168:297–302

Xiang Y, Huang Y, Xiong L (2007) Characterization of stress-responsive CIPK genes in rice for stress tolerance improvement. Plant Physiol 144:1416–1428

Xiang Y, Tang N, Du H, Ye H, Xiong L (2008) Characterization of *OsbZIP23* as a key player of the basic leucine zipper transcription factor family for conferring abscisic acid sensitivity and salinity and drought tolerance in rice. Plant Physiol 148:1938–1952

Xiong L, Yang Y (2003) Disease resistance and abiotic stress tolerance in rice are inversely modulated by an abscisic acid-inducible mitogen-activated protein kinase. Plant Cell 15: 745–759

Xu D, Duan X, Wang B, Hong B, Ho T, Wu R (1996) Expression of a late embryogenesis abundant protein gene, *HVA1*, from barley confers tolerance to water deficit and salt stress in transgenic rice. Plant Physiol 110:249–257

Xu DQ, Huang J, Guo SQ, Yang X, Bao YM, Tang HJ, Zhang HS (2008) Overexpression of a TFIIIA-type zinc finger protein gene *ZFP252* enhances drought and salt tolerance in rice (*Oryza sativa* L.). FEBS Lett 582:1037–1043

Yamamoto T, Yano M (2008) Detection and molecular cloning of genes underlying quantitative phenotypic variations in rice. In: Hirano HY, Hirai A, Sano Y, Sasaki T (eds) Rice biology in the genomics era. Springer, Berlin, pp 295–308

Yamamoto T, Yonemaru J, Yano M (2009) Towards the understanding of complex traits in rice: substantially or superficially? DNA Res 16:141–154

Yao X, Horie T, Xue S, Leung HY, Katsuhara M, Brodsky DE, Wu Y, Schroeder JI (2010) Differential sodium and potassium transport selectivities of the rice *OsHKT2;1* and *OsHKT2;2* transporters in plant cells. Plant Physiol 152:341–355

Yokotani N, Ichikawa T, Kondou Y, Matsui M, Hirochika H, Iwabuchi M, Oda K (2008) Expression of rice heat stress transcription factor *OsHsfA2e* enhances tolerance to environmental stresses in transgenic *Arabidopsis*. Planta 227:957–967

Yokotani N, Ichikawa T, Kondou Y, Matsui M, Hirochika H, Iwabuchi M, Oda K (2009) Tolerance to various environmental stresses conferred by the salt-responsive rice gene *ONAC063* in transgenic *Arabidopsis*. Planta 229:1065–1075

Zang J, Sun Y, Wang Y, Yang J, Li F, Zhou Y, Zhu L, Jessica R, Mohammadhosein F, Xu J, Li Z (2008) Dissection of genetic overlap of salt tolerance QTLs at the seedling and tillering stages using backcross introgression lines in rice. Sci China C Life Sci 51:583–591

Zapata FJ, Alejar MS, Torrizo LB, Novero AU, Singh VP, Senadhira D (1991) Field performance of anther-culture-derived lines from F_1 crosses of *indica* rices under saline and non-saline conditions. Theor Appl Genet 83:6–11

Zeng L, Shannon MC (2000) Salinity effects on seedling growth and yield components of rice. Crop Sci 40:996–1003

Zeng L, Shannon MC, Lesch SM (2001) Timing of salinity stress affects rice growth and yield components. Agric Water Manage 48:191–206

Zhang GY, Guo Y, Chen SL, Chen SY (1995) RFLP tagging of a salt-tolerance gene in rice. Plant Sci 110:227–234

Zhang ZH, Liu Q, Song HX, Rong XM, Ismail AM (2012) Responses of different rice (*Oryza sativa* L.) genotypes to salt stress and relation to carbohydrate metabolism and chlorophyll content. Afr J Agric Res 7:19–27

Zhao F, Zhang H (2006) Salt and paraquat stress tolerance results from co-expression of the *Suaeda salsa* glutathione S-transferase and catalase in transgenic rice. Plant Cell Tissue Organ Cult 86:349–358

Zhao F, Guo S, Zhang H, Zhao Y (2006a) Expression of yeast *SOD2* in transgenic rice results in increased salt tolerance. Plant Sci 170:216–224

Zhao FY, Zhang XJ, Li PH, Zhao YX, Zhang H (2006b) Co-expression of the *Suaeda salsaSs-NHX1* and *Arabidopsis AVP1* confer greater salt tolerance to transgenic rice than the single *SsNHX1*. Mol Breed 17:341–353

Zheng X, Chen B, Lu G, Han B (2009) Overexpression of a NAC transcription factor enhances rice drought and salt tolerance. Biochem Biophys Res Commun 379:985–989

Zhou W, Li Y, Zhao BC, Ge RC, Shen YZ, Wang G, Huang ZJ (2009) Overexpression of *TaSTRG* gene improves salt and drought tolerance in rice. J Plant Physiol 166:1660–1671

Zhou HK, Hayat Y, Fang1 LJ, Guo RF, JM H, HM X (2010) Analysis of genetic and genotype × environment interaction effects for agronomic traits of rice (*Oryza sativa* L.) in salt tolerance. Pak J Bot 42:3239–3246

Zhu JK (2001) Cell signaling under salt, water and cold stresses. Curr Opin Plant Biol 4:401–406

Zhu BC, Su J, Chan MC, Verma DPS, Fan YL, Wu R (1998) Overexpression of a delta(1)-pyrroline-5-carboxylate synthetase gene and analysis of tolerance to water- and salt-stress in transgenic rice. Plant Sci 139:41–48

Zou M, Guan Y, Ren H, Zhang F, Chen F (2008) A bZIP transcription factor, *OsABI5*, is involved in rice fertility and stress tolerance. Plant Mol Biol 66:675–683

Chapter 11
Approaches to Improving Salt Tolerance in Maize

Sreekala Chellamma and Bhinu V-S. Pillai

Abstract Maize is a salt-sensitive crop and is affected even by low concentrations of salt, leading to loss in crop production Changing climate conditions (environmental stress) have forced plant biologists to explore alternate strategies to make maize plants salt tolerant. Breeding for salt resistance is difficult because it is a multigenic trait. When conventional breeding fails to meet the challenges imposed by these stresses, plant scientists shifted to marker-assisted selection and transgenic approaches. Genetic transformation have been proven to be successful. It has shown that over-expression of tonoplast Na^+/H^+ antiporters in plants resulted in improved salt resistance in plants. Currently, a vast number of gene regulatory elements, including si- and mi-RNA have been identified either leading to salt tolerance or resulting from salinity stress. Engaging the right elements on a case-by-case basis may provide answers to this long-standing problem. As with most complex systems, a combination of conventional breeding, exploiting physiological knowledge, transgenic approaches and field based testing is perhaps the way forward to address salinity problem in maize.

Keywords ABA • Maize • Salt stress • Salinity tolerance • Small RNA

According to Food and Agriculture Organization (FAO) the world population is expected to reach around 9 billion by 2050, which is a 34% increase compared to today's world. As population grows rapidly, and with urbanization occurring at an accelerated pace, it is predicted that 70% of the world's population will be urban and richer, or farmland shrinkage occurs ('The State of the World's Land and Water

S. Chellamma • B.V-S. Pillai (✉)
Dow AgroSciences LLC, 9330 Zionsville Road, Indianapolis, IN 46268, USA
e-mail: bvpillai@dow.com; schellamma@dow.com

P. Ahmad et al. (eds.), *Salt Stress in Plants: Signalling, Omics and Adaptations*,
DOI 10.1007/978-1-4614-6108-1_11, © Springer Science+Business Media New York 2013

Resources for Food and Agriculture' SOLAW),' FAO (2011). Earlier predictions by the FAO indicate that annual cereal production may need to rise to approximately 3 billion tonnes from the current 2 billion tonnes while the annual meat production will need to rise by over 200 million tonnes to reach 470 million tonnes. These projections undoubtedly highlight the significance of achieving increased productivity manifold to feed the growing population. This can perhaps be accomplished by breeding for increased unit yield and/or preventing yield losses due to biotic and abiotic stresses which account for a minimum 25% yield loss.

Major abiotic factors affecting crop yield include drought, cold and salt stresses. Salinity affects 20% of the world's cultivated area and about 50% of the irrigated lands (Zhu 2001; Allakhverdiev et al. 2000; Akram et al. 2009). Of the available 14 billion hectares of land about 1 billion hectare are natural saline soils (Jamil et al. 2011). It is estimated that more than 50% of the arable land would be salinized by the year 2050, unless some type of corrections are made (Ashraf 2009; Ahmad et al. 2012). Fifty percent of existing irrigation systems globally are prone to salinization, alkalization and water logging (Szabolcs 1994). Even though only 16% of the world's croplands are irrigated, those irrigated crops produce 36% of the world's food (World Food Summit 1996).

Given the growing problem with salinization, it is essential to target measures aimed at reducing their spread and improving the salt tolerance of high yielding crops. Salinity effects are prominent in arid and semi-arid regions because they are limited by rainfall, and a combination of evapotranspiration, and high temperature. This combined with poor water and soil management magnifies the salinity problem thus giving it a global scale. Maize being a salt-sensitive crop plant (Maas 1986), poor growth followed by low yield is a major concern. In the recent decade, maize breeding programs received much attention worldwide (Bänziger and Araus 2007), among which crop improvement for stress tolerance is one of the most important areas of research both in academic and industrial research programs.

Crop Adaptations to Salinity Stress

Crop plants are sedentary and in order to cope with high salt stress, they have developmentally evolved. Research has led us to view the most obvious adaptation to salt tolerance, i.e., variation in gene expression profiles for genes involved in a wide range of biological processes in plants. Though widely debated, these developmental modifications include biochemical and physiological processes such as metabolism, signal transduction, transcription, protein biosynthesis and degradation, membrane trafficking and photosynthesis (Vinocur and Altman 2005). In addition, posttranscriptional regulations also play a role in the plant salt response (Boesani et al. 2005). Investigating salt tolerance research represents an important part of basic plant biology enhancing our understanding of salt tolerance gene regulation, signal transduction to ion transport, and mineral nutrition (Zhu et al. 2000) with the goal being genetic improvement of salt tolerance in crop plants.

In rice, for instance, drought and salinity are two major factors impacting rice production. Limited progress in releasing commercial varieties of drought tolerant or salt tolerant rice is mainly attributed to the lack of breeding efforts and also due to the complex nature of genetics and physiological processes involving drought tolerance or salt tolerance rice (Li and Xu 2007). Plant breeders are continuously working to develop high yielding and drought tolerant or salt tolerant rice cultivars for varied stress conditions of several rice growing regions. Conventional breeding efforts at International Rice Research Institute and hybrid rice have shown some promise in the development of drought tolerant or salt tolerant rice cultivars that yield high and yet indicate good levels of water use efficiency (Gregorio et al. 1997). Quantitative Trait Loci (QTL) affecting drought tolerance or salt tolerance in rice have been identified and are being employed in marker-assisted breeding programs (Lafitte and Courtois 2000). QTL pyramiding involving back crosses to develop drought tolerant or salt tolerant rice introgression lines in elite genetic backgrounds is being evaluated (Li et al. 2005).

Likewise, in soybean, drought and salinity limit soybean production and major crop loss. Soil reclamation and irrigation are not practical options for soybean cultivation especially under drought and salinity (Pathan et al. 2007). Genetic improvement for drought and salt tolerance in soybean are the most-rewarding alternatives. In soybean, conventional breeding has been successful, and the germplasm improved, leading to simple inheritance of qualitative traits that show low sensitivity to environmental variations. Quantitative traits such as yield or salinity or drought tolerance are a function of the growth environment (Tuberosa et al. 2002a, b). Most agronomic traits are quantitatively inherited and are complicated to improve via conventional breeding approaches. Marker-assisted breeding technology can identify the QTL using marker assisted technology in a shorter time span. The use of genomic tools in breeding programs will help in the development of soybean varieties with higher tolerance to drought and salt (Pathan et al. 2007).

Numerous studies show that the two other major abiotic stresses, drought and cold, are intimately linked to salt stress (Zhu et al. 1997; Kizis et al. 2001; Song and Matsuoka 2009; Hu et al. 2010; Urano et al. 2010; Jamil et al. 2011). Like drought tolerance, salt tolerance is also a complex trait involving responses to cellular osmotic and ionic stresses and their consequent secondary stresses (e.g. oxidative stress) and whole plant coordination (Ahmad and Sharma 2008; Ahmad et al. 2008, 2010a, b). The complex nature and polygenic salt stress tolerance trait impedes the breeding of salt-tolerant crop varieties (Zhu et al. 1997). The limited understanding of salt tolerance mechanisms as well as a lack of field and laboratory screening tests, availability of physiological and molecular markers are also a factor affecting breeding success. Despite the progress in molecular understanding of numerous cellular processes, we have a long and growing list of salt stress-responsive genes (Zhu et al. 1997). Two factors appear to have limited the successful elucidation of molecular mechanisms driving salt tolerance. First, the approaches used have only been correlative. Some schools propose that many salt-responsive genes do not contribute to tolerance; rather, their induction reflects salt stress damage. Second, genes or gene products have been identified based on their expression;

so genes that are important for salt tolerance that are not induced by salt stress are not captured. However, the discovery of promoter elements and transcription factors involved in the control of expression of protective proteins such as RD29A/COR78 during stress (Kasuga et al. 1999) have been valuable.

In addition to the complex biology of salt tolerance mechanisms, inferences and interpretation can lead to differences in our understanding. Salt tolerance is typically assessed as the per cent biomass production in saline versus control conditions over time. Dramatic differences are found between plant species, for instance, after a certain period of exposure to 200 mm NaCl, a salt-tolerant sugarbeet will likely have a 20% weight reduction compared to moderately tolerant cotton that could have up to 60% reduction in weight, and a sensitive species such as soybean could succumb to salinity (Greenway and Munns 1980). Understandably, halophytes grow at their optimum rate (Flowers et al. 1977, 1986). For perennial species, salt tolerance is assessed in terms of survival but for annual species, particularly for broad acre or horticultural crops, the rate of biomass production is a better measure because it correlates with yield. It is challenging to quantify differences in salt tolerance between closely related species because reduction is growth measured depends on the period of time over which the plants have been exposed to salinity. For instance, when exposed to salinity for a short time, a significant decrease in growth rate is noticed, but the decrease may be similar for species that have quite different responses for salinity tolerance. Durum wheat is known to be more salt-sensitive than bread wheat and its yield is decreased in saline conditions (Francois et al. 1986). However, short periods of exposure to salinity does not reveal any major differences between durum and bread wheat cultivars, nor between barley and triticale cultivars (Munns et al. 1995). The leaf elongation rate in the first 10 days of salinization of barley and triticale cultivars were similar indicating no physiological damage (Munns et al. 1995), including the most sensitive durum wheat and barley, the most tolerant (Rawson et al. 1988). These findings led to the experimental consideration of time scale and the various mechanisms that may be involved in controlling growth at different periods of time for plants exposed to salinity.

Salt stress tends to inhibit leaf elongation. Being a primary effect, the physiology related to leaf growth inhibition have been understood (Munns 1993; Lazof and Bernstein 1998). During an important period of leaf development in grasses, it was found that leaf elongation was confined to a basal elongation zone located near the point of leaf attachment to the node. Longitudinal expansion of the leaf was dependent on irreversible expansion of cells located at this basal region. It was found that leaf growth rate was reduced due to the shortening of the length of the leaf elongating zone and thus reducing the growth intensity in its central and distal areas (Bernstein et al. 1993a, b). Growth is synonymous with expansion as such cell tissue expansion occur by turgor-driven wall extension (Tomos and Pritchard 1994). Signaling-mediated modification of capacitance of cell walls has been proposed to be a major growth-limiting factor when plants are exposed to long-term salinity (Cramer and Bowman 1991; Neumann et al. 1994). Plants evolve and undergo physiological modifications to overcome cell-wall growth restrictions. Apoplast pH, cell-wall pH, cold temperature shock have all been shown to influence leaf elongation. Apoplast pH plays an important role in loosening cell wall and its growth (McQueen-Mason et al. 1993) and in numerous plant tissues including maize leaves higher growth rates has been associated with increased acidification of the cell wall space (Van Volkenburgh and Boyer 1985; Jahn et al. 1996; Stahlberg and Van Volkenburgh 1999). Decrease in cell wall pH promotes wall-loosening events which are necessary for cell growth (Rayle and Cleland 1992; Cosgrove 1997), and inhibition of cell wall acidification. Such double modifications may therefore keep the cell growth rate progressive. Several environmental conditions affecting growth alters apoplast acidification. For example, growth inhibition by drought is accompanied by an increase in apoplastic pH and a decrease in acidification rate (Van Volkenburgh and Boyer 1985; Hartung et al. 1988), cold temperature shock increases extracellular pH of the motor cells of *Mimosa pudica* (Kumon and Suda 1985), and gravitrophic stimulation enhances acid efflux and cell elongation along the upper surface of horizontally placed roots (Mulkey and Evans 1981). Salt stress-induced growth inhibition is associated with modifications of apoplast acidification (Neves-Piestun and Bernstein 2001). Thus changes to cellular pH seems to be a key adaptation for plants to overcome salinity stress.

Impact of Salinity on Maize Plants

Salinity is a significant factor affecting agricultural productivity largely in regions practicing irrigated farming. Salinity tolerance is a major problem in maize growing regions (Falcon and Naylor 1998). Crops have varying tolerance to salinity including, intraspecies variation, and in most cases leads to yield losses. It has been shown that above a threshold value the yield decreases as a linear function of salinity until plant death. Maize has a threshold value of 1.7 dS/m and a slope of 12% per dS/m. Maize is a moderately sensitive crop more tolerant than rice but less tolerant than barley, sorghum, bread and durum wheat (Mass and Hoffman 1977; US Salinity Laboratory 2006). Salinity threshold value can be modulated via nitrogen fertilization and evaporation (Beltrão and Asher 1997; Katerji et al. 2000); such that decreased photosynthetic ability of the leaves, and hastened leaf senescence weakens the plant. During severe stress maize plants look stunted and develop short, thick and erect stems with gray appearing foliage (Jones 2003). Even as the number of maize ears developed per plant remained unchanged, the size of their ears and kernels are reduced which translates to a reduction in yield. It is thought that maize grain yield is not as sensitive to salinity as maize forage yield (Beltrão and Asher 1997), because there may be some phloem transport of Na^+ and Cl^- to the reproductive structures (Munns 2002). According to the two-phase model of growth response to salt stress by Munns (1993), two major physiological problems limit crop performance under saline conditions. In a first phase, osmotic problems reduce extension growth so that plants show stunted growth and look dark-green in color. In a second phase, ion toxicity develops and growth of sensitive plants is severely inhibited.

Several physiological, morphological and molecular traits have been suggested for use in improving the salinity tolerance of maize. For the purpose of this article, we will briefly discuss maize-specific genetic approaches (conventional and transgenic breeding) to improve salinity tolerance in maize.

Maize has independent development of its male and female flowers and is very sensitive to stress conditions at flowering, which explains its increased susceptibility to environmental stresses including salinity stress compared to all cereal crops except rice (Grant et al. 1989). Dominance of the apical tassel could lead to protandry that gets highlighted under stress due to reduction in allocation of assimilates to ears, ovules and silks thereby limiting ear and silk growth rate, and increasing kernel and ear abortions (Edmeades et al. 1993; Westgate and Boyer 1986; Westgate and Boyer 2004). As a cross pollinating species propagated for several centuries in highly variable landrace populations (Reif et al. 2004), natural selection had minimal or little selection pressure for increased female survival or productivity under abiotic stress, given that crosses with less stressed (healthy or normal) individuals within a population could ensure survival of inferior alleles.

Improving Salinity Tolerance in Crops

Plant salt tolerance studies have shown linkages to regulatory mechanisms of ionic and osmotic homeostasis (Ashraf and Harris 2004). Apart from ionic and osmotic components, salt stress leads to oxidative stress through an increase in reactive oxygen species (ROS), such as superoxide ($O_2^{\bullet-}$), hydrogen peroxide (H_2O_2) and hydroxyl radicals ($OH^{\bullet}$) (Neill et al. 2002; Ahmad et al. 2010a, b, 2011; Ahmad and Umar 2011; Ahmad and Prasad 2012a, b; Koyro et al. 2012; Yousuf et al. 2012). ROS and redox cues, generated in the chloroplast and mitochondria, are essential for maintaining normal energy and metabolic fluxes, optimizing different cell functions, activating acclimation responses through retrograde signaling, and controlling whole-plant systemic signaling pathways. Regulation of the multiple redox and ROS signals in plants requires a high degree of coordination and balance between signaling and metabolic pathways in different cellular compartments (Ahmad et al. 2008, 2011; Ahmad and Umar 2011; Ahmad and Prasad 2012a, b). These highly reactive ROS are capable of altering normal cellular metabolism via oxidative damage to lipids, proteins and nucleic acids (Imlay 2003; Ahmad and Umar 2011; Ahmad et al. 2011; Ahmad and Prasad 2012a, b). To mitigate the ROS-mediated oxidative damage plants have developed a complex defense antioxidative system involving low-molecular weight antioxidants and antioxidative enzymes such as superoxide dismutase (SOD), catalase (CAT), ascorbate peroxidase (APX), guaiacol peroxidase (GPX) and glutathione reductase (GR) (Noctor and Foyer 1998; Ahmad and Umar 2011; Ahmad et al. 2011; Ahmad and Prasad 2012a, b). SOD is the major $O_2^{\bullet-}$ scavenger and its enzymatic action results in H_2O_2 and O_2 formation. The H_2O_2 produced is scavenged by catalase and other classes of peroxidases. Catalase is found in peroxisomes, cytosol and mitochondria and dismutates H_2O_2 into H_2O and O_2 (McKersie and Leshem 1994). Peroxidases (APX and GPX) on the other hand are distributed throughout the cell. Peroxidases catalyze the reduction of H_2O_2 to H_2O. APX uses ascorbate as electron donor in the first step of the ascorbate–glutathione cycle and is considered the most important plant peroxidase in H_2O_2 detoxification (Noctor and Foyer 1998). GPX, unlike APX has a wider range of substrates, decomposes H_2O_2 by oxidation of co-substrates such as phenolic compounds and/or ascorbate. Reduced glutathione (GSH) regenerates the ascorbate pool producing oxidized glutathione (GSSG). GR, a flavoenzyme (Edwards et al. 1988) catalyzes the reduction of NADPH-dependent GSSG, the last rate limiting step of Halliwell–Asada enzymatic pathway (Bray et al. 2000). Any increase in GR activity could increase GSH/GSSG ratio which is required for ascorbate regeneration and activation of CO_2 fixing enzymes in the chloroplasts (Crawford et al. 2000) thus ensuring the availability of electron accepting $NADP^+$ from the photosynthetic electron transport chain. Lipid oxidation results in production of malondialdehyde (MDA) which is an indicator of oxidative damage. This lipids and cell membrane stability can be utilized to differentiate salt-tolerant and salt-sensitive cultivars (Meloni et al. 2003). Some evidence suggests that resistance to oxidative stress may, at least in part, be involved in salt stress tolerance (Mittova et al. 2002; Badawi et al. 2004). However, most of these studies were performed with leaves and scarce information is available for the root, which infact is the first organ directly exposed and experiencing salt stress. The response of maize antioxidative system to abiotic stresses has been studied under anoxia (Yan et al. 1996), low temperature (Iannelli et al. 1998), drought (Aroca et al. 2003), aluminum and heavy metals toxicity (Zacchini et al. 2003) and nutritional deficiencies (Tiwari et al. 2004).

Understanding Salinity Tolerance in Maize

According to the FAO (2009) maize is the most widely grown grain crop in the Americas with over 300 M tons grown annually in the United States alone. Approximately 40% of the crop (~ 130 million tonnes) is used for corn ethanol (New York Times, 2011). Premier institutions such as CIMMYT have initiated improvement programs in tropical maize for drought and salt stress tolerance given that the two stress are important factors limiting maize production (Edmeades et al. 1999). However, increase in cropping area affected by salt stress and increasing salt levels due to poor quality of irrigation water, necessitates additional attention to salt stress research in maize.

A quick review of the stresses indicates that salinity stress is interconnected with drought. For example, many genes regulated by salt stress are also responsive to drought or cold stress (Zhu et al. 1997; Jamil et al. 2011). Because salt stress can be replicated easily and accurately, several drought stress studies in the laboratory use salt stress to study drought biology. It is interesting to note that the Hog pathway for osmotic stress perception and signaling in yeast was discovered by using NaCl stress (Brewster et al. 1993). How are these two stresses linked? Perhaps plants respond to stresses in similar ways! For instance, salinity reduces the water uptakeability of plants thus causing rapid decrease in growth rate and a series of metabolic changes identical to those caused by drought stress. It is thought that the initial reduction in shoot growth was probably due to hormonal signals generated by the roots. Also, there are salt-specific effects that could impact growth; excessive amounts of salt entering the plant could rise to toxic levels in the older transpiring leaves causing premature senescence and reduction of leaf photosynthetic area which impede growth over time. Presumably, salt-tolerant plants differ from salt-sensitive ones in having a low rate of Na^+ and Cl^- transport to leaves, and their unique ability to compartmentalize these ions in vacuoles to prevent their build-up in cytoplasm or cell walls to avoid salt toxicity (Munns 2002). In order to understand the biochemical processes that contribute to salt tolerance as distinctly as possible from tolerance of osmotic stress, and to identify genes that control the transport of salt across membranes, it is vital to avoid treatments that induce cell plasmolysis, and to design experiments that distinguish between tolerance of salt and tolerance of water stress given their connection (Munns 2002).

The genetic basis of salt tolerance in maize appears to be under the control of genes with additive and non-additive effects, with broad and narrow sense (Rao and McNeilly 1999). Studies indicate that salinity stress is connected to oxidative stress as well (Krasensky and Jonak 2012). Studies on salt signal transduction pathways revealed that salt induces oxidative stress in the form of H_2O_2. A signal transduction pathway is activated by the glutathione peroxidase (*gpx1)* promoter mediated by oxidative stress resulting in production of H_2O_2 in the intracellular space. However there is another pathway induced by extracellular H_2O_2 (Orna et al. 2004). Azevedo Neto et al. (2006) showed that SOD, APX, GPX and GR activities increased with time in the leaves of salt stressed plants compared to the controls. Comparison of salt-tolerant (BR5033) and salt-sensitive (BR5011) genotypes showed that the increase in enzyme activities was more pronounced in the salt-tolerant than in the salt-sensitive genotype (Azevedo Neto et al. 2006). Salt stress did not affect CAT activity in the salt-tolerant; the salt-sensitive lines had reduced CAT. In salt-stressed roots of the salt-tolerant genotype, SOD and CAT activities decreased while APX, GPX and GR activities remained unchanged. In roots of the salt-sensitive genotype, salinity reduced the activity of all studied enzymes. These studies proved that CAT and GPX enzymes had the greatest H_2O_2 scavenger activity in both leaves and roots of maize. Moreover, CAT, APX and GPX activities in conjunction with SOD play an essential protective role in the scavenging processes. Lipid peroxidation was enhanced only in salt-stressed leaves of the salt-sensitive genotype. These results indicate that oxidative stress plays an important role in salt-stressed maize plants and that the greater protection of BR5033 leaves and roots from salt-induced oxidative damage results, at least in part, through the maintenance and/or increase of the activity of antioxidant enzymes (Azevedo Neto et al. 2006).

Regulatory Factors Involved in Salt Stress

siRNA and miRNA in Salt Stress

Several recent studies show that small RNAs contribute to environmental stress response by modifying plant gene expression (Khraiwesh et al. 2012). Small RNAs include a group of regulatory RNAs that control a variety of biological processes. The major classes of small RNAs include microRNAs (miRNAs) and small interfering RNAs (siRNAs). siRNAs and miRNAs differ in their biogenesis. miRNAs bind to reverse complimentary sequences of cognate genes, resulting in either translational inhibition or cleavage of target transcripts. On the otherhand siRNAs most often direct DNA methylation at target sequences (Khraiwesh et al. 2012).

Some stress regulated miRNAs are proposed to be involved in cell response to salt, heat and water stress (Zhao et al. 2007; Sunkar et al. 2008; Zhou et al. 2008). Stress-responsive transcription factors (TFs) or functional genes are targets for some miRNAs (Prashanth et al. 2008; Xu 2008). In these cases, miRNA-dependent post-transcriptional regulation may contribute to plant stress response, which includes guiding target mRNAs for degradation or by repressing translation (Sunkar et al. 2007). However, there are evidences showing positive regulation by miRNAs resulting in accumulation of a set of gene products (Kim et al. 2007). Positive regulation involves accumulation of a set of gene products and negative regulation involves enhanced suppression of specific targets. It has also been observed that there are significant differences in expression levels of miRNAs in different genotypes (Mica et al. 2006) indicating genotypic differences are contributed by differences in miRNA regulation.

Based on a mi-RNA microarray hybridization experiment, Ding et al. (2009) reported a significantly altered expression of several miRNA families after salt treatment. A total of 98 miRNAs were identified in this study, of which 27 were plant miRNA families. A total of 18 miRNAs were expressed only in the salt-tolerant maize line. There were 25 miRNAs that showed a delayed regulation pattern in the salt-sensitive line. The researchers propose that similarly regulated miRNAs represent the fundamental mechanism of adapting to salt shock, whereas, the differentially regulated miRNAs represent distinct salt sensitivities between the genotypes. Members of the miR474 and miR395 families are examples of similarly regulated miRNAs between susceptible and tolerant genotypes. Members of the miR396 family were differentially regulated with respect to time indicating differential regulation during the stress process as a reason for different salt sensitivities (Ding et al. 2009). The natural antisense small interfering RNA (nat-siRNA) guided cleavage of transcripts of a salt stress-related functional gene, P5CDH [D(1)-pyrroline-5-carboxylate dehydrogenase] transcript, which is a salt stress-related functional gene was also reported (Boesani et al. 2005).

DNA Components or Genes in Salt Stress

Although there are numerous genes reported to be involved or associated with salt stress, a few key genes and interesting ones are mentioned here. Plasma membrane protein 3 (PMP3) genes include a group of small hydrophobic polypeptides that play important roles in maintenance of ion homeostasis (Fu et al. 2012). A total of eight *ZmPMP3* genes responsive to salt, drought, cold and abscisic acid were cloned from maize. Detailed studies indicated that the eight *ZmPMP3*s were membrane proteins with highly conserved sequences in trans-membrane regions. Phylogenetically these eight genes belong to three groups. Complementation experiments showed that most of these genes were capable of maintaining membrane potential, which consecutively allows for regulation of intracellular ion homeostasis, independently of the presence of Ca^{2+}. It was reported that expression levels of three ion transporter genes and four important antioxidant genes in ROS scavenging system increased significantly in transgenic plants during salt stress. This tolerance was likely achieved through diminishing oxidative stress due to the possibility of *ZmPMP3-1*'s involvement in regulation of ion homeostasis. Based on these observations, Fu et al. (2012) proposed that modulation of conserved small hydrophobic polypeptides could be an effective way to improve salt tolerance.

Mitogen-activated protein kinase (MAPK) has been investigated in detail in Arabidopsis, tobacco and rice. However, the function of MAPKs in maize (*Zea mays* L.) has not been completely documented (Gu et al. 2010). *Zea mays* salt-induced mitogen-activated protein kinase 1 (ZmSIMK1) is a known MAPK gene in maize. Over-expression of ZmSIMK1 in Arabidopsis resulted in increased resistance against salt stress and exhibited constitutive expression and upregulation of stress-responsive marker genes such as *RD29A* and *P5CS1*.

Maize MIP gene family plays are role in salt stress regulation. Salt stress is known to reduce root hydraulic conductivity and growth. Inositol phosphate genes are involved in a concomitant regulation of aquaporins. A large group of genes are involved in physiological responses related to plant stress responses and ABA is believed to play a central role (Zhang et al. 2006). Zhang et al. (2006) reported that several physiological responses that help plant survival to salt stress are mediated by ABA. In addition, the stress-related production and catabolism of ABA are themselves stress-regulated. Extensive reviews on role of ABA as a long-distance signal mediating whole plant responses to environmental stresses and control of stress related gene expression are widely available.

Improving Salinity Tolerance by Conventional Breeding Approaches

Salinity causes ionic as well as osmotic stresses in plants, in particular, osmotic stress, ionic imbalance (e.g. Na^+ versus K^+; Na^+ versus Ca^{2+}), specific ion toxicities (e.g. Na^+ and Cl^-), and developmental problems (Grattan and Grieve 1999; Munns 2002). So far the dominant effect among the two (osmotic and ionic) is unclear.

A 'two-phase growth response to salinity' model was proposed by Munns (1993) which states that water deficits inhibit growth shortly after salinisation followed by ionic effects. Maize growth under saline conditions were influenced by both osmotic and ion effects. To add, soil conditions vary between fields and also within the same field (Richards 1993). Salinity also results in changes to other soil physical and chemical properties including (but not limited to) sodicity, high pH, and boron (FAO 2000). Under such situations there will be interaction between different stress factors. This also triggers genotype-by-environment interactions, thus making breeding progress difficult and makes direct selection of superior salinity tolerant genotypes is challenging. Root and shoot growth reduction in salt sensitive plants does not appear to depend on salt levels in growing tissue but a response of osmolarity of external solution (Munns 2002). Though breeding efforts for enhanced drought tolerance in maize (Bänziger et al. 2006) are happening, the bulk of the work in maize stress has focused on genetic variation in the salinity tolerance of maize seedlings; however, some work involved modified water application procedures and use of saline irrigation water for selection purposes. It has been observed that salinity tolerance in cereal crops including maize likely increases with plant age (Flowers 2004; Yamaguchi and Blumwald 2005) because tolerance measured at maize seedling stage persists until maturity (Maiti et al. 1996). This indicates that solution culture at seedling stage is a potential tool for selecting for enhanced salinity tolerance in maize (Khan et al. 2003). Selection for salinity tolerance was seldom done methodically in maize breeding programs due to all the challenges described above. Flowers and Yeo (1995) estimated the development of about 30 salinity tolerant crop cultivars because it may not yet have influenced breeders to pursue development of salt-tolerant cultivars.

Understanding maize genetics in the context of salinity tolerance may open potential breeding approaches for incorporating salinity tolerance in maize. Populations and progenies containing increased biomass and seedling shoot length under the influence of salinity had a narrow-sense heritability of 0.54 as reported by Ashraf and McNeilly (1990), and a broad-sense heritability of 0.4 as reported by Maiti et al. (1996). Khan et al. (2003) also reported additive and non-additive effects for root length under salt stress with broad and narrow-sense heritability estimates of 0.6–0.8 and 0.4 respectively. In addition, component traits for these parameters are proposed to be controlled by complex genetic systems (Khan et al. 2003; Flowers 2004). Mechanisms of tolerance to salinity in maize is reported to be associated with differences in salt accumulation rates and leaf senescence demonstrating significance of Na^+ exclusion (Fortmeier and Schubert 1995) and varietal differences in shoot and/or root growth (Cramer et al. 1994; Mladenova 1990). It is thought that these differences are a result of response to osmolarity (Neumann 1997). For example some maize cultivars differ with respect to their response to supplemental calcium when salinized (Cramer 2002). In general, Na^+ 'exclusion' in salinity tolerant varieties is accompanied by K^+/Na^+ discrimination. Reduced loading into and removal of Na^+ from the xylem, retention of ions in the leaf sheath, tissue tolerance, and ion partitioning into leaves (Colmer et al. 2005; Flowers 2004; Munns 2002) are also common characteristics of salinity tolerance. The physiological processes involved in salinity tolerance have been an active area and several genes have been identified Munns et al. (2006).

Salinity Tolerance via Marker-Assisted Selection

Molecular markers contributed tremendously towards the identification, characterization, and comparison of quantitative trait loci (QTL) that has significant effects on plant salt tolerance during various stages of plant growth and development (Foolad et al. 2001). First attempts to apply molecular markers for QTL analysis of drought tolerance response in maize was reported by Lebreton et al. (1995), it was recently that a few applications have emerged in practical maize breeding programs. The major reasons for low popularity of this methods was complexity of the nature of the trait and its genetic basis, influence of environment and plant developmental stages on QTL effects, (Tuberosa et al. 2002), limitations for precise phenotyping of component traits, resource requirements for fine mapping QTLs, as well as understanding gene-by-gene effects (Campos et al. 2004). Ribaut et al. (1996) identified six putative drought affected QTLs for anthesis-silking interval located on chromosomes 1, 2, 5, 6, 8 and 10, collectively accounting for 47% of the phenotypic variance. Abdel-Bary et al. (2005) detected eight positive and negative RAPD markers for salinity tolerance in maize. QTLs for drought explaining 50% of the phenotypic variance of grain yield were identified on chromosomes 1, 3, 5, 6, and 8, which expressed different types of gene action (Agrama and Moussa 1996). Sanguineti et al. (1999) identified 16 of 17 QTL regions related to leaf ABA concentration. These QTLs were also related to stomatal conductance, relative leaf water content, leaf temperature, anthesis-silking interval and grain yield. Given the strong QTL-by-environment effects and low accountability for phenotypic variations by individual QTLs, Ribaut et al. (1997) concluded that marker-assisted selection should be based on best QTLs for grain yield and secondary traits. A maize line with increased drought tolerance was developed by marker assisted backcrossing based on the incorporation of five chromosome segments (Ribaut et al. 2002) as a proof of concept. In addition, 11 key regions for drought tolerance were identified by co-localization of QTLs for morphological traits, related physiological parameters, and candidate genes (Ribaut et al. 2004a). Application of this method in selecting for drought tolerance in unrelated crosses was successful (Ribaut et al. 2004b). Campos et al. (2004) hypothesized that QTLs identified for complex traits such as drought tolerance in maize are likely to be context-dependent; and these authors recommended that QTL information would have to be used selectively and based on the specific maize breeding situation to which they are to be applied.

Transgenic Approaches for Salinity Tolerance

Recent genomic approaches have contributed to our knowledge of stress adaptation and stimulated drought and salinity stress related research in a wide range of areas including osmo-protectants, stress proteins, salt shock proteins, ion/proton transporters, water status, signaling components, control of transcription, growth regulators (Cushman and Bohnert 2000). Maize has been a key commercial crop where transgenic approaches have led to enhanced drought or salinity tolerance, despite the advanced research work being conducted using model species. Jeanneau et al. (2002) successfully increased maize water use efficiency by 30% and dry weight by 20% under moderate drought conditions by over-expressing C4 phosphenolpyruvate carboxylase. Similarly, over-expression of Arabidopsis AtNHX1, a vacuolar Na^+/H^+ antiporter, resulted in enhanced salinity tolerance in transgenic maize (Yin et al. 2004). These transgenic successes are crucial to the development of salinity tolerant lines, however, no salt-resistant maize cultivars have been developed (Rozema and Flowers 2008). Quan et al. (2004) reported that transformation of maize with the *betA* gene from *Escherichia coli* encoding choline dehydrogenase resulted in higher glycine betaine accumulation, tolerance to drought stress at germination, young seedling stage and increased grain yields. This transformation with *betA* gene strengthened the cell membrane with modified catalase, peroxidase and dismutase enzymes (Saneoka et al. 1995). Shou et al. (2004) expressed a tobacco mitogen-activated protein kinase kinase kinase constitutively in maize. As a result, transgenic maize plants maintained in pot experiments had significantly higher photosynthesis rates and produced 40–60% higher kernel weights than the non-transgenic control plants. Again, the underlying protection of photosynthesis from dehydration could potentially be effective under both salinity and drought stress. More recently, information is emerging about constitutively expressed transcription factors showing yield advantage in maize exposed to drought under multi-location field conditions (Warner et al. 2005; Heard et al. 2005). Li et al. (2010) generated marker-free transgenic maize plants constitutively expressing *AtNHX1*, a Na^+/H^+ antiporter gene from Arabidopsis that conferred salt tolerance on plants, using the FLP/FRT site-specific recombination system. These transgenic plants also produced greater biomass and yields than non-transgenic plants when grown in high saline fields. Though these results are encouraging, the deployment of transgenic drought tolerant maize cultivars needs more time and research efforts.

Although drought tolerant orthologs have been identified, they may not produce the same results in a model plant and in an elite cultivar (Zhang et al. 2000), because experimental conditions between laboratory and field experiments often greatly vary, and as gene-by-crop effects may be expected between model plants and maize given its long history of improvement (Campos et al. 2004). As mentioned in previous sections, the genetics of drought and salinity tolerance is extremely complex which could mean that modulation of individual genes or even pathways may not be successful (Flowers 2004; Munns et al. 2006), as we have now known from other studies that alteration of a single process may be compensated or damped out (Sinclair and Purcell 2005), or have a minimal impact to be visible as significant phenotypic changes (Edmeades et al. 2004); tolerance mechanisms could also differ between developmental stages of a single plant (Flowers 2004; Yamaguchi and Blumwald 2005), and gene, event- and crop-specific effects may be dependent on the genotype and the target environment of deployment. Given these complexities, Sinclair and Purcell (2005) concluded that genetic improvements would have to come from the integration and concurrent improvement in several traits.

Conclusion and Future Perspective

Maize is a crop with significant commercial value in the Americas, Africa and many other regions; used as a staple or for fuel production. With increasing population, growing food demand, and constant resource shortages, it is evident that we need to achieve higher yields from minimal resource usage. Maize is considered a salt-sensitive crop despite intraspecies variability in resistance levels. Salt resistance being a multigenic trait has been a difficult goal to achieve via conventional breeding methodologies. In addition, the genetic mechanism of salinity tolerance in maize is very complex, the yield trait is strongly influenced by both genotype and environment, and the interaction of genotype-by-environment is significant. Traditional breeding programs that depend on phenotype selection are time-consuming and less efficient. At present, more powerful and efficient strategies for producing tolerant crops are possible with the development of molecular biology. For instance, the QTL pyramiding approach results from a combination of recent crop genomics and conventional breeding promotes better crop breeding.

Systematic studies to understand the physiology and genetics of salinity is needed, although there have been several discoveries of novel genes and pathways that are interconnected such as auxin, the cross-talking, gene-and event specific situations complicate characterization of salinity tolerance in plants. More knowledge on sequential steps involved in the salinity tolerance or susceptibility pathways would be valuable. While application of molecular marker techniques on traditional breeding programs can improve the efficiency of breeding for salinity-tolerant maize, alternate strategies such as the utilization of physiological knowledge combined with transgenic biology, and salt management strategies, such as exclusion of salt from plant system in combination with breeding will be valuable.

References

Abdel-Bary AA, Rashed MA, El-seoudy AA (2005) Molecular genetic studies on some maize (*Zea mays* L.) inbreds. Egypt. J Genet Cytol 34:15–27

Agrama HAS, Moussa ME (1996) Mapping QTLs in breeding for drought tolerance in maize (*Zea mays* L.). Euph 91:89–97

Ahmad P, Prasad MNV (2012a) Environmental adaptations and stress tolerance in plants in the era of climate change. Springer Science+Business Media, LLC, New York

Ahmad P, Prasad MNV (2012b) Abiotic stress responses in plants: metabolism, productivity and sustainability. Springer Science+Business Media, LLC, New York

Ahmad P, Sharma S (2008) Salt stress and phyto-biochemical responses of plants. Plant Soil Environ 54(3):89–99

Ahmad P, Umar S (2011) Oxidative stress: role of antioxidants in plants. Studium Press, New Delhi

Ahmad P, Sarwat M, Sharma S (2008) Reactive oxygen species, antioxidants and signaling in plants. J Plant Biol 51(3):167–173

Ahmad P, Jaleel CA, Salem MA, Nabi G, Sharma S (2010a) Roles of enzymatic and non-enzymatic antioxidants in plants during abiotic stress. Crit Rev Biotechnol 30(3):161–175

Ahmad P, Umar S, Sharma S (2010b) Mechanism of free radical scavenging and role of phytohormones during abiotic stress in plants. In: Ashraf M, Ozturk M, Ahmad MSA (eds) Plant adaptation and phytoremediation. Springer, Dordrecht/Heidelber/New York, pp 99–108

Ahmad P, Nabi G, Jeleel CA, Umar S (2011) Free radical production, oxidative damage and antioxidant defense mechanisms in plants under abiotic stress. In: Ahmad P, Umar S (eds) Oxidative stress: role of antioxidants in plants. Studium Press, New Delhi, pp 19–53

Ahmad P, Ashraf M, Younis M, Hu X, Kumar A, Akram NA, Al-Qurainy F (2012) Role of transgenic plants in agriculture and biopharming. Biotechnol Adv 30(3):524–540

Akram MS, Ashraf M, Akram NA (2009) Effectiveness of potassium sulfate in mitigating salt-induced adverse effects on different physio-biochemical attributes in sunflower (*Helianthus annuus* L.). Flora 204(6):471–483

Allakhverdiev SI, Sakamoto A, Nishiyama Y, Inaba M, Murata N (2000) Ionic and osmotic effects of NaCl induced inactivation of photosystems I and II in *Synechococcus* sp. Plant Physiol 123:1047–1056

Aroca R, Vernieri P, Irigoyen JJ, Sánchez-Díaz M, Tognoni F, Pardossi A (2003) Involvement of abscisic acid in leaf and root of maize (*Zea mays* L.) in avoiding chilling-induced water stress. Plant Sci 165:671–679

Ashraf M, McNeilly T (1990) Improvement of salt tolerance in maize by selection and breeding. Plant Breeding 104:101–107

Ashraf M, Harris PJC (2004) Potential biochemical indicators of salinity tolerance in plants. Plant Sci 66:3–16

Azevedo Neto AD, Prisco JT, Eneas-Filho J, Abreu CEB, Gomes-Filho E (2006) Effect of salt stress on antioxidative enzymes and lipid peroxidation in leaves and roots of salt-tolerant and salt-sensitive maize genotypes. Environ Exp Bot 56:87–94

Badawi GH, Kawano N, Yamauchi Y, Shimada E, Sasaki R, Kubo A, Tanaka K (2004) Enhanced tolerance to salt stress and water deficit by over-expressing superoxide dismutase in tobacco chloroplasts. Plant Sci 166:919–928

Bänziger M, Araus JL (2007) Recent advances in breeding maize for drought and salinity stress tolerance. In: Jenks MA et al (eds) Advances in molecular breeding toward drought and salt tolerant crops. Springer, Dordrecht, pp 587–601

Bänziger M, Setimela PS, Hodson D, Vivek B (2006) Breeding for improved drought tolerance in maize adapted to southern Africa. Agric Water Manag 80:212–224

Beltrão J, Asher JB (1997) The effect of salinity on corn yield using the CERES-maize model. Irrig Drain Syst 11:15–28

Bernstein N, Lauchli A, Silk WK (1993a) Kinematics and dynamics of sorghum (*Sorghum bicolor* L.) leaf development at various Na/Ca salinities: I. Elongation growth. Plant Physiol 103:1107–1114

Bernstein N, Silk WK, Lauchli A (1993b) Growth and development of sorghum leaves under conditions of NaCl stress: spatial and temporal aspects of leaf growth inhibition. Planta 191:433–439

Boesani O, Zhu JH, Verslues PE, Sunkar R, Zhu JK (2005) Endogenous siRNAs derived from a pair of natural cis-antisense transcripts regulate salt tolerance in Arabidopsis. Cell 123:1279–1291

Bray EA, Bailey-Serres J, Weretilnyk E (2000) Responses to abiotic stresses. In: Gruissem W, Buchannan B, Jones R (eds) Biochemistry and molecular biology of plants. American Society of Plant Physiologists, Rockville, pp 1158–1249

Brewster JL, de Valoir T, Dwyer ND, Winter E, Gustin MC (1993) An osmosensing signal transduction pathway in yeast. Science 259:1760–1763

Campos H, Cooper M, Habben JE, Edmeades GO, Schussler JR (2004) Improving drought tolerance in maize: a view from industry. Field Crop Res 90:19–34

Colmer TD, Munns R, Flowers TJ (2005) Improving salt tolerance of wheat and barley: future prospects. Aust J Exp Agric 45:1425–1443

Cosgrove DJ, Bedinger P, Durachko DM (1997) Group I allergens of grass pollen as cell wall-loosening agents. Proc Natl Acad Sci USA 94:6559–6564

Cramer GR (2002) Sodium–calcium interactions under salinity stress. In: Läuchli A, Lüttge U (eds) Salinity: environment–plants–molecules. Kluwer, Dordrecht, pp 205–227
Cramer GR, Bowman DC (1991) Kinetics of maize leaf elongation. I. Increased yield threshold limits short-term, steady-state elongation rates after exposure to salinity. J Exp Bot 42: 1417–1426
Cramer GR, Alberico GJ, Schidt C (1994) Salt tolerance is not associated with the sodium accumulation of two maize hybrids. Aust J Plant Physiol 21:675–692
Crawford NM, Kahn ML, Leustek T, Long SR (2000) Nitrogen and sulfur. In: Gruissem W, Buchannan B, Jones R (eds) Biochemistry and molecular biology of plants. American Society of Plant Physiologists, Rockville, pp 786–849
Cushman JC, Bohnert HJ (2000) Genomic approaches to plant stress tolerance. Curr Opin Plant Biol 3:117–124
Ding D, Zhang L, Wang H, Liu Z, Zhang Z, Zheng Y (2009) Differential expression of miRNAs in response to salt stress in maize roots. Ann Bot 103:29–38
Edmeades GO, Bolaños J, Hernandez M, Bello S (1993) Causes for silk delay in lowland tropical maize. Crop Sci 33:1029–1035
Edmeades GO, Bolaños J, Chapman SC, Lafitte HR, Bänziger M (1999) Selection improves drought tolerance in tropical maize populations: I. Gains in biomass, grain yield, and harvest index. Crop Sci 39:1306–1315
Edmeades GO, McMaster GS, White JW, Campos H (2004) Genomics and the physiologist: bridging the gap between genes and crop response. Field Crop Res 90:5–18
Edwards MC, Smith GN, Bowling DJF (1988) Guard-cells extrude protons prior to stomatal opening-a study using fluorescence microscopy and pH micro-electrodes. J Expt Bot 39:1541–1547
Falcon WP, Naylor RL (1998) The maize transition in Asia: unlocking the controversy. Am J Agric Econ 80:960–968
FAO (2000) Extent and causes of salt-affected soils in participating countries. FAO Rome, Italy. www.fao.org
FAO (2009) Maize, rice and wheat: area harvested, production quantity, yield. http://faostat.fao.org/site/567/DesktopDefault.aspx?PageID=567
FAO (2011) The state of the world's land and water resources for food and agriculture (SOLAW). Rome and Earthscan, London
Flowers TJ (2004) Improving crop salt tolerance. J Exp Bot 55:307–319
Flowers TJ, Yeo AR (1995) Breeding for salinity tolerance in crop plants: where next? Aust J Plant Physiol 22:875–884
Flowers TJ, Troke PF, Yeo AR (1977) The mechanism of salt tolerance in halophytes. Ann Rev Plant Physiol 28:89–121
Flowers TJ, Hajibagheri MA, Clipson NJW (1986) Halophytes. Q Rev Biol 61:313–337
Food and Agriculture Organization of the United Nations (1996) Report of the world food summit. Food and Agriculture Organization of the United Nations, Rome, 13–17 Nov 1996
Foolad MR, Zhang LP, Lin GY (2001) Identification and validation of QTLs for salt tolerance during vegetative growth in tomato by selective genotyping. Genome 44:444–454
Fortmeier R, Schubert S (1995) Salt tolerance of maize (*Zea mays* L.): the role of sodium exclusion. Plant Cell Environ 18:1041–1047
Francois LE, Maas EV, Donovan TJ, Youngs VL (1986) Effect of salinity on grain yield and quality, vegetative growth, and germination of semi-dwarf and durum wheat. Agron J 78:1053–1058
Fu J, Zhang DF, Liu YH, Ying S, Shi YS, Song YC, Li Y, Wang TY (2012) Isolation and characterization of maize PMP3 genes involved in salt stress tolerance. PLoS One 7:e31101
Grattan SR, Grieve CM (1999) Salinity – mineral nutrient relations in horticultural crops. Sci Hortic 78:127–157
Grant RF, Jackson BS, Kiniry JR, Arkin GF (1989) Water deficit timing effects on yield components in maize. Agron J 81:61–65

Greenway H, Munns R (1980) Mechanisms of salt tolerance in nonhalophytes. Annu Rev Plant Physiol 31:149–190
Gregorio GB, Senadhira D, Mendoza RD (1997) Screening rice for salinity tolerance. International Rice Research Institute, Los Banos
Gu L, Liu Y, Zong X, Liu L, Li DP, Li DQ (2010) Overexpression of maize mitogen-activated protein kinase gene, ZmSIMK1 in Arabidopsis increases tolerance to salt stress. Mol Biol Rep 37:4067–4073
Hartung W, Radin JW, Hendrix DL (1988) Abscisic acid movement into the apoplastic solution of water-stressed cotton leaves: role of apoplastic pH. Plant Physiol 86:908–913
Heard J, Adams TR, Anstrom G, Benson R, Nelson D, Warner D, Ratcliffe O, Creelman R, Dotson S (2005) Increasing yield stability in corn under drought conditions: new insights from transgenic studies. In: Abstract L 8.02 InterDrought-II of the 2nd international conference on integrated approaches to sustain and improve plant production under drought stress: final program and abstract book, Avenue media, Bologna, Italy
Hu X, Zhang Z, Xu P, Fu Z, Hu S, Song W (2010) Multifunctional genes: the cross-talk among the regulation networks of abiotic stress responses. Biol Plantarum 54:213–223
Iannelli AM, Breusegem FV, Montaguu MV, Inzé D, Massacci A (1998) Tolerance to low temperature and paraquat-mediated oxidative stress in two maize genotypes. J Expt Bot 50:523–532
Imlay JA (2003) Pathways of oxidative damage. Annu Rev Microbiol 57:395–418
Jahn T, Johansson F, Luche H, Volkmann D, Larsson C (1996) Reinvestigation of auxin and fusicoccin stimulation of the plasma-membrane H1-ATPase activity. Planta 199:359–365
Jamil M, Ashraf M, Rehman S, Ahmad M, Rha ES (2011) Salinity induced changes in cell membrane stability, protein and RNA contents. Afr J Biotechnol 11:6476–6483
Jeanneau M, Gerentes D, Foueillassar X, Zivy ML, Vidal J, Toppan A, Perez P (2002) Improvement of drought tolerance in maize: towards the functional validation of the *Zm-Asr1* gene and increase of water use efficiency by over-expressing C4–PEPC. Biochimie 84:1127–1135
Jones JB (2003) Agronomic handbook: management of crops, soils and their fertility. CRC Press, Boca Raton, p 450
Kasuga M, Liu Q, Miura S, Yamaguchi-Shinozaki K, Shinozaki K (1999) Improving plant drought, salt and freezing tolerance by gene transfer of a single stress-inducible transcription factor. Nat Biotechnol 17:287–291
Katerji N, Van Hoorn JW, Hamdy A, Mastrorilli M (2000) Salt tolerance classification of crops according to soil salinity and to water stress day index. Agric Water Manag 43:99–109
Khan AA, Rao SA, McNeilly T (2003) Assessment of salinity tolerance based upon seedling root growth response functions in maize (*Zea mays* L.). Euphytica 131:81–89
Khraiwesh B, Zhu JK, Zhu J (2012) Role of miRNAs and siRNAs in biotic and abiotic stress responses of plants. Biochim Biophys Acta 1819:137–4848
Kim YK, Kim VN (2007) Processing of intronic microRNAs. EMBO J 26:775–783
Kizis D, Lumbreras V, Pages M (2001) Role of AP2/EREBP transcription factors in gene regulation during abiotic stress. FEBS Lett 498:187–189
Koyro HW, Ahmad P, Geissler N (2012) Abiotic stress responses in plants: an overview. In: Ahmad P, Prasad MNV (eds) Environmental adaptations and stress tolerance of plants in the era of climate change. Springer, Science+business media, New York, pp 1–28
Krasensky J, Jonak C (2012) Drought, salt, and temperature stress-induced metabolic rearrangements and regulatory networks. J Exp Bot. doi: 10.1093/jxb/err460
Kumon K, Suda S (1985) Changes in the extracellular pH of the motor cells of *Mimosa pudica* L. during movement. Plant Cell Physiol 26:375–377
Lafitte HR, Courtois B (2000) Genetic variation in performance under reproductive stage water deficit in a doubled-haploid rice population in upland fields. In: Ribaut JM, Poland D (eds) Molecular approaches for the genetic improvement of cereals for stable production in water-limited environments. A strategic planning workshop held on 21–25 June 1999. CIMMYT, El Batan, pp 97–102
Lazof D, Bernstein N (1998) The NaCl-induced inhibition of shoot growth: the case for disturbed nutrition with special consideration of calcium nutrition. Adv Bot Res 29:113–189

Lebreton C, Laziejancie V, Steed A, Pekic S, Quarrie SA (1995) Identification of QTL for drought responses in maize and their testing casual relationships between traits. J Expt Bot 46:853–865

Li ZK, Xu JL (2007) Breeding for drought and salt tolerant rice (*Oryza sativa* L.): progress and perspectives. In: Jenks MA et al (eds) Advances in molecular breeding toward drought and salt tolerant crops. Springer, Dordrecht, pp 531–564

Li ZK, Fu BY, Gao YM, Xu JL, Ali J, Lafitte HR, Jiang YZ, Rey JD, Vijayakumar CHM, Maghirang R, Zheng TQ, Zhu LH (2005) Genome-wide introgression lines and a forward genetics strategy for functional genomic research of complex phenotypes in rice. Plant Mol Biol 59:33–52

Li B, Li N, Duan X, Wei A, Yang A, Zhang J (2010) Generation of marker-free transgenic maize with improved salt tolerance using the FLP/FRT recombination system. J Biotechnol 145: 206–213

Maas EV, Hoffman GJ (1977) Crop salt tolerance – current assessment. J Irrig Drain, Div Am Soc Civ Eng 103:115–134

Maas EV (1986) Salt tolerance of plants. Appl Agric Res 1:12–25

Maiti RK, Delgado LE, Amaya S, Cardona I, Dimas OAM, Rosa-Ibarra MD, Castillo HD (1996) Genotypic variability in maize cultivars (*Zea mays* L.) for resistance to drought and salinity at the seedling stage. J Plant Physiol 148:741–744

McKersie BD, Leshem YY (1994) Stress cropping in cultivated plant. Kluwer, Dordrecht

McQueen-Mason SJ, Fry SC, Durachko DM, Cosgrove DJ (1993) The relationship between xyloglucan endotransglycosylase and in vitro cell wall extension in cucumber hypocotyls. Planta 190:327–331

Meloni DA, Oliva MA, Martinez CA, Cambraia J (2003) Photosynthesis and activity of superoxide dismutase, peroxidase and glutathione reductase in cotton under salt stress. Environ Exp Bot 49:69–76

Mica E, Gianfranceschi L, Pe ME (2006) Characterization of five microRNA families in maize. J Exp Bot 57:2601–2612

Mittova V, Tal M, Volokita M, Guy M (2002) Salt stress induces up-regulation of an efficient chloroplasts antioxidant system in the salt-tolerant wild tomato species *Lycopersicon pennellii* but not in the cultivated species. Plant Physiol 115:393–400

Mladenova YI (1990) Influence of salt stress on primary metabolism of *Zea mays* L. seedlings of model genotypes. Plant Soil 123:217–222

Mulkey TJ, Evans ML (1981) Geotropism in corn roots: evidence for its mediation by differential acid efflux. Science 212:70–71

Munns R (1993) Physiological processes limiting plant growth in saline soils: some dogmas and hypotheses. Plant Cell Environ 16:15–24

Munns R (2002) Comparative physiology of salt and water stress. Plant Cell Environ 25:239–250

Munns R, Schachtman DP, Condon AG (1995) The significance of a two-phase growth response to salinity in wheat and barley. Aust J Plant Physiol 22:561–569

Munns R, James RA, Läuchli A (2006) Approaches to increasing the salt tolerance of wheat and other cereals. J Exp Bot 57:1025–1043

Neill SJ, Desikan D, Clarke A, Hancock JT (2002) Nitric oxide is a novel component of abscisic acid signaling in stomatal guard cells. Plant Physiol 128:13–16

Neumann P (1997) Salinity resistance and plant growth revisited. Plant Cell Environ 20:1193–1198

Neumann PM, Azaizeh H, Leon D (1994) Hardening of root cell walls: a growth inhibitory response to salinity stress. Plant Cell Environ 17:303–309

Neves-Piestun BG, Bernstein N (2001) Salinity-induced inhibition of leaf elongation in maize is not mediated by changes in cell wall acidification capacity. Plant Physiol 125:1419–1428

New York Times (2011) U.S. approves corn modified for ethanol. http://www.nytimes.com/2011/02/12/business/12corn.html). Accessed 11 Feb 2011

Noctor G, Foyer CH (1998) Ascorbate and glutathione: keeping active oxygen under control. Annu Rev Plant Physiol Plant Mol Biol 49:249–279

Orna AK, Yardena GD, Simcha LY, Rachel G, Gozal BH (2004) The salt-stress signal transduction pathway that activates the *gpx1* promoter is mediated by intracellular H_2O_2, different from the pathway induced by extracellular H_2O_2. Plant Physiol 135:1685–1696

Pathan MS, Lee JD, Shannon G, Nguyen HT (2007) Recent advances in breeding for drought and salt stress tolerance in soybean. In: Jenks MA et al (eds) Advances in molecular breeding toward drought and salt tolerant crops. Springer, Dordrecht, pp 531–564

Prashanth SR, Sadhasivam V, Parida A (2008) Over-expression of cytosolic copper/zinc superoxide dismutase from a mangrove plant *Avicennia marina* in indica rice var Pusa Basmati-1 confers abiotic stress tolerance. Transgenic Res 17:281–291

Quan R, Shang M, Zhang H, Zhao Y, Zhang J (2004) Engineering of enhanced glycine betaine synthesis improves drought tolerance in maize. Plant Biotechnol J 2:477

Rao SA, McNeilly T (1999) Genetic basis of variation for salt tolerance in maize (*Zea mays L*). Euphytica 108:145–150

Rawson HM, Richards RA, Munns R (1988) An examination of selection criteria for salt-tolerance in wheat, barley and triticale genotypes. Aust J Agri Res 39:759–772

Rayle DL, Cleland RE (1992) The acid growth theory of auxin-induced cell elongation is alive and well. Plant Physiol 99:1271–1274

Reif JC, Xia XC, Melchinger AE, Warburton ML, Hoisington DA, Beck D, Bohn M, Frisch M (2004) Genetic diversity determined within and among CIMMYT maize populations of tropical, subtropical, and temperate germplasm by SSR markers. Crop Sci 44:326–334

Ribaut JM, Hoisington DA, Deutsch JA, Jiang C, Gonzalez-de-Leon D (1996) Identification of quantitative trait loci under drought conditions in tropical maize. 1. Flowering parameters and the anthesis-silking interval. Theor Appl Genet 92:905–914

Ribaut JM, Jiang C, Gonzalez-de-Leon D, Edmeades GO, Hoisington DA (1997) Identification of quantitative trait loci under drought conditions in tropical maize. 2. Yield components and marker assisted selection strategies. Theor Appl Genet 94:887–896

Ribaut JM, Bänziger M, Betrán FJ, Jiang C, Edmeades GO, Dreher K, Hoisington DA (2002) Use of molecular markers in plant breeding: drought tolerance improvement in tropical maize. In: Kang MS (ed) Quantitative genetics, genomics and plant breeding. Centre for Agricultural Bioscience International, Wallingford, pp 85–99

Ribaut JM, Bänziger M, Setter T, Edmeades G, Hoisington D (2004a) Genetic dissection of drought tolerance in maize: a case study. In: Nguyen H, Blum A (eds) Physiology and biotechnology integration for plant breeding. Dekker, New York, pp 571–611

Ribaut JM, Sawkins MC, Bänziger M, Vargas M, Huerta E, Martinez C, Moreno M (2004b) Marker-assisted selection in tropical maize based on consensus map, perspectives, and limitations, p. 267–268. In Poland D, Sawkins M, Ribaut J-M, Hoisington D (eds) Resilient crops for water limited environments. Proceedings of a workshop held at Cuernavaca, CIMMYT, Mexico, 24–28 May 2004

Richards RA (1993) Should selection for yield in saline regions be made on saline or non-saline soils? Euphytica 32:431–438

Rozema J, Flowers TJ (2008) Crops for a salinized world. Science 322:1478–1480

Saneoka H, Nagasaka C, Hahn DT, Yang WJ, Premachandra GS, Joly RJ, Rhodes D (1995) Salt tolerance of glycinebetaine-defcient and containing maize lines. Plant Physiol 107: 631–638

Sanguineti MC, Tuberosa R, Landi P, Salvi S, Maccaferri M, Casarini E, Conti S (1999) QTL analysis of drought-related traits and grain yield in relation to genetic variation for leaf abscisic acid concentration in field-grown maize. J Exp Bot 50:1289–1297

Shou H, Bordallo P, Wang K (2004) Expression of the Nicotiana protein kinase (NPK1) enhanced drought tolerance in transgenic maize. J Exp Bot 55:1013–1019

Sinclair TR, Purcell LC (2005) Is a physiological perspective relevant in a 'genocentric' age? J Exp Bot 56:2777–2782

Song XJ, Matsuoka M (2009) Bar the windows: an optimized strategy to survive drought and salt adversities. Genes Dev 23:1709–1713

Stahlberg R, Van Volkenburgh E (1999) The effect of light on membrane potential, apoplastic pH and cell expansion in leaves of *Pisum sativum* L. var. *Argenteum*. Role of PM H^+ ATPase and photosynthesis. Planta 208:188–195

Sunkar R, Chinnusamy V, Zhu J, Zhu JK (2007) Small RNAs as big players in plant abiotic stress responses and nutrient deprivation. Trends Plant Sci 12:301–309

Sunkar R, Zhou X, Zheng Y, Zhang W, Zhu JK (2008) Identification of novel and candidate miRNAs in rice by high throughput sequencing. BMC Plant Biol 8:25

Szabolcs I (1994) Prospects of soil salinity for the 21st century. Paper presented at the 15th world congress of soil science, Acapulco, 10–16 July 1994, pp 123–141

Tiwari SB, Hagen G, Guilfoyle TJ (2004) Aux/IAA proteins contain a potent transcriptional repression domain. Plant Cell 16:533–543

Tomos AD, Pritchard J (1994) Biophysical and biochemical control of cell expansion in roots and leaves. J Exp Bot 45:1721–1731

Tuberosa R, Sanguineti MC, Landi P, Giuliani MM, Salvi S, Conti S (2002a) Identification of QTLs for root characteristics in maize grown in hydroponics and analysis of their overlap with QTLs for grain yield in the field at two water regimes. Plant Mol Biol 48:697–712

Tuberosa R, Gill BS, Quarrie SA (2002b) Cereal genomics: ushering in a brave new world. Plant Mol Biol 48:445–449

Urano K, Kurihara Y, Seki M, Shinozaki K (2010) 'Omics' analyses of regulatory networks in plant abiotic stress responses. Curr Opin Plant Biol 13:32–138

US Salinity Laboratory (Clyde W, 2006) Improving salt tolerance in small grain crops using physiological genomics approaches. American Society of Agronomy Meetings in Indianapolis, in 12–16 Nov 2006. Paper No. 60–6

Van Volkenburgh E, Boyer JS (1985) Inhibitory effects of water deficit on maize leaf elongation. Plant Physiol 77:190–194

Vinocur B, Altman A (2005) Recent advances in engineering plant tolerance to abiotic stress: achievements and limitations. Curr Opin Biotechnol 16:1–10

Warner DC, Heard J, Bensen R, Nelson D (2005) Development of transgenes for improvement of drought stress tolerance in maize. In: Abstracts of the international annual meetings, ASA-CSSA-SSSA, Madison, WI, 2005

Westgate ME, Boyer JS (1986) Reproduction at low silk and pollen water potentials in maize. Crop Sci 26:951–956

Westgate ME, Boyer JS (2004) Grain yields with limited water. J Exp Bot 55:2385–2394

Xu DQ, Huang J, Guo SQ (2008) Over-expression of a TFIIIA-type zinc finger protein gene ZFP252 enhances drought and salt tolerance in rice (*Oryza sativa* L.). FEBS Lett 582:1037–1043

Yamaguchi T, Blumwald E (2005) Developing salt-tolerant crop plants: challenges and opportunities. Trends Plant Sci 10:615–620

Yan B, Dai Q, Liu X, Huang S, Wang Z (1996) Flooding induced membrane damage, lipid oxidation and activated oxygen generation in corn leaves. Plant Soil 179:261–268

Yin XY, Yang AF, Zhang KW, Zhang JR (2004) Production and analysis of transgenic maize with improved salt tolerance by the introduction of *AtNHX1* gene. Acta Bot Sin 46:854–861

Yousuf PY, Hakeem KR, Chandna R, Ahmad P (2012) Role of glutathione reductase in plant abiotic stress. In: Ahmad P, Prasad MNV (eds) Abiotic stress responses in plants: metabolism, productivity and sustainability. Springer, Science + business media, New York, pp 149–158

Zacchini M, Rea E, Tullio M, Agazio M (2003) Increased antioxidative capacity in maize calli during and after oxidative stress induced by a long lead treatment. Plant Physiol Biochem 41:49–54

Zhang J, Nguyen H, Blum A (2000) Genetic analysis of osmotic adjustment in crop plants. J Exp Bot 50:291–302

Zhang J, Jia W, Yang J, Ismail AM (2006) Role of ABA in integrating plant responses to drought and salt stresses. Field Crop Res 97:111–119

Zhao B, Liang R, Ge L, Li W, Xiao H, Lin H, Ruan K, Jin Y (2007) Identification of drought-induced microRNAs in rice. Biochem Biophys Res Commun 354:585–590

Zhao B, Ge L, Liang R, Li W, Rua K, Lin H, Jin Y (2009) Members of *miR-169* family are induced by high salinity and transiently inhibit the NF-YA transcription factor. BMC Mol Biol 10:29
Zhou X, Wang G, Sutoh K, Zhu JK, Zhang W (2008) Identification of cold-inducible microRNAs in plants by transcriptome analysis. Biochim Biophys Acta 1779:780–788
Zhu JK (2000) Genetic analysis of plant salt tolerance using *Arabidopsis*. Plant Physiol 124:941–948
Zhu JK (2001) Plant salt tolerance. Trends Plant Sci 6:66–71
Zhu JK, Hasegawa PM, Bressan RA (1997) Molecular aspects of osmotic stress in plants. Crit Rev Plant Sci 16:253–277

Chapter 12
The Role of Phytochromes in Stress Tolerance

R.F. Carvalho, M.L. Campos, and R.A. Azevedo

Abstract Phytochromes, which absorb red (R)/far red (FR) light, are the most characterized photoreceptors in plants and have been shown to mediate a wide range of molecular and biochemical responses. These include responses to biotic and abiotic stresses, which are currently a major challenge in plant research. Extensive studies exploring the molecular and biochemical basis by which phytochromes modulate stresses such as drought, high light, high and low temperatures or herbivory, among other, have been carried out in many plant species. In particular, considerable efforts have also been made to unravel the involvement of phytochromes in salt stress responses. For example, many of the stress responses involving phytochrome are now understood following studies using R/FR light ratio treatments, as well as the isolation and characterization of photomorphogenic mutants and transgenic plants, which have revealed an important role of phytochromes on salt stress tolerance mechanisms. In this chapter, we will explore pieces of information available to make clear the perspectives focused on photomorphogenesis manipulation in the scope of agriculture.

Keywords Abiotic stress • Biotic stress • Photoreceptors • Phytochromes • Stress modulation

R.F. Carvalho
Departamento de Biologia Aplicada à Agropecuária, Universidade Estadual Paulista "Júlio de Mesquita Filho", Jaboticabal, SP, CEP 14884-900, Brazil

M.L. Campos
Department of Energy, Plant Research Laboratory, Michigan State University, East Lansing, MI, USA

R.A. Azevedo (✉)
Departamento de Genética, Escola Superior de Agricultura "Luiz de Queiroz", Universidade de São Paulo, Piracicaba, Brazil
e-mail: raa@usp.br

P. Ahmad et al. (eds.), *Salt Stress in Plants: Signalling, Omics and Adaptations*, DOI 10.1007/978-1-4614-6108-1_12,

Plants are sessile organisms that use light as a source of energy for key metabolic processes, growth and reproduction. Thus, it is not surprising that during evolution plants have acquired an array of photosensory mechanisms that allow them to detect light quantity, quality, direction and photoperiodicity (Alba et al. 2000b). Although plants are able to respond to different wavelengths through several photosensory mechanisms, changes in metabolism and development are controlled by three main groups of photoreceptors: phytochromes, (R; ~ 660 nm) and far red (FR; ~ 730 nm) regions of the spectrum (Essen et al. 2008; Carvalho et al. 2011a) and cryptochromes and phototropins, which perceive ultraviolet light (UV-A; 320–400 nm) (Sancar 2003). Phytochromes, the best characterized of these plant photoreceptors, are ~120 kDa peptides (termed apoproteins) with a covalently linked linear tetrapyrrole bilin chromophore (forming a complex referred to as the holoprotein). Upon excitation by R or FR light, the bilin chromophore undergoes a *Z-E* isomerization in its 15/16 double bond, converting the R-absorbing molecule (P_R) into FR-absorbing (P_{FR}) or vice versa. Importantly, in the absence of light, a thermal process known as dark conversion slowly reverts P_{FR} to the P_R form (Furuya and Schäfer 1996; Ádám et al. 2011). Since many phytochrome responses are activated by red light, P_{FR} is considered to be the active form of the phytochrome.

Phytochromes are also involved in a process called "shade avoidance". Photosynthetic pigments such as chlorophyll and carotenoids absorb light in the visible region spectrum (R, but not FR). For this reason, FR light is poorly absorbed and usually transmitted through leaves (it can also be reflected). Thus, under a leaf canopy plants face a depletion of R light, and excess of FR light (Holmes and Smith 1975; Franklin and Whitelam 2005). This low ratio of R:FR is perceived by the phytochromes of plants under the canopy, triggering responses such as an increase in elongation rate of the stem and petioles at the expend of photoassimilates. This process aims to avoid regions of shade, where there is a depletion of light available for photosynthesis, and compete for light with other plants, and is thus termed "shade avoidance" (Franklin and Whitelam 2005; Franklin 2008). It is interesting to point out, however, that non-photosynthetic organisms such as bacteria and fungi also possess phytochromes, differing slightly in their structure and spectral properties from those encountered in plant species (Davis et al. 1999).

Angiosperms are characterized as containing a small number of different apoproteins of phytochrome, which are encoded by a small *PHY* family. For example, *Arabidopsis* contains five *PHY* genes (*PHYA, PHYB, PHYC, PHYD,* and *PHYE*), all of which are expressed (Sharrock and Quail 1989; Clack et al. 1994; Winter et al. 2007). The phytochrome molecules, which can display a nuclear and/or cytoplasmic action (Rösler et al. 2010), consist of a N-terminal photosensory module containing three sub-domains: PAS (Per-ARNT-Sim), GAF (cGMP phosphodiesterase/adenyl cyclase/FhlA) and PHY (phytochrome), and a C-terminal histidine kinase-related domain (HKRD) (Rockwell et al. 2006, 2011; Rockwell and Lagarias 2010). The PAS domain is involved in protein-protein interaction and together with the HRKD appears to be involved in homo/heterodimerization of the phytochromes (Jones and Edgerton 1994; Matsushita et al. 2003). The GAF domain has a bilin-lyase activity necessary for the chromophore attachment

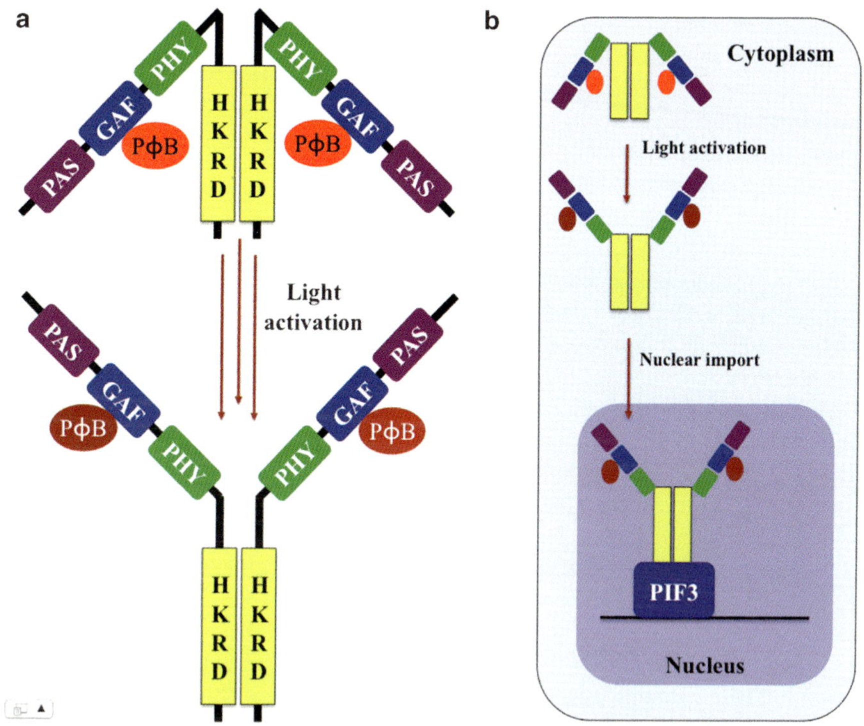

Fig. 12.1 Phytochrome model of action. (**a**) Canonical phytochromes consist of a N-terminal photosensory domain (PAS [Per-ARNT-Sim], GAF [cGMP phosphodiesterase/adenyl cyclase/FhlA] and PHY [phytochromes] sub-domains) and a C-terminal terminal histidine kinase-related (*HKRD*) regulatory domain. They act as homo/heterodimers, with the PAS and HRKD domains involved in the monomer interaction. The GAF domain has a bilin-lyase activity necessary for the chromophore attachment into the apoprotein (here the chromophore phytochromobilin – PφB – is illustrated). (**b**) In the absence of light, phytochromes are localized in the cytoplasm, however, upon light activation, the isomerization of the bilin chromophore induces a conformational change in the holoprotein (shown in "**a**") exposing a cryptic nuclear localization signal (NLS) that permits the phytochromes to move to the nucleus, where it interacts with PIF3. For a model including cytoplasmic phytochromes action, readers are suggested to see Rösler et al. (2010)

into the apoprotein (Wu and Lagarias 2000) and the PHY domain is apparently involved in the stabilization of the P_{FR} form (Montgomery and Lagarias 2002). In the absence of light, phytochromes are localized in the cytoplasm. However, upon light activation, the holoprotein translocates to the nucleus where it can physically interact with transcription factors such as PIF3 (PHYTOCHROME INTERACTING FACTOR 3 (Nagatani 2004), which is involved in light responses (Fig. 12.1).

A striking feature of phytochromes is their involvement in controlling a plethora of processes throughout plant development. From seed germination (Dechaine et al. 2009; Oh et al. 2009; Neff 2012) to flowering (Andres et al. 2009; Brock et al. 2010), phytochromes can regulate a wide range of responses. These include hypocotyl elongation (Yang et al. 2009; Kunihiro et al. 2010) and pigment synthesis (Carvalho et al. 2010; Toledo-Ortiz et al. 2010), as well as the expression of a large number of light-responsive genes (Quail 2007). Thus, it is not surprising that phytochromes are involved in the regulation of responses to diverse biotic and abiotic stresses such as those caused by salt, toxic metals, drought, high and low temperatures, high light and herbivory (Carvalho et al. 2011c; Kiyota et al. 2012). It is important to understand that phytochromes are essential to ensure that plants are able to survive, when facing a specific stressful condition. As they are sessile organisms, plants unlike animals, are unable to move away from a stress. In order to cope with a stressful condition, a plant is able to alter its own development. Thus, a molecule that can control many different steps during plant development can also play an important role during stress. Interestingly enough, phytochromes have also been the focus of attention of more applied research, since many agriculturally relevant traits are either heavily influenced or completely controlled by these molecules. These include plant architecture (Boccalandro et al. 2003; Garg et al. 2006), tuberization (Boccalandro et al. 2003; Peres et al. 2005) and fruit quality (Alba et al. 2000a). Thus, in our opinion, the study of the action of phytochrome represents a hot spot of research for global agriculture, as phytochrome is a molecule that is involved in a variety of plant stress responses and also controls many different agricultural traits.

Many of the stress responses to stress involving phytochrome are now understood following studies using R/FR light ratio treatments, as well as the isolation and characterization of photomorphogenic mutants and transgenic plants. However, since this information is scattered throughout many different research papers, the focus of the present chapter is to present the reader with the state-of-the-art view of the role of phytochrome as a key regulator of many different stress responses. We propose that analyzing this information as a whole, will provide a clear view of the dynamics and mode of action of phytochrome during the many different stressful conditions in which it is involved.

Salt Stress

Salinity lowers soil water potential and metabolic imbalances such as in the uptake of mineral nutrients and their accumulation in the plant (Munns and Tester 2008). Thus, considerable efforts have been made to unravel the genetic and biochemical mechanisms controlling plant salt tolerance. Initial evidence for the involvement of phytochromes in salt stress responses came from the seminal work of McElwain et al. (1992), who demonstrated that plants exposed to variations in the ratio of R/FR light exhibited marked alterations in the response to salt stress at the transcriptional level, which in turn lead to a change in photosynthesis biochemistry (Slocombe et al. 1993). The authors observed that *Mesembryanthemum crystallinum* switched from C_3 to CAM (Crassulacean acid metabolism) photosynthesis when exposed to salt (NaCl) stress conditions. This effect was pronounced if the plants were grown in R light rich environments, as shown by the increase in mRNA levels of PEP carboxylase, a key CAM enzyme (McElwain et al. 1992). Cockburn et al. (1996) also showed that light with a low ratio of R/FR, in contrast to light with a high ratio of R/FR, caused an induction in the formation of pinitol, a soluble carbohydrate, which accumulates in a number of plant species under stress (Guo and Oosterhuis 1997).

More recently, the biochemical and molecular mechanisms of phytochrome-mediated salt stress responses have begun to be elucidated. For example, it has been suggested that a salt tolerant protein (STO) acts as a negative regulator in phytochrome signaling. Indorf et al. (2007) showed that after R light treatment, induction of STO expression was reduced in both the *phyA* and *phyB* mutants and in the *phyA phyB* double mutant of *Arabidopsis thaliana* when compared with the wild-type. On the other hand, under FR treatment, the expression of *STO* in the wild type reached similar levels as those under R light, whereas a clear induction was observed under these conditions in the *phyB* mutant. In contrast, in the *phyA* mutant and *phyA phyB* double mutant, the transcript levels of *STO* were reduced. Although these responses reveal important interactions between both photoreceptors and STO, they appear to involve complex signaling mechanisms, which need to be further explored, since, for example, the STO interacts with key regulators of light signaling, such as ELONGATED HYPOCOTYL5 (HY5) and CONSTITUTIVELY PHOTOMORPHOGENIC1 (COP1) (Holm et al. 2001; Ma et al. 2002; Indorf et al. 2007; Datta et al. 2008).

In recent years, the role of heme oxygenase (HO) has been explored in nearly all living system including plants, animals and other organisms (Shekhawat and Verma 2010). The common role of HO is the degradation of heme, although there is a diversity of additional roles attributed to HO including iron acquisition, cellular signaling, defense against stress and biosynthesis during metabolism. Likewise, the function of HO is to provide cofactors for the photosynthetic apparatus in cyanobacteria and it also plays a major role in the synthesis of the phytochrome chromophore (Terry et al. 2002) (Fig. 12.2). HO has been associated with plant stress (Chen et al. 2009), in particular salt stress. Balestrasse et al. (2008) demonstrated in *Glycine max* leaves that the signal transduction pathways involved in oxidative stress triggered by salt stress were similar to those implicated in HO induction (Fig. 12.2). This explains the observation that upon salt stress treatments, the levels of HO gene expression and protein synthesis were significantly increased (Zilli et al. 2008). More recently, a HO mutant (*long hypocotyl* mutant: *hy1–100*) of *A. thaliana* displayed maximal sensitivity to salinity and was shown to be defective in many acclimation responses, whereas plants overexpressing *HY1* (*35S:HY1*) exhibited tolerance characteristics (Xie et al. 2011a). Nevertheless, the authors observed that the phytochrome mutants (*phyA*, *phyB* and *phyA phyB*) displayed a similar salt-sensitive phenotype to the wild-type, suggesting that HY1, rather than phytochromes A and B, mediate salt acclimation in *A. thaliana*. However, for a complete understanding of phytochrome mediated stress based on the specific roles of phytochrome chromophore or apoproteins (e.g. PHA and PHB), further studies are still required.

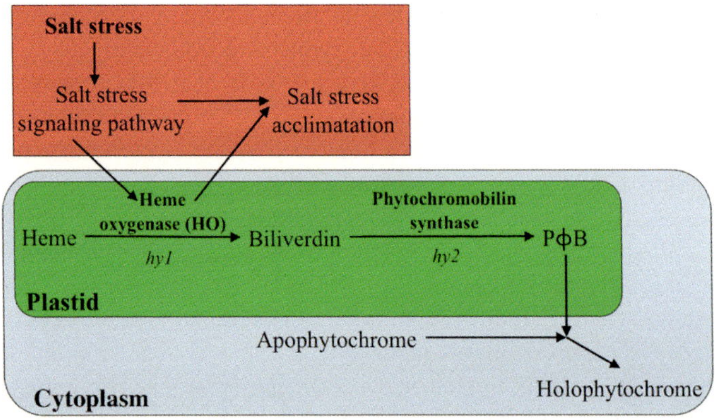

Fig. 12.2 Convergence of stalt stress signaling and the phytochrome chromophore synthesis pathways. Steps in the synthesis of the phytochrome chromophore (phytochromobilin – PϕB) involve two enzymes: The heme oxygenase enzyme (*HO*) catalyzes the stereo-specific conversion of heme to biliverdin (mutant *hy1*), which is then converted to PϕB by phytochromobilin synthase, a ferredoxin-dependent biliverdin reductase (mutant *hy2*). The PϕB is then exported from the plastid to the cytoplasm where it associates with the apophytochrome (protein component) molecules to form the holophytochrome (protein component + chromophore). Upon salt stress, the salt stress signal pathway seems to be involved in the induction of HO (Terry et al. 2002; Zilli et al. 2008). The HO then acts as a player in salt stress responses, since the *hy1* mutant is defective in many salt acclimatory processes (Xie et al. 2011a)

Drought Stress

Scarcity of water is a severe environmental constraint to plant development. Therefore, drought can possibly be classified as the single most critical threat to world food security (Farooq et al. 2009). Some observations have raised questions about a possible relationship between phytochromes and drought stress, since there is evidence that phytochromes are involved in the control of leaf transpiration (Kraepiel et al. 1994; Sokolskaya et al. 2003; Boccalandro et al. 2009; Boggs et al. 2010). In addition, phytochrome can modulate drought stress in the germination of light dormant seeds (Sanchez et al. 2002; Mollard and Insausti 2009). Moreover, phytochromes clearly interact with abscisic acid (ABA), the major signaling molecule of plants, under drought stress (Staneloni et al. 2008; Carvalho et al. 2010). Sawada et al. (2008) showed that phytochromes downregulate the expression of genes involved in the metabolism of ABA. For instance, in dehydration experiments, ABA levels were significantly higher and water was more efficiently retained in the *pew1* mutant of *Nicotiana plumbaginifolia*, which is deficient in phytochrome when compared with the wild-type (Kraepiel et al. 1994). These results could be attributed to the susceptibility of *high pigment* (*hp*) mutants of *Solanum lycopersicum*, which have been shown to have exaggerated light responses (Levin et al. 2003; Lieberman et al. 2004; Liu et al. 2004) to drought stress (Galpaz et al. 2008; Carvalho et al. 2011b). It is possible that a subset of photoreceptors including phytochromes coordinate a wide range of responses involved in light signal transduction. Moreover, *aurea* (*au*), a phytochrome-deficient mutant of *Solanum lycopersicum*, exhibited no change in transpiration rate under drought stress conditions (Biehler et al. 1997), indicating that there are variations in the specific responses of phytochrome signaling between species during drought stress.

Recently, Liu et al. (2012) provided more evidence for the role of phytochromes in drought tolerance. These authors observed that the rice *phyB* mutant exhibited decreased water loss and transpiration rate, leading to enhanced drought tolerance when compared to the wild type. Although the leaf area per plant was reduced in the *phyB* mutant, the epidermal cells were larger, leading to reduced stomatal density and stomatal size. Leaf microarray analysis lead to the conclusion that genes involved with epidermal cell expansion, such as the putative rice *ERECTA* and *EXPANSIN* homologs were upregulated in the *phyB* mutant, resulting in the reduced stomatal density in the mutant.

Molecular evidence of phytochrome-mediated drought responses can also be obtained from the work of Kidokoro et al. (2009) who showed that a member of the C-REPEAT/DEHYDRATION RESPONSIVE ELEMENT BINDING FACTOR (CBF/DREB1) transcription factor family, which associates with promoters of drought-responsive genes, was repressed by a phytochrome-interacting factor (PIF7) in *A. thaliana*. It is clear that an important part of future research will need to concentrate on widening our knowledge of how phytochrome signaling is involved in tolerance to drought stress.

High and Low Temperature Stress

Another important and severe environmental challenge to plants is variation in temperature. Different plant species vary widely in their ability to tolerate low and high temperatures. A considerable amount of research has been carried out on the interaction between phytochromes and temperature. This research has concentrated on the response of seed germination to temperature cues (both cold and warm), which are mediated by different phytochrome types, such as phyA, phyB and phyE in *A. thaliana* (Heschel et al. 2007; Dechaine et al. 2009). Seed response to temperature plays a pivotal role in the synchronization of seed germination with conditions suitable for seedling establishment and thus avoidance of stress conditions. In this way, phyB and phyD have been shown to be important for breaking cold induced dormancy, whereas phyA contributes to maintaining dormancy (Donohue et al. 2008).

In the pioneering work of Williams et al. (1972) and McKenzie et al. (1974), important information showing that phytochrome affects the cold acclimation processes of *Cornus stolonifera* was revealed. This phenomenon seems to involve molecular mechanisms, which although previously explored, are still complex. For instance, the cold-related expression of genes such as *cor14b* (Crosatti et al. 1999) and *cor15a* (Kim et al. 2002) can be controlled by phytochromes. It was observed that light can activate cold-induced gene expression through the *cis*-acting element

C/DRE (dehydration responsive element), using phytochrome B a primary photoreceptor responsible for that response (Kim et al. 2002). However, Kidokoro et al. (2009) suggested that PIF7 functions independently from the transcription factors that regulate the expression of DRE-Binding1 (DREB1) in response to low temperature. On the other hand, Foreman et al. (2011) suggested that phyB can participate through the action of PIF4 at high temperature-mediated adaptations. Although PIF4 is known to be regulated by phytochromes, this temperature-mediated control was proposed to be distinct from phytochrome activity (Koini et al. 2009). These issues are even greater when a subset of molecules, such as other photoreceptors (e.g. cryptochromes) and hormones, are considered (Leivar and Quail 2010).

High-Light Stress

Light stress is not a consequence from high light *per se*, but rather from an overload of absorbed light beyond that utilized by photosynthesis. In plants, this excess of absorbed light has the potential to damage mainly the photosynthetic apparatus, in part by the generation of reactive oxygen species (ROS) by excitation energy (e.g. UV-B) and electrons leaking from the photochemical reactions and electron transport system (Kimura et al. 2003; Foyer and Noctor 2009). Moreover, a complex crosstalk between photosensory molecules is initiated in an attempt to mitigate cell damage caused by high light (Bolink et al. 2001; Gould et al. 2010). The modulation of the harmful radiation is the most obvious response in which phytochrome signaling is involved. Extensive studies have been carried out over the last 40 years and now some molecular and biochemical mechanisms have been elucidated. In recent studies, PIFs were found to be involved in the prevention of photo-oxidation during the switch from skotomorphogenesis to photomorphogenesis in *A. thaliana* (Zhong et al. 2009). PIFs are part of phytochrome signaling and are able to interact with a subset of molecules such as kinases and hormones (Kidokoro et al. 2009; Zhong et al. 2009; Brock et al. 2010; Bu et al. 2011). PIFs are also key players in the regulation of photomorphogenesis in seedlings and they mainly act by inducing the expression of protochlorophyllide reductases (PORs), enzymes involved in the conversion of photochlorophyllide to chlorophyll. The accumulation of photochlorophyllide (by a reduction in the activity of PORs for example) can lead to ROS accumulation during photomorphogenesis and thus, photo-oxidative damage. Consequently, for instance, *pif1* mutant seedlings are more susceptible to light exposure (Huq et al. 2004; Moon et al. 2008).

In plants the major pigment group showing a wide occurrence in different tissues is the anthocyanins. These molecules may have a photo-protective function, either against light-induced photo-oxidation or against UV-B damage, which are important strategies allowing seedling survival during early development (Chalker-Scott 1999). It is probable that phytochrome plays a role in these responses (Husaineid et al. 2007; Carvalho et al. 2010), either through the induction of phenylalanine ammonia-lyase (PAL) (Lercari et al. 1986; Brödenfeldt and Mohr 1988; Alokam et al. 2002) or chalcone synthase (CHS) (Frohnmeyer et al. 1998; Wade et al. 2001), both of which play a crucial role in the anthocyanin biosynthesis pathway (Brödenfeldt and Mohr 1988). However, the mechanism by which phytochromes operate during anthocyanin induction, following irradiation with UV-B light is still not clear. In *Sinapis alba*, it was shown that phytochrome-mediated anthocyanin biosynthesis was inhibited by UV-B light (Wellmann et al. 1984; Buchholz et al. 1995), whereas continuous red light led to greatly enhanced anthocyanin synthesis in response to UV-B in *Zea mays* coleoptiles (Beggs and Wellmann 1985). Therefore, it is not surprising that the interaction between phytochrome and UV-B has been questioned. Boccalandro et al. (2001) verified that UV-B radiation enhanced the phytochrome B-mediated cotyledon opening during the de-etiolation of *A. thaliana* seedlings. In these plants, Ulm et al. (2004) also showed that UV-B caused the upregulation of *HY5* transcription, a transcription factor involved in phytochrome responses. Later on, microarray analysis showed that *HY5* is a key player in UV-B signaling, being required for *A. thaliana* survival upon UV-B radiation (Brown et al. 2005). Although these experiments have helped to elucidate the participation of phytochrome in the transduction of UV-B light signals, the genetic details of this signaling seem to be complex since UV-B alters specific sets of genes more than treatment with R and FR light (Peschke and Kretsch 2011).

Herbivory Stress

The battle between plants and phytophagous organisms has lasted for hundreds of millions of years. Under attack, plants have evolved sophisticated defense systems, using several signals to recognize the effects of herbivory (Howe and Jander 2008; Ballaré 2009; Wu and Baldwin 2009). However, plants have to thrive facing a constant dilemma: How much energy (photoassimilates) should be expended for defense to the detriment of growth or *vice-versa*? For this purpose, for example, after defoliation, allocating carbon to shoots at the expense of roots may confer tolerance mechanisms (Briske et al. 1996; Hatcher et al. 2004; Reynolds et al. 2009).

When confronted with the risks of competition (e.g. enhanced plant density: low R to FR ratio), the investment of resources into defensive mechanisms can reduce the competitive ability against neighboring plants (Herms and Mattson 1992; Taylor et al. 2004; Roberts and Paul 2006). It has been suggested that plants place their allocation priorities on maintaining their light-harvesting ability during shade avoidance (Franklin 2008), rather than on preventing biomass loss to herbivores (Ballaré 2009). Thus, increased FR light can lead to an attenuated defense response resulting in increased levels of herbivory. Moreno et al. (2009) used *Spodoptera* caterpillars to show that *A. thaliana* plants exposed to conditions of crowding and FR radiation produced an attenuated defense phenotype. The authors demonstrated the involvement of phytochromes in such responses by showing that inactivation of the molecules by FR light reduced the sensitivity of the plants to jasmonates, compounds strongly associated with responses to herbivory (see below).

Jasmonic acid and its conjugates (collectively referred as jasmonates – JA) are the main hormonal class involved with herbivory responses. For this reason, plants impaired in the synthesis of perception of JA are severely wounded by herbivores (Howe and Jander 2008). Robson et al. (2010) showed that the *coi1–16* mutant of *A. thaliana*, which is deficient in JA signaling, is also deficient in a subset of high irradiance responses to FR light. This mutant displays exaggerated shade responses to low, but not high, R/FR light ratios, suggesting a role for JA in phytochrome signaling. Strong evidence for an interaction between JA and phytochrome was shown when phytochrome chromophore deficient *hy1–100* and *hy2* mutants of *A. thaliana* also exhibited increased JA levels and constitutive expression of JA-inducible genes (Zhai et al. 2007). On the other hand, in the JA deficient mutants *hebiba* and *osjar1* of *Oryza sativa*, the growth responses to far-red light were affected (Riemann et al. 2003; Riemann and Takano 2008). Molecular analyses have revealed how JA and phytochromes are integrated during wound signaling. In *A. thaliana*, phytochrome A was required for JA- and wound-mediated degradation of JASMONATE ZIM DOMAIN1 (JAZ1) proteins (Robson et al. 2010), a family of key repressors of JA signaling. Although it is obvious that phytochromes are required for anti-herbivory through JA, a more comprehensive molecular understanding of how phytochrome interacts with JA is still not available. Certainly, it will be possible to address the regulation of JA signaling elements through phytochrome by, for example, avoiding the inhibitory effects of JA on cell growth and division (Robson et al. 2010).

Conclusion and Future Perspectives

Our aim in this chapter was to give a general overview of how phytochromes are involved in regulating plant responses to many different stresses. The information presented here delineates several features of these molecules that are presently gaining more attention. Although we have addressed salt, drought, temperature, high light and herbivory stress, it is possible that phytochromes are involved in the modulation of other environmental stresses. For example, recently, (Shen et al. 2011) showed that *Brassica napus* plants overexpressing the HO gene *BnHO-1* exhibited tolerance to mercury (Hg) toxicity. In fact, HOs may be involved in the defense against a wide range of stresses (Shekhawat and Verma 2010; Xie et al. 2011b). Although HOs were originally identified as being involved in the pathway of phytochrome chromophore biosynthesis, it is still hard to understand how HOs play a role in the stress modulation associated with phytochrome action. As seen in the salt stress section, in *A. thaliana*, the HO mutant and HO overexpresser exhibited sensitivity and tolerance, respectively, to salt stress, but *phyA, phyB* and *phyA phyB* mutants did not, indicating that, at least phytochrome A and B are not involved in salt tolerance signaling (Xie et al. 2011a). However to date, how separately or dependently HOs and phytochromes modulate stress remains largely unknown.

The induction of the antioxidant system is the most obvious response during stress (Gratão et al. 2005, 2008; Arruda and Azevedo 2009; Ghelfi et al. 2011).

Thus, in addition to the phytochrome signaling pathway, antioxidant enzymes such as peroxidases, and non-enzymatic antioxidants such as flavonoids (see light stress section) and carotenoid molecules, are antioxidants that have been found to be modulated by phytochromes under stress conditions. This and other related findings have been reported for many years (Drumm and Schopfer 1974; Thomsen et al. 1992; Zhong et al. 1997) including recent extensive recent studies on HO enzymes (Cui et al. 2011; Xu et al. 2011) and interaction with hormones (Monteiro et al. 2011; Gratão et al. 2012). Moreover, besides light signals perceived by phytochromes, coordinated responses to stress require a fully integrated signaling network (Monteiro et al. 2012), incorporating information from others photoreceptors and interactions with a wide range of molecules (we would emphasize in particular the interaction with plant hormones), this being an important emerging topic that must be addressed in future investigations. Finally, a more comprehensive view has to be considered and must necessarily include studies on gene expression, protein translation, enzyme activity, and metabolite concentrations, all considered simultaneously, wherever possible.

References

Ádám E, Hussong A, Bindics J, Wust F, Viczián A, Essing M, Medzihradszky M, Kircher S, Schafer E, Nagy F (2011) Altered dark- and photoconversion of phytochrome B mediate extreme light sensitivity and loss of photoreversibility to the phyB-401 mutant. PLoS One 6:e27250

Alba R, Cordonnier-Pratt MM, Pratt LH (2000a) Fruit-localized phytochromes regulate lycopene accumulation independently of ethylene production in tomato. Plant Physiol 123:363–370

Alba R, Kelmenson PM, Cordonnier-Pratt MM, Pratt LH (2000b) The phytochrome gene family in tomato and the rapid differential evolution of this family in angiosperms. Mol Biol Evol 17:362–373

Alokam S, Li Y, Li W, Chinnappa CC, Reid DM (2002) Photoregulation of phenylalanine ammonia-lyase (PAL) and chalcone synthase (CHS) in the accumulation of anthocyanin in alpine and prairie ecotypes of *Stellaria longipes* under varied R/FR. Physiol Plant 116:531–538

Andres F, Galbraith DW, Talon M, Domingo C (2009) Analysis of PHOTOPERIOD SENSITIVITY5 sheds light on the role of phytochromes in photoperiodic flowering in rice. Plant Physiol 151:681–690

Arruda MAZ, Azevedo RA (2009) Metallomics and chemical speciation: towards a better understanding of metal-induced stress in plants. Ann Appl Biol 155:301–307

Balestrasse KB, Zilli CG, Tomaro ML (2008) Signal transduction pathways and haem oxygenase induction in soybean leaves subjected to salt stress. Redox Rep 13:255–262

Ballaré CL (2009) Illuminated behaviour: phytochrome as a key regulator of light foraging and plant anti-herbivore defence. Plant Cell Environ 32:713–725

Beggs CJ, Wellmann E (1985) Analysis of light-controlled anthocyanin formation in coleoptiles of *Zea mays* L.: the role of UV-B, blue, red and far-red light. Photochem Photobiol 41:481–486

Biehler K, Haupt S, Beckmann J, Fock H, Becker TW (1997) Simultaneous CO_2- and $16O_2/18O_2$-gas exchange and fluorescence measurements indicate differences in light energy dissipation between the wild type and the phytochrome-deficient *aurea* mutant of tomato during water stress. J Exp Bot 48:1439–1449

Boccalandro HE, Mazza CA, Mazzella MA, Casal JJ, Ballaré CL (2001) Ultraviolet B radiation enhances a phytochrome-B-mediated photomorphogenic response in Arabidopsis. Plant Physiol 126:780–788

Boccalandro HE, Ploschuk EL, Yanovsky MJ, Sánchez RA, Gatz C, Casal JJ (2003) Increased phytochrome B alleviates density effects on tuber yield of field potato crops. Plant Physiol 133:1539–1546

Boccalandro HE, Rugnone ML, Moreno JE, Ploschuk EL, Serna L, Yanovsky MJ, Casal JJ (2009) Phytochrome B enhances photosynthesis at the expense of water-use efficiency in Arabidopsis. Plant Physiol 150:1083–1092

Boggs JZ, Loewy K, Bibee K, Heschel MS (2010) Phytochromes influence stomatal conductance plasticity in *Arabidopsis thaliana*. Plant Growth Regul 60:77–81

Bolink E, van Schalkwijk I, Posthumus F, van Hasselt P (2001) Growth under UV-B radiation increases tolerance to high-light stress in pea and bean plants. Plant Ecol 154:147–156

Briske DD, Boutton TW, Wang Z (1996) Contribution of flexible allocation priorities to herbivory tolerance in C4 perennial grasses: an evaluation with ^{13}C labeling. Oecologia 105:151–159

Brock MT, Maloof JN, Weinig C (2010) Genes underlying quantitative variation in ecologically important traits: *PIF4* (*PHYTOCHROME INTERACTING FACTOR 4*) is associated with variation in internode length, flowering time, and fruit set in *Arabidopsis thaliana*. Mol Ecol 19:1187–1199

Brödenfeldt R, Mohr H (1988) Time courses for phytochrome-induced enzyme levels in phenylpropanoid metabolism (phenylalanine ammonia-lyase, naringenin-chalcone synthase) compared with time courses for phytochrome-mediated end-product accumulation (anthocyanin, quercetin). Planta 176:383–390

Brown BA, Cloix C, Jiang GH, Kaiserli E, Herzyk P, Kliebenstein DJ, Jenkins GI (2005) A UV-B-specific signaling component orchestrates plant UV protection. Proc Natl Acad Sci USA 102:18225–18230

Bu Q, Zhu L, Dennis MD, Yu L, Lu SX, Person MD, Tobin EM, Browning KS, Huq E (2011) Phosphorylation by CK2 enhances the rapid light-induced degradation of Phytochrome Interacting Factor 1 in Arabidopsis. J Biol Chem 286:12066–12074

Buchholz G, Ehmann B, Wellmann E (1995) UV light inhibition of phytochrome-induced flavonoid biosynthesis and DNA photolyase formation in mustard cotyledons. Plant Physiol 108:227–234

Carvalho RF, Quecini V, Peres LEP (2010) Hormonal modulation of photomorphogenesis-controlled anthocyanin accumulation in tomato (*Solanum lycopersicum L.* cv Micro-Tom) hypocotyls: physiological and genetic studies. Plant Sci 178:258–264

Carvalho R, Takaki M, Azevedo R (2011a) Plant pigments: the many faces of light perception. Acta Physiol Plant 33:241–248

Carvalho RF, Aidar ST, Azevedo RA, Dodd IC, Peres LEP (2011b) Enhanced transpiration rate in the *high pigment 1* tomato mutant and its physiological significance. Plant Biol 13:546–550

Carvalho RF, Campos ML, Azevedo RA (2011c) The role of phytochrome in stress tolerance. J Integr Plant Biol 53:920–929

Chalker-Scott L (1999) Environmental significance of anthocyanins in plant stress responses. Photochem Photobiol 70:1–9

Chen X-Y, Ding X, Xu S, Wang R, Xuan W, Cao Z-Y, Chen J, Wu H-H, Ye M-B, Shen W-B (2009) Endogenous hydrogen peroxide plays a positive role in the upregulation of heme oxygenase and acclimation to oxidative stress in wheat seedling leaves. J Integr Plant Biol 51:951–960

Clack T, Mathews S, Sharrock RA (1994) The phytochrome apoprotein family in Arabidopsis is encoded by five genes: the sequences and expression of *PHYD* and *PHYE*. Plant Mol Biol 25:413–427

Cockburn W, Whitelam GC, Broad A, Smith J (1996) The participation of phytochrome in the signal transduction pathway of salt stress responses in *Mesembryanthemum crystallinum* L. J Exp Bot 47:647–653

Crosatti C, Polverino de Laureto P, Bassi R, Cattivelli L (1999) The interaction between cold and light controls the expression of the cold-regulated barley gene *cor14b* and the accumulation of the corresponding protein. Plant Physiol 119:671–680

Cui WT, Fu GQ, Wu HH, Shen WB (2011) Cadmium-induced *heme oxygenase-1* gene expression is associated with the depletion of glutathione in the roots of *Medicago sativa*. Biometals 24:93–103

Datta S, Johansson H, Hettiarachchi C, Irigoyen ML, Desai M, Rubio V, Holm M (2008) LZF1/SALT TOLERANCE HOMOLOG3, an Arabidopsis B-box protein involved in light-dependent development and gene expression, undergoes COP1-mediated ubiquitination. Plant Cell 20:2324–2338

Davis SJ, Vener AV, Vierstra RD (1999) Bacteriophytochromes: phytochrome-like photoreceptors from nonphotosynthetic Eubacteria. Science 286:2517–2520

Dechaine JM, Gardner G, Weinig C (2009) Phytochromes differentially regulate seed germination responses to light quality and temperature cues during seed maturation. Plant Cell Environ 32:1297–1309

Donohue K, Heschel MS, Butler CM, Barua D, Sharrock RA, Whitelam GC, Chiang GCK (2008) Diversification of phytochrome contributions to germination as a function of seed-maturation environment. New Phytol 177:367–379

Drumm H, Schopfer P (1974) Effect of phytochrome on development of catalase activity and isoenzyme pattern in mustard (*Sinapis alba* L.) seedlings. Planta 120:13–30

Essen LO, Mailliet J, Hughes J (2008) The structure of a complete phytochrome sensory module in the Pr ground state. Proc Natl Acad Sci USA 105:14709–14714

Farooq M, Wahid A, Kobayashi N (2009) Plant drought stress: effects, mechanisms and management. Agron Sustain Dev 29:185–212

Foreman J, Johansson H, Hornitschek P, Josse E-M, Fankhauser C, Halliday KJ (2011) Light receptor action is critical for maintaining plant biomass at warm ambient temperatures. Plant J 65:441–452

Foyer CH, Noctor G (2009) Redox regulation in photosynthetic organisms: signaling, acclimation, and practical implications. Antioxid Redox Signal 11:861–905

Franklin KA (2008) Shade avoidance. New Phytol 179:930–944

Franklin KA, Whitelam GC (2005) Phytochromes and shade-avoidance responses in plants. Ann Bot 96:169–175

Frohnmeyer H, Bowler C, Zhu JK, Yamagata H, Schafer E, Chua NH (1998) Different roles for calcium and calmodulin in phytochrome- and UV-regulated expression of chalcone synthase. Plant J 13:763–772

Furuya M, Schäfer E (1996) Photoperception and signalling of induction reactions by different phytochromes. Trends Plant Sci 1:301–307

Galpaz N, Wang Q, Menda N, Zamir D, Hirschberg J (2008) Abscisic acid deficiency in the tomato mutant *high-pigment 3* leading to increased plastid number and higher fruit lycopene content. Plant J 53:717–730

Garg A, Sawers R, Wang H, Kim J-K, Walker J, Brutnell T, Parthasarathy M, Vierstra R, Wu R (2006) Light-regulated overexpression of an Arabidopsis phytochrome A gene in rice alters plant architecture and increases grain yield. Planta 223:627–636

Ghelfi A, Gaziola SA, Cia MC, Chabregas SM, Falco MC, Kuser-Falcão PR, Azevedo RA (2011) Cloning, expression, molecular modelling and docking analysis of glutathione transferase from *Saccharum officinarum*. Ann Appl Biol 159:267–280

Gould KS, Dudle DA, Neufeld HS (2010) Why some stems are red: cauline anthocyanins shield photosystem II against high light stress. J Exp Bot 61:2707–2717

Gratão PL, Polle A, Lea PJ, Azevedo RA (2005) Making the life of heavy metal-stressed plants a little easier. Funct Plant Biol 32:481–494

Gratão PL, Monteiro MC, Peres LEP, Azevedo RA (2008) The isolation of antioxidant enzymes from mature tomato (cv. Micro-Tom) plants. HortScience 43:1608–1610

Gratão PL, Monteiro CC, Carvalho RF, Tezotto T, Pioto FA, Peres LEP, Azevedo RA (2012) Biochemical dissection of *diageotropica* and *Never ripe* tomato mutants to Cd-stressful conditions. Plant Physiol Biochem 56:79–96

Guo CX, Oosterhuis DM (1997) Effect of water-deficit stress and genotypes on pinitol occurrence in soybean plants. Environ Exp Bot 37:147–152

Hatcher PE, Moore J, Taylor JE, Tinney GW, Paul ND (2004) Phytohormones and plant-herbivore-pathogen interactions: integrating the molecular with the ecological. Ecology 85:59–69

Herms DA, Mattson WJ (1992) The dilemma of plants: to grow or defend. Q Rev Biol 67:478–478
Heschel MS, Selby J, Butler C, Whitelam GC, Sharrock RA, Donohue K (2007) A new role for phytochromes in temperature-dependent germination. New Phytol 174:735–741
Holm M, Hardtke CS, Gaudet R, Deng X-W (2001) Identification of a structural motif that confers specific interaction with the WD40 repeat domain of Arabidopsis COP1. EMBO J 20:118–127
Holmes MG, Smith H (1975) The function of phytochrome in plants growing in the natural environment. Nature 254:512–514
Howe GA, Jander G (2008) Plant immunity to insect herbivores. Annu Rev Plant Biol 59:41–66
Huq E, Al-Sady B, Hudson M, Kim CH, Apel M, Quail PH (2004) PHYTOCHROME-INTERACTING FACTOR 1 is a critical bHLH regulator of chlorophyll biosynthesis. Science 305:1937–1941
Husaineid SSH, Kok RA, Schreuder MEL, Hanumappa M, Cordonnier-Pratt MM, Pratt LH, van der Plas LHW, van der Krol AR (2007) Overexpression of homologous phytochrome genes in tomato: exploring the limits in photoperception. J Exp Bot 58:615–626
Indorf M, Cordero J, Neuhaus G, Rodriguez-Franco M (2007) Salt tolerance (STO), a stress-related protein, has a major role in light signalling. Plant J 51:563–574
Jones AM, Edgerton MD (1994) The anatomy of phytochrome, a unique photoreceptor in plants. Semin Cell Biol 5:295–302
Kidokoro S, Maruyama K, Nakashima K, Imura Y, Narusaka Y, Shinwari ZK, Osakabe Y, Fujita Y, Mizoi J, Shinozaki K, Yamaguchi-Shinozaki K (2009) The Phytochrome-Interacting Factor PIF7 negatively regulates *DREB1* expression under circadian control in Arabidopsis. Plant Physiol 151:2046–2057
Kim H-J, Kim Y-K, Park J-Y, Kim J (2002) Light signalling mediated by phytochrome plays an important role in cold-induced gene expression through the C-repeat/dehydration responsive element (C/DRE) in *Arabidopsis thaliana*. Plant J 29:693–704
Kimura M, Yamamoto YY, Seki M, Sakurai T, Sato M, Abe T, Yoshida S, Manabe K, Shinozaki K, Matsui M (2003) Identification of Arabidopsis genes regulated by high light-stress using cDNA microarray. Photochem Photobiol 77:226–233
Kiyota S, Xie X, Takano M (2012) Phytochromes A and C cooperatively regulate early and transient gene expression after red-light irradiation in rice seedlings. Plant Physiol Biochem 51:10–17
Koini MA, Alvey L, Allen T, Tilley CA, Harberd NP, Whitelam GC, Franklin KA (2009) High temperature-mediated adaptations in plant architecture require the bHLH transcription factor PIF4. Curr Biol 19:408–413
Kraepiel Y, Rousselin P, Sotta B, Kerhoas L, Einhorn J, Caboche M, Miginiac E (1994) Analysis of phytochrome- and ABA-deficient mutants suggests that ABA degradation is controlled by light in *Nicotiana plumbaginifolia*. Plant J 6:665–672
Kunihiro A, Yamashino T, Mizuno T (2010) PHYTOCHROME-INTERACTING FACTORS PIF4 and PIF5 are implicated in the regulation of hypocotyl elongation in response to blue light in *Arabidopsis thaliana*. Biosci Biotechnol Biochem 74:2538–2541
Leivar P, Quail PH (2010) PIFs: pivotal components in a cellular signaling hub. Trends Plant Sci 16:19–28
Lercari B, Sodi F, Biagioni M (1986) An analysis of UV-A effects of phytochrome-mediated induction of phenylalanine ammonia-lyase in the cotyledons of Lycopersicon esculentum mill. Environ Exp Bot 26:153–157
Levin I, Frankel P, Gilboa N, Tanny S, Lalazar A (2003) The tomato dark green mutation is a novel allele of the tomato homolog of the *DEETIOLATED1* gene. Theor Appl Genet 106:454–460
Lieberman M, Segev O, Gilboa N, Lalazar A, Levin I (2004) The tomato homolog of the gene encoding UV-damaged DNA binding protein 1 (DDB1) underlined as the gene that causes the *high-pigment* mutant phenotype. Theor Appl Genet 108:1574–1581
Liu Y, Roof S, Ye Z, Barry C, van Tuinen A, Vrebalov J, Bowler C, Giovannoni J (2004) Manipulation of light signal transduction as a means of modifying fruit nutritional quality in tomato. Proc Natl Acad Sci USA 101:9897–9902

Liu J, Zhang F, Zhou J, Chen F, Wang B, Xie X (2012) Phytochrome B control of total leaf area and stomatal density affects drought tolerance in rice. Plant Mol Biol 78:289–300

Ma L, Gao Y, Qu L, Chen Z, Li J, Zhao H, Deng XW (2002) Genomic evidence for COP1 as a repressor of light-regulated gene expression and development in Arabidopsis. Plant Cell 14:2383–2398

Matsushita T, Mochizuki N, Nagatani A (2003) Dimers of the N-terminal domain of phytochrome B are functional in the nucleus. Nature 424:571–574

McElwain EF, Bohnert HJ, Thomas JC (1992) Light moderates the induction of phosphoenolpyruvate carboxylase by NaCl and abscisic acid in *Mesembryanthemum crystallinum*. Plant Physiol 99:1261–1264

McKenzie JS, Weiser CJ, Burke MJ (1974) Effects of red and far red light on the initiation of cold acclimation in *Cornus stolonifera* Michx. Plant Physiol 53:783–789

Mollard FPO, Insausti P (2009) Soil moisture conditions affect the sensitivity of *Bromus catharticus* dormant seeds to light and the emergence pattern of seedlings. Seed Sci Res 19:81–89

Monteiro CC, Carvalho RF, Gratão PL, Carvalho G, Tezotto T, Medici LO, Peres LEP, Azevedo RA (2011) Biochemical responses of the ethylene-insensitive *Never ripe* tomato mutant subjected to cadmium and sodium stresses. Environ Exp Bot 71:306–320

Monteiro CC, Rolão MB, Franco MR, Peters LP, Cia MC, Capaldi FR, Carvalho RF, Gratão PL, Rossi ML, Martinelli AP, Peres LEP, Azevedo RA (2012) Biochemical and histological characterization of tomato mutants. An Acad Bras Cien 84:573–585

Montgomery BL, Lagarias JC (2002) Phytochrome ancestry: sensors of bilins and light. Trends Plant Sci 7:357–366

Moon J, Zhu L, Shen H, Huq E (2008) PIF1 directly and indirectly regulates chlorophyll biosynthesis to optimize the greening process in Arabidopsis. Proc Natl Acad Sci USA 105:9433–9438

Moreno JE, Tao Y, Chory J, Ballaré CL (2009) Ecological modulation of plant defense via phytochrome control of jasmonate sensitivity. Proc Natl Acad Sci USA 106:4935–4940

Munns R, Tester M (2008) Mechanisms of salinity tolerance. Annu Rev Plant Biol 59:651–681

Nagatani A (2004) Light-regulated nuclear localization of phytochromes. Curr Opin Plant Biol 7:708–711

Neff M (2012) Light-mediated seed germination: connecting phytochrome B to gibberellic acid. Dev Cell 22:687–688

Oh E, Kang H, Yamaguchi S, Park J, Lee D, Kamiya Y, Choi G (2009) Genome-wide analysis of genes targeted by PHYTOCHROME INTERACTING FACTOR 3-LIKE5 during seed germination in Arabidopsis. Plant Cell 21:403–419

Peres LEP, Carvalho RF, Zsögön A, Bermúdez-Zambrano OD, Robles WGR, Tavares S (2005) Grafting of tomato mutants onto potato rootstocks: an approach to study leaf-derived signaling on tuberization. Plant Sci 169:680–688

Peschke F, Kretsch T (2011) Genome-wide analysis of light-dependent transcript accumulation patterns during early stages of Arabidopsis seedling deetiolation. Plant Physiol 155:1353–1366

Quail PH (2007) Phytochrome-regulated gene expression. J Integr Plant Biol 49:11–20

Reynolds OL, Keeping MG, Meyer JH (2009) Silicon-augmented resistance of plants to herbivorous insects: a review. Ann Appl Biol 155:171–186

Riemann M, Takano M (2008) Rice JASMONATE RESISTANT 1 is involved in phytochrome and jasmonate signalling. Plant Cell Environ 31:783–792

Riemann M, Muller A, Korte A, Furuya M, Weiler EW, Nick P (2003) Impaired induction of the jasmonate pathway in the rice mutant hebiba. Plant Physiol 133:1820–1830

Roberts MR, Paul ND (2006) Seduced by the dark side: integrating molecular and ecological perspectives on the influence of light on plant defence against pests and pathogens. New Phytol 170:677–699

Robson F, Okamoto H, Patrick E, Harris S-R, Wasternack C, Brearley C, Turner JG (2010) Jasmonate and phytochrome A signaling in Arabidopsis wound and shade responses are integrated through JAZ1 stability. Plant Cell 22:1143–1160

Rockwell NC, Lagarias JC (2010) A brief history of phytochromes. Chem Phys Chem 11:1172–1180

Rockwell NC, Su Y-S, Lagarias JC (2006) Phytochrome structure and signaling mechanisms. Annu Rev Plant Biol 57:837–858

Rockwell NC, Martin SS, Feoktistova K, Lagarias JC (2011) Diverse two-cysteine photocycles in phytochromes and cyanobacteriochromes. Proc Natl Acad Sci USA 108:11854–11859

Rösler J, Jaedicke K, Zeidler M (2010) Cytoplasmic phytochrome action. Plant Cell Physiol 51:1248–1254

Sancar A (2003) Structure and function of DNA photolyase and cryptochrome blue-light photoreceptors. Chem Rev 103:2203–2237

Sanchez RA, de Miguel L, Lima C, de Lederkremer RM (2002) Effect of low water potential on phytochrome-induced germination, endosperm softening and cell-wall mannan degradation in *Datura ferox* seeds. Seed Sci Res 12:155–163

Sawada Y, Aoki M, Nakaminami K, Mitsuhashi W, Tatematsu K, Kushiro T, Koshiba T, Kamiya Y, Inoue Y, Nambara E, Toyomasu T (2008) Phytochrome- and gibberellin-mediated regulation of abscisic acid metabolism during germination of photoblastic lettuce seeds. Plant Physiol 146:1386–1396

Sharrock RA, Quail PH (1989) Novel phytochrome sequences in *Arabidopsis thaliana*: structure, evolution, and differential expression of a plant regulatory photoreceptor family. Genes Dev 31:745–1757

Shekhawat GS, Verma K (2010) Haem oxygenase (HO): an overlooked enzyme of plant metabolism and defence. J Exp Bot 61:2255–2270

Shen Q, Jiang M, Li H, Che LL, Yang ZM (2011) Expression of a *Brassica napus* heme oxygenase confers plant tolerance to mercury toxicity. Plant Cell Environ 34:752–763

Slocombe SP, Whitelam GC, Cockburn W (1993) Investigation of phosphoenolpyruvate carboxylase (PEP carboxylasease) in *Mesembryanthemum crvstallinum* L. in C3 and CAM photosynthetic states. Plant Cell Environ 16:403–411

Sokolskaya SV, Sveshnikova NV, Kochetova GV, Solovchenko AE, Gostimski SA, Bashtanova OB (2003) Involvement of phytochrome in regulation of transpiration: red-/far red-induced responses in the chlorophyll-deficient mutant of pea. Funct Plant Biol 30:1249–1259

Staneloni RJ, Rodriguez-Batiller MJ, Casal JJ (2008) Abscisic acid, high-light, and oxidative stress down-regulate a photosynthetic gene via a promoter motif not involved in phytochrome-mediated transcriptional regulation. Mol Plant 1:75–83

Taylor JE, Hatcher PE, Paul ND (2004) Crosstalk between plant responses to pathogens and herbivores: a view from the outside in. J Exp Bot 55:159–168

Terry MJ, Linley PJ, Kohchi T (2002) Making light of it: the role of plant haem oxygenases in phytochrome chromophore synthesis. Biochem Soc Trans 30:604–609

Thomsen B, Drumm-Herrel H, Mohr H (1992) Control of the appearance of ascorbate peroxidase (EC 1.11.1.11) in mustard seedling cotyledons by phytochrome and photooxidative treatments. Planta 186:600–608

Toledo-Ortiz G, Huq E, Rodriguez-Concepcion M (2010) Direct regulation of phytoene synthase gene expression and carotenoid biosynthesis by phytochrome-interacting factors. Proc Natl Acad Sci USA 107:11626–11631

Ulm R, Baumann A, Oravecz A, Mate Z, Adam E, Oakeley EJ, Schafer E, Nagy F (2004) Genome-wide analysis of gene expression reveals function of the bZIP transcription factor HY5 in the UV-B response of Arabidopsis. Proc Natl Acad Sci USA 101:1397–1402

Wade HK, Bibikova TN, Valentine WJ, Jenkins GI (2001) Interactions within a network of phytochrome, cryptochrome and UV-B phototransduction pathways regulate chalcone synthase gene expression in Arabidopsis leaf tissue. Plant J 25:675–685

Wellmann E, Schneider-Ziebert U, Beggs CJ (1984) UV-B inhibition of phytochrome-mediated anthocyanin formation in *Sinapis alba* L. Cotyledons: action spectrum and the role of photoreactivation. Plant Physiol 75:997–1000

Williams BJ, Pellett NE, Klein RM (1972) Phytochrome control of growth cessation and initiation of cold acclimation in selected woody plants. Plant Physiol 50:262–265

Winter D, Vinegar B, Nahal H, Ammar R, Wilson GV, Provart NJ (2007) An "eletronic fluorescent pictograph" browser for exploring and analyzing large-scale biological data sets. PLoS One 8:e718
Wu J, Baldwin IT (2009) Herbivory-induced signalling in plants: perception and action. Plant Cell Environ 32:1161–1174
Wu S-H, Lagarias JC (2000) Defining the bilin lyase domain: lessons from the extended phytochrome superfamily. Biochemistry 39:13487–13495
Xie Y-J, Xu S, Han B, Wu M-Z, Yuan X-X, Han Y, Gu Q, Xu D-K, Yang Q, Shen W-B (2011a) Evidence of Arabidopsis salt acclimation induced by up-regulation of HY1 and the regulatory role of RbohD-derived reactive oxygen species synthesis. Plant J 66:280–292
Xie Y, Cui W, Yuan X, Shen W, Yang Q (2011b) Heme oxygenase-1 is associated with wheat salinity acclimation by modulating reactive oxygen species homeostasis. J Integr Plant Biol 53:653–670
Xu S, Zhang B, Cao Z-Y, Ling T-F, Shen W-B (2011) Heme oxygenase is involved in cobalt chloride-induced lateral root development in tomato. Biometals 24:181–191
Yang SW, Jang IC, Henriques R, Chua NH (2009) FAR-RED ELONGATED HYPOCOTYL1 and FHY1-LIKE associate with the Arabidopsis transcription factors LAF1 and HFR1 to transmit phytochrome A signals for inhibition of hypocotyl elongation. Plant Cell 21:1341–1359
Zhai Q, Li C-B, Zheng W, Wu X, Zhao J, Zhou G, Jiang H, Sun J, Lou Y, Li C (2007) Phytochrome chromophore deficiency leads to overproduction of jasmonic acid and elevated expression of jasmonate-responsive genes in Arabidopsis. Plant Cell Physiol 48:1061–1071
Zhong HH, Resnick AS, Straume M, McClung CR (1997) Effects of synergistic signaling by phytochrome A and cryptochrome1 on circadian clock-regulated catalase expression. Plant Cell 9:947–955
Zhong S, Zhao M, Shi T, Shi H, An F, Zhao Q, Guo H (2009) EIN3/EIL1 cooperate with PIF1 to prevent photo-oxidation and to promote greening of Arabidopsis seedlings. Proc Natl Acad Sci USA 106:21431–21436
Zilli CG, Balestrasse KB, Yannarelli GG, Polizio AH, Santa-Cruz DM, Tomaro ML (2008) Heme oxygenase up-regulation under salt stress protects nitrogen metabolism in nodules of soybean plants. Environ Exp Bot 64:83–89

Chapter 13
Role of Arbuscular Mycorrhiza in Amelioration of Salinity

R. Hajiboland

Abstract Soil salinity is world wide problem because it negatively affect plant productivity and yield of plants particularly in arid and semi-arid regions of the world. Excessive salts decline soil water availability for plants, inhibit plants metabolism and nutrients uptake and is also responsible for osmotic imbalance. All of these changes contribute to stunted growth and less productivity of plants. Exploitation of soil microorganisms for utilizing salt affected soils is of considerable interest to plant and soil scientists. Arbuscular mycorrhizal fungi (AMF) are ubiquitous soil microorganisms inhabiting the rhizosphere and establish a symbiotic relationship with the roots of many plants. Arbuscular mycorrhizal fungi are from integral components of all natural ecosystems and are known to occur in saline soils. Symbiotic association of a plant with AMF results in higher ability for taking up the immobile nutrients in nutrient-poor soils as well as improvement of tolerance to salinity. The possible mechanisms for alleviation of salinity stress by AMF include: (1) improvement of plant nutrient uptake, particularly P, (2) elevation of K:Na ratio, (3) providing higher accumulation of osmosolutes, and (4) maintaining higher antioxidant enzymatic activities. In addition, some aquaporin genes are up-regulated in mycorrhizal plants, causing significant increase in water absorption capacity of salt-affected plants. In contrast, expression of proline biosynthetic enzymes and LEA genes as stress indicators are maintained in mycorrhizal salt stressed plants suggesting that mycorrhizal plants are less susceptible to salinity because of salinity-avoidance mechanisms.

Keywords Antioxidant defense system • Aquaporins • Arbuscular mycorrhizal fungi (AMF) • Ion homeostasis • Halophytes • K:Na ratio • Mineral nutrition • Salinity

---

R. Hajiboland (✉)
Plant Science Department, University of Tabriz, 51666-14779 Tabriz, Iran
e-mail: ehsan@tabrizu.ac.ir

P. Ahmad et al. (eds.), *Salt Stress in Plants: Signalling, Omics and Adaptations*,
DOI 10.1007/978-1-4614-6108-1_13, © Springer Science+Business Media New York 2013

Saline soils occupy nearly 8% of the earth land surface and have been considered a serious limiting factor in crop production (Pitman and Läuchli 2004). Salinity causes primarily ion toxicity because of excessive amounts of Na or Cl and reduction of osmotic potential of the soil solution resulting in reduction of plant-available water (Munns and Tester 2008). Accumulation of Na causes ionic imbalance and/or ion toxicity, leading to serious disturbances in cell metabolism and functions and significant loss of plant productivity and quality (Munns and Tester 2008).

More than 80% of all higher plants are colonized by AMF (Strack et al. 2003). The root colonization by AMF is regulated both by plant and fungus and influenced also by environmental factors (Smith and Smith 1997). Mycorrhizal fungi exploit water and mineral salts from soils more effectively than plant roots and transfer them to host (Augé 2001). Arbuscular mycorrhiza fungi occur in saline soils (Aliasgharzadeh et al. 2001). Although salinity might affect the formation and function of mycorrhizae (Juniper and Abbott 1993), AM colonization has been observed in different species grown in saline soils (Asghari et al. 2005). Many studies have demonstrated that AMF inoculation improves growth of plants under salinity (Giri and Mukerji 2004; Sannazzaro et al. 2006; Hajiboland et al. 2010). Accordingly, AMF have been considered as bio-ameliorators of salt-affected soils (Rao 1998).

The physiological and biochemical mechanisms involving in the improvement of salt tolerance in AM plants are still unknown. Although increased tolerance to salinity in AM plants can be ascribed to improved mineral nutrition, especially that of P (Giri et al. 2003, 2007), it is not the only mechanism by which salt tolerance can be induced by AMF colonization. Improvements in physiological processes such as photosynthesis and water use efficiency have also been evidenced in mycorrhizal plants grown under salinity (Sheng et al. 2008). Thus, AMF mitigate the adverse effects of excess salt in plants in various ways.

In this chapter, the most important mechanisms involved in the alleviating salt stress by AM colonization in plants are discussed. Beside this subject, the effects of soil salinity on the viability and inoculation capability of AMF and the effect of different combinations of plant genotypes and fungi species or isolates on the beneficial effect of AM association under salinity are discussed. Finally, some evidences on the changes in the expression pattern of genes involve in salinity response of plants are presented.

Structure and Function of AM

Arbuscular mycorrhizal symbiosis is formed between vascular plants and members of Glomeromycota. Arbuscular mycorrhizal fungi are obligate symbiotic fungi and their evolution is dated back 460 million years ago suggesting that AMF played a crucial role in initial colonization of terrestrial ecosystems (Smith and Read 2008). The success in the evolution of AMF association has been attributed to the role of fungi in the capture of nutrients from the soil (Bonfante and Perotto 2000).

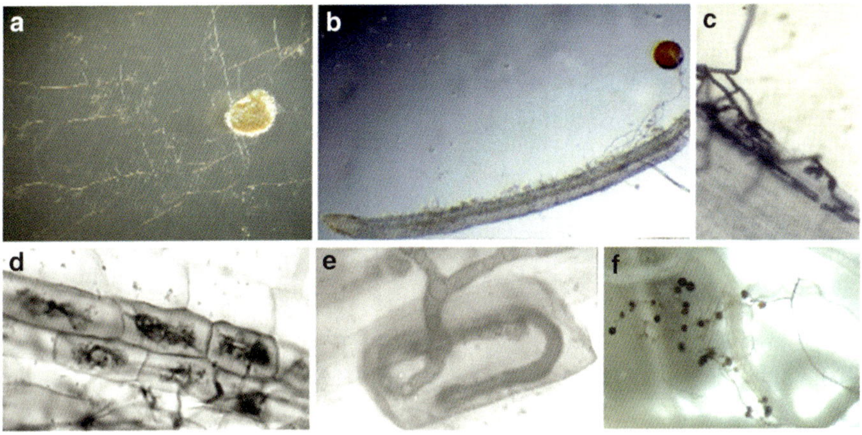

Fig. 13.1 Life cycle of an arbuscular mycorrhizal fungus. (**a**) Spore germination on water-agar. (**b**) Host recognition and pre-symbiotic growth in the proximity of a host root. (**c**) Appressoria formation on the root epidermis and colonization of the first root cortex layer. (**d**) Arbuscules in inner cortical cells. (**e**) Detail of an intracellular hypha, the so-called coil, in a cell of the root cortex. (**f**) Extraradical mycelium exploring the soil and forming the next spore generation (Retrieved from http://www.iab.kit.edu/heisenberg/286.php, 2012, Karlsruhe Institute of Technology, Germany)

Morphology, Life Cycle and Colonization of Roots

An established association between AMF and plant roots have three major components include the plant root that provides carbohydrates to the fungus, fungal structures within cortical cells (arbuscules and vesicles) that are directly interacted with the plant cytoplasm and the extraradical hyphae that are involved in the uptake of water and nutrients (Smith and Read 2008).

Arbuscules are fine and tree-like hyphal structures are separated from plant cell contents by plant plasma membrane and are responsible for the exchange of carbon and nutrients between fungi and the host cell. Vesicles found in some but not all genera of AMF and serve as carbon storage compartments for the fungi (Fig. 13.1). Formation of vesicles depends on environmental conditions (Smith and Read 2008).

Intraradical hyphae, extraradical hyphae and extraradical auxiliary cells are other important structures of AMF. The AM fungi spread its hyphae within root cortical cells by means of intraradical hyphae and form colonization units such as arbuscules and vesicles. Extraradical hyphae include branching hyphae that colonize the rhizosphere and are responsible for nutrients uptake, infective hyphae that run towards and along root surfaces and establish new entry points and reproductive hyphae that develop fertile spores after colonization of roots. Extraradical auxiliary vesicles are lipid storage compartments (Smith and Read 2008).

Under favorable environmental conditions, spores of AMF germinate and based on structural morphogenesis, show a sequence of steps include asymbiotic, presymbiotic and the symbiotic stages (Smith and Read 2008). In the asymbiotic stage, fungal spores are produced by the extraradical hyphae after establishment of symbiotic association with the host plant. In the presymbiotic stage, germinated spores grow toward the host root by production of hyphal branches that occurs before the formation of structures such as appressoria. Appressorium is an enlarged extraradical hyphal tip that attaches to the surface of the host plant root. This stage is influenced by root exudates such as organic acids, amino acids, carbohydrate monomers, phenolics, flavonoids or volatiles compounds. Plant hormones such as auxin also play a crucial role at this stage (Smith and Read 2008). The symbiotic stage refers to the penetration and development of the intraradical hyphae and the formation of arbuscles in the root cortex.

Mycorrhizal Specificity, Infectivity and Plant Responsiveness and Dependency

Arbuscular mycorrhizal fungi species are non-specific root endosymbionts and different fungal species can colonize a range of vascular plants from herbaceous to woody plants. However, these species are considerably different in both specificity and infectivity. Specificity refers to the ability of the fungus to colonize root cells of particular plant species, infectivity is the amount of colonization (Sylvia et al. 1998).

The relative performance of colonized and non-colonized plants is called 'plant responsiveness' or 'fungus effectiveness'. Responsiveness diminishes as nutrients availability becomes saturating in the soil. The inability of a plant to grow in the absence of colonization has been defined as 'dependence' and is calculated as the level of nutrient availability below which non-colonized plants cease to grow (Janos 2007).

Difference in responsiveness can be divided into dependence and non-dependence components (Sawers et al. 2008). The essential distinction between responsiveness to mycorrhizas and dependence upon mycorrhizas is that the first is the conjoint property of plant species interacting with an AMF, but the second is a constitutive property of a plant species or genotype and is used to classify plants as facultative or obligately mycotrophic. Dependence is a plant attribute, but responsiveness or effectiveness are emergent properties of the interaction between plant and fungus species (Janos 2007).

Ecological Importance of AMF

Many studies have demonstrated that AM contributes to plant growth via capture of immobile soil nutrients particularly in poor soils. Enhancement of P, Zn, Cu, Mn, and Fe uptake as well as plants growth improvement in nutrient-poor soils by AMF colonization are well documented (Smith and Read 2008). However, the benefits afforded plants from mycorrhizal symbioses are not confined to improvement of nutrients uptake. Mycorrhizal symbiosis is a key component in the adaptation of plants with adverse environmental conditions such as drought (Augé 2001; Miransari 2010), osmotic stress (Ruiz-Lozano 2003), salinity (Evelin et al. 2009; Porcel et al. 2012; Ruiz-Lozano et al. 2012) and biotic stresses (Pozo and Azcón 2007).

AMF Mediated Amelioration of Salinity Stress in Plants

Arbuscular mycorrhizas have been shown to decrease yield losses of plants in saline soils. This includes not only dry matter production, but also leaf chlorophyll content and photosynthesis rate compared with nonAM plants.

Effect of Soil Salinity on AMF Viability, Infectivity and Establishment of Symbiosis

Colonization of plant roots by AMF is influenced by various factors when plants grow under saline conditions. Reduction in AM colonization upon salinity has been reported in many plant species such as tomato (Zhi et al. 2010; Hajiboland et al. 2010), lotus (Sannazzaro et al. 2006) and acacia (Giri et al. 2003, 2007). Decline in AM colonization could be caused directly by salt-induced inhibition of spore germination (Hirrel 1981), hyphal growth and spreading (McMillen et al. 1998), reduction in the number of arbuscules (Pfeiffer and Bloss 1988); or indirectly by the effect on host plant.

Occurrence, Sporulation and Germination of AMF in Saline Soils

The abundance and distribution of AMF in different ecological regions and their relationship with soil properties have been studied by some researchers (Barrow et al. 1997; Bhardwaj et al. 1997; Cook et al. 1993; Aliasgharzadeh et al. 2001). Arbuscular mycorrhizal fungi occur naturally in saline environments (Allen and Cunningham 1983; Pond et al. 1984; Rozema et al. 1986; Ho 1987; Aliasgharzadeh et al. 2001). The AMF most commonly observed in saline soils are *Glomus* spp. (Allen and Cunningham 1983; Pond et al. 1984; Ho 1987). The occurrence of spores of almost only the *Glomus* spp. in the saline and sodic soils suggests that these fungi are the dominant root colonizers in such soils (Landwehr et al. 2002).

Formation of spores in soils depends on some physiological and ecological parameters (Redecker et al. 2000) and on the genotype of plants and fungi (Clapp et al. 1995). Published reports on the effect of soil salinity on spore production by

AMF are rare. Occurrence of relatively high spore numbers (mean of 100 per 10 g soil) has been reported in the highly saline soils (ECe ~162 dS m^{-1}) (Bhaskaran and Selvaraj 1997; Sengupta and Chaudhuri 1990; Aliasgharzadeh et al. 2001) while in other studies low or even zero spore populations were found in soils with ECe higher than 45 dS m^{-1} (Barrow et al. 1997; Hirrel 1981; Kim and Weber 1985). Stimulation of spore production under salinity as occurs in some Mucorales and *Aspergillus* species (Hirrel 1981; Tressner and Hayes 1971), or inhibition of spore germination (Juniper and Abbott 2006) both may cause accumulation of spores in saline soils. It means that the fungi may produce more spores at lower root colonization levels in severe saline conditions (Juniper and Abbott 1993).

Arbuscular mycorrhizal fungi are obligate biotrophs and it is mainly not possible to differentiate between direct and plant-mediated effects on their biology. Spore germination is the only phase of the AMF life cycle that is independent of the presence of a plant and can be studied in the absence of complex interactions with plant growth (Daniels and Graham 1976; Hepper 1979; Daniels and Trappe 1980). Germination of spores of an AMF consists of four distinct phases: hydration, activation, germ tube emergence and growth of hyphae (Tommerup 1984). The available data on the effects of salinity on germination of AMF spores indicate inhibition of spore germination by increasing concentrations of NaCl (Hirrel 1981; Estaun 1989, 1991; Juniper and Abbott 1993, 2006). It has been also demonstrated that low water potentials delay (i.e. increased the duration of the stages of germination preceding germ tube emergence) rather than prevent germination (Hirrel 1981) and the inhibitory effect of NaCl on spore germination is reversible (Hirrel 1981; Estaun 1989). Root exudates stimulate growth and alter the morphology of germ tubes (Mosse and Hepper 1975). Since root exudation is greatly influenced by soil chemistry and soil moisture availability (Rovira 1969), stimulatory effect on germ tube growth may be altered significantly under saline conditions.

In a recent work Juniper and Abbott (2006) demonstrated that, different isolates and species of AMF differ in the ability to germinate and grow in the presence of NaCl. Two isolates of *Scutellospora calospora* showed the highest germination at 300 mM NaCl, while isolates of *Acaulospora laevis* did not germinate in the presence of NaCl. The specific rate of hyphal extension was reduced by NaCl (Table 13.1). It was found that, spores of *Glomus sp.* Curragh 2 extracted from soil did not germinate at 300 mM NaCl, while intra-root propagules of this fungus germinated under these conditions. This may indicate different ecological roles for the two types of propagule. Authors stated also that overall production of hyphae was reduced in the presence of NaCl because germination was inhibited (Juniper and Abbott 2006). Increased production of the glycoprotein glomalin is related to suboptimal mycelium growth (Hammer and Rillig 2011).

The Effect of Salinity on the Formation of Mycorrhizas and Establishment of Symbiosis

Several works suggest that formation of mycorrhizas is reduced by soil salinity (Hirrel and Gerdemann 1980; Ojala et al. 1983; Poss et al. 1985; Duke et al.

Table 13.1 Percentage of maximum germination of spores of AM fungi, lengths of hyphae produced from spores (mm) and rates of hyphal extension (mm day^{-1}) from spores or pieces of colonized root in various species and isolates of AM fungi

Parameters	Germination			Length of hyphae			Rate of hyphal extension		
NaCl treatment (mM)	0	100	300	0	100	300	0	100	300
Gigaspora decipiens (WUM 6–1)	100^{a}	64^{b}	6^{c}	400^{a}	152^{b}	0	17.4	5.8	0.1
Scutellospora calospora (WUM 12–2)	97^{a}	100^{a}	97^{a}	405^{a}	148^{b}	0	19.2	11.5	3.0
Scutellospora calospora (WUM 12–3)	100^{a}	85^{a}	92^{a}	410^{a}	205^{a}	82^{c}	33.1	10.1	0.9
Glomus sp. Curragh 2 (WUM 23–1)	100^{a}	80^{a}	0	33^{a}	18^{b}	0	0.8	0.5	0.2

Data of each raw within each parameter followed by the same letter are not significantly different ($P<0.05$). From Juniper S, Abbott LK (1993) Vesicular-arbuscular mycorrhizas and soil salinity. Mycorrhiza 4:45–57, with permission

1986; Rozema et al. 1986; McMillen et al. 1998; Giri et al. 2003, 2007; Sannazzaro et al. 2006; Juniper and Abbott 2006; Sheng et al. 2008; Zhi et al. 2010; Hajiboland et al. 2010). However, other workers observed no inhibitory effect of salinity on the formation of mycorrhizas (Duke et al. 1986; Hartmond et al. 1987; Copeman et al. 1996; Yano-Melo et al. 2003). Although salinity may result in no change in the percentage of mycorrhizal roots, the total length of mycorrhizal root decreases with increasing salinity because of decreased total root growth (Poss et al. 1985). In the study on banana plants (Yano-Melo et al. 2003), salinity did not negatively influence AMF colonization expressed as percentage, but the total length of AMF-colonized roots was reduced because of decline in total root length by salinity. In this work, the increase of soil ECe stimulated root colonization by *G. clarum* and *G. etunicatum* at soil ECe up to ~5 dS m^{-1} and that by *A. scrobiculata*, up to 7.39 dS m^{-1}. Similar results were reported for tomato plants (Copeman et al. 1996).

Apart from the salt-induced changes in the hyphal morphology that affect their infective capacity (Juniper and Abbott 1993, 2006), physiological changes in the host plants caused by soil salinity may directly affect their colonization. In this respect it is important to differentiate between the primary infection, i.e. first entry into the root by the fungus, and secondary infection that occurs after ramnification of fungal hyphae from sites of initial colonization (Wilson 1984). Initial infection is dependent on germination of spores or other fungal propagules, growth of hyphae and entry into the plant root. Each of these stages can be a limiting step in the formation of mycorrhizas (Bowen 1987). Three forms of AMF propagules are responsible for colonization include AMF spores, mycorrhizal roots or root fragments and the mycelia. The relative importance of these propagules depends strongly on the environmental conditions (Juniper and Abbott 2006). Under salinity conditions, loss of viability of these propagules is a critical factor in the survival of AMF (Dixon et al. 1993).

Physiology of the host plant influences strongly the secondary infection, because for the spread of hyphae fungus needs photosynthates translocated from the plant at the arbuscules or via the internal hyphae. Apart from the toxic effects of specific ions such as sodium and chloride and physiological drought, photosynthesis *per se* is influenced negatively by salinity and reduction of the availability of photosynthates affects mycorrhizal development and function. Reduction in root growth with increasing salinity (Hirrel and Gerdemann 1980; Ojala et al. 1983; Poss et al. 1985) may also lower the probability of contact between roots and fungal hyphae and thus decrease colonization levels. Colonization of roots in the earlier growth stages is independent of root density, while after formation of initial mycorrhizas the probability of secondary hyphae encountering and infecting roots is highly dependent on root density (Abbott and Robson 1984).

With increasing concentration of toxic ions in the soil, this factor becomes more important limitation for plant growth but the availability of nutrients such as phosphorus becomes a progressively less important limitation. Under these conditions, allocation of limited photosynthates to maintain mycorrhizal association caused strong growth reduction unless there are non-nutritional benefits of AMF to the plant (Allen et al. 1981).

In some studies salt treatments were applied on plants grown on soil with a native population of AMF and, therefore, the plants were presumably colonized before the imposition of the salt stress. In other works, mycorrhizal seedlings were transplanted into saline soil (Pfeiffer and Bloss 1988). In all these cases, there was sufficient time between inoculation and imposition of salt stress, therefore, the effect of salinity on initial formation of mycorrhizas could not be evaluated. In perennial plants such as orchard trees even the age of association is important determinant on the extent to which mycorrhizal plants are influenced by salinity. This could explain contradictory results obtained by different authors (e.g. by Duke et al. 1986; Hartmond et al. 1987). In the older associations, there is much more opportunity for mycorrhizas to develop in the absence of salinity stress. Even if starting salt treatment retards hyphal growth in soil and thus formation of new entry points, colonization density increases because hyphal growth continues in the intercellular spaces of the root cortex. In addition, various AMF isolates used by different authors differ in their ability to grow under saline conditions (Estaun 1991).

Effect of AMF on Plant Growth and Performance in Saline Soils

Higher dry matter production in mycorrhizal plants under salt stress conditions has been reported by many authors in different plant species.

Plants Dry Matter Production

Root colonization by *G. mosseae* enhanced growth of maize plants irrespective to the level of P with and without NaCl (Feng et al. 2002). In tomato, salt stress

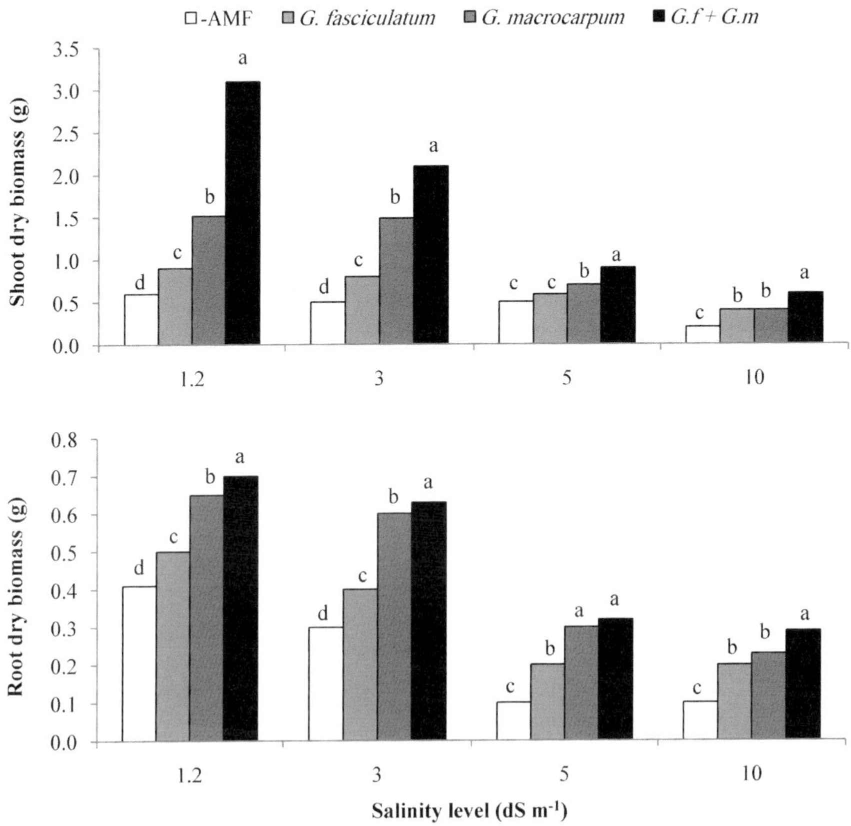

Fig. 13.2 Shoot and root dry biomass production (g) as affected by mycorrhizal colonization and salinity in acacia (*Acacia nilotica*) plants. *Bars* indicated by the same letter within each salinity level are not significantly different ($P<0.05$) (From Giri B, Kapoor R, Mukerji KG (2003) Influence of arbuscular mycorrhizal fungi and salinity on growth, biomass, and mineral nutrition of *Acacia auriculiformis*. Biol Fertil Soil 38:170–175, with permission)

significantly reduced dry matter of roots, stems, leaves, and total biomass and leaf area compared with the control treatment (Abdel-Latef and Chaoxing 2011). Effect of AMF on dry matter of tomato plants was more pronounced in the biomass of shoot than root likely due to the allocation of greater proportion of carbohydrates to the shoot compared with root in AM plants (Shokri and Maadi 2009; Abdel-Latef and Chaoxing 2011).

In acacia (*Acacia nilotica*) plants, root and shoot dry weights decreased as soil salinity increased and there was a significant influence of AM inoculation on plant growth irrespective to the salinity levels. Although high salinity levels reduced production of root and shoot biomass in both AM and nonAM plants, dry matter production was higher in AM plants (Fig. 13.2) (Giri et al. 2003, 2007). Growth improvement of *Lotus glaber* plants by *G. intraradices* was more obvious under 200 mM NaCl (Sannazzaro et al. 2006) demonstrates the capacity of the AMF to realize its symbiotic activity particularly under stress conditions.

Arbuscular mycorrhizal colonization enhanced salt tolerance of olive plants and improved its growth and nutrient acquisition (Porras-Soriano et al. 2009). Inoculating olive plantlets with *Glomus mosseae*, *Glomus intraradices* or *Glomus claroideum* increased growth and the ability to take up N, P, and K from both saline and non-saline media and improved also survival rate of plantlets after transplantation. Salt stress reduced stem diameter, number and length of shoots and nutrient content of plants, but AMF colonization alleviated all of these negative effects. Inoculation by *G. mosseae* increased shoot and root growth by 163% and 295% in the non-saline medium respectively, while the corresponding values for salinized plants were by 239% and 468% under saline conditions (Porras-Soriano et al. 2009). This finding confirmed again higher effectiveness of AMF under stress compared with non-stress conditions.

Pearl millet (*Pennisetum glaucum*) plants inoculated with AMF had significantly higher leaf number, shoot and root length under moderate level of salinity stress and both plant fresh and dry weight increased significantly in AM plants at all levels of salinity stress (Borde et al. 2011). A significant shoot dry weight increment was observed in *Theobroma cacao* inoculated with AMF (Chulan and Martin 1992). In lettuce plants and at all examined salinity levels, growth was positively influenced by AM colonization (Ruiz-Lozano and Azcón 2000). In citrus plants growth parameters of AM inoculated plants were significantly superior over non-inoculated control particularly under higher salinity level (Khalil et al. 2011).

A field experiment was under taken to study the effect of inoculation with *Glomus macrocarpum* and salinity on growth of *Sesbania aegyptiaca* and *S. grandiflora* (Giri and Mukerji 2004). It was observed that, AM seedlings had significantly higher production of root and shoot biomass compared with nonAM plants when grown under saline conditions. The number of nodules was also significantly higher in AM than nonAM plants (Giri and Mukerji 2004).

Leaf Chlorophyll Concentration

Arbuscular mycorrhizal inoculation influences positively chlorophyll concentration of leaves. Chlorophyll concentrations are usually reduced by salinity because of suppression of specific enzymes responsible for the synthesis of photosynthetic pigments (Murkute et al. 2006). Reduction in the uptake of minerals such as Mg following an antagonistic effect of Na on Mg absorption needed for chlorophyll biosynthesis also reduces the chlorophyll concentration in the leaf (Sheng et al. 2008). In the presence of mycorrhiza, the antagonistic effect of Na on Mg uptake is diminished (Giri et al. 2003) and mycorrhizal plants have shown greater absorption of Mg (Wu et al. 2010).

Mycorrhizal maize plants treated with 100 mM NaCl had 81% higher chlorophyll concentration at low P supply level and 15% at sufficient P supply than non-mycorrhizal plants (Feng et al. 2002). In acacia plants salinity decreased leaf chlorophyll concentration of nonAM plants and chlorophyll content was higher in AM than nonAM plants (Giri et al. 2003). In tomato increasing salinity caused reduction of chlorophyll content compared with control plants. Mycorrhizal colonization significantly improved chlorophyll concentration in comparison to the nonAM plants under both control and saline conditions (Abdel-Latef and Chaoxing 2011). Higher chlorophyll content in leaves of AM plants under saline conditions has been also observed by other authors (Colla et al. 2008; Kaya et al. 2009; Hajiboland et al. 2010).

Photosynthesis

Chlorophyll a fluorescence analysis reveals salt-induced structural and functional disruption of photosynthetic apparatus and damage to the PSII (Baker 2008). An improvement of photochemical parameters by AMF colonization was observed in salt affected tomato plants (Hajiboland et al. 2010). However, these parameters never were higher than nonAM plants without salt treatment (Fig. 13.3). This implies that the AMF colonization acted only for maintenance of photochemical capacity in stressed leaves and did not increase its potential for energy trapping (Hajiboland et al. 2010).

Photosynthesis is one of the primary processes to be affected by salt stress (Chaves et al. 2009). A significant reduction in the net assimilation rate, transpiration and stomatal conductance in both nonAM and AM citrus was found in salt-stressed plants (Wu et al. 2010). However, AM seedlings had significantly higher net assimilation rate, transpiration and stomatal conductance than the nonAM seedlings under salt stress. Similar results were obtained by Ruiz-Lozano et al. (1996), Sheng et al. (2008) and Hajiboland et al. (2010) under salt stress (Table 13.2). Inoculated plants under salt stress reach levels of photosynthetic capacity even superior to those of non-stressed plants (Zuccarini and Okurowska 2008). Arbuscular mycorrhizal plants have often higher CO_2 assimilation capacity because of elevated stomatal conductance. AM plants have greater sink strength of roots and this has been suggested as a reason for the often observed mycorrhizal promotion of stomatal conductance (Augé 2000). Elevated stomatal conductance due to mycorrhizal colonization of roots caused higher water loss in AM plants, however, water relations are not disrupted (Hajiboland et al. 2010). Higher shoot water potential and lower concentration of ABA in xylem sap observed in AM plants at low soil water potential (Augé et al. 2008) demonstrated an improved water uptake in AM plants because of changes in root morphology and fineness (Wu et al. 2010).

Root Growth

Roots are involved in sensing conditions in the soil environment (Davies and Zhang 1991) and are the site for exchange of carbohydrate and mineral nutrients between AMF and host plant. Root production of growth regulators that influence plant metabolism is reduced under salinity that is likely the main cause of growth inhibition in salinized plants (Amzallag 1997). Colonization by AMF improves not only

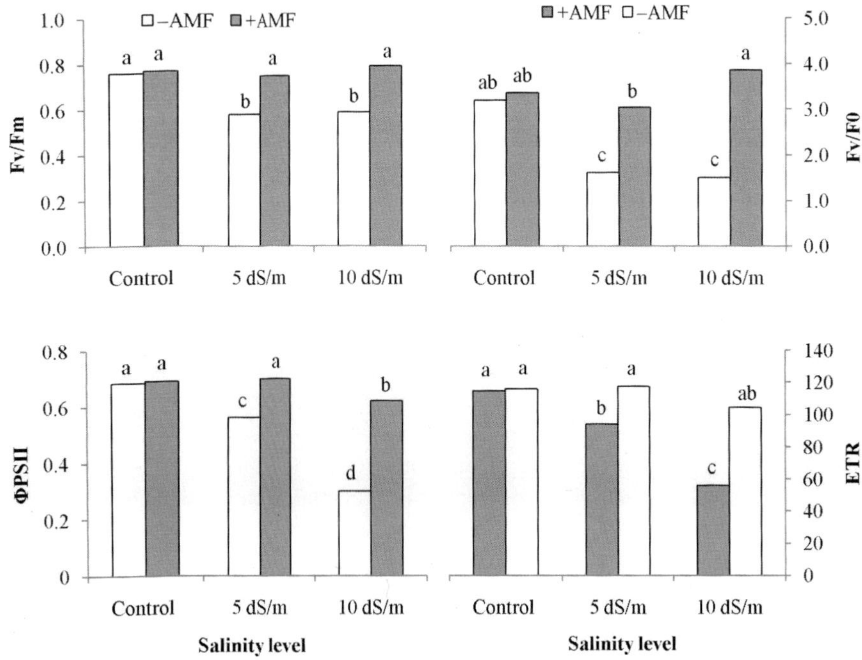

Fig. 13.3 Changes in chlorophyll fluorescence parameters (F_v/F_m: maximum quantum yield of PSII, F_v/F_0: the ratio of variable to maximum fluorescence, Φ_{PSII}: effective quantum yield of PSII and ETR: linear electron transport rate) in the leaves of tomato (*Solanum lycopersicum* cv. Behta) plants grown at different levels of salinity and inoculated with *Glomus intraradices. Bars* indicated by the same letter are not statistically different (P<0.05) (From Hajiboland R, Aliasgharzadeh N, Laiegh SF, Poschenrieder C (2010) Colonization with arbuscular mycorrhizal fungi improves salinity tolerance of tomato (*Solanum lycopersicum* L.) plants. Plant Soil 331:313–327with permission)

Table 13.2 Changes in gas exchange parameters of the leaves of tomato (*Solanum lycopersicum* cv. Behta) plants grown at different levels of salinity and inoculated with *Glomus intraradices*

		Salinity (dS m^{-1})		
		Control	5	10
Net assimilation rate (μmol m^{-2} s^{-1})	−AMF	6.51[a]	4.58[b]	2.04[c]
	+AMF	6.39[a]	6.37[a]	3.59[b]
Transpiration rate (mmol m^{-2} s^{-1})	−AMF	3.20[b]	2.39[c]	0.59[e]
	+AMF	4.16[a]	3.10[b]	1.71[d]
Stomatal conductance (mmol m^{-2} s^{-1})	−AMF	55[b]	33[c]	7[e]
	+AMF	58[b]	49[b]	19[d]

Data within each parameter indicated by the same letter are not statistically different (P<0.05). From Hajiboland R, Aliasgharzadeh N, Laiegh SF, Poschenrieder C (2010) Colonization with arbuscular mycorrhizal fungi improves salinity tolerance of tomato (*Solanum lycopersicum* L.) plants. Plant Soil 331:313–327, with permission

Table 13.3 Effect of inoculation of citrus (*Citrus tangerine*) seedlings with *Glomus mosseaes* on root morphology under control and salinity (100 mM NaCl) treatments

		Root length (cm)	Root projected area (cm^2)	Root surface area (cm^2)	Root average diameter (mm)	Root volume (cm^3)
Control	−AMF	175^c	9.59^b	30.1^b	0.55^a	0.41^b
	+AMF	221^a	12.07^a	37.9^a	0.55^a	0.52^a
Salinity	−AMF	125^d	6.58^c	20.7^c	0.54^a	0.28^c
	+AMF	188^c	9.71^b	30.5^b	0.53^a	0.40^b

Same letter within each column indicates no significant difference among treatments. From Wu QS, Zou YN, He XH (2010) Contributions of arbuscular mycorrhizal fungi to growth, photosynthesis, root morphology and ionic balance of citrus seedlings under salt stress. Acta Physiol Plant 32:297–304, with permission

growth of aerial parts but also affects root growth considerably. Arbuscular mycorrhizal association alters morphology of root in a structural, spatial, quantitative and temporal manner (Kapoor et al. 2008). It was observed that, mycorrhizal citrus seedlings had greater root length, root surface and projected area than the nonAM control seedlings under NaCl stress (Table 13.3) (Wu et al. 2010). Similar results have been observed in grapevine, apple, pepper, maize, zucchini and beach plum (Locatelli et al. 2002; Aguin et al. 2004; Schroeder and Janos 2005; Zai et al. 2007). In addition, mycorrhizal association decreases the meristematic activity of root apices and thus leads to an increase in the number of adventitious roots (Berta et al. 1993). In acacia, the number of lateral roots was significantly greater in AM than nonAM plants (Giri et al. 2003). It was also suggested thatAM colonization prolongs the life of rootlets and stimulates them to branch. This results in production of greater root system and higher total potential absorbing surface (Khalil et al. 2011). The importance of increased root surface area in enhancing nutrient uptake capacity is well known, and can be involved in stimulation of growth in AM plants. The improvement of AM root morphology is often attributed to a modified balance of growth regulators such as cytokinins and gibberellins (Berta et al. 1993). Arbuscular mycorrhizal fungi may also provide the host plant with growth hormones, including auxins, cytokinins, gibberellins and vitamins which also may affect plant growth (Barea and Azcón-Aguilar 1982).

Mycorrhizal Dependency of Plants

Although in many cases AM colonization was reduced with increasing salinity, the dependency of plants on AMF was increased (Giri and Mukerji 2004; Miranda et al. 2011). It indicates that association between AMF and plants was strengthened in saline environment once the association was established. This demonstrates the ecological importance of AM association for plants survival and growth in saline soils. Indeed under salt stress conditions, plants need mycorrhiza for both acclimatization and continued nutrient uptake during progressive growth stages (Giri and Mukerji 2004).

Effect of AMF on Mineral Nutrition of Salinized Plants

It has been widely believed that alleviation of salt stress by AMF is mainly the result of an improvement in plant P nutrition (Hirrel and Gerdemann 1980). Phosphorus concentration in plant tissues is rapidly lowered under saline conditions because phosphate ions precipitate with Ca^{2+} ions in saline soil and become unavailable to plants (Marschner 2012). Decline in plants P concentration at increased salinity levels is the result of both reduced P uptake and transport into shoot (Al-Karaki et al. 2001).

Arbuscular mycorrhizal fungi have been shown to positively influence the composition of mineral nutrients of plants under salt stress conditions (Al-Karaki and Clark 1998; Giri et al. 2003, 2007). Mycorrhizae increase particularly uptake of the nutrients that move to plant roots mainly by diffusion, especially in dry soil when nutrient diffusion rates are most limited (Marschner 2012). The enhancement of plant P uptake by AMF has been frequently reported and considered as one of the main reasons for amelioration of growth in salinized plants colonized by AMF (Al-Karaki 2000; Ruiz-Lozano and Azcón 2000). Enhanced uptake of P by AMF in salt-affected plants may reduce the negative effects of Na and Cl ions by maintaining membrane integrity, thus facilitates selective ion intake as well as compartmentalization within vacuoles and thereby, preventing ions from interference with plant metabolic pathways (Evelin et al. 2009).

The external hyphae of AMF deliver up to 80% of plants P requirement (Marschner and Dell 1994). Mycorrhizal uptake can even replace completely direct uptake by plant transporters possibly following loss of function of the direct uptake pathway in roots colonized by AMF (Smith et al. 2003). The extended network of AMF hyphae allows AM roots to explore more soil volume than nonAM plants. Indeed, AM hyphae extend beyond the depletion zones around roots and acquire nutrients that are several centimeters away from the root surface, and thus suppress the adverse effect of salinity stress (Smith and Read 2008).

In a field experiment conducted to study the effect of colonization with *Glomus macrocarpum* and salinity on growth of *Sesbania aegyptiaca* and *S. grandiflora*, AM seedlings had significantly higher P and N concentration but lower Na concentration than nonAM seedlings. It was stated that reduction in Na uptake in association with a concomitant increase in P and N absorption in AM plants are important salt-alleviating mechanisms for *S. grandiflora* plants growing in saline soil (Giri and Mukerji 2004). In *Acacia nilotica* plants, total accumulation of P, Zn, and Cu was higher in AM than in nonAM plants under both control and medium salt stress conditions (Giri et al. 2007) (Table 13.4).

Mycorrhizal colonization increased Mg uptake in two species of *Sesbania* under salinity (Giri and Mukerji 2004). Greater Mg uptake may be the reason for increasing chlorophyll concentration and hence improving photosynthetic efficiency of mycorrhizal plants. Calcium uptake was significantly lower in salt-affected tomato plants and AMF colonization enhanced significantly Ca uptake and Ca:Na ratio of both leaves and roots (Hajiboland et al. 2010). Similar results were obtained by other authors (Sharifi et al. 2007; Yano-Melo et al. 2003). An improved Ca nutrition caused maintenance of cellular ion homeostasis and improvement of plant growth under salt stress (Cramer 2004).

Table 13.4 Leaf concentration of P, Zn and Cu as affected by salinity and inoculation with *Glomus fasciculatum* in acacia (*Acacia nilotica*) plants

	1.2 dS m^{-1}		4.0 dS m^{-1}		6.5 dS m^{-1}		9.5 dS m^{-1}	
	−AMF	+AMF	−AMF	+AMF	−AMF	+AMF	−AMF	+AMF
P concentration (%)	0.6^{b}	1.2^{a}	0.5^{bc}	1.2^{a}	0.2^{c}	0.9^{b}	0.1^{c}	0.6^{b}
Zn concentration (ppm)	5.0^{c}	14^{a}	5.2^{c}	12^{a}	1.8^{d}	8.2^{b}	0.1^{d}	$3.2^{c,d}$
Cu concentration (ppm)	1.8^{c}	5.5^{a}	1.8^{c}	5.4^{a}	1.4^{c}	4.3^{b}	0.5^{a}	1.2^{c}

Data in each raw followed by the same letter are not statistically different ($P<0.05$). Data from (Giri et al. 2007)

Several mechanisms may be involved in enhancing the nutrients uptake by AMF. Mycorrhizal plants can explore a greater volume of soil beyond the zone of P, K and Zn depletion, lower the threshold concentration for absorption from soil solution, enhance root exudates and alter rhizosphere pH that increase availability of nutrients and solubilize organic P by production of phosphatase (Marschner 2012). These explanations, however, hardly apply to Ca. For this ion the AMF-mediated improvement of membrane integrity and therefore selectivity of ion uptake and transport is more likely the involving mechanism (Cramer 2004).

Activity of phosphatases increased in mycorrhizal pearl millet plants that led to increase in P uptake under saline conditions (Borde et al. 2011). The alkaline phosphatases are involved in hyphal P acquisition and one or several acid phosphatases are responsible for P transfer processes (Ezawa et al. 2002).

In summary, results of these studies indicate that AM plants have greater ability for nutrient uptake (especially P) than nonAM plants at all salinity levels. Positive influence of AMF on the mineral nutrition of plants grown under slat stress conditions could be regarded as a strategy for salt stress tolerance in plants. Moreover, improved plants growth and nutrient acquisition demonstrates the potential of AMF colonization for protecting plants against salt stress in nutrient poor soils.

Effect of AMF on Ion Homeostasis and K:Na Ratio of Salinized Plants

Ionic imbalances occur in plant cells due to salt stress. Salt stress affects mainly plant physiology and metabolism through changes in the status of ions inside the cells (Hasegawa et al. 2000; Munns et al. 2006). Ionic imbalance may be resulted from changes in nutrient availability, competitive uptake, transport or partitioning within plants (Rabie 2005). Thus, plant salt tolerance may be tightly related to its ability for regulation of ionic balance, particularly Na^{+}, K^{+}, Ca^{2+} and Mg^{2+} (Munns et al. 2006).

Elevated Na in soil solution inhibits acquisition of other nutrients by disrupting various transporters in the root plasma membrane, such as K-selective ion channels. Sodium transport is a unidirectional flow and thus results in progressive accumulation of Na in the shoot and leaf tissues with age of the plant. Detrimental effect of Na is due to its ability to compete with K for binding sites essential for various cellular functions. Potassium is one of the most important nutrients and has an important role in water balance, cell extension and solute transport in the xylem. Potassium in the plant cells is required not only for stabilizing pH in the cytoplasm, but also for increasing the osmotic potential in the vacuole (Marschner 2012). High levels of Na, or high Na:K ratios can disrupt various enzymatic processes in the cytoplasm. Maintenance of a high cytosolic K:Na ratio is a key feature of plant salt tolerance and greater K:Na ratio in AM compared with nonAM plants is the important mechanism for enhancement of salt tolerance by AMF colonization (Chinnusamy et al. 2005).

Concentration of Na was reported to be lower in AM than nonAM plants in various species regardless of salinity level (Giri et al. 2003, 2007; Giri and Mukerji 2004). This may be explained by dilution effect because of plant growth enhancement by AMF colonization. In olive plants K content was enhanced by 6.4-fold with *G. mosseae*, 3.4-fold with *G. intraradices*, and 3.7-fold with *G. claroideum* under saline conditions (Porras-Soriano et al. 2009). In *Acacia nilotica*, AM plants had lower concentration of Na in shoot tissues even at high salinity level, while Na concentration increased drastically in nonAM plants at same salinity level. In addition, AM inoculated *A. nilotica* roots accumulated more Na and thus prevented transport of Na to shoot tissues that may be considered another strategy for alleviation of detrimental effect of salinity in AM plants (Giri et al. 2007). Mycorrhizal pearl millet plants accumulated more salt in the root and prevented transport of Na into shoot (Borde et al. 2011). It was suggested that in AM plants, Na may be kept inside root cell vacuoles and intraradical fungal hyphae and by this means is prevented from being transported into the shoots (Cantrell and Linderman 2001). Increased K and Mg concentrations in citrus seedlings by AM colonization under salinity reported by Wu et al. (2010) would help the seedlings to prevent cellular Na accumulation to a toxic level and thus protect host plants against salt injury. In a field experiment, concentration of Na declined while K and Mg concentration increased under salinity when *Sesbania aegyptiaca* and *Sesbania grandiflora* inoculated with AMF (Giri and Mukerji 2004).

These results indicate that inoculation with AMF has marked influence on the acquisition and tissue concentrations of Na and K. The higher K accumulation in AM plants under salt stress conditions results in maintaining a high K:Na ratio, preventing the disruption of metabolic processes and inhibition of protein synthesis. Selective ion uptake, thus, is the main mechanism in AM plants for protection against ion imbalance caused by salinity.

It is noteworthy that reduction of Na concentration in AM plants does not necessarily mean an AM-mediated impairment of Na uptake. Lower Na accumulation in AM plants could be the result of dilution of Na in plant tissues as the consequence of growth improvement of AM compared with nonAM plants. In a pot

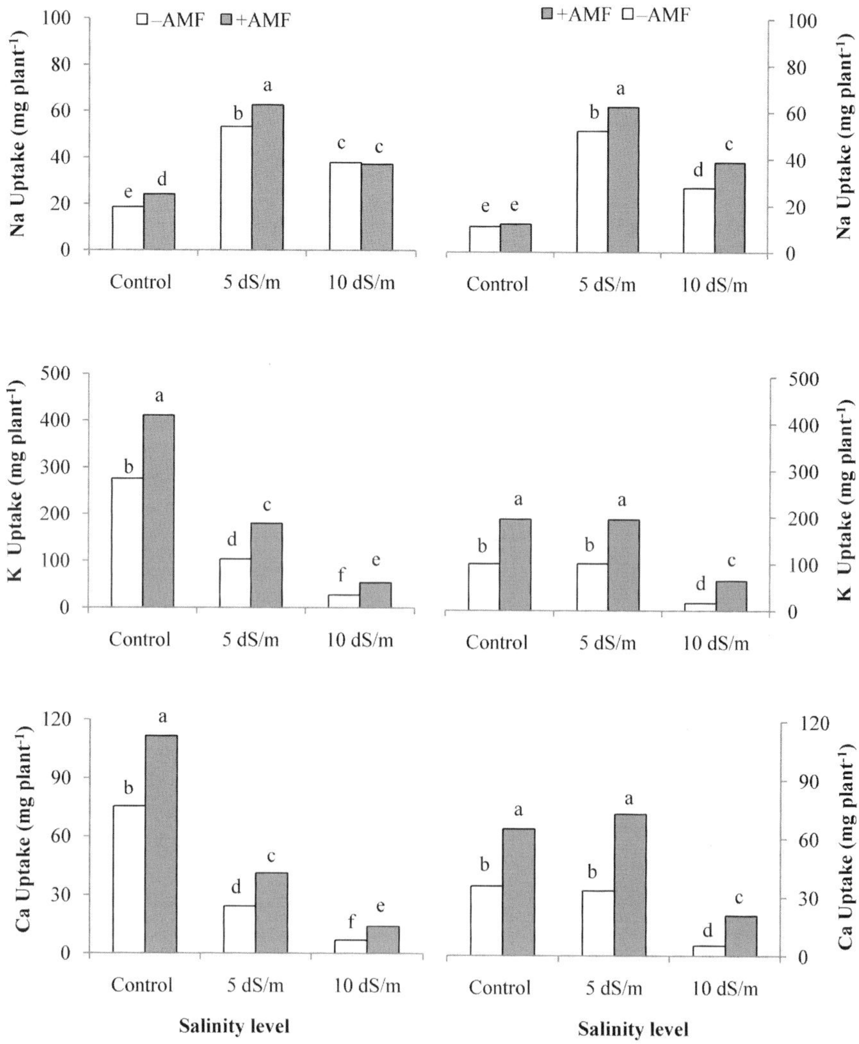

Fig. 13.4 Influence of different salinity levels on uptake of Na, K and Ca by tomato cultivars Behta and Piazar colonized (+AMF) or not (–AMF) with *Glomus intraradices*. *Bars* indicated by the same letter are not significantly different ($P<0.05$) (From Hajiboland R, Aliasgharzadeh N, Laiegh SF, Poschenrieder C (2010) Colonization with arbuscular mycorrhizal fungi improves salinity tolerance of tomato (*Solanum lycopersicum* L.) plants. Plant Soil 331:313–327 with permission)

experiment, mycorrhizal tomato plants grown in saline soil had greater Na uptake than nonAM counterparts (Hajiboland et al. 2010) (Fig. 13.4). Unfortunately, most of the authors have not provided the Na content or uptake (mg plant^{-1}) data and it is not possible to draw a precise conclusion on the influence of mycorrhization on Na uptake in plants.

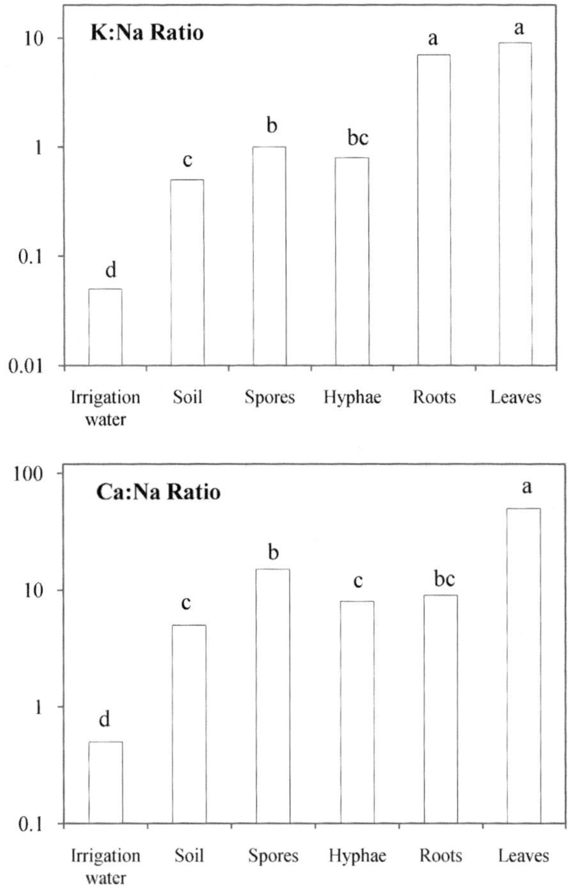

Fig. 13.5 K:Na ratio and Ca:Na ratio of the irrigation water, soil, AMF spores and hyphae and associated plant roots and leaves for samples taken from field. *Bars* indicated by the same letter are not significantly different (P<0.05)

In mycorrhizal plants, ion discrimination could take place during the uptake of nutrients from the soil by fungi or during transfer to the host plant. In general, there are morphological barriers for ion selection in plants including the root hair membrane, the casparian stript and before transfer to the shoot, the xylem membranes (Tester and Davenport 2003). In order to investigate whether AMF can prevent uptake of toxic Na in response to salinity, a pioneer work has been undertaken using proton-induced X-ray emission (PIXE) method (Hammer et al. 2011). Composition of elements in the soil solution, spores and hyphae as well as plant samples were determined in this study (Fig. 13.5). It was shown that Na ions are excluded from entering the AMF cells while K, Ca and Mg concentration were significantly elevated in the spores and hyphae compared with surrounding soil. These results revealed that mycorrhizal fungi act as the first barrier for ion selection and AMF alleviate salt stress in plants by pre-selecting nutrients and preventing toxic salt ions from entering the plant. It was also demonstrated that AMF are able to keep the internal K:Na and Ca:Na ratios within a narrow range in spite of several orders of magnitude of variation in the environment. If a significant proportion of elements acquisition by plants occurs via the AMF pathway as observed for P (Smith et al. 2003), this pre-selection mechanism in AMF could well explain the often higher K:Na ratios in AM plants (Hammer et al. 2011).

In this report, even higher K:Na and Ca:Na ratio was observed in plant tissues compared with spores and hyphae of AMF. If we consider the mycorrhizal uptake pathway as the dominant pathway for K and Na (and other ions) uptake as shown for P (Smith et al. 2003), we can hypothesize that the periarbuscular membrane is another barrier that provide a selective delivery of various ions to the host plant. Our knowledge on the function of periarbuscular membrane even for P is really limited and detailed works are needed for elucidating the selective transport of ions via periarbuscular membrane.

Effect of AMF on Water Relations and Solute Accumulation in Salinized Plants

One of the main consequences of salinity is loss of intracellular water and osmotic damages. Osmotic effects are associated with inhibited cell wall extension and cellular expansion, leading to growth reduction. Osmotic adjustment enables plants to maintain turgor potential under saline conditions (Rhodes et al. 2004; Munns and Tester 2008).

Accumulation of osmotically active organic solutes e.g. osmolytes, is a well known response to salinity in the majority of glycophytes that results from alterations in intermediary and secondary nitrogen and carbon metabolism (Hasegawa et al. 2000). This response is an important component of salinity tolerance in plants. Low molecular- weight compatible solutes like proline, glycine betaine, free amino acids, organic acids and soluble sugars accumulate to high levels without disturbing intracellular biochemistry and protect sub-cellular structures, mitigate oxidative damage, maintain the osmotic balance and protect enzymes in presence of high cytoplasmic ion concentrations (Hasegawa et al. 2000; Rhodes et al. 2004; Munns and Tester 2008).

Proline and Glycine Betaine

Free amino acids are important osmolytes contributing to osmotic adjustment in plants (Hajlaoui et al. 2010). Increasing external salt concentration causes accumulation of free amino acids in the leaves and roots (Abd-El Baki et al. 2000; Neto et al. 2009; Hajlaoui et al. 2010; Sheng et al. 2011) but this effect is less expressed in AM compared with nonAM plants (Sheng et al. 2011). Glycine betaine (N, N, N-trimethylglycine betaine) stabilizes the structure and activity of enzymes and proteins and maintain membranes integrity despite of damaging effects of excessive salt (Rhodes et al. 2004). Accumulation of betaines under salt stress was found to increase when the plant is colonized by AMF (Duke et al. 1986).

Among free amino acids, proline is much more important in osmotic adjustment of salt-stressed maize plants (Rhodes et al. 2004). Proline is synthesized by plants in response to stresses including salinity and ameliorates the abiotic stress effect. Proline also plays a role in scavenging free radicals, stabilizing subcellular structures and buffering cellular redox potential under stresses (Wang et al. 2009). The salinity stress responsive genes, containing proline responsive elements (ACTCAT) in the promoters, are induced by proline (Chinnusamy et al. 2005). In higher plants, proline is synthesized from glutamic acid via the two enzymes, pyrroline-5-carboxylate synthetase (P5CS) and pyrroline-5-carboxylate reductase (P5CR). In tobacco overexpressing the P5CS gene, proline production increased that caused enhanced salinity and drought tolerance in transgenic plants (Kishor et al. 1995).

Reports on the effect of AM symbiosis on proline accumulation are contradictory. Several authors have reported a higher proline concentration in AM plants than in nonAM plants at different salinity levels (Khaled et al. 2003; Sharifi et al. 2007). Enhanced proline accumulation in cells of AM plants can increase osmotic potential (Hajlaoui et al. 2010), thereby improve the tolerance of AM plants to salinity. On the contrary, other authors reported that AM plants accumulated significantly lower proline than nonAM plants at various salinity levels (Duke et al. 1986; Ruiz-Lozano et al. 1996; Jahromi et al. 2008; Rabie and Almadini 2005; Borde et al. 2011; Sheng et al. 2011). Lower proline content of AM plants under saline conditions compared with nonAM plants may suggest that proline accumulation in plants is an indicator of stress, and lower proline content of AM plants is a reflection of an increased salt resistance in plants upon mycorrhization, i.e. less injury. Species and genotypes with higher salt resistance accumulate lower proline under salinity conditions (Rush and Epstein 1976; Tal et al. 1979; Watad et al. 1983) that supports the view that proline accumulation in response to salt stress is an indicator of stress perception. Accordingly, it was suggested that proline is a symptom of stress in less salt-tolerant species (Wang et al. 2004; Evelin et al. 2009; Porcel et al. 2012). In citrus plants, AM seedlings accumulated less proline than nonAM seedlings under drought conditions (Wu et al. 2007). In soybean plants mycorrhization with either *G. mosseae* or *G. intraradices* did not induce the expression of the p5cs genes analyzed (Porcel et al. 2004). Under drought conditions, the levels of p5cs transcripts in AM plants were considerably lower than that in the corresponding nonAM plants, indicating that the induction of p5cs gene is not a mechanism by which the AM symbiosis protects their host plant against drought stress (Porcel et al. 2004). These results suggest that AM plants are less stressed by drought than nonAM plants due to primary drought-avoidance mechanisms. Similar salinity avoidance mechanisms are likely involved in many of species in the presence of AMF (Table 13.5).

Soluble Sugars

Accumulation of soluble sugars is a mean for lowering osmotic potential during salt stress. In a study on maize plants, concentration of soluble and reducing sugars declined by increasing salinity levels, but at a given NaCl level, sugar concentration

Table 13.5 Relative water content in shoots, proline content in roots and expression of *Lsp5cs* (pyrroline-5-carboxylate synthetase) gene probes from roots of lettuce (*Lactuca sativa*) plants subjected to 0, 50, or 100 mM NaCl and inoculated with *Glomus intraradices*

	Control		50 mM NaCl		100 mM NaCl	
	−AMF	+AMF	−AMF	+AMF	−AMF	+AMF
Relative water content (%)	93[b]	96[a]	86[d]	93[b]	91[c]	95[b]
Proline concentration (μmol g^{-1} FW)	0.1[d]	0.2[d]	6[a]	0.8[c]	5[a,b]	4.1[b]
Relative *Lsp5cs* expression	100[d]	37[e]	1354[a]	235[c]	616[b]	660[b]

Data in each raw followed by the same letter are not statistically different (P<0.05). Data from (Jahromi et al. 2008)

Table 13.6 Concentration of soluble sugars, reducing sugars, total free amino acids and total organic acids in leaves of maize plants inoculated with *Glomus mosseae* and grown at three NaCl levels

	Control		0.5 g Kg^{-1}		1.0 g Kg^{-1}	
	−AMF	+AMF	−AMF	+AMF	−AMF	+AMF
Soluble sugars (mg g^{-1} DW)	24[c]	40[a]	18[d]	32[b]	15[d]	25[c]
Reducing sugars (mg g^{-1} DW)	16[d]	38[a]	13[c]	30[b]	8[f]	20[c]
Total free amino acids (mg g^{-1} DW)	1[c]	0.5[cd]	4[b]	0.4[d]	8[a]	0.2[d]
Total organic acids (mg g^{-1} DW)	28[b,c]	40[a]	22[c]	31[b]	38[a]	40[a]

Data in each raw followed by the same letter are not statistically different (P<0.05). Data from (Sheng et al. 2011)

of AM plants was higher than nonAM plants (Table 13.6) (Sheng et al. 2011). Similar results were obtained in mungbean (Rabie 2005) and maize (Feng et al. 2002). Increased sugar concentration in the roots of AM plants was also reported in soybean subjected to drought stress (Porcel and Ruiz-Lozano 2004). However, some authors observed no role of soluble carbohydrates in the responses of AM plants to salinity (Sharifi et al. 2007).

In general, the increase in sugar content is found to be positively correlated with mycorrhization of plants. The high levels of sugars in mycorrhizal plants may be the result of an increase in photosynthetic capacity (Sheng et al. 2008; Wu et al. 2010). Symbiotic interactions in AM associations are based on the exchange of carbohydrates and mineral nutrients between the plant and the fungus. Using AM and nonAM clover plants of comparable plant size and growth rate and with similar N and P contents, it has been demonstrated that AMF colonization stimulates photosynthesis in order to compensate for the carbon requirement of the fungus and to eliminate growth reduction of plant (Wright et al. 1998).

The consumption of carbon by AMF can be up to 20% of the host photosynthates. Therefore, plant roots become a strong sink for carbohydrates when colonized by AMF and mycorrhizal sink strength influences the whole plant carbon balance. In conclusion, the requirement for carbohydrates by AMF could cause an increased allocation to and accumulation of soluble sugars in the roots (Wright et al. 1998; Lerat et al. 2003). This higher accumulation of soluble sugars in AM plant tissue, especially in roots, could make AM plants more resistant to osmotic stress induced by exposure to salt. Moreover, the increased sugar accumulation may also be due to hydrolysis of starch to sugars in the seedlings inoculated with AMF (Nemec 1981). In conclusion, AM symbiosis competes strongly for root-allocated carbon resulting in an enhanced allocation of carbohydrates to roots for AM growth and development as well as for protection of membranes and proteins.

Trehalose is a non-reducing disaccharide and the main storage carbohydrate in AMF has been found to play an important role as a stress protectant that stabilizes dehydrated enzymes and membranes and protects biological structures from desiccation damage (Paul et al. 2008). It is present in the extraradical mycelium and spores of AMF (Becard et al. 1991) but is rare in higher plants. Induction of trehalose accumulation in the roots after AMF colonization has been reported (Schubert et al. 1992). Although trehalose metabolizing enzymes showed a transient activation at 500 mM NaCl in extraradical hyphae of *Glomus intraradices*, trehalose content did not show any change (Ocón et al. 2007). In contrast, in nodulated pigeonpea plants, salinity led to an increase in trehalose-6-P synthetase and trehalose-6-P phosphatase activities resulting in increased trehalose content in nodules, which was accompanied by inhibition of trehalase activity, the catabolizing enzyme. Arbuscular mycorrhizal pigeonpea plants had lower trehalase activity under both control and saline conditions (Garg and Chandel 2011). More investigations are required for study of the accumulation of trehalose in extraradical hyphae and mycorrhizal roots in order to evaluate the potential of trehalose in protecting AM plants from salt stress.

Organic Acids

The regulation of organic acid metabolism also plays a key role in plant tolerance to saline conditions (Guo et al. 2010). It is well known that organic acids, as metabolically active solutes, play a role in osmotic adjustment, in the balance of cation excess (Hatzig et al. 2010), and in pH homeostasis (Hasegawa et al. 2000; López-Bucio et al. 2000; Yang et al. 2007; Hatzig et al. 2010). They prevent toxic chloride accumulation in cells and are important osmolytes in plant vacuoles (Guo et al. 2010). Under salt stress, increased citric acid concentration in alfalfa roots (Francoise et al. 1991), and increased salicylic acid synthesis in the leaves and roots of maize plants (Szalai and Janda 2009) have been reported. Arbuscular mycorrhizal symbiosis induces accumulation of organic acids in maize leaves (Sheng et al. 2011)

(Table 13.6). However, the effect of AM symbiosis was different depending on organic acids. Concentrations of oxalic, fumaric, acetic, malic, and citric acids increased, formic and succinic acid concentrations decreased while lactic acid concentration did not significantly affect (Sheng et al. 2011). On the other hand, AMF colonization changed the concentration of organic acids in root exudates (Zhang et al. 2003). Release of organic acids into rhizosphere of AM plants caused reduction of soil pH, EC and organic carbon, and an increase in the availability of plants to soil N, P, and K (Usha et al. 2004).

Polyamines

The three main polyamines found in plants putrescine (Put), spermidine (Spd) and spermine (Spm) are thought to play an important role in plant responses to a wide array of environmental stresses such as salinity, high osmolarity, hypoxia and oxidative stress (Bouchereau et al. 1999; Groppa and Benavides 2008). Exogenously added Spd and Spm protect plants from saline stress (Chattopadhyay et al. 2002) whereas transgenic plants overexpressing Spd and Spm biosynthetic enzymes are more tolerant to saline and hyperosmotic stress (Kasukabe et al. 2004). Similar pathways for Put synthesis to those described in plants and bacteria have been found in ectomycorhizae (Fornalé et al. 1999) and in an AM fungus (Sannazzaro et al. 2004). Information regarding polyamines in AMF or in their symbiotic interaction with plants is very limited. Free polyamines have been suggested to play an important role in the initial stages of the infection of pea roots by *Glomus intraradices* (Ghachtouli et al. 1995).

Mycorrhization changes the polyamine balance of salt-affected *Lotus glaber* plants. Colonization by *Glomus intraradices* increased (Spd+Spm)/Put ratio in lotus roots. This increment in salt stressed AM plants was even higher than those produced by salinization or AM symbiosis separately, suggesting an additive effect of both factors on the root (Spd+Spm)/Put ratio (Sannazzaro et al. 2007). It has been proposed that modulation of polyamine pools is one of the mechanisms used by AMF to improve adaptation of plants to saline soils (Sannazzaro et al. 2007).

Effect of AMF on Antioxidant Defense Capacity of Salinized Plants

One of the earliest responses of plants to salinity is the accumulation of reactive oxygen species (ROS) (Hasegawa et al. 2000; Parida and Das 2005). During salt stress, excessive generation of ROS such as superoxide radical ($O_2^{\bullet-}$), hydrogen peroxide (H_2O_2), hydroxyl radical ($^{\bullet}OH$), and singlet oxygen ($^1O_2$) occurs. These cytotoxic activated oxygen species can seriously disrupt normal metabolism through oxidative damage to membranes, proteins and nucleic acids (Apel and Hirt 2004).

Plant cells contain protection mechanisms that can minimize oxidative damage caused by ROS. The induction of ROS-scavenging enzymes, include superoxide dismutase (SOD), catalase (CAT), peroxidase (POD) and ascorbate peroxidase (APX) is the most common mechanism for detoxification of ROS synthesized during stress response. The steady-state levels of ROS in plants cells are determined by the balance between the generation of ROS and activities of scavenging enzymes (Apel and Hirt 2004).

Under salt stress, the activity of antioxidant enzymes becomes higher in order to eliminate more ROS (Parida and Das 2005). A strong correlation between the efficiency of antioxidant defense system and the level of salt tolerance has been reported in many plants (Benavides et al. 2000; Garratt et al. 2002). A constitutively high antioxidant capacity under stress conditions can prevent damages due to ROS formation (Harinasut et al. 2003). There are reports showed a greater SOD activity in salt tolerant compared with salt sensitive plants (Benavides et al. 2000). The increased POD activity in response to salinity (Harinasut et al. 2003) and higher POD activity in tolerant plants was also reported (Sreenivasulu et al. 1999). In *Solanum pennellii*, the wild salt tolerant tomato species, SOD and APX activity was higher than those in the cultivated tomato (Shalata and Tal 1998). Compared with cultivated tomato (*Solanum lycopersicum*), the better protection of *S. pennellii* root plastids from salt induced oxidative stress was correlated with increased activities of SOD, APX and POD (Mittova et al. 2002).

Apart from the salinity-induced antioxidative response, mycorrhization *per se* could also induce activation of antioxidant defense enzymes. In spite of the symbiotic nature of AM association, it represents a massive invasion of plant roots by the fungi. Induction of antioxidant enzymes observed during appressoria formation during the early stage of symbiosis development was attributed to a defense response of plants (Blilou et al. 2000). Oxidative burst and generation of superoxide radicals occur also during development of hypersensitive response in plant–pathogen interactions (Mehdy et al. 1996). Superoxide dismutase which plays a role in detoxification processes by catalyzing the conversion of free $O_2^{\bullet-}$ to O_2 and H_2O_2, very often is associated with plant-pathogen interactions (Davies and Dow 1997). Stimulation of constitutive SODs in different AM symbiosis and induction of specific SOD isoforms has been reported (Arines et al. 1994; Niki et al. 1998). Mycorrhizal clover roots exhibit two additional SOD isoforms as compared to nonAM roots: a myc-CuZn-SOD and a Mn-SOD (Palma et al. 1993). There are reports on general stimulation of CAT, POD, APX and SOD in AM compared to nonAM roots (He et al. 2007; Hajiboland et al. 2010). In bean (*Phaseolus vulgaris*) colonized by *Glomus clarum*, SOD and CAT were induced in roots at late stage of the symbiosis development under low P (Lambais et al. 2003). Cell wall bound peroxidase was measured in *Allium porrum* during root growth and development of *Glomus versiforme* (Spanu and Bonfante-Fasolo 1988). At initial stage of fungal infection, the enzyme activity was maximum and decreased in later stages when roots were highly colonized. In *Phaseolus vulgaris* inoculated with *Glomus etunicatum* peroxidase activity increased in the AM plants (Pacovsky et al. 1991).

Table 13.7 Activity of antioxidant enzymes (U mg^{-1} FW) and concentration of malondialdehyde (MDA, nmol g^{-1} FW) in the leaves of tomato (*Solanum lycopersicum* cv. Zhongzha105) plants inoculated with *Glomus mosseaes* and grown at different levels of salinity

		SOD activity	APX activity	POD activity	CAT activity	MDA content
Control	−AMF	3.2^{c}	$0.20^{d,e}$	1.8^{d}	6.3^{c}	75^{c}
	+AMF	7.1^{a}	0.34^{b}	2.1^{d}	11.3^{a}	60^{d}
50 mM	−AMF	4.0^{bc}	0.29^{d}	2.5^{cd}	9.7^{b}	110^{b}
	+AMF	6.0^{b}	$0.39^{a,b}$	4.2^{a}	11.5^{a}	99^{c}
100 mM	−AMF	1.7^{e}	0.32^{c}	3.0^{c}	4.0^{d}	160^{a}
	+AMF	2.8^{d}	0.42^{a}	3.9^{b}	$5.8^{c,d}$	100^{c}

Same letter within each column indicates no significant difference among treatments. From Abdel-Latef AAH, Chaoxing H (2011) Effect of arbuscular mycorrhizal fungi on growth, mineral nutrition, antioxidant enzymes activity and fruit yield of tomato grown under salinity stress. Sci Hort 127:228–233, with permission

Reports on the response of antioxidant defense system in AM plants under saline conditions are contradictory; increase, no change, or even decrease in the activity of enzymes have been reported in AM plants subjected to salinity. Indeed the response of the individual enzymes varies with respect to the host plant and the fungal species as well as duration and level of salinity treatment. Nevertheless, in many reports higher activity of antioxidant defense enzymes was associated with growth amelioration of AM plants under salinity. In pearl millet plants grown under salinity, enhanced SOD activity in AM plants as compared to nonAM ones supports the view that increased antioxidative enzyme activities could be involved in the beneficial effects of AM colonization on the performance of plants under saline conditions (Borde et al. 2011). In addition, gradual exposure of AMF to salinity that caused salt adaptation in the fungi enhanced its ability to colonize plant roots under saline conditions. This enhancement was correlated with induction of SOD activity in adapted fungi as compared to non adapted fungi (Borde et al. 2011). In tomato plants cultivated in soil with 0, 50 and 100 mM NaCl, AMF colonization alleviated salt induced reduction of growth and fruit yield (Abdel-Latef and Chaoxing 2011), that was accompanied by an enhancement of activity of SOD, CAT, POD and APX in leaves of both salt-affected and control plants. In addition, inoculation with AMF caused reduction in MDA content in comparison to salinized plants, indicating lower oxidative damage in the colonized plants. Similar results have been obtained by many other authors (He et al. 2007; Kohler et al. 2009; Zhi et al. 2010; Hajiboland et al. 2010) confirming that AM symbiosis really influences activity of antioxidant defense system and support the view that AMF can contribute to protect plants against salinity by alleviating the salt induced oxidative stress (Table 13.7). Enhanced antioxidant enzymes activity and lower lipid peroxidation in AM plants may contribute to better maintenance of the photochemical reactions in leaves under salinity. In addition, this mechanism improves salt tolerance via maintaining membrane integrity that would facilitate compartmentalization within vacuole and selective ion uptake. In conclusion, activation of antioxidant capacity of host plant by AMF symbiosis may be the result of common response to invasion of fungi and may also be because of a complex interaction of AMF, plant and salinity and need molecular analyses.

Effect of Different Plant Genotypes and Fungi Isolates on the Alleviating Effect of AMF in Salinized Plants

Symbiotic association between AMF and their hosts is usually believed to be nonspecific (Smith and Read 2008), however, many studies have confirmed the existence of differences in the physiological characteristics of mycorrhizal association within the species and even within isolates of the AMF (Bethlenfalvay et al. 1989, 1997; Smith and Smith 1997). Differences in fungal behavior or characteristics in interacting with hosts have attracted attention of scientists in recent years in order to improve selection of efficient isolates or to understand functional diversity or ecological plasticity of the fungi (Camprubi and Calvet 1996; Johnson et al. 1997; Douds and Millner 1999).

It has been found that under long term saline stress, although the richness of AM fungal species decreased, some species were still able to survive due to adaptation with these edaphic conditions (Copeman et al. 1996; Camprubi and Calvet 1996; del Val et al. 1999). The AMF species or isolates that are able to survive in stressed edaphic environments are considered as tolerant species/isolates and may have a higher ability to improve growth of host plants than species or isolates from non-stress edaphic condition (Tian et al. 2004).

The host plant species, cultivar and growing conditions can also influence the effectiveness of AM symbiosis in nutrient uptake (Janos 2007). Similarly, alleviation of salt stress by a given AMF species or isolates is dependent on the host plant species or genotype.

Effect of Different AMF Isolates

It has been widely accepted that AMF are able to adapt to specific edaphic conditions (Brundrett 1991; del Val et al. 1999; Copeman et al. 1996). It is expected that an isolate from saline soil would have a higher capacity to promote plant growth under saline stress. Copeman et al. (1996) suggested that differences in fungal behavior and efficiency can be due to the origin of the AMF.

In a study on cotton plants colonized by *Glomus mosseaes* isolated from non-saline soil, growth of plants were promoted under saline stress without affecting Na and Cl concentrations (Tian et al. 2004). Although isolate from saline soil increased Na and Cl concentrations in this species at higher NaCl level, it also significantly increased plant dry weight (Table 13.8) (Tian et al. 2004). This suggests that saline soil AMF isolate had a different mechanism for improving the salinity tolerance of

Table 13.8 Effect of inoculation of cotton (*Gossypium arboreum* L, cv. Xin-lu-zao No. 1) plants with two different isolates of *Glomus mosseaes* on shoot and root dry weight (DW) and concentration of P, Na and Cl under different salinity levels

		Shoot DW (g pot^{-1})	Root DW (g pot^{-1})	Shoot P (mg g^{-1})	Shoot Na (mg g^{-1})	Shoot Cl (mg g^{-1})
Control	–AMF	6.01	1.65	1.02	0.35	5.87
	Isolate from saline soil	6.46	1.51	2.27	0.44	7.06
	Isolate from non-saline soil	6.44	1.56	1.95	0.40	6.46
1 g Kg^{-1} NaCl	–AMF	5.54	1.46	1.20	0.78	9.29
	Isolate from saline soil	5.89	1.33	1.77	1.03	12.06
	Isolate from non-saline soil	6.06	1.49	1.74	0.65	9.39
2 g Kg^{-1} NaCl	–AMF	4.91	1.21	1.45	0.94	10.59
	Isolate from saline soil	4.91	1.07	1.30	2.14	16.36
	Isolate from non-saline soil	5.81	1.54	1.65	0.97	11.15
3 g Kg^{-1} NaCl	–AMF	3.09	0.99	1.54	2.45	19.82
	Isolate from saline soil	4.06	0.77	1.42	3.72	29.08
	Isolate from non-saline soil	5.19	1.21	1.71	2.49	18.95
Inoculation		***	**	***	***	***
Salt		***	***	***	***	***
Inoculation × salt		***	n.s	***	**	***

***Significant at $P<0.001$, **significant at $P<0.01$, n.s no difference

Same letter within each column indicates no significant difference among treatments. From Tian CY, Feng G, Li XL, Zhang FS (2004) Different effects of arbuscular mycorrhizal fungal isolates from saline or non-saline soil on salinity tolerance of plants. Appl Soil Ecol 26:143–148, with permission

cotton plants than the isolate from non-saline soil. Different species of AMF differentially affected growth and yield of wheat genotypes under salinity (van der Heijden et al. 1998a, b; Scheublin et al. 2004; Daei et al. 2009). The efficiency of different AMF species on enhancing plant growth under salinity was observed in the following order: *Glomus etunicatum* > *G. mosseae* > *G. intraradices* (Fig. 13.6). Higher root colonization by *Glomus etunicatum* and *G. mosseae* relative to *G. intraradices* resulted in increased nutrient uptake and less Na and Cl absorption by plant, and hence, increased plant growth under salinity (Daei et al. 2009).

The different behavior of AMF was also evidenced by Ruiz-Lozano et al. (1996) in lettuce plants subjected to salt stress and inoculated with three different AMF. In that study, the effect of *G. mosseae* and *G. fasciculatum* on salt tolerance seemed to be based on increased photosynthetic rate and water use efficiency rather than on

Fig. 13.6 Mycorrhizal responsiveness (%) calculated on the basis of shoot dry weight or grain yield in two cultivars of wheat plants inoculated with three different AMF species (From Daei G, Ardekani MR, Rejali F, Teimuri S, Miransari M (2009) Alleviation of salinity stress on wheat yield, yield components, and nutrient uptake using arbuscular mycorrhizal fungi under field conditions. J Plant Physiol 166:617–625, with permission)

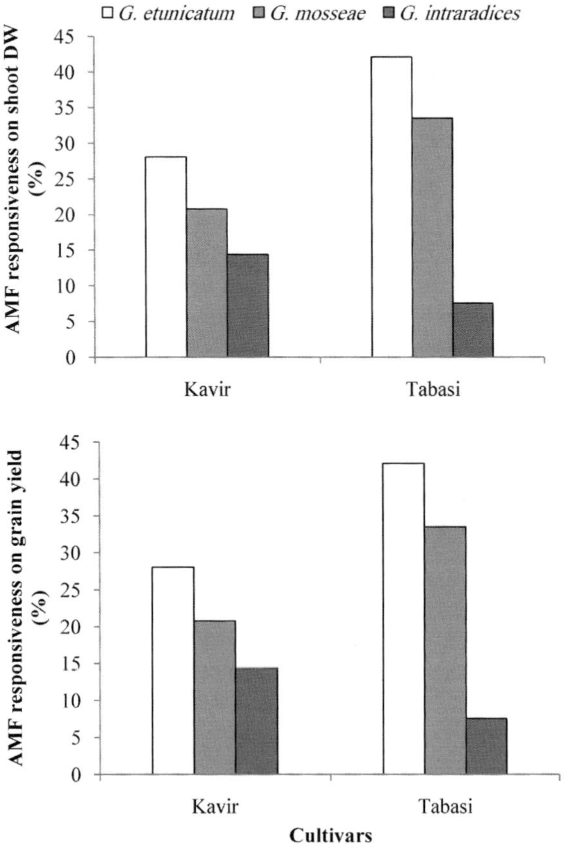

nutrient uptake (N or P). *Glomus deserticola*, in contrast, seemed to protect host plants against salt stress by increasing P uptake, in addition to the above-mentioned physiological processes. In another study on lettuce plants, *Glomus* sp. isolated from saline soils protected the host plant against salinity by stimulating root growth, while in the case of *G. deserticola,* the increase in N and P accumulation was the basis for alleviating effect under salinity (Ruiz-Lozano and Azcón 2000).

Study on *Astragalus sinicus* plants confirmed that AMF species are different in their efficiency for alleviation of salt stress. Tolerance of *A. sinicus* plants to salinity was enhanced by the three AMF species with increasing salinity levels, however, different AMF species varied in their ability to ameliorate the inhibitory effect of salt stress. This difference was reflected in plant growth response, intensity of root colonization and activity of succinate dehydrogenase and alkaline phosphatase in intraradical mycelium (Peng et al. 2011). The symbiosis formed by *G. intraradices* under increasing salinity levels were more efficient compared to that formed by *G. mosseae* or *G. claroideum* (Peng et al. 2011).

In the study on olive plants, *G. mosseae* was the most efficient AM fungus in terms of olive tree performance, and particularly in the protection offered against the detrimental effects of salinity (Porras-Soriano et al. 2009). Arbuscular mycorrhizal fungi differ in their ability to enhance nutrient uptake even when the extent of AM colonization is similar (Ruiz-Lozano and Azcón 2000). Specific mechanisms conferring functional differences among AMF could be expected from changes in fungal characteristics such as length of external mycelium, hyphae distribution, and/or nutrients translocation. In agreement with these ideas, Jakobsen et al. (1992) proposed a consistent relationship between the total length of root colonized by each fungus and the ability to alleviate the nutrient limitation effect caused by salinity.

In a study on soybean plants it was shown that exposure of the AMF to gradually increased concentrations of NaCl prior to salt stress enhanced root colonization, host plant growth, root proline and concentration of some mineral nutrients under salinity (Sharifi et al. 2007). It was hypothesized that pre-treatment of AMF by salt may result in salt acclimation that may in turn enhance the ability of the AMF to infect the roots, induction of root proline accumulation and mineral acquisition (Sharifi et al. 2007).

In contrast, other authors demonstrated that beneficial effect of AMF on host plants under saline conditions is not necessarily related to the level of salinity in the habitat of a given fungal isolate. Similar results were obtained in tomato plants inoculated with saline soil isolates of *G. mosseae* and *G. fasciculatum* (Poss et al. 1985). Copeman et al. (1996) found that an isolate from non-saline soil promoted shoot growth, but tended to increase leaf Cl concentration. Conversely, an isolate from saline soil suppressed plant growth but decreased the concentration of Cl in leaves. They suggested that although AMF originating from saline soil did not promote plant growth, reduction in leaf Cl concentration by these AMF may have beneficial implications for plant survival in saline soil (Copeman et al. 1996).

In lettuce plants, when the mycorrhizal responses were expressed per unit of mycorrhiza formed, *G. deserticola* and a native isolate from saline soil (*Glomus* sp.) differed in their symbiotic efficiency particularly under higher salinity levels (Ruiz-Lozano and Azcón 2000). The total mycorrhizal root length was highest in *Glomus* sp.-colonized plants, but plants colonized by *G. deserticola* responded more pronouncedly to inoculation. The superior ability of *G. deserticola* for improving plant growth and mineral nutrition was due to a higher rate of spread of extraradical mycelium than that of *Glomus* sp. (Ruiz-Lozano and Azcón 2000). These results indicate that specific compatibility relationships exist among symbionts, and AM symbiotic efficiency is dependent on AMF species. It implies also that, the mycorrhizae formed by *G. deserticola* was more efficient in improvement of host plants growth than that formed by *Glomus* sp. and *Glomus* sp. needed to form a higher amount of mycorrhiza than *G. deserticola* to achieve a similar symbiotic effect (Ruiz-Lozano and Azcón 2000). Authors suggested that the selection of the most suitable AMF for a specific plant genotype is of practical interest for improving the effectiveness under particular environmental conditions.

Regarding different ability of AMF to minimize stress effects and to promote plant growth (Daei et al. 2009), it has also been suggested that establishment of mixed communities by different AMF species may be more beneficial to the growth of plants than any of individual species (Koide 2000; Alkan et al. 2006; Peng et al. 2011). Root colonization of *G. mosseae* was enhanced in the presence of the other two fungi in a mix inoculation treatment particularly under high salinity conditions. There was likely a synergistic interaction between different AMF species under salt stress probably as the result of a functional complementation in P acquisition as suggested by Koide (2000). Selection of an AMF species/strain adapted to the local climate and soil conditions is the first step for a successful restoration program (Dodd and Thomson 1994).

Effect of Plant Genotypes

Salinity tolerance varies widely among species or ecotypes of the same species (Beauchamp et al. 2009). Some works also suggest that seeds collected from highly saline soils exhibit often higher establishment and performance in the presence of NaCl than seeds originated from non-saline soils (Hajiboland et al. 2010).

Study on *Lotus glaber* plants showed that *G. intraradices* established a more efficient symbiosis with the salt-tolerant than with the salt-sensitive genotype (Sannazzaro et al. 2006). Saline conditions reduced colonization rate of roots in sensitive but not in tolerant genotype, suggesting that the tolerant genotype offered the fungal partner higher protection and better chances of growth within host tissues than salt sensitive genotype. On the other hand, tolerant genotype inoculated with *G. intraradices* had not only higher root and shoot growth under saline conditions, but they also had higher leaf water and chlorophyll concentration as well as higher shoot:root ratio regardless of salinity level (Sannazzaro et al. 2006).

In tomato plants, the salt-tolerant cultivar showed higher AM colonization than the salt-sensitive cultivar (Al-Karaki et al. 2001). However, the enhancement in P, K, Zn, Cu, and Fe acquisition due to AMF inoculation was more pronounced in salt-sensitive than in salt-tolerant cultivar under saline conditions. These results suggest that although the salt-tolerant cultivar were highly infected with AMF, salt-sensitive cultivars benefited more from AMF colonization than salt-tolerant cultivar under saline soil conditions (Al-Karaki et al. 2001). In contrast, another report on tomato plants demonstrated that though similar root colonization, mycorrhizal responsiveness was greater in salt-tolerant compared with salt-sensitive cultivar likely because of greater photosynthesis rate under salinity in tolerant cultivar that could adequately provide carbohydrates for the fungi partner and result in more benefit of plants from AMF association (Hajiboland et al. 2010).

It has been stated that different wheat cultivars are able to perform differently irrespective of their mycorrhizal symbioses (Hetrick et al. 1984), however, other researchers indicated that the higher root colonization and hence, higher nutrient uptake, are the most important reasons for the greater performance of mycorrhizal tolerant varieties under salinity (Miransari 2011; Mardukhi et al. 2011). In an

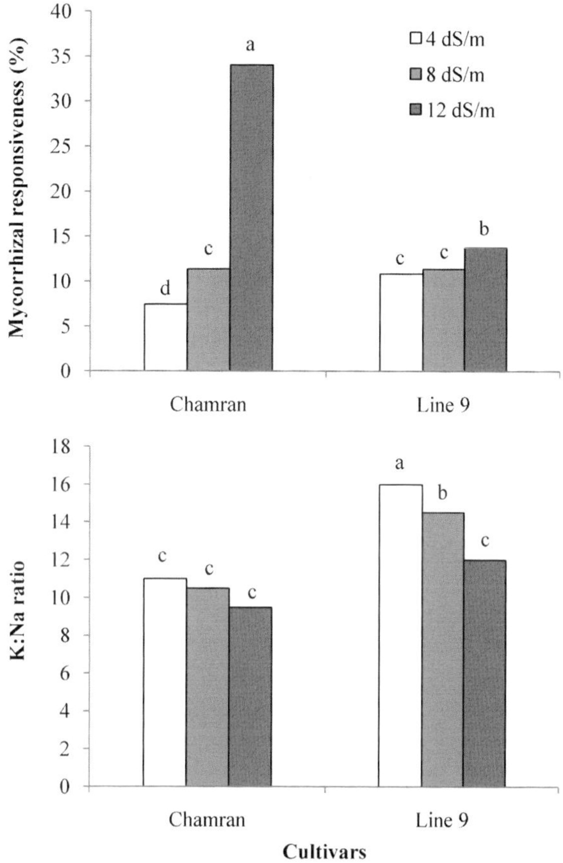

Fig. 13.7 Mycorrhizal responsiveness and K:Na ratio in two cultivars of wheat plants inoculated with AMF. Cultivar 'Chamran' is a semi- resistant and Line 9 is a resistant cultivar to salinity. *Bars* indicated by the same letter are not significantly different ($P<0.05$) (Data from (Mardukhi et al. 2011))

experiment on wheat plants grown under greenhouse conditions, two cultivars with different salinity resistance responded differently to mycorrhization under salinity (Mardukhi et al. 2011). Though a lower colonization in the roots of salinity resistant cultivar, higher nutrient uptake and K:Na ratio were observed in this cultivar. In contrast, regarding dry matter production mycorrhizal responsiveness was higher in less-tolerant cultivar (Fig. 13.7). Higher nutrient uptake in resistant cultivar under salinity despite of lower root colonization demonstrated that AMF association of salt resistant cultivar was more efficient than less-tolerant cultivar (Mardukhi et al. 2011).

Difference in symbiotic efficiency is also reflected in the mycorrhizal dependency index of plants grown under saline conditions. Mycorrhizal dependency in the salt tolerant genotype of lotus plant was similar under saline and non-saline conditions, while it was much greater under saline conditions compared with control in the salt sensitive genotype (Sannazzaro et al. 2006).

In conclusion, results of many studies demonstrated that different combinations of plant species/genotypes and AMF species/isolates can perform differently under salinity stress. Statistically significant interaction between plant genotypes and AMF species (Al-Karaki et al. 2001; Sannazzaro et al. 2006; Hajiboland et al. 2010; Mardukhi et al. 2011) emphasizes the importance of selecting the right combination of AMF species and host plant under salinity stress in order to achieve more efficient alleviation of the stress.

One of the causes for genotypic differences in the AMF responsiveness under salinity is likely different translocation of photosynthates to the roots under stress. Higher photosynthates allocation to the roots results in higher root dry weight and colonization and hence, AM symbiosis in some genotypes (Miransari and Smith 2007, 2008; Miransari et al. 2007, 2008).

On the other hand, non-dependence differences in the mycorrhizal responsiveness such as variation in the ability of genotypes to establish colonization, in the efficiency of water and nutrient uptake and exchange between fungus and host plant under salinity are very important traits for enhancement of profitability of plants-AMF interactions under saline conditions. The right combination of AMF species and host plant can partially or completely alleviate the stress of salinity, thus, knowledge on the relationship between plants and the fungi is important for successful utilization of AMF under particular conditions. Detailed studies are needed on the nature of difference between various combinations of plant species/genotypes and AMF species/isolates.

Changes in Gene Expression Patterns in Salt Stressed Plants upon Mycorrhization

The mechanisms underlying the alleviation of salt stress by AMF have not yet been elucidated at molecular level. Plant salt tolerance itself is a complex trait (Shi et al. 2000) and many different factors contribute in this process include production of compatible solutes, energy supply for the export of Na and Cl, specific transporters for the transfer of Na and Cl into the vacuole or apoplastic spaces, adequate water supply by aquaporins to maintain osmobalance (Hasegawa et al. 2000). Accordingly, any study on the impact of AMF colonization on the expression of genes with products involved in salt tolerance is faced with the multiplicity and complexity of the traits (Ouziad et al. 2006). Among the proteins functioning in salt tolerance of plants, Na/H transporters, aquaporins, proline biosynthetic enzyme and the expression of stress marker *LEA* gene have been investigated so far in AM plants at molecular level.

Na/H Transporters

Prevention of Na entry into the cell and/or sequestration of Na into the vacuole are strategies by which plants cope with high salinity. Sodium transporters contributing to Na homeostasis include Na/H antiporters in plasmamembrane (SOS1), Na/H antiporters in vacuole (NHX1) and Na uniporter in plasmamembrane (HKT1) (Zhu 2003). The Na/H antiporters mediate the transfer of Na out of the cytoplasm into either vacuole (NHX1) or apoplast (SOS1). There are six fully sequenced members of vacuolar Na/H transporters (Xia et al. 2002). Transgenic plants overexpressing vacuolar Na/H antiporters are more salt tolerant than the controls as shown for *Arabidopsis* (Apse et al. 1999; Gaxiola et al. 1999; Sottosanto et al. 2004), *Oryza sativa* (Fukuda et al. 1999) and *Brassica* (Zhang et al. 2001). Upregulation of tonoplast or plasmamembrane Na/H antiporter genes under salt stress has been reported in *Nicotiana excelsior* (Yamada et al. 1997), *Arabidopsis* (Gaxiola et al. 1999) or *Oryza sativa* (Fukuda et al. 1999), but was not observed in tomato (Ouziad et al. 2006) and in another work with *Arabidopsis* (Apse et al. 1999).

Analyzing expression of two tomato Na/H antiporter genes *LeNHX1* and *LeNHX2* showed no significant change in the expression due to colonization (Fig. 13.8) (Ouziad et al. 2006). This indicates that AMF colonization does not trigger the expression or activity of Na/H antiporter genes.

Aquaporins

Water molecules pass through the channels formed by aquaporins in the plasmalemma or the tonoplast (Maurel et al. 1997; Zeuthen 2001; Hill et al. 2004). Aquaporins belong to the major intrinsic protein (MIP) family of transmembrane channels, which permit selective membrane passage of water but not of H^+ and other ions (Weig et al. 1997; Chen et al. 2001; Hill et al. 2004). Root water uptake depends on root hydraulic conductance, which is ultimately governed by aquaporins (Luu and Maurel 2005). Plants aquaporins are divided in four groups depending on their sequence homology. These four groups are called plasma membrane intrinsic proteins (PIPs), tonoplast intrinsic proteins (TIPs), nodulin like intrinsic proteins (NIPs) and small and basic intrinsic proteins (SIPs).

There is some evidence that PIP proteins could regulate the whole water transport through plant tissues, and plants overexpressing or lacking one or more *PIP* genes have more or less root water uptake capacity, respectively (Aharon et al. 2003; Javot et al. 2003). Aquaporins are controlled at activity or transcriptional levels. As a short-term response to stresses such as drought and salinity, aquaporins activity is regulated by phosphorylation (Johansson et al. 1996; Maurel et al. 1995) while during longer time they are down-regulated by reduction of gene expression.

Overexpression of a *PIP* aquaporin in transgenic tobacco improved plant vigor under favorable growth conditions, but had not beneficial effect under salt stress and influenced even negatively plants growth under drought stress and accelerated plants wilting (Aharon et al. 2003). Similar result has been obtained in *Arabidopsis* and tobacco plants regarding two different *PIP* aquaporin genes (Jang et al. 2007).

During AM formation, the plant plasmamembrane extends to form the periarbuscular membrane. Periarbuscular membrane is closely surrounds the fungal hyphae and results in a three to tenfold increase in plant cell surface (Gianinazzi-Pearson 1996). Upregulation of aquaporins in AM plants under well-watered conditions

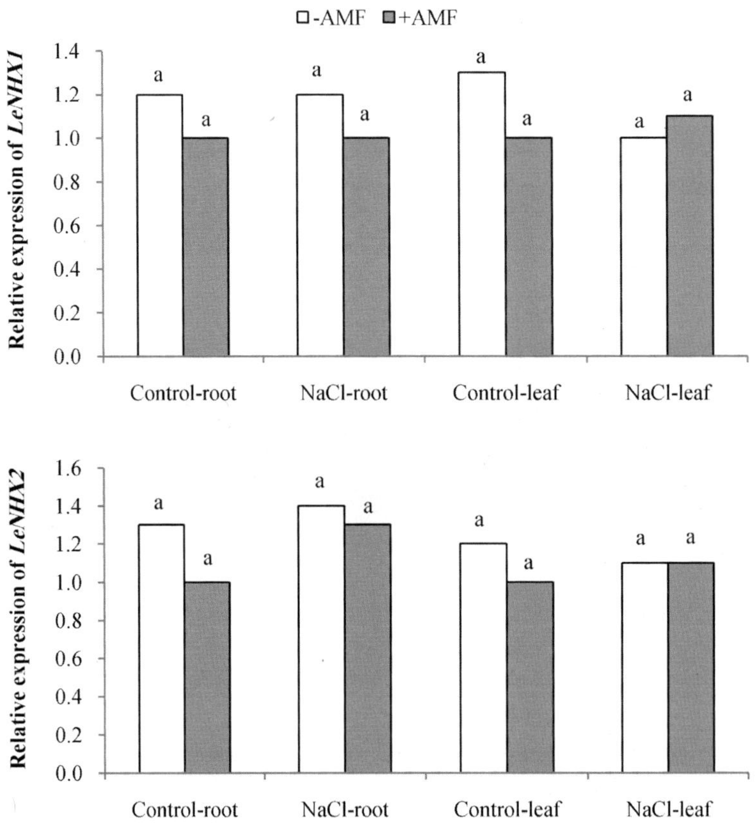

Fig. 13.8 Expression of two Na^+/H^+ antiporters (*LeNHX1* and *LeNHX2*) of tomato (*Solanum lycopersicum* cv. Tmina) plants grown under salinity and colonized with a mixture of *Glomus geosporum* and *Glomus intraradices*. *Bars* indicated by the same letter are not significantly different ($P<0.05$).(Data from (Ouziad et al. 2006))

optimizes nutrients and water exchange between two symbiotic partners (Krajiski et al. 2000).

Arbuscular mycorrhizal plants are able to take up more water from the soil than nonAM plants under water deficit conditions (Marulanda et al. 2003; Khalvati et al. 2005). In lettuce plants, *PIP2* gene expression declined by drought stress while AM inoculation by itself increased expression of the *PIP2* gene under both watering regimes (Alguacil et al. 2009). Porcel et al. (2006), in contrast, observed that under drought conditions, the AM colonization accelerated reduction of a *PIP* gene expression in roots of *Glycine max* plants and decreased the expression of two *PIP* genes in roots of *Lactuca sativa* plants. Aroca et al. (2006) reported also that, nonAM roots have higher expression of *PvPIP1;3* and *PvPIP2;1* compared with AM roots under drought conditions. In this study, colonization of *Phaseolus vulgaris* roots by AMF resulted in maintaining root hydraulic conductance under drought and prevented leaf dehydration as judged by higher relative water content of AM leaves compared with nonAM ones (Aroca et al. 2006). Lettuce plants colonized by *Glomus mosseae* showed down-regulation of *PIP* gene expression (Porcel et al. 2006). These authors concluded that down-regulation of plant aquaporins by AM symbiosis allows conservation of water in plant tissues under drought and confirms the beneficial effect of AMF in water status of host plants observed under drought conditions (Porcel et al. 2006; Aroca et al. 2006).

In a study on lettuce plants inoculated with *Glomus intraradices* and *Glomus mosseae* it was found that both fungi use different strategies to protect the host plant against drought stress (Marulanda et al. 2003). *Glomus intraradices* showed higher capacity to enhance the root water permeability, rate of root water uptake and movement by maintaining high levels of *PIP* aquaporin gene expression. *Glomus mosseae*, in contrast, protected plants against drought stress by greater conservation of the water existing in the plant and down-regulation of *PIP* genes expression. Down regulation of *PIP* genes has been suggested by other authors as a mechanism for reduction of membrane water permeability and to allow cellular water conservation (Yamada et al. 1995; Smart et al. 2001).

Because of negative water potential in saline soils, plants of saline habitats are faced with drought problem and must maintain their osmotic balance in the cytoplasm. There is a correlation between the expression or activity of aquaporins and the susceptibility to salt stress (Johansson et al. 2000).

The expression of aquaporins was strongly impaired by salt treatment in tomato plants (Ouziad et al. 2006). In ice plants, transcript levels of some aquaporins are down-regulated in the first 30 h after exposure to salt stress and recover when this stress is interrupted (Yamada et al. 1995). Salinity reduced either the activity or abundance of aquaporins in paprika pepper (Carvajal et al. 1999) and melon (Carvajal et al. 2000). Other authors, however, reported an up-regulation of aquaporin mRNA transcripts under salt stress, e.g., in *Nicotiana excelsior* (Yamada et al. 1997), *Arabidopsis thaliana* (Gaxiola et al. 1999), *Oryza sativa* (Fukuda et al. 1999) and *Beta vulgaris* (Xia et al. 2002). Such differences may be a consequence of the mode of the salt stress set, the differences between plant species and the complexity in expression pattern of different members of the large family of aquaporins (Sarda et al. 1999; Ouziad et al. 2006).

In works on *Phaseolus vulgaris* (Aroca et al. 2007) and *Lactuca sativa* (Jahromi et al. 2008), roots of mycorrhizal plants maintained or even decreased the expression of *LsPIP2* gene, whereas the expression of *LsPIP1* gene was up-regulated, particularly at 100 mM NaCl. In tomato plants, in contrast, transcript levels of both a tonoplast (*LeTIP*) and plasmamembrane (*LePIP1*) aquaporin genes in roots were reduced by AMF colonization. In contrast to the roots, in the leaves, AMF colonization resulted in a drastic increase of the mRNA of all three aquaporin genes (*LePIP1, LePIP2* and *LeTIP*) assayed under salt stress (Ouziad et al. 2006).

Inoculation with *Glomus intraradices* in the absence of salinity caused an inhibition of the expression of *LsPIP1* and *LsPIP2* genes (Jahromi et al. 2008) that was in

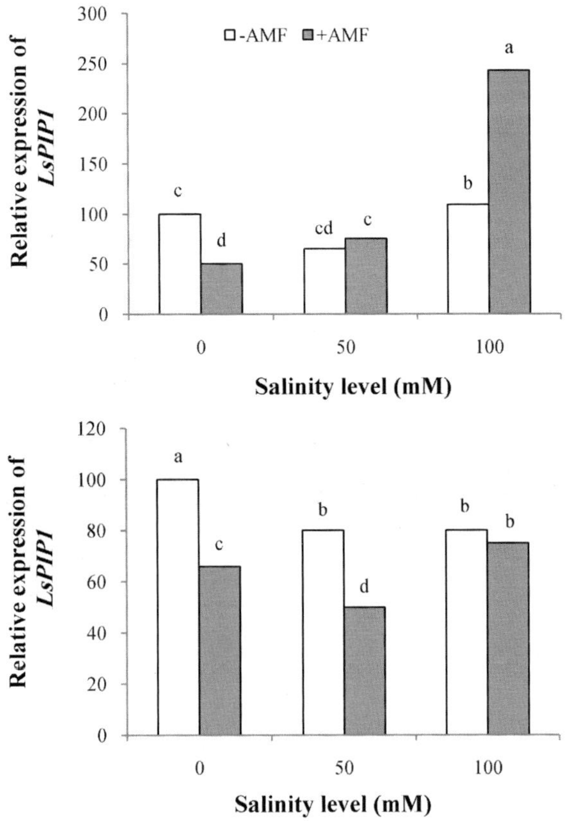

Fig. 13.9 Expression of *LsPIP1* and *LsPIP2* (plasma membrane aquaporins) gene probes from roots of lettuce (*Lactuca sativa*) plants subjected to 0, 50, or 100 mM NaCl and inoculated with Glomus intraradices. *Bars* indicated by the same letter are not significantly different (P<0.05) (Data from (Jahromi et al. 2008))

agreement with other authors (Porcel et al. 2006; Aroca et al. 2006). Under salinity conditions, however, the expression of *LsPIP2* gene maintained unaffected while *LsPIP1* gene was up-regulated (Fig. 13.9) (Jahromi et al. 2008). The latter finding was the opposite of that obtained for the same gene (*LsPIP1*) under drought conditions in plants inoculated with *Glomus mosseae* (Porcel et al. 2006). These results demonstrated that the same aquaporin gene responds differently to each AMF and the response depends also on the nature of applied osmotic stress (Ruiz-Lozano and Aroca 2010). Aroca et al. (2007) demonstrated also that, the gene *PvPIP1;2* was inhibited by three applied stresses in AM and nonAM plants in a similar way, while the gene *PvPIP1,3* showed important differences in AM and nonAM plants according to the stress imposed (Aroca et al. 2007).

The *NtAQP1* aquaporin of tobacco has been isolated and characterized as a plasma membrane intrinsic aquaporin (Biela et al. 1999). Effect of an impairment of *NtAQP1* gene expression on the AMF colonization pattern and symbiotic efficiency has been investigated (Fig. 13.10) (Porcel et al. 2005). Data from this study showed that impairment in *NtAQP1* gene expression did not influence AMF colonization ability. However, symbiotic efficiency was affected in *NtAQP1* antisense plants.

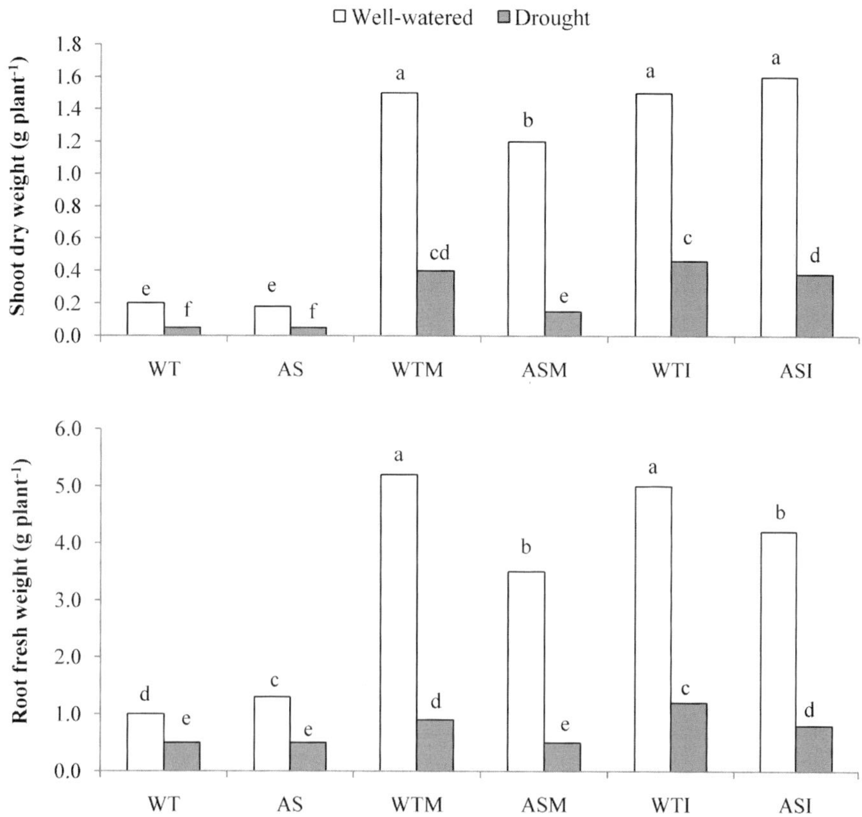

Fig. 13.10 Shoot dry weight and root fresh weight in wildtype (*WT*) or antisense (*AS*) tobacco plants inoculated or not with *Glomus mosseae* (*M*) or *Glomus intraradices* (*I*) and cultivated under well-watered conditions or subjected to drought stress. *Bars* indicated by the same letter are not significantly different ($P<0.05$) (Data from (Porcel et al. 2005))

Beneficial effect of AMF colonization was similar in wild type and antisense plants under well-watered conditions, while under drought stress mycorrhizal wild type plants grew more than mycorrhizal *NtAQP1* antisense plants. It indicates that the symbiotic efficiency of AMF was greater with wild type than with *NtAQP1* antisense plants (Porcel et al. 2005). Taken together, these results indicate that enhanced symplastic water transport via the plasma membrane aquaporin NtAQP1 is important for the efficiency of AMF symbiosis, at least under drought stress conditions.

AMF in Halophytes

In saline soils both the macrobiota (halophytes) and the microbiota (rhizosphere-constituents) usually adapt to the particular stress conditions (Ruiz-Lozano and

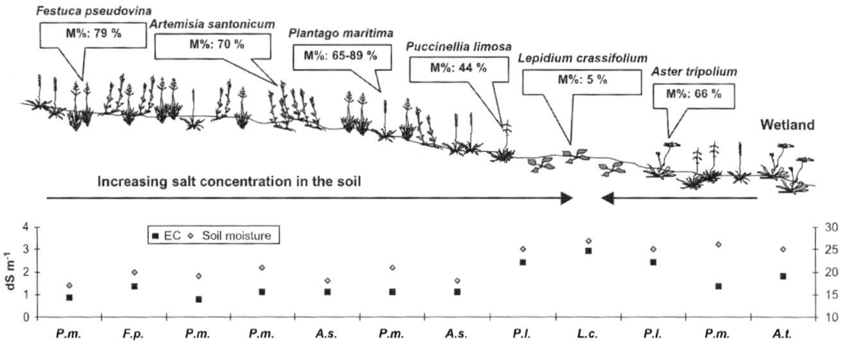

Fig. 13.11 The relative location of halophytes along the toposequence in Puszta at Apaj, Hungary. The percentage of mycorrhizal colonization was demonstrated as M% values. Soil moisture and electric conductivity were determined directly at the site in 10–20 cm soil depth close to the roots of the plants indicated. P.m., *Plantago maritima*; F.p., *Festuca pseudovina*; A.s., *Artemisia santonicum*; P.c., *Puccinellia limosa*; L.c., *Lepidium crassifolium*; A.t., *Aster tripolium* (From Füzy A, Biró B, Tóth T, Hildebrandt U, Bothe H (2008) Drought, but not salinity, determines the apparent effectiveness of halophytes colonized by arbuscular mycorrhizal fungi. J Plant Physiol 165:1181–1192, with permission)

Azcón 2000). It has been stated that high salinity in soils has adverse effects on germination and hyphal growth (Juniper and Abbott 1993, 2006) and colonization of plants by AMF (Juniper and Abbott 1993). However, there are reports from all over the world that plants of saline habitats can be colonized by AMF (Fig. 13.11) (Barrow et al. 1997; Landwehr et al. 2002; Plenchette and Duponnois 2005; Mathur et al. 2007; Asghari et al. 2008; Füzy et al. 2008).

Many halophytes belong to families like Caryophyllaceae, Chenopodiaceae and Plumbaginaceae, which are frequently reported as being nonAM (Harley and Harley 1987; Wang and Qiu 2006). Nevertheless, halophytes like *Aster tripolium* (Asteraceae), *Artemisia maritima* (Asteraceae) and *Plantago maritima* (Plantaginaceae) are intensively colonized by AMF (Harley and Harley 1987; Carvalho et al. 2001; Hildebrandt et al. 2001; Landwehr et al. 2002). Salt marsh plants such as *Spartina patens* and *Distichlis spicata* form also AMF association (Hoefnagels et al. 1993). These AMF associations could protect plants against the detrimental effects of ion toxicity and alleviate salt stress symptoms.

In a study on the biodiversity of halophytes and their mycorrhizal status in Seŏvlje salterns (Slovenia), fungal colonization was detected in the roots of *Salsola soda*, and *Salicornia europaea* and *Suaeda maritima* (Chenopodiaceae) *Plantago cornuti* (Plantaginaceae) and typical AM fungal structures were present in *Artemisia caerulescens*, *Aster tripolium* and *Inula crithmoides* (Asteraceae) and *Plantago cornuti* (Plantaginaceae) (Sonjak et al. 2009). This has been also shown for most of the halophytes collected from different salt marsh environments (Hildebrandt et al. 2001; Landwehr et al. 2002; Füzy et al. 2008), thus showing that AMF might have important roles in these extreme environments.

In a study on several salt marshes from the North and Baltic Sea and of German inland salt habitats, *Aster tripolium* and *Artemisia maritime* (Asteraceae), *Plantago maritima* and *P. coronopus* (Plantaginaceae) and *Oenanthe lachenalii* (Apiaceae) showed a high rate of AMF colonization, and low, though distinct, AMF infection was detected in samples of *Puccinellia maritime* and *P. distans* (Poaceae) and even of *Salicornia europaea* (Chenopodiaceae), at inland salt marshes, whereas other species like *Spartina anglica* (Poaceae), *Juncus gerardii* (Juncaceae), and *Triglochin maritimum* (Juncaginaceae) were nonAM (Hildebrandt et al. 2001). In a study on the two salt affected regions in Hungary, *Aster tripolium, Plantago maritime, Artemisia santonicum* and *Matricaria chamomilla* were the most heavily colonized plants (Füzy et al. 2008). In other survey on some areas of Turan Biosphere Reserve (TBR) in north east Iran that includes 1.8 million hectares of flat, semi-arid desert plains, halophyte species were studied for their mycorrhizal status (Asghari et al. 2008). Typical structures of AMF with different levels of colonization was observed in *Haloxylon aphyllum, Kochia stellaris, Halocnemum strobilaceum, Seidlitzia rosmarinus, Salsola sp.* (Chenopodiaceae) and *Zygophyllum eurypterum* and *Peganum harmala* (Zygophyllaceae). In this study, different levels of AM colonization were found in the same plant species from different locations but with the same salinity level. In addition, AM colonization in roots of halophytes existed at lower levels of salinity (<45 dSm^{-1}) while it was absent at higher salinity levels (>45–140 dSm^{-1}) (Asghari et al. 2008).

Chenopodiaceae are found in halophytic plant communities worldwide and include more halophytic species than other plant families. Although this family has been generally regarded to be nonAM (Hirrel et al. 1978; Reeves et al. 1979), there are reports on the occurrence of AM colonization in chenopods in the field (Johnson et al. 1995; Aguilera et al. 1998; O'Connor et al. 2001; He et al. 2002). A colonization rate of 60% in *Arthrocnemum indicum* with typical vesicles and arbuscules and of 48% in *Suaeda maritime* with vesicles were found in salt marshes of the Ganges delta in India (Sengupta and Chaudhuri 1990). In later works, the mycorrhizal status of numerous Chenopodiaceae, particularly the genus *Atriplex*, was confirmed (Hildebrandt et al. 2001; Plenchette and Duponnois 2005; Asghari et al. 2005), and inoculation of *Atriplex gardeneri* (Allen 1983), *A. nummularia* (Asghari et al. 2005) and *A. canescens* (Williams et al. 1974) with AMF was proven to enhance efficiently the growth and survival of this Chenopodiaceae.

Atriplex nummularia Lindl. A perennial chenopod was reported to have a relatively high level of AM colonization (10–30%) in spring and summer under field conditions (Asghari et al. 2005). In a glasshouse experiment, however, only low and patchy colonization (1–2%) was detected in inoculated *A. nummularia* plants. Despite low colonization rate, inoculation with AMF increased shoot dry weight, particularly before production of high density roots and depletion of P and other nutrients in the pots (Asghari et al. 2005). Factors characteristics of field conditions e.g. environmental factors, age and phenology of host plant (Wilson and Hartnett 1998), soil properties such as activity of hydrolytic enzymes (Mamatha et al. 2002) may be responsible for the high levels of colonization of *A. nummularia* under field versus glasshouse conditions (Asghari et al. 2005).

Colonization of halophytes is sometimes not accompanied by characteristic structures of AMF. On the other hand, results for a given species from different areas are contradictory. AMF colonization of *Salicornia europaea* (Chenopodiaceae) was reported to be usually very low, if detected (Harley and Harley 1987; Landwehr et al. 2002; Wang and Qiu 2006). No arbuscule was also detected in the roots of *Salicornia europaea* from Sečovlje salterns (Sonjak et al. 2009). In contrast, in the study of halophytes in central European salt marshes, Hildebrandt et al. (2001) demonstrated that colonization of individual specimens of this species is high and arbuscules are also present. This difference in the colonization level is likely the result of the species and subspecies composition of the genus *Salicornia* (Martinčič et al. 2007). Rare arbuscules were detected in the roots of *Arthrocnemum macrostachyum* (Chenopodiaceae) and *Limonium angustifolium* (Plumbaginaceae) and no arbuscules were seen in *Atriplex portulacoides* and *Beta vulgaris* L. subsp. Maritime (Chenopodiaceae) roots (Sonjak et al. 2009). The colonization of other halophytes from the Chenopodiaceae family including *Salicornia europaea*, *Salsola soda* and *Suaeda maritima* and *Spergularia marina* from the Caryophyllaceae family was determined only by the presence of hyphae without any arbuscules (Sonjak et al. 2009).

Indeed, there is no correlation between fungal structures frequency and fungal colonization rate of roots with efficiency of symbiosis and its influence on plant growth (Füzy et al. 2008; Smith and Read 2008; Tonin et al. 2001; Vogel-Mikuš et al. 2005). A high level of colonization is not necessarily associated with an extensive exchange of metabolites between the two symbiotic partners. A few active fungal structures in only a small number of roots may also help plants to cope with stress. On the other hand, results obtained from field sampling should be interpreted carefully because during the course of a year, the number of fungal structures particularly arbuscules (with a short half-life) vary significantly (Smith and Read 2008; Füzy et al. 2008).

Though a low diversity of AMF species, there is a high abundance of spores in inland and coastal habitats (Hildebrandt et al. 2001; Landwehr et al. 2002; Carvalho et al. 2004) indicating that saline soils are the sites where AMF thrive (Füzy et al. 2008). However, distribution of spores of AMF in saline soils and salt marshes is often patchy and highly variable from soil sample to sample (Landwehr et al. 2002).

Negative influence of soil salinity on spore germination and hyphal growth of AMF have been identified as being the most important reason for the absence of AMF colonization in halophytes (Juniper and Abbott 2006). However, in a study on some halophyte species it was shown that reducing soil salinity (0.4 dSm^{-1}) did not improve AM colonization (Asghari and Cavagnaro 2010). On the other hand, despite of low level of AM colonization, halophytes showed a positive growth response to AM inoculation in non-saline soil conditions (Asghari and Cavagnaro 2010). More studies are required to determine factors that influence AM colonization rate and responsiveness in halophytes.

Ecological Considerations and Use of AMF for Restoration of Saline Soils

Millions of hectares of land world-wide are affected by salinity. Restoration of salinized soils is a global concern. Arbuscular mycorrhizal fungi are considered as an important component of riparian ecosystem function (Beauchamp et al. 2009).

In semi-arid environments, stability of soil aggregates is important for supporting plants growth and in turn, protecting the soil against water erosion (Kohler et al. 2010). The AMF association contributes significantly to the stability of soil aggregates (Caravaca et al. 2005). Arbuscular mycorrhizal fungi influence the stability of macroaggregates (>250 mm) via hyphal enmeshment aggregates (Miller and Jastrow 2000), by deposition of organic substances (Bearden and Petersen 2000) and via production of the glycoprotein glomalin, which acts as an insoluble glue to stabilize aggregates (Rillig 2004; Gadkar and Rillig 2006). Arbuscular mycorrhizal colonization improves biochemical properties of rhizosphere and bulk soil, increases activity of catalase, neutral and alkaline phosphatases in soil, increases P solubility and decreases soil EC (Zhang et al. 2011).

The competitiveness of plants in saline soils is mainly determined by the level of salinity and drought, which favor species that can establish, grow to maturity, and reproduce under these conditions (Sonjak et al. 2009). Planting halophytes that have morphological and physiological adaptations to overcome osmotic and ionic stresses is often the only opportunity to produce forage for livestock in saline areas. The yield of such plantations could be enhanced by AM inoculation (Sonjak et al. 2009). Inoculation of sites with AMF is also needed when restoring historic floodplain areas following extended *Tamarix* occupations. *Tamarix* (Beauchamp et al. 2005) similar with some members of the Chenopodiaceae is not associated with AMF (Titus et al. 2002). Colonization rate and spore density of AMF in the rhizosphere soil of some ephemeral plant species such as *Chorispora tenella*, *Ceratocephalus testiculatus*, *Eremopyrum orientale* and *Veronica Campylopoda* growing in an area dominated by *Tamarix spp.* are significantly lower under the shrub canopies than beyond (Shi et al. 2006). During restoration process in areas with extensive and dense *Tamarix* occupation, therefore, AMF inoculation may improve the performance of native species over *Tamarix* sprouts or weeds such as *Kochia spp.* and *Salsola spp.* (Chenopodiaceae) (Johnson 1992).

Conclusion and Future Perspectives

Detailed studies are needed to elucidate biochemical, physiological and molecular mechanisms and signal transduction pathways involving in the salt alleviating effect of AMF symbiosis in plants. Results of many works suggest that AM plants are suffering from the salt stress less than nonAM plants. Higher relative water content, lower proline and ABA content and lower expression of the stress marker *LEA* gene in AM compared to nonAM plants reported for various plant species suggest that the AMF decreased salt stress injury. These results evidenced that some salt-avoidance mechanisms are likely to be improved due to mycorrhization.

The great gap in our knowledge seems to be the function of ion transporter systems operating in the symbiosis including direct pathways at the soil-root interface and the mycorrhizal pathway and the possible changes in their relative contribution under salinity. On the other hand, information on the properties of ion trafficking at arbuscule-plant cell interface is really limited. Biochemical and molecular study of Na exclusion mechanisms in the roots such as Na/H antiporters and cyclic-nucleotide-gated ion channels as well as mechanisms operating at xylem loading level may elucidate the extent to which AMF act via improvement of plants ability to exclude salts. These studies would allow understanding if the AM symbiosis affects Na uptake, distribution, compartmentation and allocation at the cellular and whole plant level.

Our knowledge of the molecular mechanisms for salt amelioration by AMF in plants is limited to the changes in the expression pattern of only a few genes. Identification of temporal and spatial patterns in the expression of genes involved in production of various antioxidant enzymes and antioxidant metabolites and enzymes controlling synthesis of various osmoregulators will provide further insights into molecular basis of the mechanisms.

Study of temporal and spatial expression patterns of different members of transporters, ion channels as well as aquaporin gene families and their localization in the shoot, root and particularly in the arbuscule-plant cell interfaces may also provide an integrative perspective on the role of specific molecules in the AM-plants interactions under salt stress. In addition of using molecular and fluorescence microscopy approaches, application of imaging techniques for localization of ions in the fungi-plant cell interaction spaces may also provide evidences on the nature of differences among the various combinations of plant genotypes/fungi isolates in the efficiency for allevation of salt stress.

References

Abbott LK, Robson AD (1984) Colonization of the root systems of subterranean clover by three species of vesicular-arbuscular mycorrhizal fungi. New Phytol 96:275–281

Abd-El Baki BGK, Siefritz F, Man HM, Weiner H, Kaldenhoff R, Kaiser WM (2000) Nitrate reductase in *Zea mays* L. under salinity. Plant Cell Environ 23:515–521

Abdel-Latef AAH, Chaoxing H (2011) Effect of arbuscular mycorrhizal fungi on growth, mineral nutrition, antioxidant enzymes activity and fruit yield of tomato grown under salinity stress. Sci Hort 127:228–233

Aguilera LE, Gutierrez JR, Moreno RJ (1998) Vesicular arbuscular mycorrhizae associated with saltbushes *Atriplex spp.* (Chenopodiaceae) in the Chilean arid zone. Rev Chil Hist Nat 71:291–302

Aguin O, Mansilla JP, Vilarino A, Sainz MJ (2004) Effects of mycorrhizal inoculation on root morphology and nursery production of three grapevine rootstocks. Am J Enol Vitic 55:108–111

Aharon R, Shahak Y, Wininger S, Bendov R, Kapulnik Y, Galili G (2003) Overexpression of plasma membrane aquaporins in transgenic tobacco improves plant vigour under favourable growth conditions but not under drought or salt stress. Plant Cell 15:439–447

Alguacil MM, Kohler J, Caravaca F, Roldán A (2009) Differential effects of *Pseudomonas mendocina* and *Glomus intraradices* on lettuce plants physiological response and aquaporin *PIP2* gene expression under elevated atmospheric CO_2 and drought. Microb Ecol 58:942–951

Aliasgharzadeh N, Saleh Rastin N, Towfighi H, Alizadeh A (2001) Occurrence of arbuscular mycorrhizal fungi in saline soils of the Tabriz Plain of Iran in relation to some physical and chemical properties of soil. Mycorrhiza 11:119–122

Alkan N, Gadkar V, Yarden O, Kapulnik Y (2006) Analysis of quantitative interactions between two species of arbuscular mycorrhizal fungi, *Glomus mosseae* and *G. intraradices*, by real-time PCR. Appl Environ Microbiol 72:4192–4199

Al-Karaki GN (2000) Growth of mycorrhizal tomato and mineral acquisition under salt stress. Mycorrhiza 10:51–54

Al-Karaki GN, Clark RB (1998) Growth, mineral acquisition, and water use by mycorrhizal wheat grown under water stress. J Plant Nutr 21:263–276

Al-Karaki GN, Hammad R, Rusan M (2001) Response of two tomato cultivars differing in salt tolerance to inoculation with mycorrhizal fungi under salt stress. Mycorrhiza 11:43–47

Allen MF (1983) Formation of vesicular-arbuscular mycorrhizae in *Atriplex gardneri* (Chenopodiaceae): seasonal response in a cold desert. Mycologia 75:773–776

Allen EB, Cunningham GL (1983) Effects of vesicular-arbuscular mycorrhizae on *Distichlis spicata* under three salinity levels. New Phytol 93:227–236

Allen MF, Smith WK, Moore TS, Christensen M (1981) Comparative water relations and photosynthesis of mycorrhizal and non-mycorrhizal *Bouteloua gracilis* H.B.K. Lag ex Steud. New Phytol 88:683–693

Amzallag GN (1997) Tolerance to salinity in plants: new concept for old problems. In: Jaiwal PK, Singh RP, Gulati A (eds) Strategies for improving salt tolerance in higher plants. Science, Enfield, pp 1–17

Apel K, Hirt H (2004) Reactive oxygen species: metabolism, oxidative stress, and signal transduction. Annu Rev Plant Biol 55:373–399

Apse M, Aharon G, Snedden W, Blumwald E (1999) Salt tolerance conferred by overexpression of a vacuolar Na^+/H^+ antiport in *Arabidopsis*. Science 285:1256–1258

Arines J, Quintela M, Vilarino A, Palma JM (1994) Protein patterns and superoxide dismutase activity in non-mycorrhizal and arbuscular-mycorrhizal *Pisum sativum* L. plants. Plant Soil 166:37–45

Aroca R, Ferrante A, Vernieri P, Chrispeels MJ (2006) Drought, abscisic acid and transpiration rate effects on the regulation of PIP aquaporin gene expression and abundance in *Phaseolus vulgaris* plants. Ann Bot 98:1301–1310

Aroca R, Porcel R, Ruiz-Lozano JM (2007) How does arbuscular mycorrhizal symbiosis regulate root hydraulic properties and plasma membrane aquaporins in *Phaseolus vulgaris* under drought, cold or salinity stresses? New Phytol 173:808–816

Asghari HR, Cavagnaro TR (2010) Mycorrhizal response of halophytes to plant growth in non-saline soil conditions. In proceedings book of the 19th world congress of soil science, soil solutions for a changing world, 1–6 Aug 2010, Brisbane

Asghari HR, Marschner P, Smith SE, Smith FA (2005) Growth response of *Atriplex nummularia* to inoculation with arbuscular mycorrhizal fungi at different salinity levels. Plant Soil 273:245–256

Asghari HR, Amerian M, Gorbani H (2008) Soil salinity affects arbuscular mycorrhizal colonization of halophytes. Pak J Biol Sci 11:1909–1915

Augé RM (2000) Stomatal behavior of arbuscular mycorrhizal plants. In: Kapulnik Y, Douds DD (eds) Arbuscular mycorrhizas: physiology and function. Kluwer, Dordrecht, pp 201–237

Augé RM (2001) Water relations, drought and vesicular-arbuscular mycorrhizal symbiosis. Mycorrhiza 11:3–42

Augé RM, Toler HD, Sams CE, Nasim G (2008) Hydraulic conductance and water potential gradients in squash leaves showing mycorrhiza-induced increases in stomatal conductance. Mycorrhiza 18:115–121

Baker NR (2008) Chlorophyll fluorescence: a probe of photosynthesis *in vivo*. Annu Rev Plant Biol 59:89–113

Barea JM, Azcón-Aguilar C (1982) Production of plant growth-regulating substances by the vesicular-arbuscular mycorrhizal fungus *Glomus mosseae*. Appl Environ Microbiol 43:810–813

Barrow JR, Havstad KM, McCaslin BD (1997) Fungal root endophytes in fourwing saltbush, *Atriplex canescens*, on arid rangelands of southwestern USA. Arid Soil Res Rehabil 1:177–185

Bearden BN, Petersen L (2000) Influence of arbuscular mycorrhizal fungi on soil structure and aggregate stability of vertisols. Plant Soil 218:173–183

Beauchamp VB, Stromberg JC, Stutz JC (2005) Interactions between *Tamarix ramosissima* (saltcedar), *Populus fremontii* (cottonwood), and mycorrhizal fungi: effects on seedling growth and plant species coexistence. Plant Soil 275:221–231

Beauchamp VB, Walz C, Shafroth PB (2009) Salinity tolerance and mycorrhizal responsiveness of native xeroriparian plants in semi-arid western USA. Appl Soil Ecol 43:175–184

Becard G, Doner LW, Rolin DB, Douds DD, Pfeffer PE (1991) Identification and quantification of trehalose in vesicular arbuscular mycorrhizal fungi *in vivo* ^{13}C NMR and HPLC analyses. New Phytol 108:547–552

Benavides MP, Marconi PL, Gallego SM, Comba ME, Tomaro ML (2000) Relationship between antioxidant defence systems and salt tolerance in *Solanum tuberosum*. Austr J Plant Physiol 27:273–278

Bentivenga SP, Bever JD, Morton JB (1997) Genetic variation of morphological characters within a single isolate of the endomycorrhizal fungus *Glomus clarum* (Glomaceae). Am J Bot 84:1211–1216

Berta G, Fusconi A, Trotta A (1993) VA mycorrhizal infection and the morphology and function of root systems. Environ Exp Bot 33:159–173

Bethlenfalvay GJ, Brown MS, Franson RL, Mihara KL (1989) The *glycine-glomus-bradyrhizobium* symbiosis. IX. Nutritional, morphological and physiological response of nodulated soybean to geographic isolates of the mycorrhizal fungus of *Glomus mosseae*. Physiol Plant 76:226–232

Bhardwaj S, Dudeja SS, Khurana AL (1997) Distribution of VAM fungi in the natural ecosystem. Fol Microbiol 42:589–594

Bhaskaran C, Selvaraj T (1997) Seasonal incidence and distribution of VAM fungi in native saline soils. J Environ Biol 18:209–212

Biela A, Grote K, Otto B, Hoth S, Hedrich R, Kaldenho R (1999) The *Nicotiana tabacum* plasma membrane aquaporin *NtAQP1* is mercury-insensitive and permeable for glycerol. Plant J 18:565–570

Blilou I, Ocampo JA, García-Garrido JM (2000) Induction of Ltp (lipid transfer protein) and Pal (phenylalanine ammonia-lyase) gene expression in rice roots colonized by the arbuscular mycorrhizal fungus *Glomus mosseae*. J Exp Bot 51:1969–1977

Bonfante P, Perotto S (2000) Outside and inside the roots: cell to cell interactions among arbuscular mycorrhizal fungi, bacteria and host plant. In: Podila GK, Douds DD (eds) Current advances in mycorrhizae research. The American Phytopathological Society, St. Paul, pp 141–155

Borde M, Dudhane M, Jite P (2011) Growth photosynthetic activity and antioxidant responses of mycorrhizal and non-mycorrhizal bajra (*Pennisetum glaucum*) crop under salinity stress condition. Crop Prot 30:265–271

Bouchereau A, Aziz A, Larher F, Martin-Tanguy T (1999) Polyamines and environmental challenges: recent development. Plant Sci 140:103–125

Bowen G (1987) The biology and physiology of infection and its development. In: Safir GR (ed) Ecophysiology of VA mycorrhizal plants. CRC, Boca Raton, pp 27–57

Brundrett M (1991) Mycorrhizas in natural ecosystems. Adv Ecol Res 21:300–313

Camprubi A, Calvet C (1996) Isolation and screening of mycorrhizal fungi from citrus nurseries and orchards and inoculation studies. Hort Sci 31:363–369

Cantrell IC, Linderman RG (2001) Preinoculation of lettuce and onion with VA mycorrhizal fungi reduces deleterious effects of soil salinity. Plant Soil 233:269–281

Caravaca F, Alguacil MM, Torres P, Roldán A (2005) Plant type mediates rhizospheric microbial activities and soil aggregation in a semiarid Mediterranean salt marsh. Geoderma 124:375–382

Carvajal M, Martínez V, Alcaraz CF (1999) Physiological function of water channels as affected by salinity in roots of paprika pepper. Physiol Plant 105:95–101

Carvajal M, Cerdá A, Martínez V (2000) Does calcium ameliorate the negative effect of NaCl on melon root water transport by regulating aquaporin activity? New Phytol 145:439–447

Carvalho LM, Cacador I, Martins-Loucão MA (2001) Temporal and spatial variation of arbuscular mycorrhizas in salt marsh plants of the Tagus estuary (Portugal). Mycorrhiza 11:303–309

Carvalho LM, Correia PM, Martins-Loucão MA (2004) Arbuscular mycorrhizal fungal propagules in a salt marsh. Mycorrhiza 14:165–170

Chattopadhyay MK, Tiwari BS, Chattopadhyay G, Bose A, Sengupta DN, Ghosh B (2002) Protective role of exogenous polyamines on salinity-stressed rice (*Oryza sativa*) plants. Physiol Plant 116:192–199

Chaves M, Flexas J, Pinheiro C (2009) Photosynthesis under drought and salt stress: regulation mechanisms from whole plant to cell. Ann Bot 103:551–560

Chen G-P, Wilson ID, Kim SH, Grierson D (2001) Inhibiting expression of a tomato ripening-associated membrane protein increases organic acids and reduces sugar levels of fruit. Planta 212:799–807

Chinnusamy V, Jagendorf A, Zhu JK (2005) Understanding and improving salt tolerance in plants. Crop Sci 45:437–448

Chulan HA, Martin K (1992) The vesicular-arbuscular (VA) mycorrhiza and its effects on growth of vegetatively propagated *Theobroma cacao* L. Plant Soil 144:227–233

Clapp JP, Young JPW, Merryweather JW, Fitter AH (1995) Diversity of fungal symbionts in arbuscular mycorrhizas from a natural community. New Phytol 130:259–265

Colla G, Rouphael Y, Cardarelli M, Tullio M, Rivera CM, Rea E (2008) Alleviation of salt stress by arbuscular mycorrhizal in zucchini plants grown at low and high phosphorus concentration. Biol Fert Soils 44:501–509

Cook JC, Butler RH, Madol T (1993) Some observations on the vertical distribution of VAM in roots of salt marsh grasses growing in saturated soils. Mycologia 85:547–550

Copeman RH, Martin CA, Stutz JC (1996) Tomato growth in response to salinity and mycorrhizal fungi from saline or nonsaline soils. Hort Sci 31:341–344

Cramer G (2004) Sodium–calcium interactions under salinity stress. In: Läuchli A, Lüttge S (eds) Salinity: environment–plants–molecules. Springer, Netherlands, pp 205–227

Daei G, Ardekani MR, Rejali F, Teimuri S, Miransari M (2009) Alleviation of salinity stress on wheat yield, yield components, and nutrient uptake using arbuscular mycorrhizal fungi under field conditions. J Plant Physiol 166:617–625

Daniels BA, Graham SO (1976) Effects of nutrition and soil extracts on germination of *Glomus mosseae* spores. Mycologia 68:108–116

Daniels BA, Trappe JM (1980) Factors affecting spore germination of the vesicular-arbuscular mycorrhizal fungus *Glomus epigaeus*. Mycologia 72:457–471

Davies HA, Dow JM (1997) Induction of extracellular matrix glycoproteins in Brassica petioles bywounding and in response to *Xanthomonas campestris*. Mol Plant-Microb Interact 10:812–820

Davies WJ, Zhang J (1991) Root signals and the regulation of growth and development of plants in drying soil. Annu Rev Plant Physiol 42:55–73

del Val C, Barea JM, Azcon-Agular C (1999) Divisity of arbuscular mycorrhizal fungus populations in heavy-metal contaminated soils. Appl Environ Microbiol 65:718–723

Dixon RK, Garg VK, Rao MV (1993) Inoculation of *Leucaena* and *Prosopis* seedlings with *Glomus* and *Rhizobium* species in saline soil: rhizosphere relations and seedling growth. Arid Soil Res Rehabil 7:133–144

Dodd JC, Thomson BD (1994) The screening and selection of inoculant arbuscular-mycorrhizal and ectomycorrhizal fungi. Plant Soil 159:149–158

Douds DD Jr, Millner PD (1999) Biodiversity of arbuscular mycorrhizal fungi in agroecosystems. Agric Ecosyst Environ 74:183–193

Duke ER, Johnson CR, Koch KE (1986) Accumulation of phosphorus, dry matter and betaine during NaC1 stress of split-root citrus seedlings colonized with vesicular-arbuscular mycorrhizal fungi on zero, one or two halves. New Phytol 104:583–590

Estaun MV (1989) Effect of sodium chloride and mannitol on germination and hyphal growth of the vesicular-arbuscular mycorrhizal fungus *Glomus mosseae*. Agric Ecosyst Environ 29:123–129

Estaun MV (1991) Effect of NaC1 and mannitol on the germination of two isolates of the vesicular-arbuscular mycorrhizal fungus *Glomus mosseae.* In: Proceedings book of 3rd European symposium on mycorrhizas. University of Sheffield, Sheffield

Evelin H, Kapoor R, Giri B (2009) Arbuscular mycorrhizal fungi in alleviation of salt stress: a review. Ann Bot 104:1263–1280

Ezawa T, Smith SE, Smith FA (2002) P metabolism and transport in AM fungi. Plant Soil 244:221–230

Feng G, Zhang FS, Li XL, Tian CY, Tang C, Rengel Z (2002) Improved tolerance of maize plants to salt stress by arbuscular mycorrhiza is related to higher accumulation of soluble sugars in roots. Mycorrhiza 12:185–190

Fornalé S, Sarjala T, Bagni N (1999) Endogenous polyamine content and metabolism in the ectomycorrhizal fungus *Paxillus involutus*. New Phytol 143:581–587

Francoise F, Daniel LR, John G (1991) Effects of salt stress on amino acid, organic acid and carbohydrate composition of root, bacteroids, and cytosol of alfalfa (*Medicago sativa* L.). Plant Physiol 96:1228–1236

Fukuda A, Nakamura A, Tanaka Y (1999) Molecular cloning and expression of the Na^+/H^+ exchanger gene in *Oryza sativa*. Biochim Biophys Acta 1446:149–155

Füzy A, Biró B, Tóth T, Hildebrandt U, Bothe H (2008) Drought, but not salinity, determines the apparent effectiveness of halophytes colonized by arbuscular mycorrhizal fungi. J Plant Physiol 165:1181–1192

Gadkar V, Rillig M (2006) The arbuscular mycorrhizal fungal protein glomalin is a putative homolog of heat shock protein 60. FEMS Microbiol Lett 263:93–101

Garg N, Chandel S (2011) The effects of salinity on nitrogen fixation and trehalose metabolism in mycorrhizal *Cajanus cajan* (L.) Millsp. plants. J Plant Growth Regul 30:490–503

Garratt LC, Janagoudar BS, Lowe KC (2002) Salinity tolerance and antioxidant status in cotton culture. Free Radic Biol Med 33:502–511

Gaxiola RA, Rao R, Sherman A, Grisafi P, Alper SL, Fink GR (1999) The *Arabidopsis thaliana* proton transporters, *AtNhx1* and *Avp1*, can function in cation detoxification in yeast. Proc Natl Acad Sci USA 96:1480–1485

Ghachtouli NE, Paynot M, Morandi D, Martintanguy J, Gianinazzi S (1995) The effect of polyamines on endomycorrhizal infection of wild-type *Pisum sativum*, cv. Frisson (nod + myc+) and 2 mutants (nod– myc + and nod–myc–). Mycorrhiza 5:189–192

Gianinazzi-Pearson V (1996) Plant cell responses to arbuscular mycorrhizal fungi: getting to the roots of the symbiosis. Plant Cell 8:1871–1883

Giri M, Mukerji KG (2004) Mycorrhizal inoculant alleviates salt stress in *Sesbania aegyptiaca* and *Sesbania grandiflora* under field conditions: evidence for reduced sodium and improved magnesium uptake. Mycorrhiza 14:307–312

Giri B, Kapoor R, Mukerji KG (2003) Influence of arbuscular mycorrhizal fungi and salinity on growth, biomass, and mineral nutrition of *Acacia auriculiformis*. Biol Fertil Soil 38:170–175

Giri B, Kapoor R, Mukerji KG (2007) Improved tolerance of Acacia nilotica to salt stress by arbuscular mycorrhiza, *Glomus fasciculatum* may be partly related to elevated K/Na ratios in root and shoot tissues. Microbiol Ecol 54:753–760

Groppa MD, Benavides MP (2008) Polyamines and abiotic stress: recent advances. Amino Acids 34:35–45

Guo LQ, Shi DC, Wang DL (2010) The key physiological response to alkali stress by the alkali-resistant halophyte *Puccinellia tenuiflora* is the accumulation of large quantities of organic acids and into the rhizosphere. J Agron Crop Sci 196:123–135
Hajiboland R, Aliasgharzadeh N, Laiegh SF, Poschenrieder C (2010) Colonization with arbuscular mycorrhizal fungi improves salinity tolerance of tomato (*Solanum lycopersicum* L.) plants. Plant Soil 331:313–327
Hajlaoui H, Ayeb NE, Garrec JP, Denden M (2010) Differential effects of salt stress on osmotic adjustment and solutes allocation on the basis of root and leaf tissue senescence of two silage maize (*Zea mays* L.) varieties. Ind Crops Prod 31:122–130
Hammer EC, Rillig MC (2011) The influence of different stresses on glomalin levels in an arbuscular mycorrhizal fungus-salinity increases glomalin content. PLoS One 6(12): e28426
Hammer EC, Nasr H, Pallon J, Olsson PA, Wallander H (2011) Elemental composition of arbuscular mycorrhizal fungi at high salinity. Mycorrhiza 21:117–129
Harinasut P, Poonsopa D, Roengmongkol K, Charoensataporn R (2003) Salinity effects on antioxidant enzymes in mulberry cultivar. Sci Asia 29:109–113
Harley JL, Harley EL (1987) A check-list of mycorrhiza in the British flora. New Phytol 10:1–102
Hartmond U, Schaesberg NV, Graham JH, Syvertsen JP (1987) Salinity and flooding stress effects on mycorrhizal and nonmycorrhizal citrus rootstock seedlings. Plant Soil 104:37–43
Hasegawa PM, Bressan RA, Zhu JK, Bohnert HJ (2000) Plant cellular and molecular responses to high salinity. Annu Rev Plant Physiol Plant Mol Biol 51:463–499
Hatzig S, Kumar A, Neubert A, Schubert S (2010) PEP-carboxylase activity: a comparison of its role in a C_4 and a C_3 species under salt stress. J Agron Crop Sci 196:185–192
He XL, Mouratov S, Steinberger Y (2002) Spatial distribution and colonization of arbuscular mycorrhizal fungi under the canopies of desert halophytes. Arid Land Res Manag 16: 149–160
He ZQ, He CX, Zhang ZB, Zou ZR, Wang HS (2007) Changes of antioxidative enzymes and cell membrane osmosis in tomato colonized by arbuscular mycorrhizae under NaCl stress. Colloids Surf B Biointerfaces 59:128–133
Hepper CM (1979) Germination and growth of *Glomus caledonius* spores: the effects of inhibitors and nutrients. Soil Biol Biochem 11:269–277
Hetrick BAD, Bockus WW, Bloom J (1984) The role of vesicular-arbuscular mycorrhizal fungi in the growth of Kansas winter wheat. Can J Bot 62:735–740
Hildebrandt U, Janetta K, Ouziad F, Renne B, Nawrath K, Bothe H (2001) Arbuscular mycorrhizal colonization of halophytes in Central European salt marshes. Mycorrhiza 10:175–183
Hill AE, Shachar-Hill B, Shachar-Hill Y (2004) What are aquaporins for? J Membr Biol 197:1–32
Hirrel MC (1981) The effect of sodium and chloride salts on the germination of *Gigaspora margarita*. Mycologia 73:610–617
Hirrel MC, Gerdemann JW (1980) Improved growth of onion and bell pepper in saline soils by two vesicular–arbuscular mycorrhizal fungi. Soil Sci Soc Am J 44:654–655
Hirrel MC, Mehravaran H, Gerdemann JW (1978) Vesiculararbuscular mycorrhiza in the Chenopodiaceae and Cruciferae: do they occur? Can J Bot 56:2813–2817
Ho I (1987) Vesicular-arbuscular mycorrhizae of halophytic grasses in the Alvard desert of Oregon. Northwest Sci 61:148–151
Hoefnagels MH, Broome SW, Shafer SR (1993) Vesicular-arbuscular mycorrhizae in salt marshes in North Carolina. Estuaries 16:851–858
Jahromi F, Aroca R, Porcel R, Ruiz-Lozano JM (2008) Influence of salinity on the *in vitro* development of *Glomus intraradices* and on the *in vivo* physiological and molecular responses of mycorrhizal lettuce plants. Microbial Ecol 55:45–53
Jakobsen I, Abbott LK, Robson AD (1992) External hyphae of vesicular-mycorrhizal fungi associated with *Trifolium subterraneum* L. New Phytol 120:371–380

Jang JY, Lee SH, Ree JY, Chung GC, Ahn SJ, Kang H (2007) Transgenic *Arabidopsis* and tobacco plants overexpressing and aquaporin respond differently to various abiotic stresses. Plant Mol Biol 64:621–632

Janos DP (2007) Plant responsiveness to mycorrhizas differs from dependence upon mycorrhizas. Mycorrhiza 17:75–91

Javot H, Lauvergeat V, Santoni V, Martin-Laurent F, Guclu J, Vinh J, Heyes J, Franck KI, Schaffner AR, Bouchez D, Maurel C (2003) Role of a single aquaporin isoform in root water uptake. Plant Cell 15:509–522

Johansson I, Larsson C, Ek B, Kjellbom P (1996) The major integral proteins of spinach leaf plasma membranes are putative aquaporins and are phosphorylated in response to Ca^{2+} and apoplastic water potential. Plant Cell 8:1181–1191

Johansson I, Karlsson M, Johanson U, Larsson C, Kjellbom P (2000) The role of aquaporins in cellular and whole plant water balance. Biochim Biophys Acta 1465:324–342

Johnson WC (1992) Dams and riparian forests: case study from the upper Missouri River. Rivers 3:229–242

Johnson GPC, Kenkel NC, Booth T (1995) The distribution and phenology of arbuscular mycorrhizae along an inland salinity gradient. Can J Bot 73:1318–1327

Johnson NC, Graham JH, Smith FA (1997) Functioning of mycorrhizal associations along the mutualism–parasitism continuum. New Phytol 135:574–586

Juniper S, Abbott LK (1993) Vesicular-arbuscular mycorrhizas and soil salinity. Mycorrhiza 4:45–57

Juniper S, Abbott LK (2006) Soil salinity delays germination and limits growth of hyphae from propagules of arbuscular mycorrhizal fungi. Mycorrhiza 16:371–379

Kapoor R, Sharma D, Bhatnagar AK (2008) Arbuscular mycorrhizae in micropropagation systems and their potential applications. Sci Hort 116:227–239

Kasukabe Y, He L, Nada K, Misawa S, Ihara I, Tachibana S (2004) Overexpression of spermidine synthase enhances tolerance to multiple environmental stresses and up-regulates the expression of various stress-regulated genes in transgenic *Arabidopsis thaliana*. Plant Cell Physiol 45:712–722

Kaya C, Ashraf M, Sonmez O, Aydemir S, Tuna LA, Cullu AM (2009) The influence of arbuscular mycorrhizal colonization on key growth parameters and fruit yield of pepper plants grown at high salinity. Sci Hort 12:1–6

Khaled LB, Gõmez AM, Ouarraqi EM, Oihabi A (2003) Physiological and biochemical responses to salt stress of mycorrhized and/or nodulated clover seedlings (*Trifolium alexandrinum* L.). Agronomie 23:571–580

Khalil HA, Eissa AM, El-Shazly SM, Aboul-Nasr AM (2011) Improved growth of salinity-stressed citrus after inoculation with mycorrhizal fungi. Sci Hort 130:624–632

Khalvati MA, Hu Y, Mozafar A, Schmidhalter U (2005) Quantification of water uptake by arbuscular mycorrhizal hyphae and its significance for leaf growth, water relations, and gas exchange of barley subjected to drought stress. Plant Biol 7:706–712

Kim CK, Weber DJ (1985) Distribution of VA mycorrhiza on halophytes on inland salt playas. Plant Soil 83:207–214

Kishor PB, Hong Z, Miao G-H (1995) Overexpression of pyrroline-5- carboxylate synthetase increases proline production and confers osmotolerance in transgenic plants. Plant Physiol 108:1387–1394

Kohler J, Hernández JA, Caravaca F, Roldán A (2009) Induction of antioxidant enzymes is involved in the greater effectiveness of a PGPR versus AM fungi with respect to increasing the tolerance of lettuce to severe salt stress. Environ Exp Bot 65:245–252

Kohler J, Caravaca F, Roldán A (2010) An AM fungus and a PGPR intensify the adverse effects of salinity on the stability of rhizosphere soil aggregates of *Lactuca sativa*. Soil Biol Biochem 42:429–434

Koide RT (2000) Functional complementarity in the arbuscular mycorrhizal symbiosis. New Phytol 147:233–235

Krajinski F, Biela A, Schubert D, Gianinazzi-Pearson V, Kaldenhoff R, Franken P (2000) Arbuscular mycorrhiza development regulates the mRNA abundance of Mtaqp1 encoding a mercury-insensitive aquaporin of *Medicago truncatula*. Planta 211:85–90

Lambais MR, Ríos-Ruiz WF, Andrade RM (2003) Antioxidant responses in bean (*Phaseolus vulgaris*) roots colonized by arbuscular mycorrhizal fungi. New Phytol 160:421–428

Landwehr M, Hildebrandt U, Wilde P, Nawrath K, Tóth T, Biró B, Bothe H (2002) The arbuscular mycorrhizal fungus *Glomus geosporum* in European saline, sodic and gypsum soils. Mycorrhiza 12:199–211

Lerat S, Lapointe L, Gutjahr S, Piché Y, Vierheilig H (2003) Carbon partitioning in a split-root system of arbuscular mycorrhizal plants is fungal and plant species dependent. New Phytol 157:589–595

Locatelli LM, Vitovski CA, Lovato PE (2002) Root architecture of apple rootstocks inoculated with arbuscular mycorrhizal fungi. Pesq Agropec Bras 37:1239–1245

López-Bucio J, Nieto-Jacobo MF, Ramírez-Rodríguez V, Herrera-Estrella L (2000) Organic acid metabolism in plants: from adaptive physiology to transgenic varieties for cultivation in extreme soils. Plant Sci 160:1–13

Luu DT, Maurel C (2005) Aquaporins in a challenging environment: molecular gears for adjusting plant water status. Plant Cell Environ 28:85–96

Mamatha G, Bagyaraj DJ, Jaganath S (2002) Inoculation of field-established mulberry and papaya with arbuscular mycorrhizal fungi and a mycorrhiza helper bacterium. Mycorrhiza 12:313–316

Mardukhi B, Rejali F, Daei G, Ardakani MR, Malakouti MJ, Miransari M (2011) Arbuscular mycorrhizas enhance nutrient uptake in different wheat genotypes at high salinity levels under field and greenhouse conditions. Com Rend Biol 334:564–571

Marschner P (2012) Rhizosphere biology. In: Marschner P (ed) Marschner's mineral nutrition of higher plants, 3rd edn. Academic Press, USA, pp 369–388

Marschner H, Dell B (1994) Nutrient uptake in mycorrhizal symbiosis. Plant Soil 159:89–102

Martinčič A, Wraber T, Jogan N, Podobnik A, Turk B, Vreš B, Ravnik V, Frajman B, Strgulc-Krajšek S, Trčak B, Bačič T, Fischer MA, Eler K, Surina B (2007) Mala Flora Slovenije: Ključ Za Določanje Praprotnic in Semenk. Tehniška založba Slovenije, Ljubljana

Marulanda A, Azcón R, Ruiz-Lozano JM (2003) Contribution of six arbuscular mycorrhizal fungal isolates to water uptake by *Lactuca sativa* plants under drought stress. Physiol Plant 119:526–533

Mathur N, Singh J, Bohra S, Vyas A (2007) Arbuscular mycorrhizal status of medicinal halophytes in saline areas of Indian Thar desert. Int J Soil Sci 2:119–127

Maurel C, Kado RT, Guern J, Chrispeels MJ (1995) Phosphorylation regulates the water channel activity of the seed-specific aquaporin alpha-TIP. EMBO J 14:3028–3035

Maurel C, Chrispeels M, Lurin C, Tacnet F, Geelen D, Ripoche P, Guern J (1997) Function and regulation of seed aquaporins. J Exp Bot 48:421–430

McMillen BG, Juniper S, Abbott LK (1998) Inhibition of hyphal growth of a VA mycorrhizal fungus in soil containing sodium chloride limits the spread of infection from spores. Soil Biol Biochem 30:1639–1646

Mehdy MC, Sharma YK, Sathasivan K, Bays NW (1996) The role of activated oxygen species in plant disease resistance. Physiol Plant 98:365–374

Miller RM, Jastrow JD (2000) Mycorrhizal fungi influence soil structure. In: Kapulnik Y, Douds DD (eds) Arbuscular mycorrhizas: molecular biology and physiology. Kluwer, Dordrecht, pp 3–18

Miranda D, Fischer G, Ulrichs C (2011) The influence of arbuscular mycorrhizal colonization on the growth parameters of cape gooseberry (*Physalis peruviana* L.) plants grown in a saline soil. J Soil Sci Plant Nutr 1:18–30

Miransari M (2010) Contribution of arbuscular mycorrhizal symbiosis to plant growth under different types of soil stress. Plant Biol 1:563–569

Miransari M (2011) Arbuscular mycorrhizal fungi and nitrogen uptake. Arch Microbiol 193:77–81

Miransari M, Smith DL (2007) Overcoming the stressful effects of salinity and acidity on soybean [*Glycine max* (L.) Merr.] nodulation and yields using signal molecule genistein under field conditions. J Plant Nutr 30:1967–1992
Miransari M, Smith DL (2008) Using signal molecule genistein to alleviate the stress of suboptimal root zone temperature on soybean-*Bradyrhizobium* symbiosis under different soil textures. J Plant Interact 3:287–295
Miransari M, Bahrami HA, Rejali F, Malakouti MJ, Torabi H (2007) Using arbuscular mycorrhiza to reduce the stressful effects of soil compaction on corn (*Zea mays* L.) growth. Soil Biol Biochem 39:2014–2026
Miransari M, Bahrami HA, Rejali F, Malakouti MJ (2008) Using arbuscular mycorrhiza to reduce the stressful effects of soil compaction on wheat (*Triticum aestivum* L.) growth. Soil Biol Biochem 40:1197–1206
Mittova V, Tal M, Volokita M, Guy M (2002) Salt stress induces up-regulation of an efficient chloroplast antioxidant system in the salt-tolerant wild tomato species *Lycopersicon pennellii* but not in the cultivated species. Physiol Plant 115:393–400
Mosse B, Hepper C (1975) Vesicular-arbuscular mycorrhizal infections in root organ cultures. Physiol Plant Pathol 5:215–223
Munns R, Tester M (2008) Mechanisms of salinity tolerance. Ann Rev Plant Biol 59:651–681
Munns R, James RA, Läuchli A (2006) Approaches to increasing the salt tolerance of wheat and other cereals. J Exp Bot 57:1025–1043
Murkute AA, Sharma S, Singh SK (2006) Studies on salt stress tolerance of citrus rootstock genotypes with arbuscular mycorrhizal fungi. Hort Sci 33:70–76
Nemec S (1981) Histochemical characterization of *Glomus etunicatum* infection of Citrus limon roots. Can J Bot 59:609–617
Neto ADA, Prisco JT, Gomes-Filho E (2009) Changes in soluble amino-N, soluble proteins and free amino acids in leaves and roots of salt-stressed maize genotypes. J Plant Interact 4:137–144
Niki T, Mitsuhara I, Ohtsubo S, Ohashi Y (1998) Antagonistic effect of salicylic acid and jasmonic acid on the expression of pathogenesis-related (PR) protein genes in wounded mature tobacco leaves. Plant Cell Physiol 39:500–507
O'Connor PJ, Smith SE, Smith FA (2001) Arbuscular mycorrhizal associations in the southern Simpson desert. Aust J Bot 49:493–499
Ocón A, Hampp R, Requena N (2007) Trehalose turnover during abiotic stress in arbuscular mycorrhizal fungi. New Phytol 174:879–891
Ojala JC, Jarrell WM, Menge JA, Johnson ELV (1983) Influence of mycorrhizal fungi on the mineral nutrition and yield of onion in saline soil. Agron J 75:255–259
Ouziad F, Wilde P, Schmelzer E, Hildebrandt U, Bothe H (2006) Analysis of expression of aquaporins and Na^+/H^+ transporters in tomato colonized by arbuscular mycorrhizal fungi and affected by salt stress. Environ Exp Bot 57:177–186
Pacovsky RS, De Silva P, Carvalho MTV, Tsai SM (1991) Growth and nutrient allocation in *Phaseolus vulgaris* L. colonized with endomycorrhizae or *Rhizobium*. Plant Soil 132:127–137
Palma JM, Longa MA, del Rio LA, Arines J (1993) Superoxide dismutase in vesicular-arbuscular red clover plants. Physiol Plant 87:77–83
Parida AK, Das AB (2005) Salt tolerance and salinity effects on plants: a review. Ecotoxicol Environ Saf 60:324–349
Paul MJ, Primavesi LF, Jhurreea D, Zhang Y (2008) Trehalose metabolism and signaling. Annu Rev Plant Biol 59:417–441
Peng J, Li Y, Shi P, Chen X, Lin H, Zhao B (2011) The differential behavior of arbuscular mycorrhizal fungi in interaction with *Astragalus sinicus* L. under salt stress. Mycorrhiza 21:27–33
Pfeiffer CM, Bloss HE (1988) Growth and nutrition of guayule *(Parthenium argentatum)* in a saline soil as influenced by vesicular- arbuscular mycorrhiza and phosphorus fertilization. New Phytol 108:315–321
Pitman M, Läuchli A (2004) Global impact of salinity and agricultural ecosystems. In: Läuchli A, Lüttge S (eds) Salinity: environment–plants–molecules. Springer, Dordrecht, pp 3–20

Plenchette C, Duponnois R (2005) Growth response of the saltbush *Atriplex nummularia* L. to inoculation with the arbuscular mycorrhizal fungus *Glomus intraradices*. J Arid Environ 61:535–540

Pond EC, Menge JA, Jarrell WM (1984) Improved growth of tomato in salinized soil by vesicular-arbuscular mycorrhizal fungi collected from saline soils. Mycologia 76:74–84

Porcel R, Ruiz-Lozano JM (2004) Arbuscular mycorrhizal influence on leaf water potential, solute accumulation and oxidative stress in soybean plants subjected to drought stress. J Exp Bot 55:1743–1750

Porcel R, Azcón R, Ruiz-Lozano JM (2004) Evaluation of the role of genes encoding for Δ^1-pyrroline-5-carboxylate synthetase (P5CS) during drought stress in arbuscular mycorrhizal *Glycine max* and *Lactuca sativa* plants. Physiol Mol Plant Pathol 65:211–221

Porcel R, Gómez M, Kaldenhoff R, Ruiz-Lozano JM (2005) Impairment of *NtAQP1* gene expression in tobacco plants does not affect root colonisation pattern by arbuscular mycorrhizal fungi but decreases their symbiotic efficiency under drought. Mycorrhiza 15:417–423

Porcel R, Aroca R, Azcón R, Ruiz-Lozano JM (2006) PIP aquaporin gene expression in arbuscular mycorrhizal Glycine max and *Lactuca sativa* plants in relation to drought stress tolerance. Plant Mol Biol 60:389–404

Porcel R, Aroca R, Ruiz-Lozano JM (2012) Salinity stress alleviation using arbuscular mycorrhizal fungi. Agron Sustain Dev 32:181–200

Porras-Soriano A, Soriano-Martín ML, Porras-Piedra A, Azcón R (2009) Arbuscular mycorrhizal fungi increased growth, nutrient uptake and tolerance to salinity in olive trees under nursery conditions. J Plant Physiol 166:1350–1359

Poss JA, Pond E, Menge JA, Harrell WM (1985) Effect of salinity on mycorrhizal onion and tomato in soil with and without additional phosphate. Plant Soil 88:307–319

Pozo MJ, Azcón C (2007) Unravelling mycorrhiza-induced resistance. Curr Opin Plant Biol 10:393–398

Rabie GH (2005) Influence of arbuscular mycorrhizal fungi and kinetin on the response of mungbean plants to irrigation with seawater. Mycorrhiza 15:225–230

Rabie GH, Almadini AM (2005) Role of bioinoculants in development of salt-tolerance of *Vicia faba* plants under salinity stress. Afr J Biotechnol 4:210–222

Rao DLN (1998) Biological amelioration of salt affected soils. In: Microbial interactions in agriculture and forestry, 1st edn. Science, Enfield, pp 21–238

Redecker D, Morton JB, Bruns TD (2000) Ancestral lineages of arbuscular mycorrhizal fungi (Glomales). Mol Phylogenet Evol 14:276–284

Reeves FB, Wagner D, Moorman T, Kiel J (1979) The role of endomycorrhizae in revegetation practices in the semi-arid west. I. A comparison of incidence of mycorrhizae in severely disturbed vs. natural environments. Am J Bot 66:6–13

Rhodes D, Nadolska-Orczyk A, Rich PJ (2004) Salinity, osmolytes and compatible solutes. In: Läuchli A, Lüttge S (eds) Salinity: environment–plants–molecules. Springer, Dordrecht, pp 181–204

Rillig MC (2004) Arbuscular mycorrhizae, glomalin, and soil aggregation. Can J Soil Sci 84:355–363

Rovira AD (1969) Plant root exudates. Bot Rev 35:35–57

Rozema J, Arp W, Van Diggelen J, Van Esbroek M, Broekma R, Punt H (1986) Occurrence and ecological significance of vesicular-arbuscular mycorrhiza in the salt marsh environment. Acta Bot Neerl 35:457–467

Ruiz-Lozano JM (2003) Arbuscular mycorrhizal symbiosis and alleviation of osmotic stress new perspectives for molecular studies. Mycorrhiza 13:309–317

Ruiz-Lozano JM, Aroca R (2010) Modulation of aquaporin genes by the arbuscular mycorrhizal symbiosis in relation to osmotic stress tolerance. In: Seckbach J, Grube M (eds) Symbioses and stress, vol 17, Cellular origin, life in extreme habitats and astrobiology. Springer, Berlin, pp 357–374

Ruiz-Lozano JM, Azcón R (2000) Symbiotic efficiency and infectivity of an autochthonous arbuscular mycorrhizal *Glomus sp.* from saline soils and *Glomus deserticola* under salinity. Mycorrhiza 10:137–143

Ruiz-Lozano JM, Azcón R, Goméz M (1996) Alleviation of salt stress by arbuscular-mycorrhizal Glomus species in *Lactuca sativa* plants. Physiol Plant 98:767–772
Ruiz-Lozano JM, Porcel R, Azcón C, Aroca R (2012) Regulation by arbuscular mycorrhizae of the integrated physiological response to salinity in plants: new challenges in physiological and molecular studies. J Exp Bot 63(11):4033–4044
Rush DW, Epstein E (1976) Genotypic responses to salinity: differences between salt-sensitive and salt-tolerant genotypes of tomato. Plant Physiol 57:162–166
Sannazzaro AI, Alvarez CL, Menéndez AB, Pieckenstain FL, Albertó EO, Ruiz OA (2004) Ornithine and arginine decarboxilase activities and effect of some polyamine biosynthesis inhibitors on *Gigaspora rosea* germinating spores. FEMS Microbiol Lett 230:115–121
Sannazzaro AI, Ruíz OA, Alberto EO, Menendez AB (2006) Alleviation of salt stress in *Lotus glaber* by *Glomus intraradices*. Plant Soil 285:279–287
Sannazzaro AI, Echeverría M, Albertó EO, Ruiz OA, Menéndez AB (2007) Modulation of polyamine balance in *Lotus glaber* by salinity and arbuscular mycorrhiza. Plant Physiol Biochem 45:39–46
Sarda X, Tousch D, Ferrare K, Cellier F, Alcon C, Dupuis JM, Casse F, Lamaze T (1999) Characterization of closely related delta- TIP genes encoding aquaporins which are differentially expressed in sunflower upon water deprivation through exposure to air. Plant Mol Biol 40:179–191
Sawers RJ, Gutjahr C, Paszkowski U (2008) Cereal mycorrhiza: an ancient symbiosis in modern agriculture. Trends Plant Sci 13:93–97
Scheublin TR, Ridgway KP, Young PW, van der Heijden MGA (2004) Non legumes, legumes, and root nodules harbor different arbuscular mycorrhizal fungal communities. Appl Environ Microbiol 70:6240–6246
Schroeder MS, Janos DP (2005) Plant growth, phosphorus nutrition, and root morphological responses to arbuscular mycorrhizas, phosphorus fertilization, and intraspecific density. Mycorrhiza 15:203–216
Schubert A, Wyss P, Wiekman A (1992) Occurrence of trehalose in vesicular-arbuscular mycorrhizal fungi and in mycorrhizal roots. J Plant Physiol 140:41–45
Sengupta A, Chaudhuri S (1990) Vesicular-arbuscular mycorrhiza in pioneer salt marsh plants of the Ganges river delta in West Bengal (India). Plant Soil 122:111–113
Shalata A, Tal M (1998) The effect of salt stress on lipid peroxidation and antioxidants in the leaf of the cultivated tomato and its wild salt-tolerant relative *Lycopersicon pennellii*. Physiol Plant 104:169–174
Sharifi M, Ghorbanli M, Ebrahimzadeh H (2007) Improved growth of salinity-stressed soybean after inoculation with salt pre-treated mycorrhizal fungi. J Plant Physiol 164:1144–1151
Sheng M, Tang M, Chen H, Yang BW, Zhang FF, Huang YH (2008) Influence of arbuscular mycorrhizae on photosynthesis and water status of maize plants under salt stress. Mycorrhiza 18:287–296
Sheng M, Tang M, Zhang F, Huang Y (2011) Influence of arbuscular mycorrhiza on organic solutes in maize leaves under salt stress. Mycorrhiza 21:423–430
Shi H, Ishitani M, Kim C, Zhu J-K (2000) The *Arabidopsis thaliana* salt tolerance gene SOS1 encodes a putative Na^+/H^+ antiporter. Proc Natl Acad Sci USA 97:6896–6901
Shi ZY, Zhang L, Feng G, Christie P, Tian CY, Li XL (2006) Diversity of arbuscular mycorrhizal fungi associated with desert ephemerals growing under and beyond the canopies of Tamarisk shrubs. Chinese Sci Bull 51:132–139
Shokri S, Maadi B (2009) Effect of arbuscular mycorrhizal fungus on the mineral nutrition and yield of *Trifolium alexandrinum* plants under salinity stress. J Agron 8:79–83
Smart LB, Moskal WA, Cameron KD, Bennett AB (2001) *MIP* genes are down-regulated under drought stress in *Nicotiana glauca*. Plant Cell Physiol 42:686–693
Smith SE, Read DJ (2008) Mycorrhizal symbiosis. Academic, San Diego
Smith FA, Smith SE (1997) Structural diversity in (vesicular)-arbuscular mycorrhizal symbioses. New Phytol 137:373–388
Smith SE, Smith FA, Jakobsen I (2003) Mycorrhizal fungi can dominate phosphate supply to plants irrespective of growth responses. Plant Physiol 133:16–20

Sonjak S, Udovič M, Wraber T, Likar M, Regvar M (2009) Diversity of halophytes and identification of arbuscular mycorrhizal fungi colonising their roots in an abandoned and sustained part of Sečovlje salterns. Soil Biol Biochem 41:1847–1856

Sottosanto JB, Gell A, Blumwald E (2004) DNA array analyses of *Arabidopsis thaliana* lacking a vacuolar Na^+/H^+ antiporter: impact of *AtNHX1* on gene expression. Plant J 40:752–771

Spanu P, Bonfante-Fasolo P (1988) Cell-wall-bound peroxidase activity in roots of mycorrhizal *Allium porrum*. New Phytol 109:119–124

Sreenivasulu N, Ramanjulu S, Ramachandra-Kini K, Prakash HS, Shekar-Shetty H, Savithri HS, Sudhakar C (1999) Total peroxidase activity and peroxidase isoforms as modified by salt stress in two cultivars of fox-tail millet with differential salt tolerance. Plant Sci 141:1–9

Strack D, Fester T, Hause B, Schliemann W, Walter MH (2003) Arbuscular mycorrhiza: biological, chemical, and molecular aspects. J Chem Ecol 29:1955–1979

Sylvia DM, Fuhrmann JJ, Hartel PG, Zuberer D (1998) Principles and applications of soil microbiology. Prentice, Upper Saddle River

Szalai G, Janda T (2009) Effect of salt stress on the salicylic acid synthesis in young maize (*Zea mays* L.) plants. J Agric Crop Sci 195:165–171

Tal M, Katz A, Heikin H, Dehan K (1979) Salt tolerance in the wild relatives of the cultivated tomato: proline accumulation in *Lycopersicon esculentum* Mill., L., Peruvianum Mill. and *Solanum pennelli* Cor. Treated with NaCl and polyethylene glycol. New Phytol 82:349–355

Tester M, Davenport R (2003) Na tolerance and Na transport in higher plants. Ann Bot 91:503–527

Tian CY, Feng G, Li XL, Zhang FS (2004) Different effects of arbuscular mycorrhizal fungal isolates from saline or non-saline soil on salinity tolerance of plants. Appl Soil Ecol 26:143–148

Titus JH, Titus PJ, Nowak RS, Smith SD (2002) Arbuscular mycorrhizae of Mojave desert plants. West North Am Naturalist 62:327–334

Tommerup IC (1984) Effect of soil water potential on spore germination by vesicular-arbuscular fungi. Trans Br Mycol Soc 83:193–202

Tonin C, Vandenkoornhuyse P, Joner E, Straczek J, Leyval C (2001) Diversity of AM fungi in the rhizosphere of *Viola calaminaria* and effect of these fungi on heavy metal uptake by clover. Mycorrhiza 10:161–168

Tressner HD, Hayes JA (1971) Sodium chloride tolerance of terrestrial fungi. Appl Microbiol 22:210–213

Usha K, Saxena A, Singh B (2004) Rhizosphere dynamics influenced by arbuscular mycorrhizal fungus (*Glomus deserticola*) and related changes in leaf nutrient status and yield of Kinnow mandarin [King (*Citrus nobilis*)×Willow Leaf (*Citrus deliciosa*)]. Aust J Agric Res 55:571–576

van der Heijden MGA, Boller T, Wiemken A, Sanders IR (1998a) Different arbuscular mycorrhizal fungal species are potential determinants of plant community structure. Ecology 79:2082–2091

van Der Heijden MGA, Klironomos JN, Ursic M, Moutoglis P, Streitwolf-Engel R, Boller T, Wiemken A, Sanders IR (1998b) Mycorrhizal fungal diversity determines plant biodiversity, ecosystem variability and productivity. Nature 396:69–72

Vogel-Mikuš K, Drobne D, Regvar M (2005) Zn, Cd and Pb accumulation and arbuscular mycorrhizal colonization of pennycress *Thlaspi praecox* Wulf Brassicaceae from the vicinity of a lead mine and smelter in Slovenia. Environ Pollut 133:232–242

Wang B, Qiu Y-L (2006) Phylogenetic distribution and evolution of mycorrhizas in land plants. Mycorrhiza 16:299–363

Wang W, Vinocur B, Shoseyov O, Altman A (2004) Role of plant heat-shock proteins and molecular chaperones in the abiotic stress response. Trends Plant Sci 9:244–252

Wang F, Zeng B, Sun ZX, Zhu C (2009) Relationship between proline and Hg^{2+}-induced oxidative stress in a tolerant rice mutant. Arch Environ Contam Toxicol 56:723–731

Watad AEA, Reinhold L, Lerner HR (1983) Comparison between a stable NaCl-selected Nicotiana cell line and the wild type: K^+, Na^+, and proline pools as a function of salinity. Plant Physiol 73:624–629

Weig A, Deswarte C, Chrispeels MJ (1997) The major intrinsic protein family of *Arabidopsis* has 23 members that form three distinct groups with functional aquaporins in each group. Plant Physiol 114:1347–1357

Williams SE, Wollum AG II, Aldon FE (1974) Growth of *Atriplex canescens* (Pursh.) Nutt. improved by formation of vesicular-arbuscular mycorrhizae. Proc Soil Sci Soc Am 38:962–965

Wilson JM (1984) Comparative development of infection by three vesicular-arbuscular mycorrhizal fungi. New Phytol 97:413–426

Wilson GWT, Hartnett DC (1998) Interspecific variation in plant responses to mycorrhizal colonization in tallgrass prairie. Am J Bot 85:1732–1738

Wright DP, Read DJ, Scholes JD (1998) Mycorrhizal sink strength influences whole-plant carbon balance of *Trifolium repens* L. Plant Cell Environ 21:881–891

Wu QS, Xia RX, Zou YN, Wang GY (2007) Osmotic solute responses of mycorrhizal citrus (*Poncirus trifoliata*) seedlings to drought stress. Acta Physiol Plant 29:543–549

Wu QS, Zou YN, He XH (2010) Contributions of arbuscular mycorrhizal fungi to growth, photosynthesis, root morphology and ionic balance of citrus seedlings under salt stress. Acta Physiol Plant 32:297–304

Xia T, Apse MP, Aharon GS, Blumwald E (2002) Identification and characterization of a NaCl-inducible vacuolar Na^+/H^+antiporter in *Beta vulgaris*. Physiol Plant 116:206–212

Yamada S, Katsuhara M, Kelly WB, Michalowski CB, Bohnert HJ (1995) A family of transcripts enpoding water channel proteins: tissue-specific expression in the common ice plant. Plant Cell 7:1129–1142

Yamada S, Komori T, Myers PN, Kuwata S, Kubo T, Imaseki H (1997) Expression of plasma membrane water channel genes under water stress in *Nicotiana excelsior*. Plant Cell Physiol 38:1226–1231

Yang CW, Chong JN, Kim CM, Li CY, Shi DC, Wang DL (2007) Osmotic adjustment and ion balance traits of an alkali resistant halophyte *Kochia sieversiana* during adaptation to salt and alkali conditions. Plant Soil 294:263–276

Yano-Melo AM, Saggin OJ, Maia LC (2003) Tolerance of mycorrhized banana (*Musa sp.* cv. Pacovan) plantlets to saline stress. Agric Ecosyst Environ 95:343–348

Zai XM, Qin P, Wan SW, Zhao FG, Wang G, Yan DL, Zhou J (2007) Effects of arbuscular mycorrhizal fungi on the rooting and growth of beach plum (*Prunus maritima*) cuttings. J Hortic Sci Biotechnol 82:863–866

Zeuthen T (2001) How water molecules pass through aquaporins. Trends Biochem Sci 26:77–81

Zhang X-H, Hodson I, Williams JP, Blumwald E (2001) Engineering salt-tolerant *Brassica* plants: characterization of yield and seed oil quality in transgenic plants with increased vacuolar sodium accumulation. Proc Natl Acad Sci USA 98:12832–12836

Zhang YF, Feng G, Li XL (2003) The effect of arbuscular mycorrhizal fungi on the components and concentrations of organic acids in the exudates of mycorrhizal red clover. Acta Ecol Sin 23:30–37

Zhang H, Wu X, Li G, Qin P (2011) Interactions between arbuscular mycorrhizal fungi and phosphate-solubilizing fungus (*Mortierella sp.*) and their effects on *Kostelelzkya virginica* growth and enzyme activities of rhizosphere and bulk soils at different salinities. Biol Fertil Soil 47:543–554

Zhi H, He CX, He ZQ, Zou ZR, Zhang ZB (2010) The effects of arbuscular mycorrhizal fungi on reactive oxyradical scavenging system of tomato under salt tolerance. Agric Sci China 9:1150–1159

Zhu JK (2003) Regulation of ion homeostasis under salt stress. Curr Opin Plant Biol 6:441–445

Zuccarini P, Okurowska P (2008) Effects of mycorrhizal colonization and fertilization on growth and photosynthesis of sweet basil under salt stress. J Plant Nutr 31:497–513

Chapter 14
Breeding Salinity Tolerance in Citrus Using Rootstocks

Maria Angeles Forner-Giner and Gema Ancillo

Abstract Citrus is a salt-sensitive crop and is influences at low concentrations of salt. Soil and/or water salinity can severely affect the growth and normal physiological processes in citrus trees. In citrus, rootstock is the most important component of the tree to assess salt tolerance or sensitivity. Hence, the choice of the appropriate rootstock plays a crucial role in yield. Numerous experiments have shown that exclusion of Cl^- and Na^+ is a hereditary/genetic function. Thus, selection of suitable parental genotypes would restrict the translocation of Cl^- or Na^+ in the grafted variety or developed hybrids.

In countries such as Spain, Australia, USA and France, one of the main goals of the citrus breeding programs is to obtain new rootstocks tolerant to salinity. In Spain during 1974, a program began at the Valencian Institute for Agricultural Research (IVIA) to breed citrus rootstocks by hybridizations, in which more than 500 hybrids were evaluated to determine their agronomic performances including tolerance to salinity. Several new commercial rootstocks have been produced in this breeding program.

Keywords Salinity stress • Citrus • Rootstock • Variety • Breeding

Salinity is a widespread problem. It is estimated that one third of the land area is affected to a due to salinity to various degrees of salinity. In nature, many organisms have adapted to saline environments for better survival (Moya 2000). When saline soils are cultivated, or saline water is used for crop irrigation, crop yield decreases.

M.A. Forner-Giner • G. Ancillo (✉)
Instituto Valenciano de Investigaciones Agrarias (IVIA), Ctra Moncada-Náquera, Km. 4.5, Valencia 46113, Spain
e-mail: ancillo@ivia.es

P. Ahmad et al. (eds.), *Salt Stress in Plants: Signalling, Omics and Adaptations*, DOI 10.1007/978-1-4614-6108-1_14, © Springer Science+Business Media New York 2013

High salinity levels may be a limiting factor for plant growth (Ahmad 2010; Ahmad et al. 2012; Katare et al. 2012), and it may seriously affect the agricultural economy of the affected regions.

Originally, salinity problems appeared in arid and semiarid areas, where rain was not enough to wash the salts of the soil. But standard cultural practices used over the years resulted in secondary salinization, which affects soil characteristics and vegetation cover. On the other hand, the need to produce more agri-horticultural produce has forced the use of marginal lands for cultivation, which are often saline (Moya 2000). In order to optimize the plant productivity criteria like improving the soil drainage to facilitate the leaching of salts, improvement of irrigation and selection of salt tolerant crops is necessary. For the selection of salt tolerant crops it is essential to know the response of the plants to salt stress and, secondly, the development of new genotypes with better adaptability to problem of salinity is some of the solutions.

Plants are classified in two groups according to their tolerance or sensitivity to salt problems:

(1) Halophytes: Those plants able to complete its life cycle in saline environments and (2) Glycophytes: Those plants that are not able to survive in saline environments.

There is no clear demarcation between the two groups, as there is a continuous spectrum ranging from the species most resistant to the most sensitive (Greenway and Munns 1980; Gorham et al. 1985).

In higher plants, the physiological mechanisms determining the sensitivity or tolerance to salinity vary with different species (Munns and Tester 2008; Kader and Lindberg 2010). In some cases, in response to high concentrations of salts in the medium, the plants tolerate the accumulation of certain ions inside. However, in most salt tolerant species, the tolerance is determined by their ability to exclude salts, restricting the absorption and/or the transport of toxic ions from the roots to the shoot (Parida and Das 2005). The glycophytes often have an inverse relationship between salt uptake and tolerance, so that exclusion is the predominant coping mechanism (Munns and Tester 2008).

Effects of Salinity in Citrus

Citrus species, as a whole, can be considered as susceptible to salinity. However, there are important differences between species and varieties belonging to the *Aurantiodea* family with regard to their capacity for absorption and translocation of different ions that may be present in the culture medium. This affects both the essential elements, such as nitrogen, phosphorus, potassium, etc., and those considered as inducers of salt problems, mainly chloride (Cl^-) and sodium (Na^+) ions. The absorption of Cl^- and Na^+ present in the culture medium, even at moderate concentrations may cause significant physiological changes in citrus, such as reduction of growth and development, necrosis of leaves and shoots, defoliation, etc. (Fig. 14.1) (Kirkpatrick and Bitters 1969; Walker et al. 1982; Behboudian et al. 1986; Syvertsen and Yelenosky 1988). Regarding

Fig. 14.1 Reduction of growth and foliar damage caused by salinity in Carrizo Citrange leaves

the mechanisms of response to salinity, there is considerable variability, with plants exhibiting high absorption of Na^+ and Cl^- exclusion (for example, Cleopatra mandarin) or the other way around (as in the case of *Poncirus trifoliata*). Moreover, some varieties present a high capacity for Cl^- absorption but with a reduced internal translocation, accumulating most of the absorbed Cl^- in the roots.

Presently, it is considered that the sensitivity to salinity in citrus species and varieties is associated with the accumulation of excessive concentrations of Cl^- in leaves, what implies a high absorption of this ion by the roots, as well as its efficient translocation (Grieve and Walker 1983). In contrast, tolerance to salinity in citrus is related to the ability to restrict the uptake and/or the transport of salt ions to the shoot (Cerdá et al. 1979; Mobaye and Milthorpe 1980; Zekri and Parsons 1992).

Taking into account that commercial citrus are usually formed by grafting a variety onto the rootstock, the response of trees to salinity depends on the individual behavior of the component parts, as well as on the possible interactions between scion and rootstock that may occur. However, as demonstrated by numerous studies and culture practices, it is evident that the component mainly involved in the tree salinity response is the rootstock. Hence, the choice of the appropriate rootstock is crucial to obtain the maximum yield in citrus crops, both in saline and other conditions (Forner-Giner et al. 2011a, b).

Effect of Salinity on Growth

Reduction of plant growth may be related to either the total concentration of soluble salts or to the root osmotic potential. Both, vegetative growth and production decrease with the increase in salt concentration (Blum and Johnson 1992). In citrus, there is a slowdown in growth and a change in water and nutritional relationships in plants undergoing saline treatments (Syvertsen and Yelenosky 1988; Nieves et al. 1991b; Bañuls and Primo-Millo 1995; Ruiz et al. 1997; Nastou et al. 1999).

It has been reported that rootstocks have a greater influence than scion on plant growth under saline conditions (Bañuls and Primo-Millo 1995; Ruiz et al. 1997;

Table 14.1 Dry weights (g) of Valencia Late orange budded on different rootstocks treated 60 days with NaCl

	Leaves	Stems	Roots	Plants
Cleopatra m.				
0 mM (NaCl)	19.74±2.39[de]	25.98±4.08[c]	20.28±1.98[f]	66.01± 6.06[f]
20 mM (NaCl)	18.68±3.80[cde]	25.10±6.26[c]	19.22±2.04[ef]	63.00±11.68[f]
40 mM (NaCl)	18.48±5.93[cde]	22.78±8.54[c]	17.97±3.29[def]	59.23±16.21[f]
60 mM (NaCl)	13.85±4.80[abc]	23.90±8.37[c]	16.95±3.93[def]	54.70±16.31[ef]
P. trifoliata				
0 mM (NaCl)	13.56±5.85[abc]	10.37±4.66[ab]	11.46±3.47[abc]	35.39±13.81[abcd]
20 mM (NaCl)	10.85±2.43[ab]	8.07±1.73[ab]	10.55±2.19[ab]	29.48± 5.66[abc]
40 mM (NaCl)	8.64±1.97[a]	5.85±1.31[a]	7.95±0.83[a]	22.45± 3.54[a]
60 mM (NaCl)	8.57±4.65[a]	6.18±2.58[a]	9.80±2.61[a]	24.56± 9.67[ab]
F&A 5				
0 mM (NaCl)	21.64±2.55[e]	20.36±2.76[c]	17.56±5.44[def]	59.56± 8.75[f]
20 mM (NaCl)	17.77±0.81[cde]	13.43±1.43[b]	12.25±0.81[abc]	43.46± 1.43[de]
40 mM (NaCl)	15.14±4.49[bcd]	13.55±4.37[b]	12.28±4.93[abc]	40.96±13.49[cde]
60 mM (NaCl)	14.94±3.86[bcd]	11.78±2.94[ab]	10.94±4.89[abc]	37.67±10.53[bcd]
F&A 13				
0 mM (NaCl)	14.77±4.40[bcd]	10.95±1.62[ab]	15.15±2.23[cde]	40.87±7.09[cde]
20 mM (NaCl)	15.30±4.17[bcd]	10.76±2.83[ab]	14.73±2.03[bcd]	40.79±7.63[cde]
40 mM (NaCl)	13.70±1.02[abc]	10.01±1.96[ab]	14.62±1.94[bcd]	38.34±3.83[bcd]
60 mM (NaCl)	8.74±1.62[a]	6.40±0.79[a]	9.32±1.61[a]	24.46±2.20[ab]

Data are means (n = 4) ± s.d. Means within a line with the same superscript letter are not significantly different (P ≥ 0.05)

Forner-Giner et al. 2011), although it has also been found an effect of the scion (Levy et al. 1992; García-Legaz et al. 1993; Ruiz et al. 1997), as well as of the grafted variety/rootstock combination (García-Lidón et al. 1998).

Salinity reduces the growth of new shoots as well as the total dry weight of sensitive rootstocks (Table 14.1), while does not affect the tolerant ones (García-Legaz et al. 1993; Lea-Cox and Syvertsen 1993; El-Hag and Sidahmed 1997; Nastou et al. 1999).

Field experiments revealed that Cl^- ion is crucial in inhibiting the growth of citrus (Chapman 1968). However, experiments accomplished in greenhouse under controlled conditions, have suggested that inhibition may be caused not only by Cl^- (Walker et al. 1982; Bañuls and Primo-Millo 1992; Bañuls et al. 1997), but also by Na^+ (Behboudian et al. 1986; Lloyd et al. 1987a, 1990; Lloyd and Howie 1989b) or by both ions (Lloyd et al. 1989).

The presence of Ca^{2+} in the medium reduces the effects of salinity on plant growth (Zekri and Parsons 1990a; Romero-Aranda et al. 1998). This phenomenon has been attributed to various effects of Ca^{2+}:

1. It prevents the absorption of toxic levels of Na^+ and allows the absorption of K^+ (Waisel 1962).
2. It maintains the selective permeability of the membranes (Munns and Tester 2008).
3. It contributes to create salt excess, which may cause flocculation of clay particles in the soil by the Na^+ (Hanson 1999).

When sour orange plants were treated with different type of salts (NaCl, KCl or SO_4Na_2) at different concentrations the growth was inhibited even at the lowest concentrations used (Zid and Grignon 1986; Salem and El-Khorieby 1989). However, the highest concentrations induced leaf necrosis and leaf fall in winter. Bañuls et al. (1997) found that the chloride salts (NaCl and KCl) significantly reduced growth in two different rootstocks (Cleopatra mandarin and *Poncirus trifoliata*) grafted with the Valencia variety. In contrast NO_3Na produced no effect.

Salinity also causes delay and reduction of seedling emergence, it reduces root and aerial biomass and it alters the mineral composition of plants, although the magnitude of these changes depends on the used rootstock (Zekri 1993; El-Desouky and Atawia 1998).

Effect of Salinity on Mineral Nutrition and Ion Interactions

Nitrogen

Nitrogen is the most essential element for plant development, conditioning on a high level the citrus production. It is a component of plant proteins and it is present in lignins and auxins. It stimulates the production of chlorophyll, the sprouting, etc.

Lea Cox and Syvertsen (1993) suggested that nutrient uptake might be affected by the osmotic effect of salt solutions, which decreases the transpiration. They found that *Citrus volkameriana* was more sensitive to Cl^- absorption than Cleopatra mandarin. Saline stress caused an equal reduction in the absorption of nitrogen in both rootstocks, with no evidence of antagonism between Cl^- and nitrate ion (NO_3^-) uptake. On the other hand, the total absorption of NO_3^- was positively correlated with transpiration in both species. Thus, it appears that the reduction in NO_3^- uptake is more probably related to the decrease of water than to the Cl^- antagonism caused by salt stress. However, other studies concluded that the inhibition of NO_3^- uptake in citrus plants exposed to salt solutions appears to be related mainly to tissue accumulation of Cl^-, being independent of the rate of transpiration (Tanner and Beevers 1990; Cerezo et al. 1997, 1999).

Cerezo et al. (1997) studied the effect of salinity and transpiration in the absorption of nitrate by Troyer citrange seedlings. Results showed that the Cl^- accumulation after saline treatment decreased NO_3^- root uptake by 80% and its accumulation between 70% and 80%. Under salt stress conditions, NO_3^- uptake and transpiration processes appear to be independent and NO_3^- uptake decrease is mainly due to salinity.

Reduction in nitrogen content in plants subjected to salinity may be due to either inhibition of nitrate absorption caused by interactions between NO_3^- and Cl^- in ion transport sites (Aslam et al. 1984; Feigin 1985; Mclure et al. 1982; Cerezo et al. 1997), or to the different ability of the rootstocks to distinguish between different available ions, showing specific preference for one or another (Bañuls et al. 1990).

Syvertsen and Yelenosky (1988) concluded that the effect of salinity on the mineral composition of tissues depends on the rootstock. In high salinity, leaves of *Poncirus trifoliata* accumulated greater amount of total nitrogen, potassium ion (K^+) and Cl^- but lower Na^+ amount compared to Pineapple sweet orange and Cleopatra mandarin. Similarly, the roots of *P. trifoliata* contained more Na^+ in high salinity conditions. By contrast, in the same conditions, leaves of Cleopatra mandarin had the lowest levels in Cl^- and K^+. Moreover, saline treatments with NaCl on sour orange plants showed, apparently, no effect on nitrogen nutrition (Zid and Grignon 1986).

Calcium

Calcium is an essential element in plant nutrition and regulates the action of growth hormones, abscission processes, senescence, maturation, secretion of cell wall, cell division and elongation, osmoregulation, role and structure of the membrane, ion transport, guard cells and photosynthesis. Some of these processes can be affected in stress conditions (salinity, drought, cold) and calcium can improve the unfavorable conditions undergoing by plants.

High amounts of Na^+, in soils or solutions, reduce plant absorption of K^+ and calcium ion (Ca^{2+}) and/or affect the internal distribution of these elements in the plant (Grattan and Grieve 1992). Zid and Grignon (1985) showed that the translocation of Na^+ to the leaves, in sour orange plants, was due to apoplasmic displacement of Ca^{2+}.

Changes in transport and accumulation of essential nutrients due to salinity result in lower concentrations of Ca^{2+} in leaves (Bañuls et al. 1990; Alva and Syvertsen; 1991; García-Lidón et al. 1998; Ruiz et al. 1999). In roots of *Citrus volkameriana*, *Citrus macrophylla* and sour orange, García-Lidón et al. (1998) found that the concentration of Ca^{2+} is hardly affected by salinity. However other authors have reported a decrease in Ca^{2+} concentration produced by salinity (Bañuls et al. 1990; Alva and Syvertsen 1991; Ruiz et al. 1999).

An important variable to be considered for the evaluation of the effects of salinity on plants is the ratio Na^+/Ca^{2+} in the saline treatment solution. Additional Ca^{2+} supply to the nutrient solution (as SO_4Ca or (NO_3) 2Ca), reduces some effects of salinity on citrus (Zekri and Parsons 1990b; Bañuls et al. 1991).

Bañuls et al. (1991) observed that growth inhibition of citrus plants produced by high NaCl concentrations could be mitigated with a supplement of Ca^{2+}. The increase in Ca^{2+} reduced Na^+ uptake in the roots and leaves of plants grafted on Troyer citrange. However, in roots of Cleopatra mandarin Na^+ concentration increased with that of Ca^{2+}. The distribution of Na^+ in plants grafted on Cleopatra mandarin, suggested that increasing concentrations of Ca^{2+} reduced Na^+ translocation to the leaves, so that its accumulation in the basal part of stem and roots is increased. The authors suggested that the Ca^{2+} reduced the transport of Na^+ from roots to leaves on Cleopatra mandarin plants what produced an accumulation of Na^+ in roots, what in addition, could caused inhibition of K^+ uptake.

Moreover, Ca^{2+} restricts Cl^- accumulation in leaves (Bañuls et al. 1991). This effect may be explained by the Ca^{2+} ability to remove Cl^- from xylem flow, particularly in the basal part of stem and roots. Bañuls et al. (1991) suggested that Ca^{2+} plays a regulatory role in membrane permeability for Cl^- and its transport.

Inhibition of Na^+ uptake under high concentrations of Ca^{2+} in the nutrient solution, have also been observed in other plant species (LaHaye and Epstein 1969; Cramer et al. 1987). Also in these cases, adequate levels of Ca^{2+} to maintain membrane integrity and K^+/Na^+ selectivity are required (LaHaye and Epstein 1969; Hanson and Kitz 1982; Kent and Laychi 1985; Cramer et al. 1986, 1987; Bañuls et al. 1997).

Potassium

Potassium, along with calcium, is the major constituent of the mineral matter in plants. It stimulates nitrate uptake and regulates the opening and closing of stomata. It has an important role in the synthesis of carbohydrates, as well as in harvest quality and quantity.

Reduction in K^+ absorption, induced by Na^+, is a well-known process (Cramer et al. 1985; Ball et al. 1987; Cerdá et al. 1995). It has been reported that K^+ deficiency induced by Na^+ is responsible for the reduction of CO_2 assimilation rate (Ball et al. 1987). Greater selectivity for K^+ over Na^+ has been associated with tolerance to salinity (Storey and Walker 1987).

Salinity reduces the concentration of K^+ in the roots (Walker 1986; Bañuls et al. 1990; Alva and Syvertsen 1991; Ruiz et al. 1999; Tozlu et al. 2000a). However, it is unclear what happens to the K^+ concentration in leaves. Some experiments revealed that K^+ concentration decreases in saline conditions (Gallasch and Dalton 1989; Bañuls et al. 1990; Alva and Syvertsen 1991), others found that the concentration did not change significantly (Nieves et al. 1990; Ruiz et al. 1999), and others showed even an increase in K^+ leaf concentration (Tozlu et al. 2000a). García-Lidón et al. (1998) observed that, in different variety/rootstock combinations under salinity conditions, foliar K^+ concentration was maintained or increased in all combinations studied while it was reduced in the rootstock. The authors attributed this effect either to a direct action of Na^+, shifting the K^+, or to a loss of K^+ in root tissues. According to Gallasch and Dalton (1989), levels of K^+ are negatively correlated with those of Na^+ and Cl^- in leaves. So that, it can be assumed that rootstocks with the ability to exclude effectively Na^+ and Cl^- would have higher levels of K^+ in leaves than the weak Na^+ and Cl^- excluders.

Magnesium

Magnesium plays an important role in plant physiology since, in addition to be present in chlorophyll and other pigments, acts as an activator of many enzymes and plays a roll in the transport of phosphorus within the plant.

We found diverse results regarding the effects of salinity on the magnesium absorption and transport in citrus plants. Several experiments indicated that salinity might reduce the foliar concentration of magnesium ions (Mg^{2+}) (Bañuls et al. 1990;

Alva and Syvertsen 1991; García-Lidón et al. 1998; Ruiz et al. 1999) while others suggest that it may have no effect (Nieves et al. 1990). In roots, it appears to exist a greater agreement as to the reported results. Mg^{2+} levels decreases in response to saline treatments (Bañuls et al. 1990; Alva and Syvertsen 1991).

Phosphorus

Phosphorus is an essential nutrient both as a part of several key plant structure compounds and as a catalyst for the conversion of numerous key biochemical reactions in plants. Phosphorus is a key nutrient, especially for its role in capturing and converting sun energy into useful plant compounds. It is a vital element of genetic inheritance (as a component of nucleic acids, DNA and RNA) and it is the basic component of ATP.

Tozlu et al. (2000) observed that, in plants subjected to saline treatments, phosphorus decreased in root tissues and in turn, increased in the leaves (Nieves et al. 1990; Tozlu et al. 2000a). However Gallasch and Dalton (1989) maintain that foliar levels of Na^{+} and Cl^{-} are not significantly associated with leaf levels of phosphorus.

Morphological, Anatomical and Histological Citrus Grown in Saline Conditions

Changes in Root Anatomy

Root system is one of the main components of the tree, being responsible for the absorption of water and nutrients. The roots have an important role in vegetative growth, in the length of the juvenile period, in fruit quality and production and in tree life cycle. Root growth is both horizontal and vertical, and it is affected by many soil factors such as texture, structure, aeration and temperature. Root growth may also be influenced by cultural practices, orchard type, etc. (Geisler 1962; Praloran 1971, and Huguet 1976; Castle et al. 1993; Tuzcu et al. 1997).

In citrus, morphology of root systems is bimorph (Castle 1978) and consists of a network of relatively shallow lateral roots and a secondary layer of small lateral and fibrous roots more or less vertically oriented. Different rootstocks differ in the lateral and vertical distribution of roots (Ford 1954; Castle and Krezdorn 1975) so that less vigorous rootstocks are those with shallower root systems.

Difference in texture between sand culture and liquid medium also appears to have a significant effect on root morphology and in the plant response to salinity (Storey 1995). Citrus rootstocks grown in liquid medium were found to accumulate much higher levels of Na^{+} and Cl^{-} in leaves than those of sand cultured plants (Storey 1995). Sand grown plants had many branched, suberized fibrous roots compared to those of plants grown in liquid medium.

Salinity can reduce the dry weight of root system and the length of fibrous roots (El-Desouky and Atawia 1998; García-Sánchez et al. 2000; Tozlu et al. 2000).

The citrus fibrous roots have an outer layer of epidermal cells covering a hypodermis consisting of a single layer of large cells as well as occasional passage cells or short cells (Walker et al. 1984; Storey and Walker 1987; Peterson and Enstone 1996). The outer periclinal and radial walls of the hypodermis are lignified and inner walls are suberized (Cossman 1940; Walker et al. 1984). How these deposits and cell wall connections between cells of the epidermis and hypodermis influence water and ion permeability is uncertain. To date, no genotypic differences in the structure and chemical composition of the root hypodermis have been reported in citrus (Walker et al. 1984; Storey and Walker 1987; García-Sánchez et al. 2000).

Root suberized hypodermis represents the first potential barrier to the mass flow of water and ions in citrus (Walker et al. 1984). Entry of water and ions into the symplast is most likely to occur at the surface of the epidermal–passage cell complex (Storey and Walker 1987), which are connected by plasmodesmata (Walker et al. 1984).

Citrus subjected to salinity produce an increase in suberization and a decrease in lateral root development (Hayward and Blair 1942). Kriedmann and Barrs (1981) suggested that highly suberized root systems might limit water supply to the grafted variety.

Suberization of endodermis walls is the second constrain to the mass flow of water and ions across the root. The structure of the primary root of citrus shows the existence of a Casparian strip in endodermis cells (Cossman 1940; Hayward and Blair 1942) and passage cells with functional plasmodesmata (Walker et al. 1984). Secondary suberization or deposition of suberin on the endodermal walls serves to block the plasmodesmata, suggesting that endodermal passage cells play an important role in the regulation of water and ion entry into the xylem (Walker et al. 1984).

In citrus, root permeability may vary along the time, because root growth is not continuous but intermittent (Bevington and Castle 1985) and the extent of suberization varies with the growth activity of the root (Castle 1978). During periods of low root growth (as winter dormancy) meristematic activity stops and suberization of fibrous roots may result in encapsulation of the whole root system, including the meristem (Castle 1978).

Lateral root emergence leads to discontinuities in the endodermis (Sanderson 1983) and, possibly too, in the hypodermis, and thus provides entry points of water in the apoplast. Highly branched fibrous roots, such as citrus ones, may have many real or potential points of water entry into the apoplast (Swietlik 1989).

Changes in Stem and Leaves

Also at this point there are controversies about the results obtained in different experiments. It has been reported that salinity may either reduce (Zekri and Parsons 1989, 1990a; Zekri 1991; Bar et al. 1998) or increase (Lea-Cox and Syvertsen 1993; Combrink et al. 1995) the ratio shoot/root in citrus plants. A relative increase in root mass or in absorption surface with respect to aerial part (sink), might likely raise the

Fig. 14.2 Foliar damage caused by NaCl treatment in Carrizo Citrange leaves

concentration of Na^+ and Cl^- in the leaf cells. It might also reduce total dry weight, stem section and total leaf area (Syvertsen et al. 1988; Aksoy et al. 2000).

Increase in salt-induced succulence in leaves has been associated with an increase in the size of spongy mesophyll cells (Cerdá et al. 1977; Zekri and Parsons 1990a; Nastou et al. 1999; Aksoy et al. 2000). It is thought that the succulence observed in leaves of plants growing in saline environments represents an adaptive plant response, as it provides the basic structure for the dilution of the accumulated salts (Romero-Aranda et al. 1998).

Saline tolerance in citrus may be associated with different Cl^- accumulation in the stem and leaf tissues (Cooper and Gorton 1952; Grieve and Walker 1983; Storey and Walker 1987). For Storey (1995), there is not evidence of citrus salinity tolerance associated with preferential accumulation of Cl^- in stem tissues. In the same way, there is little evidence of preferential accumulation of Cl^- in roots, because the levels of Cl^- were similar in the genotypes studied (Rangpur lime and Etrog citron). Figure 14.2 shows foliar damage caused by NaCl treatment in Carrizo Citrange leaves.

Effect of Salinity on Photosynthesis

Several works, in citrus plants, have shown that CO_2 assimilation rate and stomatal conductance decreased under conditions of salinity (Fig. 14.2) (Walker et al. 1982,1993; Behboudian et al. 1986; Lloyd et al. 1987a, b, 1989, 1990; Bañuls and Primo-Millo 1992, 1995; García-Legaz et al. 1993; Bañuls et al. 1997; MacHacha et al. 2000; Morinaga and Sykes 2001). Magnitude of the reduction in CO_2 assimilation and stomatal conductance is influenced by both the rootstock and the variety (Lloyd et al. 1989, 1990; García-Legaz et al. 1993; Bañuls and Primo-Millo 1995).

Possible causes of the reduction in CO_2 assimilation induced by salinity could be related to:

- Turgor reduction.
- Negative feedback regulation of the reactions of the Calvin cycle.
- Na^+-induced K^+ deficiency.
- Ion toxicity produced by Na^+ or Cl^-.

The down-regulation the Calvin cycle as a cause of the reduction of CO_2 assimilation rates in citrus grown in saline conditions, is unlikely, because foliar soluble sugars concentrations are either unaffected (Walker et al. 1993) or decreased (Lloyd and Howie 1989b). Similarly, reductions in CO_2 assimilation rates occur without a subsequent decrease in foliar K^+ concentrations (Walker et al. 1993), indicating that deficiency of K^+-induced by Na^+ is not a significant factor in reduction of CO_2 assimilation rates in citrus. While cause-effect relationships between the reduction in CO_2 assimilation rates and high levels of Cl^- and/or Na^+ in leaves has not been established, several works have found a good correlation between the reduction of CO_2 assimilation and high leaf content of Cl^- (Walker et al. 1982; Lloyd et al. 1989; Bañuls and Primo-Millo 1992; García-Legaz et al. 1993; Romero-Aranda et al. 1998) others found a correlation with high levels of Na^+ (Behboudian et al. 1986; Lloyd et al. 1987b, 1990; García-Legaz et al. 1993; Walker et al. 1993; Nastou et al. 1999), and others have concluded that neither Cl^- or Na^+ were directly responsible for the reduction in CO2 assimilation rates (Walker et al. 1982; Syvertsen et al. 1988; Bañuls and Primo-Millo 1995; Romero-Aranda and Syvertsen 1996; Bañuls et al. 1997).

It is possible that the diverse responses found in photosynthesis, could be explained by the lack of homogeneity in the experiment conditions: different graft/rootstock combinations, different ages and sizes of trees, different saline treatments, that could account for an unequal entry of Cl^- or Na^+ into leaves and subsequent charge compensation or compartmentalization of ions for normalization of physiological processes (Storey and Walker 1999).

Effect of Salinity on Transpiration

Citrus rootstocks show large differences in transpiration rates per unit leaf area per unit time. Several studies showed that transpiration rate is higher in salt-sensitive rootstocks (Graham and Syvertsen 1985; Syvertsen and Graham 1985; Walker 1986). However other studies found that transpiration rates of sensitive and tolerant rootstocks are similar (Syvertsen and Yelenosky 1988; Walker et al. 1983; Storey 1995). In general, it is accepted that salinity reduces transpiration (Walker et al. 1983; Storey 1995; Rodriguez-Gamir et al. 2012).

García-Legaz et al. (1993) observed that plants grafted on the rootstock *Citrus macrophylla* had higher transpiration rates than those grafted on sweet orange or *Citrus volkameriana*. Experiments accomplished by the same authors showed that the effect of the variety was also significant, with the result that Verna lemon plants showed the lowest transpiration rates.

Even though rootstocks with different tolerance to salinity, as Etrog citron and Rangpur lime, may have similar transpiration rates, the water flux through their roots may be notably different. In any case, a reduced water flux through the roots is not necessarily associated with a greater capacity to exclude Cl^- (Storey 1995).

Citrus Rootstock Tolerance to Salinity

It is long known that there are differences in salt tolerance between rootstocks and they are known to have a marked influence on the amount of Cl^- and/or Na^+ that accumulate in the leaves (in both grafted or non-grafted trees) (Wutscher 1979). When comparing the best and the worst rootstocks in regulating Cl^-, the range of variation in the concentration of Cl^- in leaves may be greater than ten times (Cooper and Gorton 1952; Peynado and Young 1969). However, differences in the concentration of Na^+ are usually lower than for Cl^-, being about six times (according to Kirkpatrick and Bitters 1969) or 4.5 times (according to Cooper and Gorton 1952).

Grieve and Walker (1983) suggested the existence of mechanisms, apparently different, which regulate the absorption and transport of Cl^- and Na^+ in plants undergoing salt stress.

Cooper (1961) considered the Rangpur lime (*Citrus limonia* Osb.), the Cleopatra mandarin and *Severina buxifolia* (Poir.), the most effective rootstocks in restricting the transport of Cl^- toward the scion. These results were confirmed later on with the finding that the Cleopatra mandarin (Chapman 1968; Ream and Furr 1976; Newcomb 1978; Syvertsen et al. 1988; Boman 1993; Castle et al. 1993; Bañuls et al. 1997; Levy et al. 1999) and the Rangpur lime (Walker et al. 1983) have the ability to exclude Cl^- during saline treatment, although they are unable to restrict the absorption of Na^+.

In conditions of low levels of salinity, *Poncirus trifoliata* is a weak Cl^- excluder (Peynado and Young 1969; Sykes 1985b; Castle 1987; Syvertsen and Yelenosky 1988) but a strong Na^+ excluder (Syvertsen and Yelenosky 1988; Bañuls et al. 1997). The ability of *P. trifoliata* to restrict the accumulation of Na^+ in the leaves, seems to be related to the property of reabsorbing Na^+ from the xylem sap, particularly in the region of the root near the base of the stem, but also, although to a lesser extent, in the most mature parts of the fine roots (Walker 1986).

The capacity to exclude the absorption of Cl^- and Na^+ is inherited, as evidenced by numerous studies. Thus, the ability of Rangpur lime to restrict the transport of Cl^- toward the variety can be present in their hybrids (Ream and Furr 1976; Gallasch and Dalton 1989; Sykes 1992). Similarly, the trait of *P. trifoliata* to restrict the transport of Na^+ toward the variety can be expressed in their hybrids (Fig. 14.1) (Sykes 1992).

Ream and Furr (1976) analyzed the content of Cl^- in leaves of Redblush grapefruit and lemon grafted on different rootstocks, finding that some sour oranges showed a significant difference in their absorption and transport of Cl^- in leaves, but none of them showed a tolerance comparable to that of Cleopatra mandarin. In another study, Hamlin sweet orange and lemon were grafted on 32 hybrids coming from different progenitors. Twenty out of them showed a tolerance similar to Cleopatra mandarin (that was parent of 8 of the hybrids studied; being Rangpur lime parent of 10 of them).

Troyer and Carrizo citranges take up much Cl^- compared with Rough lemon and *Citrus volkameriana* (Combrink et al. 1995). The Carrizo citrange has inherited the capacity to exclude Na^+ from *P. trifoliata* since, despite absorbing a large amount of Cl^-, it shows a very effective exclusion of Na^+ at low salt concentrations (Boman 1993). The same applies to the Troyer citrange, which restricts the transport of Na^+ toward the grafted variety (Bañuls et al. 1990).

The citrumelo 4,475 shows sensitivity to salinity less efficiently than Carrizo citrange (Boman 1993), and very similar to sweet orange. And *C. volkameriana* is more sensitive to salinity than Cleopatra mandarin (Lea-Cox and Syvertsen 1993).

According to Cooper and Gorton (1951) plants grafted on sour orange are less tolerant to salinity than when grafted on Cleopatra mandarin. Zekri (1991) attributed this to the inability of sour orange to exclude Cl^-. Similar results were obtained by Levy et al. (1999), in a field study, comparing the salt tolerance of 6 rootstocks. It was found that the accumulation of Cl^- in leaves was lower in Cleopatra mandarin, while in the sour orange the content of this ion was the highest.

Citrus Scion Tolerance to Salinity

Both the rootstock and the grafted variety can influence the accumulation of Cl^- in the leaves of the variety (Lloyd et al. 1989, 1990; Bañuls et al. 1990; Levy and Shalhevet 1990; Nieves et al. 1991a, 1992; García-Legaz et al. 1992, 1993; Bañuls and Primo-Millo 1995). However, the tests that have been accomplished to compare different combinations of rootstocks and varieties showed that the influence of the rootstock in the tree response to salinity is much higher than that exerted by the grafted variety (García-Legaz et al. 1992). The influence of the variety becomes important especially when the rootstock is a weak Cl^- excluder (Lloyd et al. 1989), although it also occurs when variety is grafted on rootstocks having different capacity to exclude Cl^- (Lloyd et al. 1990). However, rootstocks with a good capacity to exclude Cl^- have a greater influence on the level of Cl^- accumulated than the grafted variety (Cooper et al. 1952; Behboudian et al. 1986).

Bhambota and Kanwar (1969) classified some orange varieties, grafted on rough lemon, according to their tolerance to salinity, in the following descending order: Hamlin, Valencia late, Pineapple and Bloodred.

Varieties grafted on Cleopatra mandarin, accumulated less Cl^- in leaves than when grafted on Troyer citrange. However, when clementine or Navel orange are grafted on any of these rootstocks, clementine shows lower levels of Cl^- in leaves than Navel orange with the same rootstock (Bañuls and Primo-Millo 1995).

Lloyd et al. (1989) found that Prior Lisbon lemon leaves accumulated more Cl^-, and faster, than the Valencia orange, whatever they were grafted on *P. trifoliata* or on Troyer citrange. This fact may be related to an increased growth and water requirement of the lemon. Similarly, it was observed that the lemon variety also accumulated more Na^+ than orange, but the difference between them was smaller when grafted on *P. trifoliata* because this rootstock is a strong Na^+ excluder. However, the differences between varieties may vary according to the rootstock (even when the rootstock does not restrict the transport of ions).

The levels of Na^+ in leaves are also determined by both the rootstock (Bañuls et al. 1990; García-Legaz et al. 1992, 1993; Bañuls and Primo-Millo 1995), and variety (Cooper et al. 1952; Lloyd et al. 1989,1990; Levy and Shalhevet 1990; Nieves et al. 1991b). Cooper et al. (1952) observed that Shary Red grapefruit leaves accumulated significantly more Na^+ than Valencia orange, regardless the rootstock used was Cleopatra mandarin or sour orange. In contrast, the accumulation of Cl^-, which was approximately ten times greater than that of Na^+, was determined by the rootstock and not by the variety.

Bañuls et al. (1990) observed no significant differences in the Na^+ levels in leaves of clementine and Navel orange grafted on Cleopatra mandarin. These results differ from those obtained by Behboudian et al. (1986) in which Na^+ accumulation in leaves was more depending on the variety than on the rootstock.

Analysis of leaves of different varieties of lemon trees grafted on sour orange, after 3 years of saline treatments, showed that the content of Na^+ and Cl^- was higher in Primofiori variety than in Verna lemon (Cerdá et al. 1979). In contrast, Verna lemon leaves accumulated less Cl^- than Fino lemon (Nieves et al. 1992).

Assuming the hypothesis formulated by Walker (1986), in the sense that the capacity of *P. trifoliata* to exclude Na^+ is due to the reabsorption of this ion in the root (rootstock) and in the basal part of trunk (variety), Lloyd et al. (1989) argued that the rootstock should reduce the flow of Na^+ in the xylem and, consequently, the grafted variety should only have a slight effect on Na^+ accumulation in the leaves. However, Lloyd et al. (1990) found that Marsh grapefruit leaves accumulate higher levels of Na^+ than those of Valencia orange when both varieties are grafted on *P. trifoliata* and therefore they conclude that the hypothesis of Walker (1986) is just a simplification of the interaction variety/rootstock that occurs in the tree. According to that, Storey and Walker (1999) pointed out that there are some complex interactions between variety and rootstock that modulate Na^+ levels in the leaves that are still unknown.

Effect of Interstock in Salinity

In a comparative study with different salinity levels, Verna lemon was grafted directly on *Citrus macrophylla* and, using a blood orange intermediate variety, on sour orange and Cleopatra mandarin rootstocks. The blood orange intermediate variety restricted the accumulation of Cl^- in Verna lemon leaves, while Cl^- leaf levels were significantly increased in lemon on *C. macrophylla* (Cerdá et al. 1990; Nieves et al. 1992). The authors attributed the low levels of Cl^- in the leaves of Verna lemon on sour orange to the effect of the intermediate variety rather than to the rootstock, based on previous results in which variety/sour orange combinations accumulated high levels of Cl^- in leaves. They also observed that Na^+ levels hardly changed (Nieves et al. 1992). These results suggest that the use of an intermediate variety could lead to an increased resistance of citrus to salinity (Storey and Walker 1999). However, this will have to be determined in larger experiments.

Breeding Citrus Rootstocks

Early works for citrus breeding started in the spring of 1893, when the researchers WT Swingle and H.J. Webber began to accomplish multiple hybridizations with the aim of obtaining new varieties resistant to cold. Soon, other researchers, as TR Robinson, F.M. Savage, E.M. Savage, etc., followed them. Taking advantage of the power of interspecific and intergeneric hybridization held by citrus, they obtained numerous hybrids of sweet orange, mandarin, grapefruit, lemon and lime (*Citrus aurantifolia* (Christm.) Swing) among themselves and with *P. trifoliata*, a deciduous species with remarkable resistance to cold (compared with other members of the subfamily *Aurantioideae* to which it belongs). In the period 1908–1914, they obtained 2,500 hybrids with *P. trifoliata*, as well as numerous tangelos (grapefruit × mandarin) and tangors (sweet orange × mandarin) and many other types of hybrids (Cooper et al. 1962). At that time there was no specific problem with rootstocks, so that the work focused on the breeding of varieties. Later on, many of these hybrids, absolutely useless as commercial varieties, became of major importance as citrus rootstocks, as is the case of Troyer or Carrizo citrange, Swingle citrumelo, etc.

In Australia, the development of new citrus hybrid rootstocks tolerant to salinity is being accomplished through the evaluation of different citrus species selected in China and shipped to Australia for evaluation (Sykes 2011). In the same direction, in France works are under way on molecular breeding of citrus and different tetraploid citrus rootstocks are being evaluated for tolerance to salinity (Mouhaya et al. 2010).

In Spain, breeding works aimed specifically at the development of new rootstocks, are more recent. J. B. Forner initiated a rootstock breeding program through directed hybridization in the spring of 1974 at the Valencian Institute for Agricultural Research (IVIA) (Forner et al. 2003; Forner-Giner et al. 2003). This program was intended to solve some of the problems present in our citriculture and hitherto intractable with known rootstocks. One of the main objectives of the breeding program has been the development of new rootstocks tolerant to salinity (Forner-Giner et al. 2009, 2011a). As a result of this program, two new commercial rootstocks have already emerged: Forner-Alcaide 5 and Forner-Alcaide 13, both of them resistant to salinity (Table 14.1). Both of them showed better CO_2 assimilation in saline conditions than Cleopatra mandarin and *Poncirus trifoliata* (their parental genotypes) (Fig. 14.3) and less Na^+ accumulation in leaves (Fig. 14.4).

Other authors conduct the search for molecular markers to facilitate the selection of citrus rootstocks tolerant to salinity (Brumós et al. 2009, 2010).

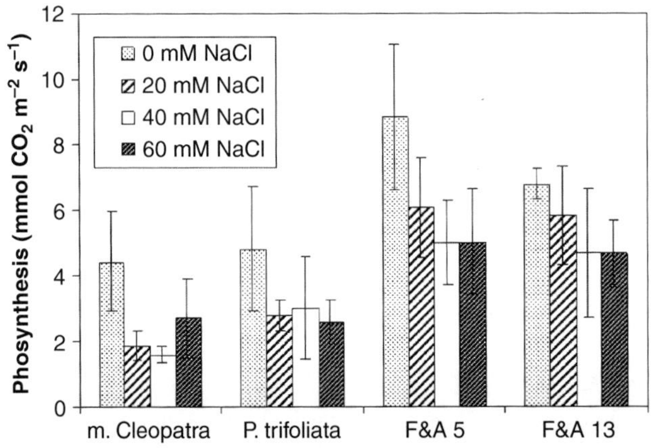

Fig. 14.3 Photosynthesis of Valencia Late orange leaves, grafted on different rootstocks and treated 60 days with NaCl. Bars indicate ± s.d (n=4)

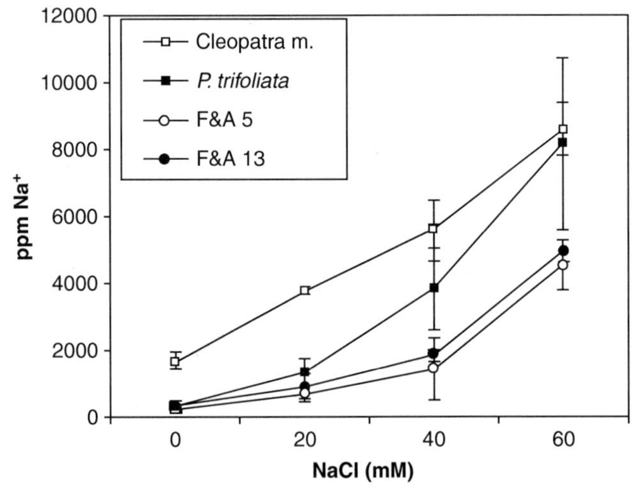

Fig. 14.4 Na^+ concentration in leaves of Valencia Late orange grafted on different rootstocks and treated 60 days with NaCl. Bars indicate ± s.d. (n=4) and are shown where larger than symbols

Conclusion and Future Perspective

Salinity is a major problem in citrus producing areas that suffer it, and the tolerant rootstocks are the only ones able to adapt to that situation. Obtaining and selecting new rootstocks tolerant to salinity is, and will be, one of the most important goals in the different breeding programs in areas with saline soils around the world.

Shortening the period of selection of new rootstocks with more effective screening methods is an essential objective for breeders. Molecular markers and physiological studies would help to understand the mechanisms underlying citrus tolerance or sensitivity to salinity and such an approach would help to evaluate the plan material in a short time.

Acknowledgments This work was financed by the Instituto Nacional de Investigación y Tecnología Agraria y Alimentaria (RTA2008-0060) and Generalitat Valenciana. G. Ancillo and M.A. Forner-Giner are recipient of a contract from Conselleria de Agricultura, Pesca y Alimentación (Generalitat Valenciana, Spain) under Proy_IVIA09/03 and Proy_IVIA09/04, respectively.

References

Ahmad P (2010) Growth and antioxidant responses in mustard (*Brassica juncea* L.) plants subjected to combined effect of gibberellic acid and salinity. Arch Agro Soil Sci 56(5):575–588

Ahmad P, Hakeem KR, Kumar A, Ashraf M, Akram NA (2012) Salt-induced changes in photosynthetic activity and oxidative defense system of three cultivars of mustard (*Brassica juncea* L.). Afri J Biotechnol 11(11):2694–2703

Aksoy U, Hepaksoy S, Can HZ, Anac S, Ul MA, Dorsan F, Anac D, Okur B, Kilic C (2000) The effect of rootstock on leaf characteristics and physiological response of Satsuma mandarins under saline conditions. In: Fokkema NJ, Beek MA, Steekelenburg N, Samyn G, Maas JL, Robinson TL, Verhoyen MNJ (ed) XXV International of agriculture congress, Part 3: culture techniques with special emphasison environmental implications. Acta-Horticulturae, vol 513. Brussels, Belgium, 2–7 August, 1998, pp 169–175

Alva AK, Syvertsen JP (1991) Soil and citrus tree nutrition are affected by salinized irrigation water. Proc Fla State Hort Soc 104:135–138

Aslam M, Huffaker RC, Rains DW (1984) Early effects of salinity on nitrate assimilation in barley seedlings. Plant Physiol 76:321–325

Ball MC, Chow WS, Anderson JM (1987) Salinity-induced potassium deficiency causes loss of functional photosystem II in leaves of the grey mangrove, *Avicennia marina*, through depletion of the atrazine-binding polypeptide. Aust J Plant Physiol 14:351–361

Bañuls J, Primo-Millo E (1992) Effects of chloride and sodium on gas exchange parameters and water relations of Citrus plants. Physiol Plant 86:115–123

Bañuls J, Primo-Millo E (1995) Effects of salinity on some Citrus scion-rootstock combinations. Ann Bot 76:97–102

Bañuls J, Legaz F, Primo-Millo E (1990) Effect of salinity on uptake and distribution of chloride and sodium in some citrus scion-rootstock combinations. J Hortic Sci 65:715–724

Bañuls J, Legaz F, Primo-Millo E (1991) Salinity-calcium interactions on growth and ionic concentration of *Citrus* plants. Plant Soil 133:39–46

Bañuls J, Serna MD, Legaz F, Primo-Millo E (1997) Growth and gas exchange parameters of Citrus plants stressed with different salts. J Plant Physiol 150:194–199

Bar Y, Apelbaum A, Kafkafi U, Goren R (1998) Ethylene association with chloride stress in citrus plants. Sci Hortic 73:99–109

Behboudian MH, Törökfalvy E, Walker RR (1986) Effects of salinity on ionic content, water relations and gas exchange parameters in some citrus scion-rootstocks combinations. Sci Hortic 28:105–116

Bevington KB, Castle WS (1985) Annual root growth pattern of young citrus trees in relation to shoot growth, soil temperature and soil water content. J Am Soc Hortic Sci 110:840–845

Bhambota JR, Kanwar JS (1969) Salinity tolerance of some rootstocks and scions of citrus species. Proc First Int Citrus Symp 3:1833–1835

Blum A, Jonson JW (1992) Transfer of water from roots into dry soil and the effect on wheat water relations and growth. Plant Soil 145:141–149

Boman JB (1993) First year response of "Ruby red" grapefruit on four rootstocks to fertilization and salinity. Proc Fla State Hort Soc 106:12–18

Brumos J, Colmenero-Flores JM, Conesa A, Izquierdo P, Sánchez G, Iglesias DJ, Lopez-Climent MF, Gomez-Cadenas A, Talon M (2009) Membrane transporters and carbon metabolism implicated in chloride homeostasis differentiate SALT stress responses in tolerant and sensitive citrus rootstocks. Funct Integr Genomics 9:293–309

Brumos J, Talon M, Bouhlal R, Colmenero-Flores JM (2010) Cl^- homeostasis in includer and excluder citrus rootstocks: transport mechanisms and identification of candidate genes. Plant Cell Environ 33:2012–2027

Castle WS (1978) Citrus root systems: their structure, function, growth and relationship to tree performance. In: Proceedings of the international society citriculture, University of Sydney, Australia, 1978, pp 62–69

Castle WS (1987) Citrus rootstocks. In: Rom RC, Carlson RF (eds) Rootstocks for fruits crops. Wiley, New York, pp 361–369

Castle WS, Krezdorn AH (1975) Effect of citrus rootstock on root distribution and leaf mineral content of Orlando Tangelo trees. J Am Soc Hortic Sci 100:1–4

Castle WS, Tucker DPH, Krezdorn AH and Youtsey CO (1993) Rootstocks for Florida Citrus. University of Florida, Florida, p 87

Cerdá A, Fernández FG, Caro M (1977) The effects of sodium chloride in the irrigation water on the suculence of leaves of citrus rootstocks. An Edafol Agrobiol 36:393–398

Cerdá A, Fernández FG, Caro M, Guillén MG (1979) Growth and mineral composition of two lemon varieties irrigated with saline waters. Agrochimica 23:387–396

Cerdá A, Nieves M, Guillén MG (1990) Salt tolerance of lemon trees as affected by rootstock. Irrig Sci 11:245–249

Cerdá A, Pardines J, Botella MA, Martínez V (1995) Effect of potassium on growth, water relations, and the inorganic and organic solute contents for two maize cultivars grown under saline conditions. J Plant Nutr 18(4):839–851

Cerezo M, García-Agustín P, Serna MD, Primo-Millo E (1997) Kinetics of nitrate uptake by *Citrus* seedlings and inhibitory effects of salinity. Plant Sci 126:105–112

Cerezo M, García-Agustín P, Primo-Millo E (1999) Influence of chloride and transpiration on net $^{15}NO_3^-$ uptake rate by Citrus roots. Ann Bot 84:117–120

Chapman HD (1968) The mineral nutrition of citrus. In: Reuther W, Batchelor LD, Webber HJ (eds) The citrus industry, vol 2. University of California Press, Davis, pp 127–289

Combrink NJJ, Labuschagne N, Barnard RO, Kotze JM (1995) The effect of chloride on four different citrus rootstocks. S Afr J Plant Soil 12:95–98

Cooper WC (1961) Toxicity and accumulation of salts in citrus trees on various rootstocks in Texas. Fla St Hort Soc 74:95–104

Cooper WC, Gorton BS (1951) Salt tolerance of various citrus rootstocks. Proc Rio Grande Valley Hortic Soc 5:46–52

Cooper WC, Gorton BS (1952) Toxicity and accumulation of chloride salts in citrus on various rootstocks. Proc Am Soc Hortic Sci 59:143–146

Cooper WC, Gorton BS, Olson EO (1952) Ionic accumulation in citrus as influenced by rootstock and scion and concentration of salts and boron in the substrate. Plant Physiol 27:191–203

Cooper WC, Reece PC, Furr JR (1962) Citrus breeding in Florida-past, present and future. Proc Fla State Hort Soc 75:5–13

Cossman KF (1940) Citrus roots: their anatomy, osmotic pressure and periodicity of growth. Palest J Bot 3:65–103

Cramer GR, Läuchi A, Polito VS (1985) Displacement of Ca^{2+} by Na^+ from the plasmalemma of root cells. Plant Physiol 79:207–211

Cramer GR, Läuchi A, Epstein E (1986) Effects of NaCl and $CaCl_2$ on ion activities in complex nutrient solutions and root growth of cotton. Plant Physiol 81:792–797

Cramer GR, Lynch J, Läuchi A, Epstein E (1987) Influx of Na^+, K^+ and Ca into roots of salt-stressed cotton seedlings. Plant Physiol 83:510–516

El-Desouky SA, Atawia AAR (1998) Growth performance of some *Citrus* rootstocks under saline conditions. Alexandria J Agr Res 43(3):231–254

El-Hag E, Agib B, Sidahmed Osman A (1997) Response of lime seedling growth and chemical composition to salinity stress. Soil Sci Plant Anal 28:1093–1101

Feigin A (1985) Fertilization management of crops irrigated with saline water. Plant Soil 89:285–299

Ford HW (1954) The influence of rootstock and tree age on root distribution of citrus. Proc Am Soc Hortic Sci 63:137–142

Forner JB, Forner-Giner MA, Alcaide A (2003) Forner-Alcaide 5 and Forner-Alcaide 13: two new citrus rootstocks released in Spain. HortScience 38:629–630

Forner-Giner MA, Alcaide A, Primo-Millo E, Forner JB (2003) Performance of "Navelina" orange on 14 rootstocks in Northern Valencia (Spain). Sci Hortic-Amsterdam 98:223–232

Forner-Giner MA, Alcaide A, Primo-Millo E, Forner JB (2009) Performance of Forner-Alcaide 5 and Forner-Alcaide 13, hybrids of Cleopatra mandarin x Poncirus trifoliata, as salinity-tolerant citrus rootstocks. J Am Pomol Soc 63:72–80

Forner-Giner MA, Legaz F, Primo-Millo E, Forner J (2011a) Nutritional responses of citrus rootstocks to salinity: performance of the new hybrids Forner-Alcaide 5 and Forner-Alcaide 13. J Plant Nut 34(9–11):1437–1452

Forner-Giner MA, Hueso JJ, Aguera JM, Legua P, Forner JB (2011b) Effects of citrus rootstocks on growth, yield and fruit quality of Navelate orange. J Food Agr Environ 9(2):400–403

Gallasch PT, Dalton GS (1989) Selecting salt-tolerant citrus rootstocks. Aust J Agric Res 40:137–144

García-Legaz MF, García-Lidón A, Porras-Castillo I and Ortiz-Marcide JM (1992) Behaviour of different scion/stock combinations of lemons (*C. Limon* (L.) Burm. F.) against Cl^- and Na^+ ions. In: Proceedings of the international society citriculture, Acireale, Italy, pp 397–399

García-Legaz MF, Ortiz JM, García-Lidon AG, Cerdá A (1993) Effect of salinity on growth, ion content and CO_2 assimilation rate in lemon varieties on different rootstocks. Physiol Plant 89:427–432

García-Lidón A, Ortiz JM, García-Legaz MF, Cerdá A (1998) Role of rootstock and scion on root and leaf ion accumulation in lemon trees grown under saline conditions. Fruits 53:89–97

García-Sánchez F, Carvajal M, Sánchez-Pina MA, Martinez V, Cerda A (2000) Salinity resistance of Citrus seedlings in relation to hydraulic conductance, plasma membrane ATPase and anatomy of the roots. J Plant Physiol 156:724–730

Geisler G (1962) The morphogenetic effect of oxygen on roots. Plant Physiol 40:85–88

Gorham J, Wyn Jones RG, McDonnell E (1985) Some mechanisms of salt tolerance in crop plants. Plant Soil 89:15–40

Graham JH, Syvertsen JP (1985) Host determinants of mycorrhizal dependency of citrus rootstock seedlings. New Phytol 101:667–676

Grattan SR, Grieve CM (1992) Mineral element acquisition and growth response of plant grown in saline environments. Agric Ecosyst Environ 38(4):275–300

Greenway H, Munns R (1980) Mechanisms of salt tolerance in nonhalophytes. Annu Rev Plant Physiol 31:149–190

Grieve AM, Walker RR (1983) Uptake and distribution of chloride, sodium and potassium ions in salt-treated citrus plants. Austr J Agric Res 34:133–143

Hanson DA, Kitz WO (1982) Metabolic responses of mesophytes to plant water deficits. Annu Rev Plant Physiol 33:163–203

Hanson B, Grattan SR, Fulton A (1999) Agricultural salinity and drainage. DANR, no 3375. University of California Irrigation Program, University of California, Davis, CA, pp 13–17

Hayward HE, Blair WM (1942) Some responses of Valencia orange seedlings to varying concentrations of chloride and hydrogen ions. Am J Bot 29:148–155

Huget JG (1976) Influence d´une irrigation localisëe sur l´enracinement de jeunes pommiers. Ann Agron 27:343–361

Kader MA, Lindberg S (2010) Cytosolic calcium and pH signaling in plants under salinity stress. Plant Signal Behav 5(3):233–238

Katare DP, Nabi G, Azooz MM, Aeri V, Ahmad P (2012) Biochemical modifications and enhancement of psoralen content in salt-stressed seedlings of *Psoralea corylifolia* Linn. J Funct Environ Bot (in press)

Kent LM, Läuchi A (1985) Germination and seedling growth of cotton: salinity-calcium interactions. Plant Cell Environ 8:155–159

Kirkpatrick JD, Bitters WP (1969) Physiological and morphological responses of various citrus rootstocks to salinity. In: Proceedings of the 1st international citrus symposium, vol 1. Ed Chapman, pp 381–389

Kriedermann PE, Barrs HD (1981) Citrus orchards. In: Kozlowski TT (ed) Water deficits and plant growth VI. Academic, New York, pp 325–417

LaHaye PA, Epstein E (1969) Salt toleration by plants: enhancement with calcium. Science 166:395–396

Lea-Cox JD, Syvertsen J (1993) Salinity reduces water use and nitrate-N-use efficiency of citrus. Ann Bot 72:47–54

Levy Y, Shalhevet J (1990) Ranking the salt tolerance of citrus rootstocks by juice analysis. Sci Hortic 45:89–98

Levy Y, Shalhevet J, Lifshitz J (1992) The effect of salinity on citrus rootstocks and scions. In: Proceedings of the international society citriculture, Acireale, Italy, pp 391–396

Levy Y, Lifshitz J, De Malach Y, David Y (1999) The responsee of several citrus genotipes to high-salinity irrigation water. Hortscience 34(5):878–881

Lloyd J, Howie H (1989) Response of orchard 'Washington navel orange', *Citrus sinensis* (L.) Osbeck to saline irrigation water I. Canopy characteristics and seasonal patterns in leaf osmotic potential, carbohydrates and ion concentrations. Aust J Agric Res 40:359–369

Lloyd J, Kriedemann PE, Syvertsen JP (1987a) Gas exchange, water relations and ion concentrations of leaves on salt-stressed Valencia orange *Citrus sinensis* (L.) Osbeck. Aust J Plant Physiol 14:387–396

Lloyd J, Syvertsen JP, Kriedemann PE (1987b) Salinity effects on leaf water relations and gas exchange of Valencia orange, *Citrus sinensis* (L.) Osbeck, on rootstock with different salt exclusion characteristics. Aust J Plant Physiol 14:605–617

Lloyd J, Kriedemann P, Aspinall D (1989) Comparative sensitivity of Prior Lisbon lemon and Valencia orange trees to foliar sodium and chloride concentrations. Plant Cell Environ 12:529–540

Lloyd J, Kriedemann P, Aspinall D (1990) Contrast between Citrus species in response to salinisation: an analysis of photosynthesis and water relations for different rootstock-scion combinations. Physiol Plant 78:236–246

MacHacha D, Pavel EW, Laker MC (2000) Rootstock/scion responses to variable salinity in irrigation water. In: Proceedings of the 6th international micro irrigation congress micro 2000, Cape Town, South Africa, 22–27, pp 1–6

Mclure PE, Valoras NY, Letey J (1982) Anion uptake in maize rotos: Interactions between clorate and nitrate. Physiologia Plantarum. 68:107–112

Mobaye RG, Milthorpe FL (1980) Response of seedling of three citrus-rootstock cultivars to salinity. Aust J Agric Res 31:117–124

Morinaga K, Sykes SR (2001) Effect of SALT and water stress on fruti quality, physiological responses, macro- and micro- element contens in leaves of Satsuma mandarin trees under greenhouse conditions. Jpn Agr Res Q 35(1):53–58

Mouhaya W, Allario T, Brumos J, Andres F, Froelicher Y, Luro F, Talon M, Ollitrault P, Morillon R (2010) Funct Plant Biol 37(7):674–685

Moya J (2000) Procesos y factores que determinan la tolerancia a la salinidad en los cítricos. Absorción y distribución de iones cloruro en los patrones mandarino Cleopatra y citrange Carrizo. Tesis Doctoral, Valencia

Munns R, Tester M (2008) Mechanism of salinity tolerance. Ann Rev Plant Biol 59:651–681

Nastou A, Chartzoulakis K, Therios I, Bosabalidis A (1999) Leaf anatomical responses, ion content and CO_2 assimilation in three lemon cultivars under NaCl salinity. Adv Hortic Sci 13(2):61–67

Newcomb DA (1978) Selection of rootstocks for salnity and disease resistence Proc Int Soc Citruculture Australia 1:117–120. Sydney, Australia
Nieves M, Martínez V, Cerdá A, Guillén MG (1990) Yield and mineral composition of Verna lemon trees as affected by salinity and rootstock combination. J Hortic Sci 65:359–366
Nieves M, García A, Cerdá A (1991a) Effects of salinity and rootstock on lemon fruit quality. J Hortic Sci 66:127–130
Nieves M, Cerdá A, Botella M (1991b) Salt tolerance of two lemon scions measured by leaf chloride and sodium accumulation. J Plant Nutr 14:623–636
Nieves M, Ruíz D, Cerdá A (1992) Influence of rootstock-scion combination in lemon trees salt tolerance. In: Proceedings of the international society citriculture, Acireale, Italy, pp 387–390
Parida AK, Das AB (2005) Salt tolerance and salinity effects on plants: a review. Ecotoxicol Environ Saf 60(3):324–349
Peterson CA, Enstone DE (1996) Functions of passage cells in the endodermis and exodermis of roots. Physiol Plant 97:592–598
Peynado A, Young R (1969) Relation of salt tolerance to cold hardiness of Redblush grapefruit and Valencia orange trees on various rootstocks. In: Proceedings of the 1st international citrus symposium, University of California, Riverside, CA, USA, pp 1793–1802
Praloran JC (1971) Les agrumes. Maisonneuve et Larose, Paris, p 565
Ream CL, Furr JR (1976) Salt tolerance of some Citrus species, relatives and hybrids tested as rootstocks. J Am Soc Hortic Sci 101:265–267
Rodriguez-Gamir J, Ancillo G, Legaz F, Primo-Millo E, Forner-Giner MA (2012) Influence of salinity on pip gene expression in citrus roots and its relationship with root hydraulic conductance, transpiration and chloride exclusion from leaves. J Exp Bot 78:163–166
Romero-Aranda R, Syvertsen JP (1996) The influence of foliar-applied urea nitrogen and saline solutions on net gas exchange of *Citrus* leaves. J Am Soc Hortic Sci 121:501–506
Romero-Aranda R, Moya JL, Tadeo FR, Legaz F, Primo-Millo E, Talon M (1998) Physiological and anatomical disturbances induced by chloride salts in sensitive and tolerant citrus: beneficial and detrimental effects of cations. Plant Cell Environ 21:1243–1253
Ruiz D, Martínez V, Cerdá A (1997) Citrus response to salinity, growth and nutrient uptake. Tree Physiol 17:141–150
Ruiz D, Martínez V, Cerdá A (1999) Demarcating specific ion (NaCl, Cl^-, Na^+) and osmotic effects in the response of two citrus rootstocks to salinity. Sci Hort 80:213–224
Salem AT, El-Khorieby MK (1989) Response of some citrus rootstocks to different types of chloride salt treatments. Ann Agric Sci (Cairo) 34:1123–1137
Sanderson J (1983) Water uptake by different regions of the barley root. Pathways of radial flow in relation to development of the endodermis. J Exp Bot 34:240–253
Storey R (1995) Salt tolerance, ion relations and the effect of root medium on the response of citrus to salinity. Aust J Plant Physiol 22:101–114
Storey R, Walker RR (1987) Some effects of root anatomy on K, Na, Cl loading of citrus roots and leaves. J Exp Bot 38:1769–1780
Storey R, Walker RR (1999) Citrus and salinity. Sci Hortic 78:39–81
Swietlik D (1989) Pressure-induced water influxes through root systems of sour orange seedlings. J Am Soc Hortic Sci 114:139–143
Sykes SR (1985b) A glasshouse screening procedure for identifying citrus hybrids which restrict chloride accumulation in shoot tissues. Aust J Agric Res 36:779–789
Sykes SR (1992) The inheritance of salt exclusion in woody perennial fruit species. Plant Soil 146:123–129
Sykes SR (2011) Chloride and sodium excluding capacities of citrus rootstock germplasm introduced to Australia from the People's Republic of China. Sci Hortic 128(4):443–449
Syvertsen JP, Graham JH (1985) Hydraulic conductivity of roots, mineral nutrition and leaf gas exchange of citrus rootstocks. J Am Soc Hortic Sci 110:865–869
Syvertsen JP, Yelenosky G (1988) Salinity can enhance tolerance of citrus rootstock seedlings by modifying growth, water relations and mineral nutrition. J Am Soc Hort Sci 113:889–893

Syvertsen JP, Lloyd J, Kriedemann PE (1988) Salinity and drought stress effects on foliar ion concentration, water relations and photosynthetic characteristics of orchard citrus. Aust J Agric Res 39:619–627

Tanner W, Beevers H (1990) Does transpiration have an essential funtion in long-distance ion transport in plants? Plant Cell Environ 13:745–750

Tozlu I, Moore GA, Guy CL (2000) Effects of increasing NaCl concentration on stem elongation, dry mass production, and macro- and micro-nutrient accumulation in *Poncirus trifoliata*. Aust J Plant Physiol 27(1):35–42

Tuzcu O, Kaplankiran M, Yesiloglu T, Cebeci Z (1997) The effects of rootstock-scion interaction on the distributions of rotos in citrus. In: Proceedings of the 5th ISCN international congress, Citrus nurseries, 75–81, 1997

Waisel Y (1962) The effect of calcium on the uptake of monovalent ions by excised barley roots. Physiol Plant 15:709–724

Walker RR (1986) Sodium exclusion and potassium-sodium selectivity in salt-treated trifoliate orange (*Poncirus trifoliata*) and Cleopatra mandarin (*Citrus reticulata*) plants. Aust J Plant Physiol 13:293–303

Walker RR, Törökfalvy E, Downton WJS (1982) Photosynthetic responses of the citrus varieties Rangpur Lime and Etrog Citron to salt treatment. Aust J Plant Physiol 9:783–790

Walker RR, Törökfalvy E, Grieve AM, Prior LD (1983) Water relations and ion concentrations of leaves on salt-treated citrus plants. Aust J Plant Physiol 10:265–277

Walker RR, Sedgley M, Blesing MA, Douglas TJ (1984) Anatomy, ultrastructure and assimilate concentrations of roots of citrus genotypes differing in ability for salt exclusion. J Exp Bot 35:1481–1494

Walker RR, Blackmore DH, Sun Q (1993) Carbon dioxide assimilation and foliar ion concentrations in leaves of lemon (*Citrus limon* L.) trees irrigated with NaCl or Na_2SO_4. Aust J Plant Physiol 20:173–185

Wutscher HK (1979) Citrus rootstocks. In: Horticultural reviews I. A.V.I, Westport, pp 237–269

Zekri M (1991) Effects of NaCl on growth and physiology of sour orange and Cleopatra mandarin seedlings. Sci Hortic 47:305–315

Zekri M (1993) Salinity and calcium effects on emergence, growth and mineral composition of seedlings of eight citrus rootstocks. J Hortic Sci 68:53–62

Zekri M, Parsons LR (1989) Growth and root hydraulic conductivity of several citrus rootstocks under salt and polyethylene glycol stresses. Physiol Plant 77:99–106

Zekri M, Parsons LR (1990a) Comparative effects of NaCl and polyethylene glycol on root distribution, growth and stomatal conductance of sour orange seedlings. Plant Soil 129:137–143

Zekri M, Parsons LR (1990b) Calcium influences growth and leaf mineral concentration of citrus under saline conditions. HortScience 25:784–786

Zekri M, Parsons LR (1992) Salinity tolerance of Citrus rootstocks: effects of salt on root and leaf mineral concentrations. Plant Soil 147:171–181

Zid, Grignon C (1985) Sodium-calcium interactions in leaves of *Citrus aurantium* grown in the presence of NaCl. Physiol Veg 23:895–903

Zid E, Grignon C (1986) Comparative effects of NaCl, KCl and Na2SO4 on the growth and mineral nutrition of young *Citrus aurantium* L. Acta-Oecologica, -Oecologica-Plantarum 7(4): 407–416

Chapter 15
Effects of Salt Stress on Photosynthesis Under Ambient and Elevated Atmospheric CO_2 Concentration

Nicole Geissler, Bernd Huchzermeyer, and Hans-Werner Koyro

Abstract Soil salinization is one of the most important factors which limit plant productivity. About 3.6 billion of the world's 5.2 billion hectares of dryland used for agriculture have already suffered erosion, soil degradation, and salinization. Global climate change caused by rising atmospheric trace gases such as CO_2 and forced migration add to the urgency of this global problem. Therefore, solutions are desperately needed, such as the improvement of drought and salinity resistance of crops or the use of (xero-) halophytes instead of glycophytic crops. As photosynthesis is a prerequisite for biomass production, this chapter focuses on information related to this essential sequence of reactions, thereby discussing the different levels of photosynthesis. At first, there are primary reactions of photosynthesis, namely absorption of light energy and (1) its conversion to redox energy, conserved in the coenzyme NADPH, and (2) energy of chemical bounds, conserved in the coenzyme ATP. On the second level, we find reactions of the Calvin cycle, nitrate and sulfate reduction as well as sugar, lipid, and amino acid metabolism. Typical reactions on the third level are trans-membrane and inter tissues transport of metabolites. The fourth level of photosynthesis relates to physiological aspects of gas exchange and water relations.

Apart from these general effects of salinity on photosynthesis, we will review the probable photosynthetic performance of salt stressed plants under future atmospheric conditions, namely under elevated CO_2 concentration. Special emphasis will be put on gas exchange and photosynthesis of C_3 and C_4 plants because these two photosynthesis types show different responses to elevated CO_2, leading to different interactions with salinity.

N. Geissler (✉) • H.-W. Koyro
Institute of Plant Ecology, Justus Liebig University Giessen,
Heinrich-Buff-Ring 26-32, D-35392 Giessen, Germany
e-mail: Nicole.Geissler@bot2.bio.uni-giessen.de

B. Huchzermeyer
Institute of Botany, Leibnitz University Hannover,
Herrenhäuser Str. 2, D-30419 Hannover, Germany

P. Ahmad et al. (eds.), *Salt Stress in Plants: Signalling, Omics and Adaptations*,
DOI 10.1007/978-1-4614-6108-1_15,

Keywords Na Cl salinity • Global change • Elevated CO_2 • Photosynthesis • Oxidative stress

Introduction

Feeding the rapidly growing human population is one of the biggest challenges we face on our planet today. Especially in arid and semiarid regions, the increasing population density is collaterally catalysed by the consequences of global climate change which is caused by anthropogenic emissions of trace gases such as CO_2 (IPCC 2007). Thus salinity – already affecting 7% of the global total land area (Glenn et al. 1998; Millennium Ecosystem Assessment 2005) and 20–50% of the global irrigated farmland (Tanji 2002; Hu and Schmidhalter 2005; Koyro et al. 2008) – will pose a more and more serious threat to agriculture and human nutrition in future. Therefore, solutions are desperately needed, such as the improvement of salt resistance of crops or the use of (xero-) halophytes instead of glycophytic crops, which in turn requires a detailed knowledge about salt and drought resistance mechanisms in plants.

The viability of plants in saline habitats depends on their ability to cope with (1) water deficit due to a low water potential of the soil, (2) restriction of CO_2 uptake, (3) ion toxicity, and (4) nutient imbalance. The first two constraints are directly related to photosynthesis which is a prerequisite for biomass and crop production, respectively. So this chapter focuses on information related to these constraints, thereby discussing the different levels of photosynthesis as described in the next chapter.

In the face of global climate change, one has also to consider the influence of elevated atmospheric CO_2 concentration on photosynthesis, which we will do in the last part of this chapter. In C_3 plants, CO_2 often improves assimilation while reducing stomatal resistance, thus increasing water use efficiency, but decreasing photorespiration and oxidative stress (Urban 2003; Kirschbaum 2004; Rogers et al. 2004). As these effects are much less pronounced in C_4 plants, it will be differentiated between these two photosynthesis types.

Effects of Salt Stress on Photosynthesis Under Ambient Atmospheric CO_2 Concentration

Levels of Photosynthesis to Be Considered

Regarding crop production and food and feed supply, limitation of plant growth by environmental factors is a matter of general concern. As photosynthesis is dominating plant growth and production of biomass, the sequence of reactions leading to the phenomenon named photosynthesis is in the focus of interest when breeding for high crop yield. In order to allow a detailed analysis of salt effects, several individual steps have to be distinguished (Fig. 15.1).

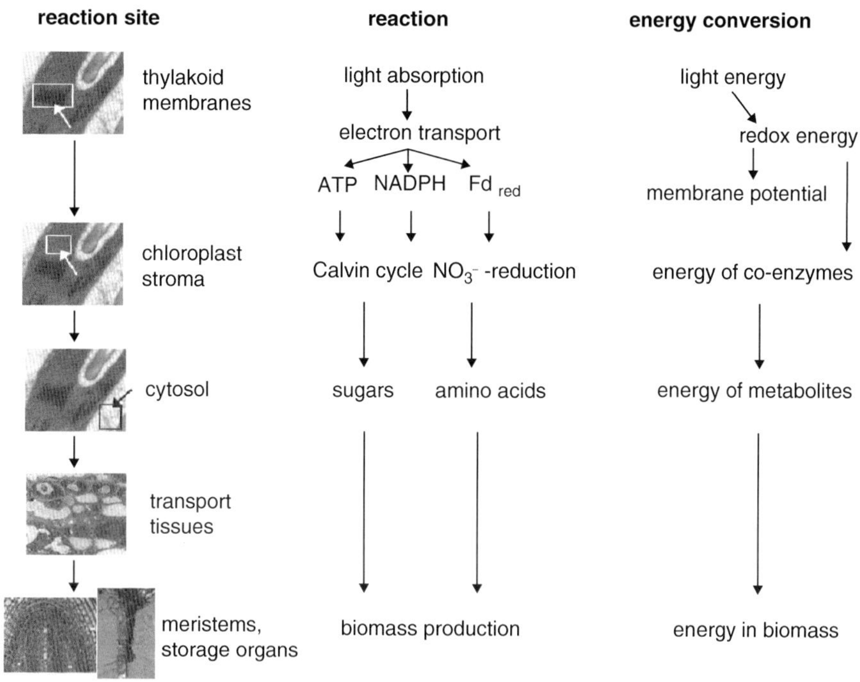

Fig. 15.1 Photosynthetic energy flow. The photosynthetic reaction sequence can be described in terms of reaction sites (*left*), sequence of reactions (*center*), and conversion of energy forms (*right*). *Fd* ferredoxin

- At first, there are primary reactions of photosynthesis, namely absorption of light energy and (1) its conversion to redox energy, conserved in the coenzyme NADPH, and (2) energy of chemical bounds, conserved in the coenzyme ATP.
- On the second level, we find reactions of the Calvin cycle, nitrate and sulfate reduction as well as sugar, lipid, and amino acid metabolism.
- Typical reactions on the third level are transmembrane and inter tissues transport of metabolites.
- The fourth level of photosynthesis relates to physiological aspects of gas exchange and water relations.

Most papers deal on salt effects on physiological aspects of biomass production rather than partial reactions of photosynthesis as defined above. As there are cross reactions as well as cell signaling involved in regulation of photosynthetic metabolic pathways, there are no strict correlations between biomass production and individual gene activities. As a consequence, predicted correlations, though they had been observed in laboratory experiments, could not be shown in subsequent field experiments. In this chapter we try a more differentiated approach and review analysis of salt stress effects at different levels of photosynthesis.

Regulation of Gas Exchange and Water Relations

According to Munns, plants show a two-phase growth response to salinity (Munns 1993, 2002; Munns et al. 2002). During the first phase of growth reduction water or osmotic stress prevails, and plant responses are presumably regulated by hormonal signals coming from the roots.

Terrestrial plants growing in saline habitats are often surrounded by low water potentials in the soil solution and atmosphere. As a part of resistance mechanisms, water flow through the soil–plant–atmosphere continuum must be ensured, so that a gradient of decreasing water potentials (Ψ) must be established. The Ψ of pure water is defined as 0 MPa; increasing salinity or concentrations of other solutes will decrease Ψ. Thus, any sharp rise in salinity could effectively hold water osmotically away from plants that lack any physiological or morphological modifications (Larcher 2003). To limit restrictions on water uptake, plants must generate increasingly lower Ψ to allow continued water flux into belowground structures (Touchette et al. 2009).

If plants cannot escape salinity (such as geophytes or pluviotherophytes), it is also important to prevent water loss by transpiration from being higher than the influx rate. This is necessary to avoid a negative tissue water balance and is only possible if the water potential remains lower in the plant than in the soil. However, experiments demonstrated that the leaf water potential of halophytes does not correlate alone as a single factor with salinity resistance. Plant species with different levels of salt resistance such as *Aster tripolium*, *Avicennia marina*, *Innula critmoides*, and *Sesuvium verrucosum* have a sufficient adjustment mechanism even at high salinities. Clearly, resistance in the form of osmotic adjustment plays an important role in halophytes residing in saline environments (Flowers and Colmer 2008). However, in addition were the osmotic potentials of all four halophytes (and many others) up to sea water salinity level sufficiently low to explain the full turgescence of the leaves (results not shown). Except for differences in concentration and type of osmotica used in plant tissues (in halophytes inorganic ions are more prevalent than organic substances), physiological responses in plants to salt stress were remarkably similar to those employed during drought (Touchette et al. 2009). For plants with limited water availability, physiological adjustments often involve avoidance and tolerance, with most plants using some combination of the two (Yue et al. 2006; Romanello et al. 2008).

Assuming there is no interruption of the water supply, water can flow passively from the root to the shoot and there seems to be no reason for growth reduction by water deficit for any of the studied species. However, by regulating the extent of apoplastic barriers and their chemical composition (long-distance response coordination), plants can effectively regulate the uptake or loss of water and solutes (by structures such as barriers in the hypo- or exodermis). This appears to be an additional or compensatory strategy of plants to acquire water and solutes (Hose et al. 2001). At the extremes of growth under saline or dry conditions even cell layers such as the exodermis can become an absolute barrier for water and ions in the strict sense (Azaizeh and Steudle 1991; North and Nobel 1991; Nublat et al. 2001).

Thus, the rate of water supply to the shoot can be restricted due to the coupling between the flows of water and solutes (Na and Cl) even if the leaf water potential is low. Therefore, the balance between water flow (sum of water accumulation and transpiration) and the decrease in the amounts of nutrients or unfavorable nutrient ratios (e.g. Na^+/K^+) are important factors for impaired leaf elongation and plant growth (Lynch et al. 1988; Munns et al. 1989; Neves-Piestun and Bernstein 2001; Cramer 2003).

In any case, plant water loss has to be minimized at low soil water potentials. However, biomass production depends mainly on the ability to keep a high net photosynthesis and a low transpiration simultaneously. In this field of tension, biomass production has always to be seen in connection with CO_2/H_2O-gas exchange, which can be estimated by the water use efficiency (WUE) of photosynthesis. It needs to be considered that anyway a critical point for the plant is reached when CO_2 fixation (apparent photosynthesis) falls below CO_2 production (compensation point).

Several halophytic plants such as *Aster tripolium*, *Beta vulgaris* ssp. *maritima*, *Chenopodium quinoa*, or *Spartina townsendii* reveal a combination of low (but positive) net photosynthesis, minimum transpiration, high stomatal resistance, and minimum internal CO_2 concentration at their threshold salinity resistance (Koyro 2000; Koyro and Huchzermeyer 2004). However, there is a big bandwidth among halophytes, especially for succulent species such as *Sesuvium portulacastrum* or *Avicennia marina*, which have alternatives if the water balance (water uptake minus water loss) is still positive and not the limiting factor for photosynthesis. In case of *S. portulacastrum*, net photosynthesis and WUE increase but stomatal resistance decreases. These results show that it is quite important to describe the regulation of gas exchange at high salinity in strong reliance with other parameters (such as water relations).

Water deficit is one major constraint at high salinity and can lead to a restriction of CO_2 uptake and to the development of reactive oxygen species (ROS). The balance between water loss and CO_2 uptake helps to find weak spots in the mechanism of adjustment (of photosynthesis) to high salinity (Badawi et al. 2004).

Primary Reactions of Photosynthesis

Photosynthetic Conversion of Energy

In plants active in photosynthesis, energy of light quanta is absorbed by chlorophyll. Just like all other pigments, activated chlorophyll can return from its activated state to the stable, non activated state by emitting heat. Other than most pigments, activated chlorophyll is sufficiently stable to allow transfer of energy to acceptor molecules of biological relevance after having absorbed energy of a red light quantum (Strasser et al. 2004; Fig. 15.2). In principle, there are two options of biological relevance in addition to a third one we can use to monitor plant performance:

1. Resonance energy transfer to activate other pigments and
2. Transfer of an electron to an acceptor. The electron acceptor can be a component of the photosynthetic electron transport chain or any other molecule having a less negative potential as compared to activated chlorophyll. This second option requires a "refill" of electrons, and it is well documented that in chloroplasts the

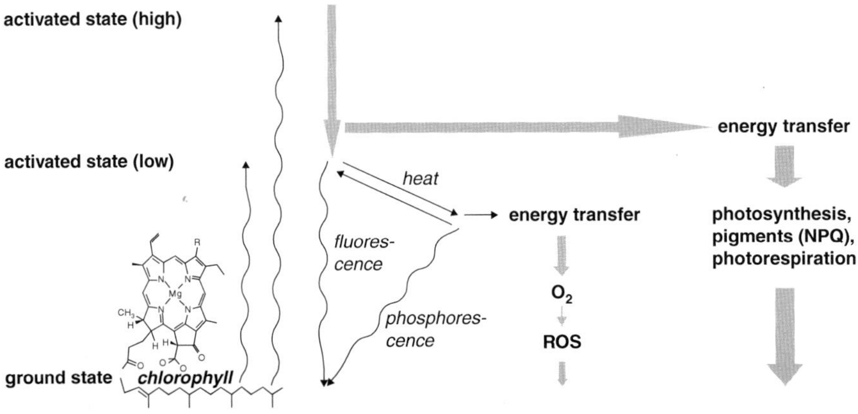

Fig. 15.2 Competition for absorbed energy of light quanta. Photosynthetic electron transport is competing with "futile reactions" for energy of light-activated chlorophyll. *ROS* reactive oxygen species, *NPQ* non-photochemical quenching

water splitting system fulfills this function, so that chlorophyll is recycled to its ground state (Strasser et al. 2004).

3. Another option, competing for energy with the already mentioned ones, is the emission of fluorescence energy (Schreiber 1997; Schreiber et al. 2002; Strasser et al. 2004). As will be discussed later, any inhibition of an individual pathway will enhance the possibility of the other pathways to occur.

In thylakoid membranes, photosystems I and II (PSI and PSII) can be distinguished from other chlorophyll containing protein complexes, namely the light-harvesting complexes. Only the special pairs of chlorophyll a located in the active center of the two photosystems are involved in electron transport. Energy transfer among other pigments occurs via resonance energy transfer. In this section, we focus on electron transfer reactions.

As known from the literature, half lifetime of activated state of chlorophyll is very short and depends on the environment of the pigment (Rees et al. 1990; Laible et al. 1994; Ma et al. 2009). It is obvious that photosynthetic efficiency depends on the probability that activated chlorophyll will transfer electrons to an acceptor of the photosynthetic electron transport chain rather than "wasting" energy by using one of the other pathways mentioned above (Ruban et al. 1993; Horton et al. 1996; Horton 2000; Allen and Forsberg 2001). In order to meet this requirement, reaction partners are arranged in ideal neighbourhood within protein complexes located in the thylakoid membranes. Moreover, it has been demonstrated that this structure undergoes permanent adjustment to match the requirement of chloroplast metabolism and to adapt to changes in the environment, i.e. changes of light quality and intensity, for instance (Allen and Bennett 1981; Allen 1992; Allen and Forsberg 2001).

By means of photosynthetic electron transport, energy of absorbed light quanta is converted to redox energy stored in the coenzyme NADPH and proton motive

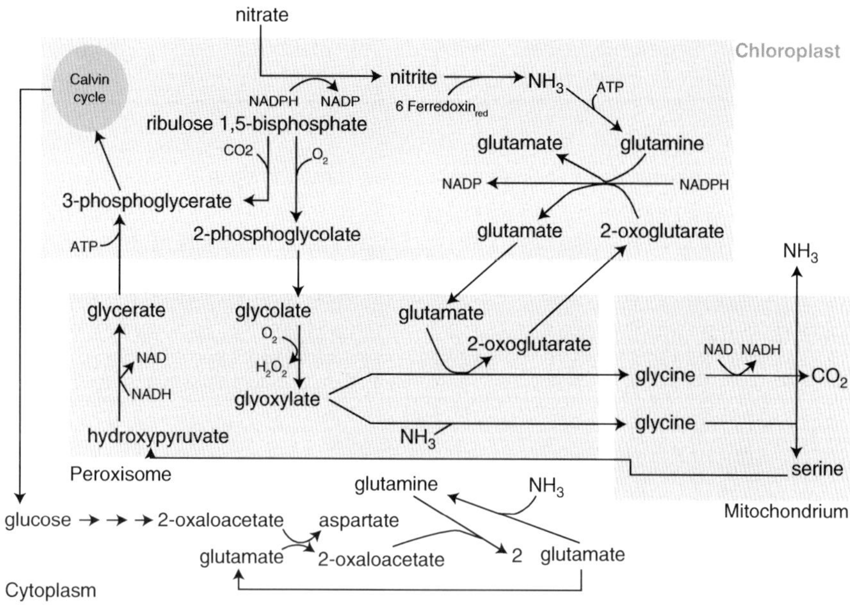

Fig. 15.3 Linking metabolic pathways of cell compartments. Biochemical pathways of chloroplasts, peroxisomes, and mitochondria are linked via shuttle systems. With respect to energy flow, photorespiration and nitrate reduction are of focal interest. Like the Calvin cycle, nitrate reduction is consuming electrons released from photosystem I. Thus, nitrate reduction is recycling the cofactors ferredoxin and $NADP^+$. Photorespiration, on the other hand, is transferring redox power to the mitochondria and is involved in shuttling of ammonia

force (Mitchell 1967) stored in a proton gradient across the thylakoid membranes. This proton gradient is the driving force for ATP synthesis catalyzed by the chloroplast F-type ATPase, called the $CFOCF_1$-complex or the chloroplast coupling factor (Strotmann et al. 1976; Huchzermeyer and Strotmann 1977; Boyer 2000). The number of coenzyme molecules ($NADP^+$/NADPH and ADP/ATP) is limited, and there is no exchange of coenzymes among cell compartments. Therefore, this machinery will work efficiently only if acceptor forms of coenzymes ($NADP^+$ and ADP, respectively) are permanently recycled by subsequent metabolic pathways. Otherwise, energy turnover by the electron transport would be inhibited, and this inhibition finally would lead to an inhibition of electron release from activated chlorophyll, enhancing the probability of alternative routes mentioned above (Fig. 15.2).

Photosynthetic CO_2 assimilation is the major consumer recycling both coenzymes in the reaction sequence of the Calvin cycle. Under physiological conditions, as a rule of thumb, in chloroplasts of non-woody plants two-thirds of the electrons from the noncyclic electron transport pathway finally will be consumed by CO_2 fixation, while one-third will be used for nitrate reduction (see top part of Fig. 15.3) (Schmidt and Jäger 1992). But it has to be mentioned here that a significant portion of the absorbed light energy will be "wasted" in futile reaction sequences rather than be used for biomass synthesis.

Salt Effects on Photosynthetic Energy Conversion

In order to understand individual photosynthetic reactions and get the principles of their interaction, thylakoid membranes and protein complexes have been isolated and analyzed with respect to their structure and function. During preparation and subsequent tests of enzyme activities, salt concentrations up to 50 mM NaCl have been applied without any inhibitory effect on individual enzyme activities (Strotmann et al. 1976). Such high salt concentrations are not found inside chloroplasts, neither under physiological conditions nor under salt stress. It therefore can be concluded that primary reactions of photosynthesis are not directly inhibited by ionic or ion specific stress (Richter et al. 2000; Huchzermeyer et al. 2004; Huchzermeyer and Koyro 2005; Koyro and Huchzermeyer 2005). However, this conclusion disagrees with an apparent inhibition of photosynthesis observed in whole plant experiments (Lawlor and Fock 1978; Lawlor 2002a). Therefore, it has to be analyzed in more detail to what extent the observed inhibition may be attributed to salt-dependent changes in thylakoid structure (Hesse et al. 1976) or to physiological effects (e.g. ion transport). One first approach to answer this question could be a detailed analysis of the kinetics of fluorescence light emission subsequent to chlorophyll activation by light pulses. Application of the pulsed amplitude modulation technique (PAM) allowed a detailed analysis of the network of primary reactions in photosynthesis (Strasser et al. 2004). It became quite obvious that salt stress does not directly inhibit primary reactions of photosynthesis, but inhibits product export and fine-tuning of primary reactions by interfering with optimal arrangement of proteins and membranes (Koyro and Huchzermeyer 1999; Huchzermeyer and Heins 2000; Huchzermeyer 2000).

Accordingly, it was observed that maximal photochemical efficiency, indicated by high F_v/F_m values of chlorophyll fluorescence, remain high under tolerable salt stress, while the growth rate of turf grass, for instance, was reduced under the same salinity level (Lee et al. 2004). This can be seen as an evidence of a reduced real photochemical efficiency.

It will be discussed in subsequent paragraphs how salt can inhibit export of products of photosynthesis and why proper function of the phosphate translocator, located in the inner envelope membrane, is essential for photophosphorylation. At this stage, we can state that inhibition of product export feeds back to primary reactions and finally will inhibit photosynthetic electron transport. One option to release energy from its activated state is blocked and chlorophyll will increase activity of fluorescence light emission, heat production, energy transfer to other pigments, and ROS production. This latter option will be discussed later in this review (see Sect. Dissipation of Surplus Energy and Production of Potential Toxic Intermediates of Photosynthesis).

Salt Effects on Chloroplast Structure and Metabolite Transfer

Chloroplasts exhibit a very high content of proteins which tend to interact and form aggregates inside the chloroplast stroma (Süss et al. 1993). Süss et al. were able to isolate such "super complexes" and found out that they contain enzymes belonging to individual metabolic pathways. Such an arrangement helps to increase substrate turnover because diffusion distances among enzymes of a pathway are close to zero. For the same reason, formation of such aggregates prevents occurrence of side reactions because intermediates are not available for other enzymes. As will be discussed in Sect. Control of Mechanisms Improving Stress Resistance: Compatible Solutes, on the one hand salt can inhibit the turnover of substrates by destroying enzyme aggregates, while on the other hand intermediates of sugar metabolism can become available for other enzymes and alternative products such as compatible solutes can be formed. Though salt effects on metabolite patterns have been analyzed in several papers, no data linking these findings to the occurrence of protein aggregates have been available up to now.

The occurrence of thylakoid grana stacks has attracted a lot of interest for years (Huchzermeyer et al. 1986; Lam and Malkin 1989; Malkin and Braun 1993; Romanowska and Albertson 1994). It was calculated that about 70% of all PSII complexes are located within stacked regions of the thylakoid membranes, while most of the PSI complexes are found in un-stacked parts of the thylakoid membranes (Schmidt and Malkin 1993; Romanowska and Albertsson 1994). From this, it was concluded that 70% of the PSII complexes must be in an inactive state because the distance between PSII and PSI would be too far to allow sufficient turnover rates of photosynthetic electron transport (Malkin and Braun 1993). Moreover, it was found that grana are unstable, permanently folding and unfolding structures. Apparently $CFOCF_1$-complexes are initiating the formation of membrane loops (Boekema et al. 1988), and grana are formed by the interaction of membrane proteins (Staehelin 1975). As active PSII has an extremely short half lifetime in the range of 20–30 min (Kuhn and Böger 1990; Trebst and Soll-Bracht 1996; Keren et al. 1997; Jansen et al. 2001), the function of this permanent modification of membrane structure may be to initiate interactions among membrane proteins and to support the formation of aggregates of protein complexes interacting in the photosynthetic electron transport chain. Such preferred neighbourhood of membrane proteins has been found, indeed (Laszlo et al. 1984; Huchzermeyer and Willms 1985). There are several publications indicating that such neighborhoods of protein complexes control the efficiency of energy conversion in primary reactions of photosynthesis (Laszlo et al. 1984; Löhr and Huchzermeyer 1985; Löhr et al. 1985; Allnutt et al. 1989) and photo-inhibition (Lu and Zhang 1999; Lu et al. 2002).

The facts presented above show that the formation of grana stacks is essential for optimal functioning and permanent repair of the photosynthetic electron transport rate. It was found that grana become destabilized if the ratio of monovalent and divalent cations is impaired (Hesse et al. 1976). Similarly, investigations on *Aster tripolium* showed that NaCl salinity leads to a decreased amount of grana stacks (Geissler et al. 2009b; Fig. 15.4). This was accompanied by an increased chlorophyll a/chlorophyll b ratio and may be an explanation for a reduced photosynthetic efficiency (Moorthy and Kathiresan 1999) because chlorophyll b is mainly located in the light harvesting complex which supplies PSII with energy and is located

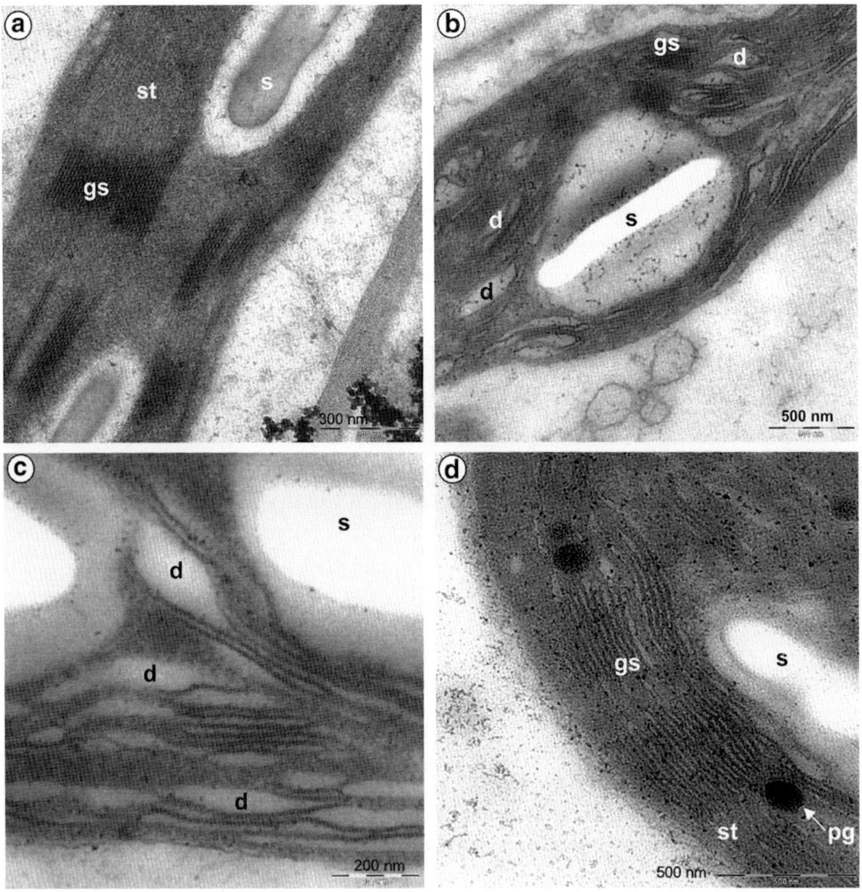

Fig. 15.4 Influence of salinity and elevated CO_2 concentration on chloroplast structure in *Aster tripolium*. (**a**) In control plants under ambient CO_2, both stroma thylakoids and grana stacks are well developed. (**b**), (**c**) In salt treatments (375 mM NaCl) under ambient CO_2, dilations of the thylakoid membranes are clearly visible, and well-developed grana stacks are almost lacking. (**d**) In salt treatments (375 mM NaCl) under elevated CO_2, the thylakoid membranes are less damaged. *st* stroma thylakoid, *gs* grana stacks, *d* dilations, *s* starch grain, *pg* plastoglobulus

mainly in the grana stacks. So the disintegration of the latter is probably correlated with the increased chlorophyll a/chlorophyll b ratio and may be an adaptation to prevent an electron surplus in the photosynthetic electron transport chain.

Last but not least it should be mentioned that in various plant species salt stress often leads to dilations of the thylakoid membranes, so that the spaces between the membranes look swollen, and undulated thylakoid areas develop (Kurkova et al. 2002; Rahman et al. 2002; Mitsuya et al. 2003; Fidalgo et al. 2004; Paramanova et al. 2004; Geissler et al. 2009b; Zhen et al. 2011; Fig. 15.4). This structural damage has been discussed by many authors as a consequence of oxidative stress

(Mitsuya et al. 2003; Fidalgo et al. 2004; Oksanen et al. 2005) and is likely to contribute to impaired photosynthetic rates under saline conditions.

Dissipation of Surplus Energy and Production of Potential Toxic Intermediates of Photosynthesis

Photorespiration

Stomatal closure and a subsequent inhibition of gas exchange is a secondary effect of salt stress, mostly brought about by ABA released from the plant roots. In the presence of light, the O_2/CO_2 ratio will increase inside the leaves and impair CO_2 fixation, especially in C_3 plants. This happens as the enzyme Rubisco can bind O_2 instead of CO_2 to its reaction center, thus catalyzing the synthesis of a C_3 plus a C_2 compound instead of two C_3 compounds in the primary reaction of the Calvin cycle (Fig. 15.3). The C_2 compound 2-phosphoglycolate will be converted to glycolate, a molecule that cannot be metabolized by chloroplasts and the concentration of which eventually would become toxic. Detoxification and recycling of the C_3 compound 3-PGA are understood to be the major function of photorespiration which takes place by metabolite transfer between chloroplasts, peroxisomes, and mitochondria.

The photorespiratory reaction cycle we know from our textbooks is a simplification allowing general estimates. We know that amino acids (glycine, serine, glutamic acid, and glutamine) may be subtracted from or fed into the cycle *in vivo*. Such reactions modify nitrogen flow among cell compartments. But, with respect to equilibrium of carbon flow, another aspect may be more important: It has been shown by Niessen et al. that mitochondrial glycolate oxidation contributes to photorespiration in higher plants as well (Niessen et al. 2007). It has to be kept in mind that photorespiration is a major source of H_2O_2 in illuminated C_3 leaves. On the other hand, H_2O_2 production and interaction with pyridine nucleotide coenzymes make photorespiration an important player in cellular redox homeostasis. Furthermore, H_2O_2 is an important second messenger controlling cell development and tuning effects of hormones like ABA, for instance. Any interference with this reaction will have effects not only on immediate energy status and phosphorylation efficiency, but also on plant hormonal responsiveness, thus on development and plant life cycle (Foyer et al. 2009).

There is evidence that glycolate can inhibit Q_A/Q_B electron transfer in PSII (Petrouleas et al. 1994). As stated above, this would lead to stimulated ROS production. This interpretation is in line with the observed bleaching of maize in presence of glycolate (s.a.). In several publications, it is suggested that the function of photorespiration is to serve as a sink to dissipate excess redox energy (Kozaki and Takebe 1996; Wingler et al. 2000). In terms of our arguments, this would mean photorespiration is recycling coenzymes functioning as acceptors of the photosynthetic electron transport chain. This would help preventing ROS production as well. But, to our understanding, this interpretation does not sufficiently take into account the compartmentation of coenzymes.

Non-photochemical Quenching of Energy

Under conditions that limit CO_2 assimilation, the potential rate of NADPH production exceeds the actual rate of consumption of reductive power. In order to be able to grow under stressful conditions, plants have to be equipped with mechanisms preventing excess reducing power. But these futile mechanisms compete with photochemistry for absorbed energy. They lead to a decrease in quantum yield of photosystem II (Genty et al. 1989; Cha-um et al. 2012).

The photosystem II antenna is highly flexible in tuning delivery of excitation energy to the photosystem II reaction center (Horton et al. 1996). The principal adaptation mechanism in photosynthesis is the control of thermal dissipation of excess energy within the photosystem II antenna, thus matching physiological needs (Johnson et al. 2009). In C_3 plants, losses by this mechanism, named non-photochemical energy quenching, may exceed the ones caused by photorespiration. Despite extensive investigations, the reaction mechanism of photochemical quenching is not yet completely understood because the turnover of intermediates is fast and the reaction depends on intact structures of protein complexes and their *in vivo* arrangement inside the thylakoid membranes; i.e. reaction partners may not be extracted and individually analyzed. But some insight was achieved by a combination of molecular biological and biophysical techniques (Johnson et al. 2009).

For experimental approaches investigating salt stress effects, it is important to know that non-photochemical quenching of excitation energy is comprised of a fast and a slow component, qE and qI, respectively. Both reactions are reversible. The trigger of qE is the ΔpH across the thylakoid membrane sensed by the PsbS subunit of the light-harvesting complex (Li et al. 2000, 2004). Full expression of qE is associated with the enzymatic de-epoxidation of violaxanthin to zeaxanthin. This reaction is part of the xanthophylls cycle (Havaux et al. 2007). Enzymes involved are pH controlled and function on the expense of NADPH (Demmig-Adams and Adams 1996). This makes the cycle a futile reversible reaction sequence on its own. The majority of photoactive xanthophylls is bound to the light-harvesting complex. In addition to the ones involved in the xanthophyll cycle, lutein and lutein epoxide are bound there as well and can be turned over in a cycle on their own (Matsubara et al. 2001). Depending on distances among pigments, these two cycles can interact with soluble xanthophylls and control non-photochemical energy quenching in a synergistic, but not well-understood way (Johnson et al. 2009).

Sensitivity of non-photochemical quenching to any experimental approach interfering with membrane structure and protein fine structure indicates that this reaction sequence of outstanding physiological importance will be highly sensitive to any reaction causing imbalance of ionic homeostasis. The threshold ion concentration resulting in significant inhibition of non-photochemical quenching will depend on availability of compatible solutes, for instance. Based on current understanding, it can be expected that such adverse effects can be monitored by measuring pigment shifts due to altered ratios of xanthophylls and by high-resolution chlorophyll fluorescence measurement. Thus, it should be possible to measure salt effects on non-photochemical quenching even in the field using noninvasive techniques.

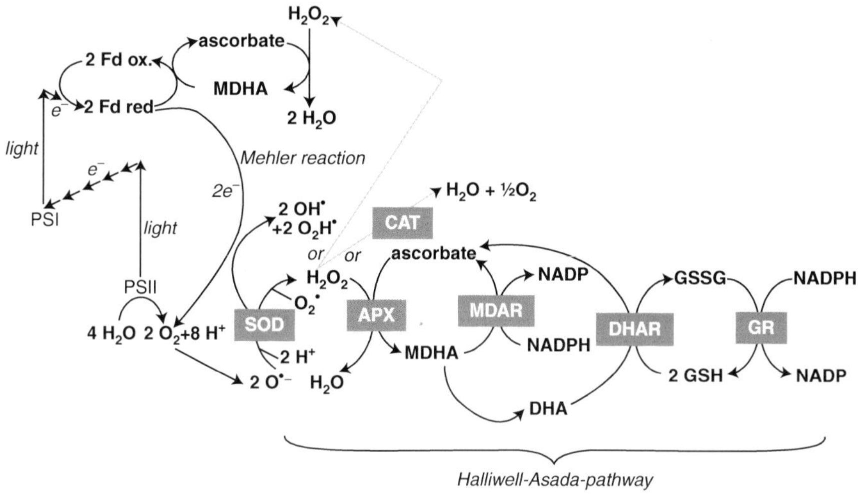

Fig. 15.5 ROS production and detoxification. Detoxification of ROS by sequential ascorbate and glutathione cycles will consume NADPH and, thus, result in a relief of $NADP^+$ shortage in high light. A prerequisite is that (1) enzymes involved are available at ample concentrations and (2) are positioned in ideal neighbourhood to allow high turnover rates. *SOD* superoxide dismutase, *Fd* ferredoxin, *CAT* catalase, *APX* ascorbate peroxidase, *MDAR* monodehydroascorbate reductase, *DHAR* dehydroascorbate reductase, *GR* glutathione reductase, *MDHA* monodehydroascorbate, *DHA* dehydroascorbate, *GSSG* oxidated glutathione, *GSH* reduced glutathione

Especially near the end of a leaf's life cycle, masking of chlorophyll by anthocyanins becomes important. It prevents photooxidative damage and allows an efficient nutrient retrieval from leaves to storage organs (Field et al. 2001). Some plants use such mechanisms regularly under saline conditions. They can be identified because leaf color will vary depending on their growth conditions like it can be observed with *Salicornia* and *Sempervivum*, for instance.

ROS Production

As described above, any reduction of the electron transport rate, especially under high light conditions, will increase the risk of ROS production (Figs. 15.2 and 15.5). The redox potential of activated chlorophyll is more negative than the one of oxygen. Therefore electron transfer from activated chlorophyll can occur unless energy is abstracted from activated chlorophyll faster than electron transfer to oxygen can occur. There are further options of ROS production as several intermediates of the photosynthetic electron transport are radicals. On the other hand, ROS can spontaneously convert or can be turned over under the control of enzymes (Fig. 15.5). Accordingly, in the literature the occurrence of various forms of ROS is described (Halliwell and Gutteridge 1986; Blokhina et al. 2002). Cytotoxicity may be attributed to oxidative damage of membrane lipids (Fridovich 1986; Wise and Naylor 1987) as well as oxidation of proteins and nucleic acids (Fridovich 1986; Imlay and Linn 1988).

In the field, salt stress results in severe damage especially in situations when its inhibitory effects occur in the presence of high light intensity. Then PSII activity will result in high oxygen concentrations, especially if stomata are closed under stress. Concomitantly, chlorophyll will remain in its active state for a prolonged period of time, as the electron transport rate is reduced by inhibited off flow of products. Thus the probability for a transfer of electrons from activated chlorophyll to molecular oxygen to form $O_2^{\bullet-}$ will increase. $O_2^{\bullet-}$ ill rapidly dismutate to yield O_2 and the less reactive ROS H_2O_2. But in the presence of some cations such as Cu and Fe, highly reactive $OH^{\bullet-}$ may be formed (Imlay and Linn 1988; Fig. 15.5).

H_2O_2 is one of the most important secondary messengers in plant tissues, modulating effects of hormones and involved in developmental control of cells and tissues (van Breusegem and Dat 2006; van Breusegem et al. 2008). It has been shown that mitogen-activated protein kinases (MAP kinases) are involved in transduction of H_2O_2 signaling on cellular level (Pitschke and Hirt 2009). Apparently, MAPK3 and MAPK6 are integrating stress signals that regulate stomatal development (Wang et al. 2008a, b).

Ascorbic acid is a major antioxidant in plants. It detoxifies reactive oxygen species and maintains photosynthetic function (Fig. 15.5). Through its ascorbate recycling function, dehydroascorbate reductase affects the level of foliar reactive oxygen species and photosynthetic activity during leaf development. As a consequence, this enzyme influences the rate of plant growth and leaf aging (Chen and Gallie 2006).

ABA-induced closure of stomata and ABA-mediated inhibition of stomata opening are two ABA effects based on different reaction sequences (Mishra et al. 2006). Nevertheless, both processes are fine-tuned by ROS signaling, and it appears to be clear that ROS signals are transduced via a cascade of MAP kinase reactions (Gudesblat et al. 2007).

ROS Detoxification

As stated above, ROS production occurs permanently at a certain probability. It is well documented that ROS as well as nitrogen radicals are involved in cell signaling (Wilken and Huchzermeyer 1999) in plant cells like in cells of most other organisms. But as high ROS concentrations are toxic, plants are equipped at varying degrees with several systems to detoxify ROS. In this respect, molecules having antioxidative potential can be discriminated from enzyme-catalyzed reaction sequences.

Alpha-tocopherol is synthesized in the chloroplast (Schultz et al. 1976) and can be found in high concentrations in thylakoid membranes. Alpha-tocopherol disrupts lipid peroxidation cascades, reacts with $O_2^{\bullet-}$, and is capable of scavenging hydroxyl-, peroxyl-, and alkoxyl-radicals (Halliwell 1987). Oxidation of alpha-tocopherol leads to the formation of an alpha-chromoxyl radical, which can be reduced by ascorbic acid.

Ascorbic acid and several redox-active tri-peptides, called glutathiones, can be found in chloroplasts in millimolar concentrations (Halliwell 1982). Several different types are known from plants (Schmidt and Jäger 1992). This finding suggests that they are involved in different pathways controlled by specific enzymes. But there is only little information available to date.

In chloroplasts, H_2O_2 can be detoxified by an ascorbate-specific peroxidase (Chen and Asada 1989) involved in the ascorbate–glutathione cycle (Halliwell and Guteridge 1986) (Fig. 15.5) while in the cytosol H_2O_2 detoxification is catalyzed in a catalase-dependent reaction. Other enzymes involved in detoxification of ROS are superoxide dismutase, which converts $O_2^{\bullet -}$ to H_2O_2, and several peroxidases (Chang et al. 1984).

Antioxidants as well as enzymes capable of detoxifying ROS are present in all plants and plant tissues. But their concentrations and catalytic activities, respectively, as well as their patterns differ a lot (Streenivasulu et al. 2000). Therefore plants differ in their capacities to immediately detoxify ROS upon their occurrence and to build up a detoxification potential under stress. If the balance between ROS production and quenching capacity of the respective tissues is upset, oxidative damage will be produced (Dhindsa and Matowe 1981; Wise and Naylor 1987; Spychalla and Desborough 1990). In experimental approaches, it was demonstrated that enzyme activities of antioxidative pathways increase as a salt stress response (Verma and Mishra 2005) and that the maximal level of salt resistance correlates with maximal respective enzyme activities (Kennedy and De Fillippis 1999; Benavides et al. 2000; Lee et al. 2001; Mittova et al. 2002, 2003; Stepien and Klobus 2005; Chang et al. 2012; Mittal et al. 2012).

Physiological Importance of ROS Production as an Electron Sink

Although O_2 uptake which could not be accounted for Rubisco oxygenation or mitochondrial respiration has been observed in the light (Gerbaud and André 1980; Osmond et al. 1997), evidence of significant extra electron transport was not always found (Genty et al. 1989; Cornic and Briantais 1991). It is thought that O_2 photoreduction increases with increasing reduction of the ferredoxin pool, thus allowing linear electron flow to continue when NADP is scarce. It has been suggested that O_2 photoreduction can assist in maintaining a high trans-thylakoid pH gradient, which in turn enhances nonradiative dissipation of light energy and protects light reactions from photodamage (Lu et al. 2003).

In this context it has to be considered that a slight alkalization of the chloroplast stroma in the light is a prerequisite for the functioning of most plastidic metabolic pathways. The phosphorylation potential of ATP, like the group transfer potential of many other metabolites, depends on pH (and cation concentrations, especially Mg), and reaction equilibria within pathways would not allow metabolism to take place in all other cell compartments because of their lower pH values compared to illuminated chloroplasts. Additionally, the stoichiometries of NADPH and ATP production and consumption in chloroplasts are different and the alternative sinks are considered necessary to enable ATP production without NADP reduction (Noctor and Foyer 1998).

Dark Reaction of Photosynthesis

Rubisco Limitation

Rubisco is the most abundant protein in the leaves, contributing to up to 50% of the total protein. It has been discussed that a part of the total Rubisco protein is not catalytically active, but functions as a CO_2 buffer. Different functions of Rubisco, CO_2 binding, and CO_2 turnover, are in accordance with the observation that for full catalytic activity in the light, Rubisco needs to be activated, which in turn requires Rubisco activase and ATP (Robinson and Portis 1988; Salvucci and Ogren 1996). Inhibitors are generally analogues of the enzyme's substrate ribulose-1,5-bis phosphate (RuBP) (Edmonson et al. 1990) and bind to the enzyme in the absence of RuBP. Activase releases tight-binding inhibitors from the Rubisco active sites, thus increasing specific activity. This reaction requires ATP (Robinson and Portis 1988), so a decreased activity and activation state of Rubisco at low relative water content may be related to inadequate ATP supply to the protein complex (Tezara et al. 1999; Parry et al. 2002).

The amount of Rubisco protein is generally little affected by moderate or severe salt stress (Flexas et al. 2002), even if experienced over a period of many days (Tezara et al. 1999; Gunasekera and Berkowitz 1993; Medrano et al. 1997). This means that specific Rubisco activity rather than protein concentration is decreased under saline conditions. Restoration of the assimilation potential by rehydration also suggests that Rubisco is not irreversibly impaired (Medrano et al. 1997; Parry et al. 2002).

Calvin Cycle Enzymes

The reduction of RuBP content at low relative water content could result from a limitation in one or more enzymes of the Calvin cycle. There is little direct evidence regarding the response of the individual enzymes of the regenerative part of the Calvin cycle to increasing cytosolic mineral content subsequent to increased salinity. There do not seem to be significant differences between enzymes from glycophytes and halophytes, respectively (Huchzermeyer and Heins 2000; Huchzermeyer 2000). Furthermore, in most cases isolated enzymes did not show a degree of salt inhibition sufficient to explain the extent of inhibition of the respective metabolic reaction in *in vivo* experiments. Therefore it is doubtful that the high salt resistance of halophytes is due to changed properties of their Calvin cycle enzymes. Thus, there is a strong evidence that the Calvin cycle per se is not the cause of the decreased assimilation rate under osmotic stress.

RuBP Concentration and Rubisco Activity

The net assimilation rate depends on the synthesis of RuBP and Rubisco activity. Therefore, the decrease in leaf Rubisco content at low relative water content is significant (Giménez et al. 1992; Gunasekera and Berkowitz 1993; Tezara et al. 1999). In stressed sunflower plants, the photosynthetic rate correlated with RuBP concentration (Giménez et al. 1992), which suggests that the assimilation potential in these experiments was determined by RuBP content, but not by CO_2.

Under salt stress, there is a general decline in Calvin cycle intermediates during the phase of stress adaptation. Decreased RuBP concentration might be caused by the general rundown of the Cavin cycle and decrease in assimilation rate. The high ratio of 3PGA/RuBP suggests a limitation in the RuBP regeneration part of the Calvin cycle, either caused by enzyme limitations or inadequate ATP, although it was interpreted by Tezara et al. (1999) as evidence of 3PGA production by mitochondria. In laboratory experiments with halophytes such as *Aster tripolium* however, the metabolite pattern recovered within less than two or three days under constant stress conditions.

Interaction between Rubisco activity and RuBP supply is well illustrated by studies on unstressed tobacco leaves with normal amounts of Rubisco (Laisk and Oja 1974; von Caemmerer 2000). The RuBP pool increased in leaves in bright light when the CO_2 concentration was transiently decreased, so that when returning to normal CO_2 concentrations, assimilation rate was much higher than the steady-state concentrations for a short time until the RuBP was consumed. Thus, under steady-state conditions, RuBP supply limited the CO_2 assimilation rate.

Limitation of RuBP Regeneration by NADPH and ATP Availability

Synthesis of RuBP depends on ATP and NADPH concentration and on the Calvin cycle activity, or more specifically on PRK activity, and concentration of the substrates ATP and ribose 5-phosphate (Lawlor 2001). In the Calvin cycle, NADPH serves as substrate of the glyceraldehyde 3-phosphate dehydrogenase. If NADPH were limiting at low relative water content, it would decrease glyceraldehyde 3-phosphate and thus ribose 5-phosphate, the same effect as decreasing the activity of an enzyme in that part of the cycle. In salt stressed leaf mesophyll cells NADPH content remained constant (Lawlor and Khanna-Chopra 1984; Tezara et al. 1999), while NADH increased (Lawlor and Khanna-Chopra 1984), indicating that the electron transport capacity is sufficient to maintain and increase the reduction state of these pyridine nucleotides. Thus, with respect to high salinity resistance of the photosynthetic electron transport system in both glycophytes and halophytes, the availability of NADPH to the Calvin cycle is unlikely to limit its capacity to form RuBP (Koyro and Huchzermeyer 1999; Huchzermeyer and Heins 2000).

The rate of ATP synthesis depends on the light reactions, generation of the transthylakoid pH gradient (ΔpH), availabiliy of ADP and P_i, and activity of the chloroplast coupling factor (CF_1) (Löhr and Huchzermeyer 1985; Löhr et al. 1985; Huchzermeyer 1988a, b; Richter et al. 2000). Inadequate ATP concentration would decrease the ability of the Calvin cycle to regenerate RuBP by PRK, so glyceraldehyde-3-phosphate would increase and RuBP decrease. This was the effect of reduced PRK activity (Paul et al. 1995), but ATP increased in the transgenics, in contrast to the case of water stress in some studies (Lawlor 2002a, b; Lawlor and Cornic 2002).

Triose Export

In the presence of light, sugars are exported as triose phosphates (DHAP and GAP) in exchange of free phosphate via the phosphate translocator (Riesmeier et al. 1993). V_{max} of sugar phosphate export is too low to keep up with sugar phosphate synthesis under average light conditions. This would cause shortage of phosphate inside the chloroplast stroma, thus inhibiting the primary reactions of photosynthesis and increasing the risk of ROS production. Starch production inside the chloroplast is a relief, because phosphate will be released and becomes available for the chloroplast coupling factor to produce ATP. V_{max} of starch production has not been measured yet as physiological sugar concentrations are too low. The capacity to produce starch is obviously high enough to turn over any sugar concentration that might become available under physiological conditions. There is an equilibrium between starch synthesis and hydrolysis, respectively. Therefore, starch will be degraded at night and allow a permanent supplementation of the cytosol of the host cell. This system allows plant cells to grow permanently on a continuous flow of incoming sugar phosphates.

In the cytosol of green leaf parenchyma cells, sugar phosphates are used to fuel cell metabolism. In a competing metabolic pathway sucrose is formed from glucose and fructose and is exported into the phloem to feed sink organs.

The description above shows that there are several alternatives how incoming salt may affect the functioning of cellular sugar metabolism. In principle, the situations in source and sink tissues are comparable. But under moderate salt stress, salt concentrations have been found to differ a lot between root, leaf, and fruit tissues. Therefore, the observed effect of salt stress also differs among tissues of plant organs.

Salt effects on sugar metabolism and sugar export from the chloroplasts into the cytosol can be explained on the basis of a competition between Cl^- and $H_2PO_4^-$ for binding sites on enzymes and receptors, respectively. This competition is due to similar diameters of these two ions when hydrated. This would mean that under salt stress the chloroplast stroma is at risk to run out of free phosphate. As mentioned above, this would inhibit ATP synthesis at the chloroplast coupling factor (Groth et al. 2000). The observed effect would be quite similar to the one observed after energy transfer inhibitors like nitrofen (Huchzermeyer and Löhr 1990). Nitrofen is a herbicide used in rice cultures and it is known to stimulate light-dependent ROS production in plants. By the way, this observation again is a proof for the tight coupling between primary reactions of photosynthesis.

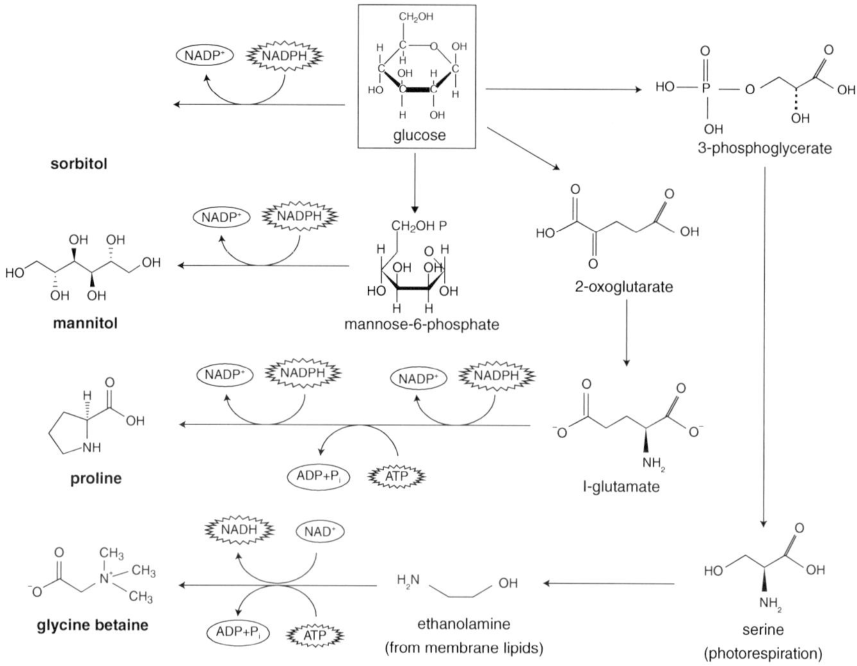

Fig. 15.6 Overview on metabolism of compatible solutes. Glucose can function as a substrate for synthesis of compatible solutes. Cytosolic synthesis often occurs at the expense of NAD(P)H. Thus, synthesis of compatible solutes is a relief under conditions, when production of redox power and glucose are exceeding consumption of electrons and export of sugar. Cytosolic pattern of compatible solutes varies among plant species depending on respective enzyme activities. In addition to regulation of respective gene expression, synthesis rates are controlled by availability of precursors

Protection of Proteins and Biomembranes from Salt Effects

Control of Mechanisms Improving Stress Resistance: Compatible Solutes

Sequestration of ions to the vacuole is a strategy leading to enhanced salt stress resistance. Some low molecular weight nontoxic compounds have been identified to significantly contribute to ionic and osmotic balance inside cells. They are called "compatible solutes" (Ford 1984; Ashihara et al. 1997; Hasegawa et al. 2000; Zhifang and Loescher 2003; Lee et al. 2008). Chemically they can be described to be poly-amines or poly-hydroxyls. In addition to their function in ionic and osmotic homeostasis, their important function is that they can replace water in its function to stabilize aggregates of soluble proteins and membrane–protein interactions, respectively (Yancey et al. 1982; Crowe et al. 1992; Ashraf and Foolad 2007).

Depending on enzyme patterns and metabolic pathways preferred in individual plants, different compatible solutes have been found in plants (Fig. 15.6).

Among the compounds described in the literature are proline (Singh et al. 2000; Kavi Kishore et al. 2005; Koca et al. 2007; Alla et al. 2012), glycine betaine (Khan et al. 1998; Wang and Nil 2000; Türkan and Demiral 2009; Zhu et al. 2011), sugars (Bohnert and Jensen 1996; Kerpesi and Galiba 2000; Gil et al. 2011), di-, tri-saccharides and other sugar derived compounds (Hagemann and Murata 2003), and polyols (Popp et al. 1985; Orthen et al. 1994; Bohnert et al. 1995; Gil et al. 2011). Overexpression of genes of metabolic pathways leading to production of compatible solutes has been shown to improve salt resistance. (Parida et al. 2002; Parida and Das 2005). Apparently, there are three ways to explain the mechanism how compatible solutes become overproduced under stress. Under optimal growth conditions, these compounds are produced at low rates, so that their concentration is found to be low. Therefore, the question is whether overproduction under stress is due to activation of already existing enzymes or de novo synthesis of such enzymes.

It was observed that many stress conditions like drought and salt stress, for instance, initially inhibit export of products of photosynthesis from their source tissues. This will result in enhanced concentrations of primary products of the Calvin cycle in leaf cells. As shown by Koyro and Huchzermeyer (2005) the most compatible solutes derive from primary products of sugar metabolism. Therefore, increased concentrations of metabolites of photosynthesis will stimulate synthesis rate of compatible solutes. Such an increase will become significant if metabolite concentrations under optimal growth conditions are below the Km values of enzymes catalyzing the initial reactions of the pathways leading to compatible solutes. This interpretation agrees with the findings of Soussi et al. (1998), who attributed enhanced concentrations of proline and carbohydrates under salt stress in chick pea to damage of metabolic pathways rather than to protective mechanisms.

A second working hypothesis is based on findings of Süss et al. (1993). They found that enzymes of individual pathways in chloroplasts tend to form clusters. Such conditions would contradict free mobility of intermediates. Thus, substrate concentrations localized to catalytic centers of enzymes may significantly differ from respective bulk phase concentrations. Calvin cycle enzymes, for instance, tend to bind to one another and thus form aggregates of proteins. This allows substrates as well as products formed to be shuttled from one enzyme to the next one of the respective pathway. Süss postulated that modulation of pathways in this model is brought about by modulation of enzyme neighbourhood. From analysis of internal signalling within cells, it is well documented that such modulations can be brought about by protein phosphorylation and de-phosphorylation, for instance. Indeed, such phosphorylation-dependent variations in protein–protein interactions have been observed by Allen and Horton, when analyzing protein localization in thylakoid membranes (Horton and Foyer 1983; Pursiheimo et al. 2001; Allen 2002).

A third model refers to the observation that salt stress response of plants has been found to be under the control of hormones (ABA, for instance) and second messengers (sugar signalling, for instance). This implies that modification of enzyme patterns will result in modified metabolic activity of cells and tissues under stress. Such argumentation would explain the above observations on the level of gene activities.

These latter arguments may suggest not to discuss three different models of regulation of metabolic pathways leading to the synthesis of compatible solutes. It rather appears to us that hormone- and secondary messenger actions explain how formation of protein aggregates may be controlled in plants.

Biochemical reaction sequences leading to improved salt resistance are likely to act synergistically (Iyengar and Reddy 1996). The reactions include the synthesis of compatible solutes, the stimulation of antioxidative enzyme activities, and modifications of the photosynthetic pathway.

Energy Consumption by N- and S-Pathways

As a rule, in non-woody plants nitrate is reduced in chloroplasts, while it is reduced in root cell plastids of trees. Nitrate competes with chloride (due to similar radii of the hydrated ions) for uptake via specific translocators. These translocators are regulated via the plant's sugar pool and their activity thus matches the photosynthetic activity of source leaves. In this chapter, we focus on coupling of nitrate reduction to photosynthesis in chloroplasts of weeds.

About one-third of the electrons released by PSII finally will be used for nitrate reduction. Therefore, nitrate reduction is a major sink for electrons and sufficient N fertilization contributes to prevention of primary reactions of photosynthesis from over-reduction in high light.

Nitrate reduction and incorporation of reduced nitrogen into glutamate is shown in Fig. 15.3. It is obvious that the first intermediates of nitrate reduction are toxic. Like in primary reactions of photosynthesis, building up of concentrations and occurrence of side reactions that eventually might be toxic for the cell are prohibited by fast turnover of products.

It has been shown in the literature that the affinity for nitrate uptake from the soil can be improved by means of a molecular biological approach (Matt et al. 2001; Wang et al. 2003; Lillo 2004). Furthermore, nitrate fertilization helps to reduce salt stress effects (Syvertsen et al. 1989; Murillo-Amador et al. 2006).

It is obvious that salt stress can have direct as well as indirect effects on nitrate reduction and amino acid biosynthesis. Direct effects mostly are due to competition among nitrate and chloride ions. Indirect effects are based on the dependence of nitrate reduction on electrons released by photosynthetic electron transport, an intact structure of enzymes and enzyme aggregates, and the dependence of amino acid biosynthesis on substrates that derive from sugar metabolism (i.e. sugar supply from photosynthesis). We have to keep in mind that sugars, sugar phosphates, and ROS are functioning as secondary messengers regulating the expression of enzymes. Therefore, there will be another level of regulatory effects. Though a lot information on sugar signaling has been analyzed and a list of participants has been outlined in pretty much detail, the complex pathway structure of second messenger signaling has not been well characterized (Gibson 2000; Sheen 2002; Baena-González and Sheen 2008).

As nitrate reduction and subsequent amino acid biosynthesis depend on photosynthetic activity, there has to be some buffer capacity for intermediates to allow continuous metabolic activity of plants. In principle, there are two storage compartments in leaf cells: plastids and the vacuole. In storage organs, plastids can differentiate to protein storage compartments. In chloroplasts, only low amounts of storage proteins typical for storage plastids are found. But in some papers it is discussed whether the huge amounts of Rubisco accumulating in chloroplasts may function as storage proteins as well. In vacuoles, free amino acids and other low molecular weight molecules are stored rather than macro molecules like proteins. It has been shown that vacuoles from tissues active in photosynthesis can actively import amino acids at final concentrations several fold exceeding the ones in the cytosol (Homeyer et al. 1989). Amino acid import occurs at the expense of a transmembrane pH gradient. Both v-Type ATPase and PPase of the tonoplast build this gradient at the expense of ATP and pyrophosphate, respectively (Homeyer et al. 1989). It is well known that the activity of these two enzymes is regulated (Taiz 1992; Davies 1997; Han et al. 2005; Park et al. 2005), and that amino acid content is affected by salt stress (Pahlich et al. 1983).

In contrast to nitrogen that is found exclusively in the reduced form in organic molecules, sulfur occurs in metabolites of biological relevance in several redox states. Only a minor portion of the photosynthetic electrons are used for the sulfate reduction pathway. As outlined by Schmidt and Jäger (1992, and citations therein), sulfate reduction pathway in chloroplasts differs from the one in bacteria that had been identified earlier.

Inhibition of sulfur metabolism has major secondary effects, because SH bounds are essential for functioning of catalytic centers of many enzymes and reduced sulfur is found in coenzymes like CoA and liponic acid, for instance. Moreover, a pool of various glutathiones forms the dominant redox buffer of plant cells, and glutathiones are involved in detoxification of heavy metals and ROS. Therefore, enhanced salt stress resistance of sulfur metabolism always goes along with improved metabolic activity and apparent resistance of other essential functions as well.

Photosynthesis of Salt Stressed Plants Under Future Elevated Atmospheric CO_2 Concentration

C_3 Plants

Elevated atmospheric CO_2 concentration leads to a higher CO_2 concentration gradient between the outside air and the intercellular spaces of the leaves, so that the diffusion of CO_2 into the leaves and the pCO_2/pO_2 ratio at the sites of photoreduction is increased (Robredo et al. 2007). Therefore, photorespiration and the rates of oxygen activation and ROS formation are usually reduced in C_3 plants due to an increased NADPH utilisation, whereas the net photosynthetic rate and thus the carbon supply is enhanced (Polle 1996; Sgherri et al. 2000; Urban 2003; Kirschbaum 2004; Long et al. 2004; Hikosaka et al. 2005; Ignatova et al. 2005; Fig. 15.7).

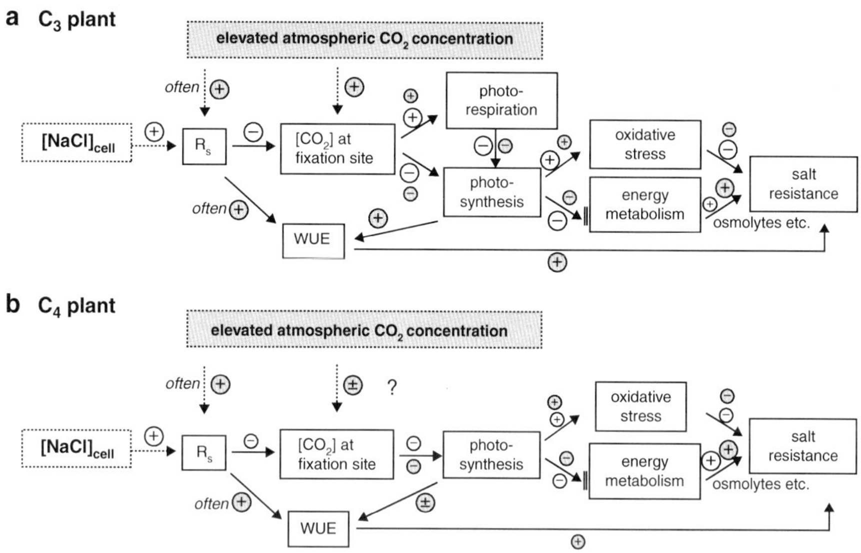

Fig. 15.7 Generalized and simplified model of the influence of NaCl salinity and elevated CO_2 concentration on the photosynthesis of (**a**) C_3 and (**b**) C_4 plants. O = influence of NaCl; O = influence of elevated CO_2 and salinity; ⊕ = positive influence; ⊖ = negative influence; O = major influence; o = minor influence. The changing quantity of a parameter is taken into account when assessing its influence on the next one (Except for energy metabolism, following ||). R_s stomatal resistance, *WUE* water use efficiency of photosynthesis

Furthermore, we often find a higher stomatal resistance (Hsiao and Jackson 1999; Sgherri et al. 2000; Li et al. 2003; Marchi et al. 2004; Rogers et al. 2004), which – together with the higher net assimilation – also leads to a better water use efficiency of photosynthesis (Amthor 1999; Morgan et al. 2001; Urban 2003). Results from free-air CO_2 enrichment (FACE) experiments showed that elevated CO_2 stimulated the light-saturated photosynthesis of C_3 plants by an average of 31% and reduced stomatal conductance by an average of 22% (Ainsworth and Rogers 2007).

However, responses of gas exchange to elevated CO_2 concentration are variable and depend on plant species and abiotic factors such as salinity and humidity (Drake et al. 1997; Ball and Munns 1992; Arp et al. 1993). So stomatal resistance of salt stressed C_3 plants may even decrease under elevated CO_2 if the maximization of photosynthesis and energy gain is given priority over a reduction of water loss in order to reduce oxidative stress. This is the case e.g. in *Aster tripolium*, which increases its WUE only by an improved photosynthesis, but not by a reduced transpiration. This is a reasonable strategy because the salt resistance of this species is mainly limited by an impaired net assimilation rate and accompanying ROS formation (Geissler et al. 2009a, b). Elevated atmospheric CO_2 concentration therefore ameliorates metabolic processes in *A. tripolium* which are typically associated with C_3 metabolism and are disadvantageous on saline habitats compared to C_4 plants (see Sect. C_4 Plants).

As a consequence of the enhanced plant water relations and/or net photosynthetic rate, there might be less need for antioxidants as elevated CO_2 ameliorates oxidative stress (Schwanz et al. 1996), such as in barley (Pérez-López et al. 2009) and *Solanum lycopersicum* (Takagi et al. 2009). On the other hand elevated CO_2 can induce carbon partitioning to various pathways associated with stress defence/response (Bokhari et al. 2007; Cseke et al. 2009; Jin et al. 2009), i.e. more energy can be provided for stress resistance mechanisms: The amount and/or activities of antioxidants can be increased, e.g. the activities of catalase (CAT), ascorbate peroxidase (APX) and superoxide dismutase (SOD) in *Quercus* (Marabottini et al. 2001; Schwanz and Polle 2001) or the expression and activities of APX, SOD and glutathione-*S*-transferase (GST) in *Aster tripolium* (Geissler et al. 2009b, 2010). Some species show also a higher amount of compatible solutes such as soluble carbohydrates (*Carthamus mareoticus*: Abdel-Nasser and Abdel-Aal 2002; *Aster tripolium*: Geissler et al. 2009a) or proline (*Aster tripolium*: Geissler et al. 2009a). Additionally, the larger number of carbon skeletons available due to an improved photosynthesis can increase the thickness of the outer epidermal cell walls and the cuticle (Tipping and Murray 1999; Tingey et al. 2003; Oksanen et al. 2005), which decreases transpiration on the one hand and light reflection and thus ROS production on the other hand (Thomas 2005; Geissler et al. 2009b). Reduced oxidative stress under elevated CO_2 concentration in turn leads to less structural damage to the thylakoid membranes of the chloroplasts (see Sect. Salt Effects on Chloroplast Structure and Metabolite Transfer) with only minor dilations of the thylakoid membranes (Oksanen et al. 2001; Geissler et al. 2009b; Fig. 15.4). This fact is of major importance because the chloroplasts – as the site of photosynthesis – fulfil a key function in adaptation to salinity, and chloroplast integrity correlates with a higher assimilation rate.

All the factors discussed above can contribute to an increased survival of salt stressed plant under elevated CO_2, which has been reported by various authors (Ball and Munns 1992; Rozema 1993; Drake et al. 1997; Fangmeier and Jäger 2001; Wullschleger et al. 2002; Urban 2003; Geissler et al. 2009a; Fig. 15.7).

C_4 Plants

While elevated CO_2 concentration generally has a positive effect on the water relations as well as on the net photosynthesis of salt stressed C_3 plants, the situation in C_4 plants is more ambiguous.

In C_4 plants, CO_2 is bound to PEP and pre-fixed by the formation of a C_4 compound (that can be reduced to malate, for instance) in the mesophyll cells which do not contain any Rubisco. The final carbon fixation step takes place in the bundle sheath cells surrounding the mesophyll cells coronally. It releases CO_2 from the C_4 compound, and assimilation will be catalyzed by Rubisco in an environment characterized by low oxygen partial pressure. This highly efficient CO_2 enrichment mechanism at the carboxylation sites in the bundle sheath cells and thus an equally efficient CO_2 fixation enable C_4 plants to maintain photosynthesis with almost closed stomata, which minimizes transpiration. Therefore they can use water more efficiently, which is of high advantage under saline conditions.

While it was proved in various studies that elevated CO_2 can ameliorate drought stress in C_4 plants similarly to C_3 plants due to its effect on stomatal conductance (Mateos-Naranjo et al. 2010; Leakey 2009; Lopes et al. 2011), it has usually only small or no positive direct effects on the photosynthesis of C_4 plants (Leakey et al. 2006; Ainsworth and Rogers 2007; Leakey 2009; Wang et al. 2012; Fig. 15.7). The latter are adapted to low atmospheric CO_2 concentrations due to their efficient CO_2 enrichment mechanism (see above), so their photosynthesis is saturated at the current atmospheric CO_2 concentration (Fangmeier and Jäger 2001). In consistence with these facts, Polley et al. (1996) predicted that the productivity of various C_4 species will probably show only a small increase under elevated CO_2 because the intrinsic water use efficiency was stimulated proportionally more by a given relative increase in CO_2 over sub-ambient than by elevated concentrations. So we expect the salt resistance of C_4 species to be less enhanced by elevated CO_2 concentration than the one of C_3 species (Fig. 15.7). This could already be confirmed by several studies which have shown that C_4 plants show less growth stimulation than C_3 species on saline soils under elevated CO_2 and/or are likely to be disadvantaged in competition (Schwarz and Gale 1984; Curtis et al. 1989; Rozema et al. 1991; Arp et al. 1993; Lenssen et al. 1993; Erickson et al. 2007; Jaggard et al. 2010).

However, this theory is not supported by Wand et al. (1999) and Ward et al. (1999), so the situation is a bit ambiguous. This is also true for the issue of photorespiration which leads to CO_2 loss and ROS production in C_3 plants (Ainsworth and Rogers 2007; Lüttge et al. 2010; Sect. ROS Production), thus limits their salt resistance and can be ameliorated by elevated CO_2. In most textbooks, it is stated that C_4 plants can overcome this problem on the expense of extra energy consumption and that the formation of significant concentrations of glycolate is not a problem of these plants. If salt stress would interfere with aggregate formation of chloroplasts, mitochondria, and peroxisomes, thus inhibiting by glycolate toxicity proper performance of C_3 plants; such effects should not be observed in C_4 species. Nevertheless, salt stress might interfere with intermediate transfer among cells and cell compartments in C_4 plants. This would inhibit photosynthetic activity but would not result in "classical" glycolate toxicity.

But again the situation is not as simple as suggested in textbooks. Several studies have pointed out that a low rate of photorespiration takes place in C_4 plants as well. In maize leaves grown at normal CO_2 concentration, photorespiration may reach 5% of the rate found in tobacco grown under identical conditions (Zelitch 1973). Though such a rate may appear to be low, glycolate oxidase activity obviously is essential for C_4 plants. Maize plants showing less than 10% of glycolate oxidase activity of the wild type could grow at high CO_2 concentrations, they became necrotic in normal air, and died within 2 weeks (Zelitch et al. 2009). When considering these facts, the salt resistance of C_4 plants may indeed be increased by elevated CO_2 concentration.

Conclusions and Future Perspectives

Abiotic stresses such as soil salinity are responsible for a decrease in yield especially in arid and semiarid regions. It is estimated that 45% of the world's agricultural land experience drought and 19.5% of the irrigated land are affected by salinization. This problem will be further catalyzed by global climate change. Salt stress directly affects the primary and secondary reactions of photosynthesis in numerous ways and consequently has also far reaching consequences on ROS production, on sugar, lipid, and amino acid metabolism, trans membrane and inter tissues transport of metabolites and on water relations and gas exchange.

Elevated atmospheric CO_2 concentration can improve photosynthesis and water relations and reduce oxidative stress in C_3 plants, so that it can enhance their salt resistance and thus their suitability as crops in a future world of climate change. In contrast, the effect of CO_2 on the salt resistance of C_4 plants is less beneficial.

The major challenge to modern plant scientists is to develop stress resistant and high-yield plants. Without doubt halophytic model plants will have to be developed as they are invaluable in precisely assessing the role which e.g. the complex photosynthetic reactions, ROS, antioxidants, or osmolytes play in the functional network which controls stress resistance. They can be of major importance not only for the breeding of halophytes as alternative crop plants, but also for increasing the salt resistance of conventional crops via genetic engineering. Halophytes as naturally salt resistant plants constitute especially valuable genetic models for understanding resistance mechanisms (Flowers and Colmer 2008). They are much more suited than classical glycophytic model plants such as *Arabidopsis thaliana* because genes which are expressed in halophytes but not in glycophytes indicate specific traits which enable survival under saline conditions.

In the face of climate change, it would be very desirable to develop model plants also regarding the interaction of salt stress with other abiotic factors associated with global change, such as elevated atmospheric CO_2 concentration. In this regard, C_3 and C_4 plants should be compared as they respond differently to elevated CO_2. Suited species for such a comparison would be *Chenopodium quinoa* (C_3) and *Atriplex spec.* (C_4) because they are related (both belong to the *Chenopodiaceae*), and show some similarities regarding their salt resistance mechanisms.

References

Abdel-Nasser LE, Abdel-Aal AE (2002) Effect of elevated CO_2 and drought on proline metabolism and growth of saf-flower (*Carthamus mareoticus* L.) seedlings without improving water status. Pak J Biol Sci 5:523–528

Ainsworth EA, Rogers A (2007) The response of photosynthesis and stomatal conductance to rising [CO_2]: mechanisms and environmental interactions. Plant Cell Environ 30:258–270

Alla MMN, Khedr A-HA, Serag MM, Abu-Alnaga AZ, Nada RM (2012) Regulation of metabolomics in *Atriplex halimus* growth under salt and drought stress. Plant Growth Regul 67:281–304

Allen JF (1992) Protein phosphorylation in regulation of photosynthesis. Biochim Biophys Acta 1098:275–335

Allen JF (2002) Plastoquinone redox control of chloroplast thylakoid protein phosphorylation and distribution of excitation energy between photosystems: discovery, background, implications. Photosynth Res 73:139–148

Allen JF, Bennett J (1981) Photosynthetic protein phosphorylation in intact chloroplasts Inhibition by DCMU and by the onset of CO_2-fixation. FEBS Lett 123:67–70

Allen JF, Forsberg J (2001) Molecular recognition in thylakoid structure and function. Trends Plant Sci 6:317–326

Allnutt FCT, Dilley RA, Kelly T (1989) Effect of high KCl concentrations on membrane localized metastable proton buffering domains in thylakoids. Photosynth Res 20:161–172

Amthor JS (1999) Increasing atmospheric CO_2 concentration, water use and water stress: scaling up from the plant to the landscape. In: Mooney HA, Luo Y (eds) Carbon dioxide and environmental stress. Academic, San Diego, pp 33–59

Arp WJ, Drake BG, Pockman WT, Curtis PS, Whigham DF (1993) Interaction between C_3 and C_4 salt marsh species during four years of exposure to elevated atmospheric CO_2. Vegetation 104(105):133–143

Ashihara H, Adachi K, Otawa M, Yasumoto E, Fukushima Y, Kato M, Sano H, Sasamoto H, Baba S (1997) Compatible solutes and inorganic ions in the mangrove plant *Avicennia marina* and their effects on activities of enzymes. Z Naturforsch 52c:433–440

Ashraf M, Foolad MR (2007) Roles of glycine betaine and proline in improving plant abiotic stress resistance. Environ Exp Bot 59:206–216

Azaizeh H, Steudle E (1991) Effects of salinity on water transport of excised maize (*Zea mays* L.) roots. Plant Physiol 97:1136–1145

Badawi GH, Yamauchi Y, Shimada E, Sasaki R, Kawano N, Tanaka K, Tanaka K (2004) Enhanced tolerance to salt stress and water deficit by overexpressing superoxide dismutase in tobacco (*Nicotiana tabacum*) chloroplasts. Plant Sci 166:919–928

Baena-González E, Sheen J (2008) Convergent energy and stress signaling. Trends Plant Sci 13:474–482

Ball MC, Munns R (1992) Plant responses to salinity under elevated atmospheric concentrations of CO_2. Aust J Bot 40:515–525

Benavides MP, Marconi PL, Gallego SM, Comba ME, Tomaro ML (2000) Relationship between antioxidant defence systems and salt tolerance in *Solanum tuberosum*. Aust J Plant Physiol 27:273–278

Blokhina O, Virolainen E, Fagerstedt KV (2002) Antioxidants, oxidative damage and oxygen deprivation stress: a review. Ann Bot 91:179–194

Boekema E, Fromme P, Gräber P (1988) On the structure of the ATP-synthase from chloroplasts. Ber Bunsenges Phys Chem 92:1031–1036

Bohnert HJ, Jensen RG (1996) Strategies for engineering water stress tolerance in plants. Trends Biotechnol 14:89–97

Bohnert HJ, Nelson DE, Jensen RG (1995) Adaptations to environmental stresses. Plant Cell 7:1099–1111

Bokhari SA, Wan X-Y, Yang YW, Zhou L, Tang WL, Liu JY (2007) Proteomic response of rice seedling leaves to elevated CO_2 levels. J Proteome Res 6:4624–4633

Boyer PD (2000) Catalytic site forms and controls in ATP synthase catalysis. Biochim Biophys Acta 1458:252–262

Chang H, Siegel BZ, Siegel SM (1984) Salinity induced changes in isoperoxidase in taro, *Colocasia esculenta*. Phytochemistry 23:233–235

Chang I-H, Cheng K-T, Huang P-C, Lin Y-Y, Cheng L-J, Cheng T-S (2012) Oxidative stress in greater duckweed (*Spirodela polyrhiza*) caused by long-term NaCl exposure. Acta Physiol Plant 34:1165–1176

Cha-um S, Chuencharoen S, Mongkolsiriwatana C, Ashraf M, Kirdmanee C (2012) Screening sugarcane (*Saccharum* sp.) genotypes for salt tolerance using multivariate cluster analysis. Plant Cell Tiss Organ Cult 110:23–33

Chen G, Asada K (1989) Ascorbate peroxidase in tea leaves: occurrence of two isoenzymes and the differences in their enzymatic and molecular properties. Plant Cell Physiol 30:987–998
Chen Z, Gallie DR (2006) Dehydroascorbate reductase affects leaf growth, development, and function. Plant Physiol 142:775–787
Cornic G, Briantais J-M (1991) Partitioning of photosynthetic electron flow between CO_2 and O_2 reduction in a C_3 leaf (*Phaseolus vulgaris*) at different CO_2 concentrations and during water stress. Planta 183:178–184
Cramer GR (2003) Differential effects of salinity on leaf elongation kinetics of three grass species. Plant Soil 253:233–244
Crowe JH, Hoekstra FA, Crowe CM (1992) Anhydrobiosis. Annu Rev Plant Physiol 54:579–599
Cseke LJ, Tsai CJ, Rogers A, Nelsen MP, White HL, Karnosky DF, Podila GK (2009) Transcriptomic comparison in the leaves of two aspen genotypes having similar carbon assimilation rates but different partitioning patterns under elevated [CO_2]. New Phytol 182:891–911
Curtis PS, Drake BG, Whigham DF (1989) Nitrogen and carbon dynamics in C_3 and C_4 estuarine marsh plants grown under elevated CO_2 in situ. Oecologia 78:297–301
Davies JM (1997) Vacuolar energization: pumps, shunts and stress. J Exp Bot 48:633–641
Demmig-Adams B, Adams WW III (1996) The role of xanthophyll cycle carotenoids in the protection of photosynthesis. Trends Plant Sci 1:21–26
Dhindsa RS, Matowe W (1981) Drought tolerance in two mosses: correlated with enzymatic defense against lipid peroxidation. J Exp Bot 32:79–91
Drake BG, Gonzalez-Meler MA, Long SP (1997) More efficient plants. a consequence of rising atmospheric CO_2? Annu Rev Plant Physiol 48:609–639
Edmonson DL, Kahne HJ, Andrews TJ (1990) Substrate isomerization inhibits ribulosebisphosphate carboxylase/oxygenase during catalysis. FEBS Lett 260:62–66
Erickson JE, Megonigal JP, Peresta G, Drake BG (2007) Salinity and sea level mediate elevated CO_2 effects on C_3–C_4 plant interactions and tissue nitrogen in a Chesapeake Bay tidal wetland. Glob Change Biol 13:202–215
Fangmeier A, Jäger HJ (2001) Wirkungen erhöhter CO_2-Konzentrationen. In: Guderian R (ed) Handbuch der Umweltveränderungen und Ökotoxikologie, vol 2a, Terrestrische Ökosysteme: Immissionsökologische Grundlagen – Wirkungen auf Boden – Wirkungen auf Pflanzen. Springer, Berlin, pp 382–433
Fidalgo F, Santos A, Santos I, Salema R (2004) Effects of long-term salt stress on antioxidant defense systems, leaf water relations and chloroplast ultrastructure of potato plants. Ann Appl Biol 145:185–192
Field TS, Lee DW, Holbrook NM (2001) Why leaves turn red in autumn: the role of anthocyanins in senescing leaves of Red-Osier Dogwood. Plant Physiol 127:566–574
Flexas J, Escalona JM, Evain S, Gulías J, Moya I, Osmond CB, Medrano H (2002) Steady-state chlorophyll fluorescence (F_s) measurements as a tool to follow variations of net CO_2 assimilation and stomatal conductance during water-stress in C_3 plants. Physiol Plant 114:231–240
Flowers TJ, Colmer TD (2008) Salinity tolerance in halophytes. New Phytol 179:945–963
Ford CW (1984) Accumulation of low molecular solutes in water stressed tropical legumes. Phytochemistry 23:1007–1015
Foyer CH, Bloom AJ, Queval G, Noctor G (2009) Photorespiratory metabolism: genes, mutants, energetics, and redox signaling. Annu Rev Plant Biol 60:455–488
Fridovich I (1986) Biological effects of the superoxide radical. Arch Biochem Biophys 247:1–11
Geissler N, Hussin S, Koyro H-W (2009a) Interactive effects of NaCl salinity and elevated atmospheric CO_2 concentration on growth, photosynthesis, water relations and chemical composition of the potential cash crop halophyte *Aster tripolium* L. Environ Exp Bot 65:220–231
Geissler N, Hussin S, Koyro H-W (2009b) Elevated atmospheric CO_2 concentration ameliorates effects of NaCl salinity on photosynthesis and leaf structure of *Aster tripolium* L. J Exp Bot 60:137–151
Geissler N, Hussin S, Koyro H-W (2010) Elevated atmospheric CO_2 concentration enhances salinity tolerance in *Aster tripolium* L. Planta 231:583–594

Genty B, Briantais JM, Baker NR (1989) The relationship between the quantum yield of photosynthetic electron-transport and quenching of chlorophyll fluorescence. Biochim Biophys Acta 990:87–92

Gerbaud A, André M (1980) Effect of CO_2, O_2 and light on photosynthesis and photorespiration in wheat. Plant Physiol 66:1032–1036

Gibson SI (2000) Plant sugar-response pathways. Part of a complex regulatory web. Plant Physiol 124:1532–1539

Gil R, Lull C, Boscaiu M, Bautista I, Lidon A, Vicente O (2011) Soluble carbohydrates as osmolytes in several halophytes from a mediterranean salt marsh. Not Bot Horti Agrobo 39:9–17

Giménez C, Mitchell VJ, Lawlor DW (1992) Regulation of photosynthetic rate of two sunflower hybrids under water stress. Plant Physiol 98:516–524

Glenn EP, Brown JJ, O'Leary JW (1998) Irrigating crops with sea water. Sci Am 279:76–81

Groth G, Mills DA, Christiansen E, Richter ML, Huchzermeyer B (2000) Characterization of a phosphate binding domain on the α subunit of chloroplast ATP synthase using the photoaffinity phosphate analog 4-azido-2-nitrophenyl phosphate. Biochemistry 39:13781–13787

Gudesblat GE, Iusem ND, Morris PC (2007) Guard cell-specific inhibition of *Arabidopsis* MAPK3 expression causes abnormal stomatal responses to abscisic acid and hydrogen peroxide. New Phytol 173:713–721

Gunasekera D, Berkowitz GA (1993) Use of transgenic plants with ribulose 1,5-bisphosphate carboxylase/oxygenase antisense DNA to evaluate the rate limitation of photosynthesis under water stress. Plant Physiol 103:629–635

Hagemann M, Murata N (2003) Glucosylglycerol, a compatible solute, sustains cell division under salt stress. Plant Physiol 131:1628–1637

Halliwell B (1982) The toxic effects of oxygen on plant tissues. In: Oberly LW (ed) Superoxide dismutase, vol I. CRC Press, Boca Raton, pp 89–123

Halliwell B (1987) Oxidative damage, lipid peroxidation, and antioxidant protection in chloroplasts. Chem Phys Lipids 44:327–340

Halliwell B, Gutteridge JMC (1986) Free radicals in biology and medicine. Oxford University Press, London

Han N, Shao Q, Lu C-M, Wang B-S (2005) The leaf tonoplast V-H^+-ATPase activity of a C_3 halophyte *Suaeda salsa* is enhanced by salt stress in a Ca-dependent mode. J Plant Physiol 162:267–274

Hasegawa PM, Bressan RA, Zhu JK, Bohnert HJ (2000) Plant cellular and molecular responses to high salinity. Annu Rev Plant Physiol Plant Mol Biol 51:463–499

Havaux M, Dall'Osto L, Bassi R (2007) Zeaxanthin has enhanced antioxidant capacity with respect to all other xanthophylls in *Arabidopsis* leaves and functions independent of binding to PSII antennae. Plant Physiol 145:1506–1520

Hesse H, Jank-Ladwig R, Strotmann H (1976) On the reconstitution of photophosphorylation in CF_1- extracted chloroplasts. Z Naturforsch 31c:445–451

Hikosaka K, Onoda Y, Kinugasa T, Nagashima H, Anten NPR, Hirose T (2005) Plant responses to elevated CO_2 concentration at different scales: leaf, whole plant, canopy, and population. Ecol Res 20:243–253

Homeyer U, Litek K, Huchzermeyer B, Schultz G (1989) Uptake of phenylalanine into isolated barley vacuoles is driven by both tonoplast adenosine triphosphatase and pyrophosphatase: evidence for a hydrophobic L-amino acid carrier system. Plant Physiol 89:1388–1393

Horton P (2000) Prospects for crop improvement through the genetic manipulation of photosynthesis: morphological and biochemical aspects of light capture. J Exp Bot 51:475–485

Horton P, Foyer C (1983) Relationship between protein phosphorylation and electron transport in the reconstituted chloroplast system. Biochem J 210:517–521

Horton P, Ruban AV, Walters RG (1996) Regulation of light harvesting in green plants. Annu Rev Plant Physiol Plant Mol Biol 47:665–684

Hose E, Clarkson DT, Steudle E, Schreiber L, Hartung W (2001) The exodermis: a variable apoplastic barrier. J Exp Bot 52:2245–2264

Hsiao TC, Jackson RB (1999) Interactive effects of water stress and elevated CO_2 on growth, photosynthesis, and water use efficiency. In: Luo Y, Mooney HA (eds) Carbon dioxide and environmental stress. Academic, San Diego, pp 3–31
Hu YC, Schmidhalter U (2005) Drought and salinity: a comparison of their effects on mineral nutrition of plants. J Plant Nutr Soil Sci 168:541–549
Huchzermeyer B (1988a) Phosphate binding to isolated chloroplast coupling factor (CF_1). Z Naturforsch 43c:213–216
Huchzermeyer B (1988b) Nucleotide binding and ATPase activity of membrane bound chloroplast coupling factor (CF_1). Z Naturforsch 43c:133–139
Huchzermeyer B (2000) Biochemical principles of salt tolerance. In: Lieth H (ed) Sustainable halophyte utilization in the mediterranean and subtropical dry regions. University of Osnabrück Publication, Osnabrück, pp 130–133
Huchzermeyer B, Heins T (2000) Energy metabolism and salt stress. In: Lieth H, Moschenko M (eds) INCO-DC annual report. University of Osnabrück Publication, Osnabrück, pp 48–73
Huchzermeyer B, Koyro H-W (2005) Salt and drought stress effects on photosynthesis. Enzyme cohesion and high turn over metabolite shuttling, essential for functioning of pathways, is impaired by changes in cytosolic water potential. In: Pessarakli M (ed) Handbook of photosynthesis, 2nd edn. CRC Press/Taylor and Francis, Boca Raton, pp 751–777
Huchzermeyer B, Löhr A (1990) Diphenylether herbicides, a tool in elucidating the mechanism of photophosphorylation. Z Naturforsch 45c:552–557
Huchzermeyer B, Strotmann H (1977) Acid/base-induced exchange of adenine nucleotides on chloroplast coupling factor (CF_1). Z Naturforsch 32c:803–809
Huchzermeyer B, Willms I (1985) Regulation of coupling factor activities by some component of chloroplast electron transport system. ICSU Short Rep 3:318–319
Huchzermeyer B, Löhr A, Willms I (1986) A direct interaction between photosystem I and the chloroplast coupling factor. Biochem J 234:217–220
Huchzermeyer B, Hausmann N, Paquet-Durant F, Koyro H-W (2004) Biochemical and physiological mechanisms leading to salt tolerance. Trop Ecol 45:141–150
Ignatova LK, Novichkova NS, Mudrik VA, Lyubimov VY, Ivanov BN, Romanova AK (2005) Growth, photosynthesis, and metabolism of sugar beet at an early stage of exposure to elevated CO_2. Russ J Plant Physiol 52:158–164
Imlay JA, Linn S (1988) DNA damage and oxygen radical toxicity. Science 240:1302–1309
IPCC (2007) Climate change 2007. The physical science basis. contribution of working group I to the fourth assessment report of the intergovernmental panel on climate change. Cambridge University Press, Cambridge\New York
Iyengar ERR, Reddy MP (1996) Photosynthesis in highly salt tolerant plants. In: Pesserakli M (ed) Handbook of photosynthesis. Marcel Dekker, New York, pp 897–909
Jaggard KW, Qi A, Ober ES (2010) Possible changes to arable crop yields by 2050. Philos T Roy Soc B 365:2835–2851
Jansen MAK, Mattoo AK, Edelmann M (2001) D1-D2 protein degradation in the chloroplast. Eur J Biochem 260:527–532
Jin CW, Du ST, Chen WW, Li GX, Zhang YS, Zheng SJ (2009) Elevated carbon dioxide improves plant iron nutrition through enhancing the iron-deficiency-induced responses under iron-limited conditions in tomato. Plant Physiol 150:272–280
Johnson MP, Pérez-Bueno M, Zia A, Horton P, Ruban AV (2009) The zeaxanthin-independent and zeaxanthin-dependent qE components of nonphotochemical quenching involve common conformational changes within the photosystem II antenna in *Arabidopsis*. Plant Physiol 149:1061–1075
Kavi Kishore PB, Sangam S, Amrutha RN, Laxmi PS, Naidu KR, Rao KRSS, Reddy KJ, Theriappan P, Sreenivasulu N (2005) Regulation of proline biosynthesis, degradation, uptake and transport in higher plants: its implications in plant growth and abiotic stress tolerance. Curr Sci 88:424–438
Kennedy BF, De Fillippis LF (1999) Physiological and oxidative response to NaCl of the salt tolerant *Grevillea ilicifolia* and the salt sensitive *Grevillea arenaria*. J Plant Physiol 155:746–754

Keren N, Berg A, Van Kan PJM, Levanon H, Ohad I (1997) Mechanism of photosystem II photoinactivation and D1 protein degradation at low light: the role of back electron flow. Proc Natl Acad Sci USA 94:1579–1584
Kerpesi I, Galiba G (2000) Osmotic and salt stress induced alteration in soluble carbohydrate content in wheat seedlings. Crop Sci 40:482–487
Khan MA, Ungar IA, Showalter AM, Dewald HD (1998) NaCl-induced accumulation of glycine-betaine in four subtropical halophytes from Pakistan. Physiol Plant 102:487–492
Kirschbaum MUF (2004) Direct and indirect climate change effects on photosynthesis and transpiration. Plant Biol 6:242–253
Koca H, Bor M, Özdemir F, Türkan I (2007) The effect of salt stress on lipid peroxidation, antioxidative enzymes and proline content of sesame cultivars. Environ Exp Bot 60:344–351
Koyro H-W (2000) Untersuchungen zur Anpassung der Wildrübe (*Beta vulgaris* ssp. *maritima*) an Trockenstreß oder NaCl-Salinität. Habilitation thesis, Justus Liebig University Giessen
Koyro H-W, Huchzermeyer B (1999) Salt and drought stress effects on metabolic regulation in maize. In: Pessarakli M (ed) Handbook of plant and crop stress, 2nd edn. Marcel Dekker, New York, pp 843–878
Koyro H-W, Huchzermeyer B (2004) Ecophysiological needs of the potential biomass crop *Spartina townsendii* GROV. Trop Ecol 45:123–139
Koyro H-W, Huchzermeyer B (2005) Recent developments in stress tolerance breeding in maize. In: Ashraf M, Harris PJC (eds) Abiotic stresses: plant resistance through breeding and molecular approaches. Food Products Press, the Haworth Press, New York, pp 545–576
Koyro H-W, Geißler N, Hussin S, Debez A, Huchzermeyer B (2008) Strategies of halophytes to survive in a salty environment. In: Khan NA, Singh S (eds) Abiotic stress and plant responses. I. K. International Publishing House, New Delhi, pp 83–104
Kozaki A, Takebe G (1996) Photorespiration protects C_3 plants from photooxidation. Nature 384:557–560
Kuhn M, Böger P (1990) Studies on the light-induced loss of the D1 protein in photosystem-II membrane fragments. Photosynth Res 23:291–296
Kurkova EB, Kalinkina LG, Baburina OK, Myasoedov NA, Naumova TG (2002) Responses of *Seidlitzia rosmarinus* to salt stress. Biol Bull 29:221–228
Laible PD, Zipfel W, Owens TG (1994) Excited state dynamics in chlorophyll-based antennae: the role of transfer equilibrium. Biophys J 66:844–860
Laisk A, Oja V (1974) Leaf photosynthesis in short pulses of CO_2. The carboxylation reaction *in vivo*. Sov Plant Physiol 21:1123–1131
Lam E, Malkin R (1989) Lateral distribution and diffusion of plastocyanin in chloroplast thylakoids. J Cell Biol 108:1397–1405
Larcher W (2003) Physiological plant ecology, 4th edn. Springer, New York
Laszlo JA, Baker GM, Dilley RA (1984) Nonequilibration of membrane-associated protons with the internal aqueous space in dark-maintained chloroplast thylakoids. J Bioenerg Biomembr 1:37–51
Lawlor DW (2001) Photosynthesis. BIOS Scientific Publishers, Oxford
Lawlor DW (2002a) Limitation to photosynthesis in water-stressed leaves: Stomata vs. metabolism and the role of ATP. Ann Bot 89:871–885
Lawlor DW (2002b) Carbon and nitrogen assimilation in relation to yield: mechanisms are the keys to understanding production systems. J Exp Bot 53:773–787
Lawlor DW, Cornic G (2002) Photosynthetic carbon assimilation and associated metabolism in relation to water deficits in higher plants. Plant Cell Environ 25:275–294
Lawlor DW, Fock H (1978) Photosynthesis, respiration and carbon assimilation in water-stressed maize at two oxygen concentrations. J Exp Bot 29:579–593
Lawlor DW, Khanna-Chopra R (1984) Regulation of photosynthesis during water stress. In: Sybesma C (ed) Advances in photosynthesis research, vol 4. Martinus-Nijhoff/Dr. W. Junk Publishers, The Hague, pp 379–382
Leakey ADB (2009) Rising atmospheric carbon dioxide concentration and the future of C_4 crops for food and fuel. Proc Roy Soc Lond B 276:2333–2343

Leakey ADB, Uribelarrea M, Ainsworth EA (2006) Photosynthesis, productivity, and yield of maize are not affected by open air elevation of CO_2 concentration in the absence of drought. Plant Physiol 140:779–790

Lee DH, Kim YS, Lee CB (2001) The inductive responses of the antioxidant enzymes by salt stress in the rice (*Oryza sativa* L.). J Plant Physiol 158:737–745

Lee G, Carrow RN, Duncan RR (2004) Photosynthetic responses to salinity stress of halophytic seashore paspalum ecotypes. Plant Sci 166:1417–1425

Lee G, Carrow RN, Duncan RR, Eiteman MA, Rieger MW (2008) Synthesis of organic osmolytes and salt tolerance mechanisms in *Paspalum vaginatum*. Environ Exp Bot 63:19–27

Lenssen GM, Lamers J, Stroetenga M, Rozema J (1993) Interactive effects of atmospheric CO_2 enrichment, salinity and flooding on growth of C_3 (*Elymus athericus*) and C_4 (*Spartina anglica*) salt marsh species. Vegetatio 104(105):379–388

Li XP, Björlmann O, Shih C, Grossmann AR, Rosenquist M, Jansson S, Niyogi KK (2000) A pigment binding protein essential for regulation of photosynthetic light harvesting. Nature 403:391–395

Li JH, Dugas WA, Hymus GJ, Johnson DP, Hinkle CR, Drake BG, Hungate BA (2003) Direct and indirect effects of elevated CO_2 on transpiration from *Quercus myrtifolia* in a scrub-oak ecosystem. Glob Change Biol 9:96–105

Li XP, Gilmore AM, Caffari S, Bassi R, Golan T, Kramer D, Niyogi KK (2004) Regulation of light harvesting involves intrathylakoid lumen pH sensing by the PsbS protein. J Biol Chem 279:22866–22874

Lillo C (2004) Light regulation of nitrate uptake, assimilation and metabolism. Plant Ecophysiol 3:149–184

Löhr A, Huchzermeyer B (1985) Regulation of photosynthetic efficiency by some component of chloroplast electron transport chain. ICSU Short Rep 3:332–333

Löhr A, Willms I, Huchzermeyer B (1985) A regulatory effect of the electron transport chain on the ATP synthase. Arch Biochem Biophys 236:832–840

Long SP, Ainsworth EA, Rogers A, Ort DR (2004) Rising atmospheric carbon dioxide: plants FACE the future. Annu Rev Plant Biol 55:591–628

Lopes MS, Araus JL, van Heerden PDR, Foyer CH (2011) Enhancing drought tolerance in C_4 crops. J Exp Bot 62:3135–3153

Lu CM, Zhang JH (1999) Effects of salt stress on PSII function and photoinhibition in the cyanobacterium Spirulina platensis. J Plant Physiol 155:740–745

Lu C, Qiu N, Lu Q, Wang B, Kuang T (2002) Does salt stress lead to increased susceptibility of photosystem II to photoinhibition and changes in photosynthetic pigment composition in the halophyte *Suaeda salsa* grown outdoors? Plant Sci 163:1063–1068

Lu Q, Wen X, Lu C, Zhang Q, Kuang T (2003) Photoinhibition and photoprotection in senescent leaves of field-grown wheat plants. Plant Physiol Biochem 41:749–754

Lüttge U, Kluge M, Thiel G (2010) Botanik. Die umfassende Biologie der Pflanzen. Wiley-VHC, Weinheim

Lynch J, Thiel G, Läuchli A (1988) Effects of salinity on the extensibility and Ca availability in the expanding region of growing barley leaves. Bot Acta 101:355–361

Ma F, Chen X-B, Sang M, Wang P, Zhang J-P, Li L-B, Kuang T-Y (2009) Singlet oxygen formation and chlorophyll a triplet excited state deactivation in cytochrome b6f complex from *Bryopsis corticulans*. Photosynth Res 100:19–28

Malkin S, Braun G (1993) The degree of functional separation between the two photosystems in isolated thylakoid membranes deduced from inhibition studies of the imbalance in photoactivities. Photosynth Res 36:89–94

Marabottini R, Schraml C, Paolacci AR, Sorgona A, Raschi A, Rennenberg H, Badiani M (2001) Foliar antioxidant status of adult Mediterranean oak species (*Quercus ilex* L. and *Q. pubescens* Willd.) exposed to permanent CO_2-enrichment and to seasonal water stress. Environ Pollut 113:413–423

Marchi S, Tognetti R, Vaccari FP, Lanini M, Kaligari M, Miglietta F, Raschi A (2004) Physiological and morphological responses of grassland species to elevated atmospheric CO_2 concentrations in FACE-systems and natural CO_2 springs. Funct Plant Biol 31:181–194

Mateos-Naranjo E, Redondo-Gómez S, Andrades-Moreno L, Davy AJ (2010) Growth and photosynthetic responses of the cordgrass *Spartina maritima* to CO_2 enrichment and salinity. Chemosphere 81:725–731
Matsubara S, Gilmore AM, Osmond CB (2001) Diurnal and acclimatory responses of violaxanthin and lutein epoxide in the Australian Mistletoe *Amyema miquelii*. Aust J Plant Physiol 28:793–800
Matt P, Geiger M, Walch-Liu P, Engels C, Krapp A, Stitt M (2001) Elevated carbon dioxide increases nitrate uptake and nitrate reductase activity when tobacco is growing on nitrate, but increases ammonium uptake and inhibits nitrate reductase activity when tobacco is growing on ammoniumnitrate. Plant Cell Environ 24:1119–1137
Medrano H, Parry MAJ, Socias X, Lawlor DW (1997) Long term water stress inactivates Rubisco in subterranean clover. Ann Appl Biol 131:491–501
Millennium Ecosystem Assessment (2005) Ecosystems and human well-being: desertification Synthesis. World Resources Institute, Washington, DC, http://www.maweb.org/documents/document. 355.aspx.pdf
Mishra G, Zhang W, Deng F, Zhao J, Wang X (2006) A bifurcating pathway directs abscisic acid effects on stomatal closure and opening in *Arabidopsis*. Science 312:264–266
Mitchell P (1967) Proton current flow in mitochondrial systems. Nature 214:1327–1328
Mitsuya S, Kawasaki M, Taniguchi M, Miyake H (2003) Relationship between salinity-induced damages and aging in rice leaf tissues. Plant Prod Sci 6:213–218
Mittal S, Kumari N, Sharma V (2012) Differential response of salt stress on *Brassica juncea*: photosynthetic performance, pigment, proline, D1 and antioxidant enzymes. Plant Physiol Biochem 54:17–26
Mittova V, Tal M, Volokita M, Guy M (2002) Salt stress induces up-regulation of an efficient chloroplast antioxidant system in the salt-tolerant wild tomato species *Lycopersicon pennellii* but not in the cultivated species. Physiol Plant 115:393–400
Mittova V, Tal M, Volokita M, Guy M (2003) Up-regulation of the leaf mitochondrial and peroxisomal antioxidative systems in response to salt-induced oxidative stress in the wild salt-tolerant tomato species *Lycopersicon pennellii*. Plant Cell Environ 26:845–856
Moorthy P, Kathiresan K (1999) Effects of UV-B radiation on photosynthesis reactions in *Rhizophora apiculata*. Plant Growth Regul 28:49–54
Morgan JA, Lecain DR, Mosier AR, Milchunas DG (2001) Elevated CO2 enhances water relations and productivity and affects gas exchange in C3 and C4 grasses of the Colorado shortgrass steppe. Glob Change Biol 7:451–466
Munns R (1993) Physiological processes limiting plant growth in saline soils: some dogmas and hypotheses. Plant Cell Environ 16:15–24
Munns R (2002) Comparative physiology of salt and water stress. Plant Cell Environ 25:239–250
Munns R, Gardner PA, Tonnet ML, Rawson HM (1989) Growth and development in NaCl-treated plants. II Do Na^+ or Cl^- concentrations in dividing or expanding tissues determine growth in barley. Aust J Plant Physiol 15:529–540
Munns R, Husain S, Rivelli AR, James RA, Condon AG, Lindsay MP, Lagudah ES, Schachtman DP, Hare RA (2002) Avenues for increasing salt tolerance of crops, and the role of physiologically based selection traits. Plant Soil 247:93–105
Murillo-Amador B, Jones HG, Kaya C, Aquilar RL, Garcia-Hermández JL, Toyo-Diéguez E, Àvilla-Serrano NY, Rueda-Puente E (2006) Effects of foliar application of calcium nitrate on growth and physiological attributes of cowpea (*Vigna unguiculata* L. Walp.) grown under salt stress. Environ Exp Bot 58:188–196
Neves-Piestun BG, Bernstein N (2001) Salinity-induced inhibition of leaf elongation in maize is not mediated by changes in cell wall acidification capacity. Plant Physiol 125:419–1428
Niessen M, Thiruveedhi K, Rosenkranz R, Kebeish R, Hirsch H-J, Kreuzaler F, Peterhänsel C (2007) Mitochondrial glycolate oxidation contributes to photorespiration in higher plants. J Exp Bot 58:2709–2715
Noctor G, Foyer CH (1998) Ascorbate and glutathione: keeping active oxygen under control. Annu Rev Plant Physiol Mol Biol 49:249–272

North GB, Nobel PS (1991) Changes in hydraulic conductivity and anatomy caused by drying and rewetting roots of *Agave desertii*, Agavaceae. Am J Bot 78:906–915

Nublat A, Desplans J, Cassea F, Berthomieua P (2001) sas1, an *Arabidopsis* mutant overaccumulating sodium in the shoot, shows deficiency in the control of the root radial transport of sodium. Plant Cell 13:125–137

Oksanen E, Sober S, Karnosky DF (2001) Impacts of elevated CO_2 and/or O_3 on leaf ultrastructure of aspen (*Populus tremuloides*) and birch (*Betula papyrifera*) in the Aspen FACE experiment. Environ Pollut 115:437–446

Oksanen E, Riikonen J, Kaakinen S, Holopainen T, Vapaavuori E (2005) Structural characteristics and chemical composition of birch (*Betula pendula*) leaves are modified by increasing CO_2 and ozone. Glob Change Biol 11:732–748

Orthen B, Popp M, Smirnoff N (1994) Hydroxyl radical scavenging properties of cyclitols. Proc R Soc Edinb Sect B 102:269–272

Osmond CB, Bagder M, Maxwell K, Björkman O, Leegood R (1997) Too many photons: photorespiration, photoinhibition and photooxidation. Trends Plant Sci 2:119–120

Pahlich E, Kerres R, Jäger H-J (1983) Influence of water stress on the vacuole/extravacuole distribution of proline in protoplasts of *Nicotiana rustica*. Plant Physiol 72:590–591

Paramanova NV, Shevyakova NI, VlV K (2004) Ultrastructure of chloroplasts and their storage inclusions in the primary leaves of *Mesembryanthemum crystallinum* affected by putrescine and NaCl. Russ J Plant Physiol 51:86–96

Parida AK, Das AB (2005) Salt tolerance and salinity effects on plants: a review. Ecotox Environ Saf 60:324–349

Parida A, Das AB, Das P (2002) NaCl stress causes changes in photosynthetic pigments, proteins and other metabolic components in the leaves of a true mangrove, *Bruguiera parviflora*, in hydroponic cultures. J Plant Biol 45:28–36

Park S, Li J, Pittman JK, Berkowitz GA, Yang H, Undurraga S, Morris J, Hirschi KD, Gaxiola RA (2005) Up-regulation of a H^+-pyrophosphatase (H^+-PPase) as a strategy to engineer drought-resistant crop plants. Proc Natl Acad Sci USA 102:18830–18835

Parry MAJ, Andolojc JP, Khan S, Lea PJ, Keys AJ (2002) Rubisco activity: effects of drought stress. Ann Bot 89:833–839

Paul MJ, Knight JS, Habash D, Parry MAJ, Lawlor DW, Barnes SA, Loynes A, Gray JC (1995) Reduction in phosphoribulokinase activity by antisense RNA in transgenic tobacco: effects on CO_2 assimilation and growth in low irradiance. Plant J 7:535–542

Pérez-López U, Robredo A, Lacuesta M, Sgherri C, Mu oz-Rueda A, Navari-Izzo F, Mena-Petite A (2009) The oxidative stress caused by salinity in two barley cultivars is mitigated by elevated CO_2. Physiol Plant 135:29–42

Petrouleas V, Deligiannakis Y, Diner BA (1994) Binding of carboxylate anions at the non-heme Fe(II) of PSII. Competition with bicarbonate and effects on the QA/QB electron transfer rate. Biochim Biophys Acta 1188:271–277

Pitschke A, Hirt H (2009) Disentangling the complexity of mitogen-activated protein kinases and reactive oxygen species signaling. Plant Physiol 149:606–615

Polle A (1996) Protection from oxidative stress in trees as affected by elevated CO_2 and environmental stress. In: Koch G, Mooney H (eds) Terrestrial ecosystem response to elevated CO_2. Academic, New York, pp 299–315

Polley HW, Johnson HB, Mayeux HS, Brown DA, White JWC (1996) Leaf and plant water use efficiency of C_4 species grown at glacial to elevated CO_2 concentrations. Int J Plant Sci 157:164–170

Popp M, Larther F, Weigel P (1985) Osmotic adaptation in Australian mangroves. Vegetation 61:247–254

Pursiheimo S, Mulo P, Rintamäki E, Aro E-M (2001) Coregulation of light-harvesting complex II phosphorylation and lhcb mRNA accumulation in winter rye. Plant J 26:317–327

Rahman MS, Miyake H, Takeoka Y (2002) Effects of exogenous glycinebetaine on growth and ultrastructure of salt-stressed rice seedlings (*Oryza sativa* L.). Plant Prod Sci 5:33–44

Rees D, Noctor GD, Horton P (1990) The effect of high-energy-state excitation quenching on maximum and dark level chlorophyll fluorescence yield. Photosynth Res 25:199–211

Richter ML, Hein R, Huchzermeyer B (2000) Important subunit interactions in chloroplast ATP synthase. Biochim Biophys Acta 1458:326–342

Riesmeier JW, Flügge U-I, Schulz B, Heineke D, Heldt H-W, Willmitzer L, Frommer WB (1993) Antisense repression of the chloroplast triose phosphate translocator affects carbon partitioning in transgenic potato plants. Proc Natl Acad Sci USA 90:6160–6164

Robinson SP, Portis AR (1988) Involvement of stromal ATP in the light activation of ribulose-1,5-bisphosphate carboxylase/oxygenase in intact chloroplasts. Plant Physiol 86:293–298

Robredo A, Pérez-López U, de la Maza HS, González-Moro B, Lacuesta M, Mena-Petite A, Mu os-Rueda A (2007) Elevated CO_2 alleviates the impact of drought on barley improving water status by lowering stomatal conductance and delaying its effects on photosynthesis. Environ Exp Bot 59:252–263

Rogers A, Allen DJ, Davey PA, Morgan PB, Ainsworth EA, Bernacchi CJ, Cornic G, Dermody O, Dohleman FG, Heaton EA, Mahoney J, Zhu XG, Delucia EH, Ort DR, Long SP (2004) Leaf photosynthesis and carbohydrate dynamics of soybeans grown throughout their life-cycle under free-air carbon dioxide enrichment. Plant Cell Environ 27:449–458

Romanello GA, Chuchra-Zbytniuk KL, Vandermer JL, Touchette BW (2008) Morphological adjustments promote drought avoidance in the wetland plant *Acorus americanus*. Aquat Bot 89:390–396

Romanowska E, Albertsson P-Å (1994) Isolation and characterization of the cytochrome bf complex from whole thylakoids, grana, and stroma lamellae vesicles from spinach chloroplasts. Plant Cell Physiol 35:557–568

Rozema J (1993) Plant responses to atmospheric carbon dioxide enrichment: interactions with some soil and atmospheric conditions. Vegetation 104(105):173–190

Rozema J, Dorel F, Janissen R, Lenssen G, Broekman R, Arp W, Drake BG (1991) Effect of elevated atmospheric CO_2 on growth, photosynthesis and water relations of salt marsh grass species. Aquat Bot 39:45–55

Ruban AV, Young AJ, Horton P (1993) Induction of non-photochemical energy dissipation and absorbance changes in leaves. Evidence for changes in the state of the light-harvesting system of photosystem II *in vivo*. Plant Physiol 102:741–750

Salvucci ME, Ogren WL (1996) The mechanism of Rubisco activase: insights from studies of the properties and structure of the enzyme. Photosynth Res 47:1–11

Schmidt A, Jäger K (1992) Open questions about sulfur metabolism in plants. Annu Rev Plant Physiol Mol Biol 43:325–349

Schmidt CL, Malkin R (1993) Low molecular weight subunits associated with the cytochrome b6f complexes from spinach and *Chlamydomonas reinhardtii*. Photosynth Res 38:73–81

Schreiber U (1997) Chlorophyll fluorescence and photosynthetic energy conversion: simple introductory experiments with the teaching-PAM chlorophyll fluorometer. H. Walz GmbH, Effeltrich

Schreiber U, Gademann R, Bird P, Ralph PJ, Larkum AWD, Kühl M (2002) Apparent light requirement for activation of photosynthesis upon rehydration of desiccated beachrock microbial mats. J Phycol 38:125–134

Schultz G, Huchzermeyer Y, Reupke B, Bickel H (1976) On the intra cellular site of biosynthesis of alpha tocopherol in Hordeum-Vulgare. Phytochemistry (Oxford) 15:1383–1386

Schwanz P, Polle A (2001) Differential stress responses of antioxidative systems to drought in pendunculate oak (*Quercus robur*) and maritime pine (*Pinus pinaster*) grown under high CO_2 concentrations. J Exp Bot 52:133–143

Schwanz P, Picon C, Vivin P, Dreyer D, Guehl J, Polle A (1996) Responses of antioxidative systems to drought stress in pendunculate oak and maritime pine as modulated by elevated CO_2. Plant Physiol 110:393–402

Schwarz M, Gale J (1984) Growth responses to salinity at high levels of carbon dioxide. J Exp Bot 35:193–196

Sgherri CLM, Salvateci P, Menconi M, Raschi A, Navari-Izzo F (2000) Interaction between drought and elevated CO_2 in the response of alfalfa plants to oxidative stress. J Plant Physiol 156:360–366
Sheen J (2002) Phosphorelay and transcription control in cytokinin signal transduction. Science 296:1650–1652
Singh SK, Sharma HC, Goswami AM, Datta SP, Singh SP (2000) In vitro growth and leaf composition of grapevine cultivars as affected by sodium chloride. Biol Plant 43:283–286
Soussi M, Lluch C, Ocana A (1998) Effects of salt stress on growth, photosynthesis and nitrogen fixation in chick pea (*Cicer arietinum* L.). J Exp Bot 49:1329–1337
Spychalla JP, Desborough SL (1990) Superoxide dismutase, catalase, and alpha-tocopherol content of stored potato tubers. Plant Physiol 94:1214–1218
Staehelin LA (1975) Chloroplast membrane structure. Intramembrane particles of different sizes make contact in stacked membrane regions. Biochim Biophys Acta 408:1–11
Stepien P, Klobus G (2005) Antioxidant defense in the leaves of C_3 and C_4 plants under salinity stress. Physiol Plant 125:31–40
Strasser RJ, Tsimilli-Michael M, Srivaslava A (2004) Analysis of the chlorophyll a fluorescence transient. In: Papageorgiou GC, Govindjee (eds) Chlorophyll Fluorescence: a signature of photosynthesis. Kluwer, Amsterdam, pp 1–42
Streenivasulu N, Grimm B, Wobus U, Weschke W (2000) Differential response of antioxidant compounds to salinity stress in salt-tolerant and salt-sensitive seedlings of fox-tail millet (*Setaria italica*). Physiol Plant 109:435–442
Strotmann H, Bickel S, Huchzermeyer B (1976) Energy-dependent release of adenine nucleotides tightly bound to chloroplast coupling factor CF1. FEBS Lett 61:194–198
Süss K-H, Arkona C, Manteuffel R, Adler K (1993) Calvin cycle multienzyme complexes are bound to chloroplast thylakoid membranes of higher plants in situ. Proc Natl Acad Sci USA 90:5514–5518
Syvertsen JP, Boman B, Tucker DPH (1989) Salinity in Florida citrus production. Proc Fla State Hort Soc 102:61–64
Taiz L (1992) The plant vacuole. J Exp Biol 172:113–122
Takagi M, El-Shemy H, Sasaki S, Toyama S, Kanai S, Saneoka H, Fujita K (2009) Elevated CO_2 concentration alleviates salinity stress in tomato plant. Acta Agr Scand B-S P 59:87–96
Tanji KK (2002) Salinity in the soil environment. In: Läuchli A, Lüttge U (eds) Salinity: environment – plants – molecules. Kluwer, Dordrecht, pp 21–51
Tezara W, Mitchell VJ, Driscoll SP, Lawlor DW (1999) Water stress inhibits plant photosynthesis by decreasing coupling factor and ATP. Nature 401:914–917
Thomas SC (2005) Increased leaf reflectance in tropical trees under elevated CO_2. Glob Change Biol 11:197–202
Tingey DT, McKane RB, Olszyk DM, Johnson MG, Rygiewicz PT, Lee H (2003) Elevated CO_2 and temperature alter nitrogen allocation in Douglas-fir. Glob Change Biol 9:1038–1050
Tipping C, Murray DR (1999) Effects of elevated atmospheric CO_2 concentration on leaf anatomy and morphology in *Panicum* species representing different photosynthetic modes. Int J Plant Sci 160:1063–1073
Touchette BW, Smith GA, Rhodes KL, Poole M (2009) Tolerance and avoidance: two contrasting physiological responses to salt stress in mature marsh halophytes *Juncus roemerianus* Scheele and *Spartina alterniflora* Loisel. J Exp Mar Biol Ecol 380:106–112
Trebst A, Soll-Bracht E (1996) Cycloheximids retards high light driven D1 protein degradation in *Chlamydomonas rheinhardtii*. Plant Sci 115:191–197
Türkan I, Demiral T (2009) Recent developments in understanding salinity tolerance. Environ Exp Bot 67:2–9
Urban O (2003) Physiological impacts of elevated CO_2 concentration ranging from molecular to whole plant responses. Photosynthetica 41:9–20
van Breusegem F, Dat JF (2006) Reactive oxygen species in plant cell death. Plant Physiol 141:384–390

van Breusegem F, Bailey-Serves J, Miller R (2008) Unraveling the tapestry of networks involving reactive oxygen species in plants. Plant Physiol 147:978–984
Verma S, Mishra SN (2005) Putrescine alleviation of growth in salt stressed *Brassica juncea* by inducing antioxidative defense system. J Plant Physiol 162:669–677
von Caemmerer S (2000) Biochemical models of leaf photosynthesis. CSIRO Publishing, Collingwood
Wand SJE, Midgley GF, Jones MH, Curtis PS (1999) Response of wild C_4 and C_3 grass (Poaceae) to elevated atmospheric CO_2 concentration: a meta-analytic test of current theories and perceptions. Glob Change Biol 5:723–741
Wang Y, Nil N (2000) Changes in chlorophyll, ribulose bisphosphate carboxylase-oxygenase, glycine betaine content, photosynthesis and transpiration in *Amaranthus tricolor* leaves during salt stress. J Hort Sci Biotechnol 75:623–627
Wang XW, Vinocur B, Altman A (2003) Plant responses to drought, salinity and extreme temperatures: towards genetic engineering for stress tolerance. Planta 218:1–14
Wang S, Assmann SM, Fedoroff NV (2008a) Characterization of the *Arabidopsis* heterotrimeric G protein. J Biol Chem 283:13913–13922
Wang H, Liu Y, Bruffett K, Lee J, Hause G, Walker JC, Zhang S (2008b) Haplo-isuffiency of *MPK3* in *MPK6* mutant background uncovers a novel function of these two MAPKs in *Arabidopsis* ovule development. Plant Cell 20:602–613
Wang D, Heckathorn SA, Wang X, Philpott SM (2012) A meta-analysis of plant physiological and growth responses to temperature and elevated CO_2. Oecologia 169:1–13
Ward JK, Tissue DT, Thomas RB, Strain BR (1999) Comparative responses of model C_3 and C_4 plants in low and elevated CO_2. Glob Change Biol 5:857–867
Wilken M, Huchzermeyer B (1999) Suppression of mycelia formation by NO produced endogenously in *Candida tropicalis*. Eur J Cell Biol 78:209–213
Wingler A, Lea PJ, Quick P, Leegood RC (2000) Photorespiration: metabolic pathways and their role in stress protection. Philos Trans R Soc Lond B Biol Sci 355:1517–1529
Wise RR, Naylor AW (1987) Chilling-enhanced photooxidation: evidence for the role of singlet oxygen and endogenous antioxidants. Plant Physiol 83:278–282
Wullschleger SD, Tschaplinski TJ, Norby RJ (2002) Plant water relations at elevated CO_2 – implications for water-limited environments. Plant Cell Environ 25:319–331
Yancey P, Clark ME, Had SC, Bowlus RD, Somero GN (1982) Living with the water stress: evolution of osmolyte system. Science 217:1214–1222
Yue B, Xue W, Xiong L, Yu X, Luo L, Cui K, Jin D, Xing Y, Zhang Q (2006) Genetic basis of drought resistance at reproductive stage in rice: separation of drought tolerance from drought avoidance. Genetics 172:1213–1228
Zelitch I (1973) Alternate pathways of glycolate synthesis in tobacco and maize and its relation to rates of photorespiration. Plant Physiol 51:299–305
Zelitch I, Schultes NP, Peterson RB, Brown P, Brutnell TP (2009) High glycolate oxidase activity is required for survival of maize in normal air. Plant Physiol 149:195–204
Zhen A, Bie Z, Huang Y, Liu Z, Lei B (2011) Effects of salt-tolerant rootstock grafting on ultrastructure, photosynthetic capacity, and H_2O_2-scavenging system in chloroplasts of cucumber seedlings under NaCl stress. Acta Physiol Plant 33:2311–2319
Zhifang G, Loescher WH (2003) Expression of a celery mannose 6-phosphate reductase in *Arabidopsis thaliana* enhances salt tolerance and induces biosynthesis of both mannitol and a glucosyl-mannitol dimer. Plant Cell Environ 26:275–283
Zhu Z, Zhang R, Liu T, Zheng H (2011) Solute accumulation and osmotic adjustment characteristics of the mangrove *Avicennia marina* under NaCl-induced salinity stress. Bot Mar 54:335–341

Chapter 16
Nitrogen-Use-Efficiency (NUE) in Plants Under NaCl Stress

Ghulam Murtaza, M.M. Azooz, Behzad Murtaza, Yawar Usman, and Muhammad Saqib

Abstract Increasing population growth rate, shortage of good arable land, use of low quality water coupled with intensive cropping system are forcing crop production into more marginal environments and thus, limiting the adaptation and productivity of food crops. In Pakistan, considerable information is available regarding nitrogen (N) requirement of plants grown in normal healthy soils. But scanty information is accessible regarding nitrogen requirement of plants grown in salt-affected soils, which cover an area of about 11.5×10^6 ha. The unfavorable conditions as well as inadequate and imbalance use of N in such soils is affecting considerable decline in crop yield. No doubt, soil salinity alters the N metabolism in plants but, use of N fertilizers affects growth dilution to alleviate detrimental effects of moderate salinity and helps to improve economic yield of crops. The concentration of soluble cations and anions in the soil solution is high enough to induce water stress, specific ion effects and nutrient imbalance which generally decrease crop growth and ultimately the harvestable yield. Generally, N fertilization increases productivity and yield of the plants at low NaCl concentration as compared to the situation where salinity is a major growth limiting factor. Various experiments have been conducted to determine appropriate N fertilizer rates for different textured soils for the protection of environment while achieving high production of crops. Since, due to different physical and chemical properties of soils, requirements of N and other mineral nutrients of crops on salt-affected soils are different than those on normal soils. Therefore, farmers will be economically benefited by the judicious use of N

G. Murtaza (✉) • B. Murtaza • Y. Usman • M. Saqib
Institute of Soil and Environmental Sciences, University of Agriculture,
Faisalabad 38040, Pakistan
e-mail: gmurtazauaf@gmail.com

M.M. Azooz
Department of Botany, Faculty of Science, South Valley University, Qena, Egypt

Department of Biological Sciences, Faculty of Science,
King Faisal University, Al Hassa, Saudi Arabia

P. Ahmad et al. (eds.), *Salt Stress in Plants: Signalling, Omics and Adaptations*,
DOI 10.1007/978-1-4614-6108-1_16, © Springer Science+Business Media New York 2013

fertilizer for optimum crop production from their marginal land resources without disturbing the biosphere equilibrium. Increased farm income will help to decrease rural poverty and migration from rural to urban areas which is another benefit of optimum use of N during amelioration of salt-affected soils by using amendments and low quality waters. This chapter presents an account of NUE in crop plants under the influenced by salt stress.

Keywords Crop improvement • Nitrogen use efficiency (NUE) • Nitrogen metabolism • Salt stress

Nitrogen (N) is the dominating element in the atmosphere that occupies central position for survival of plants. Also, it is a major essential nutrient that covers major portion of earth's surface. Nitrogen is found in the atmosphere in gaseous form as N_2 molecules, which is not directly available to the plants for their growth, development and to complete their life cycle. Although, earth's surface is a rich source of N, yet nitrogen deficiency has become a major growth limiting factor for plants worldwide, especially under salt stress.

Salinity is one of the most important abiotic stresses that is widely distributed in both irrigated and non-irrigated areas of the world (Ashraf et al. 2008; Kamal Uddin et al. 2009). On global basis salinity stress ranked second after the drought (Singh 2004). It is one of the most prevalent environmental threats to global agricultural productivity, especially in arid and semi-arid climates, where population growth, water shortage and land degradation are major concerns (Geissler et al. 2010; Munns and Tester 2008). Six percent of the world's total land area is salt-affected (approximately more than 800×10^6 ha) (Rengasamy 2006). Around 6×10^6 ha land in Pakistan is salt affected (Chatrath et al. 2007). Hence, the future of agricultural production will increasingly rely on our ability to grow plants on salt-affected lands using brackish waters (Rozema and Flowers 2008). Salinity imposes toxicities of ions like Na^+ and Cl^-, osmotic stress and ionic imbalance to the plants including soil permeability problems (Ashraf et al. 2008). Salt stress is the osmotic stress exerted on plants when they are growing under high saline conditions. In Pakistan, considerable information is available regarding N requirement of plants grown in normal healthy soils. But scanty information is accessible regarding N requirement of plants grown in salt-affected soils which cover an area of about 11.5×10^6 ha (Anonymous 2005). Salt-affected soils are characterized by high electrical conductivity (EC), sodium adsorption ratio (SAR) and pH, calcareousness, low organic matter, low biological activity and poor physical soil conditions. As chlorophyll constitutes N, reduction in plant biomass is sometimes observed under severe NaCl stress due to less concentration of N, and this is possibly because of the decrease in carbohydrate accumulation caused by reduction in carbon assimilation (Moradi and Ismail 2007; Pattanagul and Thitisaksakul 2008). No doubt, soil salinity alters the uptake of N by plants but use of fertilizer N affects growth dilution to alleviate detrimental effects of moderate salinity as well as NaCl stress and helps improve economic yield of crops. Successful crop production on moderately problematic soils demands judicious supply of N. Soil salinity and sodicity inhibit plant growth through inducing water stress, specific ion effects and nutrient imbalance resulting in deficiency of some while toxicity of others which generally decrease crop growth and ultimately the harvestable yield (Ashraf et al. 2008; Katerji et al. 2009). Exploration of halophytes for sustainable crop production is now being endorsed, particularly for monetary interests (fodder and food) or ecological reasons (CO_2-sequestration and soil desalinization) (Reddy et al. 2008).

Major impact of subsoil salinity/sodicity is expected on rooting depth and root function leading to harmful effect on water extraction, crop yields and plant growth. Subsoil NaCl salinity may also affect plants indirectly through effects on subsoil structure leading to reduced root growth. The nature and amount of N fertilizer added, its mineralization pattern, soil salinity/sodicity levels, soil type, temperature and pH determine the efficiency of N fertilizer in salt-affected soils (Chaudhry et al. 1989). The yield response to N has been mostly investigated for productive soils by different authors. Realizing the importance of crop in moderately salt-affected soils, investigations are in progress to determine the suitable application of N under such soil conditions (Murtaza 2011). Since requirements of N and other mineral nutrients for crops on salt-affected soils are different than those on normal soils due to different physical and chemical properties of soils.

Optimal rates of fertilizer application to salt-affected soils partially alleviate the adverse effects of salinity on photosynthesis and photosynthesis-related parameters and yield and yield components through mitigating the nutrient demands of salt-stressed plants (Sultana et al. 2001). Nitrogen is a limiting factor for crop growth in saline condition, since availability of nitrogenous fertilizer in salt-affected soils depends upon the nature of the fertilizer, degree of salinity and absence or presence of organic matter in soil, where N may minimize the adverse effects of salinity on plant growth and yield (Singh and Kashyap 2007; Esmaili et al. 2008; Abdelgadir et al. 2005, 2010) and help increase salt tolerance of plants (Sairam and Tyagi 2004). Salinity tolerance has been reported to be related with the ability of the plant to maintain an appropriate K^+/Na^+ ratio rather than simply maintaining low Na^+ concentrations (Shabala and Cuin 2008; Gorai et al. 2010; Bayat et al. 2011).

Nitrogen

Nitrogen in Soil

Nitrogen (N) is widely distributed throughout the lithosphere, atmosphere, hydrosphere and biosphere. Of the total, 95–99% nitrogen is found as potentially available organic form in the soil, either in plant and animal residues. This (N) is not directly available to plants, but some can be converted to available forms by microorganisms. A very minute amount of organic nitrogen may exist in soluble organic compounds, such as urea, that may be slightly available to plants. The majority of plant-available nitrogen is in the inorganic form. It is absorbed by plant roots as NO_3^- predominant under aerobic condition. In the case of rice (anaerobic condition), as NH_4^+-N is preferred. In the N-sufficient plants, its concentration varies from 1% to 5%. This nutrient is mostly deficient in soils and frequently creates serious nutritional problems under the abiotic stress like higher NaCl concentration in the soil.

Role of N in Plants

Nitrogen is an important component of chlorophyll, amino acids, proteins, nucleic acids (DNA and RNA), porphyrins, flavonoids, purines and pyrimidine nucleotides, flavin nucleotides, enzymes, co-enzymes and alkaloids. The chlorophyll in the presence of solar energy fixes atmospheric CO_2 as carbohydrates. N fertilization improves protein quality of food grains by enhancing the proportion of glutamic acid, Proline, phenylalanine, cystine, methionine and tyrosine, and decreasing the amounts of lysine, histidine, arginine, aspartic acid, threonine, glycine, valine and leucine in grains.

Nitrogen Use Efficiency (NUE) and Global Status

The NUE includes N uptake, utilization, or acquisition efficiency, expressed as a ratio of output (total plant N, grain N; biomass yield or grain yield) and input (total N, soil N or N-fertilizer applied). From one of the earliest definitions of NUE that considered the amount of plant yield in terms of either grain per unit of applied N (NUE grain) or biomass per unit of applied N (Good et al. 2004) to a recent report on the N characteristics of two rice cultivars who defined it as plant-N content expressed over the total N supplied to the plant. One of the most widely used approach measures either total biomass or grain weight (Good et al. 2004). Apart from this, efficiency of extracting N from soil is another important measure of NUE for crop plants.

Global consumption of N fertilizers has given up to 83 Mt in recent years, of which about 47 Mt is applied to cereal crops. However, significant differences exist among world regions, particularly with regard to NUE that depend on which cereal crops are grown, their attainable yield potential, soil quality, amount and form of N application, and the overall timeliness and quality of other crop management operations. In developing regions, N fertilizer use was comparatively lower in the early 1960s and increased exponentially during the course of the Green Revolution. Although the growth rate in N consumption has slowed substantially in recent years, it still averaged 1.45 Mt N $year^{-1}$ (3.2% $year^{-1}$) during the past 20 years. In developed regions, excluding Eastern Europe/Central Asia, cereal yields have continued to increase in the past 20 years without significant increases in N fertilizer use. The share of total N fertilizer consumption that is applied to cereals ranges from 32% in Northeast Asia to 71% in South East Asia. At a worldwide level, cereal production, cereal yields, and fertilizer N consumption have increased in a near-linear fashion during the past 40 years, although significant differences exist among world regions, particularly with regard to NUE. Improved NUE in case of developed countries can be attributed to investments in research and extension on crop improvement, new fertilizer products, and better management technologies by both public and private sectors, at levels that greatly exceed those currently available in the developing world. However, interventions to increase NUE and reduce N losses into environment must be accomplished at farm level through a combination of improved technologies and carefully crafted local policies that promote the adoption of improved N management practices while sustaining yield increases.

Salinity

Salinity Effects on Plants

Salinity is the major constraint to global food crop production, reducing the capacity of agriculture to sustain the growing human population increase (Flowers 2004). It is predicted that approximately 20% of all the cultivated land and nearly half of irrigated land is salt-affected, greatly reducing yield well below the genetic threshold potential. The scarcity of good quality water for irrigation purposes results in the formation of salt-affected soils (Flowers 2004; Murtaza et al. 2009). Saline conditions can influence the different steps of N metabolism such as uptake, reduction, and protein synthesis, that may be responsible, at least in part, for the lower plant growth rate observed under such conditions.

Salinity causes detrimental effects on plant growth, development, physiological and biochemical activities, which is due to osmotic stress, specific ion toxicity, nutritional imbalance, or a combination of these factors (Ashraf 2004). Sodium chloride affects plant growth and development through imposition of ion toxicity, ion imbalance, and osmotic stress, as well as by the secondary stresses, like nutritional disorders, membrane disorganization, metabolic toxicity, and inhibition of photosynthesis. Sodium chloride stress adaptation in plants occurs through organized physiological processes at cellular and molecular levels, which include ion homeostasis, osmotic adjustment, ion extrusion, and compartmentalization. Salinity adversely affects almost all stages of growth and development, flowering and fruit set, ultimately causing low economic yield and poor quality of production (Ashraf and Harris 2004).

Limited success in increasing the yield stability of crops grown on saline soils seems due, to a minimal understanding about how salinity and other abiotic stresses affect the most fundamental processes of cellular function – including cell division, differentiation and expansion – which have a substantial impact on plant growth and development (Hasegawa et al. 2000; Zhu 2001). Regardless, some degree of yield stability is achievable in salt stress environments (Flowers 2004).

Ionic and Osmotic Stresses of NaCl

Under saline conditions there is a prevalence of nonessential elements over essential elements. In the salt affected soils, plants must absorb the essential nutrients from a diluted source in the presence of highly concentrated nonessential nutrients. This requires extra energy and plants sometimes are unable to fulfill their nutritional requirements. There are two main stresses imposed by salinity on plant growth. One is water stress or osmotic stress imposed by the increase in osmotic potential of the rhizosphere as a result of high salt concentration. Another stress is the ionic stress in which toxicity of ion occurs.

Higher salt concentration in apoplast of cells generates primary and secondary effects that negatively affect survival, growth and development. Primary effects are ionic toxicity and disequilibrium, and hyper osmolality. In salt-affected soil, there are many salt contaminants, especially NaCl which readily dissolves in water to yield the toxic ions, sodium ion (Na^+) and chloride ion (Cl^-). Also, the water available in the salt-contaminated soil is restricted, inducing osmotic stress (Castillo et al. 2007; Pagter et al. 2009). The Na^+ is a small molecule that is easily absorbed into root tissues of higher plants and transported throughout plant organs, it is inhibitory to cytosolic and organellar processes which causes ion damage, osmotic stress and nutritional imbalance (Cha-um et al. 2007; Siringam et al. 2009). Root tissues are the first barrier which not only select nutrient ions but also protect against toxic ions. Excess Na^+ in plant cells directly damages membrane systems and organelles, resulting in plant growth reduction and abnormal development prior to plant death (Davenport et al. 2005; Quintero et al. 2007). In halophyte species, there are many salt-defense mechanisms including osmoregulation, ion homeostasis, hormonal regulation and antioxidant (Sairam and Tyagi 2004).

The Na^+ toxicity of many crop plants is correlated with over accumulation of Na^+ in the shoot (Tester and Davenport 2003; Møller and Tester 2007). Na^+ is taken up from the soil by the plant root system and transported to the shoot in the transpiration stream (Tester and Davenport 2003). Shoot Na^+ accumulation is the net result of distinct Na^+ transport processes occurring in different organs and cell types (Tester and Davenport 2003), and each of these processes contributes to the salinity tolerance of a plant. However, Na^+ concentration of 0.1 M is cytotoxic indicating that the ions directly affect specific biochemical and physiological processes (Serrano 1996). High concentrations of salt also impose hyperosmotic shock by lowering the water potential causing turgor reduction or loss that restricts cell expansion (Hasegawa et al. 2000; Zhu 2002). Sufficiently negative apoplastic water potential can lead to cellular water loss simulating dehydration that occurs during the episodes of severe drought.

Plant Adaptations to NaCl Stress

Ion and osmotic homeostasis (the property of a system that regulates its internal environment and tends to maintain a stable, constant condition of properties such as temperature or pH) is necessary for plants to be salt tolerant. High NaCl concentrations adversely affect the attainment of essential nutrients as Na^+ competitively inhibits K^+ and Ca^{2+} uptake, whilst Cl^- restricts anions uptake (Liu et al. 2006; Tammam et al. 2008), disturbing ion homeostasis within the plant. Moreover, salinity may create specific ion toxicity as disproportionate presence of Na^+ and Cl^- in cellular and intracellular compartments inhibits many enzymatic systems, altering a wide range of important metabolic processes that plant growth is crucially depending on (Munns 2005).

Adaptation mechanisms should therefore contribute to re-establish the homeostatic conditions needed for inward net flux of water and ion uptake. One major aspect of plant adaptation to saline environments is the utilization of massive accumulation of inorganic ions (mainly Na^+ and Cl^-) to adjust osmotically (Ottow et al. 2005). Osmotic adjustment (OA) in terms of salt accumulation is energetically efficient, but requires a combination of several tolerance and/or avoidance strategies that act in concert to avert ion toxicity and imbalance (Munns 2005; Wang et al. 2007; Koyro et al. 2011).

Intracellular ion homeostasis requires determinants that control toxic ion uptake and facilitate their compartmentalization into the vacuoles (Zhu 2003). Since the vacuole is a focal compartment for cell expansion, ion accumulation in this organelle facilitates osmotic adjustment that drives growth with minimal deleterious impact on cytosolic and organellar machinery. Controlling ion uptake at the level of plasma membrane and vacuolar compartmentalization requires intracellular coordination of transport determinants, and the regulatory molecules and system(s) are beginning to be identified (Zhu 2003). Osmotic homeostasis is accomplished by accumulation of compatible osmolytes in the cytosol for intracellular osmotic homeostasis (Hasegawa et al. 2000). Processes that function to maintain ionic and osmotic homeostatic balance in a tissue and an organismal context are less deciphered. For example, root to shoot coordination that restricts ion movement to the aerial parts of plant could minimize salt load to the shoot meristem and metabolically active cells, but mechanisms involved are not well understood.

Interactive Effects of Salinity and N Fertilization

Soil salinity is one of the major threats for agriculture in arid and semi-arid regions of the world because salinity decreases the rate of photosynthesis and plant growth to various levels. The interaction between salinity and N nutrition is very complex because it is influenced by level of salinity, plant species and genotypes within species, plant age, the concentration and composition of nutrients in the substrate and climatic conditions.

Esmaili et al. (2008) revealed in their work that soil salinity increased after application of N fertilizers to saline soils. Interactive effects of salinity and N fertilization have been studied mostly in N deficient soils. Therefore, application of N fertilizers improved growth and yield of wheat (Murtaza 2011), corn, (Absalan et al. 2011), rice (Murtaza 2011), millet, cotton (Grattan and Grieve 1999) and sorghum (Esmaili et al. 2008).

Homaee et al. (2002) indicated that yield and dry matter content of corn and cotton were generally decreased by increasing salinity but increased by N application. In salinity and N interactive studies, the form in which N is supplied seems important. Some studies indicate that increased NO_3^- in solution decreased Cl^- uptake and its accumulation (Martinez and Cerda 1989). Farmarzi et al. (2006) also showed that N as ammonium nitrate was better than urea for corn growth. Tshivhandekano and Lewis (1993) showed that NH_4^+-fed wheat and maize were more sensitive to salinity than NO_3^--fed plants when grown in solution culture. Activity of nutrient ions in soil solution is affected by high concentrations of salt ions, usually Na^+ and Cl^-, resulting in a nutritional disorder in plants (Grattan and Grieve 1999). Numerous salinity-nutrient interactions occur simultaneously that affect crop yield or quality depending upon the salinity level and composition of salts, the crop species, the nutrient status in plant tissues and a number of environmental factors (Fageria et al. 2011). Accumulation of Na^+ and Cl^- in leaves through the transpiration flow is a general and long-term process occurring in salt-stress plants (Møller and Tester 2007). The N uptake and accumulation by plants is often decreased under saline conditions as a result of competitive processes between the NO_3^- and the ambient major salt species like Cl^-.

However, this depends on the type of nutrients and composition of soil solution (Homaee et al. 2002). Although plants selectively absorb K^+ over Na^+, a Na^+-induced K^+ deficiency can develop on crops under salinity stress by Na^+ salt (Maas and Grattan 1999). Most studies related to plant nutrition and salinity interactions have been conducted in sand or solution cultures. A major difficulty in understanding plant nutrition status as affected by soil salinity is reconciling results obtained in experiments conducted in field and in solution cultures (Grattan and Grieve 1999). While application of fertilizers could improve plant nutritional status, it may also increase the salinity of soil solution.

Nitrogen Acquisition by Plants Under Saline Growth Medium

Plants acquire mineral nutrients from their native soil environments. Most crop plants are glycophytes and have evolved under conditions of low soil salinity. Consequently, they have developed mechanisms for absorbing mineral nutrients for non-saline soils. Plants which synthesize organic solutes are known as glycophytes, and they try to prevent excess salt uptake at root level because they can tolerate much lower concentrations of salt in plant tissues before cell processes are adversely affected. Even with complete osmotic adjustment, a reduction in growth may occur due to metabolic demands of maintaining osmotic adjustment.

Under saline conditions, which are characterized by low nutrient-ion activities and extreme ratio of Na^+: Ca^{2+}, Na^+: K^+, Ca^{2+}: Mg^{2+} and Cl^-: NO_3^-, nutritional disorders can develop and crop growth may be decreased. Additions of N and P generally increase the growth of plants grown in N- and P-deficient environments, provided that the plant is not experiencing severe salt stress. When salinity and nutrient deficiency are the factors limiting growth, relief of the most limiting factor will promote growth more than the relief of the less limiting factor. Therefore, addition of a limiting nutrient can increase, decrease or have no effect on relative plant tolerance to salinity, depending on the level of salt stress. Failure to account for the severity of salt stress when interpreting salinity × nutrient interactions caused considerable confusion among researchers. According to Jamil et al. (2005), however, N was severely growth limiting, salinity was found to affect the growth of some crops [broccoli (*Brassica oleracea* var. capitata), cabbage (*B. oleracea* var. botrytis)]. Conversely, when salinity severely limited growth nutritional responses of some crops decreased. Salinity did not aggravate N deficiency as judged from leaf N contents. Effects of salinity and N deficiency on other mineral constituents were highly crop specific. Salinity, however, disrupts N acquisition by plants in two ways. First, the ionic strength of the substrate, regardless of its composition, can influence nutrient uptake and translocation. The second, salinity disturb the mineral relations of plants by decreased N availability through competition with major ions (Na^+ and Cl^-) in the substrate. These interactions often lead to Na^+ induced K^+ or Ca^{2+} deficiencies or Mg^{2+} deficiency due to higher Ca^{2+}.

In salt-affected soils when pH of the soil solution exceeds 8.5, availability of some nutrients may be restricted resulting in nutrient imbalances. Bicarbonate toxicities occur primarily from decreased iron and other micronutrient availabilities at high pH while high Na^+ may lead to Ca^{2+} and Mg^{2+} deficiencies (Arshad 2008).

Reports show inhibitory and stimulatory effects on the plant N uptake under salinity stress. A considerable number of laboratory and greenhouse studies have shown that N accumulation in plants decreases under high salt-affected soils. A decrease in the NO_3^- concentration in shoots of barley, watermelon, cotton and wheat was recognized with an increase in Cl^- uptake and accumulation. In contrast to the effect of Cl^- on NO_3^- uptake, increased NO_3^- in the substrate decreased Cl^- uptake and accumulation. The possible reduction in N uptake is due to higher salt concentration which results in substitution of Cl^- for NO_3^-. For example, the N-deficiency symptom increased the Cl^- level in corn, barley, and some other crops.

Wheat is a moderately salt-tolerant crop and its yield is considerably reduced as the soil salinity level raises upto 100 mM NaCl (Munns et al. 2006). Abdul-Kadir and Paulsen (1982) reported that NaCl decreased the amount of total N in all parts of wheat plants because nitrate has an antagonistic relation with the chloride which is produced through ionization of NaCl in the growth medium. Both the chloride salts of Na^+ and K^+ inhibited the NO_3^- uptake similarly, suggesting that the process was more sensitive to anionic salinity than to cationic salinity (Aslam et al. 1984). Salinity helps to increase Nitrate Reductase Activity (NRA) in peanut plants but decreased the NRA in tomato and cucumber (*Cucumis sativus* L.) plants, the decrease in NRA in plants was probably due to inhibition of NO_3^- uptake by Cl^- in plant species (Abdul-Kadir and Paulsen 1982). In an experiment, the NH_4^+fed maize and wheat plants were more sensitive to salinity than NO_3^-fed plants grown in nutrient solution culture. Supplementation of Ca^{2+} to the growth media improved the growth rate of plants with NO_3^- treatment but not those treated with NH_4^+. Based on the results of their nutrient solution experiments, Leidi et al. (1991a) suggested that NO_3^- is a better N source than NH_4^+ for wheat grown in salt-affected soils.

Nitrogen Absorption by Different Plants Under NaCl Stress and Effect of NaCl on NUE

Excessive salinity is the most important abiotic stress that greatly affects the plant growth and nutrition of plant in arid and semi-arid regions. Salt tolerance of plants is a complex phenomenon and involves morphological and developmental changes as well as physiological and biochemical processes (Munns 2002; Ashraf and Harris 2004). On exposure to osmotic stress as a result of high salinity, plants accumulate a range of metabolically active solutes, collectively known as compatible solutes (Lauchli and Luttge 2002). Many of these osmolytes are N-containing compounds, (e.g. amino acids and amides or betaines) hence the nitrogen metabolism is of great importance under abiotic stress conditions (Läuchli and Lüttge 2002). It is significant to research the effects of NaCl stress on nitrogen metabolism in plants. It was well described that NaCl stress strongly affects nitrogen metabolism of plants (Parida and Das 2005; Munns and Tester 2008). It is necessary to osmoregulate and re-establishes the ion homeostasis in cells for plant under NaCl stress.

Salt Tolerant Grass

Very little work is found on N nutrition to turf grass under NaCl stress. Pessarakli et al. (2012) conducted an experiment in a greenhouse to determine the effect of NaCl stress on dry matter and N uptake of twelve clones of salt grass. They concluded that NaCl has no negative effect on the degree of salt tolerance of salt grass. Nitrogen uptake, growth and dry matter production of all the clones of salt grass were not affected by even at high salt concentration (EC 20 dS m^{-1}).

NaCl Stress and Its Effect on N Uptake by Legumes

Legumes are plants with seed pods that split into two halves. Edible seeds from plants in the legume family include beans, peas, lentils, soybeans, and peanuts. Legumes are excellent sources of protein, low-glycemic index carbohydrates, essential micronutrients, and fiber. Legumes are also invaluable as organic fertilizers because of their ability to fix atmospheric nitrogen. For these reasons, as well as their global economic importance, legumes have become the focus of increased interest and research activity.

Faba Bean

Faba bean (*Vicia faba* L.) is a widely cultivated legume grain, known for its great potential for yield and high protein contents (i.e., 25–40%) (Matthews and Macellos 2003). The capability of faba bean to grow under limited irrigation (Khalafallah et al. 2008; Al-Suhaibani 2009; Alderfasi and Alghamdi 2010) and moderate salinity (Khalafallah et al. 2008; Abdelhamid et al. 2010) have made it one of the most preferred crops for agricultural production in arid and semi-arid lands. However, many studies indicated that N nutritional status of faba bean possessed positive impacts on its growth, yield, yield quality (e.g., protein content) and response to salinity stress (Al-Mutawa 2003; Lopez-Bellido et al. 2003; Al-Fredan 2006). Almadini (2011) conducted an experiment to evaluate salinity (0, 100 and 200 mM NaCl) impacts on symbiosis efficacy of faba bean various rhizobium strains (Hassa-1, Hassa-2 and Hassa-3) and faba bean growth response to NaCl stress (25, 50 and 75 mM) under rhizobium inoculation (N_0), combined rhizobium with mineral N applied as urea (N_1) or urea only (N_2). He concluded that salinity stress caused a delay in the initiation of N-fixation, but it could not prevent the process of nodulation neither the potential of nodules to supply N to the shoots. However, the produced number of nodules and the total nodular activity were considerably reduced by the impacts of salinity stress that ultimately resulted in the reduction of the faba bean growth. The obtained results also depicted that changes in N and nodule activities are vital factors contributing to decreases in the biomass of grown plants under salt stress.

Broad Bean, Soybean, and Lentil

The input of mineral N in the soil mainly consists of symbiotic nitrogen fixation, transformation of organic matter, and fertilizer, whereas the output consists of N uptake by the plant and loss due to leaching beyond the root zone. The input from rainfall (10–15 kg/ha per year) and the gaseous loss to the atmosphere (0–20 kg/ha per year) may approximately be considered as negligible (Cellier et al. 1997).

Biological activity is of great importance for the N balance in the soil. Nitrogen fixation and transformation of organic matter both depend on biological activity. Fixation of atmospheric nitrogen results from symbiosis between leguminous crops and *Rhizobium* bacteria, which on one hand form nodules on the plant roots and fixes atmospheric nitrogen and on the other hand utilize carbohydrates and minerals from the host plant. The nitrogen content of the plant, expressed in g N g^{-1} dry matter, was not affected by salinity for soybean and chickpea. For broad bean it was only affected at harvest time. Lentil under saline conditions systematically showed lower values during the whole growing season.

Chickpea

NaCl stress results in the reduction of nodulation and inhibition of nitrogen fixing activity in legumes. Garg and Chandel (2011) conducted an experiment to study the interaction between mycorrhizal fungus, *Glomus mosseae*, and salinity stress in relation to nitrogen fixation, plant growth and nutrient accumulation in chickpea. Performance of two genotypes of chickpea (Pusa-329, Pusa-240) was compared under different levels of salinity with and without mycorrhizal inoculations. They concluded that N and P levels in the leaves and roots were reduced with increasing salinity in both the chickpea genotypes; the decline was more distinct in Pusa-240 than Pusa-329. Nitrogenase activity was reduced with increasing salt concentrations that ultimately resulted in reduction of nitrogen fixation, plant growth and yield.

The mineralization of organic nitrogen and the immobilization of mineral nitrogen are important for the N supply of the plant. The mineralization depends on the C: N ratio of the organic matter, the temperature, the water stress (Rodrigo et al. 1997), the water table through its effect on soil aeration (Van Hoorn 1958) and salinity, affecting the nitrification (Mengel and Kirkby 1982). Immobilization is the inversed process by which mineral N is transformed in organic nitrogen. Moreover, mineral N can be lost through denitrification and under aerobic conditions. The balance between mineralization and immobilization is estimated at a supply of about 0.5 kg/ha per day. According to Van Hoorn (1958) a lowering of the groundwater table from 0.4 m to 1.5 m depth in clay soil correspond with an increase of about 100 kg N/ha. No data are available about the salinity effect on the balance between mineralization and immobilization, but salinity could affect it directly through the nitrification or indirectly through a change in the C: N ratio or through the water stress.

The N uptake of the plant decreased with increasing salinity. The N contribution of the soil decreased stronger than the plant uptake, pointing to a salinity effect on the mineral N production by biological activity in the soil through N fixation and transformation of organic nitrogen. A salinity effect on N fixation could explain, at least partly, the salt sensitivity of grain legumes.

Effect of NaCl on N Uptake and NUE in Cereals

Cereal crops are interchangeably called grain crops. In many literatures, they are simply called grains. The world's top three cereals in 2008 ranked on the basis of economic value are rice, followed by wheat and corn (FAO Stat data).

Wheat

Wheat is the major food grain of Pakistan. It is used as staple food of the people and it has become backbone of the economy of Pakistan and plays a central role in formulation of agricultural policies. It contributes 14.4% to the value added in agriculture and 3.1% to GDP (Anonymous 2009–2010).

Botella et al. (1997a) studied the influence of N source and salinity on growth, survival and N uptake in wheat. Plants were grown in a growth chamber under controlled conditions. They applied nutrient solution contained 4 mM N, applied as either ammonium sulfate [$(NH_4)_2SO_4$] or calcium nitrate [$Ca(NO_3)_2$] or a mixture of both, and the salinity treatments consisted in two levels of sodium chloride (NaCl) (1 and 60 mM). They concluded that effect of salinity on N uptake was dependent on the N source employed, and significantly inhibited NO_3^- uptake but not NH_4^+ uptake. Salinity increased the affinity for NH_4^+ and reduced the affinity for NO_3^-, indicating a depressive effect of Cl^- on NO_3^- uptake. Several effects of Cl^- on NO_3^- uptake have been described. Chloride was found to strongly inhibit net NO_3^- uptake (Ward et al. 1986), but did not affect influx (Glass et al. 1985), but affected efflux (Deane-Drummond and Glass 1982). Nitrate reductases are molybdoenzymes that reduce NO_3^- to NO_2. Plaut (1974) reported that NaCl stress inhibits N metabolism in wheat seedlings by affecting the nitrate reductase activity. Probably, as suggested by Leidi et al. (1991b) the different interactions between these anions depend on the overall salt-tolerance of the plant species studied. Further they concluded that salinity reduces N uptake, NO_3^- uptake being inhibited to a greater extent than NH_4^+ uptake. NH_4^+ improved shoot growth rather than NO_3^-. The best N source for wheat growth seems to be a mixture of NO_3^- and NH_4^+. Under saline conditions or periods of low irradiance, the N source should be supplemented with NH_4^+ to improve plant growth since NO_3^- uptake was more affected than NH_4^+ uptake under saline conditions.

Rice

Rice (*Oryza sativa* L.) is one of the most important cereal and cash crops grown in tropical and temperate regions of the world. It is the major source of foreign exchange. It accounts for 6.4% of value added in agriculture and 1.4% in GDP (Economic survey of Pakistan 2009–2010). In many agricultural areas, especially Asia, soil salinity is major factor that decreases rice yield and productivity (Ghafoor et al. 2004). Rice has previously been reported as being salt susceptible in both the seedling and reproductive stages (Zeng et al. 2001; Moradi and Ismail 2007), leading to a reduction in yield of more than 50% in crops exposed to 6.65 dS m^{-1} EC_e (Zeng and Shannon 2000). It was reported that NaCl stress strongly affects the germination (Deng et al. 2011), the growth (Wang et al. 2011), solute accumulation (Hoai et al. 2003; Nemati et al. 2011), and gene expression (Pandit et al. 2011) of rice plants. Wang et al. (2012) studied the effects of salt stress on N metabolism and ion balance in rice plants by measuring the total amino acids, NO_3^- and inorganic ions contents in the stressed seedlings. The results indicated that when seedlings were exposed to salt stress for 4 h, in roots, salt stress strongly stimulated the accumulations of Na^+ and Cl^-, and reduced K^+ content; however, in leaves, only at 5 days these changes were observed. This confirmed that the response of root to salt stress was more sensitive than that of leaf. When seedlings were exposed to salt stress for 4 h, salt stress strongly stimulated the expression of OsAS, OsGS1;1, OsNADH-GOGAT, OsGS1;3, OsGDH1, OsGDH2, OsGDH3 in both leaves and roots of rice, after this time point their expression decreased. Namely, at 5 days most of genes involved in NH_4^+ assimilation were down regulated by salt stress, which might be the response to NO_3^- change. Salt stress did not reduce NO_3^- contents in both roots and leaves at 4 h, whereas at 5 days salt stress mightily decreased the NO_3^- contents. The deficiencies of NO_3^- in both roots and leaves can cause a large reduction in the regulation of OsNR1 followed by reduction of NH_4^+ production. This event might immediately induce the down regulations of the genes involved in NH_4^+ assimilation.

Barley

Barley (*Hordeum vulgare* L.) is considered as a highly salt-tolerant plant and is grown as a major crop under both saline and non-saline conditions. There are substantial differences in the yield and productivity of barley under these deferent conditions. This may be due to detrimental effects of salinity through the inhibition of water and nutrient uptake by plants. Even under normal (non-saline) conditions, the most common nutrient deficiency in the production of barley is N. Under saline conditions this problem may be more pronounced. Ali et al. (2001) conducted a hydroponic study under controlled conditions to investigate growth and N uptake by barley supplied with five different NH_4^+-N/NO_3^--N ratios at EC of 0 and 8 dS m^{-1}. The five NH_4^+-N/NO_3-N ratios were 0/100, 25/75, 50/50, 75/25 and 100/0, each giving a total N supply of 100 mg N L^{-1} in the root medium. The NO_3^- concentration in the plant shoots grown in saline conditions was significantly lower than that of plants grown in non-saline conditions. This shows that NaCl caused NO_3^- accumulation in plant roots and reduced NO_3^- assimilation. The most probable reason for the accumulation of NO_3^- and lower organic N synthesis in the NO_3^- treated plants is a limitation of the nitrate reductase step (Hageman and Flesher 1960). Bernstein (1963) also mentioned the possibility that increased osmotic pressure of the cell or increased concentration of specific ions may alter the activity of some enzymes. Total N contents in roots and shoots [based on mg (two plants)$^{-1}$] were significantly higher under non-saline conditions in all the treatments. The beneficial effects of fertilization in moderately saline conditions were indicated. Plant growth as well as NUE under stress conditions (salt stress and water stress) can be improved considerably by providing mixed N nutrition to the plant root medium.

Pearl Millet

Pearl millet (*Pennisetum glaucum* L.) is the most widely grown type of millet. Pearl millet is an excellent fodder for livestock. It is an important summer season fodder and grain crop of Pakistan. It can be grown throughout the country even where moisture is a limiting factor for crop growth. Its green fodder is a valuable feed for livestock which warrants developing high fodder yielding varieties of pearl millet.

Nitrogen metabolism is an essential process that determines the plant growth and productivity. The most significant factor affecting plant growth is nitrate reduction through nitrate and nitrite reductases. Application of N fertilizer has been reported to alleviate significantly the adverse effects caused by salt stress on a number of crops (Singh and Kashyap 2007; Esmaili et al. 2008). Plants supplemented with NO_3 grew better in saline media than NH_4^-fed plants, even though the latter had less costly N assimilation (Leidi et al. 1992). Different physiological responses in plants to salt stress have received extensive attention. However, little is known about the interaction effects of NaCl and NO_3 nutrition on growth and nitrate assimilation. The activities of nitrogen assimilation enzymes may be used as functional indices about N uptake and metabolism in plants under NaCl stress.

Albassam (2001) studied the influence of nitrate concentration on growth and N assimilation in NaCl-stressed pearl millet. The plants were grown in perlite and irrigated with nutrient solution containing 0, 25, 50, or 100 mM NaCl in the presence of 2 or 10 mM NO_3^-. Free amino acid content, including proline, was higher in salt-stressed plants compared to controls. Further, he found that the activities of nitrate reductase, nitrite reductase, and glutamate synthase were reduced, but the glutamine synthetase activity was less affected. High nitrate (10 mM) in the irrigation solution partially restored activities of the above enzymes and increased the soluble protein content despite the high NaCl concentration.

Stivsev et al. (1973) reported in his work that leaf dry weight, soluble proteins and total chlorophyll content of plant are decreased with NaCl stress. The decrease in the chlorophyll content under saline conditions is attributed to a salt-induced weakening of protein pigment- lipid complex and increased chlorophyllase activity.

Effect of NaCl on N Uptake and NUE in Cotton

Cotton being a non-food cash crop contributes significantly in foreign exchange earnings. Cotton is classified as a medium salt-tolerant species with a salinity threshold level of 7.7 dS m^{-1} (Maas and Hoffman 1977). Cotton growth and lint yield is significantly reduced due to high salinity (Ashraf et al. 2002). Constable and Rochester (1988) reported that ability of cotton to recover total applied N is only 30%. Due to limited supply of N from the soil, efficiency of crop must be enhanced for the fulfillment of N deficiency. For this purposes, additives of N are applied that are quick source of N to recover the N deficiency. Incorporation of additives such as N-(n-butyl), Thiophosphoric triamide (NBPT) and Dicyandiaminde (DCD) into N fertilizers are used with the purpose of increasing NUE of crops in problematic soils. Pessarakli and Tucker (1985) conducted various hydroponic experiments with cotton which indicated that N uptake was decreased due to high levels of salinity.

Alleviation of Salt Stress Through Nutrient Management

Among all the mineral elements, the N requirement of N is the largest in amount and it is constituent of many plant cell components including chlorophyll, amino acids, proteins and nucleic acids. Nitrogen use efficiency increased either by decreasing salinity or reducing N rates. An apparent increase in salt tolerance was noted when plants were fertilized with organic-N source compared to that of inorganic-N source. Investigations showed that application of fertilizers in saline soils might result in increased, decreased or unchanged plant salt tolerance. In other words, plant response to fertilizers depends on severity of salt stress in the root zone (Maas and Grattan 1999). Soils are often low in N to meet crop requirements resulting nutritional stress in field soils. Its deficiency is very common in crops and causes rapid inhibition of plant growth whether plants are growing in salt stress or non-salt stress conditions. Addition of N into N deficient soils at moderate salt stress improved growth and/or yield of chickpea (Garg and Chandel 2011), corn (*Zea mays* L.) (Absalan et al. 2011), grape (*Vitis vinifera* L.) (Taylor et al. 1987), tomato (*Lycopersicon esculentum* L.), (Maggio et al. 2007), spinach (*Spinacia oleracea* L.) (Langdale et al. 1971) and wheat (*Triticum aestivum* L.) (Murtaza 2011). In most of the cases, total N uptake (mg N plant^{-1}) decreases, but N concentration (mg N kg^{-1} dry weight) increases or remained unchanged under optimal N conditions (Hu and Schimidhater 1998). The higher shoot N concentration is not because of higher uptake, but because of concentration effect due to decreased biomass.

Under such situations, additional N application may not improve the salt tolerance of crops. On the other hand, decrease in N accumulation in plants is reported in a number of laboratory and greenhouse studies (Pessarakli et al. 2012). A decrease in N accumulation is understandable as Cl^- (saline condition) and NO_3^- (present in soil) have antagonistic effects on absorption of each other. This has been found in cucumber (Martinez and Cerda 1989), wheat (Murtaza 2011), corn, (Absalan et al. 2011), rice (Murtaza 2011), millet, cotton (Grattan and Grieve 1999), sorghum (Esmaili et al. 2008), eggplant (Savvas and Lenz 1996) tomato (Maggio et al. 2007). Salinity induced reduction in NO_3^- concentration in wheat leaves is reported without affecting the total N content and addition of NO_3^- resulted in decreased Cl^- uptake (Hu and Schmidhalter 1998). Increasing concentrations of NO_3^- linearly decreased Cl^- concentrations in plants (Kafkafi et al. 2001). An increase of 1 mmol NO_3 g^{-1} dry matter prevented the accumulation of 238 mmol Cl g^{-1} dry matter in the tomato plant (Kafkafi et al. 1992).

Accompanying cations also influenced decreased NO_3^- uptake by Cl^-. The effects of Cl^- from NaCl and KCl on inhibition of NO_3^- uptake are similar but that from $CaCl_2$ is much more pronounced at the lower salinity range (Kafkafi et al. 1992). It was found that Cl^- from $CaCl_2$ and not KCl, inhibited NO_3^- uptake in melon and tomato in a range to which the plants would likely be exposed in field conditions (i.e. up to 60 mol/m). It was only in the high concentration range (100–200 mol/m) that KCl inhibited NO_3^- uptake. Cultivars of tomato and melon having better salt-tolerance had higher NO_3^- concentration (Kafkafi et al. 1992). Increase of NO_3^- in the substrate, decreased Cl^- uptake and accumulation in several horticultural crops (Maggio et al. 2007; Garg and Chandel 2011). Plants especially fruits and vines, which are sensitive to Cl^- are likely to be benefited with NO_3^- application (Grattan and Grieve 1999). Study on avocado and citrus (both sensitive to Cl^-) showed that increasing NO_3^- in the growth media, which otherwise was sufficient for growth under non-salt stress condition decreased Cl^- concentration in their leaves with a decrease in foliage in symptoms of Cl^- toxicity and improvement in growth (Bar et al. 1997). The concentration of NO_3^- effective in decreasing Cl^- concentration in plant has to be high (molar ratio of NO_3: Cl, 0.5 and above) in soil solution. Martinez and Cerda (1989) also reported decrease in Cl^- uptake in cucumber when only NO_3^- was added to the solution but when half the NO_3^- in the solution was replaced by NH_4^+, Cl^- accumulation was enhanced. Form of N may also influence the plant sensitivity to salinity. Wheat136 and maize were more sensitive to salinity as the ratio of NH_4: NO_3 increased when grown in solution culture (Botella et al. 1997b). Similarly, NH_4 -fed maize (Lewis et al. 1989), melon (Feigin 1990) and pea (*Pisum sativum* L.) showed higher sensitivity to salinity than NO_3-fed plants when grown in solution cultures (Speer et al. 1994). All these suggest the influence of NO_3^- on Cl^- uptake. Specific cation like Ca^{2+} is also reported to play its role. As an example, added Ca^{2+} to the media improved the growth rate of the plants in the NO_3 treatment, but not those treated with NH_4^+ (Kaya et al. 2003). The role of Ca^{2+} as a second messenger in many biological systems, coupled with these observations, indicates that plants are able to adjust to high salt environments by activating a signal transduction system involving Ca^{2+} (Hasegawa et al. 2000). In presence of NO_3 as only source of N, K uptake increased in salt-stressed melon and with increase in NH_4/NO_3 ratio, the accumulation of Cl^- increased, but reverse was true for Ca^{2+} and K^+ in the leaves (Adler and Wilcox 1995; Feigin 1990). Crop response may not be same to the source of N in different growing condition. As an example, Leidi et al. (1991a), Silberbush and Lips (1991a) reported higher sensitivity of wheat to NaCl stress as the ratio of NH_4/NO_3 increased in solution and sand culture. Contrary to these, wheat grown in NaCl treated soil, showed improved salt tolerance and gave better grain yield when combination of NH_4 and NO_3 are used as N source, than NO_3 alone (Shaviv et al. 1990). Silberbush et al. (1991) also got similar results for peanut in N- NaCl study in solution cultures vs. soil on plant response. In soil cultures, the situation is more complex than solution culture, in soil culture, the NH_4^+ concentrations in the soil solution are decreased due to interaction with exchange sites.

Conclusion and Future Perspectives

Salinity is one of the most important abiotic stresses that impose ion toxicities, ionic imbalance and osmotic stress to the plants including soil permeability problems. Effects of salinity and N deficiency on other mineral constituents were highly crop specific. Salinity, however, disrupts N acquisition by plants in two ways. First, the ionic strength of the substrate, regardless of its composition, can influence nutrient uptake and translocation. The second, salinity disturb the mineral relations of plants by decreased N availability through competition with major ions (Na^+ and Cl^-) in the substrate. These interactions often lead to Na^+ induced K^+ or Ca^{2+} deficiencies. Under salt-stress conditions, when SO_4^{2-} and NO_3^- are found on an equal osmolarity concentration, Cl^- inhibited NO_3^- uptake more than SO_4^{2-}. The possible N deficiency under NaCl stress is due to substitution of Cl^- for NO_3^-. In various experiments, it is found that NO_3^- is a best N substitute for NH_4^+ for cereals especially wheat grown on salt-affected soils. Higher application of NO_3^- to soil decreased uptake of Cl^- and subsequently its accumulation in many horticultural crops. Plants especially fruits and vines, which are sensitive to Cl^- are likely to be benefited with NO_3^- application. It was also concluded that NH_4^+ concentrations in the soil solution are decreased due to interaction with exchange sites.

References

Abdelgadir EM, Oka M, Fujiyama H (2005) Characteristics of nitrate uptake by plants under salinity. J Plant Nutr 28:33–46

Abdelgadir EM, Fadul EM, Fageer EA, Ali EA (2010) Response of wheat to nitrogen fertilizer at reclaimed high terrace salt-affected soils in Sudan. J Agric Social Sci 6:43–47

Abdelhamid MT, Shoker MB, Bekhta MA (2010) Effects of induced salinity on four Vicia Faba cultivars differing in their broomrape tolerance. Paper presented at the 4th international water technology conference (IWTC), Cairo, Egypt, 421–427, 2010

Abdul-Kadir SM, Paulsen GM (1982) Effect of salinity on nitrogen metabolism in wheat. J Plant Nutr 5:1141–1151

Absalan AA, Armin M, Asghripour MR, Karimi-Yazdi S (2011) Effects of different forms of nitrogen application on yield response of corn under saline conditions. Adv Environ Biol 5(4):719–724

Adler PR, Wilcox GE (1995) Ammonium increases the net rate of sodium influx and partitioning to the leaf of Muskmelon. J Plant Nutr 18:1951–1962

Albassam BA (2001) Effect of nitrate nutrition on growth and nitrogen assimilation of pearl millet exposed to sodium chloride stress. J Plant Nutr 24:1325–1335

Alderfasi AA, Alghamdi SS (2010) Integrated water supply with nutrient requirements on growth, photosynthesis productivity, chemical status and seed yield of *faba* bean. Am Euroasian J Agron 3(1):8–17

Al-Fredan MA (2006) Effect of treated municipal waste water and Rhizobia strains on growth and nodulation of *faba* bean (*Vicia Faba* L. cv. Hassawi). Pak J Biol Sci 9(10):1960–1964

Ali A, Tucker TC, Thompson TL, Salim M (2001) Effects of salinity and mixed ammonium and nitrate nutrition on the growth and nitrogen utilization of Barley. J Agron Crop Sci 186:223–228

Almadini AM (2011) Biological and mineral nitrogen supply impacts on salinity response of faba bean (*Vicia Faba* L.). Europ J Sci Res 49(20):187–199
Al-Mutawa MM (2003) Effect of nitrogen supply on the response of faba bean to salinity. Int J Agric Biol 5(2):124–128
Al-Suhaibani NA (2009) Influence of early water deficit on seed yield and quality of faba bean under arid environment of Saudi Arabia. Am Eurasian J Agric Environ Sci 5(5):649–654
Anonymous (2005) Global network on integrated soil management for sustainable use of salt-affected soils. FAO land and plant nutrition management service, Rome, Italy. http://www.fao.org/ag/agl/agll/spush
Anonymous (2009–2010) Economic survey of Pakistan. Economic wing, Ministry of food, agriculture and livestock, Govt. Pakistan, Islamabad, Pakistan
Arshad MA (2008) Soil salinity and salinization. In: Chaseworth W (ed) Encyclopedia of soil science. Springer, Dordrecht, pp 699–703
Ashraf M (2004) Some important physiological selection criteria for salt tolerance in plants. Flora 199:361–376
Ashraf M, Harris PJC (2004) Potential biochemical indicators of salinity tolerance in plants. Plant Sci 166:3–16
Ashraf MY, Afaf R, Qureshi MS, Sarwar G, Naqvi M (2002) Salinity induced changes in α-amylase and protease activities and associated metabolism in cotton varieties during germination and early seedling growth stages. Acta Physiol Plant 24:37–44
Ashraf M, Athar HR, Harris PJC, Kwon TR (2008) Some prospective strategies for improving crop salt tolerance. Adv Agron 97:45–110
Aslam M, Huffaker RC, Rains DW (1984) Early effects of salinity on nitrate assimilation in barley seedlings. Plant Physiol 76:321–325
Bar Y, Apelbaum A, Kafkafi U, Goren R (1997) Relationship between chloride and nitrate and its effects on growth and mineral composition of avocado and citrus plants. J Plant Nutr 20:715–731
Bayat F, Shiran B, Belyaev DV (2011) Overexpression of HvNHX2, a vacuolar Na/H antiporter gene from barley, improves salt tolerance in Arabidopsis thaliana. Aust J Crop Sci 5(4):428–432
Bernstein L (1963) Osmotic adjustment of plants to saline media. II. Dynamic phase. Am J Bot 50:360–370
Botella MA, Martínez V, Nieves M, Cerdá A (1997a) Effect of salinity on the growth and nitrogen uptake by wheat seedlings. J Plant Nutr 20(6):793–804
Botella MA, Martinez V, Pardines J, Cerda A (1997b) Salinity induced potassium deficiency in maize plants. J Plant Physiol 150:200–205
Castillo EG, Tuong TP, Ismail AM, Inubushi K (2007) Response to salinity in rice: comparative effects of osmotic and ionic stresses. Plant Prod Sci 10:159–170
Cellier P, Germon JC, Henault C, Genermont S (1997) Mécanismes d'émission et quantification des flux d'azote atmosphérique. In: Lemaire et G, Nicolardot B (eds) Maıtrise de l'azote dans les agrosystèmes, INRA Editions, p 25–37
Chatrath R, Mishra B, Ferrara GO, Singh SK (2007) Challenges to wheat production in South Asia. Euphytica 157:447–456
Chaudhry MR, Khokhar IA, Shah MA (1989) Rice growth and nutrient uptake (NP) as affected by different salinity types. Pak J Agric Sci 26:348–354
Cha-um S, Vejchasarn P, Kirdmanee V (2007) An effective defensive response in Thai aromatic rice varieties (*Oryza sativa* L. spp. indica) to salinity. J Crop Sci Biotech 10:257–264
Constable G, Rochester I (1988) Nitrogen application to cotton on clay soil: timing and soil testing. Agron J 80:498–502
Davenport R, James RA, Zakrisson-Plogander A, Tester M, Munns R (2005) Control of sodium transport in durum wheat. Plant Physiol 137:807–818
Deanne-Drummond CE, Glass ADM (1982) Studies of nitrate influx into barley roots by the use of $^{36}ClO_3^-$ as a tracer for nitrate. 1. Interactions with chloride and other ions. Can J Bot 60:2148–2153

Deng XX, Zhang XQ, Song XJ, Lai KK, Guo LB, Xin YY, Wang HZ, Xue DW (2011) Response of transgenic rice at germination traits under salt and alkali stress. African J Agric Res 6:4335–4339

Esmaili E, Kapourchal SA, Malakouti MJ, Homaee M (2008) Interactive effect of salinity and two nitrogen fertilizers on growth and composition of sorghum. Plant Soil Environ 54(12):537–546

Fageria NK, Gheyi HR, Moreira A (2011) Nutrient bioavailability in salt affected soils. J Plant Nutr 34:945–962

FAOSTAT (2008) Food and Agricultural Production Commodities. http://faostat.fao.org/site/339/default.aspx. Accessed 17 Oct 2010

Faramarzi A, Jamshidi S, Siamei K (2006) The effect of amount and source of nitrogen on yield and yield components of corn. Modern Agric Sci 1(2):65–73

Feigin A (1990) Interactive effects of salinity and ammonium/nitrate ratio on growth and chemical composition of melon plants. J Plant Nutr 13:1257–1269

Flowers TJ (2004) Improving crop salt tolerance. J Exp Bot 55:307–319

Garg N, Chandel S (2011) Effect of mycorrhizal inoculation on growth, nitrogen fixation and nutrient uptake in *Cicer arietinum* L. under salt stress. Turk J Agric For 35:205–214

Geissler N, Hussin S, Koyro HW (2010) Elevated atmospheric CO_2 concentration enhances salinity tolerance in *Aster tripolium* L. Planta 231:583–594

Ghafoor A, Qadir M, Murtaza G (2004) Salt-affected soils: principles of management. Allied Book Centre, Lahore

Glass ADM, Thompson RG, Bordeleau L (1985) Regulation of NO_3 influx in barley. Plant Physiol 77:379–381

Good G, Shrawat A, Muench D (2004) Can less yield more? Is reducing nutrient input into the environment compatible with maintaining crop production? Trends Plant Sci 9:597–605

Gorai M, Ennajeh M, Khemira H, Neffati M (2010) Combined effect of NaCl-salinity and hypoxia on growth, photosynthesis, water relations and solute accumulation in Phragmites australis plants. Flora 205:462–470

Grattan SR, Grieve CM (1999) Mineral nutrient acquisition and response by plants grown in saline environments. In: Pessarakli M (ed) Handbook of plant and crop stress. Marcel Dekker, New York, pp 203–229

Hageman RH, Flesher D (1960) Nitrate reductase activity in corn seedling as affected by light and nitrate content of nutrient media. Plant Physiol 35:700–708

Hasegawa P, Bressan RA, Zhu JK, Bohnert HJ (2000) Plant cellular and molecular responses to high salinity. Annu Rev Plant Physiol Plant Mol Biol 51:463–499

Hoai NTT, Shim IS, Kobayashi K, Kenji U (2003) Accumulation of some nitrogen compounds in response to salt stress and their relationships with salt tolerance in rice (*Oryza sativa* L.) seedlings. Plant Growth Reg 41:159–164

Homaee M, Feddes RA, Dirksen C (2002) A macroscopic water extraction model for nonuniform transient salinity and water stress. Soil Sci Soc Am J 66:1764–1772

Hu Y, Oertli JJ, Schmidhalter U (1998) Interactive effects of salinity and macronutrient level on wheat. I. Growth. J Plant Nutr 20:1155–1167

Jamil M, Lee CC, Rehman S, Lee DB, Ashraf M, Rha ES (2005) Salinity (NaCl) tolerance of brassica species at germination and early seedling growth. Electron J Environ Agric Food Chem 4(4):970–976

Kafkafi U, Siddiqi MY, Ritchie RJ, Glass ADM, Ruth TJ (1992) Reduction of nitrate ($^{13}NO_3$) influx and nitrogen (^{13}N) translocation by tomato and melon varieties after short exposure to calcium and potassium chloride salts. J Plant Nutr 15:959–975

Kafkafi U, Xu G, Imas P, Magen H, Tarchitzky J (2001) Potassium and chloride in crops and soils. The role of potassium chloride fertilizer in crop nutrition. IPI Research Topics 22:220

Kamal Uddin M, Juraimi AS, Ismail MR, Rahim MA, Radziah O (2009) Growth response of eight tropical turfgrass to salinity. Afr J Biotech 8:5799–5806

Katerji N, Mastrorilli M, van Horn JW, Lahmer FZ, Hamdy A, Oweis T (2009) Durum wheat and barley productivity in saline-drought environments. Eur J Agron 3:1–9

Kaya C, Ak BE, Higg D (2003) Response of salt-stressed strawberry plants to supplementary calcium nitrate and/or potassium nitrate. J Plant Nutr 26(3):543–560
Khalafallah AA, Tawfik KM, Zinab AA (2008) Tolerance of seven faba bean varieties on drought and salt stresses. Res J Agric Biol Sci 4(2):178–186
Koyro HW, Geissler N, Seenivasan R, Huchzermeyer B (2011) Plant stress physiology; physiological and biochemical stratgies allowing to thrieve under ionic stress. In: Pessarakli M (ed) Handbook of plant and crop stress, 3th edn. CRC press, Taylor & Francis Group, pp 1051–1094
Langdale GW, Thomas JR, Littleton TG (1971) Influence of soil salinity and nitrogen fertilizer on spinach growth. J Rio Grande Valley Hort Soc 25:61–66
Läuchli A, Lüttge U (2002) Salinity and nitrogen nutrition. Boston Kluwer Academic, Boston, pp 229–248
Leidi EO, Silberbush M, Lips SH (1991a) Wheat growth as affected by nitrogen type, pH and salinity, 1. Biomass production and mineral composition. J Plant Nutr 14:235–246
Leidi EO, Silberbush M, Soares MIM, Lips SH (1991b) Salinity and nitrogen nutrition studies on peanut and cotton plants. J Plant Nutr 15(5):591–604
Leidi EO, Silberbush M, Soares MIM, Lips SH (1992) Salinity and nitrogen nutrition studies on peanut and cotton plants. J Plant Nutr 15:591–604
Lewis OAM, Leidi EO, Lips SH (1989) Effect of nitrogen source on growth response to salinity stress in maize and wheat. New Phytol 111:155–110
Liu X, Duan D, Li W, Tadano T, Khan A (2006) A comparative study on responses of growth and solute composition in halophytes Suaeda salsa and Limonium bicolor to salinity. In: Khan MA, Weber DJ (eds) Ecolphysiology of high salinity tolerant plants. Springer, Dordrecht, pp 135–143
Lopez-Bellido RJ, Lopez-Bellido L, López-Bellido FJ, Castillo JE (2003) Faba bean (*Vicia faba* L.) response to residual nitrogen in a continuous rotation with wheat (*Triticum aestivum* L.) under rain fed Mediterranean condition. Agron J 95:1253–1261
Maas EV, Grattan SR (1999) Crop yields as affected by salinity. In: Pessarakli M (ed) Handbook of plant and crop stress. Marcel Dekker, New York, pp 55–108
Maas E, Hoffman G (1977) Crop salt tolerance: current assessment. J Irrig Drain 103:115–134
Maggio A, Raimondi G, Martino A, De Pascale S (2007) Salt stress response in tomato beyond the salinity tolerance threshold. Environ Exp Bot 59:276–282
Martinez V, Cerda A (1989) Influence of N source on rate of Cl, N, Na and K uptake by cucumber seedling grown in saline condition. J Plant Nutr 12:971–983
Matthews P, Marcellos H (2003) *Faba* bean, 2nd edn. Agfact P4.2.7, Division of Plant Industries, NSW, Australia
Mengel K, Kirkby EA (1982) Nitrogen. In: Mengel K, Kirkby EA (eds) Principles of plant nutrition. International Potash Institute, Worblaufen-Bern, pp 335–368
Møller IS, Tester M (2007) Salinity tolerance of Arabidopsis: a good model for cereals? Trends Plant Sci 12:534–540
Moradi F, Ismail AM (2007) Responses of photosynthesis, chlorophyll fluorescence and ROS-Scavenging systems to salt stress during seedling and reproductive stages in rice. Ann Bot 99:1161–1173
Munns R (2002) Comparative physiology of salt and water stress. Plant Cell Environ 25:239–250
Munns R (2005) Genes and salt tolerance: bringing them together. New Phytol 167:645–663
Munns R, Tester M (2008) Mechanisms of salinity tolerance. Annu Rev Plant Biol 59:651–681
Munns R, James RA, Lauchli A (2006) Approaches to increasing the salt tolerance of wheat and other cereals. J Exp Bot 57:1025–1043
Murtaza G (2011) Optimization of nitrogen use-efficiency in rice-wheat cropping system on tile drained salt-affected soils with different textures. In: First annual technical progress report for higher education commission (HEC) funded project (July 2010 to June 2013). Inst. Soil Environ Sci Univ Agric, Faisalabad, Pakistan
Murtaza G, Ghafoor A, Owens G, Qadir M, Kahlon UZ (2009) Environmental and economic benefits of saline-sodic soil reclamation using low-quality water and soil amendments in conjunction with a rice-wheat cropping system. J Agron Crop Sci 195:124–136

Nemati I, Moradi F, Gholizadeh S, Esmaeili MA, Bihamta MR (2011) The effect of salinity stress on ions and soluble sugars distribution in leaves, leaf sheaths and roots of rice (*Oryza sativa* L.) seedlings. Plant Soil Environ 57:26–33

Ottow EA, Brinker M, Teichmann T, Fritz E, Kaiser W, Brosché M, Kangasjärvi J, Jiang X, Polle A (2005) *Populus euphratica* displays apoplastic sodium accumulation, osmotic adjustment by decreases in calcium and soluble carbohydrates, and develops leaf succulence under salt stress. Plant Physiol 139(4):1762–1772

Pagter M, Bragato C, Malagori M, Brix H (2009) Osmotic and ionic effects of NaCl and Na_2SO_4 salinity on *Phragmites australis*. Aquat Bot 90:43–51

Pandit A, Rai V, Sharma TR, Sharma PC, Singh NK (2011) Differentially expressed genes in sensitive and tolerant rice varieties in response to salt-stress. J Plant Biochem Biotechnol 20:149–154

Parida AK, Das AB (2005) Salt tolerance and salinity effects on plants: a review. Ecotoxicol Environ Safety 60(3):324–349

Pattanagul W, Thitisaksakul M (2008) Effects of salinity stress on growth and carbohydrate metabolism in three rice (*Oryza sativa* L.) cultivars differing in salinity tolerance. Ind J Exp Biol 46:736–742

Pessarakli M, Tucker TC (1985) Uptake of nitrogen-15 by cotton under salt stress. Soil Sci Soc Am J 49:149–152

Pessarakli M, Harivandi MA, Kopec DM, Ray DT (2012) Growth responses and nitrogen uptake by saltgrass (*Distichlis spicata* L.), a halophytic plant species, under salt stress, using the ^{15}N technique. Int J Agron 2012:1–9

Plaut Z (1974) Nitrate reductase activity of wheat seedlings during exposure to and recovery from water stress and salinity. Physiol Plant 30:212–217

Quintero JM, Fournier JM, Benlloch M (2007) Na^+ accumulation in shoot is related to water transport in K^+-starved sunflower plants but not in plants with a normal K^+ status. J Plant Physiol 164:60–67

Reddy MP, Shah MT, Patolia JS (2008) *Salvadora persica*, a potential species for industrial oil production in semiarid saline and alkali soils. Ind Crop Prod 28:273–278

Rengasamy P (2006) World salinization with emphasis on Australia. J Exp Bot 57:1017–1023

Rodrigo AS, Recous Néel C, Mary B (1997) Modelling temperature and moisture effects on C:N transformations in soils: comparison of nine models. Ecol Model 102:325–339

Rozema J, Flowers TJ (2008) Crops for a salinized world. Science 322:1478–1480

Sairam RK, Tyagi A (2004) Physiology and molecular biology of salinity stress tolerance in plants. Curr Sci 86:407–421

Savvas D, Lenz F (1996) Influence of NaCl concentration in the nutrient solution on mineral composition of eggplants grown in sand culture. Angew Bot 70:124–127

Serrano R (1996) Salt tolerance in plants and microorganisms: toxicity targets and defense responses. Int Rev Cytol 165:1–52

Shabala S, Cuin TA (2008) Potassium transport and plant salt tolerance. Physiol Plantarum 133:651–669

Shaviv A, Hazan O, Neumann PM, Hagin J (1990) Increasing salt tolerance of wheat by mixed ammonium nitrate nutrition. J Plant Nutr 13:1227–1239

Silberbush M, Lips SH (1991) Potassium, nitrogen, ammonium/nitrate ratio, and sodium chloride effects on wheat growth. I. Shoot and root growth and mineral composition. J Plant Nutr 14:751–764

Silberbush M, Ben-Asher J, Lips SH (1991) Nutrient-salt interaction and peanut growth. Adv Desert Arid Land Tech Dev 5:631–645

Singh BD (2004) Plant breeding, 4th edn. Kalyani Publishers, Banaras

Singh JS, Kashyap AK (2007) Variations in soil N-mineralization and nitrification in seasonally dry tropical forest and savanna ecosystems in Vindhyan region, India. Trop Ecol 48(1):27–35

Siringam K, Juntawong N, Cha-um S, Kirdmanee C (2009) Relationships between sodium ion accumulation and physiological characteristics in rice (*Oryza sativa* L. spp. indica) seedlings grown under iso-osmotic salinity stress. Pak J Bot 41:1837–1850

Speer M, Brune A, Kaiser WM (1994) Replacement of nitrate by ammonium as the nitrogen sources increases the salt sensitivity of pea plants. I. Ion concentrations in roots and leaves. Plant Cell Environ 17:1215–1221

Stivsev MV, Ponnamoreva SA, Kuznestova EA (1973) Effect of salinization and herbicides on chlorophyllase activity in tomato leaves. Fiziol Rast 20:62–65

Sultana N, Ikeda T, Kashem MA (2001) Effect of foliar spray of nutrient solutions on photosynthesis, dry matter accumulation and yield in seawater-stressed rice. Environ Expt Bot 46:129–140

Tammam AA, Alhamd MFA, Hemeda MM (2008) Study of salt tolerance in wheat (*Triticum aestivum* L.) cultivar Banysoif 1. Aust J Crop Sci 1(3):115–125

Taylor RM, Fenn LB, Petty CA (1987) ^{15}Nitrogen uptake by grapes with divided roots growing in differently salinised soils. Hort Sci 22:664

Tester M, Davenport RJ (2003) Na^+ tolerance and Na^+ transport in higher plants. Annu Bot (Lond) 91:503–527

Tshivhandekano TR, Lewis OAM (1993) Differences in response between nitrate and ammonium-fed maize to salinity stress and its amelioration by potassium. S Afr J Bot 59:597–601

Van Hoorn JW (1958) Results of a groundwater level experimental field with arable crops on clay soil. Neth J Agric Sci 6(1):1–10

Wang H, Wu Z, Chen Y, Yang C, Shi D (2011) Effects of salt and alkali stresses on growth and ion balance in rice (*Oryza sativa* L.). Plant Soil Environ 57:286–294

Wang H, Wu ZY, Zhou Han J, Shi D (2012) Effects of salt stress on ion balance and nitrogen metabolism in rice. Plant Soil Environ 58(2):62–67

Wang YC, Yang CP, Liu GF, Zang GD, Ban QY (2007) Microarray and suppression subtractive hybridization analyses of gene expression in Puccinellia tenuiflora after exposure to NaHCO3. Plant Sci 173:309–320

Ward MR, Aslam M, Huffaker RC (1986) Enhancement of nitrate uptake and growth of barley seedlings by calcium under saline conditions. Plant Physiol 80:520–523

Zang L, Shannon MC, Lesch SM (2001) Timing of salinity stress affects rice growth and yield components. Agr Water Manage 48:191–206

Zeng L, Shannon MC (2000) Salinity effects on seedling growth and yield components of rice. Crop Sci 40:996–1003

Zhu JK (2001) Plant salt tolerance. Trends Plant Sci 6:66–71

Zhu JK (2002) Salt and drought stress signal transduction in plants. Annu Rev Plant Physiol Plant Mol Biol 53:247–273

Zhu JK (2003) Regulation of ion homeostasis under salt stress. Curr Opin Plant Biol 6:441–445

Chapter 17
The Responses of Salt-Affected Plants to Cadmium

Gabrijel Ondrasek

Abstract Salinisation and contamination by trace elements are expected to be some of the most critical environmental and sustainability issues in the forthcoming era of global warming and human population growth. Agro-ecosystems could be increasingly influenced by salinity, given that exploitation of saline pedo/hydroresources (>20 mM in water or soil saturation extract) will have to increase across many irrigated as well as rain-fed areas, despite poor yield and low food stuff quality. Using experimental and computational (modelling) approaches it has been well established that one of crucial plant responses under salinity is increased Cd phytoextraction. In salt-affected rhizosphere environments excessive concentrations of dissolved ions can impact Cd biogeochemistry through complexation and/or competition reactions with inorganic and organic ligands. For ecologically relevant conditions (e.g. Cd-contaminated and organically-depleted salinised soil) it was estimated that under low salinity (<30 mM) free Cd^{2+} ion predominates across a wide range of pHs (3.5–9), whereas in moderate-to-strong salinity (135–270 mM) Cd-chloro/sulphate complexes prevail (mostly comprising Cd-Cls). Although Cd-Cl interactions are still under intensive investigation and not fully understood, complexation is likely to be one of the main mechanisms for increasing Cd transfer to plants from the salt-affected rhizosphere. Also, there is evidence that Cd uptake by plants is underpinned by additive effects of salt and Cd stresses. Great efforts have been made in elucidating Cd biochemistry after uptake, (re)translocation and its deposition in plants differing in salt/metal tolerance; it is highly likely that the role of Cd-Cl complexation in plants is of negligible importance vs. that in the rhizosphere. Sharing similar or the same routes for crossing plant membranes with essential elements (Zn, Cu, Fe, Ca), Cd-organo complexes (with S, O and N radicals) relatively easily reach transpirational tissues via the xylem, and thereafter the depositing tissues dominantly via the phloem. Genetic engineering is a promising strategy

G. Ondrasek (✉)
University of Zagreb Faculty of Agriculture, Svetosimunska 25, Zagreb, Croatia
e-mail: gondrasek@agr.hr

P. Ahmad et al. (eds.), *Salt Stress in Plants: Signalling, Omics and Adaptations*,
DOI 10.1007/978-1-4614-6108-1_17, © Springer Science+Business Media New York 2013

in (1) increasing plant resistance to excessive soil salinity, and (2) producing genotypes for enhanced phytoremediation of areas overloaded by metals. However, under ecologically-relevant conditions (e.g. rhizosphere soil with poor metal-buffering capacity) Cd soil-plant transfer can be enhanced by soil salinity, simultaneously interfering with nutrient (Cu, Zn, Fe) extraction. So, if salinity resistance (as a multi-gene trait) is closely associated with gene loci responsible for Cd extraction, care needs to be taken that genetically improved (e.g. salt-resistant) genotypes do not impair crops foodstuff safety/security via increased Cd accumulation and/or micronutrient deficiency.

Keywords Salinity • Metal contamination • Cd complexation • Cd soil-plant transfer

Introduction

Salinity is the most widespread abiotic constraint to higher plants, affecting metabolism of cultured and/or native vegetation on ~ 1 billion hectares of Earth's surface. Naturally- (sea water intrusion, groundwater fluctuation) and/or anthropogenically induced (unsustainable agricultural management practices, excessive groundwater withdrawal) environmental salinisation processes in the era of climate change and global warming may be especially pronounced and unpredictable. Recently, WMO (World Meteorological Organisation) confirmed that the last decade (2000–2009) was recorded as the warmest, and some of the recent years (e.g. 2009) are among the warmest for the past 160 years, i.e. since the beginning of official meteorological measurements in 1850 (WMO 2009). Under such conditions of global warming, salinity is expected to induce more severe phytoeffects, principally to cultivated glycophytes (crop food production), but also in long-term, exacerbate a certain interrelations (soil buffering potential to retain contaminants, provoke further soil degradation, decrease the quality of hydro-resources, reduce biodiversity etc.) in terrestrial ecosystems.

In the short-term, salinity impairs crop production and food/feed quality as a consequence of a wide range of physiological disorders known as salt stress effects (Zhu 2001; Chinnusamy et al. 2005; Zhang et al. 2012). In the long-term, particularly under specific geological (parent rock material), pedological (lower hydraulic permeability, excessive soil alkalinity and sodicity), atmospheric (high evapotranspiration demands, low precipitation) and/or hydrospheric (shallow ground water level) surroundings, salinity leads to permanent land degradation, i.e. desertification (Ondrasek et al. 2011). Topsoil crusting and salt crystallisation, dispersion of secondary clay minerals (soil aggregates), subsoil sodicity and/or toxicity of specific elements (boron, aluminium, chloride) are some of the most important long-term salinity impacts on the pedosphere.

Salt-affected areas are often situated on land used for agriculture or industrial purposes. For instance, a large part of some marine estuaries, such as Nile (Egypt) (Rashad and Dultz 2007), Murray-Darling (Australia) (Biggs et al. 2010), Ebro

(Spain) (Romani et al. 2011), Neretva (Croatia) (Romic et al. 2012), developed on floodplains (alluvial soils) and used for agriculture over a long time, are subject to salinisation. Large infrastructural projects (power generation, water flow regulation, irrigation) may substantially change the hydrology of an area (e.g. decrease downstream water refreshing) and exacerbate salinisation. In the modern era, estuarine ecosystems have been shown to be some of the most vulnerable regarding environmental protection and food safety and security. Due to their naturally low topsoil fertility (e.g. light sandy soils usually with < 1% organic matter) or human-induced activities (e.g. emission of industrial/municipal effluents, intensive cultivation) aimed at retaining or even improving agricultural production, alluvial soils may gradually become overloaded by various (in)organic substances, i.e. potential contaminants. Estuarine environment thus may (1) represent an important sink for hazardous materials as a result of fluvial deposition of potentially-contaminated sediment material, and (2) become more susceptible to salinisation as a consequence of altered fresh water hydrology (river/aquifer level decreasing) and global warming (sea level rising).

Contamination of biosphere by trace elements (TEs) is expected to be one of the most critical environmental and sustainability issues of this century. Among TEs, cadmium (Cd) and its biogeochemical interactions are frequently elaborated regarding soil (rhizosphere) ecology and transfer to higher plants, i.e. into the food chain (McLaughlin and Singh 1999 and references therein; Adriano et al. 2004). Summarising the most of soil-, nutrient solution- and plant-based studies yields a several key statements for Cd: (1) its rhizosphere biogeochemistry is highly pH-dependent; (2) it is relatively easily bioavailable, especially from acidic environments; (3) it has a substantial potential for complexation/chemosorption with inorganic (salts) and organic ligands; (4) it does not any known essential biological function in plants (animals/humans), although shares some uptake/translocation mechanisms with particular essential elements; (5) it is preferentially accumulated in underground plant tissues; (6) it enters the human/animal food chain via food/feed consumption; and (7) it is toxic to a wide range of organisms, including plants and humans.

Salinisation of soil environment may substantially change Cd biogeochemistry and consequently its dynamics in the soil-plant-human continuum. For instance, by adding one of the most used agro-chemicals such as phosphatic fertilizers (e.g. diammonium phosphate, potassium dihydrogen orthophosphate) and liming salts (e.g. calcium carbonates, calcium hydroxides) can increase soil Cd adsorption, i.e. reduce its phytoavailability and/or phytotoxicity (Basta et al. 2001; Bolan et al. 2003a, b). In contrast, NaCl as the widespread salt in the nature can improve Cd mobilization by the formation of soluble inorganic chloride complexes (see Sect. 3.2.1) and thus enhance Cd phytoaccumulation, i.e. food/feed contamination. In the alkaline sodic soils Cd mobility and uptake can be enhanced due to facilitated organo-complexation with certain DOC (dissolved organic carbon) fractions (Harter and Naidu 1995). Using artificial seawater (15 g salts/L), Lores and Pennock (1998) obtained 100% desorption of Cd from dissolved organic complexes at relatively low DOC concentration (10 mg humic acid/L). Therefore, it is of great importance to elucidate Cd interactions under different salinity given that some outcomes could be highly important for improving management strategies to minimise Cd contamination and achieve food safety and security.

Plant Responses to Increased Environmental Salinity

Increased sodicity (Na) and chloride (Cl) concentration are two the most naturally widespread and intensively studied environmental types of salinity affecting plant populations worldwide. Unfortunately, in agricultural food/feed production those salinities are often accompanied by numerous environmental constrains such as: (1) water-logging or water scarcity, (2) high evapotranspirational demands, (3) shallow groundwater salinity, (4) soil acidity/alkalinity, (5) soil organic matter depletion, (6) ion toxicity (B, Al), (7) increased exchangeable sodium percentage (ESP) and/or sodium adsorption ratio (SAR), and (8) disturbed soil structure (Ondrasek et al. 2011). Complex interrelations among these variables make it almost impossible to detect which constraint is the major limitation factor in plant growth (e.g. Dang et al. 2008); hence, land management of salt-affected areas is exceedingly complicated with usually high cost-to-benefit ratio for grown crops.

In a majority of cultivated plant species, the important economic responses to increased environmental salinity (e.g. a decline in yield quality/quantity) occur even at relatively low salinity in either (1) soil (e.g. threshold salinity of ~10 mM reduced bean and onion yield by ~20%) (Chinnusamy et al. 2005) or (2) irrigation water (e.g. sprinkler irrigation with ~10 and ~30 mM salinised solution reduced watermelon yield by 50 and 100%, respectively, of the yield achieved with drip irrigation using the same salt concentrations) (Romic et al. 2008).

Many other plant metabolic functions (e.g. root/leaf membrane selectivity, photosynthesis, transpiration) (e.g. Grattan and Grieve 1999) usually are seriously compromised under excessive salinity. Moreover, using a variety of experimental approaches and many plant genotypes, it has been well established that one of crucial plant responses to increased salinity could be reduced ion selectivity resulting in enhanced Cd phytoextraction (see Sect. 3.2.1).

Cadmium in the Pedosphere

Origin of Soil Cadmium

Under natural pedosphere conditions, native soil Cd mostly has geogenic origin. Some of the most Cd-enriched natural materials are sedimentary rocks (e.g. phosphorites up to 980 and marine shales up to 219 mg Cd/kg) and sulphide minerals (e.g. sphalerite up to 50,000 and metacinnabar up to 117,000 mg Cd/kg) (Krishnamurti et al. 2005). These minerals are fairly rare in the surface lithosphere, which is the main reason for Cd naturally occurring in trace amounts, usually at several tenths of mg per kg of dry soil. Generally, soils with <0.5 mg Cd/kg (Zovko and Romic 2011), i.e. soil solution with <0.3 μM (Sanitá di Toppi and Gabbrielli 1999) can be considered as non-contaminated, whereas in contaminated environment Cd concentrations may be several magnitudes higher.

The main contributors of high Cd emission to environment during recent past have been industry, urbanisation and agriculture. Given certain physical properties of Cd (melting point 321 °C), during the ore smelting and emissions from metallurgic plants, Cd relatively easily reaches atmosphere and can cause long-distance environmental contamination. From the anthropogenic Cd emission annually (~30 million kg), about 8 million kg (27%) is emitted to the atmosphere mostly as a consequence of ore smelting and related heavy industry (metallurgic) activities, whereas a dominant portion (74%) of Cd emission (~22 million kg) is deposited/applied to pedosphere, mostly as industrial wastes/by-products (9.5 million kg), urban wastes/effluents (4.4 million kg) and agricultural wastes/fertilisers (2.4 million kg) (adopted from Sanitá di Toppi and Gabbrielli 1999).

On salt-affected soils, which are usually deficient in phytonutrients and/or organic matter, a common practice for improving crop production is application of mineral (inorganic) and organic fertilisers/amendments. As mentioned above, sedimentary phosphorus-(P) and Cd-enriched minerals like phosphorites, as a main raw material for P fertilisers, are some of the most ubiquitous sources of Cd contamination in modern agricultural production. An application of P fertilisers may be totally avoided (extensive pastures), but on the other side additions of >500 kg/ha (double cropping in intensive horticulture) (Romic et al. 2008) may in a long-term increase soil Cd from 0.3 up to >4 g/ha annually (e.g. Singh 1994). This is in a line with recent findings by Romic et al. (2012) who in the horticultural topsoils observed high association of bioavailable P and total Cd content probably due to overuse of mineral P fertilisers.

Most soil organic amendments from industrial/urban (biosolids, composts) or agricultural (slurries, manures) sectors used for improving soil productivity are usually enriched (contaminated) by Cd even more than P-fertilisers. Allochthonous soil Cd, irrespective of whether it enters the pedosphere via previously mentioned antropogenic activities or natural processes (precipitation, flooding), is firstly deposited in the surface horizons and thereafter is transferred by soil management practices (e.g. ploughing, discing) into the rhizosphere zone of cultivated crops. However, even on such a relatively small scale (rhizosphere vs. bulk soil, soil solution vs. particulate soil), certain soil properties can be markedly different, thus variably affecting Cd chemical forms/distribution, i.e. its environmental biogeochemistry (Ondrasek and Rengel 2012).

Out of the total soil Cd content, in most pedospheres and climates, almost negligible Cd portion (<1%) (e.g. Christensen and Huang 1999) is bioavailable. However, this fact does not diminish ecological importance of Cd highlighted in recent reviews by Kabata-Pendias (2004) and Clemens (2006).

Chemical Speciation of Cd in Salinized Environment

According to Florence (1982), chemical speciation of Cd would be a function of concentrations of its various chemical forms (species) that together make up the total Cd concentration (Cd_{TOT}) in a given sample. However, mobility, bioavailability and thus toxicity of Cd in soil media in most cases cannot be related only to Cd_{TOT}. Particular forms in the Cd_{TOT} pool differ in their physico-chemical properties, from the readily available to the highly unavailable, and thus have varied environmental impacts. Accordingly, Cd_{TOT} content comprises three main physico-chemical Cd forms: (1) dissolved in soil solution (the most bioavailable and potentially the most toxic), (2) adsorbed onto soil in/organic particles (less bioavailable but still potentially toxic), and (3) Cd precipitates (inactive, i.e. non-toxic).

Chemical speciation as well as chemisorption (adsorption) and complexation of Cd with particular inorganic (chlorides, carbonates, Al/Fe/Mn hydroxides and organic (low-/high-molecular weight) ligands in soil environment can be detected, i.e. predicted (modelled) by appropriate analytical/computational procedures (Fig. 17.1). Formation of Cd precipitates (salts) is very rare in most naturally-occurring environmental conditions; however, it may be induced in controlled experiments and simulated by modelling. For instance, speciation-solubility geochemical modelling is a convenient concept for studying Cd (TEs) mobility and bioavailability, i.e. assessing a potential risk from metal contamination under a wide range (pH, texture, redox potential) of natural conditions. Certain methods/techniques for controlled (e.g. rhizoboxes, soil columns, chelate-buffered nutrient solution cultures) or non-controlled field (e.g. lysimeters) conditions can be useful tools for quantifying Cd in the soil solution pool and its mobility to other surrounding interfaces, as well to plants (food) (Fig. 17.1).

The soil (rhizosphere) solution generally contains most of the readily bioavailable (dissolved) and potentially the most toxic Cd forms such as free ionic hydrated Cd^{2+}. Consequently, the potential Cd toxicity is mostly related to specific Cd^{2+} concentrations and physico-chemical properties of aqueous soil media, particular pH as a soil master variable (e.g. Rengel 2002). Even though dispersed inorganic (e.g. metal hydroxioxides) and organic (e.g. humic acids) colloids cannot be technically considered soil solution constituents (Helmke 1999), their active interfaces (can be altered by sampling techniques and filtration) are highly relevant in biogeochemistry of the dissolved Cd pool (Fig. 17.1). In salt-affected environments, excessive concentrations of dissolved salts, particularly anions (Cl^-, SO_4^{2-}, HCO_3^-) have a large influence on Cd chemical speciation through complexation (McLaughlin and Singh 1999). Also, presence of cationic metals (Na^+, Mg^{2+}, Ca^{2+}) is important for (1) competition with in/organic rhizosphere matrix constituents (Ondrasek and Rengel 2012), thus increasing Cd solubility and mobility in the rhizosphere solution, and (2) possibly influencing pH reaction of the media (e.g. Lores and Pennock 1998).

The predominant salt forms in salinised agro-exploited natural resources (e.g. soils, water used for irrigation) are those of sodium (Na^+), chloride (Cl^-) and sulphate (SO_4^{2-}) (Ondrasek et al. 2011), and they may influence Cd bioavailability in

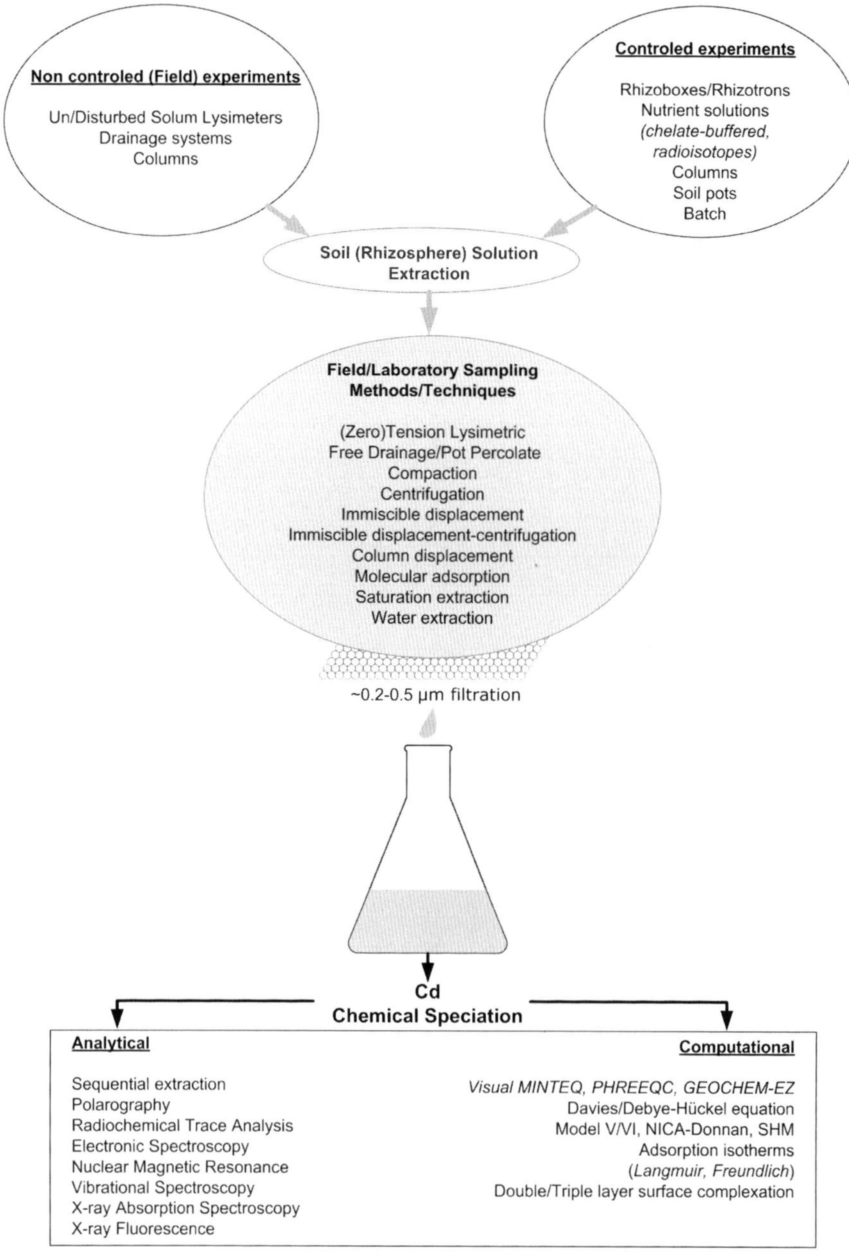

Fig. 17.1 Some analytical and computational approaches for determination/evaluation of Cd chemical speciation under field as well as controlled-experiment conditions

the rhizosphere. In the next section, an emphasis will be on Cd biogeochemistry and Cd transfer into food/feed plants under salinity caused by Cl^-, Na^+ and SO_4^{2-}. To create a supporting material for further discussion, a Cd "pre-screening modelling" (i.e. Cd speciation assessment) was performed for realistic salt-affected aqueous environments (e.g. soil solutions of salt-affected sandy soils, salinised channel waters, sea-intruded aquifers) over a wide range of pH levels. For instance, the modelled pH range 3.5–9.5 covers the most naturally-occurring conditions (e.g. Tipping 2005), whereas under strong acidity (pH < 3.5) or alkalinity (pH > 10.5) it is almost impossible to characterise in full the proton-affinity distribution because of a lack of reliable data (Kinniburgh et al. 1999). Inorganic speciation modelling in the Visual MINTEQ framework (Gustafsson 2006) was done using a default thermodynamic database and settings (e.g. Davies equation at 25°C), whereas organic complexation was performed by incorporated NICA-Donnan model (Kinniburgh et al. 1999) as one of the most advanced models for competitive Cd binding to dissolved high-molecular-weight organics (DOC) (Weng et al. 2001). Assumed constant concentrations of Cd (1 μM) and DOC (1 mg/L) in all models correspond well to Cd-contaminated and organically-depleted salinised soil environments.

Free Cd^{2+} pool was estimated from the modell to predominate in most of tested pHs but only under low salinity, whereas in moderate-to-strong salinity conditions the most prevalent pool was Cd-chloro-/sulphate complex (Cd-Cl/Cd-SO_4) (Fig. 17.2). Inorganic Cd-complexes mostly comprised $CdCl^+$, $CdCl_2$(aq), $Cd(SO_4)_2^{-2}$ and $CdSO_4$(aq) species (data not shown), although concentration of Cd in the chloro-complex (vs. Cd-SO_4) was multi-fold higher (Fig. 17.2). In all three models, inorganically-complexed and free Cd^{2+} pools decreased with pH rising, contrary to organically-complexed pool (Cd-ORG) whose concentration positively responded to increasing pH, but predominated only under high alkalinity (pH > 9.0) and low salinity (Fig. 17.2).

Cl-induced Salinity and Cd Soil (Rhizosphere) Solution-plant Transfer

Over the last several decades it has been well documented that Cl salinity may be of crucial importance for food/feed safety and security (quality) with respect to Cd (trace metals) contamination, because of Cl effects on Cd mobility (Fig. 17.2) and thus soil-plant transfer (Table 17.1). In a survey study with 124 paired topsoil and durum wheat grain samples collected under highly variable field conditions (e.g. pH 5.95–8.07; Cl^{2-}1696 mg/kg; Cd_{DTPA} 0.064–0.155 mg/kg) Norvell et al. (2000) confirmed that Cd phytoaccumulation in grains (Cd_{grain}) was strongly positively correlated ($P < 0.01$) with water-extractable soil Cl (Cl_w). Although Cd-Cl interactions are still under intensive investigation and not fully understood, complexation is likely to be one of the main mechanisms for enhancing Cd uptake from Cl-affected rhizospheres and deposition in plant tissues (Table 17.1). Generally, it is accepted that cadmium free cationic form (Cd^{2+}) is the most efficient in crossing the root

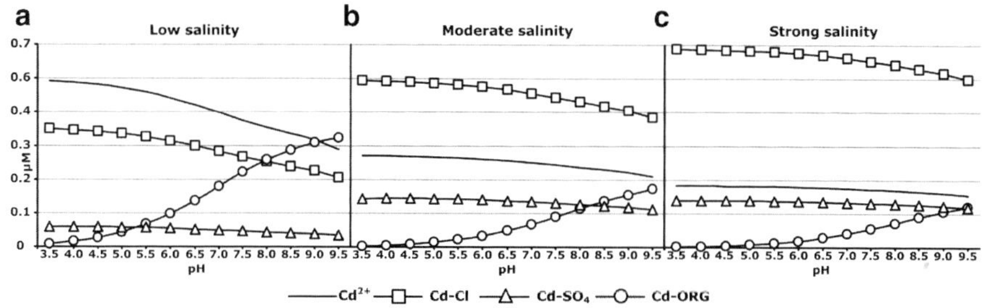

Fig. 17.2 Concentration distribution of Cd species in particular pools (uncomplexed cationic Cd^{2+}; chloro-complexed Cd-Cl; sulphate-complexed Cd-SO_4 and organo-complexed Cd-ORG) modelled by Visual MINTEQ under different pHs and salinities (in mM): (**a**) low (Na^+ 15, Cl^- 12, SO_4^{2-} 1.5), (**b**) moderate (Na^+ 75, Cl^- 45, SO_4^{2-} 15) and (**c**) strong (Na^+ 150, Cl^- 90, SO_4^{2-} 30). In all models, concentrations of Cd (1 µM) and DOC (1 mg/L) were constant

membranes (Sect. 3.3). Most studies listed in Table 17.1 as well as others (Bingham et al. 1984; Boukhars and Rada 2000; Li et al. 1994; McLaughlin et al. 1994, 1997; Smolders and McLaughlin 1996a, b; Smolders et al. 1998; Weggler-Beaton et al. 2000; López-Chuken and Young 2005; López-Chuken et al. 2012) indicated possible root extraction of certain Cd forms from the Cd-Cl pool, although with low efficiency (in comparison with Cd^{2+}). Based on various experimental approaches (e.g. nutrient solution, soils) and a wide range of plant species (differing in salinity and Cd tolerance) and conditions (pH, Cl concentrations/length of exposure), it was suggested that certain Cd-Cl forms could be phytoavailable, i.e. enter roots (1) directly as a Cd-Cl-complex and/or (2) serving as a proxy form for facilitated diffusion to the root-binding sites, and afterwards dissociating and entering cells as uncomplexed Cd^{2+} (Cl^-) (McLaughlin et al. 1994, 1997; Smolders and McLaughlin 1996a, b) (Fig. 17.3 in Sect. 3.3).

In interpreting results from Table 17.1, it is impossible to consider all variables and their influence to Cd phytoavailability. Soils (especially clayic, organic) usually have a complex matrix, composed of different organic (inorganic) metal sorbents that interact with Cd and decrease its concentration (activity) in the dissolved pool (Ondrasek et al. 2009a; Ondrasek et al. 2012). Similarly to some other metals (e.g. Cu), Cd has a strong potential for complexation with organic ligands (Romic et al. 2012). Also, a soil background and applied Cd concentrations varied considerably, from inorganic (e.g. $CdCl_2$, $Cd(NO_3)_2$, fertilisers) to organically-enriched sources (histosols, biosolids) (Table 17.1). Phytoavailability of applied Cd is generally higher from inorganic vs. organic sources, particularly under salinity (NaCl) (Weggler et al. 2004). For instance, biosolids and some other organically-enriched materials (ORG) (e.g. peats, composts) are frequently used for improving fertility of organic matter- and nutrient-depleted soils, although simultaneously may induce contamination by Cd (TEs). Due to a substantial portion of negatively-charged reactive interfaces, ORG substances strongly compete with Cl^- (SO_4^{2-}) ligands in

Table 17.1 Soil (rhizosphere) solution and plant responses to induced salinity and Cd contamination in nutrient solution (NS) or soil experiments. When present, type of letters (small/capital) indicates inside of which groups treatments were compared. Treatments with the same letters are not significantly different. Approximate values (~) are taken from graphical presentation in cited references

EXPERIMENTAL CONDITIONS							SOIL SOLUTION AND PLANT RESPONSES		
	Medium characteristics						Cd content		Biomass Parameters on dry or fresh weight (DW or FW) basis
Plant spp.	Type	Texture	pH	Salinity Type mM	Salinity duration	Cd content mg/kg	In soil solution µg/L	In plant tissue	
1. Cereal	**Soil** Weggler et al. 2004							Shoot (% of total Cd in soil solution)	Shoot DW g/pot
Triticum aestivum L. cv. Halberd		Alfisol 95% sand	6.3	0	30 days	0.12	n.d.	n.d.	1.32a
		Alfisol 95% sand		27.4 NaCl		0.12	0.23	0.383	1.11a
		Alfisol+8% biosolid		27.4 NaCl		0.72	1.6	0.444	1.10a
		Alfisol+16% biosolid		27.4 NaCl		1.32	0.75	0.114	1.32a
2. Cereal	**Soil** Khoshgoftar et al. 2004							Shoot mg/kg	Shoot DW g/plant
Triticum aestivum L. cv. Rushan		Calcigypsids 54% clay	7.8	0	35 days	3.8	10a	0.90a	~15a
				120 NaCl			130b	1.27c	~11b
				120 $NaNO_3$			20a	1.01b	~11b
3. Cereal	**Soil** Khoshgoftarmanesh et al. 2006							Shoot mg/kg	
Triticum aestivum L. cvs. Rushan, Kavir, Cross and Falat *Triticum turgidum* L. cv. Durum		Calcigypsids 46% clay	7.6	0	35 days	3.2	*14a	**0.01a	
				180 NaCl			*96b	**0.05b	

4. Cereal	**Soil** Ozkutlu et al. 2007						Grain mg/kg		Grain yield g/plant
Triticum turgidum L. cv. Balcali-2000	Clayic	8.1	0	140 seconds	1.27		~2,8A		3.5A
			167 $NaCl^{x}$				~3,8B		3.2A
			0	140 seconds	0.27		0.05a		2.52a
			167 $NaCl^{y}$ +8.8 CdCl2				0.55c		2.79a
			167 $NaCl^{y}$ +8.8 CdSO4				0.34b		2.40a
			167 $NaCl^{y}$ +8.8$Cd(NO_3)_2$				0.49bc		2.57a
5. Vegetables	**Soil** McLaughlin et al. 1998b						Shoot nmol/g	Total uptake nmol/pot	Shoot DW g/plant
Beta vulgaris L. cv. Fordhook Giant	Alfisol 81% sand	5.4	0	19 days	0.31 (EDTA-extractable)	2.61a	~11.6A	18.2A	~0.187A
			120 Na_2SO_4			7.9c	~20.84B	12.6B	~0.079B
			0			2.63a	~12.5a	18.2a	~0.181a
			120 $NaNO_3$			5.0b	~16.22a	7.3b	~0.061b
6. Vegetables	**Soil** Ondrasek et al. 2009a, 2009b						Leaves mg/kg	Pulp mg/kg	Yield kg/plant
*Cucumis melo L. cv*Early Dawn	Histosol >90% OM	5.7	0	50 [1](60-73) days	5.4^{3n}	18.4^{3n}	10a	[1](0.4a)	[1](3.72a)
			60 NaCl				15b	[1](0.4a)	[1](1.94b)
							Leaves mg/kg	Hypocotyls mg/kg	Yield g/m2
Raphanus sativus L. var. sativus, cv. Tarzan			0	34 days	2.6^{3n}	22.4^{3n}	27.5A	6.4A	659A
			60 NaCl				33.1B	5.9A	426B
7. Vegetables	**Soil** Oporto et al. 2009						Root uptake flux pmol/cm^{2}h	Shoot mg/kg	Shoot DW mg/plant
Spinacia oleracea L. cv. Viking		6.7	0	18 days	0.4		~0.1a	~9a	448
			120 NaCl				~0.6b	~55b	160
			120 $NaNO_3$				~0.2a	~25c	114

(continued)

Table 17.1 (continued)

			0		10.5		~10A		~700A	150		
			120 NaCl				~10A		~2000B	13		
			120 $NaNO_3$				~10A		~1000B	41		
8. Legume	**Soil** Helal et al. 1999						Root	Steam	Leaf	DW g/plant	Root length density cm/L	
							µg/g					
Leucaena leucocephala L.	Sandy 73% sand	7.9	[2]Control	134 days	0.76	24a	3.96a	2.17a	0.38a	47a	141a	
			[2]Control + 10 NaCl			35b	3.75a	2.04a	0.89b	57b	128b	
9. Submersed spp.	[3]**NS** Fritioff et al. 2005				In NS µg/L		In all plant biomass µg/g			FW g/L	DW g/L	Length cm
Elodea canadensis L.		5.0	0	48 hours	112.4		269a			8.7a	0.45a	78.5a
			0.5 NaCl				244b			8.7a	0.44a	78.8a
			5.0 NaCl				212c			8.3a	0.41a	78.2a
Potamogeton natans Michx.			0				151A			11.7A	1.35A	-
			0.5 NaCl				100B			11.7A	1.39A	-
			5.0 NaCl				63C			10.3A	1.19A	-
10. Cereal	**NS** Mühling and Läuchli 2003						Root g/FW		Leaf µM	Root DW	Shoot DW	
										g/pot		
Triticum aestivum L. cv. Chinese Spring		6.0	1 NaCl	3 weeks	0		0		0a	~5.8a	~32a	
			1 NaCl		22.5 (2 weeks) 1124 (1 week)		~440a		23ab	~5.0a	~20b	
			75 NaCl				~490a		33b	~4.1a	~13b	
11. Vegetables	**NS** Smolders and Mclaughlin 1996a						Root		Shoot	Root DW	Shoot DW	
							per kg DW			mg		
Beta vulgaris L., cv. Fordhook Giant unbuffered		6.1	0.01 NaCl	11 days	5.6		~250 mg[4n]		~33 mg[4n]	7.0	52.4	
			120 NaCl				~115 mg[2n]		~17 mg[2n]	8.2	55.9	
NTA-buffered			0.01				~100 mg[2n]		~10 mg[2n]	5.0	51.6	
			120 NaCl				~115 mg		~15 mg	5.2	51.4	
EGTA-buffered			0.01				~240 µg		~60 µg	11.1	92.5	
			120 NaCl				~310 µg		~60 µg	12.6	106.6	
12. Vegetables	**NS** McLaughlin et al. 1998a						Root		Shoot	Root DW	Shoot DW	
							mmol/kg			g/pot		
Beta vulgaris L., cv. Fordhook Giant		6.1	8 Na_2SO_4	11 days	5.6		~1.4a		~0.16a	~0.20a	~1.8a	
			58 Na_2SO_4				~1.6a		~0.15a	~0.19a	~1.9a	

[*]Mean value of four bread and one durum wheat genotypes
[**]Mean effect of NaCl salinity on shoot Cd concentration in four bread and one durum wheat genotypes
[x]Salinity treatment was applied (nearly after 80 and 110 days of growth) by immersion of leaf to NaCl solution
[y]Salinity and Cd treatment were applied (nearly after 80 and 110 days of growth) by immersion of leaf to NaCl- and Cd-containing solution
[1]Fruit harvesting was obtained in between 60 and 73 days of salinity treatment and values in brackets are referred to that period
[2]Control had an average molar ratio: Na 3.1, Cl 4.6, Mg 1, Ca 1.2, K 1.4, SO_4 1.4
[3]Values are refereed for 20°C
[2n]Mean values of two replicates
[3n]Mean values of three Cd treatments
[4n]Mean values of four replicates

complexing positively-charged Cd forms, and therefore affect Cd availability and phytoextraction (Ondrasek and Rengel 2012). Bolan et al. (2003c) indicated that the addition of ORG (biosolid compost) to soil decreased the concentration of the soluble/exchangeable Cd pool at the expense of the Cd-ORG pool. As a consequence, the same authors observed decreased Cd phytoaccumulation (i.e. alleviation of Cd phytotoxicity) by ORG addition. Similar results were reported by others (e.g. Shuman et al. 2002; Pinto et al. 2004) and are mostly due to the formation of poorly phytoextractable Cd forms in the Cd-ORG pool.

However, ORG application (up to 18% of biosolids) to sandy soil (Weggler et al. 2004) and increasing Cl salinity in predominantly (>90%) ORG soil (Ondrasek et al. 2009a) resulted in (1) increased Cd (Cd-Cls) concentration in soil solutions, (2) enhanced Cd phytoaccumulation in wheat shoots or muskmelon leaves (Table 17.1), and (3) closer correlation of Cd concentration in wheat shoots with $CdCl^+$ than with Cd^{2+}. Both studies indicated that even in the presence of relatively high ORG content, Cd-Cl complexation was of crucial importance for the soil-plant Cd transfer.

Norvell et al. (2000) obtained a curvilinear relationship among Cd_{grain} and soil Cl, with an increased rate of Cd phytoaccumulation being most pronounced at marginal Cl concentrations (~10 mM) and relatively low Cd concentration in the topsoil (Cd_{DTPA} ranged from 0.064 to 0.155 mg/kg). Though specific mechanisms responsible for Cd acquisition under low Cl_w salinity from uncontaminated soil are still unclear, possible influence of particular pedosphere and/or plant variables was tested. Fig. 17.2a (low Cl salinity/Cd) shows that one of possible explanation could be Cd biogeochemistry in the soil solution pool, whereby in the pH range ~6–8 (similar to Norvell et al. 2000), the most bioavailable Cd^{2+} and thereafter Cd-Cl forms predominated. However, in study by Norvell et al. (2000), topsoils (0–15 cm) were quite enriched with dissolved ORGs (i.e. forming poorly available Cd-ORG pool), given that organic C was ~20–40 g/kg. Another explanation for better Cd than CdCl uptake by plants at relatively low soil background concentration was suggested by Oporto et al. (2009), whereby Cd-Cl-complexation was of negligible importance for plant uptake under higher Cd supply.

By employing the technique of diffusive gradients in thin-films (DGT; in which a Chelex resin induces a diffusive flux of Cd^{2+} from a labile complexes such as dissolved Cd-Cls and/or sorbed forms on soil solids) (see Zhang and Davison 1995, 2001), fluxes of Cd (in soil/soil solution) were quantified together with root Cd uptake and shoot Cd concentration at variable Cl (0–120 mM) and Cd supply (0.4–10.5 mg/kg). They observed that (1) rising Cl salinity significantly increased shoot Cd at all soil Cd levels, but it was relatively more pronounced at the low (background) than at high soil Cd, (2) Cd uptake flux into root increased significantly (>5-fold) due to addition of Cl at the lowest (but not at the highest) soil Cd concentration (Table 17.1), whereas DGT fluxes increased to the same extent but at all Cd levels. Correlation between the fluxes measured based on DGT and fluxes calculated based on root uptake at the background soil Cd suggested that at low Cd supply Cd phytoextraction was controlled (limited) by diffusion of Cd^{2+} and its replenishment from labile (e.g. Cd-Cl) pools (e.g. Oporto et al. 2009). This is in a line with recent observations by López-Chuken et al. (2012), whereby $CdCl^+$ complexes appeared to saturate root sorption sites even at low activities and, therefore, $CdCl^+$ activities greater than this saturation level do not cause any increase in Cd phytoextraction. With high concentration of Cd in the contaminated rhizosphere environment, the kinetics of other Cd-borne forms should also be considered, given that precipitation-dissolution reactions may significantly control element speciation (e.g. Khoshgoftar et al. 2004) and thus uptake.

With respect to possible genotypic influence on the Cd-Cl interaction, it is well known that durum is more effective in Cd_{grain} accumulation than bread wheat (Norvell et al. 2000). Increasing Cl-salinity exacerbated Cd accumulation in durum wheat shoots more than in four bread wheat cultivars (Khoshgoftarmanesh et al. 2006). More recently, Ozkutlu et al. (2007) confirmed that even short (<2.5 min) foliar application of Cl may be important for soil Cd extraction and/or Cd (re) mobilisation in durum wheat, i.e. Cd_{grain} deposition. In experiment 1, where soil was spiked with 1.0 mg Cd kg (i.e. with a background concentration totally contained 1.27 mg Cd/kg) they observed increases in Cd_{grain} concentrations by up to 41% with 167 mM Cl foilar application (vs. H_2O treatment). In experiment 2 (with unspiked soil) foliar application of Cd-contaminated solution (8.8 mM with different Cd salts) significantly enhanced Cd_{grain} concentration but it was even more pronounced (~45% in case with $CdCl_2$) by applying of Cl salinity (Table 17.1). Above studies suggesting that durum (vs. bread wheat) genotypes could be more effective, not only in Cd root extraction, but also in Cd root-shoot (i.e. leaf-fruit) translocation under exposure to Cl salinity.

Although in many studies (Table 17.1), Cd phytoaccumulation positively correlates with salinity, there are certain exceptions. Fritioff et al. (2005) detected that accumulation of Cd in submergent spp. (*Elodea canadensis* L. and *Potamogeton natans* Michx.) significantly decreased with NaCl, without any influence of salinity on growth (Table 17.1). However, they used relatively low NaCl (up to 5 mM), corresponding well to the modelled situation in Fig. 17.2a where the proportion of Cd^{2+} appeared to increase with increasing salinity.

Na-induced Salinity

Na is the most frequent causative agents of naturally-induced soil alkalinity and is accompanied with salinity induced by Cl. As a consequence, similar to Cl, Na might also correlate well with phytoextracted Cd, although it seems unlikely to play any significant role in enhancing Cd uptake (e.g. Smolders et al. 1998; Mühling and Läuchli 2003); such relationships are attributed mostly to multi-colinearity with Cl (Norvell et al. 2000). Indirectly, it is possible that excessive Na^+ substantially impacts Cd biogeochemistry in the rhizosphere via (1) competition with Cd cationic free/complexed species for (in)organic exchange interfaces, root transport mechanisms, and thereafter (re)translocation/deposition in the plant (Sect. 3.3), (2) deprotonation from (in)organic soil matrix (e.g. humics) and thus acidification of surrounded (unbuffered) solution (e.g. Lores and Pennock 1998), and/or (3) induced

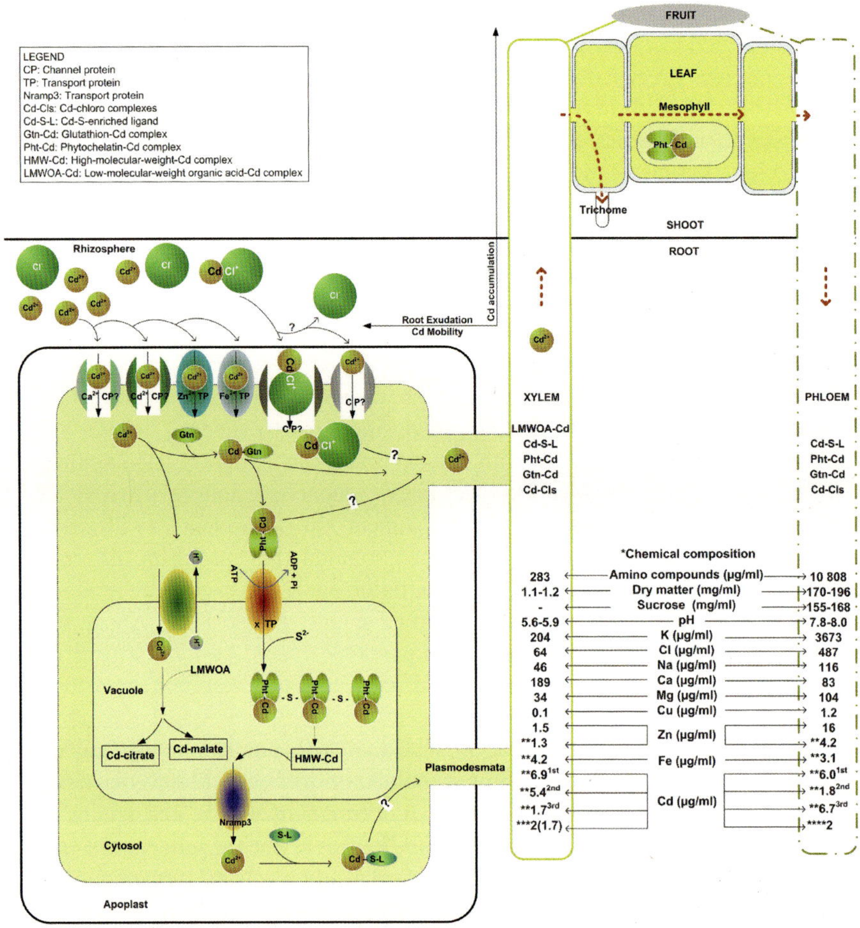

Fig. 17.3 A conceptual representation of root uptake and translocation of Cd and characterisation of transport systems principally for plants with aerial fruiting bodies (adopted from references cited in the text). *Chemical composition of xylem/phloem sap in wild tobacco (*Nicotiana glauca* Grah.) (Hocking 1980), **Cd concentration in xylem/phloem sap collected during 3 stages [1st (at 10th), 2nd (at 14th) and 3rd (at the early grain-filling) stage] from rice (*Oryza sativa* L.) grown in flooded-soil containing 1.55 mg Cd/kg (extracted with 0.1 M HCl). **Fe and **Zn concentration represent their average concentration obtained during 3 stages (Yoneyama et al. 2010). ***Cd concentration in xylem sap of Indian mustard (*Brassica juncea* L.) exposed to 0.6 μg/mL Cd for 10 h or 7 days (in parenthesis) (Salt et al. 1995), ****Cd concentration in phloem sap of rice (*Oryza sativa* L.) treated with a nutrient solution containing 10 μM Cd for 2 days (Tanaka et al. 2007)

osmotic effects in the rhizosphere solution (e.g. water stress, Na toxicity, disruption of integrity/selectivity of root plasma membrane and the plant homeostasis) (reviewed by Tester and Davenport 2003; Munns and Tester 2008). Na^+ may become harmful for many plants if its cytosolic concentration exceeds 10 mM (e.g. Munns and Tester 2008). This is relatively easily achievable in natural saline environments, causing common plant responses to Na-toxicity such as growth inhibition (Table 17.1) and in extreme cases early senescence and plant mortality (e.g. Romic et al. 2008).

When associated with anions other than Cl^-, Na salts may induce relatively moderate plant (environmental) responses. An addition of $NaNO_3$ can be sevenfold less effective than NaCl in increasing Cd_{TOT} content (Table 17.1) and without influence on Cd^{2+} and $CdCl^+$ concentrations in soil solution (Khoshgoftar et al. 2004). In comparison to non-salinised control, $NaNO_3$ salinity did not increase concentration of soil Cd_{TOT}, as in case of NaCl, but it significantly reduced wheat growth/yield, and marginally (but significantly at $P<0.05$) improved shoot Cd accumulation (Table 17.1). Similarly, the relative increase in Cd_{TOT} in soil solution was fivefold at 120 mM $NaNO_3$ and 20-fold at 120 mM NaCl, which was attributed to formation of Cd-Cls (Oporto et al. 2009). The twofold greater Cd accumulation in shoots was obtained with NaCl than $NaNO_3$ application at the same molar rates (Table 17.1), probably as a result of the reduced plant yield at increasing ionic strength (Oporto et al. 2009). Both salinity types imposed similar negative effects on shoot yield at low but not at high Cd supply, where NaCl depressed the yield more severely (12-fold) than $NaNO_3$ (4-fold) compared to non-salinised control (Table 17.1), probably due to additive NaCl/Cd toxicities, i.e. stresses (Sect. 3.3).

SO_4-Induced Salinity

Sulphate is a relatively abundant anion in agro-exploited pedo/hydrosphere resources affected by salinity (Rhoades et al. 1999). In such circumstances, and in sufficient magnitude of Cd^{2+} pool, Cd-sulphate complexes were common, although orders of magnitude less abundant (Fig. 17.2) and less stable compared to Cd-Cls (cf. Norvell et al. 2000). The Cd-sulphate complexes might also influence Cd chemistry in the rhizosphere (e.g. McLaughlin et al. 1998b). From negatively-charged $Cd(SO_4)_2^{2-}$ to neutral $CdSO_4^0$ complexes, these complexes can enhance soil-plant Cd transport by either (1) diffusive transport of $CdSO_4^0$ through the lipophilic plasma membrane bilayers as shown for $HgCl_2^0$ by Gutknecht (1981), and suggested for H_2BO species by Welch (1995) and recently in durum wheat leaf epiderms for $CdCl_2^0$ by Ozkutlu et al. (2007), or (2) through facilitated diffusive transport of charged Cd-SO_4 complexes to absorption sites at the root interface as suggested above for Cd-Cl (Fig. 17.3). In support of the first assumption, McLaughlin at al. (1998a) did not observe better plant uptake of Cd by increasing dissolved calcium (Ca) concentration in solution to compensate for $CaSO_4$ complexation, confirming that $CdSO_4^0$ could be phytoextracted as efficiently as Cd^{2+}. Direct crossing of negative Cd-SO_4 forms across the plasma membrane via channel/transport proteins seems less likely given possible size restriction of their ionic radii (see discussion by Smolders and McLaughlin 1996a).

Compared to studies focused on Cd-Cl interactions and their plant responses, those including Cd-SO_4 are rare and less instructive. Cd-SO_4 complexation in nutrient solutions (up to 58 mM SO_4) has been shown as having an insignificant effect on Cd phytoextraction (McLaughlin et al. 1998a), whereas in approximately double SO_4-concentrated soil solution, Cd shoot deposition increased marginally but significantly ($P<0.05$) (McLaughlin et al. 1998b). Comparing some other studies that used low Cl salinity and other crop species (Table 17.1), and also earlier similar approaches and conditions (e.g. salinity duration, chemical environment in the rhizosphere) with the same tested plant species (Smolders and McLaughlin 1996a; Smolders et al. 1998), obviously a substantially lower effects of Cd-SO_4 complexation may be expected in Cd transfer to crops than in Cd-Cl complexation. Norvell and associates (2000) also obtained (1) highly positive correlation among concentrations of Cd_{grain} and water-extractable sulphates (SO_{4w}) (Pearson correlation coefficient was even higher then with Cl_w), and (2) explained high variability (around 60%) in Cd_{grain} by attributing such a relationship mostly to SO_4^{2-} association (i.e. multi-colinearity) with Cl, rather than to direct causality.

Soil-Plant Transfer of Cd from Saline Environment

As one of the most soluble and bioavailable trace metals in the rhizosphere, and due to low selectivity for root extraction in majority of plant species, Cd relatively easily overcomes the soil-plant barriers. Although concentration as well as distribution of Cd decreases with the upstream transpirational flow (hence, it is higher in belowground than aboveground plant parts, Table 17.1, Fig. 17.3), consumption of crops produced in Cd-enriched environment is the principal route of Cd "intoxication" for humans. Relatively enhanced soil-plant transfer of Cd and its biotoxicity are main causes for strictly defined acceptable limits of Cd intake from foodstuff (e.g. 1 μg Cd per kg body weight per day) (WHO 1992). In Australia and New Zealand for example, Cd is recognised as the most common trace metal reaching the food chain through animal transfer in pastoral agriculture (Bolan et al. 2003a). Also, for the same populations consumption of crop foodstuffs represents predominant route (>80%) in total daily Cd intake of ~9 μg, with >60% originating from potatoes (~4.3 μg) and wheat (~1.5 μg) consumption (Ondrasek and Rengel 2012), whereas ~50% of the total Cd intake by Japanese comes from white rice (Yoneyama et al. 2010). A similar situation could be expected across other Cd-noncontaminated areas worldwide, given that potatoes, wheat and rice are among the top five agricultural crop commodities in the world (FAO 2009). In irrigated potato production, but under highly contaminated (e.g. Cd in soil up to 80 mg/kg and in irrigated water up to 240 μg/L) and salinised conditions [e.g. in soil solution (mM) 107 Cl^-, 90 Na^+ and 42 SO_4^{2-}], it was estimated that daily Cd intake from potato consumption could reach ~100 μg (Oporto et al. 2007).

Similarly to other metals, the Cd species most capable of entering root is soluble Cd^{2+}, though under certain conditions Cd uptake is better explained by the activity of the Cd-Cl complexes than Cd^{2+} (López-Chuken et al. 2012). Evidence on Cd-complexes being extracted by plant root is relatively scarce and based mostly on assumptions. Besides, in the presence of Cd-Cl (Table 17.1), increased Cd uptake was confirmed under increasing concentrations of low- (Chiang et al. 2011) and high-molecular-weight organic acid (L/HMWOA) ligands (Evangelou et al. 2004; Bandiera et al. 2009). Undoubtedly, metal complexation with dissolved ligands would enhance Cd desorption from the (in)organic solid interfaces in the rhizosphere, and improve the transfer of Cd complexes into plants, although complete mechanisms are not fully explained. Also, Cd-Cl$^+$ complexes (vs. Cd^{2+}) are less strongly sorbed onto rhizosphere interfaces, and thus formation of Cl-complexes tends to shift Cd from solid to dissolved phase, thereby enhancing its solubility/mobility (López-Chuken et al. 2012). The most logical explanations of Cd uptake from the rhizosphere would rely on well characterised Cd^{2+} and $CdCl^+$ forms (Fig. 17.3).

Due to many similarities between Cd and some divalent metal phytonutrients in physical (e.g. ionic radii in pm: Cd^{2+} 97, Ca^{2+} 99, Mn^{2+} 80, Fe^{2+} and Zn^{2+} 74) or chemical (e.g. redox-activity, Lewis acidity) properties, non-essential Cd would most probably enter roots via routes specific for similar essential elements. Cadmium uptake across the root cell plasma membrane was shown to occur via a concentration-dependent mechanism exhibiting saturable kinetics (Salt et al. 1995; Hart et al. 1998), with confirmed Cd-Zn; Cd-Mn and Cd-Cu competition (Hart et al. 2002; Salt et al. 1995) or Fe-Cd complementarity (Nakanishi et al. 2006). Cadmium uptake by plants could be facilitated under salt- (Na, Cl) (Ondrasek and Rengel 2012; Ondrasek et al. 2012) and/or Cd-stressed (e.g. Salt et al. 1995; López-Chuken and Young 2005; Ahmad et al. 2011) conditions. For instance, Mühling and Läuchli (2003) showed that a combined NaCl/Cd stress (vs. their separate influence) can enhance Cd phytoextraction, probably by exacerbating oxidative stress (i.e. production of O_2^- radicals and H_2O_2) which could increase plasma membrane permeability and/or inhibit activities of antioxidant enzymes involved in the oxidative defense mechanism (Ahmad et al. 2011). Also, Mühling and Läuchli (2003) highlighted that salt-sensitive genotype exposed to salinity showed higher Cd accumulation than salt-resistant ones. Similarly, López-Chuken and Young (2005) confirmed Cl-enhanced uptake of Cd (at 100 mM NaCl) from highly contaminated soil (58 mg Cd/kg) by salt-tolerant genotypes (*Brassica juncea* G32192 and *Zea mays* hybrid W23/L317), with some seed not germinating probably due to salt and/or Cd toxicity. Both studies indicate large challenges faced by genetic engineering for improved food safety and security (e.g. Ahmad et al. 2012), especially if salinity tolerance (as multi-gene trait) is closely associated with gene loci responsible for Cd phytoextraction.

The likely Cd uptake routes from the rhizosphere are via channel (CP) and/or transport proteins (TP) (e.g. Ca^{2+}CP; Zn^{2+}TP; Fe^{2+}TP), and less probably via specific Cd^{2+}CP or Cd^{2+}TP embedded into the plasma membrane of root cells (Clemens 2006; Lee and An 2009; Yoneyama et al. 2010) (Fig. 17.3). It remains unclear whether Cd enters the roots as $CdCl^+$ and is then translocated in the same form via the xylem to shoot (discussed in above sections).

To moderate high reactivity/toxicity of Cd^{2+} in the symplast caused by its strong complexation with cytosolic metabolites (e.g. organic acids, amino acids, peptides etc.; Kato et al. 2010 and refs therein), Cd must be immobilised. Salt et al. (1995) and recently Karlsson et al. (2005) confirmed by EXAFS (Extended X-ray Absorption Fine Structure) experiments that reduced organic S ligands could be involved in Cd complexation in the root tissues. After entering root cytoplasm, possible initial Cd complexation could be with glutathion (Gtn) to form Cd-Gtn complex as a precursor of Pht-Cd (Phytochelatin-Cd complex) (Fig. 17.3). Both Gtn and Pht are LMW cistein-enriched peptides involved in maintenance of low activity of free forms of cytoplasmic metals. As a Pht-Cd complex, Cd can cross the tonoplast (via certain TPs) and create Phytochelatin-based HMW-Cd complexes in the vacuole (Salt and Rauser 1995); however, it remains unclear whether the same form can be loaded into the xylem. To some extent free cytosoilic Cd can be anti-ported into the vacuole (Salt and Wagner 1993) and sequestered to LMWOA-Cd compounds, whereas a fraction of Cd from the LMW vacuolar pool may be remobilised into the cytosol (over AtNramp3 proteins) and chelated with S-enriched ligands (Cd-S-L) (Fig. 17.3). The physiological mechanisms responsible for differential movement of Cd in roots and from roots to shoots (i.e. Cd xylem/phloem loading) still remain unclear (Hart et al. 2006; Yoneyama et al. 2010).

Due to slight acidity of the xylem sap (pH ~ 6) (Fig. 17.3) there is a possibility that a large proportion of Cd is complexed. It is presumed that the same or similar transport and deposition routes (as in root) of Cd complexation occur in the aboveground tissues, as shown in tobacco cell leaves, where cytoplasmic Pht-Cd is directly transported to vacuole (Vögeli-Lange and Wagner 1990). Cadmium deposition could specifically be targeted to specific leaf cells such as trichomes, i.e. leaf hair/gland cells derived from specialized epidermal layer on the leaf or stem surfaces. Compared to concentration within leaves (without trichomes), Salt et al. (1995) observed a 43-fold greater accumulation of Cd in trichomes of Indian mustard leaves after 24-h exposure to 0.1 μg Cd/mL, explaining one of possible detoxification strategies in plants.

Under slightly basic conditions and prevalence of organic assimilates (e.g. S-, O-, N-rich compounds) that exist in the phloem sap, dominant Cd forms in the phloem could be those from Cd-ORG pool, but Cd-Cl complexation cannot be excluded because of Cl abundance (Fig. 17.3). During last decades, biochemistry of Cd transport in phloem was studied in potato tubers (Dunbar et al. 2003; Reid et al. 2003), peanuts seeds (Popelka et al. 1996), soybeans (Yada et al. 2004), wheat (Harris and Taylor 2001) and rice (Tanaka et al. 2007; Yoneyama et al. 2010); however, it is likely that Cd-Cl complexation inside the vascular system is less relevant than in the rhizosphere. Recently, Kato et al. (2010) reported that major Cd chelators in phloem rice sap are proteinous/S-rich (e.g. sulfhydryl) substances of ~ 13 kDa, which distinguishes it from similar metals (Fe, Zn, Cu, Mn, Ni, Co), probably complexed with other organics (e.g. nicotianamine, 2'-deoxymugineic acid, citrate and histidine). Although in both vascular systems (xylem and phloem) Cd has similar and relatively high mobility (Reid et al. 2003; Dunbar et al. 2003), and pose similar fluctuation pattern during vegetation period (from 10th leaf to early grain rice filling stage) in Cd concentration (Yoneyama et al. 2010), retranslocation and redistribution of Cd from leaves to fruits in cereals could predominantly (>90%) occur in phloem (Tanaka et al. 2007).

Conclusions and Future Perspective

It is expected that in the forthcoming era of global warming and human population growth agro-exploited resources could be dramatically impacted by salinity and metal contamination. During last decades a significant research effort with different genotypes has been devoted to elucidation of Cd biogeochemistry in the rhizosphere, its (re)translocation and deposition in plants. The importance of Cd complexation with inorganic salt ligands (particularly Cl) in the rhizosphere was established. Cadmium taken up is transported (notably in organo-complexed forms) to the transpirational tissues via the xylem, and thereafter to the depositing tissues dominantly via the phloem. Recently, genetic engineering combined with traditional breeding has been highlighted as a promising strategy for improving (1) agri-food production in saline environment, and (2) phytoremediation of metal-contaminated areas. Soil salinity can increase the soil-plant transfer of Cd under ecologically-relevant conditions (e.g. soils with poor metal-buffering capacity), but can also interfere with phytoextraction of nutrients (Zn, Cu, Fe). Thus, growing salt-resistant crops might impact food safety and security by exacerbating Cd accumulation and decreasing micronutrient content, especially if salinity tolerance (as a multi-gene trait) is closely associated with gene loci responsible for Cd phytoextraction.

Acknowledgments The author is grateful to Winthrop Professor Zed Rengel (University of Western Australia) for valuable discussion, comments and text improvement.

References

Adriano DC, Wenzel WW, Vangronsveld J, Bolan NS (2004) Role of assisted natural remediation in environmental cleanup. Geoderma 122:121–142

Ahmad P, Nabi G, Ashraf M (2011) Cadmium-induced oxidative damage in mustard [Brassica juncea (L.) Czern. & Coss.] plants can be alleviated by salicylic acid. S Afr J Bot 77:36–44

Ahmad P, Ashraf M, Younis M, Hu X, Kumar K, Akram NA, Al-Qurainy F (2012) Role of transgenic plants in agriculture and biopharming. Biotechnol Adv 30(3):524–540

Bandiera M, Moscaa G, Vamerali T (2009) Humic acids affect root characteristics of fodder radish (*Raphanus sativus* L. var. *oleiformis* Pers.) in metal-polluted wastes. Desalination 246:78–91

Basta NT, Gradwohl R, Snethen KL, Schroder JL (2001) Chemical immobilisation of lead, zinc and cadmium in smelter contaminated soils using biosolids and rock phosphate. J Environ Qual 30:1222–1230

Biggs AJW, Watling KM, Cupples N, Minehan K (2010) Salinity risk assessment for the Queensland Murray-Darling region. Queensland Department of Environment and Resource Management, Toowoomba

Bingham FT, Sposito G, Strong JE (1984) The effect of chloride on the availability of cadmium. J Environ Qual 13:71–74

Bolan NS, Adriano DC, Mani P, Duraisamy A, Arulmozhiselvan S (2003a) Immobilization and phytoavailability of cadmium in variable charge soils: I. Effect of phosphate addition. Plant Soil 250:83–94

Bolan NS, Adriano DC, Mani P, Duraisamy A, Arulmozhiselvan S (2003b) Immobilization and phytoavailability of cadmium in variable charge soils: II. Effect of lime addition. Plant Soil 250:187–198

Bolan NS, Adriano DC, Duraisamy P, Mani A (2003c) Immobilization and phytoavailability of cadmium in variable charge soils III. Effect of biosolid compost addition. Plant Soil 256:231–241

Boukhars L, Rada A (2000) Plant exposure to cadmium in Moroccan calcareous saline soils treated with sewage sludge and waste waters. Agrochimica 44:641–652

Chiang PN, Chiu CY, Wang MK, Chen BT (2011) Low-molecular-weight organic acids exuded by Millet (*Setaria italica* (L.) Beauv) roots and their effect on the remediation of cadmium-contaminated soil. Soil Sci 176:33–38

Chinnusamy V, Jagendorf A, Zhu JK (2005) Understanding and improving salt tolerance in plants. Crop Sci 45:437–448

Christensen TH, Huang PM (1999) Solid phase cadmium and the reactions of aqueous cadmium with soil surfaces. In: McLaughlin MJ, Singh BR (eds) Cadmium in soils and plants. Kluwer Academic Publ, Dordrecht\The Netherlands, pp 65–96

Clemens S (2006) Toxic metal accumulation, responses to exposure and mechanisms of tolerance in plants. Biochimie 88:1707–1719

Dang YP, Dalal RC, Mayer DG, McDonald M, Routley R, Schwenke GD, Buck SR, Daniells IG, Singh DK, Manning W, Ferguson N (2008) High subsoil chloride concentrations reduce soil water extraction and crop yield on Vertosols in north-eastern Australia. Aust J Agric Res 59:321–330

Dunbar KR, McLaughlin MJ, Reid RJ (2003) The uptake and partitioning of cadmium in two cultivars of potato (*Solanum tuberosum* L.). J Exp Bot 54:349–354

Evangelou MWH, Dagan H, Schaeffer A (2004) The influence of humic acids on the phytoextraction of cadmium from soil. Chemosphere 57:207–213

FAO (2009) http://faostat.fao.org/site/339/default.aspx. Accessed 07 Jan 2012

Florence TM (1982) The speciation of trace elements in waters. Talanta 29(5):345–364

Fritioff A, Kautsky L, Greger M (2005) Influence of temperature and salinity on heavy metal uptake by submersed plants. Environ Pollut 133:265–274

Grattan SR, Grieve CM (1999) Salinity-mineral nutrient relations in horticultural crops. Sci Hortic-Amsterdam 78(1):127–157

Gustafsson JP (2006) Arsenate adsorption to soils: modelling the competition from humic substances. Geoderma 136:320–330

Gutknecht J (1981) Inorganic mercury (Hg2+) transport through lipid bilayer membranes. J Membr Biol 61:61–66

Harris NS, Taylor GJ (2001) Remobilization of Cd in maturing shoots of near isogenic lines of durum wheat that differ in grain Cd accumulation. J Exp Bot 52:1473–1481

Hart JJ, Welch RM, Norvell WA, Sullivan LA, Kochian LV (1998) Characterization of cadmium binding, uptake, and translocation in intact seedlings of bread and durum wheat cultivars. Plant Physiol 116:1413–1420

Hart JJ, Welch RM, Norvell WA, Kochian LV (2002) Transport interactions between cadmium and zinc in roots of bread and durum wheat seedlings. Physiol Plantarum 116:73–78

Hart JJ, Welch RM, Norvell WA, Kochian LV (2006) Characterization of cadmium uptake, translocation and storage in near-isogenic lines of durum wheat that differ in grain cadmium concentration. New Phytol 172:261–271

Harter RDR, Naidu R (1995) Role of metal-organic complexation in metal sorption by soils. Adv Agron 55:219–264

Helal HM, Upenov A, Issa GJ (1999) Growth and uptake of Cd and Zn by Leucaena leucocephala in reclaimed soils as affected by NaCl salinity. J Plant Nut Soil Sci 162:589–592

Helmke PA (1999) Chemistry of cadmium in soil solution. In: McLaughlin MJ, Singh BR (eds) Cadmium in soils and plants. Kluwer Academic Publ, Dordrecht\The Netherlands, pp 39–64

Hocking PJ (1980) The composition of phloem exudates and xylem sap from tree tobacco (*Nicotiana glauca* Grah). Ann Bot-London 45:633–643

Kabata-Pendias A (2004) Soil-plant transfer of trace elements – an environmental issue. Geoderma 122:143–149

Karlsson T, Persson P, Skyllberg U (2005) Extended X-ray absorption fine structure spectroscopy evidence for the complexation of cadmium by reduced sulfur groups in natural organic matter. Environ Sci Tech 39:3048–3055

Kato M, Ishikawa S, Inagaki K, Chiba K, Hayashi H, Yanagisawa S, Yoneyama T (2010) Possible chemical forms of cadmium and varietal differences in cadmium concentrations in the phloem sap of rice plants (*Oryza sativa* L.). Soil Sci Plant Nutr 56:839–847

Khoshgoftar AH, Shariatmadari H, Karimian N, Kalbasi M, van der Seatm Z, Parker DR (2004) Salinity and Zn application effects on phytoavailability of Cd and Zn. Soil Sci Soc Am J 68:1885–1889

Khoshgoftarmanesh AH, Shariatmadari H, Karimian N, Kalbasi M, van der Seatm Z (2006) Cadmium and zinc in saline soil solutions and their concentrations in wheat. Soil Sci Soc Am J 70:582–589

Kinniburgh DG, van Riemsdijk WH, Koopal LK, Borkovec M, Benedetti MF, Avena MJ (1999) Ion binding to natural organic matter: competition, heterogeneity, stoichiometry and thermodynamic consistency. Colloid Surface A151:147–166

Krishnamurti GSR, McArthur DFE, Wang MK, Kozak LM, Huang PM (2005) Biogeochemistry of soil cadmium and the impact on terrestrial food chain contamination. In: Huang PM, Gobran GR (eds) Biogeochemistry of trace elements in the rhizosphere. Elsevier, Amsterdam, pp 197–257

Lee S, An G (2009) Over-expression of OsIRT1 leads to increased iron and zinc accumulation in rice. Plant Cell Environ 32:408–416

Li YM, Chaney L, Schneiter AA (1994) Effect of soil chloride level on cadmium concentration in sunflower kernels. Plant Soil 167:275–280

López-Chuken UJ, Young SD (2005) Plant screening of halophyte species for cadmium phytoremediation. Z Naturforsch 60:236–243

López-Chuken UJ, López-Domínguez U, Parra-Saldivar R, Moreno-Jimánez E, Hinojosa-Reyes L, Guzmán-Mar JL, Olivares-Sáenz E (2012) Implications of chloride-enhanced cadmium uptake in saline agriculture: modeling cadmium uptake by maize and tobacco. Int J Environ Sci Technol 9:69–77

Lores EM, Pennock JR (1998) The effect of salinity on binding of Cd, Cr, Cu and Zn to dissolved organic matter. Chemosphere 37:861–874

McLaughlin MJ, Singh BR (1999) Cadmium in soils and plants. Kluwer Academic Publishers, Dordrecht\The Netherlands

McLaughlin MJ, Palmer LT, Tiller KG, Beech TA, Smart MK (1994) Increased soil salinity causes elevated cadmium concentrations in field grown potato tubers. J Environ Qual 23:1013–1018

McLaughlin MJ, Tiller KG, Smart MK (1997) Speciation of cadmium in soil solutions of saline/sodic soils and relationships with cadmium concentrations in potato tubers (*Solanum tuberosum* L). Aust J Soil Res 35:183–198

McLaughlin MJ, Andrews SJ, Smart MK, Smolders E (1998a) Effects of sulfate on cadmium uptake by Swiss chard I. Effects of complexation and calcium competition in nutrient solutions. Plant Soil 202:211–216

McLaughlin MJ, Lambrechts RM, Smolders E, Smart MK (1998b) Effects of sulfate on cadmium uptake by Swiss chard: II. Effects due to sulfate addition to soil. Plant Soil 202:217–222

Mühling KH, Läuchli A (2003) Interaction of NaCl and Cd stress on compartmentation pattern of cations, antioxidant enzymes and proteins in leaves of two wheat genotypes differing in salt tolerance. Plant Soil 253:219–231

Munns R, Tester M (2008) Mechanisms of salinity tolerance. Annu Rev Plant Biol 59:651–681

Nakanishi H, Ogawa I, Ishimaru Y, Mori S, Nishizawa NK (2006) Iron deficiency enhances cadmium uptake and translocation mediated by the Fe2+ transporters OsIRT1 and OsIRT2 in rice. Soil Sci Plant Nutr 52:464–469

Norvell WA, Wu J, Hopkins DG, Welch RM (2000) Association of cadmium in durum wheat grain with soil chloride and chelateextractable soil cadmium. Soil Sci Soc Am J 64:2162–2168

Ondrasek G, Rengel Z (2012) The role of soil organic matter in trace elements bioavailability and toxicity. In: Ahmad P, Prasad MNV (eds) Abiotic stress responses in plants: metabolism productivity and sustainability. Springer, New York, pp 403–423

Ondrasek G, Romic D, Rengel Z, Romic M, Zovko M (2009a) Cadmium accumulation by muskmelon under salt stress in contaminated organic soil. Sci Tot Environ 407:2175–2182

Ondrasek G, Rengel Z, Romic D, Poljak M, Romic M (2009b) Accumulation of non/essential elements in radish plants grown in salt-affected and cadmium-contaminated environment. Cereal Res Commun 37(1):9–12

Ondrasek G, Rengel Z, Veres S (2011) Soil salinisation and salt stress in crop production. In: Shanker AK and Venkateswarlu B (eds.) Abiotic stress in plants: mechanisms and adaptations.

In Tech, Rijeka, pp 171–190. Available from http://www.intechopen.com/articles/show/title/soil-salinisation-and-salt-stress-in-crop-production. Accessed on 20 Dec, 2012

Ondrasek G, Rengel Z, Romic D, Savic R (2012) Salinity decreases dissolved organic carbon in the rhizosphere and increases trace elements phyto-accumulation. Eur J Soil Sci. doi:10.1111/j.1365-2389.2012.01463.x

Oporto C, Vandecasteele C, Smolders E (2007) Elevated cadmium concentrations in potato tubers due to irrigation with river water contaminated by mining in Potosí, Bolivia. J Environ Qual 36:1181–1186

Oporto C, Smolders E, Degryse F, Verheyen L, Vandecasteele C (2009) DGT-measured fluxes explain the chloride-enhanced cadmium uptake by plants at low but not at high Cd supply. Plant Soil 318:127–135

Ozkutlu F, Ozturk L, Erdem H, McLaughlin MJ, Cakmak I (2007) Leaf-applied sodium chloride promotes cadmium accumulation in durum wheat grain. Plant Soil 290:323–331

Pinto AP, Mota AM, de Varennes A, Pinto FC (2004) Influence of organic matter on the uptake of cadmium, zinc, copper and iron by sorghum plants. Sci Tot Environ 326:239–247

Popelka JC, Schubert S, Schulz R, Hansen AP (1996) Cadmium uptake and translocation during reproductive development of peanut (*Arachis hypogoea* L.). Angew Bot 70:140–143

Rashad M, Dultz S (2007) Decision factors of clay dispersion in alluvial oils of the Nile River Delta – a study on surface charge properties. American-Eurasian J Agric Environ Sci 2(3):213–219

Reid RJ, Dunbar KR, McLaughlin MJ (2003) Cadmium loading into potato tubers: the roles of the periderm, xylem and phloem. Plant Cell Environ 26:201–206

Rengel Z (2002) Role of pH in availability of ions in soil. In: Rengel Z (ed) Handbook of plant growth. pH as the master variable. Marcel Dekker, Inc, New York, pp 323–350

Rhoades JD, Chanduvi F, Lesch S (1999) Soil salinity assessment. Methods and interpretation of electrical conductivity measurements. FAO Irrig and Drain Paper 57, Rome, p 165

Romani AM, Sabater S, Munoz I (2011) The physical framework and historic human influences in the Ebro River. In: Barcelo D, Petrovic M (eds) The Ebro River basin, Hdb. Env. Chem. 13. Springer, New York, pp 1–20

Romic D, Ondrasek G, Romic M, Borosic J, Vranjes M, Petosic D (2008) Salinity and irrigation method affect crop yield and soil quality in watermelon (*Citrullus lanatus* L.) Growing. Irrig Drain 57:463–469

Romic D, Romic M, Rengel Z, Zovko M, Bakic H, Ondrasek G (2012) Trace metals in the coastal soils developed from estuarine floodplain sediments in the Croatian Mediterranean region. Environ Geochem Health 34:399–416. doi:10.1007/s10653-012-9449-z

Salt DE, Rauser WE (1995) MgATP-dependent transport of phytochelatins across the tonoplast of oat roots. Plant Physiol 107:1293–1301

Salt DE, Wagner GJ (1993) Cadmium transport across tonoplast of vesicles from oat roots. Evidence for a Cd2+/H+antiport activity. J Biol Chem 268:12297–12302

Salt DE, Prince RC, Pickering IJ, Raskin I (1995) Mechanisms of cadmium mobility and accumulation in Indian Mustard. Plant Physiol 109:1427–1433

Sanitá di Toppi L, Gabbrielli R (1999) Response to cadmium in higher plants. Environ Exp Bot 41:105–130

Shuman LM, Dudka S, Das K (2002) Cadmium forms and plant availability in compost-amended soil. Commun Soil Sci Plant Anal 33:737–748

Singh BR (1994) Trace element availability to plants in agricultural soils, with special emphasis on fertilizer inputs. Environ Rev 2:133–146

Smolders E, McLaughlin MJ (1996a) Effect of Cl and Cd uptake by Swiss chard in nutrient solution. Plant Soil 179:57–64

Smolders E, McLaughlin MJ (1996b) Chloride increases Cd uptake in Swiss chard in a resin-buffered nutrient solution. Soil Sci Soc Am J 60:1443–1447

Smolders E, Lambregts RM, McLaughlin MJ, Tiller KG (1998) Effect of soil solution chloride on cadmium availability to Swiss chard. J Environ Qual 27:426–431

Tanaka K, Fujimaki S, Fujiwara T, Yoneyama T, Hayashi H (2007) Quantitative estimation of the contribution of the phloem in cadmium transport to grains in rice plants (*Oryza sativa* L.). Soil Sci Plant Nutr 53:72–77

Tester M, Davenport R (2003) Na^+ tolerance and Na^+ transport in higher plants. Ann Bot-London 91(5):503–527

Tipping E (2005) Modelling Al competition for heavy metal binding by dissolved organic matter in soil and surface waters of acid and neutral pH. Geoderma 127:293–304

Vögeli-Lange R, Wagner GJ (1990) Subcellular localization of cadmium and cadmium-binding peptides in tobacco leaves. Implication of a transport function for cadmium-binding peptides. Plant Physiol 92:1086–1093

Weggler K, McLaughlin MJ, Graham RD (2004) Effect of chloride in soil solution on the plant availability of biosolid-borne cadmium. J Enviro Qual 33:496–504

Weggler-Beaton K, McLaughlin MJ, Graham RD (2000) Salinity increases cadmium uptake by wheat and Swiss chard from soil amended with biosolids. Aust J Soil Res 38:37–45

Welch RM (1995) Micronutrient nutrition of plants. Crit Rev Plant Sci 14:49–82

Weng L, Temminghoff EJM, van Riemsdijk WH (2001) Contribution of individual sorbents to the control of heavy metal activity in sandy soil. Environ Sci Techn 35:4436–4443

World Health Organisation (WHO) (1992) Cadmium. Environmental health criteria 134. World Health Organization, Geneva

World Meteorological Organization (WMO) (2009) Statement on the status of the global climate in 2009, Press No 869. World Meteorological Organization, Geneva

Yada S, Oda H, Kawasaki A (2004) Uptake and transport of Cd supplied at early growth stage in hydroponically cultured soybean plants. Biomed Res Trace Elements 15:292–294

Yoneyama T, Gosho T, Kato M, Goto S, Hayashi H (2010) Xylem and phloem transport of Cd, Zn, and Fe into the grains of rice plants (*Oryza sativa* L.) grown in continuously flooded Cd-contaminated soil. Soil Sci Plant Nutr 56:445–453

Zhang H, Davison W (1995) Performance-characteristics of diffusion gradients in thin-films for the in-situ measurement of trace-metals in aqueous-solution. Anal Chem 67:3391–3400

Zhang H, Davison W (2001) A new method to measure effective soil solution concentration predicts copper availability to plants. Environ Sci Technol 35:2602–2607

Zhang H, Han B, Wang T, Chen SX, Li HY, Zhang YH, Dai SJ (2012) Mechanisms of plant salt response: insights from proteomics. J Proteome Res 11(1):49–67

Zhu JK (2001) Plant salt tolerance. Trends Plant Sci 6(2):66–71

Zovko M, Romic M (2011) Soil contamination by trace metals: geochemical behaviour as an element of risk assessment. In: Ahmad DI and Ahmad DM (eds.) Earth and environmental sciences. InTech Rijeka, pp 437–456. Available from http://www.intechopen.com/articles/show/title/soil-contamination-by-trace-metals-geochemical-behaviour-as-an-element-of-risk-assessment. Accessed on 20 Dec, 2012

Chapter 18
Plant Tissue Culture: A Useful Measure for the Screening of Salt Tolerance in Plants

Mukhtar A. Bhat, Sumiya Jamshieed, Abdul Mujib, M.M. Azooz, Mahmooduzzafar, Junaid Aslam, and Parvaiz Ahmad

Abstract Soil salinity is one of the most important problems worldwide, which has decreased crop production to a great extent. The major or deleterious effects of salinity on plant growth and development are associated with low osmotic potential of soil solution, nutritional imbalance etc. Consequently these can ultimately lead to plant death because of growth arrest and molecular damage. Salt stress affects all the major processes such as photosynthesis, protein synthesis, lipid metabolism etc. The use of plant cell and tissue culture offers a means to focus on those physiological and biochemical processes inherent to cell which contribute to the adaptation to salt stress. The response depends on the species and the genotype, the length and the severity of the salt stress, the age and stage of development, the organ and cell type. In this article, various in vitro strategies have been made for salt tolerance. Effect of salinity on biochemical and antioxidant properties of plants have also been highlighted. The chapter also covers the role of genetic transformation for the development of salt tolerance in plants.

M.A. Bhat
Department of Botany, Degree College Beerwah, University of Kashmir, Srinagar, India

S. Jamshieed • A. Mujib • Mahmooduzzafar
Department of Botany, Jamia Hamdard (Hamdard University), New Delhi, India

M.M. Azooz
Department of Botany, Faculty of Science, South Valley University, Qena, Egypt

Department of Biological Sciences, Faculty of Science, King Faisal University, Al Hassa, Saudi Arabia

J. Aslam
Department of Biotechnology, Jamia Hamdard (Hamdard University), New Delhi, India

P. Ahmad (✉)
Department of Botany, A.S. College, University of Kashmir, Srinagar, India
e-mail: parvaizbot@yahoo.com

P. Ahmad et al. (eds.), *Salt Stress in Plants: Signalling, Omics and Adaptations*, DOI 10.1007/978-1-4614-6108-1_18,

Keywords Plant tissue culture • Salt tolerance • Photosynthesis • Protein synthesis • Genotypes

Crop production is restricted due to increasing area of saline soil through out the world by naturally saline soil. Salinity inhibits growth of plants by affecting both water absorption and biochemical processes, such as nitrogen assimilation and protein biosynthesis (Dubey 1994). Under saline conditions, the plants fail to maintain the required balance of organic constituents leading to suppressed growth and yield. In developing countries, the limited supply of good quality water in many arid and semi-arid regions necessitates the use of saline water where available for crop production. This, in turn, requires the screening of crop plant varieties for their tolerance to salinity. Screening for improved salt tolerance is difficult in the field because of lateral, vertical and temporal variability in salt distribution within the soil profile. In addition, plant salt tolerance varies with ontogeny, the growth parameter measured and environmental factors. Salinity has been a threat to agriculture in various parts of the world for over 3,000 years and this threat is growing enormously (Ahmad et al. 2008, 2010, 2011, 2012). As the world population continues to increase, more food needs to be grown to feed the people. This can be achieved by an increase in cultivated land and by an increase in crop productivity per area.

Tissue culture techniques have been applied to the plant species in an attempt to produce new clones and cultivars with improved characteristics. In this respect, number of researchers have suggested that cultured tissues and cells may prove useful both in selections of the salt-tolerant plants and in studies of the physiological basis for salinity tolerance (Chen et al. 1980; Umiel et al. 1980). Easy manipulation of salt mixture concentrations in media, especially in suspension culture, also permits uniform and direct treatment on cell growth with a given salt stress level. Most efforts at selecting salt resistant cell lines have involved direct selection for capacity to grow on inhibitory levels of NaCl (Dix 1985; Tal 1983; Bhat et al. 2008; Benderradji et al. 2012).

Plant Tissue Culture

Plant tissue culture is a collection of techniques used to maintain or grow plant cells, tissues or organs under sterile conditions on a nutrient culture medium of known composition. Plant tissue culture is widely used to produce clones of a plant in a method known as micropropagation. Plant tissue culture relies on the fact that many plant cells have the ability to regenerate a whole plant (totipotency) (Davey et al. 2005). In vitro cell and tissue- based system have tremendous potential in fundamental research and for commercial applications such as clonal propagation, genetic engineering and production of valuable metabolites (Neelakandan and Wang 2012).

Single cells and protoplasts, pieces of leaves or roots can often be used to generate a new plant on culture media given the required nutrients and plant hormones (Wang et al. 2011). Different techniques in plant tissue culture may offer certain advantages over traditional methods of propagation, including:

1. The production of exact copies of plants that produce particularly good flowers, fruits, or have other desirable traits.
2. Quick production of mature plants.
3. The production of multiples of plants in the absence of seeds or necessary pollinators.
4. The regeneration of whole plants from plant cells that have been genetically modified.
5. The production of plants in sterile containers that allows them to be moved with greatly reduced chances of transmitting diseases, pests and pathogens.
6. To clean particular plants with viral and other infections and to quickly multiply these plants as cleaned stock for agriculture.

Development of tissue culture is closely linked to improvement in techniques of protoplast, cell, tissue and organ culture, followed by the success achieved in regenerating whole plants from cultured plant materials (Mustafa et al. 2011). The technique has advanced rapidly over the years due to extensive investigations into problems related to basic and applied aspects of plants. Knowledge of tissue culture has contributed greatly to the understanding of the factors responsible for growth, metabolism, differentiation and morphogenesis of plant cells (Kärkönen et al. 2011). Plant tissue culture is of great interest to molecular biologists, plant breeders and industrialists. The methods of tissue culture have been used as an important aid to conventional methods of plant improvement (Germanà 2011; Bhatti and Jha 2010). They have been used for the production of genetically modified superior clones, ex-situ conservation of germplasm, pathogen free plants as well as in the synthesis of many important secondary compounds (including pharmaceuticals) has been very significant (Loyola-Vargas and Ochoa-Alejo 2012; Sharry et al. 2011). The advantages offered by tissue culture in agriculture and general plant biotechnology have well been witnessed by many research labs and industries.

Tissue Culture and Secondary Metabolites

Plant, cell, tissue and organ culture techniques have emerged as an inescapable tool with possibilities for complementing the conventional methods in plant breeding and also in the study of biosynthetic pathways and plant genetics (Hussain et al. 2012; Ono and Tian 2011). These techniques have proved successful and are now being used globally for the ex-situ conservation of the plants including medicinal herbs (Taheri et al. 2011; Gasparetto et al. 2012). The endeavour is to adopt the method to multiply the medicinal herbs and monitor their secondary metabolites.

Success achieved in enhancing the synthesis of secondary metabolites needs to be exploited commercially.

Plant cell cultures have advantages in metabolite production over intact plants due to:

- High rate of cell growth and biosynthesis in cultures initiated from a very small amount of plant material and the final product may be produced in a considerably short period of time.
- Continuous yield of metabolites.
- More effective mechanism of incorporating precursors into cells in suspension cultures.
- Recovering new routes of synthesis from deviant and mutant cell lines, which can lead to production of novel compounds.
- More economical for those plants which take long periods to achieve maturity.
- Some cell cultures have the capacity for biotransformation of specific substances to more valuable products by means of single or multiple step enzyme activity.

Examples of some economically important plant derived drugs and intermediates that are still obtained commercially from plant sources are presented in Table 18.1. Medicinal plant biotechnology has grown from cell technology, specifically plant tissue culture. The powerful techniques of plant tissue culture, and recombinant DNA and bioprocessing technologies etc., coupled with sophisticated analytical tools such as NMR, HPLC, GC-MS, LC-MS etc., have offered mankind a great opportunity to exploit the totipotent, biosynthetic and biotransformation capabilities of plant cells under in vitro conditions. HPTLC analysis of *Nothapodytes foetida* cultures have shown the presence of a number of flavonoids with antimicrobial activity including camptothecin (Namdeo et al. 2010). The scope for in vitro germplasm preservation and large-scale production of plant secondary metabolites has brightened due to these improved technologies (Volk 2010; Hussain et al. 2012). Tissue culture technique is becoming popular because of its well-known ability to enhance the content of secondary metabolites in plants (Satheesan et al. 2012). Plants yield a number of secondary products, but the yield depends upon many factors like the stage of development, environmental conditions, and the plant part being analyzed (Chatterjee et al. 2003; Ghasemzadeh and Jaafar 2011). Plant tissue culture can help to enhance the yield of these active constituents via change in chemical mileu and hormone concentrations (Zare et al. 2011) Flavonoids showing antimicrobial activity have been isolated from callus cultures of *Tridax procumbens* (Jindal and Kumar 2011). Cakar et al. (2012) have reported antioxidative and antitumor properties in broccoli (*Brassica oleracea* var. italic).

Tissue Culture and Herbal Medicine

Medicinal plants are no new introduction in the modern times. They were known to ancient people who more than us depended on plants directly or indirectly for their day to day needs. Use of indigenous drugs from plant origin forms a major part of

Table 18.1 List of some economically important plant derived drugs and intermediates that are obtained commercially from plant sources

Drug/chemical	Plant source	Therapeutic use/action
Hormones (derived from diosgenin, lecogenin)	*Dioscorea spp.*	Oral contraceptives, other steroid hormones
Glycosides (digoxin, digitoxin, digitalin)	*Digitalis purpurea, D. lanata*	Cardiotonic
Atropine	*Atropa belladonna*	Anticholinergic
Hyoscyamine	*Hyoscyamus niger*	Analgesic, sedative, smooth muscle relaxant
Scopolamine	*Datura stromonium*	
Codeine, Morphine	*Papaver somniferum*	
Reserpine, Ajmalicine	*Rauwolfia serpentina*	Antihypertensive, psychotropic, vasodilator
Vincristine, Vinblastin	*Catharanthus roseus*	Anticancer
Physostigmine	*Physostigma venenosum*	Cholinergic
Pilocarpine	*Pilocarpus jaborandi*	Cholinergic, parasympathomimetic
Quinine, quinidine	*Cinchona spp.*	Antimalarial, antipyretic
Colchicine	*Colchicum autumnale*	Antigout, antitumor
Cocaine	*Erythroxylon coca*	Local anaesthetic
Tubocurarine	*Chondodendron tomentosum*	Skeletal muscle relaxant
Berberin	*Berberis vulgaris*	Bacillary dysentery
Camptothesin	*Camtotheia accuminata*	Anticancerous
Ephedrin	*Ephedra spp.*	Sympathomimetic, antihistamine
Podophyllotoxin	*Podophyllum peltatum*	Antitumor, anticancer
Curcumin	*Curcuma longa*	Chloretic
Taxol	*Taxus baccata*	Antitumor

complementary and traditional medicine. The world market for herbal medicine, including herbal products and raw materials has been estimated to have an annual growth rate between 5% and 15% and the total global herbal drug market is estimated to be US $ 5 trillion by the year 2050 (Narula et al. 2004). In India, the knowledge of herbal medicine is widespread- ranging from tribal folklore use to age old practices and closely guarded recipes handed down from generation to generation, to highly evolved systems of medicines like Ayurveda, Sidha and Unani. Herbal medicinal products may vary in composition and properties, unlike conventional pharmaceutical products, which are usually prepared from synthetic, chemically pure materials by means of reproducible manufacturing techniques and procedures (Nile et al. 2012).

With the advent of sulfa drugs and antibiotics these plants and traditional systems of medicine were relegated to the back seat. However, the toxicity and harmful side effects of synthetic drugs and antibiotics brought medicinal plants back to the forefront. There is global renaissance of research on medicinal plants (Calixto 2000). During the last three decades the globe has been swept with green waves. This has resulted in the emergence of a large number of health food shops.

The trend is increasing and seems to continue. Therefore, the plants that serve as raw materials in herbal drug formulations are now required in much larger quantities, strategies have to be developed for the mass propagation of several precious medicinal herbs (Lucchesini and Mensuali-Sodi 2010). The medicinal property of a plant is due to the presence of a particular secondary metabolite. Thus, along with mass multiplication, enhancement in yield of active constituents is also desirable. Since secondary metabolites are produced in low amounts, substantial quantity of raw plant material is required to extract only a few grams of the useful product (Croteau et al. 2000). Novel methods therefore, have to be adopted to enhance the biosynthesis of secondary products that would help to prevent further loss of such useful plants. Anticancerous activity has been reported from aqueous extracts of cultures of *Brassica javanica* on cervical and several other cancer cells (Gao et al. 2011). The results of this study imply that it may lead to the development of novel-anticancer drugs. The root of *Salvia miltiorrhiza* is an important herb in Chinese medicine used for the treatment of cardiovascular diseases (Yang et al. 2012). Diterpenoids tanshinons are the active constituent of this drug. Plant tissue culture is the major biotechnological processes for rapid production of tanshinones and other bioactive components in the herb (Yan et al. 2011). Various in vitro cultures of *S. miltiorrhiza* have been established, including cell suspension cultures, adventitious root, and hairy root cultures (Wu et al. 2007). Tanshinone production in cell and hairy root cultures has been dramatically enhanced using different strategies of tissue culture (Wang and Wu 2010). Chen et al. (2007) have investigated the effects of *Astragalus* polysaccharides on myocardial chymase in diabetic hamsters and found that APS can inhibit the local chymase-Ang II system in diabetic cardiomyopathy.

Tissue Culture and Abiotic Stress Tolerance

Abiotic stress is defined as the negative impact of non-living factors on the living organism in a specific environment. The non living variable must influence the environment beyond its normal range of variation to adversely affect the population performance or individual physiology of the organism in a significant way. Abiotic stress is the most harmful factor concerning the growth and productivity of crops worldwide (Gao and Ji-Ping 2007). Research has also shown that abiotic stressors are at their most harmful effects when they occur together i.e., in combinations of abiotic stress factors (Mittler and Ron 2006). This facilitation will not go so far as to protect an entire species. Plants are extremely sensitive to such changes, and don't generally adapt quickly (Lane and Jarvis 2007). Plants also adapt very differently from one another, even from the same area. When a group of different plant species was prompted by a variety of different stress signals, such as drought or cold, each plant responded uniquely. Hardly any of the responses were similar, even though the plants had become accustomed to exactly the same home environment (Mittler and Ron 2006).

Plants are sessile and prone to multiple stresses in the changing environmental conditions. A plant's first line of defense against abiotic stress is in its roots. If the soil holding the plant is healthy and biologically diverse, the plant will have higher chance of surviving stressful conditions (Brussaard et al. 2007). Facilitation, or the positive interactions between different species of plants, is an intricate web of association in a natural environment. It is how plants work together. In areas of high stress, the level of facilitation is especially high as well. This could possibly be because the plants need a stronger network to survive in a harsher environment, so their interactions between species, such as cross-pollination or mutualistic actions, become more common to cope with the severity of their habitat (Maestre et al. 2007). Of the several strategies adopted by plants to counteract the adverse effects of abiotic stress, phytohormones provide signals to allow plants to survive under stress conditions. They are one of the key systems integrating metabolic and developmental events in the whole plant and the response of plants to the external factors and are essential for many processes throughout the life of a plant and influence the yield and quality of crops.

Nitric oxide and hydrogen peroxide function as signaling molecules in plants under abiotic stresses. Callus from *Populus euphratica*, which show salt tolerance, were used to study the interaction of Nitric oxide (NO) and hydrogen peroxide (H_2O_2) in plant adaptation to salt resistance (Zhang et al. 2007). The results indicate that NO and H_2O_2 served as intermediate molecules in inducing salt resistance in the calluses from *P. euphratica* under salt stress by increasing the K/Na ratio (Zhang et al. 2007).

Evidences in literature have shown that plants treated with hydrogen peroxide exogenously acquire abiotic stress tolerance potential, without substantial disturbances in the endogenous hydrogen peroxide pool. Low temperature is one of the major abiotic stresses limiting the productivity and geographical distribution of many important crops. Schützendübel and Polle (2002) enhanced endogenous hydrogen peroxide content of *Nicotiana tabaccum* plants by the constitutive expression of glucose oxidase gene and studied the cold tolerance level. Most of the transgenic lines showed tolerance to cold treatment (Zhang et al. 2007). To identify proteins associated with chilling stress in *Nicotiana tabacum* cv. bright yellow cell suspension culture, electrophoresis was carried out to compare proteins from samples of treated with or without chilling treatment and a protein specifically more abundant in treated sample was identified and designated as NtLEA7-3 (Gai et al. 2011). All of these, taken together, suggest that NtLEA7-3 is worthwhile to elucidate the contribution of the proteins to the tolerance mechanism to chilling stress, and can be considered as a potential target for crop genetic improvement in the future (Gai et al. 2011).

Other Uses of Tissue Culture

Plant tissue culture involves the culture of all types of plant cells, tissues and organs under aseptic conditions. This definition also extends to the culture of excised embryos and to protoplast culture. Plant tissue culture now has direct commercial applications as well as value in basic research into cell biology, genetics and biochemistry. The techniques include culture of cells, anthers, ovules and embryos on experimental to industrial scales, protoplast isolation and fusion, cell selection and meristem and bud culture. Various applications of tissue culture include:

1. Micropropagation using meristem and shoot culture to produce large numbers of identical individuals.
2. Screening programmes of cells rather than plants for advantageous characters.
3. Large scale growth of plant cells in liquid culture as a source of secondary products.
4. Crossing distantly related species by protoplast fusion and regeneration of the novel hybrid.
5. Production of dihaploid plants from haploid cultures to achieve homozygous lines more rapidly in breeding programmes.
6. As a tissue for transformation, followed by either short-term testing of genetic constructs or regeneration of transgenic plants.
7. Removal of viruses by propagation from meristematic tissues.

The use of tissue culture in plant breeding is not new. Plant tissue culture techniques are used by breeders to generate new varieties of crops (Gai et al. 2011). The in vitro culture of crops has been used routinely by plant breeders for many decades and has resulted in numerous varities that are grown and eaten as food all over the world (Gai et al. 2011) these varities have worked well in farmers' field, and have not caused any health problems. The amount of genetic variation coming from tissue culture is modest and manageable, and mutations are an integral part of crop breeding (Gai et al. 2011). They occur all the time, and can be either positive or detrimental in terms of the value of the crop. It is a plant breeder's job to choose for favourable combinations of traits and eliminate deleterious mutations. The phenomenon of somaclonal variation is a valuable tool in the plant breeder's kit (Kaeppler 1998, 2000; Larkin 2004; Dan et al. 2010).

There are many varities of conventionally bred plants that have been produced using tissue culture. Plants produced through tissue culture have frequently been introduced into our food supply in the past and don't cause any health problems. Commercial cultivars with improvement through tissue culture are documented in barley, wheat, potato, blackberry, flax, celery, rice (Larkin 2004). The technique has been used to make celery and tomato varieties that are resistant to fungal attack (Heath-Pagliuso and Rappaport 1990). It has also been used to breed linseed flax, and commercial canola varities used to produce vegetable oils (Obert et al. 2009). When plant tissue culture is used in making transgenic plants the appearance of mutations is minimized by optimizing the plantlet regeneration method and rejecting any regenerated plantlets that have unwanted changes to them (Lakshmanan 2006). The plants derived from tissue culture are commonly used in commercial conventional breeding operations is the best evidence that the technique doesn't cause problems.

Salinity

Environmental stresses such as, low and high temperatures, drought, alkalinity, salinity etc. are potentially harmful to the plants (Breusegem et al. 2001). Soil salinity problem is as old as human civilization and is probably a cause of the breakdown of the ancient Sumerian civilization (Jacobson and Adams1958). Salt stress in soil or water is one of the major stresses especially in arid and semiarid regions which severely limit plant growth and productivity (Shannon 1998; Allakhverdiev et al. 2000). Mostly earth water contains about 30 g of NaCl (Flowers 2004; Bhat et al. 2008; Benderradji et al. 2012) and in plants, salt concentration is usually higher in root zone than that of irrigation water. Salts are concentrated due to evaporation and plant transpiration, which selectivity removes water, leaving salts in soil. Salinity of the soil is characterized by toxic levels of chlorides and sulphates of sodium. The problem of soil salinity is increasing owing to:

- Use of poor quality water for irrigation.
- Improper drainage in canal irrigated wetland agro-ecosystems.
- Entry of sea water during cyclones in coastal areas and
- Salt accumulation in root zone in arid and semiarid regions due to high evaporative demand and insufficient leaching of ions as the fall is inadequate.

The salinity causes atmospheric pollution by inducing increased rate of degradation of agricultural lands and by destroying key water resources ultimately causing alteration in the natural (biological and physical) diversity within the agricultural area of the world. NaCl is the main salt responsible for world salinity. Plants are classified on the basis of growth in salt medium as glycophytes or halophytes (Sairam and Tyagi 2004). The deleterious effects of salinity on plant growth are associated with low osmotic potential of soil solution, nutritional imbalance and specific ion effect (Ashraf 1994; Marschner 1995). The consequence of all these can lead to plant death as a result of growth arrest and molecular damage.

Properties of Na^+ and Chlorine (Cl^-)

Sodium Na^+ is a soft sliver white metal with melting point 97.9°C and boiling point 828°C. It is extremely reactive metal and occurs in combined state as sodium chloride (NaCl) (rock salt), sodium carbonate (Na_2CO_3), sodium aluminate fluride (Na_3AlF_6), sodium nitrate $NaNO_3$ sodium sesquicarbonate ($Na_2CO_3.2H_2O$), borax ($Na_2B_4O_7.10H_2O$), sodium sulphate ($Na_2SO_4.10H_2O$, glaubers salt) and sodium aluminum silicate ($NaAlS_3O_8$, feldspar). It is sufficiently malleable, ductile and good conductor of heat and electricity.

For higher plants, Cl is an essential micronutrient and a minimal requirement of 1 g/kg dry weight has been suggested for crop growth (Marschner 1995). Plants can generally obtain this quantity through rainfall. However, high Cl concentration in tissues can be toxic to the plant and may restrict the agriculture of saline regions (Xu et al. 2000). Plants acquire most of their Cl from the soil as Cl^- anion to support plant growth.

Chloride Distribution Within the Plant

Ability of plants to accumulate Cl^- varied considerably (Greenway and Munns 1980). Halophytes generate turgor by accumulating high Cl^- concentrations in plant tissue ($[Cl^-]_{tissue}$; 340–475 mM), and the fluctuations in osmotic pressure during their growth, or in response to environmental changes are usually affected by changes in NaCl concentration. In contrast, glycophytes growing in natural habitats have much lower $[Cl^-]_{tissue}$ (7–70 mM), and Cl^- is generally only a minor component of their cell-sap osmotic pressure.

Chloride Transport into the Plant

Plants acquire most of their Cl from the soil as the Cl^- anion. To support plant growth, Cl^- is loaded into the xylem and thereby delivered to the shoot. In principle, there are two pathways termed as the symplastic (cytoplasmic) and the apoplastic (extracellular) pathways. Anions, following the symplastic pathway enter the root cells across their plasma membranes, are transferred from cell to cell through plasmodesmatal connections and are exported across the plasma membrane which could control the selectivity and magnitude of Cl^- fluxes to the shoot. Alternatively, anions may travel extracellularly through the cell walls and water-filled spaces to reach the stele (White and Broadley 2001). During the growth of a plant, Cl^- is translocated from the root to shoot via xylem and is redistributed between tissues via the phloem.

Salt Transport Inside Plants

Transpirational flux, which results the movement of salt into roots and shoots, is required to maintain water status of the plant (Flowers and Yeo 1992). Toxic levels of ion accumulation can result from unregulated transpiration in the aerial parts of the plant. However, because of the water potential difference between the atmosphere and the leaf cells, and because of the need for carbon fixation, this is an untenable long-term strategy of salt tolerance (Munns and Termaat 1986).

Plants regulate ion movements into tissues so as to protect the actively growing and metabolizing cells (Munns 1993). One mode by which plant controls salt flux to the shoot is the entry of the ions into the xylem stream. Still debated is the extent to which the symplastic ion transport through the epidermal and cortical cells contributes to a reduction in Na^+ that is delivered to the xylem (Flower and Yeo 1992). However, at the endodermis, radial movement of solutes must be via a symplastic pathway, as the casparian strips constitute a physical barrier to the apoplastic transport.

It has been observed that there is an accumulation of large quantities of ions in mature and old leaves under salt stress (Flowers and Yeo 1992; Munns 1993). The meristematic cells, which are not directly connected to the vasculature, are less exposed to ion delivered through transpirational stream, and the smaller vacuolar space of these cells is not conductive to ion storage. Due to this fact the solute content of those tissues containing cells with little vacuolation (e.g. meristematic region) is predominated by the organic osmolytes while that of tissues with highly vacuolated cells by ions (Binzel et al. 1988).

In many plant species, salt sensitivity is associated with the accumulation of sodium (Na^+) in photosynthetic tissues. Na^+ uptake to leaves involves a series of transport steps and for which only few candidates' genes have been so far characterized. Benderradji et al. (2011) studied at physiological and molecular level of two Algerian bread wheat varieties (*Triticum aestivum* L.), Mahon-Demias (MD) a salt sensitive and Hidhab (HD) a salt tolerant varieties. The comparative analysis of Na^+ transport revealed two major differences between the two genotype (1) a lower rate of transfer from the root to the shoot (xylem loading) in the salt tolerant genotype, and (2) A higher capacity of the leaf sheath in the tolerant genotype to extract and sequester Na^+ as it entered the leaf. In addition, an enhanced uptake of K^+ in leaves of Hidhab compared to Mahon-Demias resulting in a higher K^+/Na^+ ratio in leaf blades and hence improving cellular homeostasis in the tolerant variety. Moreover, correlation was observed between the expression patterns of the transcripts encoding the plasma membrane Na^+/H^+ antiporter (TaSOS1), two members of the HKT transporters family (HKT1;5 and HKT2;2) and the Na+fluxes from roots to leaves which help to understand the differential salt stress tolerance between Hidhab and Mahon-Demias wheat varieties (Benderradji et al. 2011).

Effect of Salinity on Plant Growth

Plant growth and yield is affected by salt stress in glycophytes (Bernstein 1975; Greenway and Munns 1980). A variation occurs in number, size, and biomass etc. of the plant. Grapevines were grown under varied salinity stresses with 25, 50, 75, 125 mM NaCl (Downtown 1977). He noted that shoot growth was maintained at a rate similar to the control for the first week of experiment at the 25 mM NaCl concentration while as growth rate of treated plants declined from days 12 onwards. Fresh weight of roots and shoots were more severely affected than that of the leaves. Shoot: root ratio was increased by salinity treated plant and root growth was more adversely affected compared to shoot growth. Shoot length is one of the most reliable response indicators of salt stress in a wide range of tomato genotypes (Cruz and

Cuartero 1990). The reduction in shoot dry weight is due to decline in vegetative growth and is most widely used index in studies on salt tolerance in tomato. Bolarin et al. (1993) also noted the significant reduction in fresh and dry weights of tomato shoots under salt stress.

Peter et al. (1986) cultivated *Hibiscus cannabinus* in 37 and 75 mM NaCl and noted reduction in dry weight at 75 mM, leaf area was affected at both 37 and 75 mM. A significant decline in leaf area but no decrease in the net carbon assimilation rate was observed under salt stress. Plants grown in 37 mM showed increased net photosynthesis and unchanged with 75 mM NaCl. Zidan et al. (1990) reported that high levels of salinity (100 mm NaCl) was accompanied by reduction in the length of the root tip elongation zone, in the length of the epidermal cells and in the apparent rate of cell production.

A considerable decrease in growth of *Phaseolus aconitifolius* in terms of both vegetative and reproductive plant parts at high salinity stress (100 mM) was studied by Kulkarni and Karadge (1991). With increasing Na^+ supply, dry matter production was decreased in Lettuce (Hamid and Talibuddeen 1976).

Roots exposed to 30 or 100 mM of NaCl showed decreased dry weight by 40% in 30 mM NaCl and by 93% in 100 mM (Gersani et al. 1993). The dry weight as well as fresh weight production was enhanced in rice cultivars at 180 mol m^{-3} NaCl while the same was reduced by 17% in other cultivars at the same level of NaCl (Heuer and Plaut 1989).

Biochemical Response Under Salt Stress

Pigments

The green pigments of the plant (chlorophyll) are present in the mesophyll cells specialized in light absorption. Salinity affects their contents leading to decreased absorption of light required for optimum photosynthesis and hence indirectly impairing photosynthesis. Ziska et al. (1990), Nieves et al. (1991) reported reduction in chl. 'a', chl. 'b' as well as total chl. by salinity to be determined by the absolute concentration of chloride and/or sodium in the leaves. Rice cultivars grown at salinity levels ranging from 10 mmhos/cm to 15 mmhos/cm, showed differential response in terms of leaf chlorophyll contents (Pandey and Srivastava 1987). The chlorophyll content was reduced by about 8–33% in some varieties, while increased (5–15%) in others.

The chl. 'a' and chl. 'b' contents increased in salt tolerant varieties but decreased in the susceptible varieties. Changes were also noted in the carotenoids under saline conditions. Carotene content increased and neoxanthin content decreased in the resistant varieties, while the neoxanthin increased, but the carotene content did not change in susceptible varieties (Sinel'nickova et al. 1988).

Protein Content

Proteins that accumulate in plants grown under saline conditions may provide a storage form of nitrogen that is re-utilized when stress is over (Singh et al. 1987) and may play a role in osmotic adjustment. Proteins may be synthesized *de novo* in response to salt stress or may be present constitutively at low concentration and increase when plants are exposed to salt stress (Pareek et al. 1997). A higher content of soluble proteins has been observed in salt tolerant barley than in salt sensitive cultivars (Hurkman et al. 1989), sunflower (Ashraf and Tufail 1995), finger millet (Uma et al. 1995), and in rice (Pareek et al. 1997; Rains 1989; Lutts et al. 1996).

In higher plants, osmotic stress induces several proteins in vegetative tissues, which are related to late-embryogenesis-abundant (LEA) proteins. The correlation between LEA protein accumulation in vegetative tissues and stress tolerance in various plant species indicates its protective role under dehydration stress (Ingram and Bartels 1996). Engineered rice plants over expressing a barley *LEA* gene, *HVA1*, under control of the rice actin 1 promoter showed better stress tolerance under 200 mM NaCl and drought stress than did the wild type (Xu et al. 1996).

Carbohydrates

Carbohydrates such as sugars (glucose, fructose, sucrose, fructans etc.) and starch accumulate under salt stress (Parida et al. 2002). Their major functions are osmoprotection, osmotic adjustment, carbon storage, and radical scavenging. Salt stress increases reducing sugars (glucose, fructose), sucrose, and fructans in a number of plants (Singh et al. 2000). In *Vicia faba*, salinity decreases soluble and hydrolysable sugars (Gadallah 1999). Sugar content increases in some genotypes of rice but also decreases in some genotypes (Alamgir and Ali 1999). Under salinity, the starch content in roots of rice plant declines and remains unchanged in shoots. A decrease in starch content and an increase in both reducing and nonreducing sugar have been reported in leaves of *Bruguiera parviflora* (Parida et al. 2002). In leaves of tomato the contents of soluble sugars and total saccharides are significantly increased, but the starch content is not affected significantly by NaCl treatment (Khavarinejad and Mostofi 1998). It has also been reported that polyphenol level increases in leaves of *B. parviflora* by salt stress (Parida et al. 2002).

Trehalose, a disaccharide, accumulates in many organisms under various abiotic stresses and has been reported to be both an osmolyte and an osmoprotectant (Hounsa et al. 1998). It protects membranes and proteins in cells exposed to stresses that cause water deficit (Goddijin and Van Dun 1999) and reduces aggregation of denatured proteins (Singer and Lindquist 1998). Recently, Yamada et al. (2003) have reported that trehalose has a suppressive effect on apoptotic cell death. There is now conclusive evidence that trehalose is present in trace amounts in vascular plants, including major crops, but the actual role of this osmolyte in metabolism of these plants is still unclear. In alfalfa under salt stress, trehalose accumulated in only very low concentrations in roots and nodules and thus the role of this sugar in osmoregulation was thought to be negligible (Fougere et al. 1991).

Amino Acids

Amino acids have been reported to accumulate in higher plants under salinity stress (Mansour 2000). The important amino acids include alanine, arginine, glycine, serine, leucine, and valine, together with the imino acid, proline, and the non-protein amino acids, citrulline and ornithine (Mansour 2000). Amides such as glutamine and asparagine have also been reported to accumulate in plants subject to salt stress (Mansour 2000). Total free amino acids in the leaves are reported to be higher in salt tolerant than in salt sensitive lines of sunflower (Ashraf and Tufail 1995), safflower (Ashraf and Fatima 1995), *Eruca sativa* (Ashraf 1994) and *Lens culinaris* (Hurkman et al. 1991).

Proline

Proline an imino acid (a secondary amino acid), is an uncharged organic molecule that does not affect protein function, protein synthesis and accumulation in the cytoplasm, may have contribution (a) towards the osmotic balance when osmolytes are lower in cytoplasm than in the vacuoles (Stewart and Lee 1974), and (b) towards the protection of enzymes in the presence of high electrolytes in the cytoplasm (Polard and Wyn Jones 1979).

Proline, which occurs widely in higher plants, accumulates in larger amounts than other amino acids in salt stressed plants (Ashraf 1994; Abraham et al. 2003). Proline accumulation is one of the common characteristics in many monocotyledons under saline conditions. Proline regulates the accumulation of usable N, is osmotically very active (Ashraf 1994), contributes to membrane stability and mitigates the effect of NaCl on cell membrane disruption. Even at supra-optimal levels, proline does not suppress enzyme activity. Maggaio et al. (2002) are of the view that proline may act as a signalling/regulatory molecule, able to activate multiple responses that are component of the adaptation process. Proline concentration in many salt tolerant plants has been found to be higher than that in salt sensitive ones.

The increase in proline is due to the capacity of some plants to accumulate organic and inorganic compounds in the cytoplasm to reduce the water potential and change the osmotic gradient, assuring the water flow to the plant. Accumulation of proline acts as an osmotic regulator. It is assumed that proline contributes to osmoregulation in plants by accumulating in cytoplasm as a compatible osmolytes.

Oxidative Stress

Oxidative stress occurs when plants are exposed to various forms of stress (Krause 1994). Reactive oxygen species (ROS) namely super oxide radicals ($O_2^{-\bullet}$), hydrogen peroxide (H_2O_2) and hydroxyl radicals ($OH^\bullet$) are produced in aerobic cellular processes such as mitochondrial and chloroplast electron transport or oxidation of glycolate (photorespiration), xanthine, and glucose under stress conditions due to metabolic disturbances. ROS production increases under abiotic stresses including salinity (Gomez et al. 1999; Hernandez 2001).

Reactive Oxygen Species (ROS)

As a consequence of ion imbalance and hyperosmotic stresses, which are primary effects of salt stress, secondary stresses such as oxidative damage may occur under abiotic stresses. $O_2^{-\bullet}$ production enhances from 240 uM S^{-1} to 720 uM S^{-1} and H_2O_2 production in chloroplast enhances from 5 uM to 15 uM (Polle 2001; Mittler 2002).

ROS causes damage to lipids, proteins and DNA. Peroxidation of membrane lipids occur, when ROS react with unsaturated fatty acids which leads to leakage of cellular contents, rapid desiccation and cell death. The ability of plant tissues to mobilize enzymatic defense against uncontrolled lipid peroxidation may be an important facet of their tolerance (Srivalli et al. 2003). The harmful effects of ROS is primarily due to their ability to initiate a variety of antioxidative chain reactions and unsaturated fatty acids (Smirnoff 1993). In presence of Fe, the chain reactions are propagated and amplified. Oxidative attack on proteins results in site specific amino acid modifications, fragmentation of the peptide chain, aggregation of cross linked reaction products and increased susceptibility to proteolysis. ROS can also induce numerous lesions in DNA that causes deletions, mutations and other lethal genetic effects (Srivalli et al. 2003).

Antioxidant Defense

Studies carried out with whole plants (Gossett et al. 1992, 1993) and callus tissues (Gossett et al. 1996; Csiszar et al. 2004) exposed to NaCl have revealed significant increase in activities of antioxidant enzymes. The activities of the antioxidative enzymes such as catalase (CAT), ascorbate peroxidase (APX), guaicol peroxidase (POD), glutathione reductase (GR), and superoxide dismutase (SOD) increases under salt stress in plants and a correlation of these enzyme levels and salt tolerance exists (Lee et al. 2001; Mittova et al. 2002). SOD activity in plant leaves of barley and H^+-ATPase activity in the plant roots increases by salt stress, where as malondialdehyde (MDA) concentration in plant leaves decreases (Liang 1999).

Rodriguez Rosales et al. (1999) has reported that SOD activity of tomato increases under NaCl stress. Hernandez et al. (1999) have reported that in pea (*Pisum sativum* cv. Puget), higher concentration of NaCl (110–130 mM) enhances the activities of cytosolic CuZn-SOD II, chloroplastic CuZn-SOD II and mitochondrial and/or proximal Mn-SOD. SOD activity decreased during water stress in rice (Boo and Jung 1999) but increased under salt stress in a tolerant cultivar of pea (Hernandez et al. 1995). Shalata et al. (2001) found that SOD and CAT activities decreased in roots of a salt sensitive tomato cultivar but increased in roots of salt tolerant tomato cultivar under salt stress.

Mechanism of Salt Tolerance

The ability of plants to tolerate excess salt in their habitat without any significant impairment of their vital function is termed as salt resistance. It is a complex combination of various mechanisms, and therefore, certainly not controlled by a single gene (Munns 1993). Plants can achieve resistance to salt stress either by tolerating the stress or by avoiding it. Resistance is the more general term, including mechanism. High salinity causes hyperosmotic stress and ion imbalance that produces adverse effects on plants (Zhu 2001). Resistance to environmental stress occurs when a plant withstands the imposed stress and may arise from either tolerance or a mechanism that permits avoidance of the situation. Whole plant mechanism can contribute to the avoidance of stress during the plants life cycle and avoidance can also occur at the cellular level. Plants are either dormant during the salt episode or there must be cellular adjustment to tolerate the saline environment (Yokoi et al. 2002). The response depends on the species and the genotypes, the length and severity of the salt stress, the age and stage of development, the organ and the cell type and the sub-cellular compartment. An example of avoidance at the cellular level is the process of osmotic adjustment, where the osmotic potential of the cell is lowered in order to maintain the water potential gradient to favour water uptake and maintenance of turgor. (Bray 1997). Conventional selection and breeding techniques have been used in recent decades to improve salinity tolerance in crop plants (Ashraf 1994; Shannon 1998). The agronomical parameters used by scientists for salt tolerance are yield, survival, plant height (Noble and Rogers 2002), leaf area (Franco et al. 1993) leaf injury (Munns 1993), relative growth rate (Cramer et al. 1990) and relative growth reduction (He and Cramer 1992).

Many scientists are of the view that plant species should possess distinctive indicators of salt tolerance to whole plant, tissue or cellular level (Greenway and Munns 1980; Munns 2002). The adaptive mechanisms utilized by the plants to survive under salt stress are still not well understood. Nonetheless, parallels have been drawn between different biochemical indicators and plant tolerance to salt stress. For example, there is strong evidence that glycinebetaine (quaternary ammonium compound) and proline (imino acid) play an adaptive role in mediating osmotic adjustment and protecting the subcellular structures in stressed plants. In many studies a positive correlation between the accumulation of these two osmolytes and stress tolerance in plants has been found (Yamada et al. 2003; Yang et al. 2003). Similarly, trehalose accumulation was found to be correlated with cellular stress resistance to heat and dessication in yeast and desert resurrection plants such as *Selaginella lepidophylla* and *Myrothamnus flabellifolius* (Muller et al. 1995) and in transgenic rice plants (Garg et al. 2002). While determining the role of various antioxidants in the salt tolerance of tomato, Mittova et al. (2002) found that higher salt tolerance of wild tomato (*Lycopersicon pennelli*) as compared to cultivated tomato (*L. esculentum*) was correlated with increased activities of SOD (superoxide dismutase), APX (ascorbate peroxidase), and POD (guiacol peroxidase).

The other factors affecting the plant tolerance mechanisms are salt concentration, type of salt, length of exposure etc. In most of the studies, NaCl was used as the selective agent. NaCl selection is likely to produce genotypes with resistance to Na^+ and Cl^-ions or osmotic stress, but not to the other ions contributing to salinity in certain agricultural situations (Rains et al. 1986). However some authors have also reported the selection of cell lines tolerant to salts other than NaCl such as Na_2SO_4, K_2SO_4, KCl, $MgCl_2$ and $MgSO_4$ (Muralitharan et al. 1992).

Enzymes in the cytoplasm are directly affected by salt stress. In salt exposed plants, higher concentrations of salt ions like Na^+ and Cl^- have been found in the cytosol. Under such conditions the cytoplasmic enzymes have to function in the presence of salt ions. This was studied with the enzyme phosphoenol pyruvate carboxylase (PEP), extracted from the halophytes *Sanacda monoica* and *Chloris gayana* (Shomer-Iian et al. 1985). It is the key enzyme in the process of CO_2 fixation in leaves. The in vitro salt tolerance of PEP carboxylase depends on the enzyme pretreatment as well as on the substrate (PEP) concentration in the medium. Enzymes are stabilized by the addition of PEP to the extraction and storage medium. A low PEP level of the assay medium, NaCl inhibited the pretreatment enzymes, but at high PEP levels NaCl activated the enzymes. Kinetic properties of the enzymes are changed by salt ions and are suggested to function as allosteric effectors.

Solmon et al. (1994) found a mechanism of enzyme protection with ribulose-1, 5-bisphosphate carboxylase/oxygenase (rubisco) of the woody halophyte *Tamarix jordanis*, NaCl inhibited the function of this key enzyme. The salt tolerance of *Tamarix jordanis* seems to be based on two mechanisms, the increase of Rubisco content and the formation of compatible solutes like N-methyl-L-proline. Such solutes enable the Rubisco to function at high rates in the presence of salt ions in the cytosol.

In Vitro Assessment of Salinity

Nearly 800 ppm salinity has been recognized as a threshold level for agricultural damage. The salinity values are much higher than this value in the problematic areas. The average damage accounts to more than Rs. 224 million (Nabors and Dykes 1985). So there is a need to modify the environmental conditions in such areas to suit the plant growth and development. But the economic impacts of modifying the environment always limit the wide applicability for practices like soil amelioration, irrigation, agronomical managements etc.

Assessment of plants growing under in vitro condition has been carried out through a numbers of parameters, which includes: morphogenetic response of in vitro growing plantlets, on the basis of quantification of secondary metabolites, transformation of salinity specific genes in those crops which growth and development is directly affected by the presence of salt or preparation of salt stress tolerance varieties for future. In addition to this, salinity and drought resistant mutant were produced by applying abiotic stress under in vitro conditions. It was assumed that they may show a different morphological and physiological response and also show genetic diversity among the tolerant plantlets. This investigation was aimed to study the effect of abiotic stress viz. salinity and drought on the morphology, physiology and biochemical constituents like proline, protein and chlorophyll content in the in vitro grown plantlets and at the same time analyze the molecular diversity using random amplified polymorphic DNA (RAPD) molecular marker analysis among the stress grown plantlets in comparison with the control plants. The assessment of genetic identity and purity in the salt tolerant plantlets can help us in further identification of the stress related genes.

In Vitro Morphogenetic Response of Plants Growing Under Salt Stress

Plant tissue culture technology is one of the alternative methods which modify the plant to suit their environments with little initial inputs and long lasting benefits. Plant tissue culture with cell genetic manipulation is providing a very useful tool for effectively implementing the plant breeding programme (Croughan et al. 1981). These methods are now becoming integrated part of the successful breeding procedures to produce high yield and tolerant cultivars at a more rapid pace (Sibi and Fakiri 2000; Alvarez et al. 2003). Nabors et al. (1975) reported that salt tolerance can be selected at increasing salt concentrations in callus or suspension cultures, and that the enhanced tolerance is both stable and passed to progeny from regenerated plants. It has been shown with the work on *Solanum pennelli, Lycopersicon peruvlanum* and *L. esculentum* (Tal et al. 1978) and *Hordeum juvatum, H. vulgare* (Orton 1980) that plants could be ranked for selective salt tolerance through tissue culture.

The advantages of plant tissue culture for the production of salt tolerant plants and to understand the mechanism of salt tolerance are as follows:

- Evaluation of large number of genotypes and selection in the laboratory using relatively small space.
- The environment and nutrient conditions can be uniformly and precisely controlled.

- Complications due to differences in morphology and developmental stages can be reduced by growing cells uniformly in culture medium.
- Time between generations can be reduced.
- During culture conditions, variations in genotypes can be generated.
- Traits selected at the cell level of somaclonal variability for that trait can be evaluated in regenerated plants and their progeny.
- Physiological and biochemical processes which regulate the salt tolerance (Rains et al. 1986) can be studied by using isolated protoplasts, cell and suspension cultures.

Dix and Street (1975) first isolated the salt tolerant cell lines from *Capsicum annum*. Since then such cell lines have been isolated from many plants. Screening and selection for salt tolerance of many plants have been proposed (Pecetti and Gorham 1997). Most screening experiments for salt tolerant genotypes have been conducted either in vitro or controlled environmental conditions (Houshmand et al. 2005; Ahmad et al. 2006). Benderradji et al. (2012) evaluated the response of two genotypes of bread wheat (*Triticum aestivum*), Mahon-Demias (MD) and Hidhab (HD1220), to mature embryo culture, callus production, and in vitro salt and heat tolerance. For assessment of genotypes to salt and heat tolerance, growing morphogenic calli were exposed to different concentrations of NaCl (0, 5, 10, and 15 $g \cdot L^{-1}$) and under different thermal stress intensities (25°C, 30°C, 35°C, and 40°C). Comparison of the two genotypes was reported for callus induction efficiency from mature embryo. While, for salt and heat tolerance, the proliferation efficiency, embryonic efficiency, and regeneration efficiency were used. The results show significant medium and genotype effects for the embryogenesis capacity of calluses induction and plantlets regeneration under saline and thermal stresses. Mahon-Demias showed good callus induction and ability to proliferate and regenerate seedling under heat and salt stress conditions compared to Hidhab. No sizeable differences were observed between the two genotypes at higher salt stress rates. This study will serve as a base line for in vitro screening of several elite wheat cultivars for their ability to induce callus and regenerate plants from mature embryos, and to start selection for tolerance to salinity.

Somaclonal Variant

Somaclonal variant strains from callus cultures are supplementary tools to traditional breeding for production of salt stressed resistant plants (Larkin and Scowcroft 1981; Dix 1993; Ashraf 1994). The introduction of a given genotype in in vitro selection depends on its aptitude to in vivo cultures, particularly to callus induction and embryogenic callus production. Studies have shown in many species that genotype affects plant in in vitro culture response. Following strategies have been used for the isolation of salt tolerant cell lines.

Selection of Salt Tolerance Cell Lines

Direct Selection

Salt tolerant cell lines have been isolated by exposing explants mainly callus cultures, cell suspension cultures, protoplasts and even microspores to a sublethal concentrations of salts, once or few times or by gradual stepwise increase in salt(s) at each subcultures (Anju and Pawan 1997). One-step selection strategy is more effective than stepwise method for the selection of salt tolerant mutants (Mc Hughen and Swartz 1984). A better understanding of the phenomenon of adaptation to salt stress may enable a distinction between mutant and adopted cells and thus may help to improve the efficiency of selection of salt tolerant mutant cells (Tal 1993).

Indirect Selection

In some of the plant species salt tolerant cell lines have been isolated by selecting the cells which accumulate higher levels of osmoregulatory compounds e.g. proline, quaternary ammonium compounds, polyols, or by selecting for salt tolerant to osmotic stress component of salinity (Benderradji et al. 2012).

Genetic Transformation to Develop Salt Stress Transgenic and Their Assessment

There are numbers of established protocol for most of the valuable plants including medicinal, crops and ornamental through the organogenesis and somatic embryogenesis procedure. But, in present scenario, scientist has change their thought keeping the view of global warming. The growth and the developmental process of plants being directly affecting due to the emission of carbon, incorporation of salt in soil and other toxic gases become a threat for agriculture (Ahmad et al. 2010, 2011, 2012). Therefore, it has been reported that, due to the salinity stress or other abiotic stress, propagation of plants is directly effecting or they failed to grow properly in salinity soil. The problems of soil salinity and improving the salt tolerance of cultivated plants are of particular urgency where soils have been already either been partially salinized or can become saline because of irrigation. Salinity can affect any process in the life cycle of a plant resulting in loss of yield. Salt stress (NaCl) was shown to have greater toxic effects than $CaCl_2$ on the growth and metabolism in number of plant system which functions as growth inhibitor. *Osmotin* is a 24 kDa stress responsive multifunctional protein that accumulates during adaptation of cells to high osmotic stress including salt or drought (Singh et al. 1987). The protective abilities of osmotin against abiotic stresses have been confirmed by over expression studies of osmotin gene in most angiosperm plant like potato, tomato, tobacco and strawberry. Constitutive expression of an *osmotin* gene in transgenic tobacco improved their tolerance to salinity and drought stress (Barthakur et al. 2001). Chemicals such as acetosyringone are recommended in most of the crops transformation protocols (Ishida et al. 1996) for *vir* induction. Southern hybridization pattern of selected transgenic plants confirmed single as well as multiple gene insertion. Studies have shown that it is desirable to have single gene insertion in transgenic plants as multiple copies of T-DNA adversely influence the expression of the introduce gene.

Salt tolerance transgenic plants perform significantly better in terms of chlorophyll, soluble protein, and proline accumulation than wild type plants, when expose to different concentrations of NaCl. Similarly, the over expression of *osmotin* in tobacco enabled the transgenics to perform significantly better than the wild type plants when subjected to either salt or water stress (Babu and Bansal 1998; Barthakur et al. 2001). Barthakur et al. (2001) hypothesized that *osmotin* induce substantial increase in free proline accumulation in transgenic tobacco plant with or without stress as compared to wild type plants is most likely the basis of improved performance. However, the mechanism governing the increased proline accumulation in *osmotin* transgenic remains to be unrevealed. Relatively higher proline content was also detected in transgenic tomato plants ecotopically expressing *Arabidopsis* CBFI transcription factor gene (Hseih et al. 2002) and in transgenic *Arabidopsis* over expressing DREBIA (CBF3) gene (Gilmour et al. 2000). It appears that similar to CBF1 and CBF3, *osmotin* plays a role directly or indirectly in inducing the expression of gene encoding enzymes of protein biosynthesis by activating the proline-5'carboxylate synthetase (P5C5) enzyme those catalyses the rate limiting step in protein biosynthesis expressing the proline catabolic pathway (Igarashi et al. 1997; Savoure et al. 1995; Yoshiba et al. 1995; Kavi Kishore et al. 1995; Delauney and Verma 1993) or by suppressing feedback inhibition of proline biosynthesis (Phutela et al. 2003). Further, it may be pointed out that *osmotin* is a proline rich protein and degradation could also possible lead to increased accumulation of proline at least under conditions where the protein is over produced. It has also been shown in developing grapevine fruits that proline could accumulate in plant cell due to degeneration of proline rich proteins independently and not associated with either increase in steady state levels of P5CS mRNA or proteins or a decreased in steady state proline dehydrogenase protein (Stines et al. 1999). The high chlorophyll content suggesting transgenic overexpressing *osmotin* are probably due to *the* osmotic adjustment effect of proline in addition to other unknown effects of *osmotin* and its role in conferring stress tolerance in plants. Proline has been earlier shown to act as a compatible osmolyte and its increased production confers osmotolerance in transgenic plants (Kavi Kishore et al. 1995; Nanjo et al. 1999). Sarin et al. (2004) also observed that *Vigna mungo* upon over-expression of the Glyoxalase 1 (Gly 1) gene exhibit tolerance to NaCl and methylglyoxal induced stress and retained more chlorophyll as compared to the untransformed controls under short term stress conditions. The production of transgenic sweetpotato (cv. Xushu 18) plants exhibiting enhanced salt tolerance using salt overlysensitive (SOS) genes was achieved through *Agrobacterium* tumefaciens-mediated transformation. *A. tumefaciens* strainEHA105 harbors a binary vector pCAMBIA3301 with SOS genes

(SOS1, SOS2 and SOS3) and bar gene. Selection culture was conducted using 0.3 mg L^{-1} phosphinothricin (PPT). A total of 40 plants were produced from the inoculated 170 cell aggregates via somatic embryogenesis. PCR analysis showed that 37 of the 40 regenerated plants were transgenic plants. The in vitro assay demonstrated that superoxide dismutase (SOD) and proline were significantly more accumulated and malonaldehyde (MDA) was significantly less accumulated in 21 transgenic plants than in control plants when they were exposed to 86 mmol L^{-1} NaCl. Salt tolerance of these 21 plants was further evaluated with Hoagland solution containing 0, 51, 86, and 120 mmol L^{-1} NaCl in the greenhouse. The results indicated that six of them had significantly better growth and rooting ability than the remaining 15 transgenic plants and control plants. Expression of SOS genes in the six salt-tolerant transgenic plants was demonstrated by RT-PCR analysis (Shang et al. 2012).

Conclusion and Future Perspective

Accumulation of high levels of salts in the soil is characteristic of arid and semi-arid regions. All salts can alter plant growth but not all inhibit growth. In addition, salts do not act alone in soil. They interact in their effects on plants; some of these interactions are simple (e.g., interaction between Na^+ and Ca^{2+}), whereas some are complex (e.g. carbonates and their effects via increased soil pH). Although different curative and management measures are being used to render salt-affected soils fit for agriculture, they are extremely expensive and do not provide permanent solutions to overcome the salinity problem. In contrast, a biotic approach for overcoming salinity stress has gained considerable recognition within the past few decades in view of the vast experimental evidence from what has happened in nature concerning the evolution of highly salt-tolerant ecotypes of different plant species, and also from the remarkable achievements that have been made in improving different agronomic traits.

One of the important applications of modern biotechnology in agriculture is use of tissue culture techniques. In vitro micropropagation techniques are increasingly being applied to large scale production of quality planting materials especially in agricultural and horticultural crops. The goal of tissue culture is to mass produce genetically identical, physiologically uniform, developmentally normal and pathogen free plantlets which can be acclimatized in a reduced time period and at a lower cost. Plant cell/tissue culture is a rapidly developing technology which holds promise of restructuring agricultural, horticultural and forestry practices. Cultured explants undergo frequent genetic changes which are expressed at biochemical or molecular level. The genetic variability expressed in regenerated plants can be transmitted to the progeny through sexual or vegetative propagation. Salt resistant plant variants have been developed in vitro to reclaim salt affected wastelands. Somaclonal variation may be an additional tool for crop improvement during salt stress.

References

Abraham E, Rigo G, Szekely G, Nagy R, Koncz C, Szabados L (2003) Light-dependent induction of proline biosynthesis by abscisic acid and salt stress is inhibited by brassinosteroid in *Arabidopsis*. Plant Mol Biol 51:363–372

Ahmad P, Sharma S, Srivastava PS (2006) Differential physio-biochemical responses of high yielding varieties of Mulberry (*Morus alba*) under alkalinity (Na_2CO_3) stress in vitro. Physiol Mol Biol Plants 12(1):59–66

Ahmad P, Sarwat M, Sharma S (2008) Reactive oxygen species, antioxidants and signaling in plants. J Plant Biol 51(3):167–173

Ahmad P, Jaleel CA, Salem MA, Nabi G, Sharma S (2010) Roles of Enzymatic and non-enzymatic antioxidants in plants during abiotic stress. Crit Rev Biotechnol 30(3):161–175

Ahmad P, Nabi G, Jeleel CA, Umar S (2011) Free radical production, oxidative damage and antioxidant defense mechanisms in plants under abiotic stress. In: Ahmad P, Umar S (eds) Oxidative stress: role of antioxidants in plants. Studium Press, New Delhi, pp 19–53

Ahmad P, Hakeem KR, Kumar A, Ashraf M, Akram NA (2012a) Salt-induced changes in photosynthetic activity and oxidative defense system of three cultivars of mustard (*Brassica juncea* L.). Afr J Biotechnol 11(11):2694–2703

Alamgir ANM, Ali MY (1999) Effect of salinity on leaf pigments, sugar and protein concentrations and chloroplast ATPAase activity of rice (*Oryza sativa* L.). Bangladesh J Bot 28:145–149

Allakhverdiev SI, Sakamoto A, Nishiyama Y, Inaba M, Murata N (2000) Ionic and osmotic effects of NaCl-induced inactivation of photosystems I and II in *Synechococcus* sp. Plant Physiol 123:1047–1056

Alvarez I, Tomaro LM, Benavides PM (2003) Changes in polyamines, proline and ethylene in sunflower calluses treated with NaCl. Plant Cell Tissu Org Cult 00:1–9

Anju G, Pawan J (1997) The potential of plant tissue culture and related techniques for the improvement of salt tolerance in higher plants. In: Jaiswal PK, Singh RP, Gulati A (eds) Strategies for improving salt tolerance in higher plants. Oxford & IBH Publishing, New Delhi, pp 321–363

Ashraf M (1994) Organic substances responsible for salt tolerance in *Eurica sativa*. Biol Plant 36:255–259

Ashraf M, Fatima H (1995) Responses of some salt tolerant and salt sensitive lines of safflower (*Carthamus tinctorius* L.). Acta Physiol Plant 17:61–71

Ashraf M, Tufail M (1995) Variation in salinity tolerance in sunflower (*Helianthus annuus* L.). J Agron Soil Sci 174:351–362

Babu V, Bansal KC (1998) *Osmotin* overexpression in transgenic potato plants provides protection against osmotic stress. In: Proceedings of the 5th international symposium on molecular biology of the Potato, Bogensee, Germany

Barthakur S, Babu V, Bansal KC (2001) Over expression of *osmotin* induces proline accumulation and confers tolerance to osmotic stress in transgenic tobacco. J Plant Biochem Biotechnol 10:31–37

Benderradj L, Brini F, Amar SB, Kellou K, Azaza J, Khaled Masmoudi K, Bouzerzou H, Hanin M (2011) Sodium transport in the seedlings of two bread wheat (*Triticum aestivum* L.) genotypes showing contrasting salt stress tolerance. AJCS 5(3):233–241

Benderradji L, Brini F, Kellou K, Ykhlef N, Djekoun A, Masmoudi K, Bouzerzour H (2012) Callus induction, proliferation, and plantlets regeneration of two bread wheat (*Triticum aestivum* L.) genotypes under saline and heat stress conditions. ISRN Agron 2012:1–8

Bernstein N (1975) Effect of salinity and sodicity on plant growth. Annu Rev Phytopathol 13:295–312

Bhat MA, Ahmad S, Aslam J, Mujib A, Mahmooduzzfar (2008) Salinity stress enhances production of solasodine in *Solanum nigrum* L. Chem Pharm Bull 56(1):17–21

Bhatti S, Jha G (2010) Current trends and future prospects of biotechnological interventions through tissue culture in apple. Plant Cell Rep 29:1215–1225

Binzel ML, Hess FD, Bressan RA, Hasegawa PM (1988) Intracellular compartmentation of ions in salt adapted tobacco cells. Plant Physiol 86:607–614

Bolarin MC, Periz-Alfocea F, Cano EA, Estan MT, Caro M (1993) Growth, fruit yield and ion concentration in tomato genotype after pre-emergence and post emergence salt treatments. J Am Soc Hortic Sci 118:655–660

Boo YC, Jung J (1999) Water deficit induced oxidative stress and antioxidative defence in rice plants. J Plant Physiol 51:255–261

Bray EA (1997) Plant responses to water deficit. Trends Plant Sci 2:48–54

Breusegem FV, Vranova E, Dat JF, Inze D (2001) The role of active oxygen species in plant signal transduction. Plant Sci 161:405–414

Brussaard L, De Ruiter PC, Brown GG (2007) Soil biodiversity for agricultural sustainability. Agricul Ecosyst Environ 121:233–244

Cakar J, Parick A, Maksimovic M, Bajrovic K (2012) Antioxidative and antitumor properties of in vitro-cultivated broccoli (*Brassica oleracea* var. italica). Pharm Biol 50:175–181

Calixto JB (2000) Efficacy, safety, quality control, marketing and regulatory guidelines for herbal medicines (phytotherapeutic agents). Braz J Med Biol Res 33:179–189

Chatterjee A, Çui Y, Yang H, Collmer A, Alfano JR, Chatterjee AK (2003) GacA, the response regulator of a two-component system, acts as a master regulator in *Pseudomonas syringae* pv. tomato DC3000 by controlling regulatory RNA, transcriptional activators, and alternate sigma factors. Mol Plant Microbe Interact 16:1106–11017

Chen Y, Zahavi E, Barek P, Umiel N (1980) Effect of salinity stresses on tobacco 1. The growth of *Nicotiana tabacum* callus cultures under water, NaCl and mannitol stress. Z Pflanzenphysiol 98:141–143

Chen W, Li YM, Yu MH (2007) Effects of Astragalus polysaccharides on chymase, angiotensin-converting enzyme and angiotensin II in diabetic cardiomyopathy in hamsters. J Int Med Res 35:873–877

Cramer GR, Epstein E, Lauchli A (1990) Effects of sodium, potassium and calcium on salt stressed barley. Physiol Plant 80:83–88

Croteau R, Kutchan TM, Lewis NG (2000) Natural products (secondary metabolites). In: Buchanan B, Gruissem W, Jone W, Jone R (eds) Biochemistry and molecular biology of plants. American Society of Plant Physiologists, Rockville, pp 1250–1318

Croughan TP, Stavared SJ, Rains DW (1981) In vitro development of salt resistant plants. Environ Exp Bot 21:317–324

Cruz V, Cuartero J (1990) Effect on salinity at several developmental stages of six genotype of tomato (*Lycopersicon* sp.) In: Cuartero J, Gomez-Guillamon ML, Fernandez-Munoz (eds) Proceedings of the X1. Eucarpic meeting on tomato genetics and breeding, Spain, pp 81–86

Csisar J, Szabo M, Erdei L, Marton L, Horvath F, Tari I (2004) Auxin autotrophic tobacco callus tissues resist oxidative stress: the importance of glutathione S-transferase and glutathione peroxidase activities in auxin heterotrophic and autotrophic calli. J Plant Physiol 161:691–699

Dan Y, Baxter A, Zhang S, Pantazis CJ, Veilleux RE (2010) Development of efficient plant regeneration and transformation system for impatiens using *Agrobacterium tumefaciens* and multiple bud cultures as explants. BMC Plant Biol 10:165, Aug 9

Davey MR, Anthony P, Power JB, Lowe KC (2005) Plant protoplasts: status and biotechnological perspectives. Biotechnol Adv 23:131–171

Delauney AJ, Hu C-AA, Kavi Kishore PB, Verma DPS (1993) Cloning of ornithine δ-aminotransferase cDNA from *Vigna acontifolia* by trans complementation in *Escherichia coli* and regulation of proline biosynthesis. J Biol Chem 268:18673–18678

Dix PJ (1985) Cell line selection. In: Yeoman MM (ed) Plant cell culture technology. Blackwell Scientific, Oxford, pp 141–199

Dix PG (1993) The role of mutant cell lines in studies on environmental stress tolerance: an assessment. Plant J 3:309–313

Dix PJ, Street HE (1975) Sodium chloride-resistant cultured cell lines from *Nicotiana sylvestris* and *Capsicum annum*. Plant Sci Lett 5:231–237

Downtwon WSJ (1977) Photosynthesis I salt-stressed grapevines. Aust J Plant Physiol 4:183–192
Dubey R (1994) Protein synthesis by plants under stressful conditions. In: Mohammed P (ed) Handbook of plant and crop stress. Marcel Dekker, New York, pp 277–299
Flowers TJ (2004) Improving crop salt tolerance. J Expt Bot 55:1–13
Flowers TJ, Yeo AR (1992) Solute transport in plants. Chapman and Hall, Landon
Fougere F, Le Rudulier D, Streeter JG (1991) Effects of salt stress on amino acid, organic acid, and carbohydrate composition of roots, bacteroids, and cytosol of alfalfa (*Medicago sativa* L.). Plant Physiol 96:1228–1236
Franco JA, Esteban C, Rodriguez C (1993) Effects of salinity on various growth stages of musk-melon cv. Revigal J Hort Sci 68:899–904
Gadallah MAA (1999) Effect of proline and glycinebetaine on *Vicia faba* responses to salt stress. Biol Plant 42:249–257
Gai YP, Ji XL, Lu W, Han XJ, Yang GD, Zheng CC (2011) A novel late embryogenesis abundant like protein associated with chilling stress in *Nicotiana tabacum* cv. bright yellow-2 cell suspension culture. Mol Cell Proteom 10:M111.010363
Gao JP (2007) Understanding abiotic stress tolerance mechanisms: recent studies on stress response in rice. J Integ Plant Biol 49:742–750
Gao H, Lamusta J, Zhang WF, Salmonsen R, Liu Y, O'Connell E, Evans JE, Burstein S, Chen JJ (2011) Tumor cell selective cytotoxicity and apoptosis induction by an herbal preparation from *Brucea javanica*. North Am J Med Sci 4:62–66
Garg AK, Kim JK, Owens TG, Ranwala AP, DoChoi Y, Kochian LV, Wu RJ (2002) Trehalose accumulation in rice plants confers high tolerance levels to different abiotic stresses. Proc Natl Acad Sci USA 99:15898–15903
Gasparetto JC, Martins CA, Hayashi SS, Otuky MF, Pontarolo R (2012) Ethnobotanical and scientific aspects of *Malva sylvestris* L.: a millennial herbal medicine. J Pharm Pharmacol 64:172–189
Germanà MA (2011) Gametic embryogenesis and haploid technology as valuable support to plant breeding. Plant Cell Rep 30:839–857
Gersani M, Gorham EA, Nobel S (1993) Growth responses on individual roots of *Opuntia ficus* indica to salinity. Plant Cell Environ 16:827–834
Ghasemzadeh A, Jaafar HZ (2011) Effect of CO(2) enrichment on synthesis of some primary and secondary metabolites in ginger (*Zingiber officinale* Roscoe). Int J Mol Sci 12:1101–1114
Gilmour SJ, Sebolt AM, Salazar MP, Everard JD, Thomashow MF (2000) Overexpression of the *Arabidopsis* CBF3 transcriptional activator mimics multiple biochemical changes associated with cold acclimation. Plant Physiol 124:1854–1865
Goddijn OJM, Van Dun K (1999) Trehalose metabolism in plants. Trends Plant Sci 4:315–319
Gomez JM, Hernandez JA, Jimenez A, del Rio LA, Sevilla F (1999) Differential response of anti-oxidative enzymes of chloroplast and mitochondria to long term NaCl stress of pea plants. Free Radic Res 31:11–18
Gossett DR, Lucas MC, Millhollon EP, Caldwell WD, Barclay A (1992) Isozyme variation among salt tolerant and salt sensitive varieties of cotton. In: Herber DG (ed) Beltwide cotton production research conferences, Nashville, 6–10 Jan 1992. Cotton Council of America, Memphis, pp 1036–1039
Gossett DR, Millhollon EP, Lucas MC, Marney MM (1993) Antioxidant enzymes and salt tolerance in cotton. In: Herber DG (ed) Beltwide cotton production conferences, New Orleans LA, 10–14 Jan 1993. National cotton council of America. Memphis TN, pp 1262–1266
Gossett DR, Banks SW, Millhollon EP, Lucas MC (1996) Antioxidant response to NaCl stress in a control and a NaCl-tolerant cotton line grown in the presence of paraquat, buthionine sul-phoxime, and exogenous glutathione. Plant Physiol 112:803–809
Greenway H, Munns R (1980) Mechanism of salt tolerance in non-halophytes. Ann Rev Plant Physiol 31:149–190
Hamid A, Talibuddeen C (1976) Effect of sodium on the growth and ion uptake by barley sugar beet and broad beans. J Agric Sci 86:49–56

He T, Cramer GR (1992) Growth and mineral nutrition of six rapid-cycling *Brassica* species in response to sea water salinity. Plant Soil 139:285–294

Heath-Pagliuso S, Rappaport L (1990) Somaclonal variant UC-T3: the expression of *Fusarium* wilt resistance in progeny arrays of celery, *Apium graveolens* L. Theor Appl Genet 80:390–394

Hernandez JA, Olmos E, Corpas FJ, Sevilla F, del Rio LA (1995) Salt-induced oxidative stress in chloroplasts of pea plants. Plant Sci 105:151–167

Hernandez JA, Campillo A, Jimenez A, Alacon JJ, Sevilla F (1999) Response of antioxidant systems and leaf water relations to NaCl stress in pea plants. New Phytol 141:241–251

Hernandez JA, Ferrer MA, Jimenez A, Barcelo AR, Sevilla F (2001) Antioxidant system and $O_2^{\cdot-}$/H_2O_2 production in the apoplast of pea leaves. Its relation with salt induced necrotic lesions in minor veins. Plant Physiol 127:817–831

Heuer B, Plaut Z (1989) Photosynthesis and osmotic adjustment of two sugarbeet cultivars grown under saline conditions. J Exp Bot 40:437–440

Hounsa CG, Brandt EV, Thevelein J, Hohmann S, Prior BA (1998) Role of trehalose in survival of *Saccharomyces cerevisiae* under osmotic stress. Microbiology 144:671–680

Houshmand S, Arzani A, Maibody SAM, Feizi M (2005) Evaluation of salt-tolerant genotypes of drum wheat derived from in vitro and field experiments. Field Crop Res 91:345–354

Hseih TH, Lee JT, Charng YY, Chan MT (2002) Tomato plant ectopically expressing *Arabidopsis* CBF1 show enhanced resistance to water deficit stress. Plant Physiol 130:618–626

Hurkman WJ, Fornari CS, Tanaka CK (1989) A comparison of the effect of salt on the polypeptide and translatable mRNA in roots of the salt tolerant and salt sensitive cultivar of barley. Plant Physiol 90:1444–1456

Hurkman WJ, Rao HP, Tanaka CK (1991) Germinlike polypeptides increase in barley roots during salt stress. Plant Physiol 97:366–374

Hussain MS, Fareed S, Ansari S, Rahman MA, Ahmad IZ, Saeed M (2012) Current approaches toward production of secondary plant metabolites. J Pharm Bioallied Sci 4:10–20

Igarashi Y, Yoshiba Y, Sanada Y, Yamaguchi-Shinozaki K, Wada K, Shinozaki K (1997) Characterization of gene for σ -pyrroline-5-carboxylate synthetase and correlation between the expression of the gene and salt tolerance in *Oryza sativa* L. Plant Mol Biol 33:857–865

Ingram J, Bartels D (1996) The molecular basis of dehydration tolerance in plants. Ann Rev Plant Physiol Plant Mol Biol 47:377–403

Ishida Y, Saito H, Ohta S, Hiei Y, Komari T, Kumashiro T (1996) High efficiency transformation of maize (*Zea mays* L.) mediated by *Agrobacterium tumefaciens*. Nat Biotechnol 14:745–750

Jacobsen T, Adams RM (1958) Salt and silt in ancient mesopotamian agriculture. Science 128:1251–1258

Jindal A, Kumar P (2011) Antimicrobial flavonoids from Tridax procumbens. Nat Prod Res 26:2072–2077

Kaeppler SM (1998) Molecular basis of heritable tissue culture-induced variation in plants. In: Jain DSBJSM, Ahloowalia BS (eds) Somaclonal variation and induced mutations in crop improvement. Curr Plant Sci Biotech Agr, vol 32. Kluwer Academic, Dordrecht, The Netherlands, pp 465–484

Kaeppler SM (2000) Epigenetic aspects of somaclonal variation in plants. Plant Mol Biol 43:179–188

Kärkönen A, Santanen A, Iwamoto K, Fukuda H (2011) Plant tissue cultures. Methods Mol Biol 715:1–20

Kavi Kishore PB, Hong Z, Miao GH, Hu CAA, Verma DPS (1995) Overexpression of σ^1–pyrroline-5-carboxylate synthetase increases proline production and confers osmotolerance in transgenic plants. Plant Physiol 108:1387–1394

Khavarinejad RA, Mostofi Y (1998) Effects of NaCl on photosynthetic pigments, saccharides, and chloroplast ultrastructure in leaves of tomato cultivars. Photosynthetica 35:151–154

Krause GH (1994) The role of oxygen in photoinhibition in photosynthesis. In: Foyer CH, Mullineux PM (eds) Cause of photoxidative stress and amelioration of defense system in plants. CRC, Boca Raton, pp 43–76

Kulkarni H, Karadge BA (1991) Growth and mineral nutrition of mothbean (*Phaseolus aconitifolius* Jacb) under saline condition. Indian J Plant Physiol 14:14–24

Lakshmanan P, Geijskes RJ, Wang L, Elliott A, Grof CP, Berding N, Smith GR (2006) Developmental and hormonal regulation of direct shoot organogenesis and somatic embryogenesis in sugarcane (*Saccharum* spp. interspecific hybrids) leaf culture. Plant Cell Rep 25:1007–1015

Lane A, Jarvis A (2007) Changes in climate will modify the geography of crop suitability: agricultural biodiversity can help with adaptation. SAT e J 4(1):1–12

Larkin PJ (2004) Somaclonal variation: origins and causes. In: Goodman RM (ed) Encyclopedia of plant and crop science. Marcel Dekker, New York, pp 1185–1191

Larkin PJ, Scowcroft WR (1981) Somaclonal variation- a novel source of variability from cell culture from plant improvement. Theor Appl Genet 60:197–214

Lee DH, Kim YS, Lee CB (2001) The inductive responses of the antioxidant enzymes by salt stress in the rice (*Oryza sativa* L.). J Plant Physiol 158:737–745

Liang YC (1999) Effects of silicon on enzyme activity and sodium, potassium and calcium concentration in barley under salt stress. Plant Soil 209:217–224

Loyola-Vargas VM, Ochoa-Alejo N (2012) An introduction to plant cell culture: the future ahead. Methods Mol Biol 877:1–8

Lucchesini M, Mensuali-Sodi A (2010) Plant tissue culture–an opportunity for the production of nutraceuticals. Adv Exp Med Biol 698:185–202

Lutts S, Kinet JM, Bouharmont J (1996) Effects of salt stress on growth, mineral nutrition and proline accumulation in relation to osmotic adjustment in rice (*Oryza sativa* L.) cultivars differing in salinity tolerance. Plant Growth Regul 19:207–218

Maestre FT, Jordi C, Susana B (2007) Mechanisms underlying the interaction between *Pinus halepensis* and the native late-successional shrub *Pistacia lentiscus* in a semi-arid plantation. Ecography 27:776–786

Maggio A, Miyazaki S, Veronese P, Fujita T, Ibeas JI, Damsz B, Narasimhan ML, Hasegawa PM, Joly RJ, Bressan RA (2002) Does proline accumulation play an active role in stress induced growth reduction? Plant J 31:699–712

Mansour MMF (2000) Nitrogen containing compounds and adaptation of plants to salinity stress. Biol Plant 43:491–500

Marschner H (1995) Mineral nutrition of higher plants, 2nd edn. Academic, London

McHughen A, Swartz M (1984) A tissue culture derived salt tolerant line of flax (*Linum ustatissimum*). J Plant Physiol 117:109–117

Mittler R (2002) Oxidation stress, antioxidants and stress tolerance. Trends Plant Sci 7:405–410

Mittler R, Ron V (2006) Abiotic stress, the field environment and stress combination. Trends Plant Sci 11:15–19

Mittova V, Tal M, Volokita M, Guy M (2002) Salt stress induces up-regulation of an efficient chloroplast antioxidant system in the salt-tolerant wild tomato species *Lycopersicon pennellii* but not in the cultivated species. Physiol Plant 115:393–400

Muller J, Boller T, Wiemken A (1995) Trehalose and trehalase in plants: recent developments. Plant Sci 112:1–9

Munns R (1993) Physiological process limiting plant growth in saline soils: some dogmas and hypothesis. Plant Cell Environ 16:15–24

Munns R (2002) Comparative physiology of salt and water stress. Plant Cell Environ 25: 239–250

Munns R, Termaat A (1986) Whole plant responses to salinity. Aust J Plant Physiol 13:143–160

Muralitharan MS, Chandler SF, Van Steveninck RFM (1992) Effect of Na_2So_4, K_2SO_4 and KCl on growth and ion uptake of callus cultures of *Vaccinium carymbosum* L. cv blue crop. Ann Bot 69:459–465

Mustafa NR, de Winter W, van Iren F, Verpoorte R (2011) Initiation, growth and cryopreservation of plant cell suspension cultures. Nat Protoc 6:715–742

Nabores MW, Daniels A, Nadolmy L, Brown C (1975) Sodium chloride tolerant lines of tobacco cells. Plant Sci Lett 4:155–159

Nabors MW, Dykes TA (1985) Tissue culture of cereal cultivars with increased salt drought and acid tolerance. In: Biotechnology in international research. IRRI, Manila, pp 121–138

Namdeo AG, Sharma A, Sathiyanarayanan L, Fulzele D, Mahadik KR (2010) HPTLC densitometric evaluation of tissue culture extracts of *Nothapodytes foetida* compared to conventional extracts for camptothecin content and antimicrobial activity. Planta Med 76:474–480

Nanjo T, Kobayashi M, Yoshiba Y, Sanada Y, Wada K, Tsukaya H, Kakubari Y, Yamaguchi-Shinozaki K, Shinozaki K (1999) Antisence suppression of proline degradation improves tolerance to freezing and salinity in *Arabidopsis thaliana*. FEBS Lett 461:205–210

Narula A, Kumar S, Bansal KC, Srivastava PS (2004) Biotechnological approaches towards improvement of medicinal plants. In: Srivastava PS, Narula A, Srivastava S (eds) Plant biotechnology and molecular markers. Anamaya, New Delhi

Neelakandan AK, Wang K (2012) Recent progress in the understanding of tissue culture-induced genome level changes in plants and potential applications. Plant Cell Rep 4:597–620

Nieves M, Cerda A, Botalla M (1991) Salt tolerance of two lemon scions measured by leaf chloride and sodium accumulation. J Plant Nutr 14:623–636

Nile SH, Kobragde CN, Park SW (2012) Optimized and comparative antioxidant assays and its application in herbal and synthetic drug analysis as an antioxidants. Mini Rev Med Chem 22:115–121

Noble CL, Rogers ME (2002) Arguments for the use of physiological criteria for improving the salt tolerance in crops. Plant Soil 146:99–107

Obert B, Zácková Z, Samaj J, Pretová A (2009) Doubled haploid production in flax (*Linum usitatissimum* L.). Biotechnol Adv 27:371–375

Ono NN, Tian L (2011) The multiplicity of hairy root cultures: prolific possibilities. Plant Sci 80:439–446

Orton TJ (1980) Comparison of salt tolerance between *Hordeum vulgare* and *H. jubatum* in whole plants and callus cultures. Z Pflanzenphysiol 98:105

Pandey UK, Srivastava RDL (1987) Salinity index in relation to nitrate reductase activity and proline accumulation in paddy genotypes. Indian J Plant Physiol 32:175–177

Pareek A, Singla SL, Grover A (1997) Salt responsive proteins/genes in crop plants. In: Jaiwal PK, Singh RP, Gulati A (eds) Strategies for improving salt tolerance in higher plants. Oxford and IBH Publication Co, New Delhi, p 365

Parida A, Das AB, Das P (2002) NaCl stress causes changes in photosynthetic pigments, proteins and other metabolic components in the leaves of a true mangrove, *Bruguiera parviflora*, in hydroponic cultures. J Plant Biol 45:28–36

Pecetti L, Gorham J (1997) Screening of drum wheat germplasm for 22Na uptake under moderate salinity. Cereal Res Commun 25:923–930

Peter S, Curtis PS, Luchli A (1986) The role of leaf area development and photosynthetic capacity in determining growth of Kenaf under moderate salt stress. Aust J Plant Physiol 13:553–565

Phutela A, Jain V, Dhawan K, Nainawatee HS (2003) Proline metabolism and growth of Brassica juncea seedlings under water deficit stress. Indian J Agric Biochem 16:29–32

Polard A, Wyn Jones RG (1979) Enzymes activities in concentrated solutions of glycinebetain and other solutes. Planta 144:291–298

Polle A (2001) Dissecting the superoxide dismutase-ascorbate-glutathione-pathway in chloroplasts by metabolic modeling. Computer simulations as a step towards flux analysis. Plant Physiol 126:445–462

Rains DW (1989) Plant tissue and protoplast culture: application to stress physiology and biochemistry. In: Jones HG, Flowers TJ, Jones MB (eds) Plants under stress. Biochemistry, physiology and ecology and their application to plant improvement. Cambridge University Press, Cambridge, pp 181–196

Rains DW, Croughan SS, Croughan TP (1986) Isolation and characterization of mutant cell lines and plants: salt tolerance. In: Vasil IK (ed) Cell culture and somatic cell genetics of plants. Academic, Orlando, pp 537–547

Rodríguez-Rosales MP, Kerkeb L, Bueno P, Donaire JP (1999) Changes induced by NaCl in lipid content and composition, lipoxygenase, plasma membrane H^+-ATPase and antioxidant enzyme activities of tomato (*Lycopersicon esculentum.* Mill) calli. Plant Sci 143:143–150
Sairam RK, Tyagi A (2004) Physiology and molecular biology of salinity stress tolerance in plants. Curr Sci 86:407–421
Sarin N, Bhomkar P, Debroy S, Sharma N, Saxena M, Chandrama P, Upadhyaya, Muthusamy A, Pooggin M, Hohn T (2004) Transformation of *Vigna mungo* (blackgram) for abiotic stress tolerance using marker free approach. In: Proceedings of the 4th international crop science congress, Brisbane, Australia, 2004
Satheesan J, Narayanan AK, Sakunthala M (2012) Induction of root colonization by Piriformospora indica leads to enhanced asiaticoside production in *Centella asiatica*. Mycorrhiza 22: 195–202
Savoure A, Jaoua S, Hua XJ, Van Ardiles W, Montagu M, Verbruggen N (1995) Isolation characterization and chromosomal location of a gene encoding pyrroline –5-carboxylate synthetase in *Arabidopsis thaliana*. FEBS Lett 372:13–19
Schützendübel A, Polle A (2002) Plant responses to abiotic stresses: heavy metal-induced oxidative stress and protection by mycorrhization. J Exp Bot 53:1351–1365
Shalata A, Mittova V, Volokita M, Guy M, Tal M (2001) Response of the cultivated tomato and its wild salt- tolerant relative *Lycopersicon pennellii* to salt-dependent oxidative stress: the root antioxidant system. Physiol Plant 112:487–494
Shang G, Hong Z, Shao-zhen H, Qing-chang L (2012) Overexpression of SOS genes enhanced salt tolerance in sweet potato. J Integ Agricul 11:378–386
Shannon MC (1998) Adaptation of plants to salinity. Adv Agron 60:75–119
Sharry S, Adema M, Basiglio Cordal MA, Villarreal B, Nikoloff N, Briones V, Abedini W (2011) Propagation and conservation of native forest genetic resources of medicinal use by means of in vitro and ex vitro techniques. Nat Prod Commun 6:985–988
Shomer-Ilan A, Beno DM, Waisel Y (1985) Effect of NaCl on the properties of phosphoenol pyruvate carboxylase from *Suaeda monoica* and *Chloris gayana*. Physiol Plant 65:72–78
Sibi ML, Fakiri M (2000) Androgenese et gynogenese, sources de vitrovariation et de tolerance a la salinite chez l'orge *Hordeum vulgare*. Secheresse 11(2):125–132
Sinel'nikova VN, Bazhanov IA, Kosareva IA (1988) Effect of chloride salinity on functional changes in the photosynthetic apparatus of tomato cultivars. Plant Breed Abst 60:849–1990
Singer MA, Lindquist S (1998) Multiple effects of trehalose on protein folding in vitro and in vivo. Mol Cell 1:639–648
Singh NK, Bracken CA, Hasegawa PM, Handa AK, Buckel S, Hermodson MA, Pfankoch F, Regnier FE, Bressan RA (1987a) Characterization of osmotin. A thaumatin-like protein associated with osmotic adjustment in plant cells. Plant Physiol 85:529–536
Singh NK, Bracker CA, Hasegawa PM, Handa AK, Buckel S, Harmodson MA, Pfankoch E, Regnier FE, Bressan RA (1987b) Characterization of *Osmotin*. Plant Physiol 85:529–536
Singh SK, Sharma HC, Goswami AM, Datta SP, Singh SP (2000) In vitro growth and leaf composition of grapevine cultivars as affected by sodium chloride. Biol Plant 43(2):283–286
Smirnoff N (1993) The role of active oxygen in the response of plants to water deficit and dessication. New Phytol 125:27–58
Solmon A, Beer S, Waisel Y, Jones GP, Paleg LG (1994) Effect of NaCl on the carboxylating activity of Rubisco from *Tamarrix Jordans* in the presence and absence of proline related compatible solutes. Plant Physiol 90:198–204
Srivalli B, Chinnusamy V, Khanna-Chopra R (2003) Antioxidant defense in response to abiotic stresses in plants. J Plant Biol 30:121–139
Stewart GR, Lee JA (1974) The role of proline accumulation in halophytes. Planta 120:279–289
Stines AP, Naylor DJ, Hoj PB, Heeswijck RV (1999) Proline accumulation in developing grapevine fruit occurs independently of changes in the level of σ^1 –pyrroline-5-carboxylate synthetase mRNA or protein. Plant Physiol 120:923–993
Taheri JB, Azimi S, Rafieian N, Zanjani HA (2011) Herbs in dentistry. Int Dent J 61:287–296

Tal M (1983) Selection for stress tolerance. In: Evans DA, Sharp WR, Ammirato PV, Yamada Y (eds) Handbook of plant cell culture. Macmillan, New York, pp 481–488, vol 1

Tal M (1993) In vitro methodology for increasing salt tolerance in crop plants. Acta Hort 336:69–78

Tal M, Heikin H, Dehan K (1978) Salt tolerance in wild relatives of the cultivated tomato: responses of callus tissue of *L. esculentum, L. peruvianum* and *S. pennellii* to high salinity. Z Pflanzenphysiol 86:231–240

Uma S, Prasad TG, Kumar MU (1995) Genetic variability in recovery growth and synthesis of stress proteins in response to polyethylene glycol and salt stress in finger millet. Ann Bot 76:43–49

Umiel N, Zahavi E, Chen Y (1980) Effect of salinity stresses on tobacco, 2. Short term kinetics of Na^+ and K^+ uptake by callus cultures grown on media containing NaCl. Z Pflanzenphysiol 100:363–367

Volk GM (2010) Application of functional genomics and proteomics to plant cryopreservation. Curr Genomics 11:24–29

Wang JW, Wu JY (2010) Tanshinone biosynthesis in *Salvia miltiorrhiza* and production in plant tissue cultures. Appl Microbiol Biotechnol 88:437–449

Wang XD, Nolan KE, Irwanto RR, Sheahan MB, Rose RJ (2011) Ontogeny of embryogenic callus in Medicago truncatula: the fate of the pluripotent and totipotent stem cells. Ann Bot 107:599–609

White PJ, Broadley MR (2001) Chloride in soils and its uptake and movement within the plant: a review. Ann Bot 88:967–988

Wu JY, Ng J, Shi M, Wu SJ (2007) Enhanced secondary metabolite (tanshinone) production of *Salvia miltiorrhiza* hairy roots in a novel root-bacteria co-culture process. Appl Microbiol Biotechnol 77:543–550

Xu D, Duan X, Wang B, Hong B, Ho TD, Wu R (1996) Expression of late embryogenesis abundant protein gene, *HVA1*, from barley confers tolerance to water deficit and salt stress in transgenic rice. Plant Physiol 110:249–257

Xu G, Magan H, Tarchitzky J, Kafkafi U (2000) Advances in chloride nutrition. Adv Agron 68:96–150

Yamada T, Takatsu Y, Manabe T, Kasumi M, Marubashi W (2003) Suppressive effects of trehalose on apoptotic cell death leading to petal senescence in ethylene-insentive flowers of gladiolus. Plant Sci 164:213–221

Yan Q, Wu J, Liu R (2011) Modeling of tanshinone synthesis and phase distribution under the combined effect of elicitation and in situ adsorption in *Salvia miltiorrhiza* hairy root cultures. Biotechnol Lett 33:813–819

Yang WJ, Rich PJ, Axtell JD, Wood KV, Bonham CC, Ejeta G, Mickelbart MV, Rhodes D (2003) Genotypic variation for glycinebetaine in sorghum. Crop Sci 43:162–169

Yang D, Ma P, Liang X, Wei Z, Liang Z, Liu Y, Liu F (2012) PEG and ABA trigger methyl jasmonate accumulation to induce the MEP pathway and increase tanshinone production in *Salvia miltiorrhiza* hairy roots. Physiol Plant 22:1399–3054

Yokoi S, Quintero FJ, Cubero B, Ruiz MT, Bressan RA, Hasegawa PM, Pardo JM (2002) Differential expression and function of *Arabidopsis thaliana* NHX Na^+/H^+ antiporters in the salt stress response. Plant J 30:529–539

Yoshiba Y, Kiyosue T, Katagiri T, Ueda H, Mizoguchi T, Yamaguchi-Shinozaki K, Wada K, Harada Y, Shinozaki K (1995) Correlation between the induction of a gene for σ^1-pyrroline-5-carboxylate synthetase and the accumulation of proline in *Arabidopsis thaliana* under osmotic stress. Plant J 7:751–760

Zare K, Khosrowshahli M, Nazemiyeh H, Movafeghi A, Azar AM, Omidi Y (2011) Callus culture of *Echium italicum* L. towards production of a shikonin derivative. Nat Prod Res 25:1480–1487

Zhang F, Wang Y, Yang Y, Wu H, Wang D, Liu J (2007) Involvement of hydrogen peroxide and nitric oxide in salt resistance in the calluses from *Populus euphratica*. Plant Cell Environ 30:775–785

Zhu JK (2001) Plant salt tolerance. Trends Plant Sci 6:66–71
Zidan I, Aziazeh H, Neumann PM (1990) Does salinity reduce growth in maize root epidermal cells by inhibiting their capacity for cell wall acidification? Plant Physiol 93:7–11
Ziska LH, Seemann JR, DeJong TM (1990) Salinity induced limitations on photosynthesis in *Prunus salica*, a deciduous tree species. Plant Physiol 93:864–870

Index

P. Ahmad et al. (eds.), *Salt Stress in Plants: Signalling, Omics and Adaptations*,
DOI 10.1007/978-1-4614-6108-1,

B

C

T

Printed by Printforce, the Netherlands